Excel

ADVANCED-LEVEL MATHEMATICS

STUDY GUIDE
YEARS 9–10

Get the Results You Want!

John Compton & Allyn Jones

Reprinted 2017, 2019, 2021, 2022, 2024

ISBN 978 1 74125 241 5

Pascal Press
PO Box 250
Glebe NSW 2037
(02) 9198 1748
www.pascalpress.com.au

Publisher: Vivienne Joannou
Project Editors: Leanne Poll and Karen Enkelaar
Edited by Karen Enkelaar
Indexed by Puddingburn Publishing Services
Answers checked by Peter Little
Typeset by Nikki M Group Pty Ltd
Page design by Larissa Petryca
Cover by DiZign Pty Ltd
Printed by Vivar Printing/Green Giant Press

Students
All care has been taken in compiling this study guide, but please check with your teacher about the exact requirements of the course as these can change from year to year.

Table of Contents

How to use this book

- This book covers **the topics in the Year 9, 10 and 10A Australian Curriculum (Mathematics)**, as well as the New South Wales Mathematics Stage 5.1, Stage 5.2 and Stage 5.3 syllabuses. It is best to work through it bit by bit over the whole two years. However, you can also use it for revision before tests or examinations. You will also find it useful for finding out about particular topics.
- Try to work through the book in **chapter order** (Chapter 1 first, then Chapter 2, Chapter 3, and so on). Your teacher may cover topics in a different order. If you are not sure which chapter of the book relates to your classroom work, then ask your teacher for assistance.
- If you want to practise a particular topic, then look up that topic in the Table of contents (the pages before this one) or the index (at the back of the book).

Chapter by Chapter ...

- If you want to work through a whole chapter start at the contents page. Here you will find **an overview** of the topics covered in that chapter.

Keywords

Coefficient	Literal
Elimination	Quadratic
Equation	Simultaneous
Exact	Solution
Factorise	Subject
Formula	Substitution
Formulae	Surd
Inequality	Variable

Keywords from Chapter 4: Equations.

- You should also study the **keywords** at the start of the chapter. Look out for these important terms as you work through the chapter.

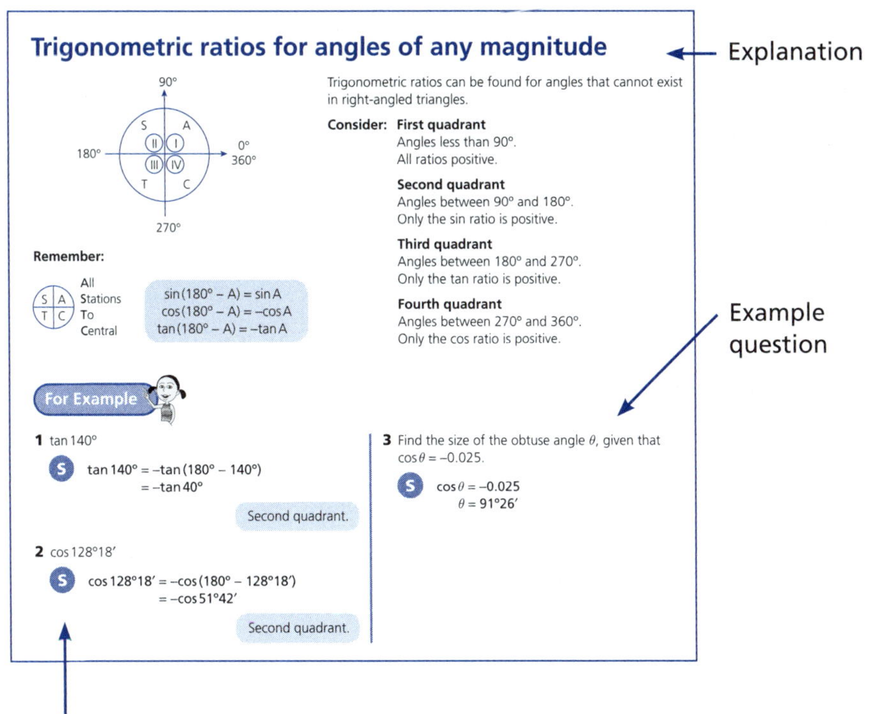

Trigonometric ratios for angles of any magnitude

Trigonometric ratios can be found for angles that cannot exist in right-angled triangles.

Consider: **First quadrant**
Angles less than 90°.
All ratios positive.

Second quadrant
Angles between 90° and 180°.
Only the sin ratio is positive.

Third quadrant
Angles between 180° and 270°.
Only the tan ratio is positive.

Fourth quadrant
Angles between 270° and 360°.
Only the cos ratio is positive.

Remember:

All Stations To Central

$\sin(180° - A) = \sin A$
$\cos(180° - A) = -\cos A$
$\tan(180° - A) = -\tan A$

For Example

1 $\tan 140°$

S $\tan 140° = -\tan(180° - 140°)$
$= -\tan 40°$

Second quadrant.

2 $\cos 128°18'$

S $\cos 128°18' = -\cos(180° - 128°18')$
$= -\cos 51°42'$

Second quadrant.

3 Find the size of the obtuse angle θ, given that $\cos\theta = -0.025$.

S $\cos\theta = -0.025$
$\theta = 91°26'$

Explanation

Example question

Solution

Extract from Chapter 8: Pythagoras and Trigonometry

- Each concept is explained using both an **explanation** and an **example question**. Read the explanation and then work through the example question. The worked solution to the **example question** follows and is signalled by the S solution icon. Make sure you understand how the answer was found. If you are not sure, read through the topic again or ask your teacher for help.
- Use an exercise book to make notes as you go along. This will help you remember what you have learned.

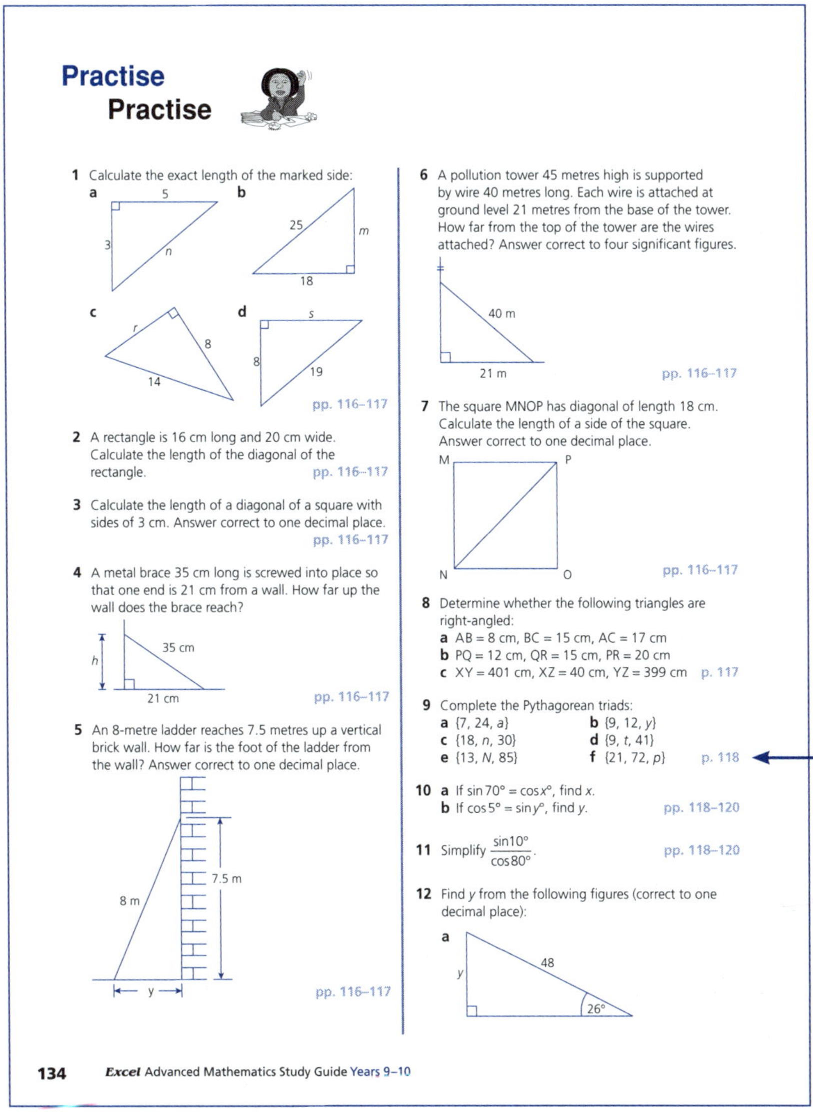

Practise Practise

1 Calculate the exact length of the marked side:

a (5, 3, n) **b** (25, 18, m)

c (r, 8, 14) **d** (s, 8, 19)

pp. 116–117

2 A rectangle is 16 cm long and 20 cm wide. Calculate the length of the diagonal of the rectangle. pp. 116–117

3 Calculate the length of a diagonal of a square with sides of 3 cm. Answer correct to one decimal place. pp. 116–117

4 A metal brace 35 cm long is screwed into place so that one end is 21 cm from a wall. How far up the wall does the brace reach?

(h, 35 cm, 21 cm)

pp. 116–117

5 An 8-metre ladder reaches 7.5 metres up a vertical brick wall. How far is the foot of the ladder from the wall? Answer correct to one decimal place.

(8 m, 7.5 m, y)

pp. 116–117

6 A pollution tower 45 metres high is supported by wire 40 metres long. Each wire is attached at ground level 21 metres from the base of the tower. How far from the top of the tower are the wires attached? Answer correct to four significant figures.

(40 m, 21 m)

pp. 116–117

7 The square MNOP has diagonal of length 18 cm. Calculate the length of a side of the square. Answer correct to one decimal place.

(M, P, N, O)

pp. 116–117

8 Determine whether the following triangles are right-angled:
a AB = 8 cm, BC = 15 cm, AC = 17 cm
b PQ = 12 cm, QR = 15 cm, PR = 20 cm
c XY = 401 cm, XZ = 40 cm, YZ = 399 cm p. 117

9 Complete the Pythagorean triads:
a $\{7, 24, a\}$ **b** $\{9, 12, y\}$
c $\{18, n, 30\}$ **d** $\{9, t, 41\}$
e $\{13, N, 85\}$ **f** $\{21, 72, p\}$ p. 118

10 **a** If $\sin 70° = \cos x°$, find x.
b If $\cos 5° = \sin y°$, find y. pp. 118–120

11 Simplify $\frac{\sin 10°}{\cos 80°}$. pp. 118–120

12 Find y from the following figures (correct to one decimal place):

a (y, 48, 26°)

134 *Excel* Advanced Mathematics Study Guide Years 9–10

- When you have worked through all topics in the chapter, try the **Practise Practise** section. Quick answers and worked solutions for this section may be found at the back of the book, with page references to answers and solutions at the end of the section. The page reference on the right of each question tells you where each concept is explained in the chapter.

Page reference to explanations in the chapter

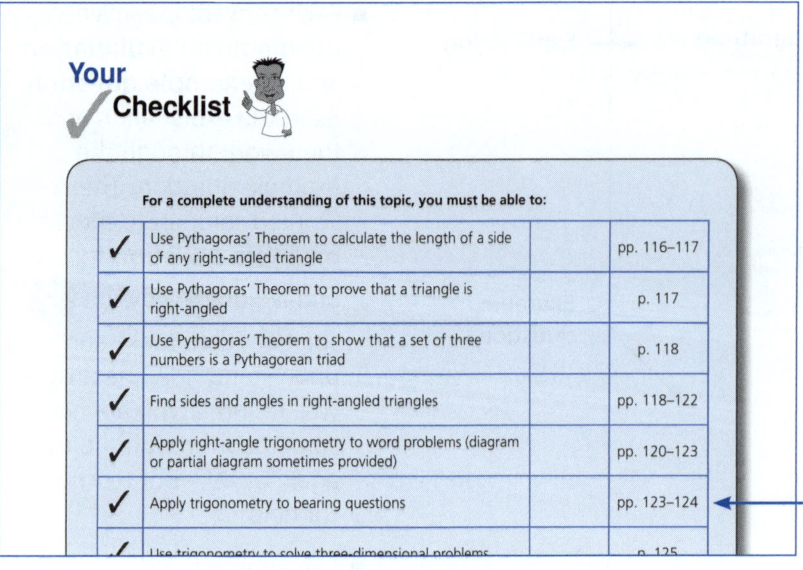

Your Checklist

For a complete understanding of this topic, you must be able to:

✓	Use Pythagoras' Theorem to calculate the length of a side of any right-angled triangle		pp. 116–117
✓	Use Pythagoras' Theorem to prove that a triangle is right-angled		p. 117
✓	Use Pythagoras' Theorem to show that a set of three numbers is a Pythagorean triad		p. 118
✓	Find sides and angles in right-angled triangles		pp. 118–122
✓	Apply right-angle trigonometry to word problems (diagram or partial diagram sometimes provided)		pp. 120–123
✓	Apply trigonometry to bearing questions		pp. 123–124
✓	Use trigonometry to solve three-dimensional problems		p. 125

- Go through the **Checklist** near the end of the chapter to ensure that you understand the key concepts of the chapter. Put a tick next to each checklist item that you have covered. You need to be able to tick every item to be able to do the tests. If you have missed any items, go through the chapter to ensure that you cover the relevant topics. The page reference next to each item tells you where the concept is explained.

Testing times ...

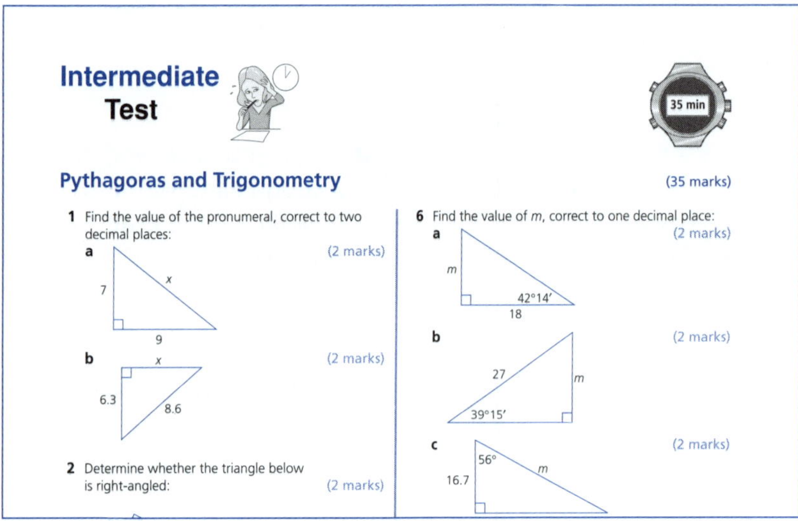

Intermediate Test

35 min

Pythagoras and Trigonometry (35 marks)

1 Find the value of the pronumeral, correct to two decimal places:

a (2 marks)

b (2 marks)

2 Determine whether the triangle below is right-angled: (2 marks)

6 Find the value of m, correct to one decimal place:

a (2 marks)

b (2 marks)

c (2 marks)

- Next, attempt both **Tests**. The tests are of two levels of difficulty: the intermediate tests are of average difficulty and the advanced tests are above average. Start with the intermediate test. If you find it easy, it will still be good practice before you attempt the advanced test. Complete all the questions you can—even if you find them difficult. Because there are worked solutions to each question at the back of the book, you will always be able to find out how to obtain each answer.

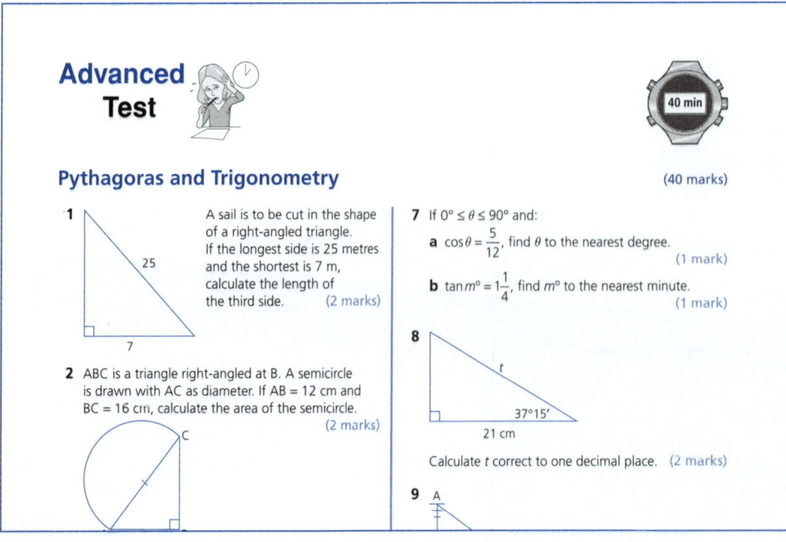

Advanced Test

40 min

Pythagoras and Trigonometry (40 marks)

1 A sail is to be cut in the shape of a right-angled triangle. If the longest side is 25 metres and the shortest is 7 m, calculate the length of the third side. (2 marks)

2 ABC is a triangle right-angled at B. A semicircle is drawn with AC as diameter. If AB = 12 cm and BC = 16 cm, calculate the area of the semicircle. (2 marks)

7 If $0° \le \theta \le 90°$ and:

a $\cos\theta = \frac{5}{12}$, find θ to the nearest degree. (1 mark)

b $\tan m° = 1\frac{1}{4}$, find $m°$ to the nearest minute. (1 mark)

8

Calculate t correct to one decimal place. (2 marks)

9

- The tests are designed to **prepare you for your school tests or examinations** so try to keep to the time on the stopwatch at the start of the test. Avoid looking at your textbook or notes while doing the test—this will help you develop your examination technique.
- **Marks** are allocated for each question. These are similar to the marks you will be trying for in your school tests and exams. Spend more time on the questions that are worth the most marks. For example, if there are 20 marks in total, and 20 minutes have been allocated for completion of the test, then spend about one minute on each mark.

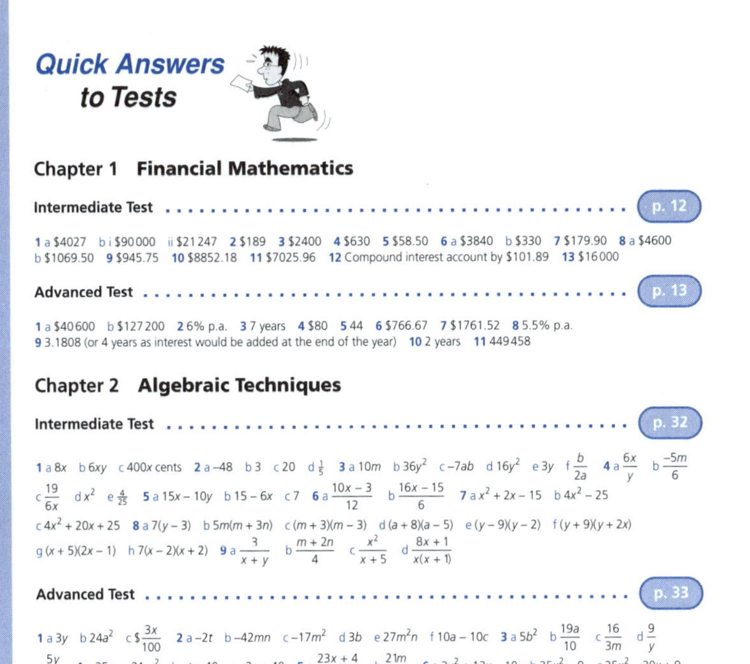

Quick Answers to Tests

Chapter 1 Financial Mathematics

Intermediate Test .... p. 12

1 a $4027 b i $90000 ii $21247 2 $189 3 $2400 4 $630 5 $58.50 6 a $3840 b $330 7 $179.90 8 a $4600 b $1069.50 9 $945.75 10 $8852.18 11 $7025.96 12 Compound interest account by $101.89 13 $16000

Advanced Test .... p. 13

1 a $40600 b $127200 2 6% p.a. 3 7 years 4 $80 5 44 6 $766.67 7 $1761.52 8 5.5% p.a. 9 3.1808 (or 4 years as interest would be added at the end of the year) 10 2 years 11 449458

Chapter 2 Algebraic Techniques

Intermediate Test .... p. 32

1 a $8x$ b $6xy$ c $400x$ cents 2 a -48 b 3 c 20 d $\frac{1}{3}$ 3 a $10m$ b $36y^2$ c $-7ab$ d $16y^2$ e $3y$ f $\frac{b}{2a}$ 4 a $\frac{6x}{y}$ b $\frac{-5m}{6}$ c $\frac{19}{6x}$ d x^2 e $\frac{4}{25}$ 5 a $15x-10y$ b $15-6x$ c 7 6 a $\frac{10x-3}{12}$ b $\frac{16x-15}{6}$ 7 a $x^2+2x-15$ b $4x^2-25$ c $4x^2+20x+25$ 8 a $7(y-3)$ b $5m(m+3n)$ c $(m+3)(m-3)$ d $(a+8)(a-5)$ e $(y-9)(y-2)$ f $(y+9)(y+2x)$ g $(x+5)(2x-1)$ h $7(x-2)(x+2)$ 9 a $\frac{3}{x+y}$ b $\frac{m+2n}{4}$ c $\frac{x^2}{x+5}$ d $\frac{8x+1}{x(x+1)}$

Advanced Test .... p. 33

1 a $3y$ b $24a^2$ c $\$\frac{3x}{100}$ 2 a $-2t$ b $-42mn$ c $-17m^2$ d $3b$ e $27m^2n$ f $10a-10c$ 3 a $5b^2$ b $\frac{19a}{10}$ c $\frac{16}{3m}$ d $\frac{9}{y}$ e $\frac{5y}{...}$ 4 a $35m-21m^2$ b $-t+40$ c $-3a+40$ 5 a $\frac{23x+4}{...}$ b $\frac{21m}{...}$ 6 a $3x^2+13x-10$ b $25x^2-9$ c $25x^2-30x+9$

Worked Solutions to Tests

Chapter 1 Financial Mathematics

Intermediate Test .... p. 12

1 a Tax = 3572 + 0.325 × (38400 − 37000)
= 4027
∴ Tax of $4027. ✓

b i Taxable income = 100000 − 10000
= 90000
∴ Taxable income of $90000. ✓

ii Tax = 17547 + 0.37 × (90000 − 80000)
= 21247
∴ Tax of $21247. ✓

2 $I = Prn$
= 700 × 0.09 × 3
= 189
∴ Interest of $189. ✓

3 $I = Prn$
= 6400 × 0.075 × 5
= 2400
∴ Interest of $2400. ✓

4 $I = Prn$
= 4200 × 0.05 × 3
= 2400
∴ Interest of $630.
4200 + 630 = 4380
∴ Amount repaid is $4380. ✓

5 $I = Prn$
= 650 × 0.0075 × 12 ✓
= 58.5
∴ Interest of $58.50. ✓

6 a $I = Prn$
= 12000 × 0.08 × 4
= 3840
∴ Interest of $3840. ✓

b Total repayments = 12000 + 3840
= 15840
Monthly repayments = 15840 ÷ 48
= 330
∴ Repayments of $330 each month. ✓

7 New price = 249 × 0.85 × 0.85 ✓
= 179.9025
= 179.90 (2 dec. pl.)
∴ New price is $179.90. ✓

8 a Deposit = 0.1 × 46000
= 4600
∴ Deposit is $4600. ✓

b Balance owing = $41400
Interest = 41400 × 0.06 × 4
= 9936 ✓
Total repayments = 41400 + 9936
= 51336 ✓
Monthly repayment = 51336 ÷ 48
= 1069.5
∴ Instalment of $1069.50. ✓

9 $A = P(1+r)^n$
= 6000(1.05)^3
= 6945.75 ✓
Interest = 6945.75 − 6000
= 945.75
∴ Compound interest of $945.75. ✓

10 $A = P(1+r)^n$
= 4500(1.07)^{10}
= 8852.181108...
= 8852.18 (2 dec. pl.)
∴ Compound interest of $8852.18. ✓

11 $A = P(1-r)^n$
= 20000(0.84)^6
= 7025.960632...
= 7025.96 (2 dec. pl.)
∴ Value of car is $7025.96. ✓

12 $I = Prn$
= 2000 × 0.05 × 15
= 1500 ✓
$A = P(1+r)^n$
= 2000(1.04)^{15}
= 3601.887011...
= 3601.89
Comp. int. = 3601.89 − 2000
= 1601.89 ✓
Difference = 1601.89 − 1500
= 101.89
∴ Compound interest account by $101.89. ✓

13 $A = P(1-r)^n$
8352.1 = P(0.85)^4 ✓
$P = \frac{8352.1}{0.85^4}$
= 16000
∴ The jet ski cost $16000. ✓

Advanced Test .... p. 13

1 a $4742 tax means the taxable income is between $37000 and $80000.
Amount over 3572 = 4742 − 3572
= 1170
As 1170 ÷ 0.325 = 3600, the income is $3600 more than $37000.
∴ Taxable income is $40600. ✓

b $35011 tax means the taxable income is between $80000 and $180000.
Amount over 17547 = 35011 − 17547
= 17464
As 17464 ÷ 0.37 = 47200, the income is $47200 more than $80000.
∴ Taxable income is $127200. ✓

2 $I = Prn$
1536 = 6400 × r × 4 ✓
= 25600r
$r = \frac{1536}{25600}$
= 0.06
∴ Interest rate of 6% p.a. ✓

3 $I = Prn$
1071 = 3600 × 0.0425 × n ✓
= 153n
$n = \frac{1071}{153}$
= 7
∴ Olivia invested for 7 years. ✓

4 New price = 200 × 0.75 × 0.8
= 120 ✓
Total discount = 200 − 120
= 80
∴ Total discount of $80. ✓

5 As 0.7 × 0.8 = 0.56, and 1 − 0.46 = 0.44, the discounts is 44%, ∴ P is 44 ✓✓
OR: Let original price be $100
New price = 100 × 0.7 × 0.8
= 56 ✓
Total discount = 100 − 56
= 44
Total % discount = 44%
∴ P is 44. ✓

6 Deposit = 0.25 × 32000
= $8000 ✓
Balance owing = $24000
Interest = 24000 × 0.05 × 3
= 3600 ✓
Total repayments = 24000 + 3600
= 27600
Monthly repayment = 27600 ÷ 36
= 766.666666...
= 766.67 (2 dec. pl.)
∴ Instalment of $766.67. ✓

7 4% p.a. is 1% per quarter, for 20 quarters
$A = P(1+r)^n$
= 8000(1.01)^{20}
= 9761.52032...
= 9761.52 (2 dec. pl.) ✓
Interest = 9761.52 − 8000
= 1761.52
∴ Interest of $1761.52. ✓

Worked Solutions to Tests 361

- **Mark your work** by referring to the answers at the back of the book. The **Quick answers** section allow you to see straight away if you have the right answer—only the answers are shown. The **Worked solutions** section is also found at the back of the book and sets out the solutions to the question in full, so look at this section if your answer was incorrect. The **ticks** in worked solutions indicate those parts of the working which receive marks. Therefore, even if your answer is wrong, you may be entitled to some marks for the question. Compare the worked solution to your own working to find out if you are entitled to any marks for the question.
- Write your own score in the **Your Feedback** space at the end of the test. From this you will be able to work out your percentage score. If your score for the test was less than 50%, you will need to work through the chapter again. If you are still having trouble, ask your teacher to explain those topics that you can't understand.

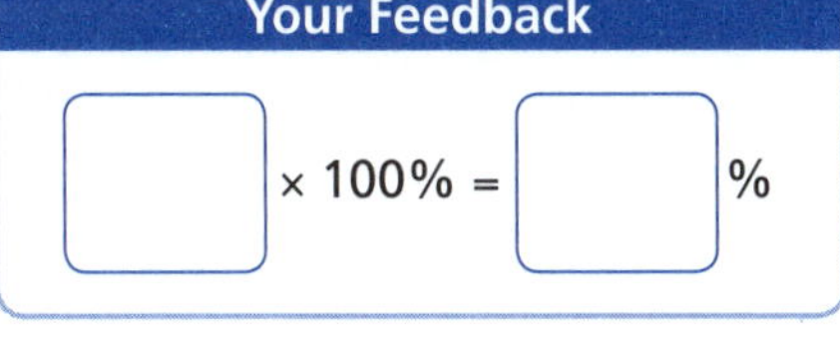

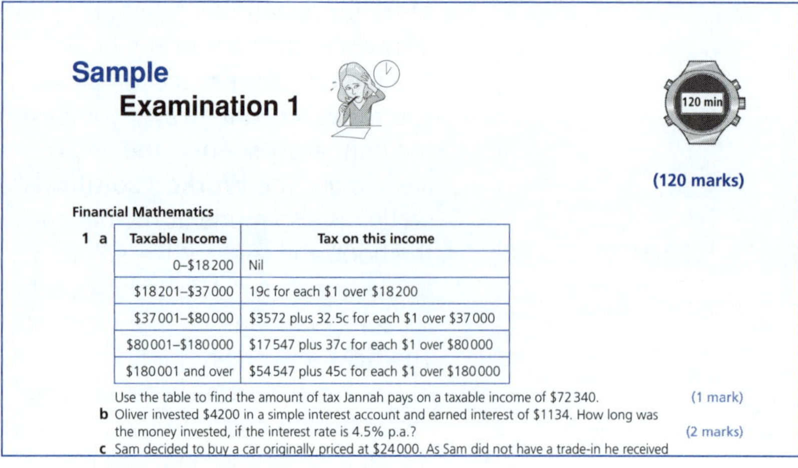

Sample Examination 1

120 min

(120 marks)

Financial Mathematics

1 a

Taxable Income	Tax on this income
0–$18 200	Nil
$18 201–$37 000	19c for each $1 over $18 200
$37 001–$80 000	$3572 plus 32.5c for each $1 over $37 000
$80 001–$180 000	$17 547 plus 37c for each $1 over $80 000
$180 001 and over	$54 547 plus 45c for each $1 over $180 000

Use the table to find the amount of tax Jannah pays on a taxable income of $72 340. (1 mark)

b Oliver invested $4200 in a simple interest account and earned interest of $1134. How long was the money invested, if the interest rate is 4.5% p.a.? (2 marks)

c Sam decided to buy a car originally priced at $24 000. As Sam did not have a trade-in he received

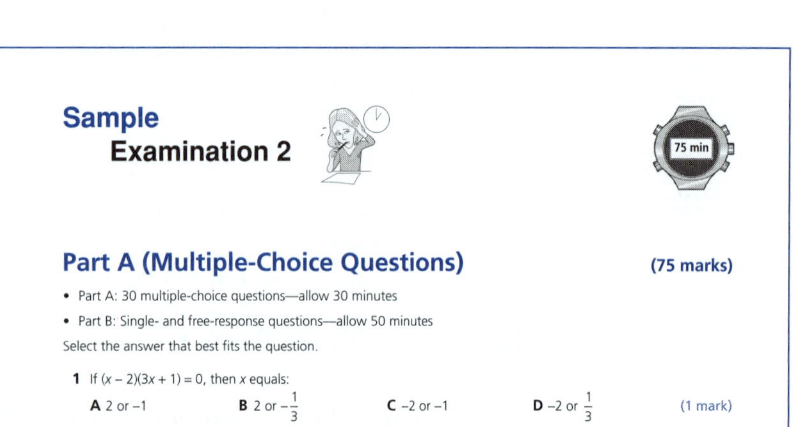

Sample Examination 2

75 min

Part A (Multiple-Choice Questions)

(75 marks)

- Part A: 30 multiple-choice questions—allow 30 minutes
- Part B: Single- and free-response questions—allow 50 minutes

Select the answer that best fits the question.

1 If $(x - 2)(3x + 1) = 0$, then x equals:

A 2 or -1 **B** 2 or $-\frac{1}{3}$ **C** -2 or -1 **D** -2 or $\frac{1}{3}$ (1 mark)

- When you have completed all of the chapters you can try the **Sample Examination Papers**. The Sample Examination Papers cover the entire course's work, so you should only attempt the papers if you have covered all the topics in this book. (Ask your teacher if you are not sure.) The Sample Examination Papers will provide you with good practice for your final examination. Set aside the time allowed for each paper and complete them under exam conditions.

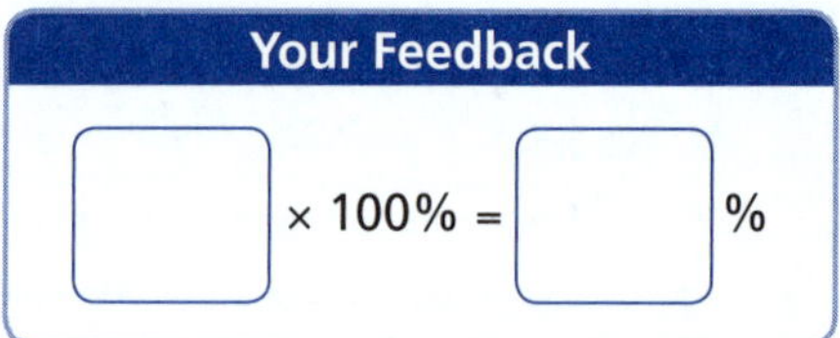

- Write your own score in the box at the end of the exam paper. From this you will be able to work out your percentage score.

Some general advice on succeeding in Mathematics

- Mathematics is a subject that **builds on knowledge**. Often you need to master one topic in order to be able to understand a later topic. This is why you should work through the book from beginning to end.
- Many students fail to **read questions carefully**. This accounts for a large number of the mistakes in exams. Although your time is limited, you must still take enough time to read each question carefully. If necessary, read the question a second (or even a third) time. Don't start your answer until you are sure you understand the question. For the longer questions, take a moment or two to **plan** your answer.
- Take care with your working. Write all answers down the page and avoid skipping any steps, as this often leads to errors. Your teachers like to see all your working, so it is wise to get into good habits. For questions that require a lot of working (such as equations), **write down each step of the solution on a separate line, lining up any equal signs** (=). This helps your teachers follow your working and will also help you check your work.
- Set out all of your working, because in Mathematics you may get some marks for your working, even if your answer is wrong.
- Always work through the solutions to any questions you answer incorrectly. This helps you learn. Remember: if you have attempted the questions, you are entitled to look at the answers—this is not cheating. If you don't get feedback, you won't be able to improve!
- You cannot expect to score 100% all the time. Remember that this book has been designed to help you identify your strengths as well as your weaknesses, so it is OK to make mistakes. The key to success is to learn from those mistakes.

Chapter 1
Financial Mathematics

Income tax

A tax table provided by the Australian Taxation Office (ATO) is used by workers to calculate the tax payable on their taxable incomes. It contains *tax brackets*, which represent different ranges of incomes.

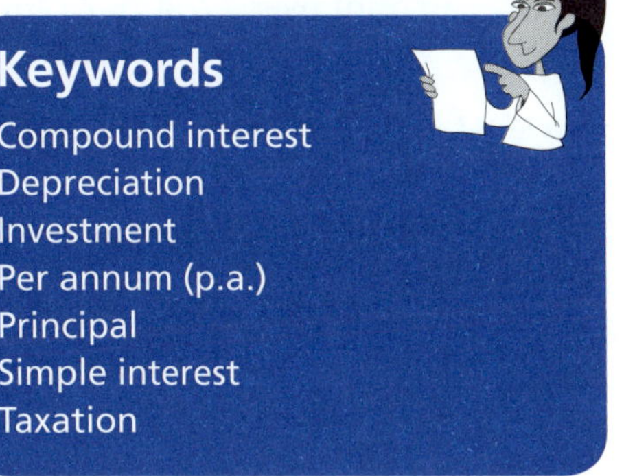

The table below allows taxpayers to work out their tax payable for the financial year:

Taxable income	Tax on this income
\$0–\$18 200	Nil
\$18 201–\$37 000	19c for each \$1 over \$18 200
\$37 001–\$80 000	\$3572 plus 32.5c for each \$1 over \$37 000
\$80 001–\$180 000	\$17 547 plus 37c for each \$1 over \$80 000
\$180 001 and over	\$54 547 plus 45c for each \$1 over \$180 000

1 Paul's taxable income is \$43 500. Use the above table to calculate the amount of tax he must pay.

S Tax payable $= 3572 + 0.325 \times 6500$
$= 5684.5$

$\therefore$ Paul pays \$5684.50.

2 Susan's gross income is \$89 750. Her total deductions are \$3450.

a Find her taxable income.

S Taxable income = Gross income − Deductions
$= 89\,750 - 3450$
$= 86\,300$

$\therefore$ \$86 300

b Calculate the amount of tax she must pay.

S Tax payable
$= 17\,547 + 0.37 \times (86\,300 - 80\,000)$
$= 19\,878$

$\therefore$ Susan pays \$19 878.

c If she pays \$385 per week in tax, how much refund will she receive for the year?

S Annual tax paid $= 385 \times 52$
$= 20\,020$

Refund $= 20\,020 - 19\,878$
$= 142$

$\therefore$ Susan receives \$142 back from tax office.

3 Sooni pays \$1482 tax. What was his taxable income?

S From the table, Sooni must have a taxable income between \$18 201 and \$37 000.

As $1482 \div 0.19 = 7800$, Sooni earns \$7800 more than \$18 200.

$\therefore$ He earns \$26 000.

4 If Sarah paid \$23 467 in tax, what was her taxable income?

S \$23 467 tax means the taxable income is between \$80 001 and \$180 000.

Amount over 17 547 = 23 467 − 17 547
= 5920

As 5920 ÷ 0.37 = 16 000, the income is \$16 000 more than \$80 000.

∴ Sarah's taxable income is \$96 000.

5 Kerrie pays twice as much tax as Terry. If Terry's taxable income is \$70 000, find Kerrie's taxable income.

S Terry: Tax payable
= 3572 + 0.325 × (70 000 − 37 000)
= 14 297

∴ Terry pays \$14 297 tax.

∴ Kerry pays \$28 594 tax.

Kerrie: From the table, she must have a taxable income between \$80 001 and \$180 000.

Firstly, 28 594 − 17 547 = 11 047

Also, 11 047 ÷ 0.37 = 29 856.76 (2 dec. pl.)

∴ Kerrie earns \$29 856.76 more than \$80 000.

∴ She earns \$109 856.76.

Simple interest

Interest earned when investing at a simple (or flat) interest rate is found by the formula:

$I = Prn$

where I = interest charged (\$)
P = principal (amount invested) (\$)
r = percentage interest rate, expressed as a decimal
n = number of periods.

When borrowing money, the same formula can be used to find the extra money paid for the loan (the interest charged on the loan).

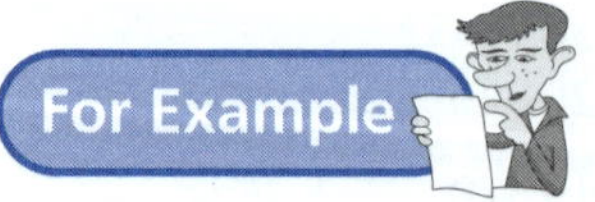

1 Calculate the amount of interest earned if \$5000 is invested for 3 years at $8\frac{1}{2}\%$ p.a.

S $I = Prn$ where $P = 5000$, $r = 0.085$, $n = 3$
= 5000 × 0.085 × 3
= 1275
∴ \$1275

2 Tom borrows \$3500 from a bank to buy a car. He has to pay it back, plus 8% p.a. simple interest over 36 months. How much interest does the bank charge Tom?

S $I = Prn$ where $P = 3500$, $r = 0.08$, $n = 3$
= 3500 × 0.08 × 3
= 840
∴ \$840

Therefore, the bank charges Tom \$840 interest over 36 months.

3 Determine the interest rate if \$2300 invested at simple interest for 2 years earns \$402.50 interest.

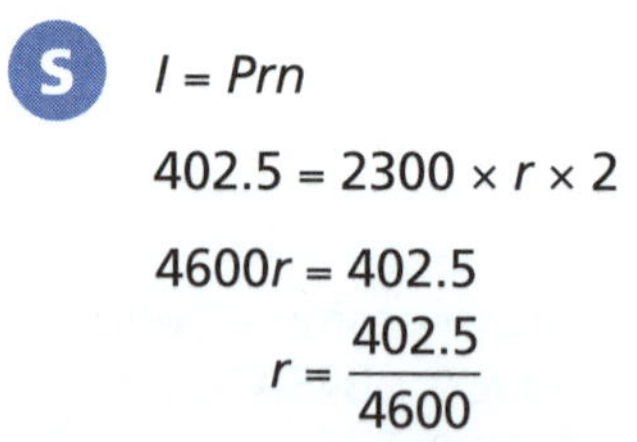

S $I = Prn$

$402.5 = 2300 \times r \times 2$

$4600r = 402.5$

$r = \frac{402.5}{4600}$

$r = 0.0875$

∴ The interest rate is 8.75% p.a.

4 How long is $7000 invested at 6.5% p.a. to earn simple interest of $1365?

S $I = Prn$

$1365 = 7000 \times 0.065 \times n$

$455n = 1365$

$n = \frac{1365}{455}$

$n = 3$

∴ The investment is for 3 years.

Loan repayments

A loan repayment is found by first subtracting the **deposit** from the **cash price** to find the **balance owing**. Interest is then calculated over the length of the loan and added to the balance owing to find the **total repayments**. This is then divided by the number of repayments to find the size of each repayment.

1 Susan borrows $5400 from a bank to buy a car. She has to pay it back, plus $9\frac{1}{2}$% p.a. simple interest over a 5-year period.

a How much extra money does Susan have to pay the bank?

S The extra amount she has to repay equals the interest charged by the bank.

$I = Prn$ $\quad P = 5400$

$= 5400 \times 0.095 \times 5$ $\quad r = 0.095$

$= 2565$ $\quad n = 5$

Therefore, Susan has to repay $2565 extra to the bank.

b How much does she have to repay altogether?

S The total amount to be repaid

= Amount borrowed + Interest

= 5400 + 2565

= 7965

∴ $7965

c If Susan repays the loan in monthly repayments over the 5-year period, how much does she pay per month?

Note: 5 years = 60 months

S Monthly repayment

= Total amount to be repaid ÷ Number of months

= 7965 ÷ 60

= 132.75

Therefore, Susan pays an instalment of $132.75 per month for 5 years to pay back the loan.

2 Albert borrows $12 000 from a building society to buy a boat. He is charged 7.25% p.a. simple interest over a 4-year period. What will be his monthly repayment?

S Interest = $12\,000 \times 0.0725 \times 4$

= 3480

Total repayment = 12 000 + 3480

= 15 480

Monthly repayment = 15 480 ÷ 48

= 322.5

∴ The monthly repayment will be $322.50.

Time payment

Time payment is used to buy goods without paying the cost immediately. Interest is charged on the balance owing after a deposit (or trade-in) is taken off the cost price. Repayments are usually weekly or monthly, spread over a period of time.

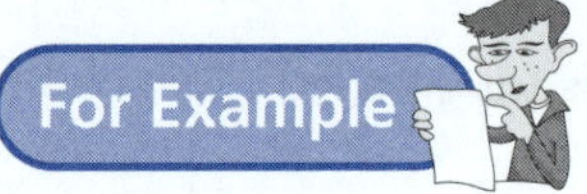

1 A lounge suite worth \$1850 is bought under a time-payment agreement. The terms required are a deposit of \$110 and payment of 24 equal monthly instalments of \$96.

a Find the total cost of the lounge suite.

S Total price = Deposit + Instalments
= $110 + 24 \times 96$
= $110 + 2304$
= 2414

The total cost of the lounge suite is \$2414.

Note: the extra amount paid is the interest charged.

b How much do you save by paying cash?

S Interest = Total price − Cash price
= $2414 - 1850$
= 564

Therefore, if you pay cash you will save \$564.

2 A car can be obtained for \$25 840 cash, or through a time-payment agreement involving a deposit of 15% of the cash price and 36 monthly instalments of \$875.50. Find the total cost of the car using time payments.

S Deposit = 15% of 25 840
$= \frac{15}{100} \times 25\,840$
$= 3876$

Total price = Deposit + Instalments
= $3876 + 36 \times 875.50$
= $35\,394$

The total cost of the car using time payments is \$35 394.

Successive discounts

Sometimes retailers may offer discounts on items that have already been discounted in price. If an item is discounted in price by 15% it means the purchaser pays 85% of the original price. A second discount would not be added to the original percentage discount, but instead would be applied to the discounted price.

1 A car is discounted by 20%, but after a hailstorm damaged the car it dropped in price by another 30%. If the car was originally priced at \$28 000, what is the new price?

S New price = $0.80 \times 0.70 \times 28\,000$
= $15\,680$

∴ The car's new price is \$15 680.

2 All power tools have been dropped in price by 15%. If a further $7\frac{1}{2}$% discount is given for cash, how much will be saved when purchasing a drill originally priced at \$96?

S New price = $0.85 \times 0.925 \times 96$
= 75.48

∴ Savings = $96 - 75.48$
= 20.52

∴ Savings of \$20.52.

3 A television is discounted by 25% and then by a further 20%. What is the total discount, expressed as a percentage?

S Let the price of the television be \$100.

New price = $0.75 \times 0.80 \times 100$
= 60

∴ Price is \$60, total discount is \$40, so the discount is 40%.

4 The price of a motorbike drops by 20% and a further discount of 10% is available for customers who pay cash. If the final price is \$14 400, what was the original price?

S Original price = $14400 \div 0.80 \div 0.90$
= 20 000

∴ Motorbike originally priced at \$20 000.

Compound interest

Compound interest is more realistic than simple interest, and assumes depositors will gain interest on their interest as their money lies untouched in their account, rather than withdrawing interest on the day it is earned.

For example, \$1000 gaining compound interest of 12% per annum over 3 years can be calculated the following way:

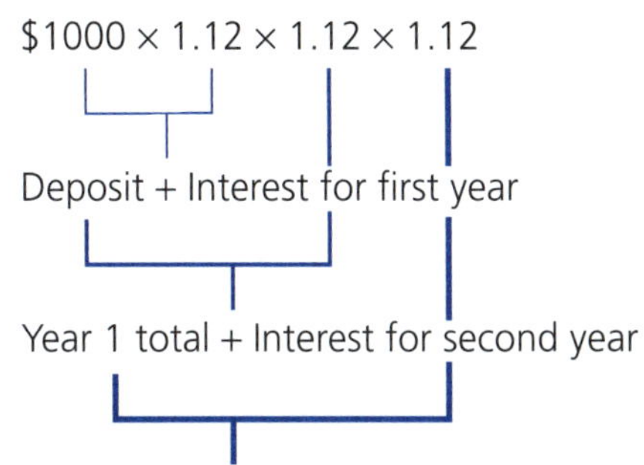

Year 2 total + Interest for third year.

Remember: to increase \$1000 by 12% you could find 12% of \$1000 and add to \$1000, but it is simply easier to multiply \$1000 × 1.12.

That is, total after 3 years = $1000 \times 1.12 \times 1.12 \times 1.12$
= 1000×1.123
= 1404.93 (to the nearest cent)

Compound interest = 1404.93 − 1000 ← \$1000 is the original deposit.
= 404.93 ∴ \$404.93

We can use the formula:

$A = P(1 + r)^n$

where A = accumulated balance
P = principal (amount invested)
r = percentage interest rate per period, expressed as a decimal
n = number of periods.

1 Terry invested \$42 000 in a savings account that attracted an 8.5% interest rate compounded annually.

Find out how much money Terry has in the account after 4 years.

$A = P(1 + r)^n$
$= 42000\,(1 + 0.085)^4$
$= 42000 \times 1.085^4$
= 58 206.07 (to the nearest cent)

Terry has \$58 206.07 in his savings account.

2 Mercia works as a manager of a clothing factory. Her pay conditions involve an increase of 6% every year. If this year she has a salary of \$48 320, what will her salary be in 7 years' time?

S
$$\begin{aligned} A &= P(1+r)^n \\ &= 48\,320(1+0.06)^7 \\ &= 48\,320 \times 1.06^7 \\ &= 72\,655.41 \text{ (to the nearest cent)} \end{aligned}$$

Mercia will be paid a salary of \$72 655.41.

3 Megan can choose between two accounts: the first offering simple interest at a rate of 12% and the second compound interest at a rate of 11%. If she had \$400 to invest for 3 years, which account should she use to gain the greater amount of interest?

S
$$\begin{aligned} \text{Simple interest} &= 400 \times 0.12 \times 3 \\ &= 144 \quad \therefore \$144 \end{aligned}$$

For compound interest:

$$\begin{aligned} A &= 400(1.11)^3 \\ &= 547.05 \text{ (to 2 dec. pl.)} \quad \therefore \$547.05 \end{aligned}$$

$$\begin{aligned} \text{Compound interest} &= \$547.05 - \$400 \\ &= \$147.05 \end{aligned}$$

Megan would gain more interest by using the simple interest account.

4 Noel deposits \$1000 in an account that offers interest at 6% per annum *compounded monthly*. He invests the money for 2 years. Find his balance after 2 years.

S As the interest is determined monthly, there will be 24 payments over the 2 years. Also, the rate is 6% per annum, that is:

$$\frac{6}{12} = 0.5\% \text{ per month}$$

Note: 0.5% = 0.005

$$\begin{aligned} A &= P(1+r)^n \\ &= 1000(1+0.005)^{24} \\ &= 1000(1.005)^{24} \\ &= 1127.16 \end{aligned}$$

Noel will have \$1127.16 in his account after 2 years.

5 Oliver invested a sum of money in an account earning compound interest of 4.5% per annum. After 8 years the balance in the account was \$16 396. How much did Oliver originally invest, to the nearest dollar?

S
$$\begin{aligned} A &= P(1+r)^n \\ 16\,396 &= P(1.045)^8 \\ P &= \frac{16\,395}{1.045^8} \\ &= 11\,529.423\,34\ldots \\ &= 11\,529 \text{ (nearest whole)} \end{aligned}$$

$\therefore$ Oliver originally invested \$11 529.

6 Lucas earned compound interest of \$807 on an investment of \$5400 over a period of 7 years. What was the rate of interest? Give your answer to the nearest percentage.

S
$$\begin{aligned} A &= 5400 + 807 \\ &= 6207 \end{aligned}$$

$\therefore$ The compounded amount is \$6207.

$$\begin{aligned} A &= P(1+r)^n \\ 6207 &= 5400(1+r)^7 \\ (1+r)^7 &= \frac{6207}{5400} \\ &= \frac{6207}{5400}^{\frac{1}{7}} \\ 1+r &= 1.020\,096\,226\ldots \\ r &= 0.020\,096\,226\ldots \\ &= 0.02 \text{ (to 2 dec. pl.)} \end{aligned}$$

$\therefore$ The rate of interest was 2%.

7 Mia invested \$4000 in an account earning compound interest of 5% per annum. After a period of time the balance in the account was \$5360.38. How long did Mia invest her money? (Answer to nearest year.)

S
$$\begin{aligned} A &= P(1+r)^n \\ 5360.38 &= 4000(1.05)^n \\ 1.05^n &= \frac{5360.38}{4000} \\ &= 1.340\,095 \end{aligned}$$

$n = 6$ (using a 'guess and check' strategy)

$\therefore$ Mia has invested for 6 years.

Depreciation

Whereas compound interest is calculating values that are increasing, depreciation is the opposite—that is, calculating values that are *decreasing*.

We can use the formula:

$A = P(1 - r)^n$

where A = final value
P = initial value
r = percentage interest rate as a decimal
n = number of time periods.

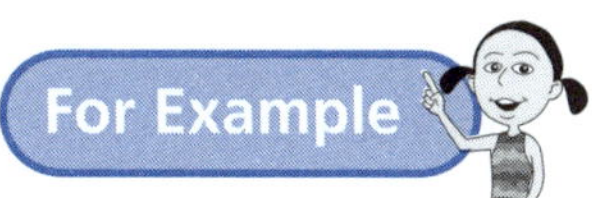

1 Lloyd buys a second-hand car for \$14400. If it depreciates in value at 13% per annum, find its value after 4 years.

S
$$\begin{aligned} A &= P(1 - r)^n \\ &= 14400(1 - 0.13)^4 \\ &= 14400(0.87)^4 \\ &= 8249.73 \text{ (to the nearest cent)} \end{aligned}$$

Lloyd's car is valued at \$8249.73.

2 Christine purchases a computer valued at \$2100. If it depreciates at 20% per annum, how much will it depreciate in its third year?

S $A = P(1 - r)^n$

Value after 2 years:
$$2100(1 - 0.2)^2 = 2100(0.8)^2 = 1344$$

Value after 3 years:
$$2100(1 - 0.2)^3 = 2100(0.8)^3 = 1075.20$$

Amount of depreciation in third year
= 1344 − 1075.20
= 268.80

The computer depreciates \$268.80 in its third year.

3 What is the annual rate of depreciation if a motorbike drops in price from \$12500 to \$5800 in 3 years? Give your answer to one decimal place.

S
$$\begin{aligned} A &= P(1 - r)^n \\ 5800 &= 12500(1 - r)^3 \\ (1 - r)^3 &= \frac{5800}{12500} \\ &= 0.464 \\ 1 - r &= \sqrt[3]{0.464} \\ &= 0.774175328 \\ r &= 1 - 0.774175328 \\ &= 0.225824671 \end{aligned}$$

∴ The rate of depreciation is 22.6%.

4 The value of a caravan depreciates at the rate of 25% per annum. The caravan was originally purchased on 1 January 2014. In what year will the value of the caravan drop to one-tenth of its original cost?

S
$$\begin{aligned} A &= P(1 - r)^n \\ 1 &= 10(1 - 0.25)^n \\ 0.75^n &= 0.1 \\ n &= 8 \quad \text{(using a 'guess and check' strategy)} \end{aligned}$$

∴ It will be one-tenth in 2022.

Practise Practise

1

Taxable income	Tax on this income
\$0–\$18 200	Nil
\$18 201–\$37 000	19c for each \$1 over \$18 200
\$37 001–\$80 000	\$3572 plus 32.5c for each \$1 over \$37 000
\$80 001–\$180 000	\$17 547 plus 37c for each \$1 over \$80 000
\$180 001 and over	\$54 547 plus 45c for each \$1 over \$180 000

Find the tax payable on a taxable income of:
- **a** \$17 305
- **b** \$29 430
- **c** \$68 495
- **d** \$132 427
- **e** \$193 483 p. 1

2 Find the taxable income if the tax paid was:
- **a** \$912
- **b** \$24 947
- **c** \$198 547 pp. 1–2

3 Emma pays twice as much tax as Aiden. If Aiden's taxable income is \$59 950, find Emma's taxable income. pp. 1–2

4 Jacob's taxable income for the year was \$72 000.
- **a** Calculate his payable tax.
- **b** If Jacob was given a bonus of \$5000, how much of his bonus is paid in tax? pp. 1–2

5 Find the simple interest on:
- **a** \$1800 for 5 years at 7% p.a.
- **b** \$450 for 3 years at $8\frac{1}{2}$% p.a.
- **c** \$1200 for 18 months at 9% p.a.
- **d** \$680 for 3 months at 0.5% per month
- **e** \$10 000 for a year at 1.2% per quarter. p. 2

6 Find the annual interest rate of the following investments:
- **a** \$3500 over 4 years earns simple interest of \$840
- **b** \$6100 over 7 years earns simple interest of \$1921.50
- **c** \$10 000 over 6 years earns simple interest of \$2250
- **d** \$2700 over 4 years earns simple interest of \$410.40. p. 2

7 Find the amount of money invested if it earns simple interest of:
- **a** \$500 at 4% p.a. over 5 years
- **b** \$680.80 at 3.7% p.a. over 8 years
- **c** \$3570 at 4.25% p.a. over 10 years
- **d** \$280 at 0.4% per month for 7 months. p. 2

8 Find the length of time of the following investments:
- **a** \$2800 at 4% p.a. earning simple interest of \$336
- **b** \$5700 at 2.75% p.a. earning simple interest of \$940.50
- **c** \$3600 at $5\frac{1}{2}$% p.a. earning simple interest of \$1782
- **d** \$5000 at 0.3% per month earning simple interest of \$225. p. 3

9 Jack invests an amount of money at a simple interest rate of 5% per annum. How long will it take to double his money? p. 3

10 The Great Building Society offers the following interest rates for investments:
- 9% p.a. for a 1-year term
- 11% p.a. for a 2-year term
- $12\frac{1}{2}$% p.a. for a 3-year term.

If Jill invests \$1200 for a 2-year term, how much simple interest will she earn on her investment? p. 3

11 The cash price for a sofa was marked at \$1080. The Cox family bought it on time payment. They paid no deposit and \$49 per month for 2 years. What was the total amount they paid? p. 3

12 Vicky borrows \$20 000. She has to pay it back, plus 9% p.a. simple interest over 4 years.
- **a** How much must she repay altogether?
- **b** What is her monthly repayment? p. 3

13 Mr Smith borrowed \$280 000 to buy a house. The terms of his loan are \$1400 per month for 21 years.
- **a** Calculate the total amount Mr Smith will pay.
- **b** Calculate the total interest he will pay over the 21 years. p. 3

14 A bed valued at \$1495 is bought on terms of \$150 deposit and \$11.50 per week for 4 years.
- **a** What does the bed cost if bought on terms?
- **b** How much interest was charged? p. 4

15 The following terms are offered by four different retailers for the purchase of goods with a cash price of \$2400. Which one is the cheapest buy?
- **A** No deposit and \$250 per month for a year.
- **B** \$1500 deposit and \$100 per month for a year.
- **C** \$1000 deposit and \$150 per month for a year.
- **D** 20% deposit on the cash price and \$210 per month for a year. p. 4

16 The cash price for a bike is $240. Paul bought it on term payment over a 1-year period. A deposit of $60 was paid and 12% p.a. interest was charged on the balance owing. Calculate the:
a actual interest charged
b total amount to be repaid
c amount of each monthly instalment. p. 4

17 A television (cash price $670) is bought for a deposit of 20% of the cash price and monthly repayments to be made over $1\frac{1}{2}$ years. Interest of 12% is charged on the balance owing. Calculate the amount of each payment. p. 4

18 A furniture store held a '20% off everything' weekend sale. Shareholders in the store receive a further 10% discount. Find the price paid by a shareholder for a television originally priced at $1450. pp. 4–5

19 Which is the larger discount?
A Successive discounts of 15% and 25%.
B One discount of 35%. pp. 4–5

20 Topside steak usually costs $11.90/kg. It is discounted by 15%. Jake gains a further 10% discount when buying more than 5 kg of any product. Find the cost for Jake of 7.5 kg of topside steak. pp. 4–5

21 A travel website discounted all cruises by 10%. A further 10% reduction in price was offered to all repeat customers. Find the total discount. pp. 4–5

22 The price of a bracelet is discounted by 25%. If the new price is $150, what was the original price? pp. 4–5

23 The price of an outdoor setting was reduced by 5% but was not sold. A month later the store owner took another 5% off the marked price. If the final price was $902.50, what was the original price? pp. 4–5

24 Yesterday a sofa was priced at $1200. Today, and each day, the price is to be discounted 2% per day. What will be the price of the sofa on the seventh day, to the nearest dollar? pp. 4–5

25 The price of a car was reduced in early February by 18% and a further 12% was taken from the price in early March. If the car is valued at $25 977.60 in late March, what was the value in January? pp. 4–5

26 Find the compound interest on:
a $6400 at 6% p.a. over 3 years
b $4900 at $8\frac{1}{2}$ p.a. over 5 years
c $2600 at 6% p.a. over 2 years, compounded monthly
d $14 000 at 7% p.a. over 4 years, compounded 6 monthly. pp. 5–6

27 If $9400 is invested earning 8% p.a. compounded quarterly, how much is in the account after 10 years? pp. 5–6

28 The town of Singleton had a population in 1980 of 11 240 and it increased at the rate of 3% per year. Find the population in 2005. pp. 5–6

29 Find the original investment, to the nearest dollar, if the balance in an account is:
a $2920 after 5 years earning compound interest of 4% p.a.
b $5926 after 8 years earning compound interest of 3.5% p.a.
c $39 763 after 10 years earning compound interest of $4\frac{3}{4}$% p.a. pp. 5–6

30 Ava invested a sum of money in an account earning interest of 4.8% per annum compounded monthly. After 6 years the balance in the account was $26 660. How much did she originally invest, to the nearest dollar? pp. 5–6

31 Find the annual rate of interest, to one decimal place, if an original investment of:
a $2400 after 6 years earns $466 compound interest
b $8000 after 10 years earns $4424 compound interest
c $25 000 after 8 years earns $12 646 compound interest. pp. 5–6

32 Cameron invested $20 000 in an account for 15 years at an interest rate of 6% p.a. compounded monthly. However, after 10 years, the interest rate changed to 9% p.a. What was the total amount of interest earned in the 15 years? pp. 5–6

33 Brianna earned compound interest of $3223 on an investment of $9000 over a period of 8 years. What was the rate of interest? Give your answer to two significant figures. pp. 5–6

34 How long is $5000 invested if it earns:
a $2036 compound interest at 5% p.a.?
b $955 compound interest at 6% p.a.?
c $10 861 compound interest at 8% p.a.? pp. 5–6

35 Ben invested \$7500 in an account earning interest of 6% per annum, compounded monthly. After a period of time the balance in the account was \$33 487. How long did Ben invest his money, to nearest year? pp. 5–6

36 A computer is purchased for \$1600. If the value decreases by 15% each year, what is its value after 3 years? p. 7

37 A motorbike is purchased in 2010 for \$16 300. If it depreciates by 18% per annum, find its value in 2015. p. 7

38 What is the annual rate of depreciation if a taxi drops in price from \$42 500 to \$17 930 in 3 years? Give your answer to nearest percentage. p. 7

39 In 4 years a photocopier depreciates in value from \$21 000 to \$5644. What is the annual rate of depreciation, to nearest percentage? p. 7

40 The value of a motor home depreciates at the rate of 20% per annum. The caravan was originally purchased on 1 January 2015. In what year will the value of the caravan drop to one-quarter of its original cost? p. 7

Go to p. 299 for **Quick Answers** or to pp. 310–312 for **Worked Solutions**

For a complete understanding of this topic, you must be able to:

✓	Use a tax table to calculate the tax payable on a given taxable income		pp. 1–2
✓	Calculate simple interest and apply to problems involving investing money		pp. 2–3
✓	Solve problems involving loans and repayments		pp. 3–4
✓	Calculate the results of successive discounts		pp. 4–5
✓	Use the formula for compound interest and apply it to problems involving investing money		pp. 5–6
✓	Calculate depreciation by modifying the compound interest formula.		p. 7

Now you are ready to do the tests!

Intermediate Test

Financial Mathematics

(25 marks)

1

Taxable income	Tax on this income
\$0–\$18 200	Nil
\$18 201–\$37 000	19c for each \$1 over \$18 200
\$37 001–\$80 000	\$3572 plus 32.5c for each \$1 over \$37 000
\$80 001–\$180 000	\$17 547 plus 37c for each \$1 over \$80 000
\$180 001 and over	\$54 547 plus 45c for each \$1 over \$180 000

Use the table to answer the following questions:

a Tyler has a taxable income of \$38 400. How much tax will he pay? (1 mark)

b Regan is paid \$100 000 per year. His deductions total \$10 000.

  i What is his taxable income? (1 mark)

  ii How much tax will Regan have to pay? (1 mark)

2 Find the simple interest on \$700 at 9% p.a. over 3 years. (1 mark)

3 Find the amount of simple interest on \$6400 at $7\frac{1}{2}$% p.a. over 5 years. (1 mark)

4 Guy borrows \$4200 from his father to buy a secondhand motorbike. He was charged simple interest of 5% per annum over a period of 3 years. How much will be repaid to his father? (1 mark)

5 Find the amount of simple interest on \$650 at 0.75% per month for 1 year. (1 mark)

6 Natasha borrows \$12 000 from a credit union. She is charged a flat rate (simple) interest rate of 8% p.a. over a period of 4 years.

a How much interest is Natasha charged by the credit union? (1 mark)

b If Natasha repays the loan in monthly repayments, how much will she pay each month? (1 mark)

7 A heater was priced at \$249. It was discounted by 15% and then a further 10% discount during a weekend sale. What was the new price of the heater? (2 marks)

8 A car priced at \$46 000 is purchased by paying a deposit of 10% and the balance is paid on monthly instalments for 4 years at 6% p.a. simple interest. Find the:

a deposit (1 mark)

b amount of each monthly instalment. (2 marks)

9 Find the amount of interest on \$6000 at 5% p.a. compounded annually for 3 years. (2 marks)

10 What is the compounded amount if \$4500 is invested for 10 years at 7% p.a. compound interest? (2 marks)

11 Find the value of a car in 6 years' time if it is purchased for \$20 000 and loses its value at a rate of 16% per annum. (2 marks)

12 Teghan invests \$2000 in two different savings accounts for 15 years. The first account offers 5% per annum simple interest and the second account offers 4% compound interest. Which account provided more interest and by how much? (3 marks)

13 The value of a jet-ski depreciates at the rate of 15% per annum. After 4 years its value is \$8352.10. What was the original value of the jet-ski? (2 marks)

Your Feedback

☐ × 100% = ☐ %

QA PAGE 304

WS PAGE 361

Financial Mathematics

(25 marks)

1

Taxable income	Tax on this income
\$0–\$18 200	Nil
\$18 201–\$37 000	19c for each \$1 over \$18 200
\$37 001–\$80 000	\$3572 plus 32.5c for each \$1 over \$37 000
\$80 001–\$180 000	\$17 547 plus 37c for each \$1 over \$80 000
\$180 001 and over	\$54 547 plus 45c for each \$1 over \$180 000

Use the table to answer the following questions:

a Sophia paid \$4742 tax. What was her taxable income? (1 mark)

b If James paid \$35 011 in tax, what was his taxable income? (1 mark)

2 Find the interest rate if \$6400 is invested at simple interest for 4 years and earns \$1536. (2 marks)

3 Olivia invested \$3600 in a simple interest account and earned interest of \$1071. How long was the money invested, if the interest rate is 4.25% p.a.? (2 marks)

4 Sam decided to buy a bracelet originally priced at \$200. The shop owner reduced the price by 25% and then by another 20% when Sam offered cash. What was the total discount on the bracelet? (2 marks)

5 Sheridan found that successive discounts of 30% and 20% is the same as a single discount of P%. Find the value of P. (2 marks)

6 Travis purchased a \$32 000 car by paying a deposit of 25% and a simple interest charge of 5% p.a. on the balance. If he made monthly repayments for 3 years, how much did he pay each month? (3 marks)

7 What is the amount of compound interest earned on an investment of \$8000 for 5 years at 4% per annum, compounded quarterly? (2 marks)

8 If \$12 000 earns compound interest of \$6416 when invested for 8 years, what is the annual interest rate for the investment? (2 marks)

9 Kaylee and Joseph both inherited \$20 000 each from their grandmother's estate. Kaylee invested her money in an account offering 6% p.a. compound interest for 10 years. Joseph chose a different account offering 6% p.a. simple interest. How much longer will it take Joseph to earn the same amount of interest as Kaylee? (4 marks)

10 A lathe is purchased for \$360 000 and depreciates at the rate of 18% per annum. How long will it take to reduce in value by \$117 936? (2 marks)

11 The population of a city in 1995 was 500 000. In each of the following 10 years the population increased annually by 2%. In the decade from 2005, the population dropped annually by 3%. What is the population of the city in 2015? (2 marks)

Your Feedback

☐ × 100% = ☐ %

QA PAGE 304
WS PAGE 361

Chapter 2
Algebraic Techniques

Generalised arithmetic

To find the number of days in p hours, it may be helpful to replace temporarily the pronumerals with numbers: alter the question to ask how many days in 48 hours? The answer is $\frac{48}{24} = 2$, so that the answer to the original question is $\frac{p}{24}$.

Keywords

Algebraic	Factorise
Binomial	Monic
Common	Quadratic
Expand	Simplify
Expansion	Substitution
Expression	Trinomial
Factor	

For Example

1 The sum of x and y is: $x + y$

Sum: add
Product: multiply
Difference: subtract
Quotient: divide

2 The average of a, b and c is: $\frac{a + b + c}{3}$

$$\text{Average} = \text{Mean} = \frac{\text{Sum of scores}}{\text{No. of scores}}$$

3 The number 4 more than c is: $c + 4$

4 The next three consecutive whole numbers after x are: $x + 1$, $x + 2$ and $x + 3$.

Note: in generalised arithmetic, it can be helpful to substitute numbers for the pronumerals.

5 If y is odd, find the next three consecutive odd numbers.

S $y + 2, y + 4, y + 6$

All odd, and even, numbers are separated by two.

6 Convert:

a \$$y$ to cents

S $100 \times y = 100y$
\$$y = 100y$ cents

Try \$7
$\therefore 7 \times 100$
i.e. \$7 = 700c

b p litres to mL

S $p \times 1000 = 1000p$
p litres $= 1000p$ mL

Try 8 litres.
$\therefore 8 \times 1000 = 8000$ mL

c y minutes to hours.

S $y \div 60 = \frac{y}{60}$

y minutes $= \frac{y}{60}$ hours

Try 120 minutes.
$\therefore \frac{120}{60} = 2$ hours

7 Find the area of the rectangle:

3 cm

(2x + 4) cm

S $A = (2x + 4) \times 3$
$= 3(2x + 4)$

Area is $3(2x + 4)$ cm^2.

8 Find the change from \$5 if y cakes are purchased at k cents each.

S Cost $= y \times k$ or $k \times y$
$= yk$ cents or ky cents.

Change $= 500 - ky$
Change is $(500 - ky)$ cents.

Substitution into algebraic expressions

The pronumerals in algebraic expressions are replaced with numbers.

1 If $a = 3$, $b = 4$ and $c = -5$, evaluate:

a $ab + c$

S $ab + c = 3 \times 4 + (-5)$
$= 12 - 5$
$= 7$

b $b - c$

S $b - c = 4 - (-5)$
$= 4 + 5$
$= 9$

c $\dfrac{bc - 1}{a}$

S $\dfrac{bc - 1}{a} = \dfrac{4 \times -5 - 1}{3}$
$= \dfrac{-20 - 1}{3}$
$= \dfrac{-21}{3}$
$= -7$

d $c(a - b)$

S $c(a - b) = -5(3 - 4)$
$= -5(-1)$
$= 5$

e $b^2 + c^2$

S $b^2 + c^2 = (4)^2 + (-5)^2$
$= 16 + 25$
$= 41$

f $2c^2 - (2c)^2$

S $2c^2 - (2c)^2 = 2(-5)^2 - (2 \times -5)^2$
$= 2(25) - (-10)^2$
$= 50 - 100$
$= -50$

g $\sqrt{c^2 - b^2}$

S $\sqrt{c^2 - b^2} = \sqrt{(-5)^2 - 4^2}$
$= \sqrt{25 - 16}$
$= \sqrt{9}$
$= 3$

h $\dfrac{a}{b} - \dfrac{b}{c}$

S $\dfrac{a}{b} - \dfrac{b}{c} = \dfrac{3}{4} - \dfrac{4}{-5}$
$= \dfrac{3}{4} + \dfrac{4}{5}$
$= \dfrac{15 + 16}{20}$
$= \dfrac{31}{20}$
$= 1\dfrac{11}{20}$

Simplifying algebraic expressions

Like terms can be added or subtracted—unlike terms cannot.

For Example

Simplify the following expressions:

1 $3x + 5x + 12x$

S $3x + 5x + 12x = 20x$

2 $4x - 2y + 3x + 4y$

S $4x - 2y + 3x + 4y = 7x + 2y$

3 $12xy - 3yx$

S $12xy - 3yx = 12xy - 3xy$
$= 9xy$

Remember: $ab = ba$

4 $4a \times (-2b)$

S $4a \times (-2b) = -8ab$

5 $(-14a) \div 7$

S $(-14a) \div 7 = -2a$

6 $(-6y)^2$

S $(-6y)^2 = -6y \times -6y$
$= 36y^2$

7 $cd \div c$

S
$$cd \div c = \frac{cd}{c} = \frac{{}^{1}\not{c} \times d}{\not{c}_{1}} = \frac{d}{1} = d$$

8 $12ab \div 3a$

S
$$12ab \div 3a = \frac{12ab}{3a} = \frac{{}^{4}\not{12} \times \not{a}^{1} \times b}{{}_{1}\not{3} \times \not{a}_{1}} = \frac{4 \times b}{1} = 4b$$

9 $5pq \div p^2q$

S
$$5pq \div p^2q = \frac{5pq}{p^2q} = \frac{5 \times \not{p}^{1} \times \not{q}^{1}}{{}_{1}\not{p} \times p \times \not{q}_{1}} = \frac{5 \times 1}{1 \times p} = \frac{5}{p}$$

10 $\frac{12a + 3a}{5}$

S
$$\frac{12a + 3a}{5} = \frac{15a}{5} = \frac{{}^{3}\not{15} \times a}{\not{5}_{1}} = 3a$$

11 $\frac{6a - 2a}{10a - 8a}$

S
$$\frac{6a - 2a}{10a - 8a} = \frac{{}^{2}\not{4}\not{a}^{1}}{{}_{1}\not{2}\not{a}_{1}} = 2$$

Simple algebraic fractions

Addition and subtraction

Find a common denominator and then add or subtract the numerators.

For Example

Add or subtract the following fractions:

1 $\frac{5y}{2} + \frac{y}{3}$

S $\frac{5y}{2} + \frac{y}{3} = \frac{15y}{6} + \frac{2y}{6}$

$= \frac{17y}{6}$

- Lowest common denominator of 2 and 3 is 6
- Two times 3 is 6
  ∴ $5y$ times $3 = 15y$ and so on ...

2 $\frac{3x}{4} - \frac{x}{2}$

S $\frac{3x}{4} - \frac{x}{2} = \frac{3x}{4} - \frac{2x}{4}$

$= \frac{x}{4}$

3 $\frac{4}{x} - \frac{3}{2x}$

S $\frac{4}{x} - \frac{3}{2x} = \frac{8}{2x} - \frac{3}{2x}$

$= \frac{5}{2x}$

4 $\frac{5}{2y} + \frac{8}{5y}$

S $\frac{5}{2y} + \frac{8}{5y} = \frac{25}{10y} + \frac{16}{10y}$

$= \frac{41}{10y}$

Multiplication and division

To multiply: cancel and then multiply numerators and denominators.
To divide: find the **reciprocal** of the second fraction (that is, turn it upside down) and then multiply.

For Example

Multiply or divide the following fractions:

1 $\frac{x}{3} \times \frac{9}{2x}$

S $\frac{x}{3} \times \frac{9}{2x} = \frac{{}^{1}\cancel{x}}{\cancel{3}_{1}} \times \frac{\cancel{9}^{3}}{2\cancel{x}_{1}}$

$= \frac{3}{2}$

$= 1\frac{1}{2}$

2 $\frac{4y}{7} \times \frac{21}{6y}$

S $\frac{4y}{7} \times \frac{21}{6y} = \frac{{}^{2}\cancel{4}\cancel{y}^{1}}{\cancel{7}_{1}} \times \frac{\cancel{21}^{\cancel{3}1}}{{}_{1}\cancel{3}\cancel{6}\cancel{y}_{1}}$

$= \frac{2}{1}$

$= 2$

3 $\frac{x+2}{6} \times \frac{18}{x+2}$

S $\frac{x+2}{6} \times \frac{18}{x+2} = \frac{{}^{1}\cancel{x+2}}{\cancel{6}_{1}} \times \frac{\cancel{18}^{3}}{\cancel{x+2}_{1}}$

$= 3$

4 $\frac{ab}{4} \surd \frac{a}{6}$

S $\frac{ab}{4} \div \frac{a}{6} = \frac{{}^{1}\cancel{a}b}{{}_{2}\cancel{4}} \times \frac{\cancel{6}^{3}}{\cancel{a}_{1}}$

$= \frac{3b}{2}$

5 $\frac{4x}{5ab} \surd \frac{12}{10b}$

S $\frac{4x}{5ab} \div \frac{12}{10b} = \frac{{}^{1}\cancel{4}x}{{}_{1}\cancel{5}a\cancel{b}_{1}} \times \frac{{}^{2}\cancel{10}\,\cancel{b}^{1}}{\cancel{12}_{3}}$

$= \frac{2x}{3a}$

Removing grouping symbols

The term outside the grouping symbols multiplies the contents of the grouping symbols.

For Example

1 Expand and simplify:

a $3(2x + 5y)$

S $3(2x + 5y) = 3 \times 2x + 3 \times 5y$
$= 6x + 15y$

b $a(2a - 7)$

S $a(2a - 7) = a \times 2a - a \times 7$
$= 2a^2 - 7a$

c $-(3 - 4y)$

S $-(3 - 4y) = -1(3 - 4y)$
$= -1 \times 3 - (-1) \times 4y$
$= -3 + 4y$

Note: the – sign before the grouping symbols has the effect of negating the contents of grouping symbols.

d $5(2x - 4) - 3(5 - x)$

S $5(2x - 4) - 3(5 - x) = 10x - 20 - 15 + 3x$
$= 13x - 35$

e $x(x + 3) - 2(x + 3)$

S $x(x + 3) - 2(x + 3) = x^2 + 3x - 2x - 6$
$= x^2 + x - 6$

2 Simplifying fractions:

a $\frac{x+2}{3} + \frac{x-4}{4}$

S $\frac{x+2}{3} + \frac{x-4}{4} = \frac{4(x+2)}{12} + \frac{3(x-4)}{12}$

$= \frac{4(x+2) + 3(x-4)}{12}$

$= \frac{4x + 8 + 3x - 12}{12}$

$= \frac{7x - 4}{12}$

Note: good to include the second step (in **2a**) with grouping symbols.

b $\frac{x}{5} - \frac{x+1}{2}$

S $\frac{x}{5} - \frac{x+1}{2} = \frac{2x}{10} - \frac{5(x+1)}{10}$

$= \frac{2x - 5(x+1)}{10}$

$= \frac{2x - 5x - 5}{10}$

$= \frac{-3x - 5}{10}$

3 Find the difference between $7x^2 - 4x$ and $3x + 5x^2$.

S $7x^2 - 4x - (3x + 5x^2) = 7x^2 - 4x - 3x - 5x^2$
$= 2x^2 - 7x$

Binomial products

A binomial expression has two terms, for example $2x + 1$, so a binomial product is the result of multiplying two binomial expressions.

Each term in the first binomial expression multiplies each term in the second expression.

For Example

1 Expand and simplify:

a $(x + 2)(x + 4)$

S $(x + 2)(x + 4) = x(x + 4) + 2(x + 4)$
$= x^2 + 4x + 2x + 8$
$= x^2 + 6x + 8$

b $(x - 2y)(x - 3y)$

S $(x - 2y)(x - 3y) = x(x - 3y) - 2y(x - 3y)$
$= x^2 - 3xy - 2yx + 6y^2$
$= x^2 - 5xy + 6y^2$

c $(a - b)(a + b)$

S $(a - b)(a + b) = a(a + b) - b(a + b)$
$= a^2 + ab - ba - b^2$
$= a^2 - b^2$

This method can be shortened of course, or other methods used, such as:

- The **Robin Hood method** (with arrows …).

For Example

1 Expand and simplify:
$(x + 4)(x - 3)$

S $(x + 4)(x - 3) = x^2 - 3x + 4x - 12$
$= x^2 + x - 12$

- The **FOIL method** (First, Outside, Inside, Last).

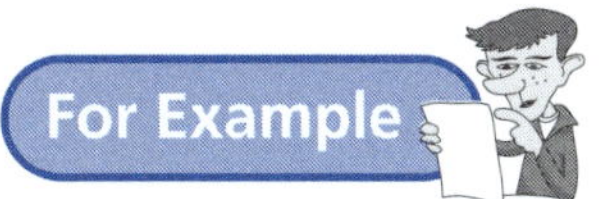

1 Expand and simplify:
$(3x + 1)(2x + 5)$

S $(3x + 1)(2x + 5)$
$= (3x)\ (2x) + (3x)\ (5) + (1)\ (2x) + (1)\ (5)$
first, outside, inside, last
$= 6x^2 + 15x + 2x + 5$
$= 6x^2 + 17x + 5$

Special products

The following results are very important and must be known for success in Year 10:

$$(a + b)^2 = a^2 + 2ab + b^2$$
$$(a - b)^2 = a^2 - 2ab + b^2$$
$$(a - b)(a + b) = a^2 - b^2$$

For Example

1 Expand and simplify:

a $(x + 5)^2$

S $(x + 5)^2 = (x)^2 + 2(x)(5) + (5)^2$
$= x^2 + 10x + 25$

b $(a - 2)^2$

S $(a - 2)^2 = (a)^2 - 2(a)(2) + (2)^2$
$= a^2 - 4a + 4$

c $(3x - 4)^2$

S $(3x - 4)^2 = (3x)^2 - 2(3x)(4) + (4)^2$
$= 9x^2 - 24x + 16$

d $(d - 2)(d + 2)$

S $(d - 2)(d + 2) = (d)^2 - (2)^2$
$= d^2 - 4$

e $(5y + 3)(5y - 3)$

S $(5y + 3)(5y - 3) = (5y)^2 - (3)^2$
$= 25y^2 - 9$

f $(x^2 - y^2)(x^2 + y^2)$

S $(x^2 - y^2)(x^2 + y^2) = (x^2)^2 - (y^2)^2$
$= x^4 - y^4$

Using these special products

The special products can help in numerical problems.

For Example

1 Use the special products to evaluate:

a 101^2

S $101^2 = (100 + 1)^2$
$= 100^2 + 2(100)(1) + 1^2$

$(a + b)^2 = a^2 + 2ab + b^2$

$= 10\,000 + 200 + 1$
$= 10\,201$

b 98^2

S $98^2 = (100 - 2)^2$
$= 100^2 - 2(100)(2) + 2^2$

$(a - b)^2 = a^2 - 2ab + b^2$

$= 10\,000 - 400 + 4$
$= 9604$

More-difficult expansions

The special products can make complicated questions easier.

Expand and simplify:

1 $(x - 3)(x + 3) - (x - 4)^2$

S $(x - 3)(x + 3) - (x - 4)^2$
$= x^2 - 9 - (x^2 - 8x + 16)$
$= x^2 - 9 - x^2 + 8x - 16$
$= 8x - 25$

2 $2(x + 6)^2 - (4 - x)(x + 4)$

S $2(x + 6)^2 - (4 - x)(x + 4)$
$= 2(x^2 + 12x + 36) - (16 - x^2)$
$= 2x^2 + 24x + 72 - 16 + x^2$
$= 3x^2 + 24x + 56$

Common factors

We look for the highest, or largest, factor common to the terms in the expression—this is the opposite to expanding.

Factorise:

1 $4x - 6$

S $4x - 6 = 2(2x - 3)$

Check by expanding.

2 $xy - 3x$

S $xy - 3x = x(y - 3)$

3 $12ab - 14a$

S $12ab - 14a = 2a(6b - 7)$

You must take both 2 and a as common factors.

4 $3x^2 - 6x - 3$

S $3x^2 - 6x - 3 = 3(x^2 - 2x - 1)$

5 $-12x - 4$

S $-12x - 4 = -4(3x + 1)$

6 $9a^2 + 12a^2b$

S $9a^2 + 12a^2b = 3a^2(3 + 4b)$

7 $4x(x + y) + 3(x + y)$

S $4x(x + y) + 3(x + y) = (x + y)[4x + 3]$
$= (x + y)(4x + 3)$

Factorising by grouping in pairs

We can factorise four-term expressions by grouping in pairs.

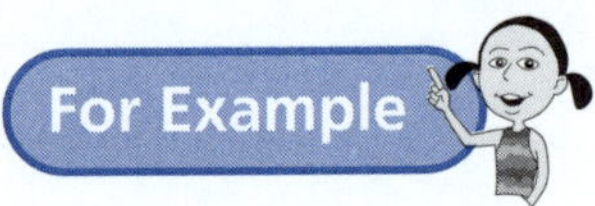

Factorise:

1 $ax + bx + ay + by$

S $ax + bx + ay + by = x(a + b) + y(a + b)$
$= (a + b)[x + y]$
$= (a + b)(x + y)$

2 $xy + y^2 - x - y$

S $xy + y^2 - x - y = y(x + y) - 1(x + y)$
$= (x + y)(y - 1)$

3 $p^2 + mq + pq + mp$

S $p^2 + mq + pq + mp = p^2 + pq + mq + mp$
$= p(p + q) + m(p + q)$
$= (p + q)(p + m)$

Difference of two squares

We can reverse an earlier rule: $a^2 - b^2 = (a - b)(a + b)$

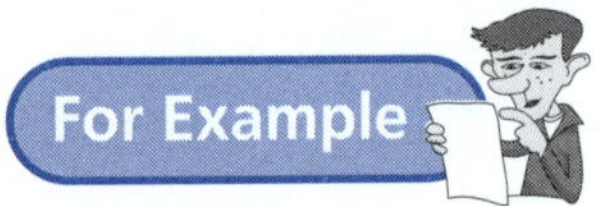

Factorise:

1 $c^2 - 9$

S $c^2 - 9 = (c)^2 - (3)^2$
$= (c - 3)(c + 3)$

Note: it does not matter if your answer has this order: $(c + 3)(c - 3)$.

The order of factors is not important, that is, $2 \times 3 = 3 \times 2$.

2 $49 - 4a^2$

S $49 - 4a^2 = (7)^2 - (2a)^2$
$= (7 - 2a)(7 + 2a)$

3 $x^4 - 1$

S $x^4 - 1 = (x^2)^2 - (1)^2$
$= (x^2 - 1)(x^2 + 1)$
$= (x - 1)(x + 1)(x^2 + 1)$

The monic quadratic trinomial

An expression with three terms is called a **trinomial**.

A trinomial with a highest power of 2 is called a **quadratic**.

If the coefficient (number in front of) the term with the power of 2 is 1, the quadratic trinomial is **monic**, $\therefore x^2 + 5x + 6$ is a monic quadratic trinomial.

As $(x + a)(x + b) = x^2 + bx + ax + ab$
$= x^2 + (a + b)x + ab$

Then, factorising, $x^2 + (a + b)x + ab = (x + a)(x + b)$

Hence, to factorise $x^2 + 5x + 6$ we are looking for two numbers that add together to give 5 (that is, $a + b = 5$) and multiply together to give 6 (that is, $ab = 6$).

Factorise:

1 $x^2 + 5x + 4$

S $x^2 + 5x + 4$
(that is, $a + b = 5$, $ab = 4$)
$= (x + 4)(x + 1)$

2 $x^2 - 5x + 6$

S $x^2 - 5x + 6$
(that is, $a + b = -5$, $ab = 6$)
$= (x - 3)(x - 2)$

3 $x^2 - 3x - 4$

S $x^2 - 3x - 4$
(that is, $a + b = -3$, $ab = -4$)
$= (x - 4)(x + 1)$

4 $x^2 + 5x - 14$

S $x^2 + 5x - 14$
(that is, $a + b = 5$, $ab = -14$)
$= (x + 7)(x - 2)$

Other helpful rules are:

- For $x^2 + 5x + 4$
  ↑ ↑ positive here, we say 'both'
  'this sign', that is, **both positive**.
- For $x^2 - 3x - 4$
  ↑ ↑ negative here, we say 'the bigger number is'
  'this sign', that is, **bigger one is negative**.

Factorise:

1 $x^2 + 7x + 12$

S $x^2 + 7x + 12 = (x + 4)(x + 3)$
↑
both positive

2 $x^2 - 8x + 15$

S $x^2 - 8x + 15 = (x - 5)(x - 3)$
↑
both negative

3 $x^2 - 4x - 12$

S $x^2 - 4x - 12 = (x - 6)(x + 2)$

The 'bigger number' is negative.

4 $x^2 + x - 20$

S $x^2 + x - 20 = (x + 5)(x - 4)$

The 'bigger number' is positive.

Non-monic quadratic trinomial

Most of the hints given above no longer apply, although:

- the two numbers a and b still multiply together (ab) to give our constant term
- if this constant term is positive, we can still say 'both positive/negative' (see above).

In $ax^2 + bx + c$, the c is the constant term and is said to be independent of x.

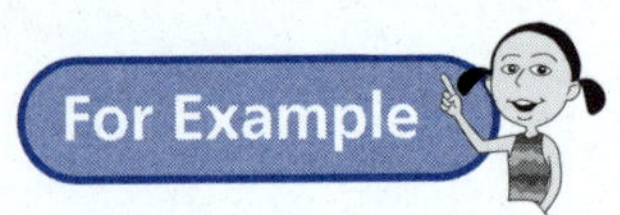

1 Factorise $2x^2 + 13x + 21$.

S $2x^2 + 13x + 21 = (2x \quad)(x \quad)$

We can see they are both positive,

$\therefore\ 2x^2 + 13x + 21 = (2x + \quad)(x + \quad)$

Now, it's simply trial and error, and checking by expanding the answer to get back to the expression.

$\therefore\ 2x^2 + 13x + 21 = (2x + 7)(x + 3)$

2 Factorise $3x^2 - 2x - 8$.

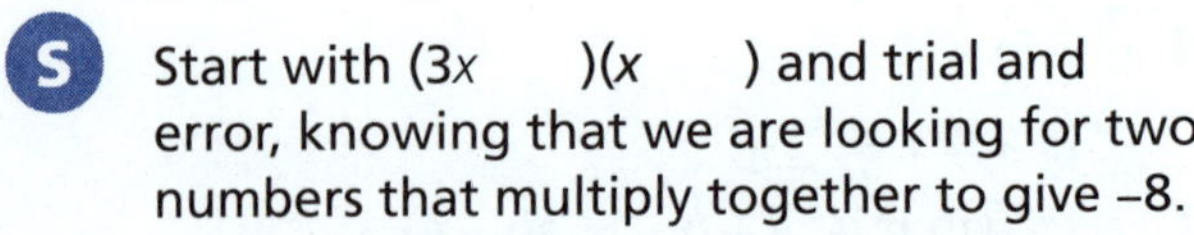

S Start with $(3x \quad)(x \quad)$ and trial and error, knowing that we are looking for two numbers that multiply together to give -8.

$\therefore\ 3x^2 - 2x - 8 = (3x + 4)(x - 2)$.

Now try a second method: This is called the **cross method**, where the guess and verify stage occurs around a cross.

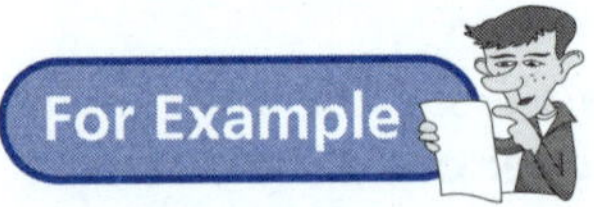

1 Factorise $3x^2 + 10x - 8$.

S

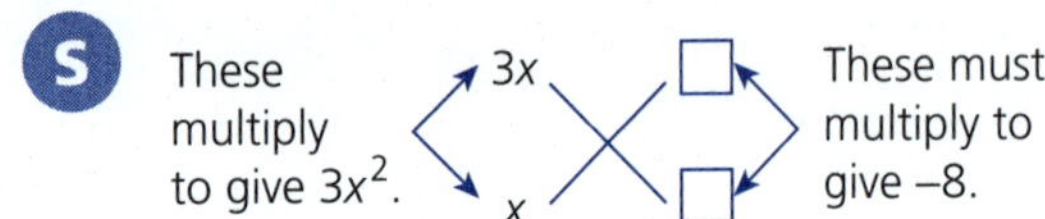

We then multiply across the diagonals and add terms.

Thus:

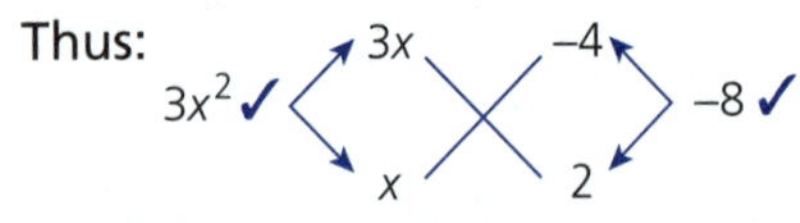

$\therefore\ 3x \times 2 + x \times -4$
$= 6x - 4x$
$= 2x$ No! It must be $10x$.

Try again:

$3x^2$ ✓ $3x$, x ; -2, 4 -8 ✓

$\therefore\ 3x \times 4 + x \times -2$
$= 12x - 2x$
$= 10x$ Yes

$\therefore\ (3x - 2)(x + 4)$

2 Factorise $4x^2 - 16x + 15$.

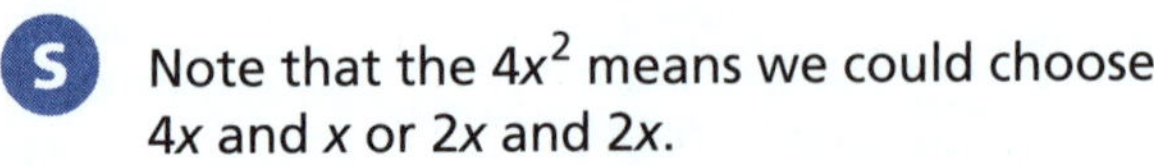

S Note that the $4x^2$ means we could choose $4x$ and x or $2x$ and $2x$.

Once again, by trial and error:

$4x^2$ $2x$, $2x$; -5, -3 15

$2x \times -3 + 2x \times -5 = -6x - 10x$
$= -16x$

$\therefore\ 4x^2 - 16x + 15 = (2x - 5)(2x - 3)$

3 Factorise $6x^2 - 10x - 4$.

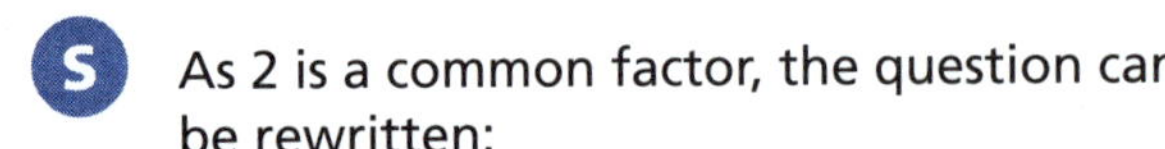

S As 2 is a common factor, the question can be rewritten:

$2(3x^2 - 5x - 2)$

For $3x^2$ we choose $3x$ and x, and for -2, use trial and error:

$3x^2$ $3x$, x ; $+1$, -2 -2

$3x \times -2 + x \times 1 = -6x + x$
$= -5x$

$\therefore\ 6x^2 - 10x - 4 = 2(3x + 1)(x - 2)$

Another method of factorising non-monic quadratic trinomial is as follows: if the trinomial is of the form $ax^2 + bx + c$, we rewrite bx as two separate terms, which allows us to factorise pairs of terms.

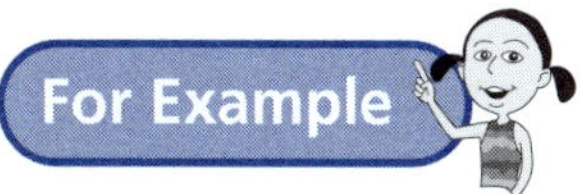

1 Factorise $2x^2 + 5x - 3$.

- We multiply $2x^2$ by -3, that is, $-6x^2$.
- We now look for two factors that add together to give 5 (from the $5x$) and multiply together to give -6 (from the $-6x^2$).

Thus the factors are 6 and -1.

That is:

$$\begin{aligned} 2x^2 + 5x - 3 &= 2x^2 + 6x - x - 3 \\ &= 2x(x + 3) - 1(x + 3) \\ &= (x + 3)(2x - 1) \\ &= (2x - 1)(x + 3) \end{aligned}$$

2 Factorise $8a^2 + 10a - 3$.

$8a^2 \times -3 = -24a^2$

Therefore look for factors that add to give 10 and multiply to give -24.

That is, 12 and -2.

Therefore:

$$\begin{aligned} 8a^2 + 10a - 3 &= 8a^2 - 2a + 12a - 3 \\ &= 2a(4a - 1) + 3(4a - 1) \\ &= (4a - 1)(2a + 3) \end{aligned}$$

A further method is as follows:

For Example

1 Factorise $4a^2 - 13a + 9$.

- The coefficient of a^2 and the constant term are multiplied … here $4 \times 9 = 36$.
- The coefficient of a^2 (here it is 4) is put in both brackets and as a denominator.

Thus:

$$4a^2 - 13a + 9 = \frac{(4a \quad)(4a \quad)}{4} \qquad \text{(36)}$$

- We now look for two numbers that add together to get -13 (from $-13a$), and multiply together to get 36.

That is, -9 and -4.

Therefore:

$$4a^2 - 13a + 9 = \frac{(4a - 9)(4a - 4)}{4}$$

- We then cancel … here $4a - 4$ is divided by 4 to get $a - 1$.

Thus $4a^2 - 13a + 9 = (4a - 9)(a - 1)$

2 Factorise $6x^2 + 11x - 10$.

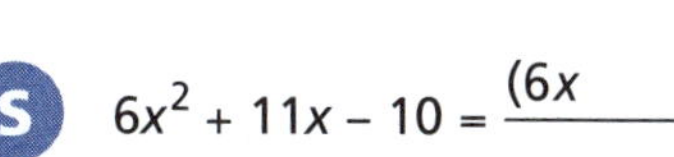

$$6x^2 + 11x - 10 = \frac{(6x \quad)(6x \quad)}{6} \qquad \text{(−60)}$$

Now $6 \times -10 = -60$

$\therefore$ Add to get 11 and multiply to get -60, that is, 15, -4.

$$= \frac{(6x + 15)(6x - 4)}{6}$$

$$= \frac{{}^{1}\cancel{3}(2x + 5) \times {}^{1}\cancel{2}(3x - 2)}{\cancel{6}\,\cancel{2}_1}$$

$$= (2x + 5)(3x - 2)$$

Thus $6x^2 + 11x - 10 = (2x + 5)(3x - 2)$

Further examples

Some expressions require two steps of factorisations.

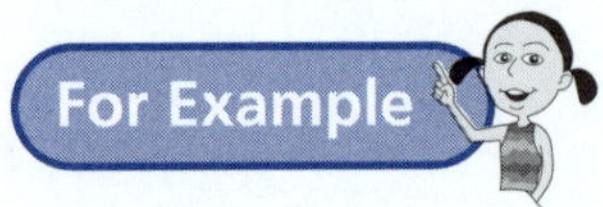

Factorising:

1 $x^3 - x$

S $x^3 - x = x(x^2 - 1)$
$= x(x - 1)(x + 1)$

2 $y^8 - 1$

S $y^8 - 1$
$= (y^4 - 1)(y^4 + 1)$ Now factor $y^4 - 1$
$= (y^2 - 1)(y^2 + 1)(y^4 + 1)$ Now factor $y^2 - 1$
$= (y - 1)(y + 1)(y^2 + 1)(y^4 + 1)$

3 $6x^3 + 13x^2 - 5x$

S $6x^3 + 13x^2 - 5x$
$= x(6x^2 + 13x - 5)$
$= x(3x - 1)(2x + 5)$

Always look first for common factors.

4 $(x - 3)^2 - (y + 2)^2$

S $(x - 3)^2 - (y + 2)^2$
$= [(x - 3) + (y + 2)][(x - 3) - (y + 2)]$
$= (x - 3 + y + 2)(x - 3 - y - 2)$
$= (x + y - 1)(x - y - 5)$

5 $a^4 + 2a^2 + 1$

S $a^4 + 2a^2 + 1$
$= (a^2 + 1)(a^2 + 1)$
$= (a^2 + 1)^2$

6 $c^6 + 3c^3 - 4$

S $c^6 + 3c^3 - 4$
$= (c^3 + 4)(c^3 - 1)$

Note: in Year 11 you might learn to factorise $x^3 - 1$: a difference of two cubes.

Complex algebraic fractions

Factorising numerators and denominators

Numerators and denominators of fractions should be factorised if possible.

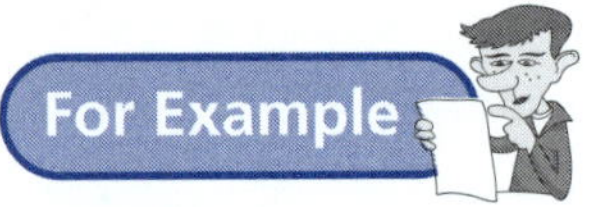

Simplify:

1 $\dfrac{3x + 12}{3}$

S $\dfrac{3x + 12}{3} = \dfrac{{}^1\not{3}(x + 4)}{\not{3}_1}$
$= x + 4$

2 $\dfrac{5x + 15y}{2x + 6y}$

S $\dfrac{5x + 15y}{2x + 6y} = \dfrac{5\cancel{(x + 3y)}^1}{2\cancel{(x + 3y)}_1}$
$= \dfrac{5}{2}$
$= 2\dfrac{1}{2}$

3 $\dfrac{4}{4x+8}$

S $\dfrac{4}{4x+8} = \dfrac{^1\cancel{4}}{_1\cancel{4}(x+2)}$
$= \dfrac{1}{x+2}$

Don't forget the 1 as the numerator.

4 $\dfrac{2a-2b}{b-a}$

S $\dfrac{2a-2b}{b-a} = \dfrac{2(a-b)}{-1(-b+a)}$
$= \dfrac{2\cancel{(a-b)}^1}{-1\cancel{(-b+a)}_1}$
$= \dfrac{2}{-1}$
$= -2$

Remember: $x - y = -(y - x)$

5 $\dfrac{x^2+5x-6}{x^2-36}$

S $\dfrac{x^2+5x-6}{x^2-36} = \dfrac{^1\cancel{(x+6)}(x-1)}{(x-6)\cancel{(x+6)}_1}$
$= \dfrac{x-1}{x-6}$

6 $\dfrac{ax+ay+2x+2y}{x^2-y^2}$

S $\dfrac{ax+ay+2x+2y}{x^2-y^2} = \dfrac{a(x+y)+2(x+y)}{(x+y)(x-y)}$
$= \dfrac{^1\cancel{(x+y)}(a+2)}{_1\cancel{(x+y)}(x-y)}$
$= \dfrac{a+2}{x-y}$

Algebraic fractions involving multiplication and division

In some questions, factorise before simplifying.

For Example

Simplify:

1 $\dfrac{a^2-9}{a-3} \times \dfrac{5}{a+3}$

S $\dfrac{a^2-9}{a-3} \times \dfrac{5}{a+3}$
$= \dfrac{^1\cancel{(a-3)}\cancel{(a+3)}^1}{\cancel{a-3}_1} \times \dfrac{5}{\cancel{a+3}^1}$
$= 5$

2 $\dfrac{y^2+7y+10}{y^2-8y+12} \times \dfrac{y-6}{y+5}$

S $\dfrac{y^2+7y+10}{y^2-8y+12} \times \dfrac{y-6}{y+5}$
$= \dfrac{^1\cancel{(y+5)}(y+2)}{_1\cancel{(y-6)}(y-2)} \times \dfrac{\cancel{y-6}^1}{\cancel{y+5}_1}$
$= \dfrac{y+2}{y-2}$

3 $\dfrac{3b^2-27}{ab^2-16a} \div \dfrac{b+3}{2b+8}$

We multiply by the reciprocal …

S $\dfrac{3b^2-27}{ab^2-16a} \div \dfrac{b+3}{2b+8}$
$= \dfrac{3(b-3)\cancel{(b+3)}^1}{a(b-4)\cancel{(b+4)}_1} \times \dfrac{2\cancel{(b+4)}^1}{\cancel{b+3}_1}$
$= \dfrac{6(b-3)}{a(b-4)}$

4 $\dfrac{x^2-4x+3}{x^2-x-6} \div \dfrac{x^2+3x-4}{x^2+x-12}$

S $\dfrac{x^2-4x+3}{x^2-x-6} \div \dfrac{x^2+3x-4}{x^2+x-12}$
$= \dfrac{^1\cancel{(x-3)}\cancel{(x-1)}^1}{_1\cancel{(x-3)}(x+2)} \times \dfrac{^1\cancel{(x+4)}(x-3)}{_1\cancel{(x+4)}\cancel{(x-1)}_1}$
$= \dfrac{x-3}{x+2}$

Algebraic fractions involving addition and subtraction

First find a common denominator—often the product of the denominators will give the denominator.

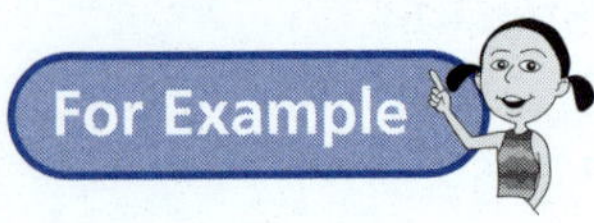

Simplify:

1 $\dfrac{3}{x+1} - \dfrac{4}{x+3}$

S
$$\frac{3}{x+1} - \frac{4}{x+3} = \frac{3(x+3) - 4(x+1)}{(x+1)(x+3)}$$
$$= \frac{3x + 9 - 4x - 4}{(x+1)(x+3)}$$
$$= \frac{-x + 5}{(x+1)(x+3)}$$

2 $\dfrac{2}{x-1} + \dfrac{3}{x^2-1}$

S
$$\frac{2}{x-1} + \frac{3}{x^2-1} = \frac{2}{x-1} + \frac{3}{(x+1)(x-1)}$$
$$= \frac{2(x+1) + 3}{(x+1)(x-1)}$$
$$= \frac{2x + 2 + 3}{(x+1)(x-1)}$$
$$= \frac{2x + 5}{(x+1)(x-1)}$$

3 $\dfrac{4}{a^2+a} - \dfrac{3}{a^2-1}$

S
$$\frac{4}{a^2+a} - \frac{3}{a^2-1} = \frac{4}{a(a+1)} - \frac{3}{(a+1)(a-1)}$$
$$= \frac{4(a-1) - 3a}{a(a+1)(a-1)}$$
$$= \frac{4a - 4 - 3a}{a(a+1)(a-1)}$$
$$= \frac{a - 4}{a(a+1)(a-1)}$$

4 $\dfrac{2}{x^2-y^2} + \dfrac{1}{x^2-xy} - \dfrac{3}{xy+y^2}$

S
$$\frac{2}{x^2-y^2} + \frac{1}{x^2-xy} - \frac{3}{xy+y^2}$$
$$= \frac{2}{(x-y)(x+y)} + \frac{1}{x(x-y)} - \frac{3}{y(x+y)}$$
$$= \frac{2xy + y(x+y) - 3x(x-y)}{xy(x-y)(x+y)}$$
$$= \frac{2xy + xy + y^2 - 3x^2 + 3xy}{xy(x-y)(x+y)}$$
$$= \frac{6xy + y^2 - 3x^2}{xy(x-y)(x+y)}$$

Practise Practise

1 The maximum speed of a car is x kilometres per hour, but when it has a trailer, it is slowed up by y kilometres per hour. What would be its maximum speed when towing the trailer? pp. 14–15

2 A boy is now 14 years old.
- **a** How old will he be in x years' time?
- **b** How old was he y years ago?
- **c** If his mother was p years old when he was born, how old is she now?
- **d** How old will his mother be in q years' time? pp. 14–15

3 A brick wall x metres long and y metres high has r windows, each m metres high and q metres wide. Find the:
- **a** overall area of the wall
- **b** area of one window
- **c** area of r windows
- **d** area of the wall that is completely brick. pp. 14–15

4 Convert:
- **a** k cm to mm
- **b** p days to minutes
- **c** k kilolitres to litres
- **d** r cents to dollars
- **e** y grams to kilograms. pp. 14–15

5 A barrel of sugar contains 16.4 kg. If x kilograms is removed but then y grams is put back, find the amount of sugar, in grams, now in the barrel. pp. 14–15

6 Find the number of cents in x dollars and y cents. pp. 14–15

7 If 20 books cost \$$c$, find the cost of three books. pp. 14–15

8 In a triangle, two angles are found to be $p°$ and $q°$. Find the size of the third angle. pp. 14–15

9
- **a** How far will Sharon walk at s km/h in t hours?
- **b** What is Bob's average speed if his car travels k km in p hours?
- **c** How long will it take Pat to cycle b km if she travels at c km/h? pp. 14–15

10 Evaluate the following expressions, if $a = -2$, $b = 3$, $c = -5$:
- **a** abc
- **b** $bc - a$
- **c** $\frac{1}{a} - \frac{1}{b}$
- **d** $\frac{ab}{c - 1}$
- **e** $a^2 - b^2$
- **f** $c^2 - 2a^2$ p. 15

11 Simplify:
- **a** $12y - 3x - 6y + 7x$
- **b** $5y^2 - 2y + 3y^2 - 4y$
- **c** $12xy - 4xy + xy$
- **d** $3a \times (-2b)$
- **e** $(-5y)^2$
- **f** $7a \times a \times a$
- **g** $12xy \times -2x$
- **h** $14ab \div 7a$
- **i** $3xy \div -x$
- **j** $(4x + 6x) \div 5x$ p. 16

12 Simplify:
- **a** $\frac{2x}{3} + \frac{x}{2}$
- **b** $\frac{5a}{4} - \frac{a}{2}$
- **c** $\frac{12x}{13} - \frac{x}{2}$
- **d** $\frac{3}{x} - \frac{1}{x}$
- **e** $\frac{5}{2a} - \frac{3}{a}$
- **f** $\frac{a}{b} - \frac{c}{d}$
- **g** $\frac{2a}{3b} - \frac{4a}{5b}$
- **h** $\frac{6}{y} \times \frac{2y}{3}$
- **i** $\frac{6m}{5} \times \frac{10}{18m^2}$
- **j** $\frac{3}{2m} \div \frac{6}{4m}$
- **k** $\frac{6a}{5} \div \frac{4a}{25}$
- **l** $\frac{3a}{8bc} \times \frac{b}{9a} \times 4c$
- **m** $\frac{4}{x} \div \frac{3}{x} \times 1\frac{1}{3}$ p. 16–17

13 Expand and simplify:
- **a** $4(2x - 7)$
- **b** $-3(5a - 12)$
- **c** $2 + 3(4y - 1)$
- **d** $7 - 5(x - 1)$
- **e** $-(3x - 4)$
- **f** $4(x - 1) + 2(x - 5)$
- **g** $5(2a - 5) - 6(4 - a)$
- **h** $x(x - 7) - 2(x - 7)$ p. 18

14 Simplify:
- **a** $\frac{3x - 4}{2} + \frac{x - 1}{4}$
- **b** $\frac{5x - 4}{2} - \frac{3 - x}{3}$
- **c** $\frac{4a - 2}{2x} - \frac{3a - 4}{3x}$ p. 18

15 Find the difference between:
- **a** $4x - 2$ and $x + 5$
- **b** $12y - 4$ and $5 - 2y$
- **c** $3x - x^2$ and $7x^2 - 5x + 2$ p. 19

16 Expand and simplify:

a $(a-2)(a+4)$ **b** $(2x-1)(x+3)$
c $(4x-7)(x+8)$ **d** $(4-x)(3x+1)$
e $(x-4)^2$ **f** $(4x+1)^2$
g $(2y-1)(2y+1)$ **h** $(5y-7)(5y+7)$ p. 19

17 Find the area of:

a

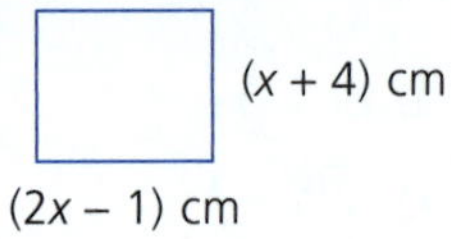

b

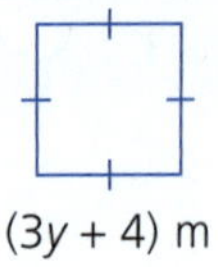

p. 19

18 Expand and simplify:

a $x^2-(x-4)^2$
b $(x-1)^2-(x+1)^2$
c $\left(x+\frac{1}{x}\right)^2$
d $(a-b)(b-a)-a^2-b^2$
e $3(x-4)^2$
f $2(a-1)^2-2(a-1)(a+1)$ pp. 20-21

19 Evaluate, using the special products:

a 41^2 **b** 29^2 pp. 20–21

20 Factorise:

a $5x-10y$ **b** $8a-6$
c $12xy-8y$ **d** $15y^2-10y$
e $-10a-18$ **f** $4y^2z-6yz$
g $3x^2-9x-48$ **h** $\pi r^2-2\pi r$
i $2l+2b$ **j** $3(x+y)+x(x+y)$
k $a(a-7)-b(a-7)$ p. 21

21 Factorise:

a $2x+2y+ax+ay$ **b** $-2a+2b-ca+cb$
c $6a-ct+ac-6t$ **d** g^3-3g^2-2g+6
e $xy-y^2-7x+7y$ p. 22

22 Factorise:

a u^2-36 **b** $1-25p^2$
c $4a^2-9b^2$ **d** $25k^2-144m^2$
e $16x^4-1$ **f** $1-x^4y^4$
g $2a^2-2$ **h** $5a^2-125$ p. 22

23 Factorise:

a x^2+2x-8 **b** $k^2-12k+11$
c $h^2+13h-48$ **d** $b^2-3b-88$
e $x^2-6xy+8y^2$ **f** $x^2+3xy-4y^2$
pp. 22–23

24 Factorise:

a $2x^2-x-1$ **b** $3p^2+11p+8$
c $3c^2+5c-12$ **d** $9x^2+17x+8$
e $8v^2-v-7$ **f** $3b^2+16bc+5c^2$
pp. 23–26

25 Simplify:

a $\dfrac{8}{2a-6}$ **b** $\dfrac{3y-4}{9y-12}$
c $\dfrac{a^2-4a}{6a}$ **d** $\dfrac{a^2-6a+9}{a^2-9}$
e $\dfrac{x^2+7x+12}{x^2-16}$ **f** $\dfrac{p^2+4p+pq+4q}{p^2-q^2}$
pp. 26–27

26 Simplify:

a $\dfrac{y-6}{y^2-36}\times\dfrac{y+6}{2}$
b $\dfrac{x^2-1}{x-1}\div\dfrac{x+1}{5}$
c $\dfrac{a^2-6a-7}{a^2-1}\times\dfrac{a^2+a}{a^2}$ p. 27

27 Simplify:

a $\dfrac{4}{x+3}+\dfrac{5}{x+7}$
b $\dfrac{2}{x-3}+\dfrac{4}{x-5}$
c $\dfrac{x-3}{x^2-5x+6}-\dfrac{3x}{x^2+4x}$
d $\dfrac{2}{c^2-49}+\dfrac{4}{c^2-4c-21}$
e $\dfrac{y+2}{y^2+2y-3}-\dfrac{y}{y^2+7y+12}$ p. 28

Go to p. 299 for **Quick Answers** or to pp. 312–314 for **Worked Solutions**

For a complete understanding of this topic, you must be able to:

✓	Complete arithmetic questions algebraically		pp. 14–15
✓	Substitute into algebraic expressions		p. 15
✓	Simplify algebraic expressions, including algebraic fractions		pp. 16–18
✓	Remove grouping symbols and simplify resulting expressions		pp. 18–19
✓	Expand binomial products, including difference of two squares and perfect squares		pp. 19–21
✓	Factorise algebraic expressions		pp. 21–26
✓	Simplify algebraic fractions, including factoring.		pp. 26–28

Now you are ready to do the tests!

Intermediate Test

Algebraic Techniques

(50 marks)

1 Write down the:

a sum of $3x$ and $5x$ (1 mark)

b product of $6x$ and y (1 mark)

c cents in $\$4x$. (1 mark)

2 If $m = -6$ and $n = 4$, evaluate:

a $2mn$ (1 mark)

b $5 + m + n$ (1 mark)

c $m^2 - n^2$ (1 mark)

d $\dfrac{m+n}{m-n}$ (1 mark)

3 Simplify the expressions:

a $7m - m + 4m$ (1 mark)

b $9y \times 4y$ (1 mark)

c $5ab - 4a \times 3b$ (1 mark)

d $(-2y) \times (-8y)$ (1 mark)

e $21xy \div 7x$ (1 mark)

f $9ab \div 18a^2$ (1 mark)

4 Simplify the fractions:

a $\dfrac{42xy}{7xy^2}$ (1 mark)

b $\dfrac{2m}{3} - \dfrac{3m}{2}$ (1 mark)

c $\dfrac{11}{2x} - \dfrac{7}{3x}$ (1 mark)

d $\dfrac{2x}{5} \times \dfrac{10x}{4}$ (1 mark)

e $\dfrac{2x}{5} \surd \dfrac{10x}{4}$ (1 mark)

5 Expand and simplify:

a $5(3x - 2y)$ (1 mark)

b $5 - 2(3x - 5)$ (2 marks)

c $3(4x - 3) - 4(3x - 4)$ (2 marks)

6 Simplify:

a $\dfrac{x}{3} + \dfrac{2x-1}{4}$ (2 marks)

b $\dfrac{4x-3}{2} + \dfrac{2x-3}{3}$ (2 marks)

7 Expand and simplify:

a $(x + 5)(x - 3)$ (1 mark)

b $(2x - 5)(2x + 5)$ (1 mark)

c $(2x + 5)^2$ (1 mark)

8 Factorise the following completely:

a $7y - 21$ (1 mark)

b $5m^2 + 15mn$ (1 mark)

c $m^2 - 9$ (1 mark)

d $a^2 + 3a - 40$ (1 mark)

e $y^2 - 11y + 18$ (1 mark)

f $y^2 + 2xy + 9y + 18x$ (2 marks)

g $2x^2 + 9x - 5$ (2 marks)

h $7x^2 - 28$ (2 marks)

9 Simplify:

a $\dfrac{27}{9x + 9y}$ (1 mark)

b $\dfrac{m^2 - 4n^2}{4m - 8n}$ (2 marks)

c $\dfrac{5x}{x^2 - 25} \times \dfrac{5x^2 - 25x}{25}$ (3 marks)

d $\dfrac{5}{x+1} + \dfrac{3x+1}{x^2 + x}$ (3 marks)

Your Feedback

☐ × 100% = ☐ %

QA PAGE 304

WS PAGE 362

Advanced Test

Algebraic Techniques

(50 marks)

1 Write down the:

a difference between $8y$ and $5y$ (1 mark)
b product of $8a$ and $3a$ (1 mark)
c dollars in $3x$ cents. (1 mark)

2 Simplify the expressions:

a $7t - 4t - 5t$ (1 mark)
b $7m \times (-6n)$ (1 mark)
c $18m^2 - 7m \times 5m$ (1 mark)
d $42ab \div 14a$ (1 mark)
e $(-3m) \times (-9mn)$ (1 mark)
f $8a - 6c + 2a - 4c$ (1 mark)

3 Simplify the fractions:

a $\dfrac{35ab^3}{7ab}$ (1 mark)
b $\dfrac{5a}{2} - \dfrac{3a}{5}$ (1 mark)
c $\dfrac{7}{m} - \dfrac{5}{3m}$ (1 mark)
d $\dfrac{15}{xy} \times \dfrac{3x}{5}$ (1 mark)
e $\dfrac{25y^2}{14x} \surd \dfrac{5xy}{7x^2}$ (1 mark)

4 Expand and simplify:

a $7m(5 - 3m)$ (1 mark)
b $15t - 8(2t - 5)$ (2 marks)
c $6(2a + 5) - 5(3a - 2)$ (2 marks)

5 Simplify:

a $\dfrac{13x}{4} - \dfrac{4x - 1}{3}$ (2 marks)
b $\dfrac{5m - 2}{2} - \dfrac{2m - 5}{5}$ (2 marks)

6 Expand and simplify:

a $(3x - 2)(x + 5)$ (1 mark)
b $(5x + 3)(5x - 3)$ (1 mark)
c $(5x - 3)^2$ (1 mark)
d $(2x + 3)(x - 2) - 2x^2 + 6$ (2 marks)
e $(2x - 1)^2 - (2x + 1)^2$ (2 marks)

7 Factorise the following completely:

a $54 - 9a$ (1 mark)
b $m^2 - 64$ (1 mark)
c $27y^2 - 9y$ (1 mark)
d $p^2 + p - 42$ (1 mark)
e $p^2 - p - 72$ (1 mark)
f $6x^2 - x - 2$ (1 mark)
g $m^2 - 8m - 7tm + 56t$ (2 marks)
h $a^4 - b^4$ (2 marks)

8 Simplify:

a $\dfrac{42x - 35}{6x^2 - 5x}$ (2 marks)
b $\dfrac{5}{2x + 1} - \dfrac{5}{2x - 1}$ (2 marks)
c $\dfrac{15ab}{9a^2 - 1} \times \dfrac{9a + 3}{5b^2}$ (2 marks)
d $\dfrac{3y - 2}{8y} \div \dfrac{9y^2 - 4}{6y + 4}$ (2 marks)
e $\dfrac{2}{x^2 - 16} - \dfrac{1}{x^2 - 4x}$ (2 marks)

Your Feedback

☐ × 100% = ☐ %

QA PAGE 304
WS PAGE 362

Chapter 3
Surds and Indices

Surds

Surds are numerical expressions that involve irrational numbers.

Irrational numbers are numbers that cannot be written in the form $\frac{a}{b}$, where a and b are integers.

Keywords
Index
Indices
Rationalise
Recurring
Scientific notation
Surd
Terminating

Approximation of surds

The calculator can be used to approximate surds.

For Example

1 Arrange the following numbers in order, from smallest to largest:

$3, \sqrt{5}, \sqrt{7}, 2, 4$

S $3 = \sqrt{9}, \sqrt{5}, \sqrt{7}, 2 = \sqrt{4}, 4 = \sqrt{16}$

In order: $\sqrt{4}, \sqrt{5}, \sqrt{7}, \sqrt{9}, \sqrt{16}$

That is, $2, \sqrt{5}, \sqrt{7}, 3, 4$.

Rules for surds

If $a > 0$, $b > 0$, then:

- $\sqrt{ab} = \sqrt{a} \times \sqrt{b}$
- $\sqrt{\frac{a}{b}} = \frac{\sqrt{a}}{\sqrt{b}}$
- $(\sqrt{a})^2 = a$

For Example

1 Simplify the following:

a $\sqrt{48}$

S $\sqrt{48} = \sqrt{16} \times \sqrt{3}$
$= 4 \times \sqrt{3}$
$= 4\sqrt{3}$

Note: we look for two factors, the first of which is a perfect square.

b $3\sqrt{12}$

S $3\sqrt{12} = 3 \times \sqrt{4} \times \sqrt{3}$
$= 3 \times 2 \times \sqrt{3}$
$= 6\sqrt{3}$

c $\sqrt{\frac{49}{64}}$

S $\sqrt{\frac{49}{64}} = \frac{\sqrt{49}}{\sqrt{64}}$
$= \frac{7}{8}$

d $(\sqrt{16})^2$

S $(\sqrt{16})^2 = 4^2$
$= 16$

Addition and subtraction of surds

We can add or subtract only like terms.

In algebra, like terms are $3a$, $7a$, $2a$, etc. Like surds are $\sqrt{3}$, $3\sqrt{3}$, $-7\sqrt{3}$, etc.

For Example

1 Simplify:

a $4\sqrt{3} + 2\sqrt{3} - \sqrt{3}$

S $4\sqrt{3} + 2\sqrt{3} - \sqrt{3}$
$= 5\sqrt{3}$

b $2\sqrt{5} + \sqrt{20}$

S $2\sqrt{5} + \sqrt{20}$
$= 2\sqrt{5} + \sqrt{4} \times \sqrt{5}$
$= 2\sqrt{5} + 2\sqrt{5}$
$= 4\sqrt{5}$

We may need to simplify surds before adding or subtracting.

c $5\sqrt{7} - \sqrt{63} + 2\sqrt{28}$

S $5\sqrt{7} - \sqrt{63} + 2\sqrt{28}$
$= 5\sqrt{7} - \sqrt{9} \times \sqrt{7} + 2 \times \sqrt{4} \times \sqrt{7}$
$= 5\sqrt{7} - 3\sqrt{7} + 4\sqrt{7}$
$= 6\sqrt{7}$

d $\sqrt{a^3} + 3a\sqrt{a}$

S $\sqrt{a^3} + 3a\sqrt{a}$
$= \sqrt{a^2} \times \sqrt{a} + 3a\sqrt{a}$
$= a\sqrt{a} + 3a\sqrt{a}$
$= 4a\sqrt{a}$

Multiplication and division of surds

We reverse our surd rules: $\sqrt{a} \times \sqrt{b} = \sqrt{ab}$

$$\frac{\sqrt{a}}{\sqrt{b}} = \sqrt{\frac{a}{b}}$$

For Example

1 Simplify:

a $\sqrt{5} \times \sqrt{5}$

S $\sqrt{5} \times \sqrt{5} = \sqrt{25} = 5$

b $\sqrt{7} \times \sqrt{3}$

S $\sqrt{7} \times \sqrt{3} = \sqrt{21}$

c $2\sqrt{3} \times \sqrt{3}$

S $2\sqrt{3} \times \sqrt{3} = 2 \times 3 = 6$

d $3\sqrt{5} \times 2\sqrt{7}$

S $3\sqrt{5} \times 2\sqrt{7} = 6\sqrt{35}$

e $(3\sqrt{2})^2$

S $(3\sqrt{2})^2 = 3\sqrt{2} \times 3\sqrt{2}$
$= 9\sqrt{4}$
$= 9 \times 2$
$= 18$

f $\sqrt{4a + 4}$

S $\sqrt{4a + 4} = \sqrt{4(a + 1)}$
$= \sqrt{4} \times \sqrt{a + 1}$
$= 2\sqrt{a + 1}$

g $\sqrt{54}\ \surd\ \sqrt{18}$

S $\sqrt{54}\ \surd\ \sqrt{18} = \sqrt{\frac{54}{18}}$
$= \sqrt{3}$

h $\dfrac{3\sqrt{2} \times \sqrt{6}}{\sqrt{3}}$

S $\dfrac{3\sqrt{2} \times \sqrt{6}}{\sqrt{3}} = \dfrac{3\sqrt{12}}{\sqrt{3}}$

$= \dfrac{3 \times \sqrt{4} \times \cancel{\sqrt{3}}^{1}}{\cancel{\sqrt{3}}_{1}}$

$= 3 \times 2$

$= 6$

2 Expand and simplify:

a $\sqrt{3}(2 - \sqrt{3})$

S $\sqrt{3}(2 - \sqrt{3}) = \sqrt{3} \times 2 - \sqrt{3} \times \sqrt{3}$

$= 2\sqrt{3} - \sqrt{9}$

$= 2\sqrt{3} - 3$

b $4\sqrt{2}(\sqrt{2} - 1)$

S $4\sqrt{2}(\sqrt{2} - 1) = 4\sqrt{2} \times \sqrt{2} - 4\sqrt{2} \times 1$

$= 4 \times \sqrt{4} - 4\sqrt{2}$

$= 4 \times 2 - 4\sqrt{2}$

$= 8 - 4\sqrt{2}$

3 Simplify:

a $\dfrac{2 + 4\sqrt{3}}{2}$

S $\dfrac{2 + 4\sqrt{3}}{2} = \dfrac{2(1 + 2\sqrt{3})}{2}$

$= 1 + 2\sqrt{3}$

b $\dfrac{\sqrt{12} + 4\sqrt{3}}{\sqrt{3}}$

S $\dfrac{\sqrt{12} + 4\sqrt{3}}{\sqrt{3}} = \dfrac{2\sqrt{3} + 4\sqrt{3}}{\sqrt{3}}$

$= \dfrac{6\sqrt{3}}{\sqrt{3}}$

$= 6$

Binomial products

We can use the following three rules:

$$(a + b)^2 = a^2 + 2ab + b^2$$
$$(a - b)^2 = a^2 - 2ab + b^2$$
$$(a - b)(a + b) = a^2 - b^2$$

For Example

1 Expand and simplify:

a $(\sqrt{2} + 1)^2$

S $(\sqrt{2} + 1)^2 = (\sqrt{2})^2 + 2(\sqrt{2})(1) + 1^2$

$= 2 + 2\sqrt{2} + 1$

$= 3 + 2\sqrt{2}$

b $(\sqrt{3} - \sqrt{5})^2$

S $(\sqrt{3} - \sqrt{5})^2 = (\sqrt{3})^2 - 2(\sqrt{3})(\sqrt{5}) + (\sqrt{5})^2$

$= 3 - 2\sqrt{15} + 5$

$= 8 - 2\sqrt{15}$

c $(\sqrt{2} - 1)(\sqrt{2} + 1)$

S $(\sqrt{2} - 1)(\sqrt{2} + 1) = (\sqrt{2})^2 - 1^2$

$= 2 - 1$

$= 1$

d $(2\sqrt{3} - 1)(\sqrt{5} - \sqrt{2})$

S $(2\sqrt{3} - 1)(\sqrt{5} - \sqrt{2})$

$= 2\sqrt{3}(\sqrt{5} - \sqrt{2}) - 1(\sqrt{5} - \sqrt{2})$

$= 2\sqrt{15} - 2\sqrt{6} - \sqrt{5} + \sqrt{2}$

$= 2\sqrt{3}(\sqrt{5} - \sqrt{2})$

Rationalising the denominator

If the denominator of a fraction is irrational and a single term, we can multiply 'top and bottom' by the same surd to rationalise the denominator.

If the denominator is irrational and is of the form $a + \sqrt{b}$, we can multiply 'top and bottom' by the conjugate (i.e. $a - \sqrt{b}$).

For Example

1 Rationalise the denominator:

a $\dfrac{\sqrt{3} + 1}{\sqrt{2}}$

S
$$\frac{\sqrt{3}+1}{\sqrt{2}} = \frac{\sqrt{3}+1}{\sqrt{2}} \times \frac{\sqrt{2}}{\sqrt{2}}$$
$$= \frac{\sqrt{2}(\sqrt{3}+1)}{\sqrt{4}}$$
$$= \frac{\sqrt{6}+\sqrt{2}}{2}$$

b $\dfrac{\sqrt{2} - 1}{5\sqrt{3}}$

S
$$\frac{\sqrt{2}-1}{5\sqrt{3}} = \frac{\sqrt{2}-1}{5\sqrt{3}} \times \frac{\sqrt{3}}{\sqrt{3}}$$
$$= \frac{\sqrt{3}(\sqrt{2}-1)}{5\sqrt{9}}$$
$$= \frac{\sqrt{6}-\sqrt{3}}{15}$$

Note: we do not need to multiply by $5\sqrt{3}$; $\sqrt{3}$ will do.

c $\dfrac{2}{\sqrt{3} - 1}$

S
$$\frac{2}{\sqrt{3}-1} = \frac{2}{\sqrt{3}-1} \times \frac{\sqrt{3}+1}{\sqrt{3}+1}$$
$$= \frac{2\sqrt{3}+2}{\sqrt{9}-1}$$
$$= \frac{2\sqrt{3}+2}{3-1}$$
$$= \frac{2\sqrt{3}+2}{2}$$
$$= \frac{{}^{1}\cancel{2}(\sqrt{3}+1)}{\cancel{2}_{1}}$$
$$= \sqrt{3}+1$$

d $\dfrac{\sqrt{3} - \sqrt{2}}{\sqrt{3} + \sqrt{2}}$

S
$$\frac{\sqrt{3}-\sqrt{2}}{\sqrt{3}+\sqrt{2}} = \frac{\sqrt{3}-\sqrt{2}}{\sqrt{3}+\sqrt{2}} \times \frac{\sqrt{3}-\sqrt{2}}{\sqrt{3}-\sqrt{2}}$$
$$= \frac{(\sqrt{3}-\sqrt{2})^2}{(\sqrt{3}+\sqrt{2})(\sqrt{3}-\sqrt{2})}$$
$$= \frac{(\sqrt{3})^2 - 2(\sqrt{3})(\sqrt{2}) + (\sqrt{2})^2}{(\sqrt{3})^2 - (\sqrt{2})^2}$$
$$= \frac{3 - 2\sqrt{6} + 2}{3-2}$$
$$= 5 - 2\sqrt{6}$$

Note: we multiply by the conjugate of the denominator—the same expression but with the opposite sign between the terms.

Indices

For x^n, x is the base, and n is the index.

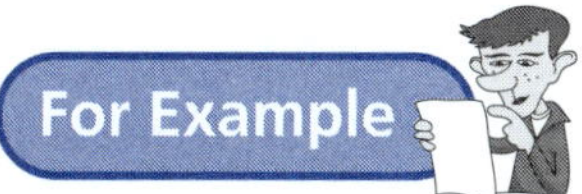

1 Use your calculator to evaluate 3^7.

S We use our x^y button. To find $3^7 = 2187$, key:
3 [x^y] 7 [=]

Index rules

We can use the following four rules:

$x^a \times x^b = x^{a+b}$ $\quad (x^a)^b = x^{ab}$

$x^a \div x^b = x^{a-b}$ $\quad x^0 = 1$ (**Note:** $x \neq 0$)

1 Simplify:

a $p^4 \times p^2$

S $p^4 \times p^2 = p^{(4+2)}$
$= p^6$

b $p^6 \div p$

S $p^6 \div p = p^6 \div p^1$
$= p^{6-1}$
$= p^5$

c $4^5 \div 4^3$

S $4^5 \div 4^3 = 4^{5-3}$
$= 4^2$
$= 16$

d $12a^4b \div 3a^3$

S $12a^4b \div 3a^3 = 4ab$

e $(x^2)^4$

S $(x^2)^4 = x^{2 \times 4}$
$= x^8$

f $(a^{\frac{1}{2}})^3$

S $(a^{\frac{1}{2}})^3 = a^{\frac{3}{2}}$

g $(4y^2)^3$

S $(4y^2)^3 = 4^3 \times (y^2)^3$
$= 64y^6$

h $4x^0$

S $4x^0 = 4 \times x^0$
$= 4 \times 1$
$= 4$

i $(3a)^0 + 3a^0$

S $(3a)^0 + 3a^0 = 1 + 3 \times 1$
$= 1 + 3$
$= 4$

Negative powers

We can use the following rule: $x^{-a} = \dfrac{1}{x^a}$

For Example

1 Simplify:

a 4^{-2}

S $4^{-2} = \dfrac{1}{4^2} = \dfrac{1}{16}$

Note: the negative power is the reciprocal.

b $\dfrac{1}{2}^{-3}$

S $\dfrac{1}{2}^{-3} = \dfrac{1}{\left(\frac{1}{2}\right)^3} = \dfrac{1}{\frac{1}{8}}$
$= 1 \div \dfrac{1}{8}$
$= 1 \times \dfrac{8}{1}$
$= 8$

c $3x^{-2}$

S $3x^{-2} = 3 \times x^{-2}$
$= 3 \times \frac{1}{x^2}$
$= \frac{3}{x^2}$

d $(5x^3)^{-2}$

S $(5x^3)^{-2} = \frac{1}{(5x^3)^2}$
$= \frac{1}{25x^6}$

e $\frac{x^2}{x^5}$

S $\frac{x^2}{x^5} = x^2 \div x^5$
$= x^{-3}$
$= \frac{1}{x^3}$

Fractional powers

The following two rules apply:

- $x^{\frac{1}{2}} = \sqrt[2]{x}$

  That is, $x^{\frac{1}{2}} = \sqrt{x}$, $x^{\frac{1}{3}} = \sqrt[3]{x}$, $x^{\frac{1}{n}} =$ *n*th root of x
- $x^{\frac{a}{b}} = \sqrt[b]{x^a}$

1 Simplify:

a $4^{\frac{1}{2}}$

S $4^{\frac{1}{2}} = \sqrt{4}$
$= 2$

b $8^{\frac{1}{3}}$

S $8^{\frac{1}{3}} = \sqrt[3]{8}$
$= 2$

c $4a^{\frac{1}{3}}$

S $4a^{\frac{1}{3}} = 4\sqrt[3]{a}$

d $(16y)^{\frac{1}{2}}$

S $(16y)^{\frac{1}{2}} = 16^{\frac{1}{2}} \times y^{\frac{1}{2}}$
$= \sqrt{16} \times \sqrt{y}$
$= 4\sqrt{y}$

e $16^{\frac{3}{2}}$

S $16^{\frac{3}{2}} = (16^{\frac{1}{2}})^3$
$= 4^3$
$= 64$

To keep numbers small it is best to take the root first and then calculate the power.

f $16^{-\frac{3}{4}}$

S $16^{-\frac{3}{4}} = (2^4)^{-\frac{3}{4}}$
$= 2^{-3}$
$= \frac{1}{8}$

g $x\sqrt{x}$

S $x\sqrt{x} = x \times x^{\frac{1}{2}}$
$= x^1 \times x^{\frac{1}{2}}$
$= x^{1\frac{1}{2}}$
$= x^{\frac{3}{2}}$
$= \sqrt{x^3}$

h $\frac{x}{\sqrt{x}}$

S $\frac{x}{\sqrt{x}} = x^1 \div x^{\frac{1}{2}}$
$= x^{\frac{1}{2}}$
$= \sqrt{x}$

Algebraic expressions involving negative and fractional powers

Some expressions involving pronumerals can be expressed with negative and/or fractional powers.

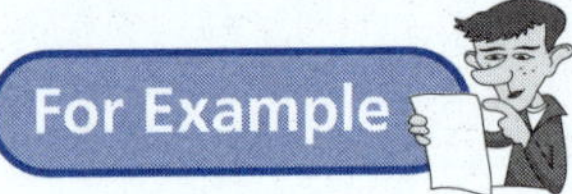

1 Rewrite the following using negative and/or fractional powers:

a $\sqrt{3a+4}$

S $\sqrt{3a+4} = (3a+4)^{\frac{1}{2}}$

b $\dfrac{3}{\sqrt{2x-1}}$

S $\dfrac{3}{\sqrt{2x-1}} = 3(2x-1)^{-\frac{1}{2}}$

c $\dfrac{5a}{\sqrt[3]{(1-2a)^2}}$

S $\dfrac{5a}{\sqrt[3]{(1-2a)^2}} = 5a(1-2a)^{-\frac{2}{3}}$

The calculator and indices

Calculators can be used to evaluate difficult indices.

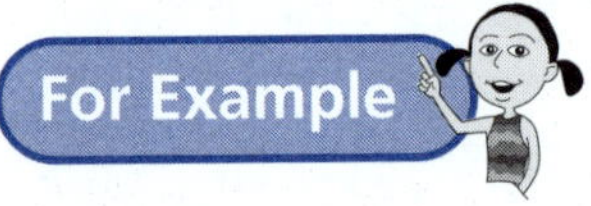

1 Evaluate:

a $256^{\frac{1}{8}}$

S $256^{\frac{1}{8}} = 2$

b $\dfrac{1}{16}^{-\frac{3}{2}}$

S $\dfrac{1}{16}^{-\frac{3}{2}} = 64$

Use your x^y key and enter the power as a fraction.

Solving algebraic problems involving indices

Equations can be solved by equating the indices.

1 Solve for x:

a $2^x = 64$

① Express both sides as a power of the same base number.

② Then the indices must be equal.

S $2^x = 2^6$

$\therefore x = 6$

Both as powers of 2. Indices are equal.

b $3^{x+2} = 81$

S $3^{x+2} = 3^4$

$\therefore x + 2 = 4$

$x = 2$

Both as powers of 3. Indices are equal.

c $25^{2x-1} = 3125$

S $25^{2x-1} = 3125$

$(5^2)^{2x-1} = 5^5$

$5^{4x-2} = 5^5$

$\therefore 4x - 2 = 5$

$4x = 7$

$x = \dfrac{7}{4}$

Powers of 5.

Conversions involving recurring decimals

Decimals can be **terminating** or **recurring** (repeating).

A number is rational if it can be expressed in the form $\frac{p}{q}$ where p and q are integers, and $q \neq 0$.

Examples of rational numbers are:

$\frac{3}{5}, -\frac{3}{4}, 2\frac{1}{3} = \frac{7}{3}$, $24\% = \frac{6}{25}$, $2 = \frac{2}{1}$

$\sqrt{16} = 4 = \frac{4}{1}$, $4:7 = \frac{4}{7}$, $0.\dot{3} = \frac{1}{3}$, etc.

For Example

1 Convert the following to decimals:

a $\frac{5}{8}$

S $\frac{5}{8} = 0.625$

Use calculator: 5 [÷] 8 [=] 0.625

This is a terminating decimal.

b $\frac{4}{9}$

S $\frac{4}{9} = 0.4444\ldots$
$= 0.\dot{4}$

This is a recurring decimal.

c $\frac{8}{11}$

S $\frac{8}{11} = 0.727\,272\ldots$
$= 0.\dot{7}\dot{2}$

d 23%

S $23\% = \frac{23}{100}$
$= 0.23$

Simple recurring decimals can be expressed as rational numbers (fractions) by solving equations.

For Example

1 Express as rationals, that is, in the form $\frac{p}{q}$.

a $0.\dot{7}$

S Let $x = 0.\dot{7}$

$\therefore x = 0.7777\ldots$ (1)

(1) × 10 $\quad 10x = 7.7777\ldots$ (2)

(2) − (1) $\quad 9x = 7$

$x = \frac{7}{9}$

$\therefore 0.\dot{7} = \frac{7}{9}$

b $2.\dot{6}$

S Let $x = 2.\dot{6}$

$\therefore x = 2.666\ldots$ (1)

(1) × 10 $\quad 10x = 26.666\ldots$ (2)

(2) − (1) $\quad 9x = 24$

$x = \frac{24}{9}$

$x = 2\frac{2}{3}$

$\therefore 2.\dot{6} = 2\frac{2}{3}$

c $0.\dot{1}\dot{2}$

S Let $x = 0.\dot{1}\dot{2}$

$\therefore x = 0.1212\ldots \quad (1)$

$(1) \times 100 \quad 100x = 12.1212\ldots \quad (2)$

$(2) - (1) \quad 99x = 12$

$x = \frac{12}{99}$

$x = \frac{4}{33}$

$\therefore 0.\dot{1}\dot{2} = \frac{4}{33}$

Note: all recurring decimals can be expressed in the form $\frac{p}{q}$. In other words, they are all rational numbers.

Scientific notation

Very large or very small numbers are written as a product of a number between 1 and 10, and a power of ten. This is sometimes referred to as **standard notation**.

1 Express in scientific notation:

a 181.4

S $181.4 = 1.814 \times 100$
$= 1.814 \times 10^2$ ← $\therefore$ 2 places

The number of decimal places moved by the decimal point equals the power of ten.

b 2 000 000

S $2\,000\,000 = 2 \times 1\,000\,000$
$= 2 \times 10^6$ ← $\therefore$ 6 places

c 0.0215

S $0.0215 = 2.15 \div 100$
$= 2.15 \times \frac{1}{100}$
$= 2.15 \times 10^{-2}$ ← $\therefore$ 2 places

'Negative power' means 'reciprocal' (or move decimal to the left).

d 0.0001

S $0.0001 = 1 \div 10\,000$
$= 1 \times \frac{1}{10\,000}$
$= 1 \times 10^{-4}$ ← $\therefore$ 4 places

2 Express in ordinary, or decimal, notation:

a 4.271×10^2

S $4.271 \times 10^2 = 4.271 \times 100$
$= 427.1$

The power of ten equals the number of decimal places moved by the decimal point.

b 3.08×10^4

S $3.08 \times 10^4 = 3.08 \times 10\,000$ ← $\therefore$ 4 places
$= 30\,800$

Add some zeros to help.

c 1.56×10^{-2}

S $1.56 \times 10^{-2} = 1.56 \times \frac{1}{100}$
$= 1.56 \div 100$ ← $\therefore$ 2 places
$= 0.0156$

d 5×10^{-3}

S $5 \times 10^{-3} = 5 \times \frac{1}{1000}$
$= 5 \div 1000$ ← $\therefore$ 3 places
$= 0.005$

Scientific notation and the calculator

The calculator expresses very large and very small numbers in scientific notation.

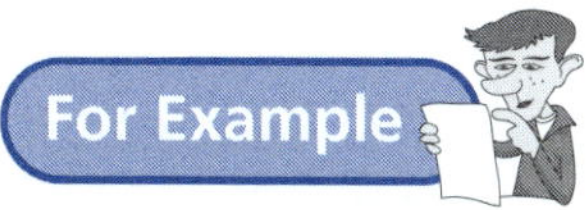

1 Calculate the following expressions, leaving your answer correct to three significant figures.

a $37\,649 \times 28\,547 \times 486$

S $37\,649 \times 28\,547 \times 486$

= 5.223362775 11 ← Calculator display.

$= 5.223\,362\,775 \times 10^{11}$

$= 5.22 \times 10^{11}$

(correct to three significant figures)

b $(0.0023)^4$

S $(0.0023)^4 =$ 2.79841 −11

$= 2.798\,41 \times 10^{-11}$

$= 0.000\,000\,000\,027\,9841$

$= 0.000\,000\,000\,0280$

(correct to three significant figures)

Or

$= 2.80 \times 10^{-11}$

(correct to three significant figures)

c 5^{15}

S $5^{15} =$ 3.051 757 813 10

$= 3.051\,757\,813 \times 10^{10}$

$= 30\,517\,578\,130$

$= 30\,500\,000\,000$

(correct to three significant figures)

Or

$= 3.05 \times 10^{10}$

(correct to three significant figures)

Entering numbers in scientific notation in the calculator

EXP exponent key—this key is used to enter numbers in scientific notation.

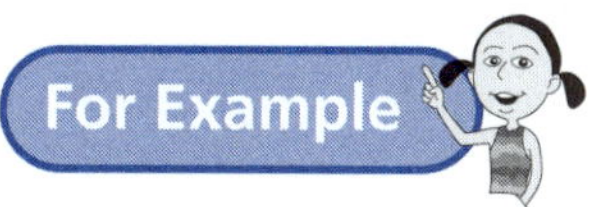

1 Enter the number 2.34×10^5 in the calculator.

S Calculator steps:

2.34 EXP 5

2 Calculate $\sqrt{5.21 \times 10^6}$, giving the answer in scientific notation correct to three significant figures.

S Calculator steps:

√ 5.21 EXP 6

$\therefore 2.28 \times 10^3$

Practise Practise

1 Arrange in ascending order:

a $\sqrt{11}, \sqrt{7}, \sqrt{12}, 3, 5$

b $6, \sqrt{33}, \sqrt{24}, 5, \sqrt{40}$

p. 34

2 Arrange in descending order:

a $\sqrt{15}, \sqrt{8}, \sqrt{17}, 3, 4$

b $8, \sqrt{70}, \sqrt{56}, 7, \sqrt{50}$

p. 34

3 Simplify:

a $\sqrt{8}$ **b** $\sqrt{27}$

c $\sqrt{128}$ **d** $\sqrt{80}$

e $2\sqrt{50}$ **f** $4\sqrt{300}$

g $\sqrt{\frac{4}{9}}$ **h** $\sqrt{\frac{81}{100}}$

i $\sqrt{\frac{3}{4}}$ **j** $(\sqrt{3})^2$

k $(2\sqrt{5})^2$ **l** $(4\sqrt{3})^2$

p. 34

4 Simplify:

a $2\sqrt{3} - \sqrt{3}$ **b** $12\sqrt{5} + \sqrt{5} - 7\sqrt{5}$

c $\sqrt{2} - \sqrt{3} + 3\sqrt{2}$ **d** $\sqrt{63} - \sqrt{28}$

e $2\sqrt{2} + \sqrt{128}$ **f** $x\sqrt{x} + \sqrt{x^3} - 4\sqrt{x}$

p. 35

5 Simplify:

a $\sqrt{3} \times \sqrt{5}$ **b** $\sqrt{2} \times \sqrt{2}$

c $2\sqrt{2} \times 4\sqrt{7}$ **d** $\sqrt{5} \times 2\sqrt{5}$

e $3\sqrt{2} \times -2\sqrt{2}$ **f** $(2\sqrt{3})^2$

pp. 35–36

6 Simplify:

a $\sqrt{9y - 18}$ **b** $\sqrt{8a + 4}$

c $\sqrt{x^3 - x^2}$

pp. 35–36

7 Simplify:

a $\sqrt{\frac{48}{3}}$ **b** $\frac{\sqrt{24}}{\sqrt{6}}$

c $\frac{2\sqrt{5}}{\sqrt{45}}$ **d** $\frac{3\sqrt{10}}{6\sqrt{5}}$

e $\frac{4\sqrt{3} \times \sqrt{2}}{\sqrt{6}}$ **f** $\frac{2\sqrt{3} \times \sqrt{2}}{\sqrt{8}}$

pp. 35–36

8 Expand:

a $\sqrt{2}(\sqrt{2} - 1)$ **b** $\sqrt{5}(3\sqrt{5} + 2)$

c $4\sqrt{3}(2 - \sqrt{3})$ **d** $\sqrt{7}(2\sqrt{7} - \sqrt{2})$

e $2\sqrt{3}(\sqrt{3} + 5\sqrt{2})$ **f** $\sqrt{x}(x - x\sqrt{x})$

pp. 35–36

9 Simplify:

a $\frac{3 + 3\sqrt{3}}{3}$ **b** $\frac{4 - 6\sqrt{2}}{4}$

c $\frac{12 + 3\sqrt{5}}{9}$ **d** $\frac{12 + \sqrt{8}}{8}$

e $\frac{\sqrt{12} - \sqrt{24}}{6}$ **f** $\frac{\sqrt{48} + \sqrt{128}}{4}$

pp. 35–36

10 Expand and simplify:

a $(\sqrt{3} + 2)^2$ **b** $(2\sqrt{2} - 1)^2$

c $(3\sqrt{2} + \sqrt{3})^2$ **d** $(\sqrt{5} - 2)(\sqrt{5} + 2)$

e $(3 - 2\sqrt{2})(3 + 2\sqrt{2})$ **f** $(\sqrt{2} - 1)(\sqrt{3} + 2)$

g $(\sqrt{x} - 1)^2$ **h** $(x\sqrt{x} + 1)^2$

i $(\sqrt{x} - \frac{1}{\sqrt{x}})^2$

p. 36

11 Rationalise the denominator:

a $\frac{1}{\sqrt{6}}$ **b** $\frac{3}{2\sqrt{5}}$

c $\frac{\sqrt{2}}{\sqrt{3}}$ **d** $\frac{\sqrt{2} - 1}{\sqrt{7}}$

e $\frac{1}{2 + \sqrt{3}}$ **f** $\frac{1}{3 - \sqrt{8}}$

g $\frac{12}{\sqrt{7} + \sqrt{3}}$ **h** $\frac{\sqrt{5} + \sqrt{2}}{2\sqrt{3}}$

i $\frac{\sqrt{3} - \sqrt{2}}{\sqrt{3} + \sqrt{2}}$

p. 37

12 Simplify:

a $y^4 \times y^3$ **b** $x^2 \times x^4 \times x$

c $4q^2 \times (-3q^3)$ **d** $a^7 \div a^2 \div a$

e $\frac{m^5}{m^3}$ **f** $16b^6 \div 2b^3$

g $3^2 \times 3$ **h** $5^8 \div 5^7$

i $6^4 \div 6^2 \times 6$ **j** $(x^7)^2$

k $(2a^3)^3$ **i** $(a^4)^{\frac{1}{2}}$

m $3a^0$ **n** $(4a^3 + 2a^3)^0$

o $(2y)^0 + 2y^0$

pp. 37–38

13 Simplify:

a 3^{-2}
b 2^{-3}
c $\left(\frac{1}{2}\right)^{-2}$
d $\left(\frac{3}{5}\right)^{-2}$
e $4x^{-2}$
f $(4x)^{-2}$
g $(2a^4)^{-2}$
h $x^{-2} \times x^{-3}$
i $x^{-2} \div x^{-3}$
j $\frac{y^3}{y^7}$
k $\frac{a^2b}{ab^3}$
l $\frac{10x^5}{5x^7}$

pp. 38–39

14 Simplify:

a $25^{\frac{1}{2}}$
b $27^{\frac{1}{3}}$
c $\left(8^{\frac{1}{3}}\right)^2$
d $1000^{\frac{2}{3}}$
e $4^{\frac{3}{2}}$
f $32^{-\frac{1}{5}}$
g $\left(\frac{1}{4}\right)^{-\frac{1}{2}}$
h $(25a)^{\frac{1}{2}}$
i $\left(64y^6\right)^{\frac{1}{3}}$
j $\left(\frac{1}{9x^4}\right)^{-\frac{1}{2}}$

p. 39

15 Rewrite using negative or fractional powers:

a $\sqrt{2y-1}$
b $\sqrt[3]{x+5}$
c $\frac{1}{5y+2}$
d $\frac{2a}{(3a-1)^2}$
e $\frac{1}{\sqrt{7b-3}}$
f $\frac{1}{\sqrt[3]{1-8p}}$
g $\frac{2y+3}{\sqrt{6y-5}}$
h $\frac{1}{x\sqrt{x}}$
i $\frac{1}{\sqrt[3]{(8y-7)^2}}$

p. 40

16 Calculate:

a $361^{\frac{1}{2}}$
b $81^{\frac{3}{2}}$
c $\left(\frac{4}{9}\right)^{-\frac{1}{2}}$

p. 40

17 Solve, by equating powers:

a $3^x = 81$
b $2^{x+3} = 1024$
c $5^{2x-3} = 125$
d $8^{2x-1} = 512$
e $9^{x+1} = 3^{4x}$
f $64^{\frac{x}{2}} = 4^{x+1}$

p. 40

18 Convert the following to decimals:

a $\frac{3}{7}$
b $\frac{6}{11}$
c $\frac{31}{99}$

p. 41

19 Express as rationals:

a $0.\dot{9}\dot{7}$
b $0.\dot{3}8\dot{1}$
c $1.2\dot{5}$

pp. 41–42

20 Express the following in scientific notation:

a 38 230
b 5 000 000 000
c 0.237
d 0.000 000 3
e 0.09
f 0.001 007

p. 42

21 Rewrite in ordinary form:

a 4.2×10^4
b 1.07×10^{-2}
c 8×10^{-4}

p. 42

22 Calculate, expressing your answer in scientific notation, correct to three significant figures:

a $\frac{125.6 \times 1.09}{0.0032}$
b $\left(\frac{8.038}{0.003 + 0.1392}\right)^3$
c $\frac{\sqrt{10.857 - 9.8}}{3.92^3}$

p. 43

23 Calculate, expressing your answer in scientific notation, correct to three significant figures:

a $(3.76 \times 10^3) \times (1.982 \times 10^5)$
b $\frac{2.35 \times 10^5}{1.0089 \times 10^3}$
c $\sqrt{2.9812 \times 10^{-3}}$

p. 43

Go to pp. 299–300 for **Quick Answers** or to pp. 314–316 for **Worked Solutions**

For a complete understanding of this topic, you must be able to:

✓	Simplify surds		p. 34
✓	Add and subtract surds		p. 35
✓	Multiply and divide surds		pp. 35–36
✓	Expand binomial products involving surds		p. 36
✓	Rationalise the denominator of a surd		p. 37
✓	Simplify expressions involving fractional and negative powers		pp. 37–39
✓	Rewrite algebra expression using fractional and negative powers		p. 40
✓	Solve equations involving fractional and negative powers		p. 40
✓	Convert recurring decimals to the form $\frac{p}{q}$		pp. 41–42
✓	Express numbers in scientific notation.		pp. 42–43

Now you are ready to do the tests!

Intermediate Test

Surds and Indices

(20 marks)

1 Simplify:

a $\sqrt{\dfrac{9}{25}}$ (1 mark)

b $\dfrac{\sqrt{12}}{\sqrt{3}}$ (1 mark)

2 Expand and simplify:

a $3\sqrt{2}(\sqrt{2} - 3)$ (1 mark)

b $(\sqrt{2} - 1)^2$ (2 marks)

3 If $\sqrt{18} + \sqrt{32} = a\sqrt{2}$, find the value of a. (2 marks)

4 If $x = 2 + \sqrt{3}$, write $\dfrac{1}{x}$ with a rational denominator. (2 marks)

5 Simplify: $\dfrac{4\sqrt{3} - \sqrt{2}}{3\sqrt{3}}$ (2 marks)

6 Simplify, $\dfrac{4x^4 \times 6x^2}{12x^9}$, leaving your answer with positive indices. (2 marks)

7 Evaluate $27^{-\frac{2}{3}}$, leaving your answer in the form $\dfrac{p}{q}$. (1 mark)

8 Solve $3^{x-1} = 27$. (2 marks)

9 Simplify:

a $\sqrt{x^4y^6}$ (1 mark)

b $(8x)^{\frac{1}{3}}$ (1 mark)

10 Rewrite $0.\dot{4}\dot{7}$ in the form $\dfrac{p}{q}$. (1 mark)

11 Evaluate $\sqrt{\dfrac{3.9812 - 0.0012}{0.0063^2}}$, expressing your answer in scientific notation correct to three significant figures. (1 mark)

Your Feedback

☐ × 100% = ☐ %

QA PAGE 304

WS PAGE 363

Advanced Test

Surds and Indices

(20 marks)

1 Simplify $\sqrt{a^3} + a\sqrt{a}$. (2 marks)

2 If $y = 2\sqrt{3}$, evaluate $y^4 - 2y^2 + 1$. (2 marks)

3 If $\frac{3 - 2\sqrt{2}}{3 + 2\sqrt{2}} = a + b\sqrt{2}$, find a, b. (2 marks)

4 Find the value of x and y such that:
$\frac{6}{\sqrt{5} + \sqrt{2}} = \sqrt{x} - \sqrt{y}$ (2 marks)

5 Expand and simplify:
$(2\sqrt{3} - \sqrt{11})(2\sqrt{3} + 11)(2\sqrt{2} - 3)(2\sqrt{2} + 3)$ (2 marks)

6 Simplify $8^{\frac{2}{3}} \times 9^{-\frac{3}{2}}$, writing your answer in the form $\frac{a}{b}$. (2 marks)

7 Simplify, leaving your answer with positive indices, $\frac{4a^{-2} \times 3a^{-4}}{12a}$. (1 mark)

8 Solve:
a $4^{x+2} = 32$ (2 marks)
b $3^x - 1 = 0$ (1 mark)

9 If $(2x^m)^3 \times (2x)^{m-2} = nx^2$, find the value of m and n. (3 marks)

10 Rewrite $\frac{5}{\sqrt[3]{1 - 3x + x^2}}$ using fractional and negative indices. (1 mark)

Your Feedback

☐ × 100% = ☐ %

QA PAGE 304
WS PAGE 363

Chapter 4
Equations

Simple equations and inequations

Solutions of equations

The main rules to remember are:

- Pronumeral on one side and number on the other side.
- To *change sides* means we have to *change signs*.

Keywords

Coefficient	Literal
Elimination	Quadratic
Equation	Simultaneous
Exact	Solution
Factorise	Subject
Formula	Substitution
Formulae	Surd
Inequality	Variable

For Example

1 Solve:

a $3x - 1 = x + 2$

this is $+3x$; this is -1; this is $+x$; this is $+2$

S

$3x - 1 = x + 2$

$3x - x = 2 + 1$

$2x = 3$

$\frac{{}^1\cancel{2}x}{\cancel{2}_1} = \frac{3}{2}$ ← We divide by the coefficient (number in front of *x*).

$x = 1\frac{1}{2}$

b $5y - 7 = 3(2y + 4)$ ← First we expand the bracket.

S

$5y - 7 = 3(2y + 4)$

$5y - 7 = 6y + 12$

$5y - 6y = 12 + 7$

$-y = 19$

$\frac{-y}{-1} = \frac{19}{-1}$ ← The coefficient of *y* is -1

$y = -19$

c $\frac{3a - 4}{2} = 8$

S

$\frac{3a - 4}{2} = 8$

We have to get rid of denominators, so we multiply both sides by 2.

${}^1\cancel{2}\ \frac{3a - 4}{\cancel{2}_1} = 8 \times 2$

$3a - 4 = 16$

$3a = 16 + 4$

$3a = 20$

$\frac{{}^1\cancel{3}a}{\cancel{3}_1} = \frac{20}{3}$

$a = 6\frac{2}{3}$

We can leave this as a mixed numeral or $6.\dot{6}$.

d $\frac{5a - 1}{2} = \frac{4a + 8}{3}$

S

$\frac{5a - 1}{2} = \frac{4a + 8}{3}$

Best to multiply by lowest common denominator.

${}^3\cancel{6}\ \frac{5a - 1}{\cancel{2}_1} = {}^2\cancel{6}\ \frac{4a + 8}{\cancel{3}_1}$

$3(5a - 1) = 2(4a + 8)$

$15a - 3 = 8a + 16$

$15a - 8a = 16 + 3$

$7a = 19$

$\frac{{}^1\cancel{7}a}{\cancel{7}_1} = \frac{19}{7}$

$a = 2\frac{5}{7}$

Note that we could have jumped to this step by cross-multiplying: ⤫

e $\frac{5}{2a} - \frac{3}{4a} = \frac{1}{2}$

S

$$\frac{5}{2a} - \frac{3}{4a} = \frac{1}{2}$$

$$^{2}\not{4}a\,\frac{5}{\not{2}a_1} - {^{1}\not{4}a}\,\frac{3}{\not{4}a_1} = {^{2}\not{4}a}\,\frac{1}{\not{2}_1}$$

$$2(5) - 1(3) = 2a(1)$$

$$10 - 3 = 2a$$

$$7 = 2a$$

$$2a = 7$$

Simply swap sides around.

$$\frac{^{1}\not{2}a}{\not{2}_1} = \frac{7}{2}$$

$$a = 3\frac{1}{2}$$

f $\frac{5}{y} - \frac{2}{y-1} = 0$

S

$$\frac{5}{y} - \frac{2}{y-1} = 0$$

$$\frac{5}{y} = \frac{2}{y-1}$$

$$5(y - 1) = 2y$$

$$5y - 5 = 2y$$

$$5y - 2y = 5$$

$$3y = 5$$

$$\frac{^{1}\not{3}y}{\not{3}_1} = \frac{5}{3}$$

$$y = 1\frac{2}{3}$$

By cross-multiplying:

Solution of inequalities

Algebraic expressions with an inequality symbol have a range of solutions. An important rule is:

- When multiplying or dividing an inequality by a negative, we must reverse the inequality symbol.

1 Solve:

a $3x - 4 > x + 7$

S

$$3x - 4 > x + 7$$

$$3x - x > 7 + 4$$

$$2x > 11$$

$$\frac{2x}{2} > \frac{11}{2}$$

$$x > 5\frac{1}{2}$$

b $x + 7 \geq 3x - 4$

S

$$x + 7 \geq 3x - 4$$

$$x - 3x \geq -4 - 7$$

$$-2x \geq -11$$

$$\frac{-2x}{-2} \leq \frac{-11}{-2}$$

$$x \leq 5\frac{1}{2}$$

When dividing by a negative we reverse the inequality.

c $7 - 5x > 12$

S

$$7 - 5x > 12$$

$$-5x > 12 - 7$$

$$-5x > 5$$

$$\frac{-5x}{-5} < \frac{5}{-5}$$

$$x < -1$$

d $4(3 - 2x) \geq \frac{2x - 5}{5}$

S

$$4(3 - 2x) \geq \frac{2x - 5}{5}$$

$$5[4(3 - 2x)] \geq {^{1}\not{5}}\,\frac{2x - 5}{\not{5}_1}$$

$$20(3 - 2x) \geq 2x - 5$$

$$60 - 40x \geq 2x - 5$$

$$-40x - 2x \geq -5 - 60$$

$$-42x \geq -65$$

$$\frac{-{^{1}\not{42}}x}{-\not{42}_1} \leq \frac{-65}{-42}$$

$$x \leq 1\frac{23}{42}$$

Graphing on the number line

When graphing a solution on a number line, a closed circle is used to denote inclusion (e.g. $x \geq 3$), while an open circle is used to denote an exclusion (e.g. $x > 3$).

For Example 

1 Graph on separate number lines:

a $x \geq 2$

S

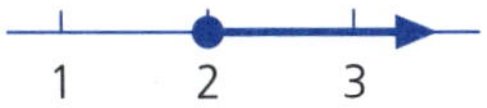

We use a closed circle, to denote the inclusion of 2.

Hint: the arrow-head looks like the inequality symbol.

b $x < -1$

S

We use an open circle to denote the *exclusion* of –1.

c $-2 < x \leq 1$

S

We want all values of x greater than –2 but less than or equal to 1.

2 Write down the inequality that corresponds to the graph.

a

S $x < 2$

b

S $-2 < x \leq 4$

c

S $x < 0, x > 3$

3 Solve, and graph your solution on a number line:

a $2x - 4 > x - 1$

S

$$2x - 4 > x - 1$$
$$2x - x > -1 + 4$$
$$x > 3$$

b $4 \leq 3(4 - 5x)$

S

$$4 \leq 3(4 - 5x)$$
$$4 \leq 12 - 15x$$
$$15x \leq 12 - 4$$
$$15x \leq 8$$
$$\frac{\cancel{15}x}{\cancel{15}} \leq \frac{8}{15}$$
$$x \leq \frac{8}{15}$$

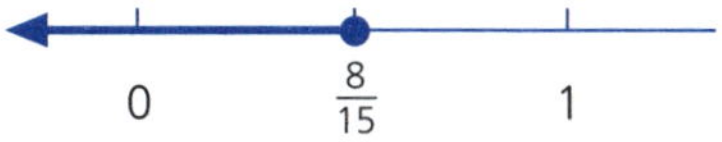

Formulae

Formulae are widely used and usually represent physical quantities. Unlike equations, formulae involve more than one pronumeral. The subject of a formula is the pronumeral on the left-hand side of the equals sign (the pronumeral by itself). A formula can also be called a literal equation.

Substitution into a formula

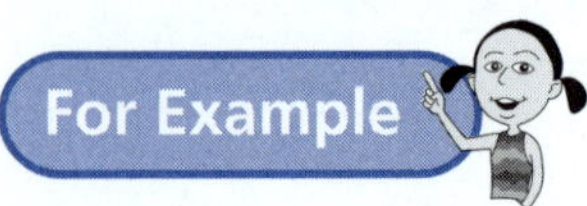

1 If $S = n(n + 1)$, find S when $n = 16$.

S
$$\begin{aligned} S &= n(n + 1) \\ &= 16(16 + 1) \\ &= 16(17) \\ &= 272 \end{aligned}$$

2 If $A = P\left(1 + \frac{r}{100}\right)^n$, find A when:
$P = 10\,000$, $r = 20$, $n = 2$.

S
$$\begin{aligned} A &= P\left(1 + \frac{r}{100}\right)^n \\ &= 10\,000\left(1 + \frac{20}{100}\right)^2 \\ &= 10\,000\,(1.2)^2 \\ &= 14\,400 \end{aligned}$$

3 If $T = \frac{n}{2}[2a + (n - 1)d]$, find T if:
$a = 6$, $d = 3$, $n = 10$.

S
$$\begin{aligned} T &= \frac{n}{2}[2a + (n - 1)d] \\ &= \frac{10}{2}[2(6) + (10 - 1) \times 3] \\ &= 5(12 + 9 \times 3) \\ &= 5(12 + 27) \\ &= 5(39) \\ &= 195 \end{aligned}$$

4 If $M = \frac{1}{M_1} + \frac{1}{M_2}$, find M when $M_1 = 1.6$ and $M_2 = 0.5$.

S
$$\begin{aligned} M &= \frac{1}{M_1} + \frac{1}{M_2} \\ &= \frac{1}{1.6} + \frac{1}{0.5} \\ &= 0.625 + 2 \\ &= 2.625 \end{aligned}$$

Calculator

1.6 [1/x] [+] 0.5 [1/x] [=]

Note: your calculator may require: [INV] [1/x]

Constructing formulae

A formula relating two or more variables can be developed from given information.

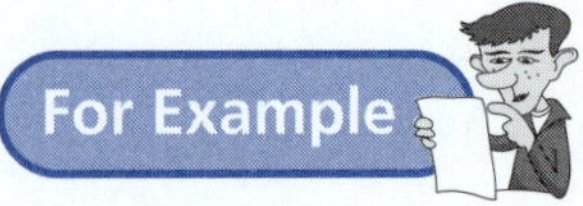

1 Write down a formula for the speed, in kilometres/hour if a vehicle has travelled D kilometres in T hours.

S $S = \frac{D}{T}$

2 The table shows the relation between speed S km/h and the stopping distance D metres, of a passenger van.

S	40	60	80	100
D	16	36	64	100

a Find a formula linking D with S, that is, expressing D in terms of S.

S We can see that D is found by dividing S by 10 and then squaring:

That is $D = \left(\frac{S}{10}\right)^2$

Therefore $D = \frac{S^2}{100}$

b Find D when $S = 70$.

S $D = \frac{70^2}{100}$
$= \frac{4900}{100}$
$= 49$

c What would be the stopping distance of a passenger van travelling at 120 km/h?

S $D = \frac{S^2}{100}$
$= \frac{(120)^2}{100}$
$= \frac{14400}{100}$
$= 144$

The distance is 144 metres.

3 Consider a square garden of side x metres. A path p metres wide surrounds a square area of lawn with side y metres, as shown in the diagram.

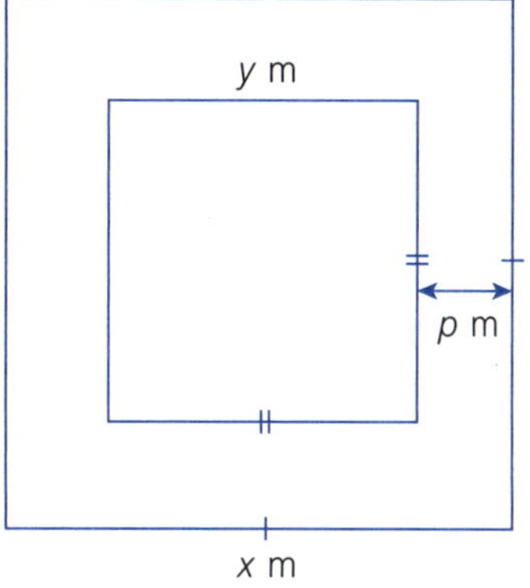

Write down a formula for:

a y in terms of x and p

S $2p = x - y$
$y = x - 2p$

Path on both sides ∴ $2p$.

b the area A of the path in terms of x and y

S $A = x^2 - y^2$

Area of larger square minus area of smaller square.

c the area A of the path in terms of x and p.

S $A = x^2 - y^2$
But $y = x - 2p$ (From **a**)
Therefore $A = x^2 - (x - 2p)^2$
$= x^2 - (x^2 - 4xp + 4p^2)$
$= x^2 - x^2 + 4xp - 4p^2$
$= 4xp - 4p^2$
$= 4p(x - p)$

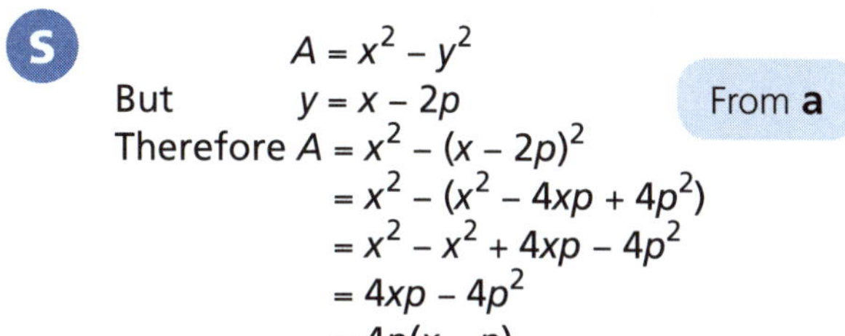

Changing the subject of a formula (solving literal equations)

A formula such as $v^2 = u^2 - 2as$ has v^2 as the subject and we can easily find v, if we know u, a and s, by simple substitution (and taking the square root). However to find u, a or s it is best to change the subject to the unknown pronumeral. This is known as **solving a literal equation**.

1 Make y the subject:

a $2x + y = 7$

S $y = 7 - 2x$

b $3x - y = 5$

S $3x - 5 = y$

$y = 3x - 5$

Taking y to the RHS changes its sign. Then we just swap sides.

Note: we could have left the y on the LHS as $-y$ and then divide through by -1.

c $4x - 3y = 12$

S $4x - 12 = 3y$

$3y = 4x - 12$

$\frac{\cancel{3}y}{\cancel{3}} = \frac{4x - 12}{3}$

$y = \frac{4x - 12}{3}$

d $xy - 7 = 3a + 2b$

S $xy = 3a + 2b + 7$

$\frac{\cancel{x}y}{\cancel{x}} = \frac{3a + 2b + 7}{x}$

$y = \frac{3a + 2b + 7}{x}$

Divide both sides by x.

e $3(4x - 2y) = 13x + 1$

S $12x - 6y = 13x + 1$

$12x - 13x - 1 = 6y$

$6y = -x - 1$

$\frac{\cancel{6}y}{\cancel{6}} = \frac{-x - 1}{6}$

$y = \frac{-x - 1}{6}$

Expand the bracket first.

f $A = \frac{3xy}{7}$

S $7[A] = \frac{3xy}{\cancel{7}_1}\ \cancel{7}_1$

$7A = 3xy$

$3xy = 7A$

$\frac{3\cancel{x}y}{3\cancel{x}} = \frac{7A}{3x}$

$y = \frac{7A}{3x}$

Multiply both sides by 7.

g $P = \frac{3x - 4}{y}$

S $y[P] = \frac{3x - 4}{\cancel{y}_1}\ \cancel{y}_1$

$Py = 3x - 4$

$\frac{\cancel{P}y}{\cancel{P}} = \frac{3x - 4}{P}$

$y = \frac{3x - 4}{P}$

h $x = \sqrt{\frac{A}{y}}$

S $x^2 = \frac{A}{y}$

$x^2y = A$

$y = \frac{A}{x^2}$

Squaring both sides.

i $3xy = 7x - y$

S $3xy + y = 7x$

$y(3x + 1) = 7x$

$\frac{y(\cancel{3x + 1})}{\cancel{3x + 1}} = \frac{7x}{3x + 1}$

$y = \frac{7x}{3x + 1}$

Take y as a common factor.

j $x^2 = y^2 - 4x$

S $y^2 = x^2 + 4x$

$y = \pm\sqrt{x^2 + 4x}$

Don't forget both cases.

k $x = \frac{y}{y + 3}$

S $x(y + 3) = y$

$xy + 3x = y$

$xy - y = -3x$

$y(x - 1) = -3x$

$\frac{y(\cancel{x - 1})}{\cancel{x - 1}} = \frac{-3x}{x - 1}$

$y = \frac{-3x}{x - 1}$

Cross-multiplying

This could be written as $\frac{\cancel{-}3x}{\cancel{-}(1 - x)} = \frac{3x}{1 - x}$

Solving problems involving formulae

A word problem can be solved by forming and then solving an equation.

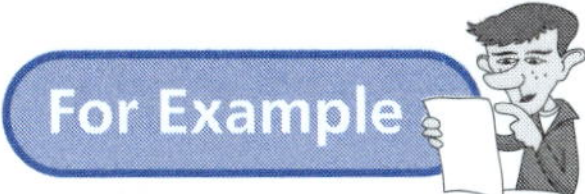

1 Write down the formula for the curved surface area A of a cylinder:

$A = 2\pi rh$

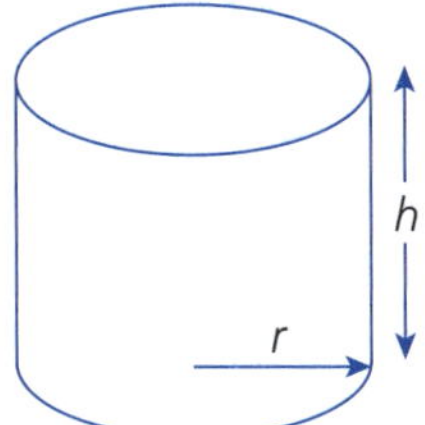

a Find the area in terms of π if the radius is 12 cm and the height is 10 cm.

S $A = 2\pi rh$

$A = 2 \times \pi \times 12 \times 10$
$= 240\pi$

The area is 240π cm^2.

b Find the height h if the area A of the curved surface is 132π cm^2 and the radius is 6 cm.

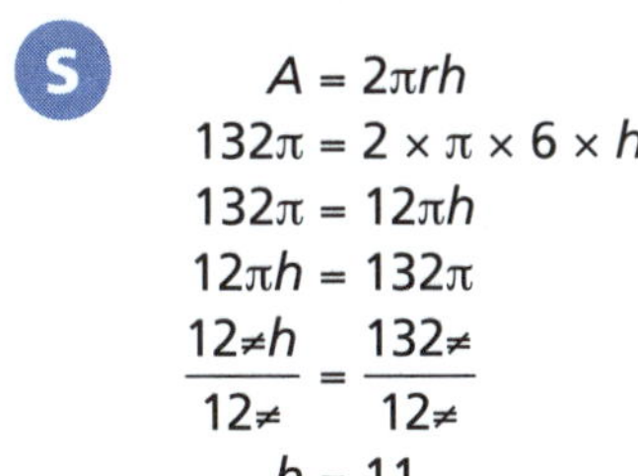

S

$$A = 2\pi rh$$
$$132\pi = 2 \times \pi \times 6 \times h$$
$$132\pi = 12\pi h$$
$$12\pi h = 132\pi$$
$$\frac{12\not\pi h}{12\not\pi} = \frac{132\not\pi}{12\not\pi}$$
$$h = 11$$

The height is 11 cm.

2 Write down the formula for the area A of the annulus (right) if the radius of the larger circle is R while the radius of the smaller circle is r.

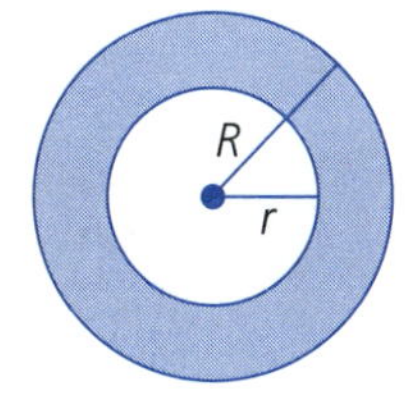

That is,
$A = \pi R^2 - \pi r^2$
$= \pi(R^2 - r^2)$.

a Find the area of an annulus where $R = 6.2$ cm and $r = 4.7$ cm, correct to three significant figures.

S

$$A = \pi(R^2 - r^2)$$
$$= \pi(6.2^2 - 4.7^2)$$
$$= \pi(38.44 - 22.09)$$
$$= \pi(16.35)$$
$$= 51.365\,039\,89$$
$$= 51.4 \text{ (to three significant figures)}$$

The area is 51.4 cm^2.

b Find the radius of the smaller circle if the radius of the larger circle is 10 cm and the area of the annulus is 36π cm^2.

S

$$A = \pi(R^2 - r^2)$$
$$36\pi = \pi(10^2 - r^2)$$
$$36\pi = \pi(100 - r^2)$$

Divide by π:

$$36 = 100 - r^2$$
$$r^2 = 100 - 36$$
$$r^2 = 64$$
$$r = 8 \; (r > 0)$$

The radius of the smaller circle is 8 cm.

Simultaneous equations

If an equation has two unknowns, for example $x + y = 4$, there is an infinite number of solutions. If there is a pair of equations with two unknowns we may be able to find a solution to satisfy both equations at the same time, that is, simultaneously.

There are three methods of solution: graphically, algebraically through *substitution*, and algebraically through *elimination*. The latter two are treated below.

The substitution method

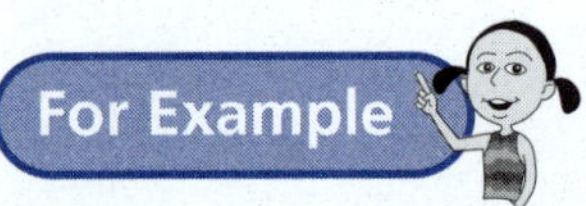

1 Solve the simultaneous equations:

a $2x + y = 5$ (1)
$5x - 3y = 7$ (2)

We number our equations to help.

From (1), make y the subject:
$y = 5 - 2x$ (3)

It is better here to choose y as it is a monic term (i.e. has a coefficient of 1).

Now substitute into (2):

$$5x - 3(5 - 2x) = 7$$
$$5x - 15 + 6x = 7$$
$$11x - 15 = 7$$
$$11x = 7 + 15$$
$$11x = 22$$
$$x = 2$$

Now we substitute $x = 2$ in (3).

Equation (3) is 'simpler'.

$$y = 5 - 2(2)$$
$$y = 5 - 4$$
$$\therefore y = 1$$

Therefore $x = 2$, $y = 1$.
[*Or* (2, 1) as an ordered pair.]

It is good to write the complete solution here.

Note: we could check this answer by substituting into both equations, (1) and (2).

b $3x - 4y = 11$
$2x + 3y = -4$

S $3x - 4y = 11$ (1)
$2x + 3y = -4$ (2)

From (2),

$$2x = -4 - 3y$$
$$x = \frac{-4 - 3y}{2}$$

Not as easy as Example 1a; we choose the term with smallest coefficient.

Substituting in (1):

$$3\ \frac{-4 - 3y}{2} - 4y = 11$$
$$3(-4 - 3y) - 8y = 22$$

By multiplying through by 2.

$$-12 - 9y - 8y = 22$$
$$-12 - 17y = 22$$
$$-17y = 22 + 12$$
$$-17y = 34$$
$$y = -2$$

By dividing through by −17.

Substituting $y = -2$ in (2):

$$2x + 3(-2) = -4$$
$$2x - 6 = -4$$
$$2x = -4 + 6$$
$$2x = 2$$
$$x = 1$$

Therefore, $x = 1$, $y = -2$.

The elimination method

The second example above could have been solved in a simpler way by using the elimination method. One of the pronumerals is eliminated.

1 Solve:

a $2x - 3y = 3$ (1)
$4x + 3y = 15$ (2)

S Add equations (1) and (2):
$6x = 18$
$x = 3$

The pronumeral y has been eliminated by adding.

From here the method is the same as the substitution method.

Substituting in (2):
$4(3) + 3y = 15$
$12 + 3y = 15$
$3y = 15 - 12$
$3y = 3$
$y = 1$

Remember that we can substitute into either equation—choose the simpler.

Therefore, $x = 3$ and $y = 1$.

b $3x - 4y = 11$ (1)
$2x + 3y = -4$ (2)

S $2 \times (1)$ $6x - 8y = 22$ (3)
$3 \times (2)$ $6x + 9y = -12$ (4)

We chose to eliminate the pronumeral x by using the lowest common multiple, 6.

Subtracting (4) from (3):
$-17y = 34$
$y = -2$

Substituting in (2):
$2x + 3(-2) = -4$
$2x - 6 = -4$
$2x = -4 + 6$
$2x = 2$
$x = 1$

Therefore $x = 1$, $y = -2$.

Using equations to solve problems

Problems can be solved by introducing pronumerals to form equations.

1 The sum of two consecutive even numbers is 22. Find the numbers.

S Let the numbers be x and $x + 2$.
Then $x + x + 2 = 22$
$2x + 2 = 22$
$2x = 20$
$x = 10$

Therefore, $x = 10$, $x + 2 = 12$, and so the numbers are 10 and 12.

2 Find two numbers such that their sum is 20, while half their difference is 1.

S Let the numbers be x and y.
Then $x + y = 20$ (1)
$\frac{1}{2}(x - y) = 1$ (2)
$2 \times (2)$ $x - y = 2$ (3)
$[(1) - (3)]$ $2y = 18$
$y = 9$

Substituting $y = 9$ in (1):
$x + 9 = 20$
$x = 20 - 9$
$x = 11$

Therefore, $x = 11$, $y = 9$, and so the numbers are 11 and 9.

3 If the sides of an isosceles triangle are $(x + 20)$ cm, $(3x - 16)$ cm and $(x + 2)$ cm, what are the possible values of x?

S Two of the three sides are equal. We form equations and solve them to find value(s) of x, but there is no point in setting $x + 20 = x + 2$ as this is nonsense.

$$\begin{aligned} 3x - 16 &= x + 20 \\ 3x - x &= 20 + 16 \\ 2x &= 36 \\ x &= 18 \end{aligned} \qquad \text{or} \qquad \begin{aligned} 3x - 16 &= x + 2 \\ 3x - x &= 2 + 16 \\ 2x &= 18 \\ x &= 9 \end{aligned}$$

But $x \neq 9$ as the sides are then 11 cm, 11 cm and 29 cm, ∴ not a triangle.

Therefore, $x = 18$ is the only solution.

4 John is 10 years older than Peter, but 25 years ago, John was twice Peter's age. Find Peter's present age.

S Let Peter's age $= x$
John's age $= x + 10$

Twenty-five years ago:
Peter $= x - 25$
John $= x + 10 - 25 = x - 15$

Therefore 25 years ago:

$$\begin{aligned} 2(x - 25) &= x - 15 \\ 2x - 50 &= x - 15 \\ 2x - x &= 50 - 15 \\ x &= 35 \end{aligned}$$

Therefore Peter's present age is 35 years.

5 Daniel has 22 coins, each being either a one-dollar coin or a two-dollar coin. He has a total of \$35. How many of each coin does he have?

S Let x be the number of \$1 coins, and y be the number of \$2 coins.

The equations are:
$x + y = 22$ [he has 22 coins] (1)
$x + 2y = 35$ [as \$1 and \$2] (2)
[(2) – (1)] $y = 13$

Substituting $y = 13$ in (1):
$x + 13 = 22$
$x = 9$

Therefore he has nine \$1 coins and thirteen \$2 coins.

6 Find the value of x and y for the rectangle:

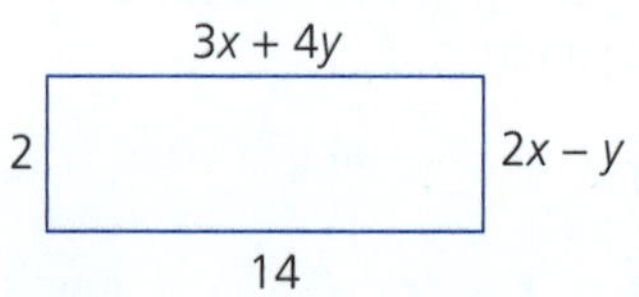

S

$$\begin{aligned} 3x + 4y &= 14 && (1) \\ 2x - y &= 2 && (2) \\ 4 \times (2) \therefore\ 8x - 4y &= 8 && (3) \\ [(1) + (3)] \quad 11x &= 22 \\ x &= 2 \end{aligned}$$

Substituting $x = 2$ in (2)

$$\begin{aligned} 4 - y &= 2 \\ -y &= 2 \\ -y &= -2 \\ y &= 2 \end{aligned}$$

Therefore $x = 2$ and $y = 2$.

7 Three peaches and four plums cost \$2.75, while five peaches and two plums cost \$2.95. Find the price of each.

S Let a = price of a peach, b = price of a plum.

$$\begin{aligned} \therefore \quad 3a + 4b &= 275 && (1) \\ 5a + 2b &= 295 && (2) \\ 2 \times (2) \quad 10a + 4b &= 590 && (3) \\ \therefore\ [(3) - (1)] \quad 7a &= 315 \\ \therefore \quad a &= 45 \end{aligned}$$

Substituting $a = 45$ in (1)

$$\begin{aligned} \therefore \quad 135 + 4b &= 275 \\ \therefore \quad 4b &= 275 - 135 \\ 4b &= 140 \\ b &= 35 \end{aligned}$$

∴ Peach costs 45c, plum costs 35c.

8 Tickets for a concert were \$8 for adults and \$5 for children. If 200 tickets were sold and \$1420 in ticket sales was collected, how many adult tickets were sold?

S Let a be the number of adult tickets and c be the number of children's tickets.

$$\begin{aligned} \therefore\ a + c &= 200 && (1) \\ 8a + 5c &= 1420 && (2) \\ \text{From (1)} \quad c &= 200 - a && (3) \end{aligned}$$

Subs. (3) in (2)

$$\begin{aligned} 8a + 5(200 - a) &= 1420 \\ 8a + 1000 - 5a &= 1420 \\ 3a &= 420 \\ a &= 140 \end{aligned}$$

∴ There were 140 adult tickets sold.

Quadratic equations

A quadratic equation is in the form $ax^2 + bx + c = 0$, where $a \neq 0$.

Using the result that if $ab = 0$, then either $a = 0$ or $b = 0$ (or $a = b = 0$), we attempt to factorise our quadratic expression and then solve the equation.

In this book we will focus on solving quadratic equations by first factorising.

Note: students should be confident with Chapter 2 work on factorising quadratic trinomials.

For Example

1 Solve:

a $x^2 - 5x - 6 = 0$

S $(x - 6)(x + 1) = 0$
$x = 6, -1$

We could check by substitution.

This means that $x = 6$ or -1.

b $x^2 - 4x = -4$

S $x^2 - 4x + 4 = 0$
$(x - 2)(x - 2) = 0$
$x = 2$

We must have the quadratic equalling zero.

c $x^2 = 5x$

S $x^2 - 5x = 0$
$x(x - 5) = 0$
$x = 0, 5$

d $6 + x - x^2 = 0$

S Multiply through by -1.

$x^2 - x - 6 = 0$
$(x - 3)(x + 2) = 0$
$x = 3, -2$

e $2x^2 + 5x - 3 = 0$

S $(2x - 1)(x + 3) = 0$
$2x - 1 = 0 \quad x + 3 = 0$
$2x = 1 \quad x = -3$
$x = \frac{1}{2}$

Therefore, $x = \frac{1}{2}, -3$.

f $3x^2 - 8x + 5 = 0$

S $(3x - 5)(x - 1) = 0$
$3x = 5 \quad x = 1$
$x = \frac{5}{3}$

Therefore, $x = 1\frac{2}{3}, 1$.

g $x^2 - 5x + 6 = 2$

S $x^2 - 5x + 6 - 2 = 0$
$x^2 - 5x + 4 = 0$
$(x - 4)(x - 1) = 0$
$x = 4, 1$

We must always make one side equal to zero.

h $2(x - 3)(x - 2) = x^2 - 3x + 2$

S $2(x^2 - 5x + 6) = x^2 - 3x + 2$
$2x^2 - 10x + 12 = x^2 - 3x + 2$
$2x^2 - 10x + 12 - x^2 + 3x - 2 = 0$
$x^2 - 7x + 10 = 0$
$(x - 5)(x - 2) = 0$
$x = 5, 2$

i $x = \frac{x + 12}{x}$

S $x^2 = x + 12$
$x^2 - x - 12 = 0$
$(x - 4)(x + 3) = 0$
$x = 4, -3$

Multiply through by x.

j $(x - 5)^2 = 9$

S $x - 5 = \pm\sqrt{9}$
$x - 5 = \pm 3$

Take the square root of both sides.

Therefore, $x = 8, 2$.

Solving quadratics by completing the square

A quadratic equation can be solved by adding a constant to both sides of the equation so as to form a perfect square.

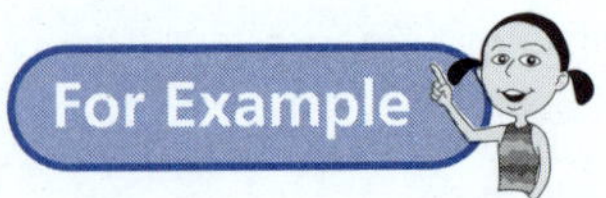

1 Solve these quadratic equations by completing the square. Write your answer correct to one decimal place if necessary:

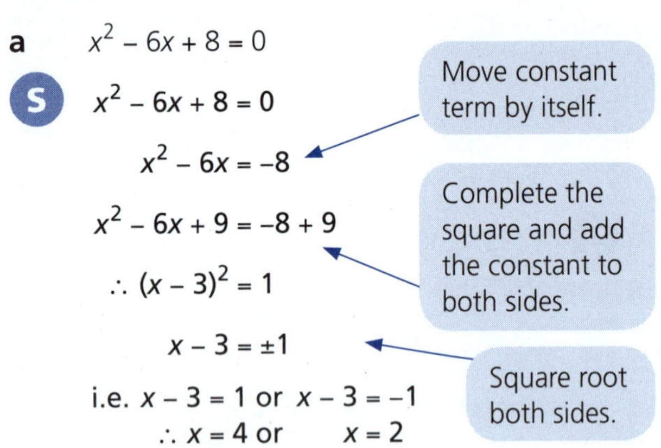

a $x^2 - 6x + 8 = 0$

S $x^2 - 6x + 8 = 0$

$x^2 - 6x = -8$ ← Move constant term by itself.

$x^2 - 6x + 9 = -8 + 9$ ← Complete the square and add the constant to both sides.

$\therefore (x - 3)^2 = 1$

$x - 3 = \pm 1$ ← Square root both sides.

i.e. $x - 3 = 1$ or $x - 3 = -1$

$\therefore x = 4$ or $x = 2$

b $y^2 - 6y + 7 = 0$

S $y^2 - 6y + 7 = 0$

$y^2 - 6y = -7$ ← Move constant term by itself.

$y^2 - 6y + 9 = -7 + 9$ ← Complete the square and add the constant to both sides.

$(y - 3)^2 = 2$

$y - 3 = \pm\sqrt{2}$ ← Square root both sides.

$y - 3 = \sqrt{2}$ or $y - 3 = -\sqrt{2}$

$y = 3 + \sqrt{2}$ or $3 - \sqrt{2}$

$y = 4.4$ or 1.6

The quadratic formula

When we cannot factorise the quadratic, we can use the quadratic formula:

For $ax^2 + bx + c = 0,\ a \neq 0$

$$x = \frac{-b \pm \sqrt{b^2 - 4ac}}{2a}$$

Note: the algebraic expression 'housed under' the square root has a great bearing on the final solution—it determines whether the solutions are *surds*, or there is only *one solution*, or there is *no real solution*.

This expression, $b^2 - 4ac$, is called the *discriminant*.

1 Solve the equations, leaving answers as simplified surds:

a $x^2 - 5x + 2 = 0$

S Here $a = 1$, $b = -5$, $c = 2$.

Note: it is a good idea to write down the values for *a*, *b* and *c* before substituting in the formula.

$$x = \frac{-b \pm \sqrt{b^2 - 4ac}}{2a}$$

$$= \frac{-(-5) \pm \sqrt{(-5)^2 - 4(1)(2)}}{2(1)}$$

$$= \frac{5 \pm \sqrt{25 - 8}}{2}$$

$$= \frac{5 \pm \sqrt{17}}{2}$$

This is often called the *exact form*.

b $3x^2 - 4x - 5 = 0$

S $x = \dfrac{-b \pm \sqrt{b^2 - 4ac}}{2a}$ $(a = 3,\ b = -4,\ c = -5)$

$$x = \frac{-(-4) \pm \sqrt{(-4)^2 - 4(3)(-5)}}{2(3)}$$

$$= \frac{4 \pm \sqrt{16 + 60}}{6}$$

$$= \frac{4 \pm \sqrt{76}}{6}$$

$$= \frac{4 \pm 2\sqrt{19}}{6}$$

$$= \frac{{}^{1}\not{2}(2 \pm \sqrt{19})}{\not{6}_{3}}$$

$$= \frac{2 \pm \sqrt{19}}{3}$$

2 Solve, correct to two decimal places if necessary:

a $2x^2 + 3x + 8 = 0$

S $x = \dfrac{-b \pm \sqrt{b^2 - 4ac}}{2a}$ $(a = 2, b = 3, c = 8)$

$x = \dfrac{-3 \pm \sqrt{3^2 - 4(2)(8)}}{2(2)}$

$= \dfrac{-3 \pm \sqrt{9 - 64}}{4}$

$= \dfrac{-3 \pm \sqrt{-55}}{4}$

No real solution.

Cannot find $\sqrt{-55}$ in 'real' number system.

b $2 - 4x - 5x^2 = 0$

S $x = \dfrac{-b \pm \sqrt{b^2 - 4ac}}{2a}$ $(a = -5, b = -4, c = 2)$

$= \dfrac{-(-4) \pm \sqrt{(-4)^2 - 4(-5)(2)}}{2a}$

$= \dfrac{4 \pm \sqrt{16 + 40}}{-10}$

$= \dfrac{4 \pm \sqrt{56}}{-10}$

$= \dfrac{4 + \sqrt{56}}{-10}$ or $\dfrac{4 - \sqrt{56}}{-10}$

$= -1.15, 0.35$

c $2x^2 - 5x - 1 = 0$

S $x = \dfrac{-b \pm \sqrt{b^2 - 4ac}}{2a}$ $(a = 2, b = -5, c = -1)$

$= \dfrac{-(-5) \pm \sqrt{(-5)^2 - 4(2)(-1)}}{2(2)}$

$= \dfrac{5 \pm \sqrt{25 + 8}}{4}$

$= \dfrac{5 \pm \sqrt{33}}{4}$

$= \dfrac{5 + \sqrt{33}}{4}$ or $\dfrac{5 - \sqrt{33}}{4}$

$= 2.69$ or -0.19

d $5x^2 = 2x + 1$

S $5x^2 - 2x - 1 = 0$

$x = \dfrac{-b \pm \sqrt{b^2 - 4ac}}{2a}$

$= \dfrac{-(-2) \pm \sqrt{(-2)^2 - 4(5)(-1)}}{2(5)}$

$= \dfrac{2 \pm \sqrt{4 + 20}}{10}$

$= \dfrac{2 \pm \sqrt{24}}{10}$

$= \dfrac{2 + \sqrt{24}}{10}$ or $\dfrac{2 - \sqrt{24}}{10}$

$= 0.69, -0.29$

Simultaneous equations leading to quadratic equations

When quadratics are involved in simultaneous equations there will usually be more than one solution.

For Example

1 Solve $2x + y = 9$ (1)

$xy = 4$ (2)

S From (1), $y = 9 - 2x$

Substituting in (2):

$x(9 - 2x) = 4$

$9x - 2x^2 = 4$

$2x^2 - 9x + 4 = 0$

$(2x - 1)(x - 4) = 0$

Therefore, $x = \frac{1}{2}, 4$.

Now, there are two values of x to substitute in (1).

Substitute $x = \frac{1}{2}$:

$2\;\frac{1}{2} + y = 9$

$1 + y = 9$

$y = 9 - 1$

$= 8$

Therefore, $x = \frac{1}{2}, y = 8$.

Also, $x = 4$:

$2(4) + y = 9$

$8 + y = 9$

$y = 9 - 8$

$y = 1$

Therefore, $x = 4, y = 1$.

The answer is:

$x = \frac{1}{2}, y = 8$, or $x = 4, y = 1$

Use of equations to solve problems

A complicated problem can be solved using a simultaneous equation.

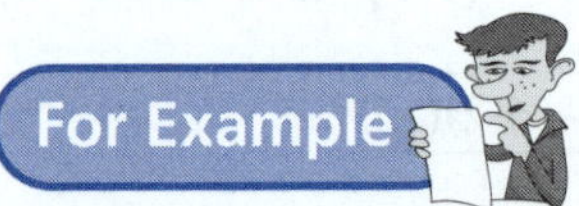

1 The perimeter of the rectangle is 40 cm and the area is 96 cm^2. Find its dimensions.

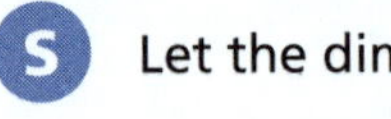

S Let the dimensions be x cm by y cm.

$2x + 2y = 40$ (1)

$xy = 96$ (2)

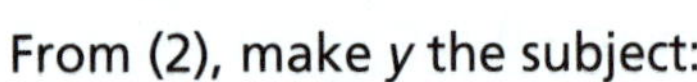

From (2), make y the subject:

$y = \dfrac{96}{x}$ (3)

Substituting (3) in (1):

$$2x + 2\frac{96}{x} = 40$$
$$2x + \frac{192}{x} = 40$$
$$2x^2 + 192 = 40x$$
$$2x^2 - 40x + 192 = 0$$
$$x^2 - 20x + 96 = 0$$
$$(x - 12)(x - 8) = 0$$
$$x = 12, 8$$

If $x = 12$, $y = 8$.
If $x = 8$, $y = 12$.
$\therefore$ Dimensions of rectangle are 12 cm by 8 cm.

Cubic equations

A simple cubic equation is of the form $ax^3 = k$.

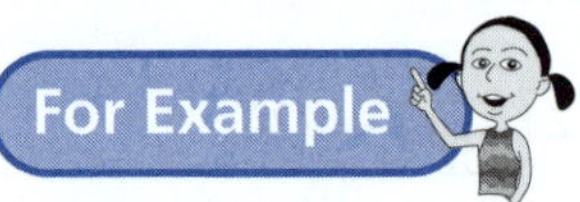

1 Solve $2x^3 = 16$.

S $x^3 = 8$
$x = 2$

2 Solve $4x^3 + 108 = 0$.

S $4x^3 = -108$
$x^3 = -27$
$x = -3$

Variable substitution

Occasionally in algebra a variable is substituted into an expression or equation. This is generally done to simplify the question.

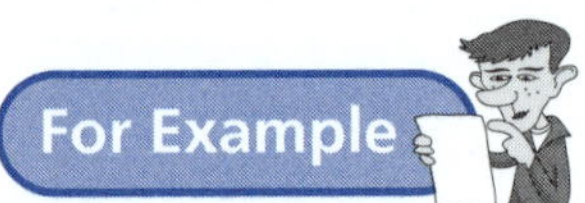

1 By substituting $m = x^2$ into $x^4 - 6x^2 + 8 = 0$, find all solutions.

$x^4 - 6x^2 + 8 = 0$

$(x^2)^2 - 6x^2 + 8 = 0$ — Rewrite equation in terms of x^2.

$m^2 - 6m + 8 = 0$ — Substitute m for x^2.

$(m - 4)(m - 2) = 0$ — Factor and solve the quadratic for m.

$\therefore m = 4$ or 2

$x^2 = 4$ or $x^2 = 2$ — Replace x^2 for m.

$x = 2$ or -2

or $x = \sqrt{2}$ or $-\sqrt{2}$ — Solve equation for x by taking square root of both sides.

$\therefore x = 2, \sqrt{2}, -2, -\sqrt{2}$

Note: the equation must be solved for the original variable.

Practise Practise

1 Solve for the given variable:

a $4x - 12 = 3x + 6$
b $5y + 2 = 3y + 18$
c $7(2y - 4) = 3(2y - 11)$
d $\frac{4x - 2}{3} = 5$
e $\frac{4 - 2y}{4} = 3y + 1$
f $\frac{7a - 12}{2} - \frac{a + 1}{3} = 5$
g $\frac{12x - 4}{3} = \frac{4x + 11}{2}$
h $\frac{5}{2x} = \frac{3}{4}$
i $\frac{6}{3y + 1} = \frac{2}{2y - 1}$
j $\frac{3}{2y - 1} + 4 = 0$
k $\frac{5a - 3}{4} - \frac{a - 1}{2} = \frac{3}{5}$ pp. 49–50

2 Solve the following inequations:

a $5a - 3 > 2a + 12$
b $7y - 4 \le 9y + 12$
c $4 - 2y \ge 11$
d $3 \le 4a + 6$
e $4 - 2(x - 4) \ge 12$
f $\frac{3 - 4x}{2} \le 3$ pp. 49–50

3 Graph on separate number lines:

a $-2 < x \le 3$
b $3 \le x < 4$ p. 50

4 Write down the inequality that corresponds to the graph:

a

−2 4

b

−4 −2

p. 51

5 Solve, and graph your solution on a number line:

a $3x - 5 \ge 2x - 10$
b $\frac{4 - 5x}{2} < 1$ pp. 50–51

6 Solve and graph your solution on a number line:

$\frac{4x - 1}{3} \le \frac{2 + 5x}{4}$ pp. 50–51

7 If $P = m(v - u)$, find P when $m = 16$, $v = 11$, $u = 2$. p. 52

8 If $S = \frac{a(r^n - 1)}{r - 1}$, find S when $a = 100$, $r = 2$, $n = 4$. p. 52

9 If $I = \frac{Prn}{100}$, find I if $P = 160$, $r = 11$, $n = 9$. p. 52

10 Examine each pattern of numbers in each table, and find a formula for y in terms of x.

a

x	0	1	2	3
y	2	4	6	8

b

x	1	2	3	4
y	5	8	13	20

pp. 52–53

11 The following table shows the cost ($$C$) of catching a taxi over a variety of distances (t km):

t	0	1	2	3	4
C	2	3.5	5	6.5	8

a Find a formula for C in terms of t.
b Find the cost of catching a taxi and travelling 6 kilometres. pp. 52–53

12 Find a formula for A, the area, of the following two shapes:

a

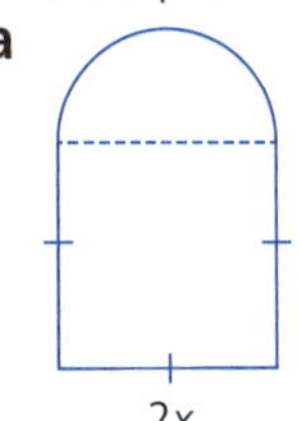

b

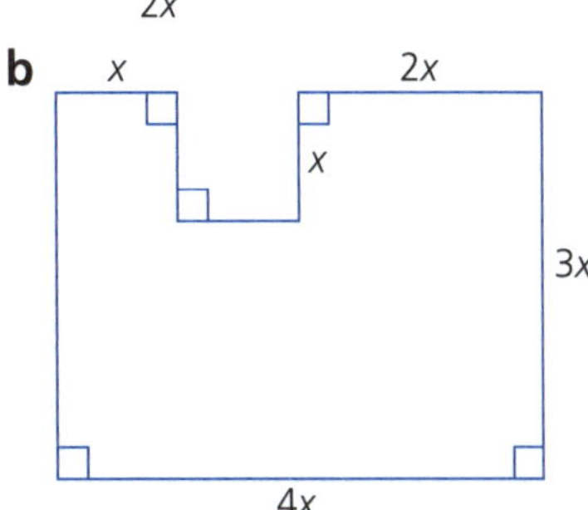

pp. 52–53

13 A 20 cm × 8 cm print is to be surrounded with a t cm-wide frame. Find a formula for A, the area of the entire picture (print + frame) in terms of t.

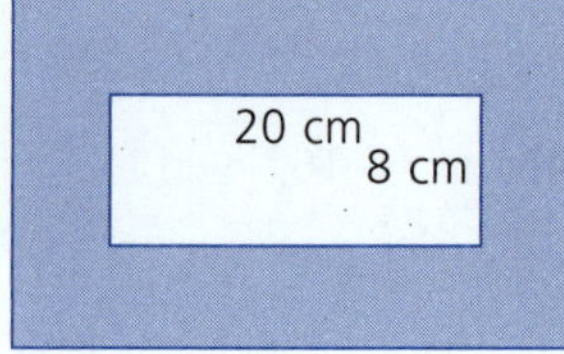

pp. 52–53

14 Make y the subject of:

a $4x + 2y = 16$ **b** $5x - 3y = 11$
c $2xy = 11 - x$ **d** $5y - 2xy = 7$
e $A = \dfrac{4x}{y}$ **f** $P = \dfrac{\sqrt{xy}}{2}$
g $k = \dfrac{y-1}{y}$ **h** $x^2 = \dfrac{y}{y-2}$ pp. 53–54

15 a If $V = \pi r^2 h$, find h if $V = 770$, $r = 7$, $\pi = \dfrac{22}{7}$.

b If $s = ut + \dfrac{1}{2}at^2$, find u if $s = 345.5$, $t = 5$, $a = 10$. pp. 53–54

16 An object dropped from rest is acted upon by gravity. It will have a velocity of v km/h after falling d metres (ignoring air resistance) according to the formula $v = 8.4\sqrt{d}$.

a Make d the subject of the formula.
b Hence, or otherwise, find how far the object will have to fall before its velocity is 42 km/h.
p. 53–54

17 Solve the following pairs of simultaneous equations:

a $x + 2y = 4$
$3x + 14y = 36$
b $3x - 8y = 2$
$4x + 10y = 44$
c $3x = 5 - 2y$
$4x = 1 + 3y$
d $x - y = 8$
$\dfrac{x}{2} + \dfrac{y}{3} = -1$ pp. 55–57

18 a If 5 is subtracted from twice a number, the result is the same as when 4 is added to the number. Find the number.

b The sum of two numbers is 12 and their difference is 2. Find the numbers.

c Find the values of a and b.

i

$2a + b$
$3a - 2b$
7

ii

$3a - 1$
$5a - 7$
$b + 4$
$3 - 2b$

d Six pens and five pencils cost \$2.50, while three pens and two pencils cost \$1.15. What will be the cost of a pen and four pencils? pp. 57–58

19 Solve the following quadratic equations for x:

a $(3x - 1)(x + 2) = 0$
b $x^2 - 5x = 0$
c $3x = x^2$
d $x^2 - 4x + 4 = 0$
e $5x - 6 - x^2 = 0$
f $x^2 = 49$
g $4x^2 = 36$
h $5x^2 - 2x - 3 = 0$
i $9x^2 = 2(9x + 8)$
j $4(x^2 - 3) + 13x = 0$
k $3x^2 - 8x + 6 = x^2 - 3x + 4$
l $(x - 3)^2 = 16$
m $\dfrac{3x + 4}{x} = x$ p. 59

20 Solve the following equations, using the 'complete the square' method:

a $x^2 + 4x + 1 = 0$
b $x^2 - 6x + 2 = 0$ p. 60

21 Solve the following equations, leaving your answer as a surd:

a $x^2 - 4x - 1 = 0$
b $x^2 - 8x - 5 = 0$ pp. 60–61

22 Solve the following equations, writing your answer correct to three decimal places:

a $3x^2 + 9x + 5 = 0$
b $5x^2 = 2x + 4$ pp. 60–61

23 **a** The product of two consecutive numbers is 30. By solving a quadratic equation, find the numbers.

b The sum of two numbers is 10 and the sum of their squares is 52. Find the numbers.

c

For the above rectangle, the length is known to be 8 cm longer than the width.

- **i** If the width is x cm, find the length of the rectangle in terms of x.
- **ii** If its area is 240 cm^2, find the dimensions of the rectangle.

d

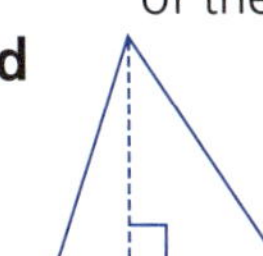

The base of a triangle is 5 cm longer than the perpendicular height. If the area of the triangle is 18 cm^2, find the length of the base.

e

$(4x + 12)°$

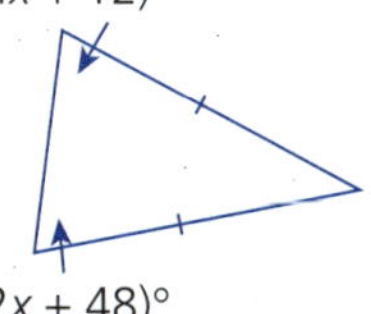

$(2x + 48)°$

The base angles of an isosceles triangle are given above. Find the size of all the angles of the triangle.

f Find the points of intersection of $y = x^2 + 6x - 2$ and $y = 2x + 3$.

g What positive integer, when increased by 20, will be equal to its square? pp. 61–62

24 Solve:

a $x + y = 3$
$xy = 2$

b $y = 2x^2 + x - 3$
$y = x^2 - 2x + 7$

c $y = \frac{2}{x}$
$y = x + 1$ p. 61

25 Find the points of intersection of $y = x^2 + 6x - 2$ and $y = 2x + 3$. pp. 61–62

26 The perimeter of a rectangle is 24 cm and its area is 32 cm^2. Find its dimensions. pp. 61–62

27 The difference between two numbers is 4 and the product is 21. Find the two numbers. pp. 61–62

28 Solve the following:

a $4x^3 = 4$

b $8x^3 = 64$

c $2x^3 + 250 = 0$ p. 62

29 **a** Use the substitution $y = x^2$ to solve the equation $x^4 - 10x^2 + 9 = 0$.

b Use the substitution $w = \frac{1}{x}$ to solve the equation $\frac{1}{x^2} + \frac{3}{x} - 10 = 0$. p. 62

Go to p. 300 for **Quick Answers** or to pp. 317–320 for **Worked Solutions**

For a complete understanding of this topic, you must be able to:

✓	Solve simple equations		pp. 49–50
✓	Solve inequations and graph the solution on the number line		pp. 50–51
✓	Substitute into formulae, construct formulae, solve equations resulting from substitution		pp. 52–53
✓	Solve literal equations (changing the subject)		pp. 53–54
✓	Solve problems by involving formulae		p. 55
✓	Solve simultaneous equations		pp. 55–57
✓	Use equations to solve problems		pp. 57–58
✓	Solve quadratic equations using a variety of methods		pp. 59–61
✓	Solve simultaneous equations involving quadratic equations		pp. 61–62
✓	Solve cubic equations		p. 62
✓	Substitute a variable into an expression or equation.		p. 62

Now you are ready to do the tests!

Intermediate Test

Equations

(35 marks)

1 Solve the equations:

a $4x - 11 = x + 17$ (2 marks)

b $\frac{x - 2}{5} = 8$ (2 marks)

2 Write down the set of values represented by the graph: (1 mark)

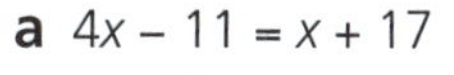

3 Solve and graph the solutions of $1 - 4x > 9$. (3 marks)

4 For the formula $V = ut + \frac{1}{2}at^2$, calculate the value of V where $u = 7$, $t = 4$ and $a = 5$. (2 marks)

5 Make y the subject of $7x - 2y = 9$. (2 marks)

6 Solve $4x^3 = 32$. (1 mark)

7 Solve the simultaneous equations:

$2x - y = 11$

$3x + y = 4$ (3 marks)

8 Solve the quadratic equations:

a $x^2 - 7x - 8 = 0$ (2 marks)

b $5x^2 = 4x$ (2 marks)

9 Use the quadratic formula to find the exact roots of $x^2 - 5x - 2 = 0$. (2 marks)

10 By completing the square, find the roots of $x^2 - 8x + 12 = 0$. (3 marks)

11 Find the points of intersection of the line $2x + y = 14$ and the curve $y = x^2 + 3x$. (4 marks)

12

A rectangle has a length three times larger than twice the width. If the width is x cm and the area of the rectangle is 2 cm^2, find the dimensions of the rectangle. (6 marks)

Your Feedback

☐ × 100% = ☐ %

QA PAGE 305

WS PAGE 364

Advanced Test

Equations

(35 marks)

1 Solve the equations:

a $7 - 4x = 13 - 7x$ (2 marks)

b $\frac{7x}{4} - \frac{3x - 5}{3} = 2$ (3 marks)

2 On a number line sketch the set of values such that $-1 \leq x < 3$. (1 mark)

3 Solve the inequation $3(2x - 5) - 8(x - 4) < 13$. (2 marks)

4 For the formula $b^2 - 4ac = (2ax + b)^2$, find the value of x where $a = 2$, $b = 5$, $c = 2$. (3 marks)

5 Decide whether the roots of the following equation are real or not real:

a $y = x^2 - 5x + 2$ (1 mark)

b $y = 3 + x + x^2$ (1 mark)

6 Make y the subject:

a $M = x(y - b)$ (2 marks)

b $P = \frac{x + y}{m} + \frac{x - y}{n}$ (3 marks)

7 Solve the simultaneous equations:

$3x - 2y = 1$

$x + 3y = 15$ (3 marks)

8 Solve the equations:

a $6x^2 - 11x = 10$ (2 marks)

b $x - 2 = \frac{35}{x}$ (2 marks)

9 Find the roots of the equation $y^2 - 5y + 1 = 0$ correct to two decimal places. (2 marks)

10 Find any points of intersection of the line $y = 2x - 1$ with the curve $y = \frac{3}{x}$. (4 marks)

11 The sum of the squares of two consecutive positive integers in 221. By first forming an equation, find the two integers. (4 marks)

Your Feedback

$\square \times 100\% = \square\ \%$

QA PAGE 305

WS PAGE 364

Chapter 5
Linear Relationships

Important formulae

Three very important formulae used in coordinate geometry are:

1 The distance formula:

$$d = \sqrt{(x_2 - x_1)^2 + (y_2 - y_1)^2}$$

(Based on Pythagoras' Theorem)

2 The midpoint formula:

$$MP = \frac{x_1 + x_2}{2}, \frac{y_1 + y_2}{2}$$

3 The gradient formula:

$$m = \frac{y_2 - y_1}{x_2 - x_1}$$

Keywords

Axes	Median
Axis	Midpoint
Bisector	Negative
Collinear	Origin
Collinearity	Parallel
Concurrence	Perpendicular
Concurrent	Positive
Coordinate	Reciprocal
Equation	Simultaneously
Gradient	Substitute
Horizontal	Tangent
Intercept	Vertex
Intersection	Vertical
Locus	

The distance formula is used to calculate the straight-line distance between points (x_1, y_1) and (x_2, y_2). The midpoint formula is used to find the coordinates of the midpoint of the interval joining (x_1, y_1) and (x_2, y_2), while the gradient formula is used to find the gradient (or slope) of the line containing the points (x_1, y_1) and (x_2, y_2).

For Example

1 Consider the points A(–2, –1) and B(3, 3).

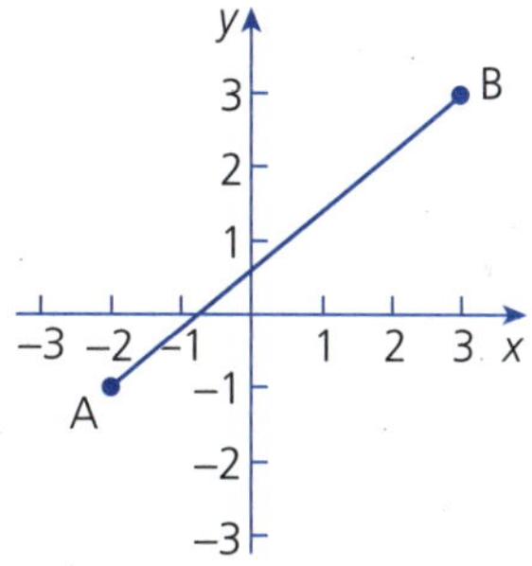

Put $(x_1, y_1) = (-2, -1)$; $(x_2, y_2) = (3, 3)$.

$x_1 = -2, x_2 = 3$

$y_1 = -1, y_2 = 3$

a Calculate distance AB.

S

$$\begin{aligned} d &= \sqrt{(x_2 - x_1)^2 + (y_2 - y_1)^2} \\ &= \sqrt{(3 + 2)^2 + (3 + 1)^2} \\ &= \sqrt{5^2 + 4^2} \\ &= \sqrt{25 + 16} \\ &= \sqrt{41} \end{aligned}$$

AB is $\sqrt{41}$ units long.

Note: this is simply an application of Pythagoras' Theorem.

Note: this is the exact distance AB.

b Calculate the midpoint of AB.

S $MP = \dfrac{x_1 + x_2}{2}, \dfrac{y_1 + y_2}{2}$

$= \dfrac{-2 + 3}{2}, \dfrac{-1 + 3}{2}$

$= \dfrac{1}{2}, 1$

Midpoint is $\dfrac{1}{2}, 1$.

Average of x-coordinates, average of y-coordinates.

c Calculate the gradient of AB.

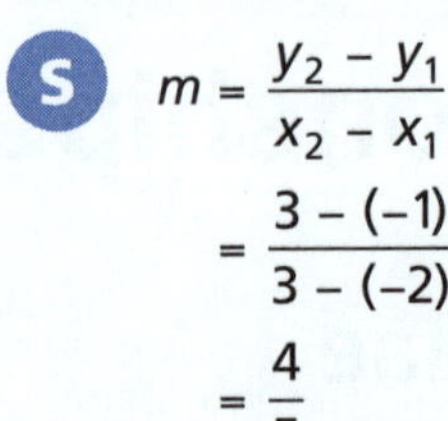

S $m = \dfrac{y_2 - y_1}{x_2 - x_1}$

$= \dfrac{3 - (-1)}{3 - (-2)}$

$= \dfrac{4}{5}$

Gradient is $\dfrac{4}{5}$.

$\dfrac{\text{Difference in } y}{\text{Difference in } x}$

Note: the gradient is always left as a simple fraction. Do not convert either to a decimal or to a mixed number.

A note on gradient

The gradient or slope of a line can be thought of as $\dfrac{\text{Vertical rise}}{\text{Horizontal run}}$ *or* $\dfrac{\text{Rise}}{\text{Run}}$.

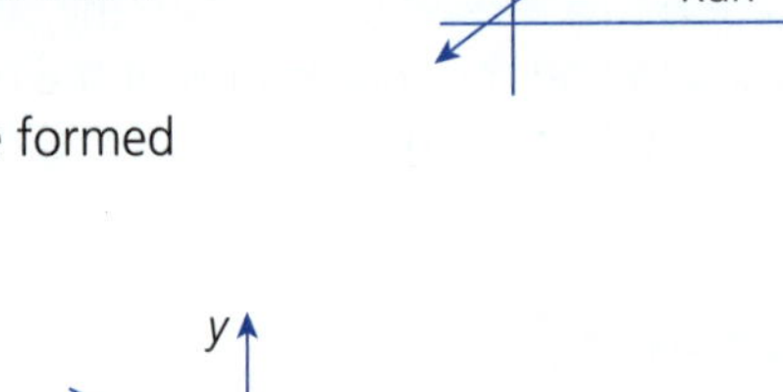

This is the reason why the y-values are on top in the formula.

The gradient of a line can be either positive or negative. A line sloping upwards to the right has a positive gradient; one sloping downwards to the right has a negative gradient.

The gradient can be found from the diagram by looking at the triangle formed by the line and the coordinate axes.

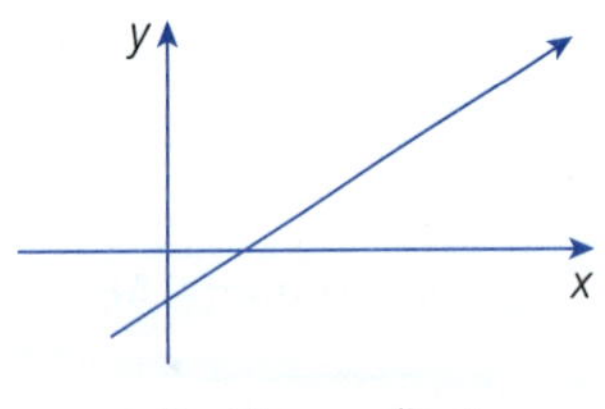

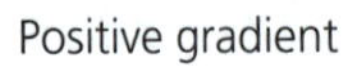

Positive gradient

Negative gradient

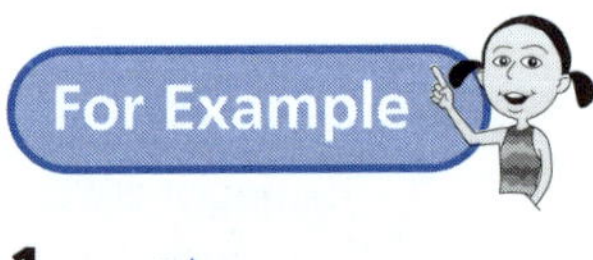

1

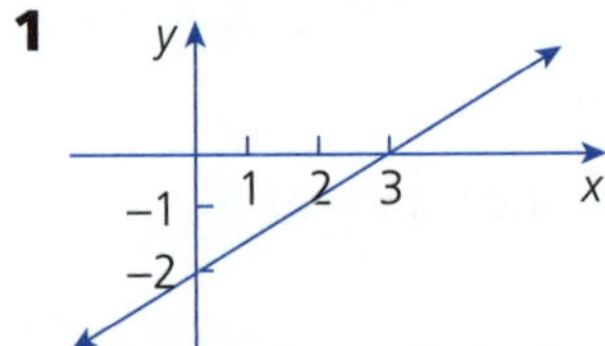

S Rise = 2 ↑

Run = 3 →

Direction is positive ↗

Gradient $= \dfrac{\text{Rise}}{\text{Run}}$

$= \dfrac{2}{3}$

2

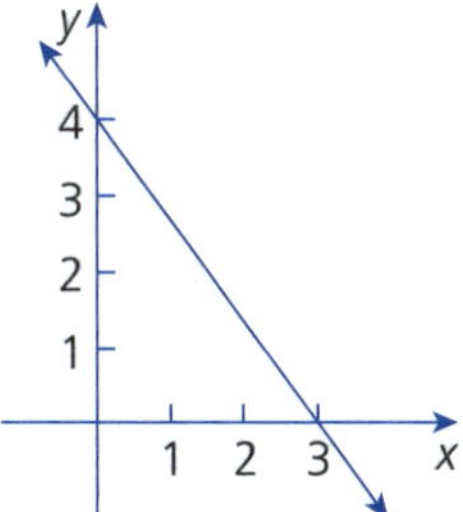

S Rise = 4 ↓

Run = 3 →

But, direction is negative ↘

$$\text{Gradient} = \frac{\text{Rise}}{\text{Run}}$$
$$= -\frac{4}{3}$$

3

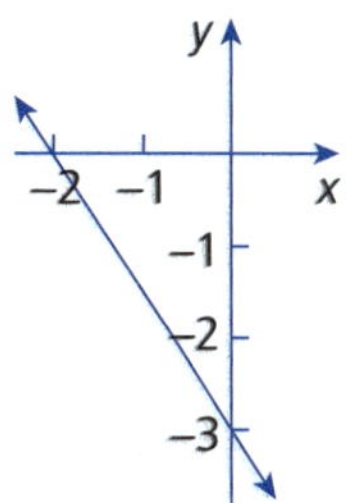

Rise = 3

Run = 2

But, direction is negative ↘

$$\textbf{Gradient} = \frac{\textbf{Rise}}{\textbf{Run}}$$
$$= -\frac{3}{2}$$

Use of the formulae

The distance formula can be applied to number plane problems.

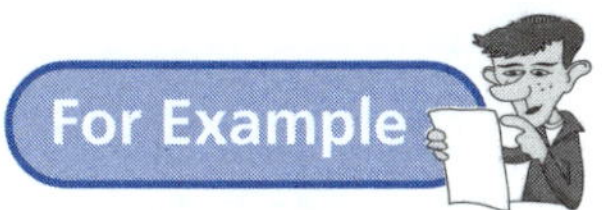

1 Plot the points A(–2, –2), B(2, –2) and C(0, 3) on a number plane.

Draw a clear sketch.

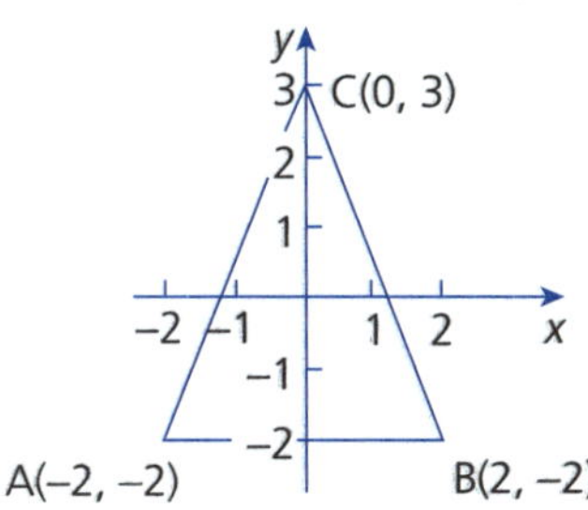

a Prove that length AC is $\sqrt{29}$ units.

S $d = \sqrt{(x_2 - x_1)^2 + (y_2 - y_1)^2}$

$= \sqrt{(0 + 2)^2 + (3 + 2)^2}$

$= \sqrt{4 + 25}$

$= \sqrt{29}$ units

AC is $\sqrt{29}$ units.

b Prove that △ABC is isosceles.

S **For △ABC to be isosceles, two sides must have the same length. Check the length of BC.**

$BC = \sqrt{(x_2 - x_1)^2 + (y_2 - y_1)^2}$

$= \sqrt{(2 - 0)^2 + (-2 - 3)^2}$

$= \sqrt{4 + 25}$

$= \sqrt{29}$ units

∴ AC = BC (both $\sqrt{29}$).
That is, △ABC is isosceles.

c Prove that the midpoint of AB lies on the *y*-axis.

S **Midpoint** $= \frac{x_1 + x_2}{2}, \frac{y_1 + y_2}{2}$

$= \frac{-2 + 2}{2}, \frac{-2 + -2}{2}$

$= (0, -2)$

The point (0, –2) lies on the *y*-axis as the *x*-coordinate is 0.

2 The points A(4, 8), B(−1, 3), C(−3, −11) and D(2, −6) are the vertices of a parallelogram. Draw a clear sketch.

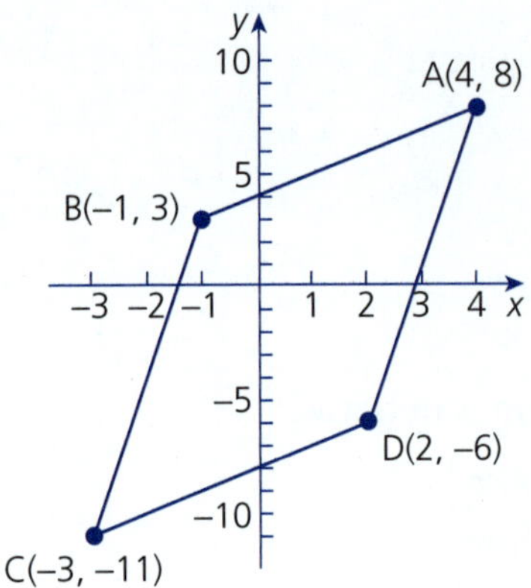

a Show that the opposite sides are equal.

S Using the distance formula

$d = \sqrt{(x_2 - x_1)^2 + (y_2 - y_1)^2}$ in each case:

$AB = \sqrt{(-1-4)^2 + (3-8)^2} = \sqrt{50}$

$CD = \sqrt{(-3-2)^2 + (-11+6)^2} = \sqrt{50}$

$\therefore AB = CD$

$AD = \sqrt{(4-2)^2 + (8+6)^2} = \sqrt{200}$

$BC = \sqrt{(-1+3)^2 + (3+11)^2} = \sqrt{200}$

$\therefore AD = BC$

The opposite sides are equal.

b Show that the diagonals bisect each other.

S This example requires the midpoint of each diagonal, AC and BD. If they are the same, the diagonals must bisect each other. Using:

$MP = \frac{x_1 + x_2}{2}, \frac{y_1 + y_2}{2}$ in each case:

$$MP_{AC} = \frac{-3+4}{2}, \frac{-11+8}{2}$$

$$= \frac{1}{2}, -\frac{3}{2}$$

$$MP_{BD} = \frac{-1+2}{2}, \frac{3-6}{2}$$

$$= \frac{1}{2}, -\frac{3}{2}$$

Therefore the diagonals intersect at $\frac{1}{2}, -\frac{3}{2}$, the midpoint of each diagonal.

c Show that the opposite sides have equal gradients.

S Using in each case $m = \frac{y_2 - y_1}{x_2 - x_1}$:

$$m_{AB} = \frac{8-3}{4-(-1)} = \frac{5}{5} = 1$$

$$m_{CD} = \frac{-6-(-11)}{2-(-3)} = \frac{5}{5} = 1$$

$$m_{AD} = \frac{8-(-6)}{4-2} = \frac{14}{2} = 7$$

$$m_{BC} = \frac{3-(-11)}{-1-(-3)} = \frac{14}{2} = 7$$

The opposite sides have equal gradients.

The gradient form

This form of the straight-line equation can be used either to find the gradient of a given line or to write the equation of a line, given the gradient and y-intercept.

$y = mx + b$

↑ gradient (m) ↑ y-intercept (b)

1 For the line with equation $2x + 3y = 5$, write down the gradient.

S First, express $2x + 3y = 5$ with y as subject.

$$2x + 3y = 5$$
$$3y = -2x + 5$$
$$y = -\frac{2}{3}x + \frac{5}{3}$$

Then the gradient is the coefficient of x, that is, $m = -\frac{2}{3}$.

2 A straight line has gradient $-\frac{1}{4}$ and cuts the y-axis at 2. Find the equation of the line.

S The equation has the form $y = mx + b$, where $m = -\frac{1}{4}$, $b = 2$, that is, $y = -\frac{1}{4}x + 2$.

Now multiply all terms by 4 to clear fractions:

$4y = -x + 8$

Move all terms to the side that will make the x-term positive:

$x + 4y - 8 = 0$

3 Find the equation of the line AB, indicated in the diagram.

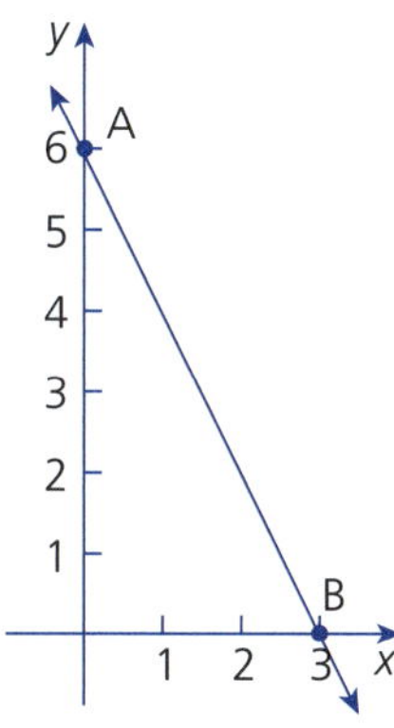

S Rise = 6, run = 3.

Gradient is negative. ↘

$$\text{Gradient} = \frac{\text{Rise}}{\text{Run}}$$
$$= -\frac{6}{3}$$
$$= -2$$

y-intercept = 6

The equation is $y = mx + b$, that is, $y = -2x + 6$.

Therefore the equation is $2x + y - 6 = 0$.

4 Sketch the line $y = 3x - 4$ by first finding the gradient and y-intercept.

S The equation $y = 3x - 4$ is in the form $y = mx + b$.

$\therefore$ $m = 3$ and y-intercept = -4

- Mark on y-axis the intercept of -4.
- Gradient = $\frac{3}{1}$, that is, line rises 3 units and goes across 1 unit. (Note ↗ positive gradient.)
- Draw right-angle triangle with rise = 3, run = 1, hypotenuse ↗ .
- Required line runs along the hypotenuse.

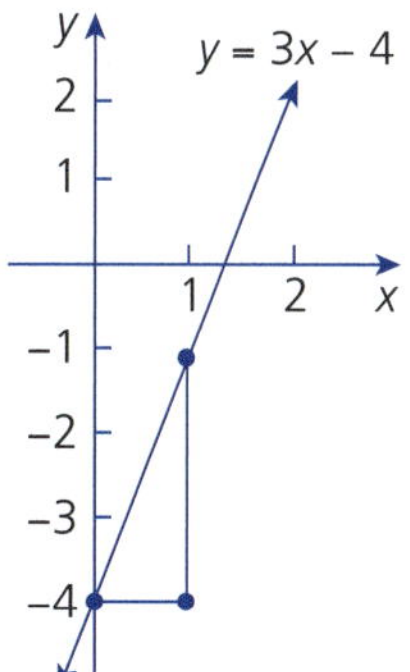

Note: usually the equation of the line is written in **general form**, that is, $\mathbf{Ax + By + C = 0}$, where $A \geq 0$, and A, B and C are integers.

The point–gradient formula

The point–gradient form of the straight-line equation is $\mathbf{y - y_1 = m(x - x_1)}$, where (x_1, y_1) is the given point and m is the gradient of the line.

For Example

1 Write down the equation of the line through $(2, -3)$ with gradient $\frac{3}{4}$.

S The equation is of the form:

$$y - y_1 = m(x - x_1) \qquad (x_1, y_1) = (2, -3)$$

$$y - (-3) = \frac{3}{4}(x - 2) \qquad m = \frac{3}{4}$$

$$y + 3 = \frac{3}{4}(x - 2)$$

← Multiply both sides by 4.

Therefore $\quad 4(y + 3) = 3(x - 2)$

That is $\quad 4y + 12 = 3x - 6$

Therefore, $3x - 4y - 18 = 0$. General form.

2 Find the equation of the line with gradient $-\frac{2}{3}$ making an intercept of -6 on the x-axis.

S The equation is of the form:

$$y - y_1 = m(x - x_1)$$

$$y - 0 = -\frac{2}{3}(x - [-6])$$

Intercept of -6 means that the line passes through $(-6, 0)$.

$$y = -\frac{2}{3}(x + 6)$$

Multiply both sides by 3.

$$3y = -2(x + 6)$$

$$= -2x - 12$$

Therefore the equation is

$2x + 3y + 12 = 0$. General form.

The two-point formula

The two-point formula for finding the equation of a line is: $\dfrac{\mathbf{y - y_1}}{\mathbf{x - x_1}} = \dfrac{\mathbf{y_2 - y_1}}{\mathbf{x_2 - x_1}}$

where (x_1, y_1) and (x_2, y_2) are two points through which the line passes.

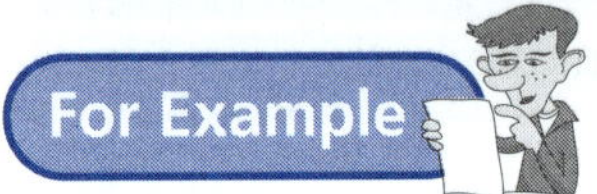

1 Find the equation of the line passing through the points $(-1, 4)$ and $(6, 2)$.

S The equation of the line is of the form:

$$\frac{y - y_1}{x - x_1} = \frac{y_2 - y_1}{x_2 - x_1}, \qquad (x_1, y_1) = (-1, 4)$$
$$(x_2, y_2) = (6, 2)$$

$$\frac{y - 4}{x - (-1)} = \frac{2 - 4}{6 - (-1)}$$

$$\frac{y - 4}{x + 1} = \frac{-2}{7}$$

Cross multiply

$$7(y - 4) = -2(x + 1)$$

$$7y - 28 = -2x - 2$$

That is, $2x + 7y - 26 = 0$. General form.

2 Find the equation of the line making intercepts -3 and 2 on the x- and y-axes, respectively.

S

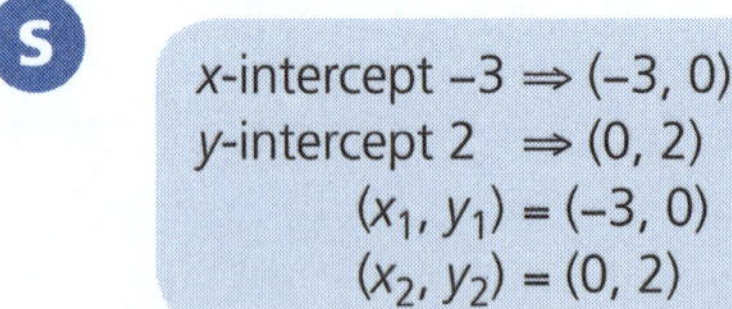

The equation is of the form:

$$\frac{y - y_1}{x - x_1} = \frac{y_2 - y_1}{x_2 - x_1}$$

$$\frac{y - 0}{x - (-3)} = \frac{2 - 0}{0 - (-3)}$$

$$\frac{y}{x + 3} = \frac{2}{3}$$

Cross multiply

$$3y = 2(x + 3)$$

$$3y = 2x + 6$$

That is, $2y - 3y + 6 = 0$.

The condition for a point to lie on a line

A point lies on a line (or a curve) if the coordinates of the point satisfy the equation of the line (or curve).

1 Show that $(2, -1)$ lies on the line $y = 2x - 5$.

S Substitute $x = 2$ and $y = -1$ into the equation.

LHS $= -1$ RHS $= 2 \times 2 - 5$
$= -1$

Therefore $(2, -1)$ satisfies the equation and $(2, -1)$ lies on the line $y = 2x - 5$.

2 Determine whether the point $(-5, 17)$ lies on the line $4x + y = -1$.

S Substitute $x = -5$ and $y = 17$ into the equation.

LHS $= 4 \times (-5) + 17$ RHS $= -1$
$= -20 + 17$
$= -3$

$(-5, 17)$ does not satisfy the equation. Therefore $(-5, 17)$ does *not* lie on line $4x + y = -1$.

3 The point $(-3, k)$ lies on the line $x + 2y = 5$. Find the value of k.

S Substitute $x = -3$, $y = k$ into
$x + 2y = 5$

Thus $-3 + 2k = 5$
$2k = 8$
$k = 4$

Solve the simple equation for k.

4 The points $(2, 1)$ and $(-1, -1)$ lie on the line $2ax - by = 5$. Find a and b and hence the equation of the line.

S The point $(2, 1)$ lies on $2ax - by = 5$

Substitute $x = 2, y = 1$

$\therefore 4a - b = 5$ (1)

The point $(-1, -1)$ lies on $2ax - by = 5$

Substitute $x = -1, y = -1$

$\therefore -2a + b = 5$ (2)

Add (1) + (2): $2a = 10$
$\therefore a = 5$
Substitute $a = 5$ in (2)
$-10 + b = 5$
$\therefore b = 15$

Then $a = 5$, $b = 15$.
Therefore the equation is
$10x - 15y = 5$, that is, $2x - 3y = 1$.

Intercepts cut from the axes

At any point on the x-axis: $y = 0$

At any point on the y-axis: $x = 0$

This is useful in two ways:

1 It enables you to find where any line or curve cuts the coordinate axes.

2 It allows you to quickly sketch the graphs of straight lines.

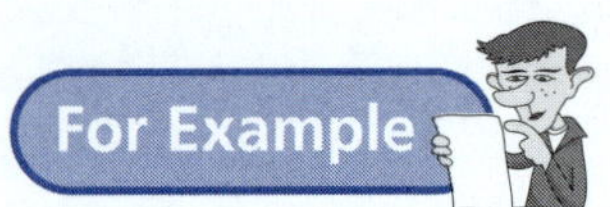

1 By finding the x and y intercepts sketch the following:

a $x + y = 4$

S $x + y = 4$

x	0	4
y	4	0

← Abbreviated table to find intercepts.

Intercepts are $x = 4$
$y = 4$

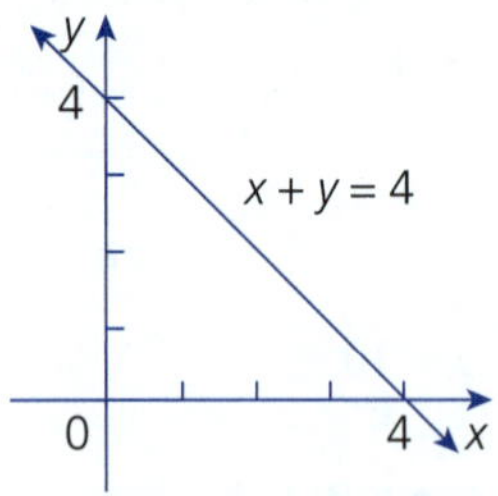

b Sketch $2x - y = 6$.

S $2x - y = 6$

x	0	3
y	−6	0

$x = 0, \quad -y = 6$
$\therefore \quad y = -6$
$y = 0, \quad 2x = 6$
$x = 3$

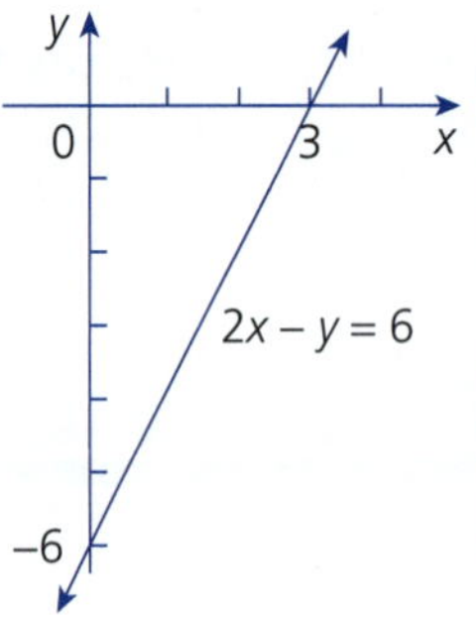

2 Find the area of the triangle contained between the line $2x - y = 5$ and the coordinate axes.

S $2x - y = 5$

x	0	$2\frac{1}{2}$
y	−5	0

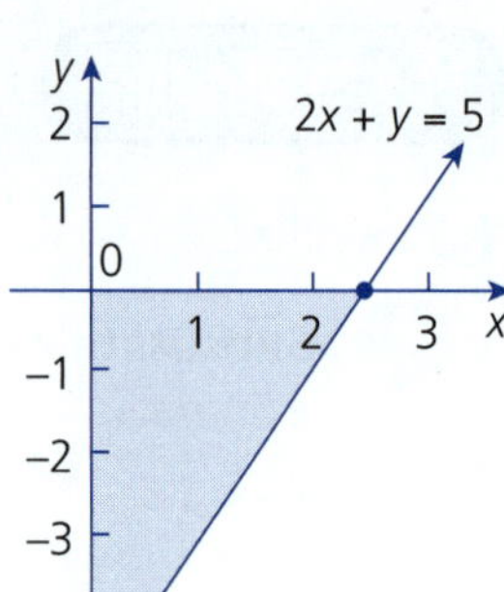

$$A = \frac{1}{2}bh$$
$$= \frac{1}{2} \times 2\frac{1}{2} \times 5$$
$$= \frac{1}{2} \times \frac{5}{2} \times 5$$
$$= \frac{25}{4}$$
$$\text{Area} = 6\frac{1}{4} \text{ units}^2$$

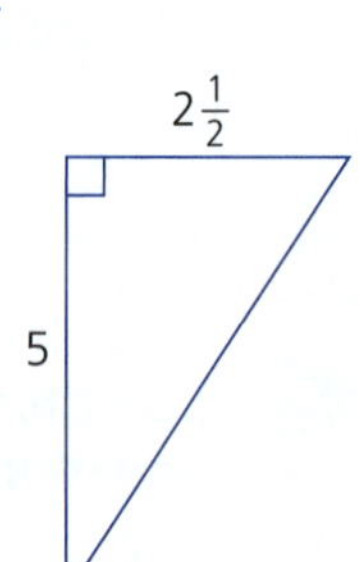

Parallel and perpendicular lines

If two lines have gradients m_1 and m_2 respectively, the condition for the lines to be:

1 parallel || is: $m_1 = m_2$

2 perpendicular ⊥ is: $m_1 = -\frac{1}{m_2}$ or $m_1 m_2 = -1$ (Negative reciprocal)

1 Show that l_1, $2x - y = 7$ is parallel to l_2, $y = 2x + 1$, but perpendicular to l_3, $x + 2y - 3 = 0$.

S Let m_1, m_2, m_3 be the gradients of lines l_1, l_2, l_3 respectively.

Line 1 (l_1)	Line 2 (l_2)
$2x - y = 7$	$y = 2x + 1$
$\Rightarrow \quad y = 2x - 7$	$\therefore m_2 = 2$
$\therefore m_1 = 2$	

Therefore $2x - y = 7$ is parallel to $y = 2x + 1$. (Both have a gradient of 2.)

Each equation must be expressed in the form $y = mx + b$ in order to find gradients. The gradients are then compared using the above criteria.

Line 3 (l_3)

$$x + 2y - 3 = 0$$
$$2y = -x + 3$$
$$y = -\frac{1}{2}x + \frac{3}{2}$$
$$\therefore m_3 = -\frac{1}{2}$$

y is made the subject of the equation in each case.

Compare m_1 and m_3.

$$2 \times -\frac{1}{2} = -1 \quad (m_1 \times m_3 = -1)$$

Therefore $2x - y = 7$ is perpendicular to $x + 2y - 3 = 0$.

2 Show that $\triangle$ABC is right-angled, given that the coordinates of A, B and C are (–3, 4), (2, 1) and (–1, –4) respectively. Which angle is the right-angle?

S This question could be answered using Pythagoras' Theorem but it is easier to consider the gradients.

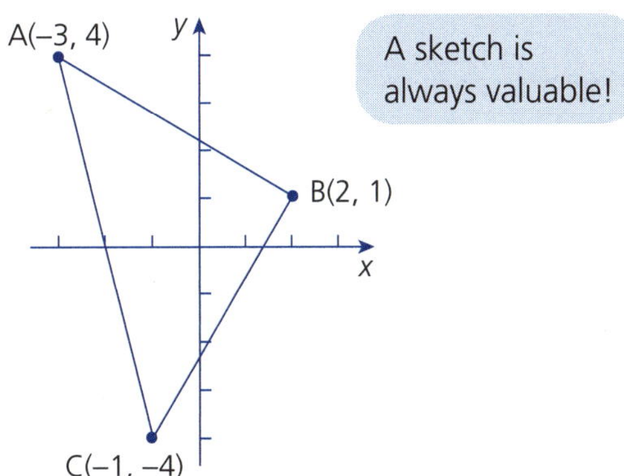

A sketch is always valuable!

Using $m = \dfrac{y_2 - y_1}{x_2 - x_1}$ in each case:

$$\text{Gradient AB} = \frac{4 - 1}{-3 - 2} = \frac{3}{-5} = -\frac{3}{5}$$

$$\text{Gradient BC} = \frac{1 - (-4)}{2 - (-1)} = \frac{1 + 4}{2 + 1} = \frac{5}{3}$$

$$\text{Gradient AC} = \frac{4 - (-4)}{-3 - (-1)} = \frac{4 + 4}{-3 + 1} = \frac{8}{-2} = -4$$

Now compare the gradients:

$$m_{AB} \times m_{BC} = -\frac{3}{5} \times \frac{5}{3} = -1$$

$\therefore$ AB $\perp$ BC, that is, the triangle is right-angled at B.

3 Prove that the line $2x + y = 1$ is the perpendicular bisector of the interval joining A(–3, 2) to B(1, 4).

S

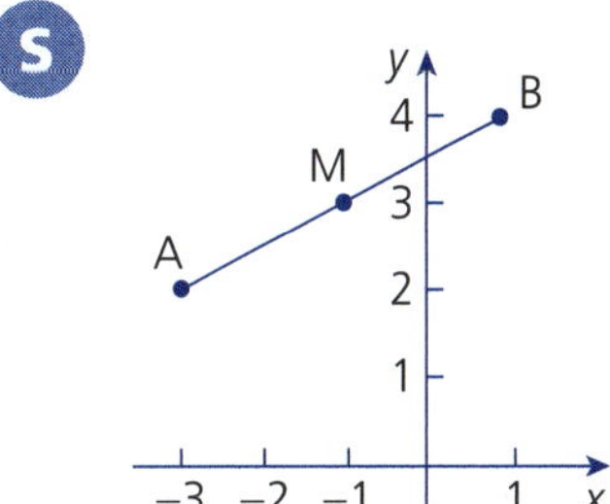

Take M as the midpoint of AB.

$$M = \left(\frac{x_1 + x_2}{2}, \frac{y_1 + y_2}{2}\right)$$
$$= \left(\frac{-3 + 1}{2}, \frac{2 + 4}{2}\right)$$
$$= (-1, 3)$$

We need the midpoint of AB and the gradient of AB. If the line is to be the perpendicular bisector it must pass through the midpoint of AB and the gradient must be the negative reciprocal of the gradient of AB.

$$\text{The gradient of AB} = \frac{y_2 - y_1}{x_2 - x_1}$$
$$= \frac{4 - 2}{1 - (-3)}$$
$$= \frac{2}{4}$$
$$= \frac{1}{2}$$

∴ Gradient of perpendicular is –2.

$\therefore y - 3 = -2(x + 1)$

$\therefore y - 3 = -2x - 2$

$2x + y = 1$

Thus the line $2x + y = 1$ passes through the midpoint of AB and is perpendicular to AB, that is, $2x + y = 1$ is the perpendicular bisector of AB.

4 On a number plane, plot the points A(–1, 3), B(4, 4), C(2, –1) and D(–3, –2). By showing that both pairs of opposite sides are parallel, prove that ABCD is a parallelogram.

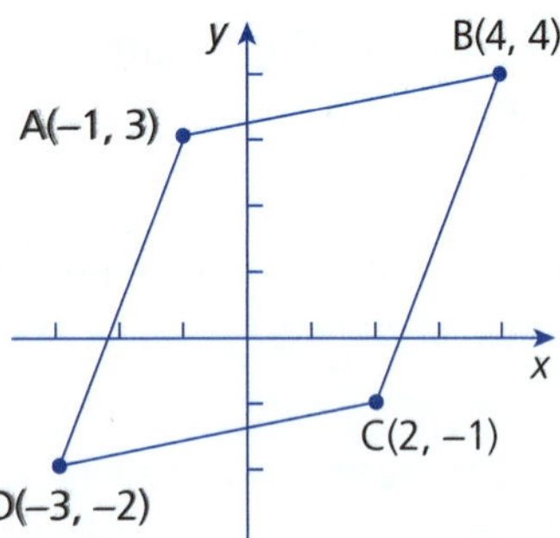

Using $m = \dfrac{y_2 - y_1}{x_2 - x_1}$ in each case:

Gradient AB $= \dfrac{4 - 3}{4 - (-1)} = \dfrac{1}{5}$

Gradient CD $= \dfrac{-1 - (-2)}{2 - (-3)} = \dfrac{1}{5}$

Gradient BC $= \dfrac{4 - (-1)}{4 - 2} = \dfrac{5}{2}$

Gradient DA $= \dfrac{-2 - 3}{-3 - (-1)} = \dfrac{-5}{-2} = \dfrac{5}{2}$

Then $m_{AB} = m_{CD} = \dfrac{1}{5}$

and $m_{BC} = m_{DA} = \dfrac{5}{2}$

That is, AB ∥ DC (Same gradient.)

BC ∥ AD (Same gradient.)

Hence ABCD is a parallelogram.

Both pairs of opposite sides are parallel.

5 Derive the equation of the line **parallel** to $2x - 3y - 5 = 0$ passing through (–2, 1).

We need the gradient of $2x - 3y - 5 = 0$, so rearrange the equation with y as subject, that is, $y = mx + b$, with gradient m.

$2x - 3y - 5 = 0$

$3y = 2x - 5$

$y = \dfrac{2}{3}x - \dfrac{5}{3}$

∴ Gradient $= \dfrac{2}{3}$

Any parallel line will have a gradient of $\dfrac{2}{3}$. This one passes through (–2, 1).

The equation is of the form:

$y - y_1 = m(x - x_1)$ ($-(-2) = +2$)

That is, $y - 1 = \dfrac{2}{3}(x + 2)$ (Multiply both sides by 3.)

$3(y - 1) = 2(x + 2)$

$3y - 3 = 2x + 4$

$2x - 3y + 7 = 0$

Therefore $2x - 3y + 7 = 0$ is the required equation.

6 Derive the equation of the line **perpendicular** to $2x - 3y - 5 = 0$ passing through (–2, 1).

From (5), gradient $= \dfrac{2}{3}$, therefore the gradient of the perpendicular $= -\dfrac{3}{2}$.

The equation is of the form:

$y - y_1 = m(x - x_1)$

$y - 1 = -\dfrac{3}{2}(x + 2)$ (Multiply both sides by 2.)

$2(y - 1) = -3(x + 2)$

$2y - 2 = -3x - 6$

$3x + 2y + 4 = 0$

Therefore $3x + 2y + 4 = 0$ is the required equation.

7 Prove that the line passing through A(–1, 1) and B(2, 4) is perpendicular to the line passing through C(1, 2) and D(4, –1).

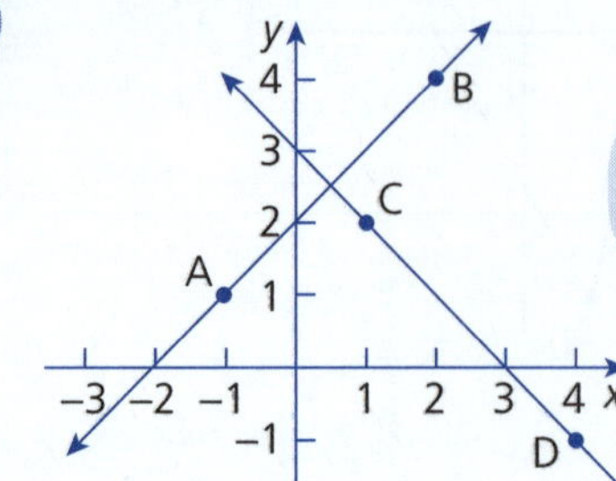

A sketch is always helpful.

Using the formula $m = \frac{y_2 - y_1}{x_2 - x_1}$:

Gradient of AB $= \frac{1-4}{-1-2} = \frac{-3}{-3} = 1$

Gradient of CD $= \frac{2-(-1)}{1-4} = \frac{2+1}{-3} = -1$

Compare the gradients of AB and CD.

$m_{AB} \times m_{CD} = 1 \times -1 = -1$

Therefore AB ⊥ CD.

Product of gradients = −1

Collinearity and concurrence

Collinearity

Three or more points lying on the same line are **collinear points**:

For example, the points A(0, 5), B(2.5, 0) and C(1, 3) all lie on the line $2x + y = 5$. (Check by substitution!) The points A, B and C are collinear.

1 Show that the points P(−1, 4), Q(2, −2) and R(3, −4) are collinear.

S One method of answering this question is to find the equation of the line through P and Q (two-point formula) and then show that R lies on that line. However, a quicker method is to consider the gradients of PQ and QR.

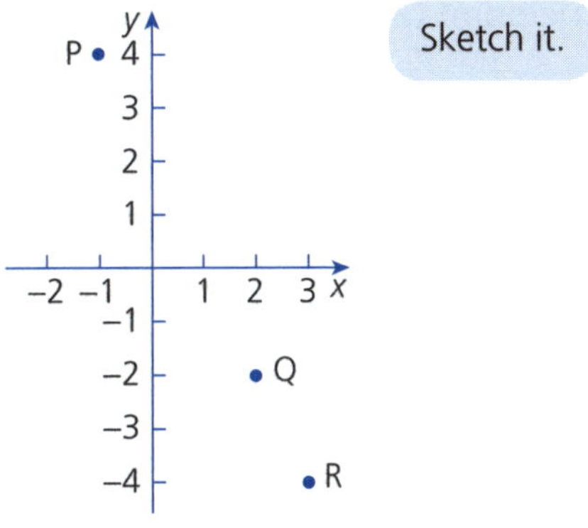

Sketch it.

Using $m = \frac{y_2 - y_1}{x_2 - x_1}$:

Gradient of PQ $= \frac{4-(-2)}{-1-2} = \frac{4+2}{-3} = \frac{6}{-3} = -2$

Gradient of QR $= \frac{-2-(4)}{2-3} = \frac{2}{-1} = -2$

Both line segments have the same gradient and Q lies on both, hence P, Q and R lie on the same line, that is, P, Q and R are collinear.

Alternative method

The equation of PQ is of the form:

$$\frac{y - y_1}{x - x_1} = \frac{y_2 - y_1}{x_2 - x_1}$$

$(x_1, y_1) = (-1, 4)$
$(x_2, y_2) = (2, -2)$

Thus, $\frac{y-4}{x-(-1)} = \frac{-2-4}{2-(-1)}$

That is,

$$\frac{y-4}{x+1} = \frac{-6}{3} = \frac{-2}{1}$$

$$y - 4 = -2(x+1)$$

$$y - 4 = -2x - 2,$$

$$2x + y - 2 = 0$$

Cross multiply

Therefore $2x + y = 2$ is the equation of PQ, that is, we have the line through P and Q.

Now check R(3, −4) in the equation, that is:

Put $x = 3$, $y = -4$ in $2x + y = 2$

LHS = 6 − 4 RHS = 2
= 2

Therefore LHS = RHS.

The point R(3, −4) lies on the line through P and Q, that is, P, Q and R are collinear.

Concurrence

Three or more lines passing through the same point are **concurrent lines**.

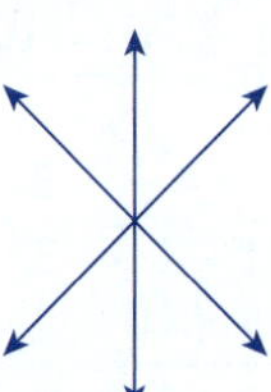

For Example

1 Show that the lines $x + y = 2$, $2x - y = 7$ and $x - 2y = 5$ are concurrent.

Consider the simplest pair of equations and solve them simultaneously to obtain the point of intersection.

Show that this point lies on the third line.

Consider $x + y = 2$ (1)

$2x - y = 7$ (2)

Add equations (1) and (2).

$3x = 9$

$x = 3$

Substitute $x = 3$ into equation (1).

$3 + y = 2$

$y = -1$

Therefore the point of intersection is (3, –1).

Now check whether (3, –1) lies on $x - 2y = 5$.

LHS = 3 + 2 RHS = 5

= 5

LHS = RHS

The point (3, –1) satisfies the equation $x - 2y = 5$, therefore it also lies on $x - 2y = 5$.

Thus all three lines pass through (3, –1), that is, the lines are concurrent.

2 Show that the lines $y = 3x - 2$, $x + 2y = 10$ and $4x - 5y + 12 = 0$ are concurrent.

Consider $y = 3x - 2$ (1)

$x + 2y = 10$ (2)

Substitute (1) in (2)

$x + 2(3x - 2) = 10$

$7x - 4 = 10$

$7x = 14$

$x = 2$

Substitute $x = 2$ in (1)

$y = 3(2) - 2$

$y = 4$

Now check whether (2, 4) lies on $4x - 5y + 12 = 0$.

LHS = 4(2) – 5(4) + 12

= 8 – 20 + 12

= 0

= RHS

The point (2, 4) lies on the line $4x - 5y + 12 = 0$.

The point lies on all three lines, so the lines are concurrent.

Practise Practise

1 Use the distance formula to calculate the length AB (leave your answer in surd from if necessary), given that:

a A(–1, 4), B(1, –4)
b A(2, –1), B(4, –1)
c A(–3, –2), C(1, 1)
d A(0, 4), B(–3, 7) p. 69

2 Find the midpoint of the interval PQ given that:

a P(2, –1), Q(4, –3)
b P(–4, 2), Q(1, –4)
c P(5, 3), Q(–3, 5)
d P(7, –2), Q(–3, –2) pp. 69–70

3 Find the gradient of the line containing R and S given that:

a R(4, 1), S(–4, –1)
b R(1, 4), S(–1, 1)
c R(–3, –2), S(–2, –3)
d R(4, –1), S(3, –1) pp. 69–70

4 From the following diagrams, find the gradients of the lines:

a

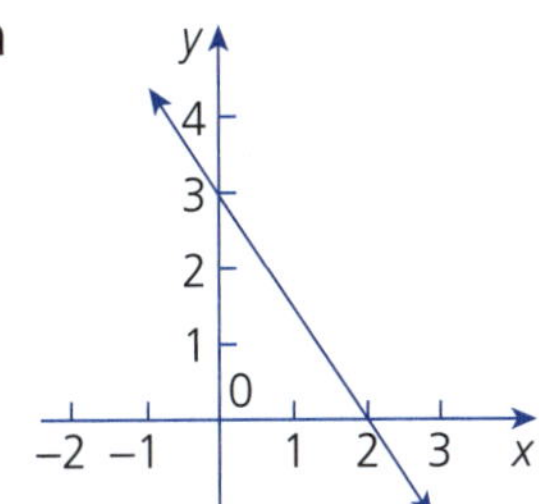

b

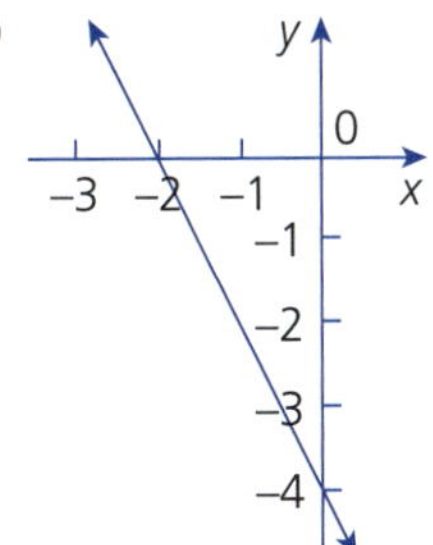

c

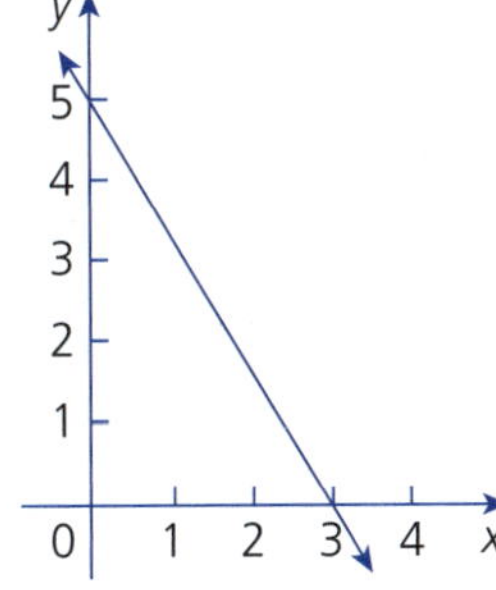

d

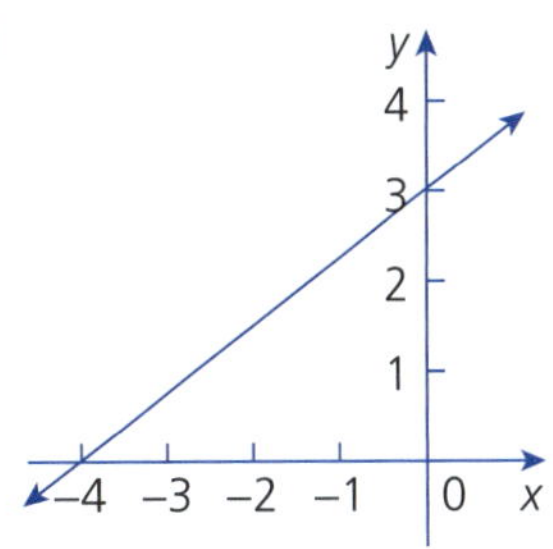

e

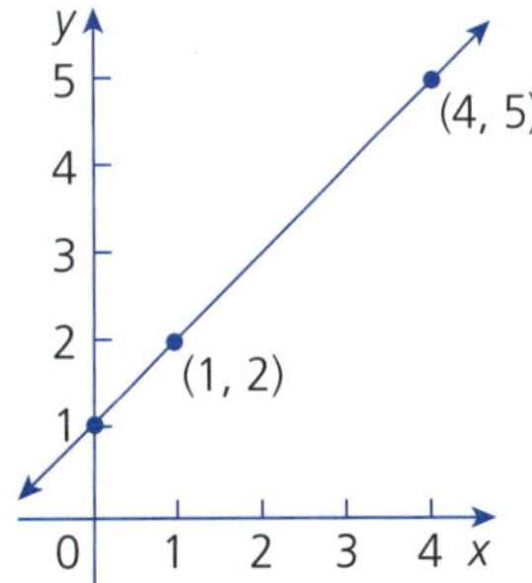

pp. 70–71

5 The points A(–2, 4), B(4, 2) and C(0, –2) are plotted on a number plane. Show that the:

a △ABC is isosceles
b gradient of AB is $-\frac{1}{3}$
c midpoint of AC is (–1, 1)
d line joining the midpoints of AC and BC is parallel to the side AB and half its length. pp. 71–72

6

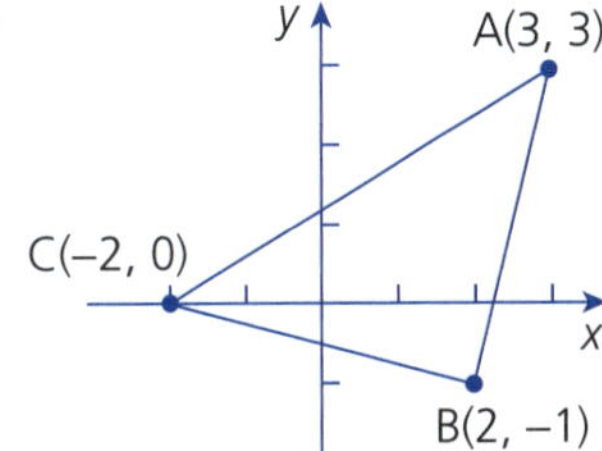

A, B and C are the three vertices of the triangle with coordinates (3, 3), (2, –1) and (–2, 0) respectively.

a By considering gradients, show that △ABC is right-angled.
b Prove that $AC^2 = AB^2 + BC^2$.
c Calculate the area of the ABC.
d Show that △ABC is also isosceles. pp. 71–72

7 Find the gradient of the lines with equations:

a $y = -3x + 4$
b $4y = x + 2$
c $3x - y = 7$
d $x + 5y - 8 = 0$ pp. 72–73

8 Write down the equation of the line with gradient $\frac{1}{2}$ and y-intercept -2. pp. 72–73

9 Find the equation of the line passing through $(0, 4)$ with gradient -3. pp. 72–73

10 Find the equation of the lines with:
a gradient -2 passing through $(-1, 1)$
b gradient $\frac{1}{2}$ passing through $(4, 2)$
c gradient $\frac{3}{4}$ passing through $(-3, 1)$. p. 73

11 Find the equation of the line through $(2, -5)$ with gradient $\frac{2}{3}$. p. 73

12 Find the equation of the line passing through the points:
a $(1, -1), (2, 4)$
b $(0, 3), (-3, 0)$
c $(-2, 1), (4, 2)$
d $(1, -5), (-1, -5)$ p. 74

13 Which of the points A$(-3, 7)$, B$(3, -7)$ or C$(-11, 23)$ lie on the line $2x + y = 1$? p. 75

14 By first finding the intercepts, sketch the following lines:
a $y = x - 2$
b $2x + y = 6$
c $x - 3y - 9 = 0$ pp. 75–76

15 State whether the following pairs of lines are parallel, perpendicular or otherwise:
a $2x - 3y = 7$ and $6x - 12y - 5 = 0$
b $x + 2y - 3 = 0$ and $y = 2x + 1$
c $2x + 3y = 6$ and $4y = 6x - 5$ pp. 76–79

16 Find the equation of the line that is:
a parallel to $y = 4x - 7$ and passing through $(0, -4)$
b parallel to $2x - 3y = 5$ and passing through $(-1, -1)$
c perpendicular to $x + 3y - 6 = 0$ and passing through $(4, 1)$
d perpendicular to $2y = 5 - x$ and passing through $(-2, 3)$. pp. 76–79

17 If $2x - ky = 5$ is parallel to $3x + 4y - 7 = 0$, find the value of k. pp. 76–79

18 Find the equation of the line that is the perpendicular bisector of the interval AB, given that A is $(-3, 4)$ and B is $(1, -2)$. pp. 76–79

19

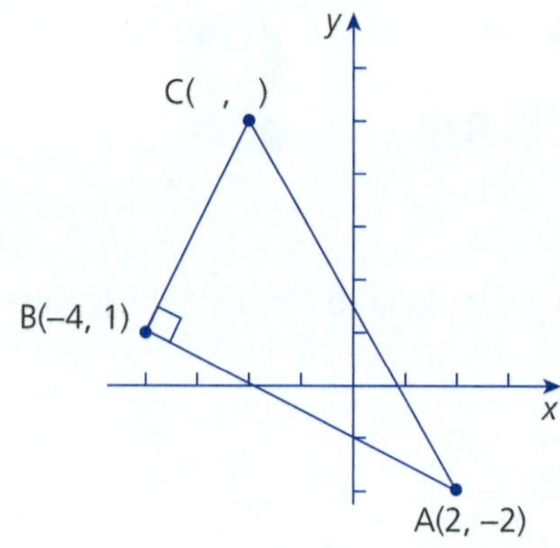

A$(2, -2)$, B$(-4, 1)$ and C are vertices of a right-angled triangle, with the right angle at B.
a Find the equation of the line AB.
b Find the gradient of the line BC.
c Find the equation of the line BC.
d If C has coordinates $(-2, 5)$, calculate the length of BC.
e Find the area of $\triangle ABC$.
f Find the equation of the median drawn from the midpoint of AB to the vertex C. pp. 76–79

20

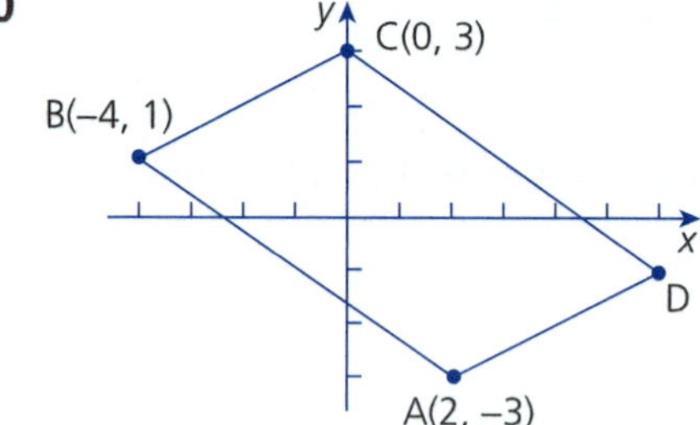

A$(2, -3)$, B$(-4, 1)$ C$(0, 3)$ and D are the vertices of the parallelogram ABCD with AC as a diagonal.
a Calculate the gradient of AB and hence find its equation.
b Find the equation of CD.
c Write down the coordinates of D, the fourth vertex.
d Show that the interval joining the midpoints of AB and CD is parallel to BC. pp. 76–79

21 Find the equation of the line passing through $(2, 1)$ and parallel to the line $2x - y + 3 = 0$. pp. 76–79

22 Find the equation of the line passing through $(-1, 3)$ and perpendicular to the line $4x + y + 5 = 0$. pp. 76–79

23 Show that the point X$(-2, 0)$ lies on the line passing through P$(7, -3)$ and Q$(-5, 1)$. pp. 76–79

24 The vertices of the triangle XYZ below are X(–2, 6), Y(4, 2) and Z(0, 0). Let M, N and P be the midpoints of XY, YZ and ZX respectively.

a Find the coordinates of M, N and P.
b Find the equations of the medians MZ, NX and PY.
c Show that the medians are concurrent.
d Prove that the product of the gradients of XY, YZ and ZX is 1.
e Find the equation of the *altitude* of the triangle through Z. pp. 76–79

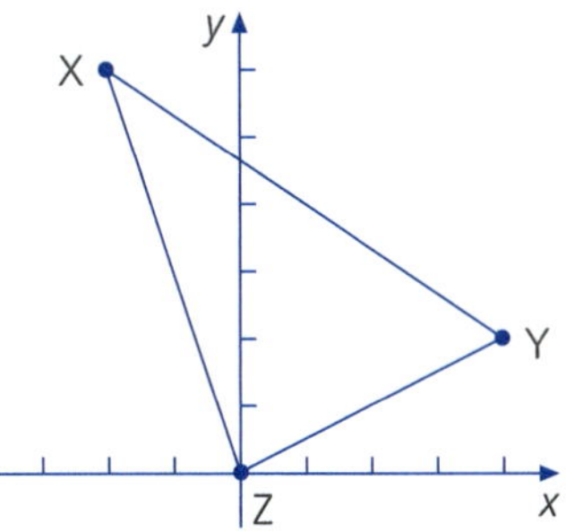

Hint: the altitude of a triangle is the line perpendicular to one side and passing through the opposite vertex.

25 The point $(2d, 5)$ lies on the line $2x - y - 8 = 0$. Find d. pp. 76–79

26 Show that the quadrilateral with vertices P(–2, 3), Q(1, 5), R(5, –1) and S(2, –3) is a rectangle. pp. 76–79

27 Find the equation of the line perpendicular to AB, where A is (–3, 5), B is (2, 4) and passing through the origin. pp. 76–79

28 By first plotting the points A(2, 2), B(4, 8) and C(–2, 5) find the:

a distance BC
b equation of the line through B and C
c equation of the line parallel to BC through A
d equation of the line perpendicular to BC through A
e midpoint of AB
f equation of the median CL, where L is the midpoint of AB. pp. 76–79

29 Show that the points (–1, 3), (1, 1) and (4, –2) are collinear. p. 79

30 Prove that $2x - 3y = 3$, $x + 2y = 5$ and $3x - y = 8$ are concurrent lines. p. 80

Go to p. 300 for **Quick Answers** or to pp. 320–324 for **Worked Solutions**

Your Checklist

For a complete understanding of this topic, you must be able to:

✓	Find the distance between two points, the midpoint of two points and the gradient of a line joining two points using the appropriate formulae		pp. 69–70
✓	Find the gradient of any line from a diagram		pp. 70–71
✓	Apply formulae to number plane problems		pp. 71–72
✓	Apply the gradient form of the equation of a straight line		pp. 72–73
✓	Find the equation of a straight line using its gradient and a point on the line		p. 74
✓	Find the equation of a straight line using any two points on the line		p. 74
✓	Show that a point lies on a line algebraically		p. 75
✓	Find intercepts cut from the axes by a line		pp. 75–76
✓	Find gradients of lines parallel or perpendicular to a given line		pp. 76–79
✓	Show concurrence and collinearity.		pp. 79–80

Now you are ready to do the tests!

Intermediate Test

Linear Relationships

(30 marks)

1 For the points A(–3, 1) and B(5, –4) find the:

a distance AB (2 marks)

b midpoint of AC (2 marks)

c gradient of BC. (2 marks)

2

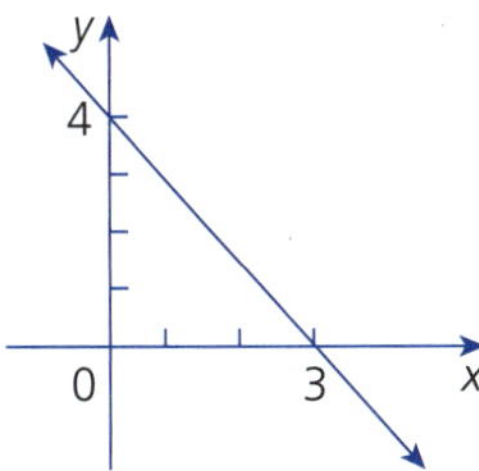

From the diagram write down the gradient of the line and hence write down its equation. (3 marks)

3 From the equation $4y = 2x - 5$, write down the gradient of the line and the y-intercept. (2 marks)

4 Derive the equation of the line passing through (0, –4) with gradient 2. (2 marks)

5 Derive the equation of the line passing through the points (–2, 5) and (5, –2). (2 marks)

6 Show that the points (3, –5), (9, 7) and (–2, –15) are collinear. (2 marks)

7 Derive the equation of the line that is parallel to $y = \frac{x - 5}{2}$ and passes through (0, 3). (2 marks)

8

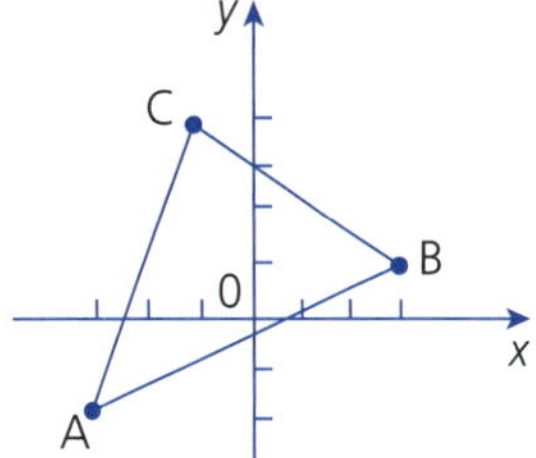

A(–3, –2), B(3, 1) and C(–1, 4) are points on the number plane. Find the:

a distance AB (1 mark)

b gradient AC (1 mark)

c midpoint AC (1 mark)

d equation of the line AC (2 marks)

e equation of the perpendicular bisector of AC (2 marks)

f equation of the line parallel to OC that passes through the midpoint of AB. (2 marks)

9 If $(2, k)$ lies on the line $x + 3y = 8$, find the value of k. (2 marks)

Your Feedback

☐ × 100% = ☐ %

QA PAGE 305

WS PAGE 365

Advanced Test

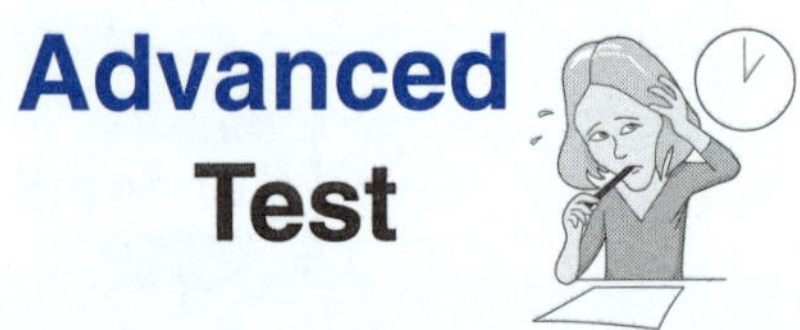

40 min

Linear Relationships

(40 marks)

1 For the points A(–3, 1), B(4, –3) and C(–2, –5), find the:

a distance AB (2 marks)
b gradient AC (2 marks)
c midpoint BC (2 marks)
d the coordinates of the point one quarter of the way between C and B respectively. (2 marks)

2

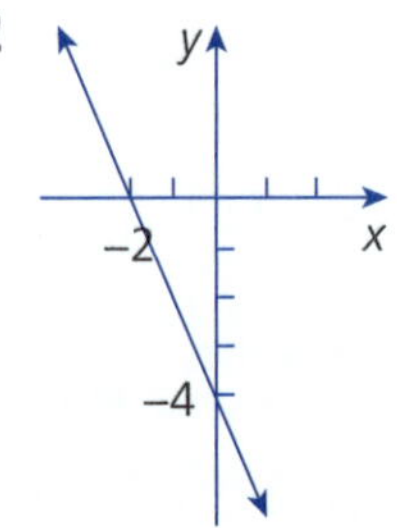

From the diagram write down the gradient of the line and hence write down its equation. (3 marks)

3 For the equation $6x = 2y - 7$, find the gradient of the line and the y-intercept. (2 marks)

4 Derive the equation of the line passing through (2, –3) with gradient $\frac{3}{4}$. (2 marks)

5 Derive the equation of the line passing through (–3, 1) and (3, –5). (2 marks)

6 Derive the equation of the line perpendicular to $x + 2y - 5 = 0$ and passing through (–1, 4). (3 marks)

7 The point $(3m, -5)$ lies on the line $2x - 3y - 3 = 0$. Find the value of m. (2 marks)

8 X(–4, 3) and Y(p, q) are the end points of a diameter of a circle. The centre of this circle is (–1, –1). Find:

a p and q (2 marks)
b the radius of the circle. (1 mark)

9 Show that the lines $x + 2y - 6 = 0$, $x = 2y - 2$ and $y = 3x - 4$ are concurrent. (3 marks)

10

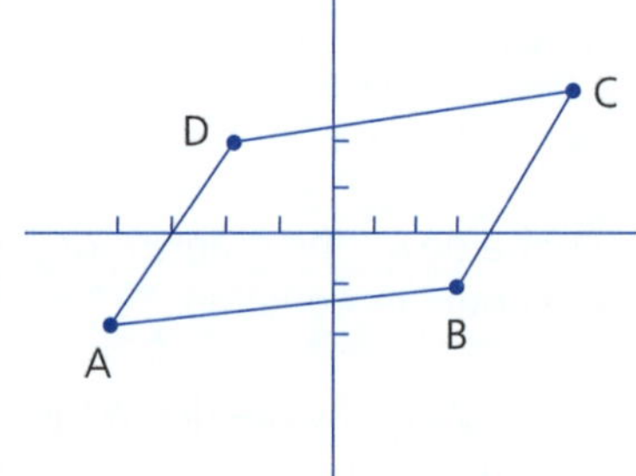

A(–4, –2), B(3, –1), C and D(–2, 2) are the vertices of the parallelogram ABCD, with AB || DC.
Find the:

a distance AB (1 mark)
b gradient of AB (1 mark)
c equation of AB (2 marks)
d equation of DC (2 marks)
e midpoint of AD (1 mark)
f coordinates of C (1 mark)
g distance CD. (1 mark)

11 A(1, 1), B(3, 1), C(3, –1), D(1, –1) are the vertices of a square. Show that the diagonals of the square bisect each other at right-angles. (3 mark)

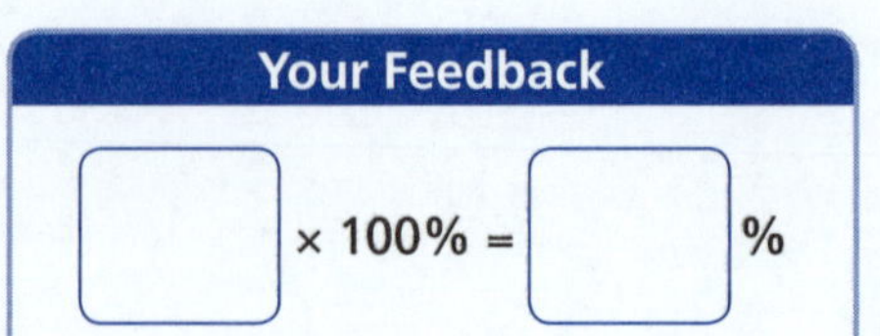

QA PAGE 305

WS PAGE 366

Chapter 6
Non-linear Relationships

The parabola

The graph of $y = x^2$

To graph any equation which results in a curve we must plot points to make a clear shape.

Keywords

Asymptote
Axis of symmetry
Coefficient
Concavity
Cubic
Exponential
Half-plane
Hyperbola
Parabola
Region
Vertex
x-intercept
y-intercept

1 For $y = x^2$, complete the table and graph the curve:

x	−3	−2	−1	0	1	2	3
y							

S Our points are (−3, 9), (−2, 4), (−1, 1), (0, 0), (1, 1), (2, 4), (3, 9)

The y-axis is the axis of symmetry. The curve is symmetrical about the y-axis.

The curve is always above the x-axis (or passes through the origin) because both positives and negatives squared give positive answers (except at [0, 0]).

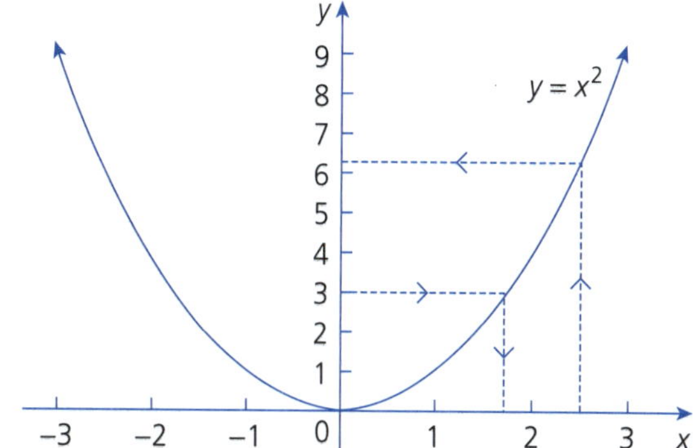

The graph can be used to estimate:

$2.5^2 \approx 6.2$

$\sqrt{3} \approx 1.7$

Graphing $y = ax^2 + c$, $y = (x - k)^2$, $y = (x - m)(x - n)$, etc.

The basic shape of $y = x^2$ can be adjusted to graph other parabolas.

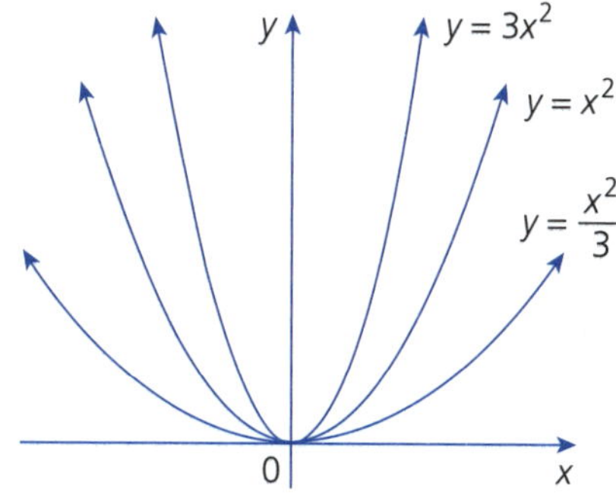

$y = \dfrac{x^2}{3}$ (or $y = \dfrac{1}{3}x^2$) is 'outside' $y = x^2$

$y = 3x^2$ is 'inside' $y = x^2$

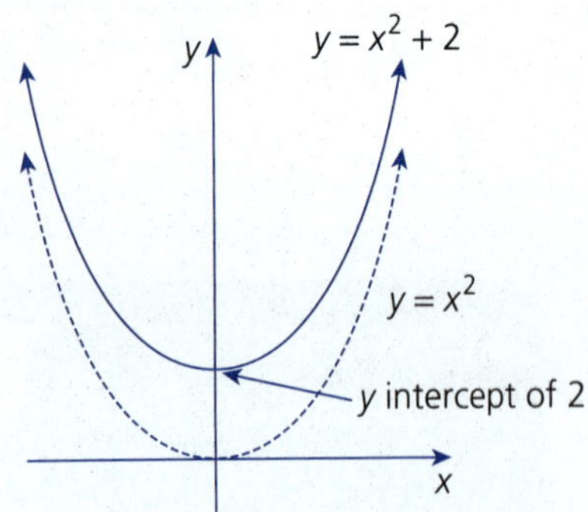

$y = x^2 + 2$ is 'shifted up' two units from $y = x^2$

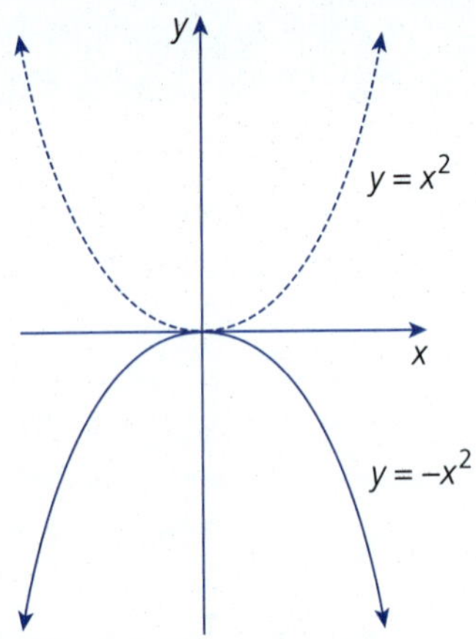

$y = -x^2$ is an 'upside down' view of $y = x^2$, that is, the coefficient of x^2 is negative, therefore concave down, not concave up

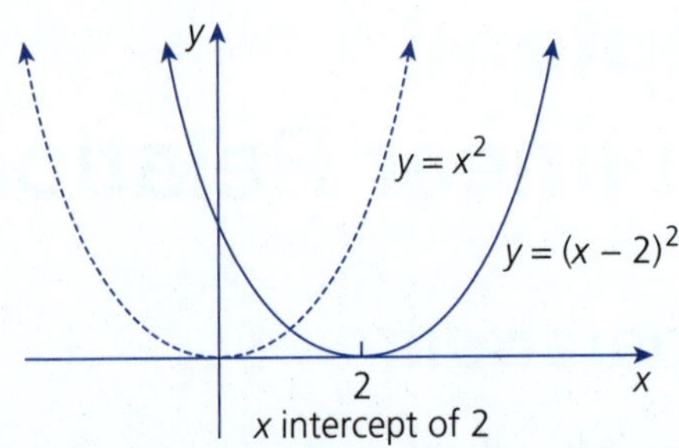

$y = (x - 2)^2$ is shifted to the right two units from $y = x^2$

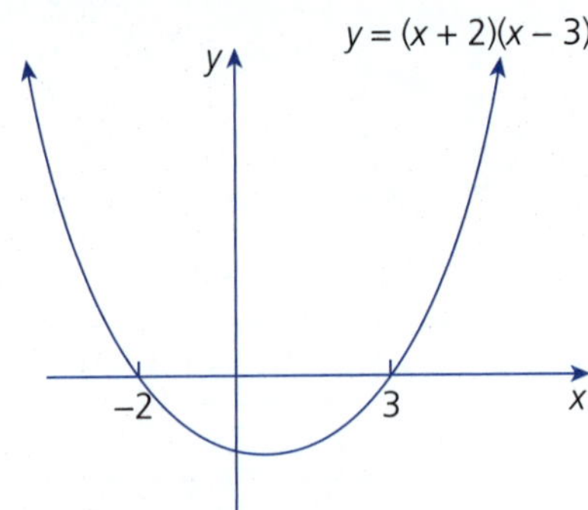

When $y = 0$, $x = -2$ and $x = 3$.

For Example

1 Graph the following parabolas on the same number plane:

a $y = x^2$ and $y = 2 - x^2$

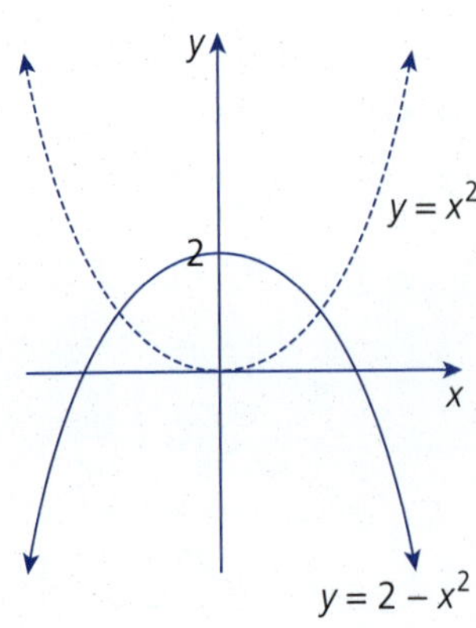

b $y = x^2 + 3$ and $y = (x + 3)^2$

S

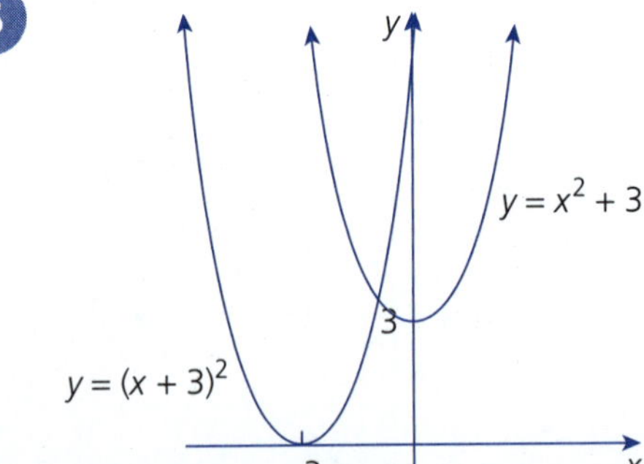

c $y = x^2 - 4x + 4$ and $y = x^2 + 4x + 4$

S Now, $y = x^2 - 4x + 4$ $y = x^2 + 4x + 4$

$\therefore y = (x - 2)^2$ $y = (x + 2)^2$

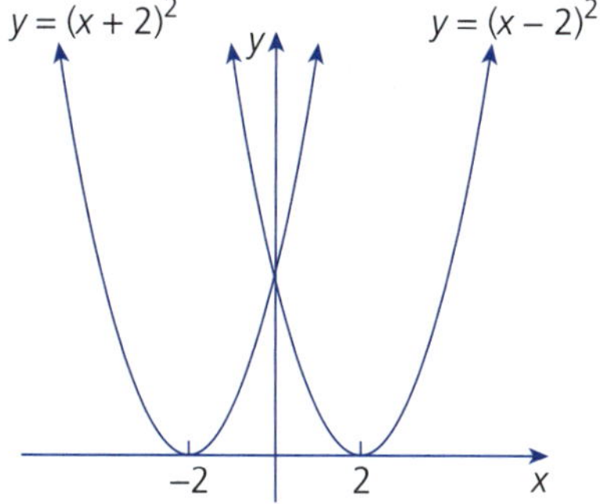

d $y = (x + 2)(x - 3)$ and $y = x^2 - 7x$

S For $y = (x + 2)(x - 3)$

to find x-intercepts, let $y = 0$

$\therefore (x + 2)(x - 3) = 0$

$x = -2, 3$

For $y = x^2 - 7x$, let $y = 0$

$\therefore x(x - 7) = 0$

$x = 0, 7$

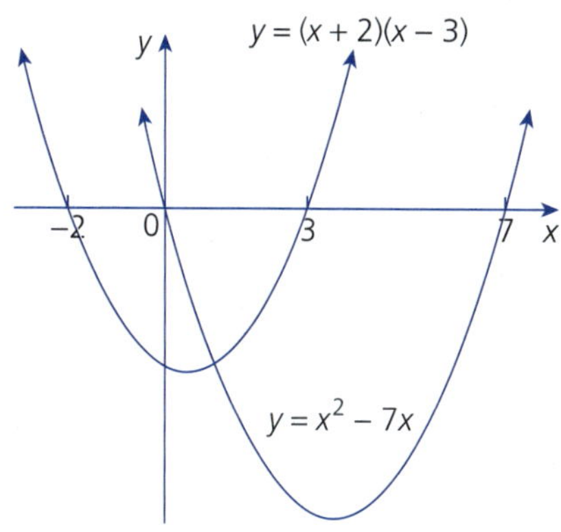

e $y = 2x^2 - x - 4$

S Cannot factorise, so must use table of values:

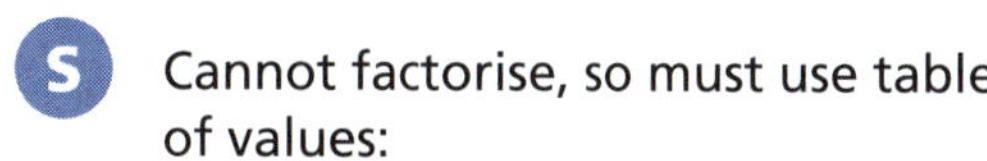

x	−3	−2	−1	0	1	2	3
y	17	6	−1	−4	−3	2	11

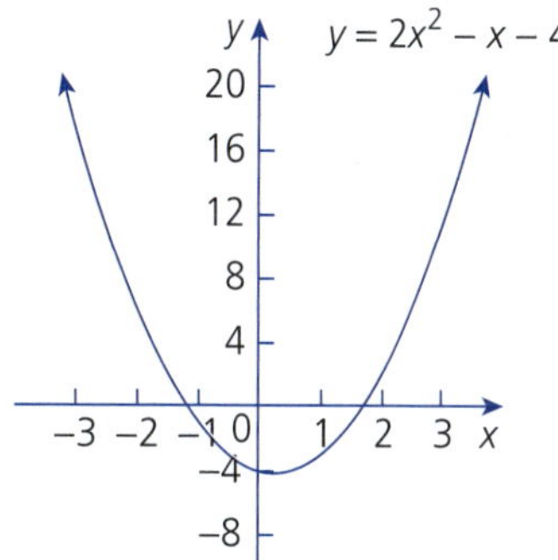

The vertex of a parabola

Three methods can be used to find the vertex of a parabola:

- Midpoint of x-intercepts
- Completing the square
- Using $x = \frac{-b}{2a}$ (axis of symmetry)

1 Using midpoint of x-intercepts, find the vertex of $y = (x + 3)(x - 1)$.

S x-intercept when $y = 0$

$\therefore (x + 3)(x - 1) = 0$

$x = -3, 1$

$\therefore$ Middle of −3 and 1

$= \frac{-3 + 1}{2}$

$= -1$

Now substitute in y

That is, $y = (-1 + 3)(-1 - 1)$

$= (2)(-2)$

$= -4$

$\therefore$ Vertex is $(-1, -4)$.

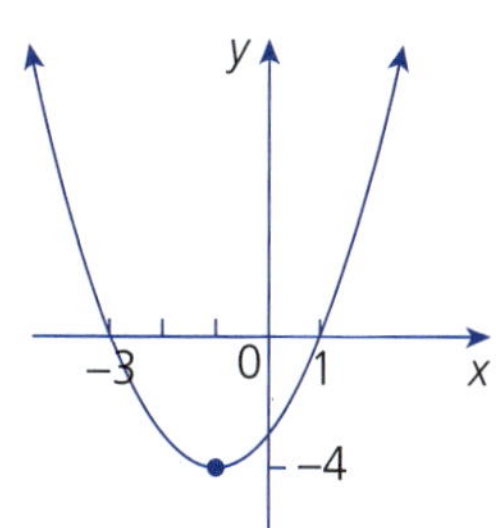

The parabola has minimum value (as $a > 0$) of −4.

For Example

1 By completing the square, find the vertex of: $y = x^2 - 4x + 3$.

$$y = x^2 - 4x + 3$$
$$\therefore\ y = x^2 - 4x + 3$$
$$\therefore\ y = x^2 - 4x + 4 + 3 - 4$$

We have added a 4, so subtract 4.

$$\therefore\ y = (x - 2)^2 - 1$$

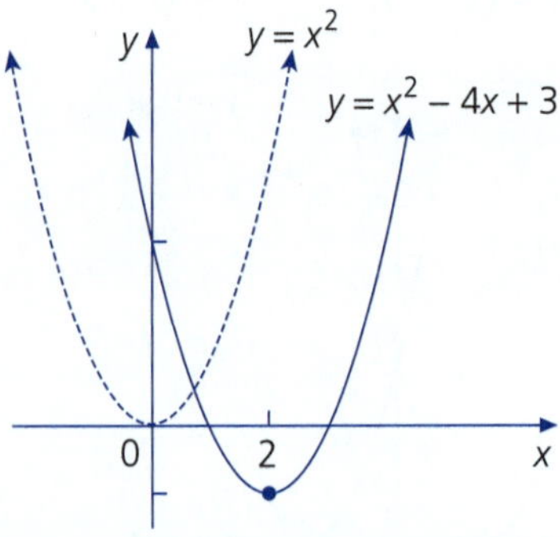

$y = x^2$ moved across 2 units and down 1 unit. Vertex is (2, −1). The parabola has minimum value of −1.

For Example

1 Using $x = \frac{-b}{2a}$, find the vertex of: $y = -x^2 - 3x + 7$

As $y = -x^2 - 3x + 7$ is in the form of $y = ax^2 + bx + c$ $\therefore\ a = -1, b = -3$

$$\therefore\ x = \frac{-b}{2a}$$
$$= \frac{3}{-2}$$
$$= -1\frac{1}{2}$$

$\therefore$ Axis of symmetry is $x = -1\frac{1}{2}$.

Substitute in $x = -1\frac{1}{2}$ in

$$y = -x^2 - 3x + 7$$
$$\therefore\ y = -\left(-1\frac{1}{2}\right)^2 - 3\left(-1\frac{1}{2}\right) + 7$$
$$= -2\frac{1}{4} + 4\frac{1}{2} + 7$$
$$= 9\frac{1}{4}$$

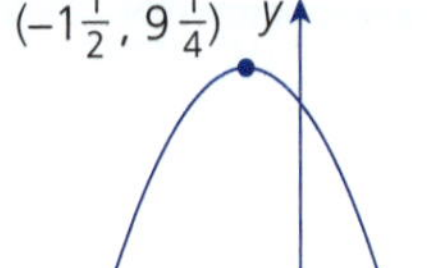

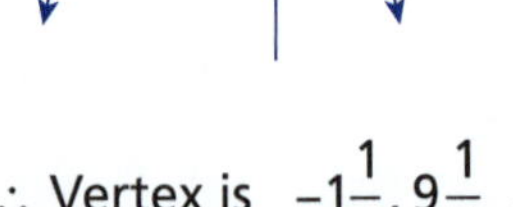

$\therefore$ Vertex is $\left(-1\frac{1}{2}, 9\frac{1}{4}\right)$.

The parabola has a maximum value (as $a < 0$) of $9\frac{1}{4}$.

Determining the equation of the parabola

The equation of the parabola can be determined using points that lie on the parabola.

For Example

1 Find the equation of:

a

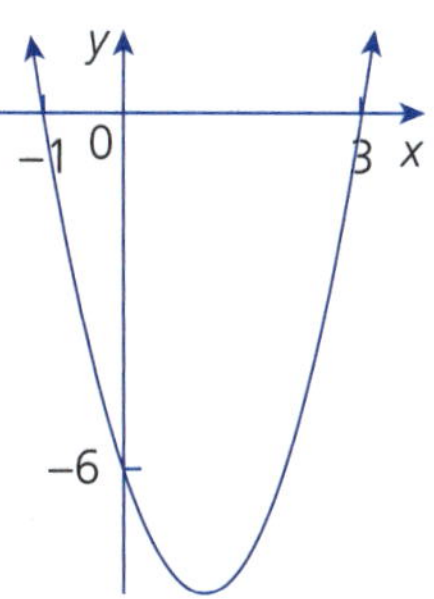

S x-intercepts of –1 and 3

$\therefore\ y = a(x + 1)(x - 3)$

Now substitute in (0, –6)

$\therefore\ -6 = a(0 + 1)(0 - 3)$

$-3a = -6$

$a = 2$

$\therefore\ y = 2(x + 1)(x - 3)$

b

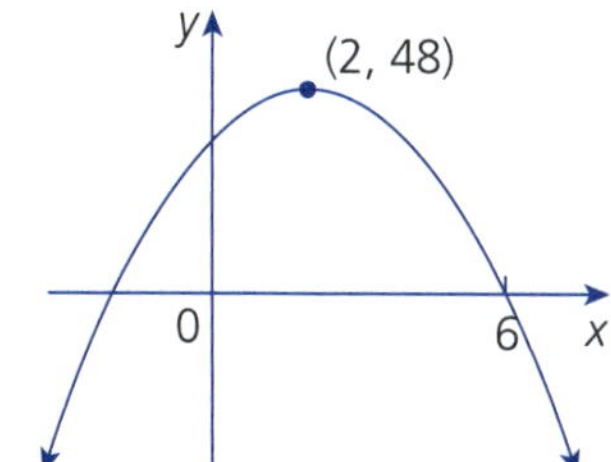

S As axis of symmetry at $x = 2$, and one x-intercept of 6, then the other x-intercept will be at –2.

$\therefore\ y = a(x + 2)(x - 6)$

Substitute in (2, 48)

$\therefore\ 48 = a(2 + 2)(2 - 6)$

$-16a = 48$

$a = -3$

$\therefore\ y = -3(x + 2)(x - 6)$

The cubic (for example $y = x^3$)

A positive number cubed gives a positive. A negative number cubed gives a negative.

For Example

1 Graph the following values for $y = x^3$:

x	–2	–2	$-\frac{1}{2}$	0	$\frac{1}{2}$	1	2
y	–8	–1	$-\frac{1}{8}$	0	$\frac{1}{8}$	1	8

S

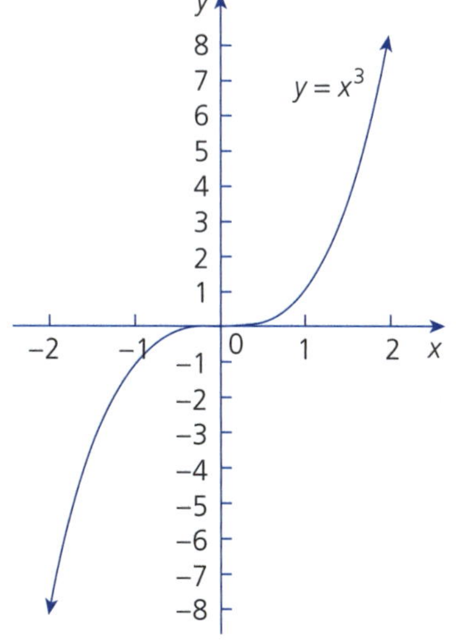

2 On the same number plane, graph $y = x^3 + 4$, $y = -x^3$.

S

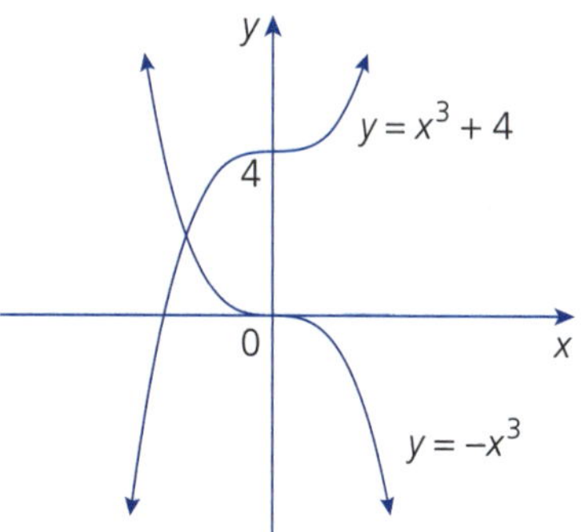

The hyperbola ($y = \frac{a}{x}$, or $xy = a$)

For the hyperbola, the curve gets closer and closer to the axes without touching. The axes are called the *asymptotes* of the curve.

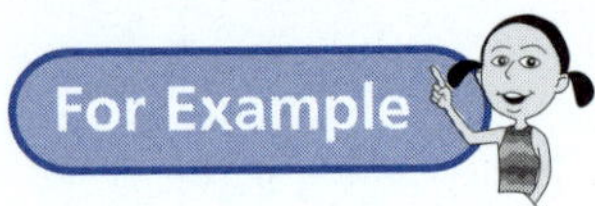

1 Graph $y = \frac{2}{x}$.

x	−4	−2	−1	$-\frac{1}{2}$	0	$\frac{1}{2}$	1	2	4
y	$-\frac{1}{2}$	−1	−2	−4	–	4	2	1	$\frac{1}{2}$

S

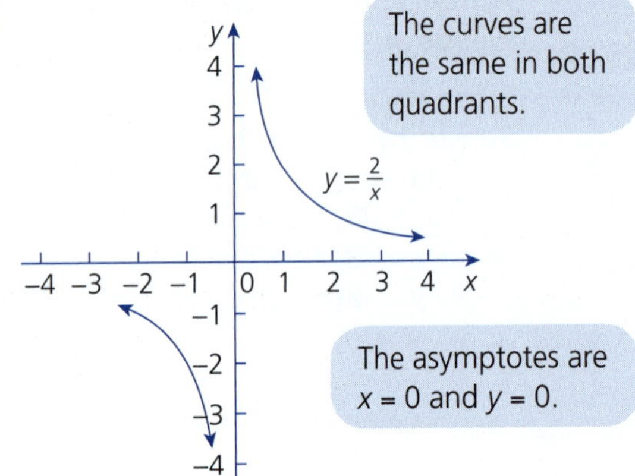

Note that there is a problem for $x = 0$ as $y = \frac{2}{0}$ is indeterminate. Therefore the graph has a discontinuity, that is, it has two parts.

If $a < 0$, for example, $y = -\frac{2}{x}$, we would graph it as follows:

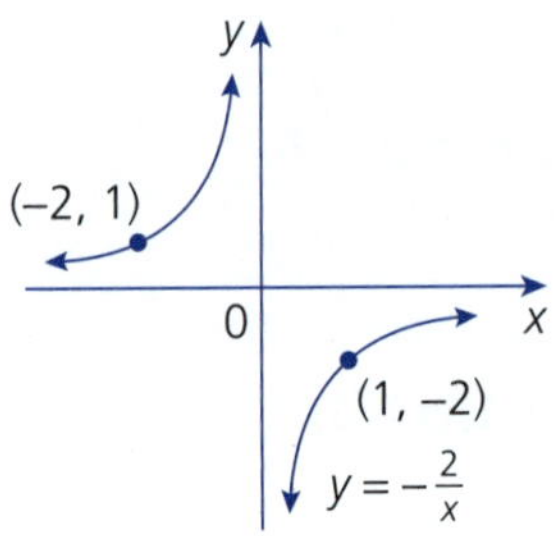

The exponential (for example, $y = 2^x$)

The curve of an exponential always passes through (0, 1) and is always positive.

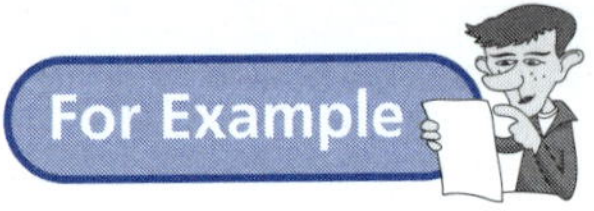

1 Complete the table for $y = 2^x$ and graph $y = 2^x$ on a number plane.

x	−3	−2	−1	0	1	2	3
y	0.125	0.25	0.5	1	2	4	8

S To find 2^{-3} by calculator we use:

2 [x^y] [+/−] 3 [=]

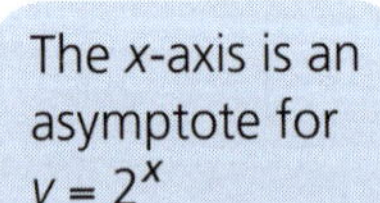

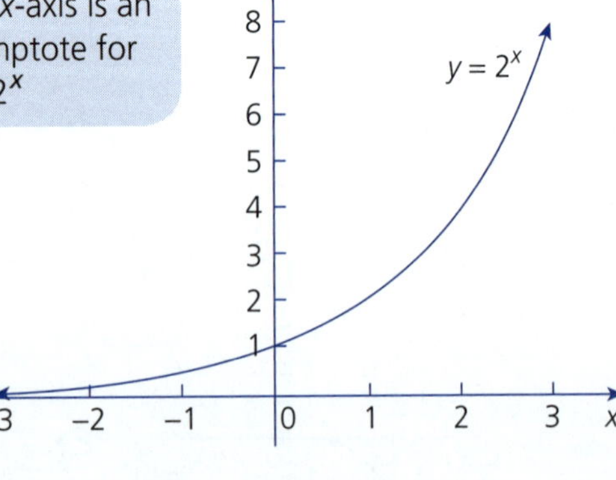

We can also graph $y = 2^{-x}$, that is, $y = \frac{1}{2^x}$, the reciprocal of $y = 2^x$. →

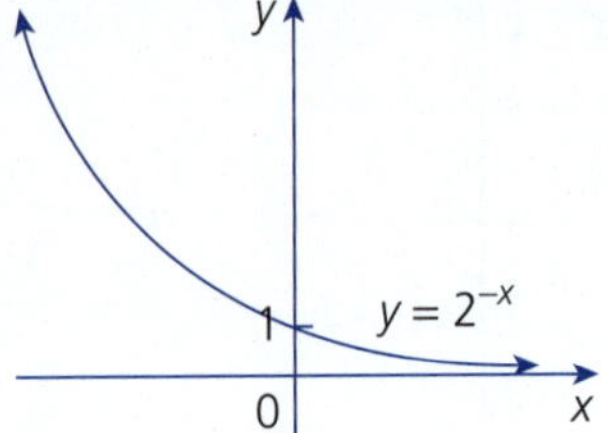

Graphs of the form $y = x^n$

These first examples are of the graph of $y = x^n$ when n is even.

For Example

1 On separate number planes graph the following:

a $y = x^4$

S

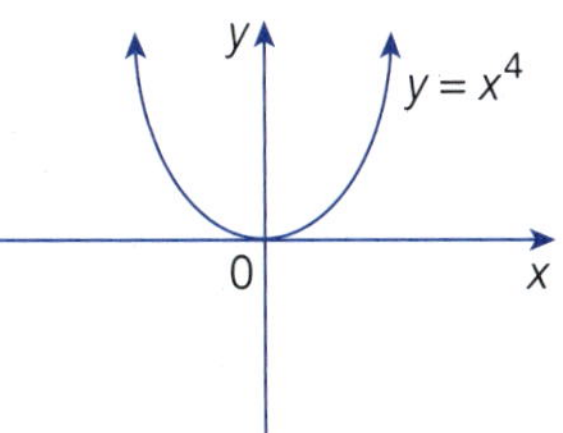

Similar to $y = x^2$ but steeper.

b $y = x^4 + 2$

S

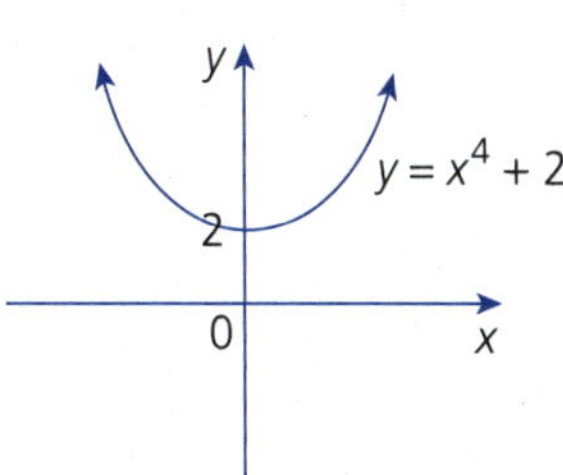

$y = x^4$ is translated vertically by 2 units

c $y = (x + 2)^4$

S

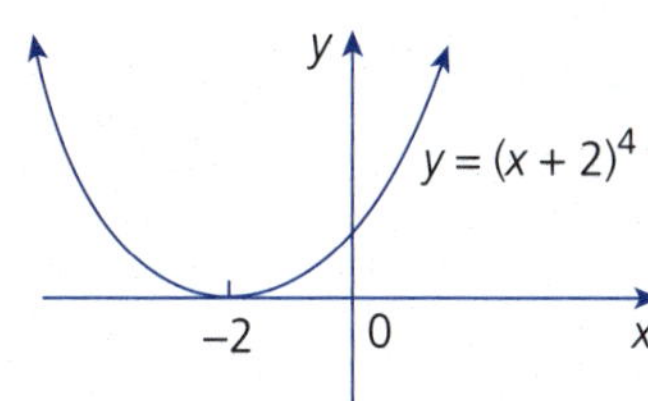

$y = x^4$ is translated horizontally to left 2 units

The next examples are of the graph of $y = x^n$ when n is odd (where $n > 1$).

For Example

1 On separate number planes graph of the following:

a $y = x^5$

S

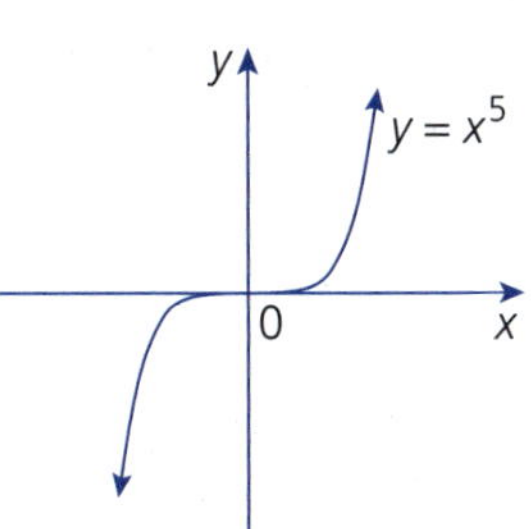

Similar to $y = x^3$ but steeper.

b $y = x^5 - 3$

S

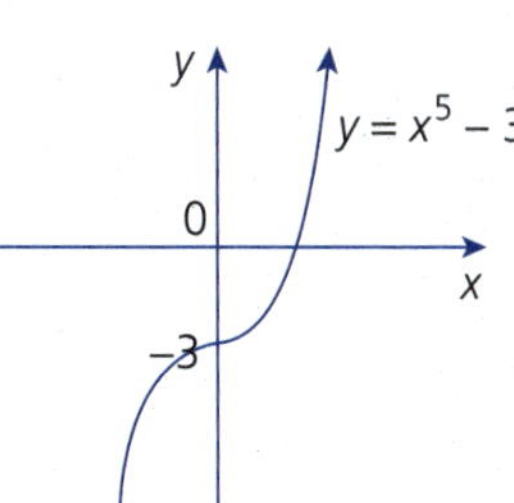

$y = x^5$ is translated vertically down 3 units

c $y = (x - 3)^5$

S

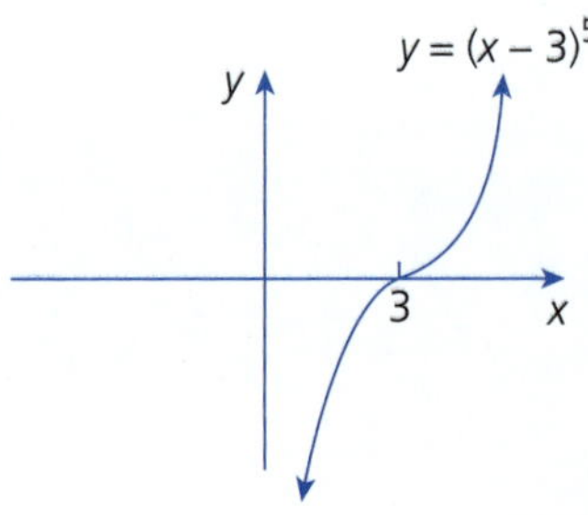

$y = x^5$ is translated horizontally to right 3 units.

d $y = 1 - x^5$

S

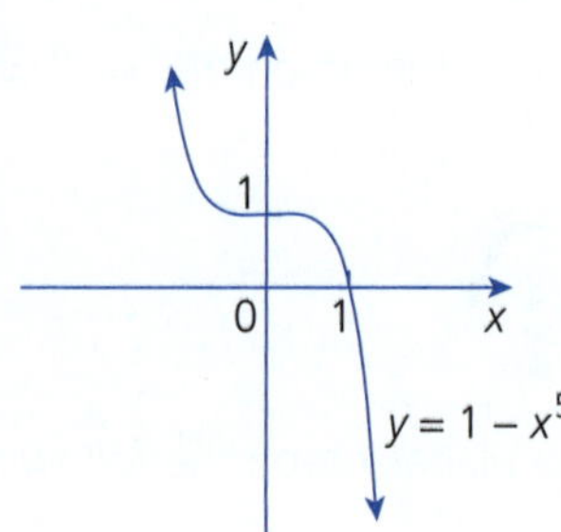

Graphs of the form $y = a(x - r)(x - s)(x - t)$

The next special graph is the graph of $y = a(x - r)(x - s)(x - t)$.

This graph has x-intercepts of r, s, t and a y-intercept of $-a \times r \times s \times t$.

For Example

1 Graph the following:

a $y = (x + 2)(x - 3)(x + 1)$

S

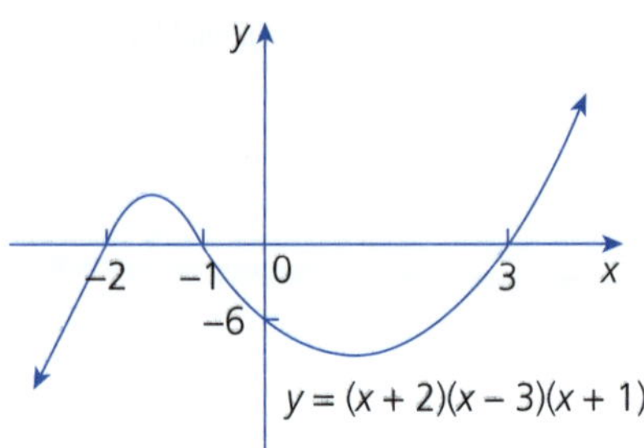

$y = (x + 2)(x - 3)(x + 1)$

Let $y = 0$ $\quad \therefore (x + 2)(x - 3)(x + 1) = 0$

$\therefore x = -2, 3, -1$

$\therefore$ x-intercepts of $-2, 3, -1$

Let $x = 0$ $\quad \therefore y = (0 + 2)(0 - 3)(0 + 1) = -6$

$\therefore$ y-intercept of -6

Finally, choose value of x beyond $x = 3$ to check whether y is positive/negative:

Let $x = 4$ $\quad \therefore y = (4 + 2)(4 - 3)(4 + 1)$

$= 6 \times 1 \times 5 > 0$

b $y = 2(3 - x)(x + 1)(x - 4)$

S

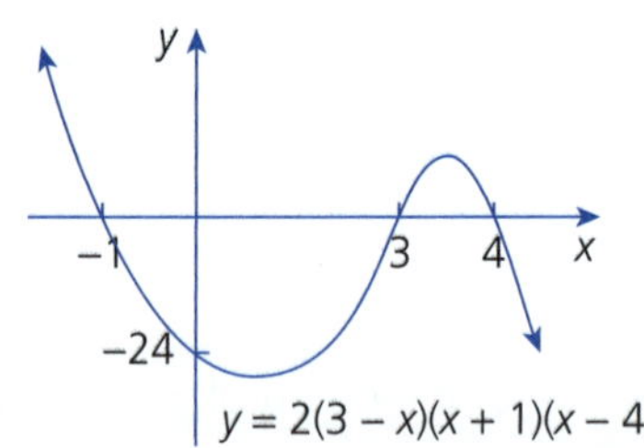

$y = 2(3 - x)(x + 1)(x - 4)$

Let $y = 0$ $\quad \therefore 2(3 - x)(x + 1)(x - 4) = 0$

$\therefore x = 3, -1, 4$

$\therefore$ x-intercepts of $3, -1, 4$

Let $x = 0$ $\quad \therefore y = 2(3 - 0)(0 + 1)(0 - 4) = -24$

$\therefore$ y-intercept of -24

Check y when $x = 5$:

$\therefore y = 2(3 - 5)(5 + 1)(5 - 4) < 0$

c $y = x(x - 1)^2$

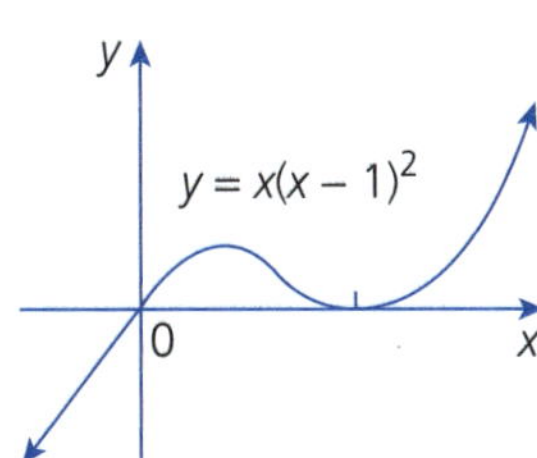

$y = x(x - 1)^2$

Let $y = 0$ $\quad \therefore x(x - 1)^2 = 0$

$\therefore x = 0, 1$

$\therefore$ x-intercepts at 0 and 1

Let $x = 0$ $\quad \therefore y = 0(0 - 1)^2 = 0$

$\therefore$ y-intercept of 0

Check y when $x = 2$:

$\therefore y = 2(2 - 1)^2 > 0$

d $y = x(x + 1)^2(x - 2)^3$

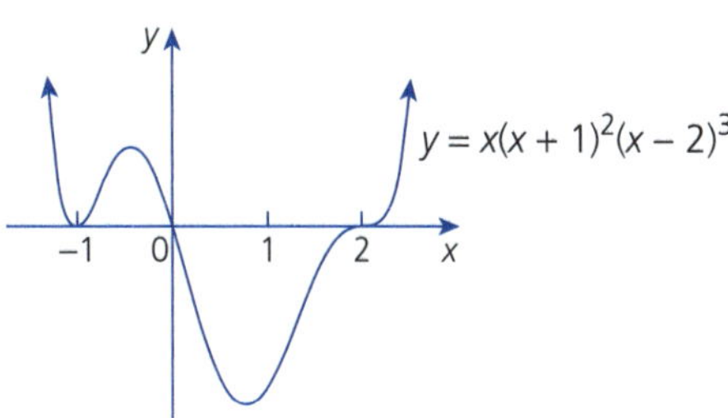

Let $y = 0$ $\quad \therefore x(x + 1)^2(x - 2)^3 = 0$

$\therefore x = 0, -1, 2$

$\therefore$ x-intercepts at 0, −1, 2

Let $x = 0$ $\quad \therefore y = 0(0 + 1)^2(0 - 2)^3 = 0$

$\therefore$ y-intercept of 0

Check y when $x = 3$:

$\therefore y = 3(3 + 1)^2(3 - 2)^3 > 0$

The circle

Equations of the form $\mathbf{x^2 + y^2 = r^2}$ are circles with centre (0, 0) and radius r units.

1 Write the equation of the circles:

a

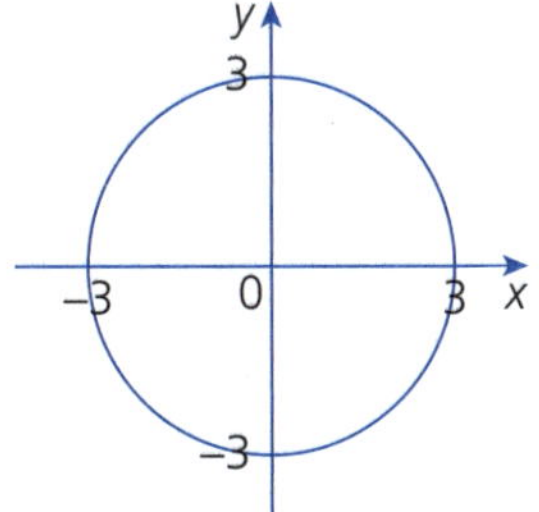

Circle centre (0, 0), radius = 3 units

$\therefore x^2 + y^2 = 9$

b

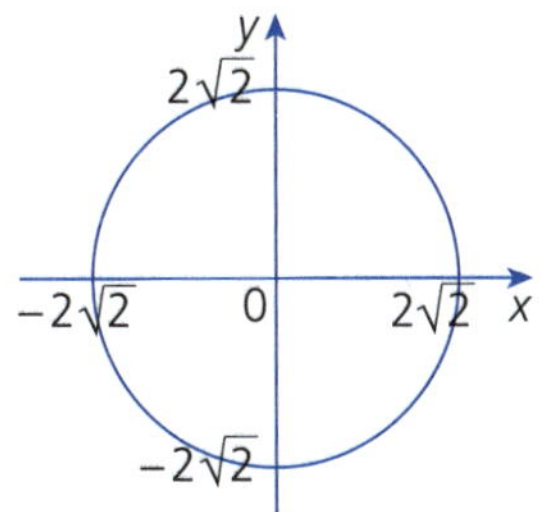

Circle, centre (0, 0), radius $2\sqrt{2}$ units

Now, $(2\sqrt{2})^2 = 8$

$\therefore x^2 + y^2 = 8$

2 For the circles, find the radius:

a $x^2 + y^2 = 49$

$x^2 + y^2 = 49$

Radius $= \sqrt{49}$

$= 7 \quad \therefore$ 7 units

b $x^2 + y^2 = 48$

$x^2 + y^2 = 48$

Radius $= \sqrt{48}$

$= 4\sqrt{3} \quad \therefore 4\sqrt{3}$ units

3 Draw the graph of a circle with centre (0, 0) and diameter 10 units. Write the equation of the graph.

As diameter is 10 units, then radius is 5 units.

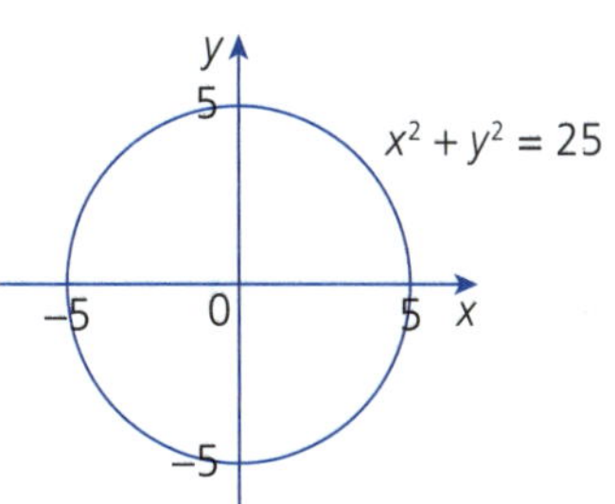

Equation is $x^2 + y = 25$.

Circles of the form $(x - h)^2 + (y - k)^2 = r^2$

The diagram on the right illustrates circles of this form, with circle centre (h, k), radius r units.

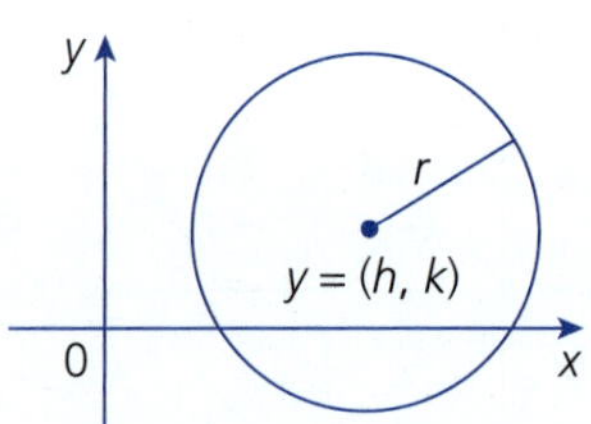

For Example

1 Write the equation of the circle with the following values:

a $C(4, 3)$, radius 3 units

S $(x - h)^2 + (y - k)^2 = r^2$
$(x - 4)^2 + (y - 3)^2 = 9$

b $C(-5, 1)$, radius $\sqrt{5}$ units

S $(x - h)^2 + (y - k)^2 = r^2$
$(x + 5)^2 + (y - 1)^2 = 5$

c $C(0, -4)$, radius $3\sqrt{2}$ units

S $(x - h)^2 + (y - k)^2 = r^2$
$x^2 + (y + 4)^2 = 18$
(as $[3\sqrt{2}]^2 = 18$)

2 Find the coordinates of the centre and radius of the following circles:

a $(x - 2)^2 + (y + 3)^2 = 25$

S $(x - 2)^2 + (y + 3)^2 = 25$
$C(2, -3)$, radius 5 units

b $x^2 + (y - 1)^2 = 18$

S $x^2 + (y - 1)^2 = 18$
$C(0, 1)$, radius = $\sqrt{18}$ units
i.e. $3\sqrt{2}$ units

c $x^2 + 8x + y^2 - 6y = 0$

S $x^2 + 8x + y^2 - 6y = 0$
$x^2 + 8x + 16 + y^2 - 6y + 9 = 16 + 9$

Complete the square by halving coefficient of x (or y) and squaring.

$(x + 4)^2 + (y - 3)^2 = 25$
$\therefore C(-4, 3)$, radius = 5 units

d $x^2 + x + y^2 + 10y = 4$

S $x^2 + x + y^2 + 10y = 4$

$$x^2 + x + \frac{1}{4} + y^2 + 10y + 25 = 4 + \frac{1}{4} + 25$$

$$\left(x + \frac{1}{2}\right)^2 + (y + 5)^2 = 29\frac{1}{4}$$

$$\left(x + \frac{1}{2}\right)^2 + (y + 5)^2 = \sqrt{\frac{117}{4}}$$

$$\therefore C\left(-\frac{1}{2}, -5\right), \text{ radius} = \frac{\sqrt{117}}{2}$$

3 Sketch the graph of the following equations:

a $(x - 2)^2 + (y - 2)^2 = 4$

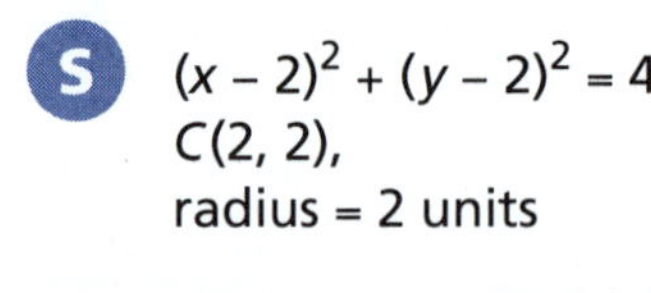

S $(x - 2)^2 + (y - 2)^2 = 4$
$C(2, 2)$,
radius = 2 units

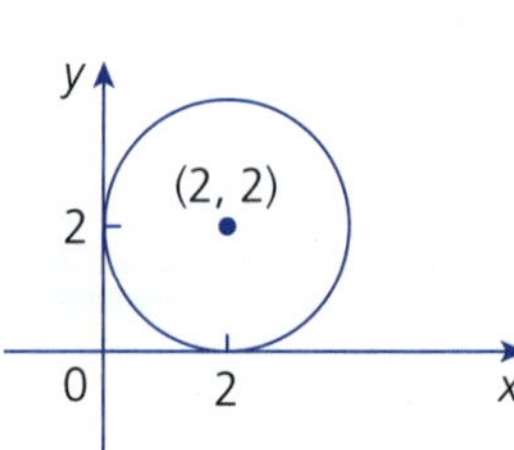

b $(x + 1)^2 + (y + 3)^2 = 9$

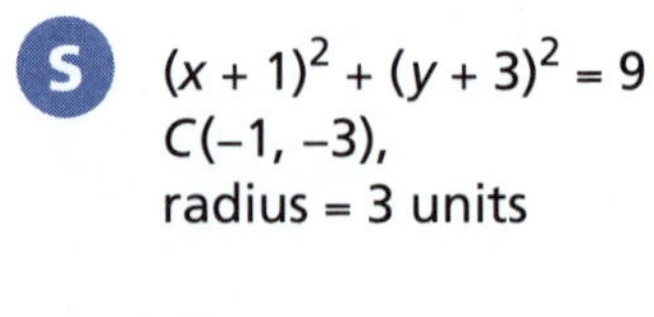

S $(x + 1)^2 + (y + 3)^2 = 9$
$C(-1, -3)$,
radius = 3 units

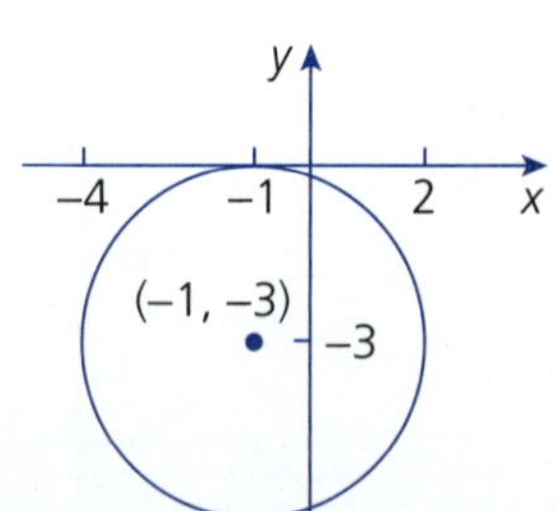

Intersection of graphs

To find the point(s) of intersection of graphs we can solve simultaneously, either algebraically or graphically.

For Example

1 Find the point(s) of intersection of the following:

a $y = x^2$ and $y = x + 2$

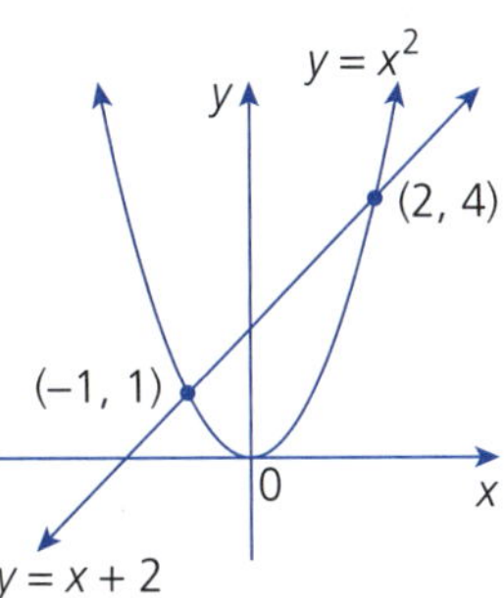

$y = x^2 \quad (1)$

$y = x + 2 \quad (2)$

Let (1) = (2) $\quad x^2 = x + 2$

$x^2 - x - 2 = 0$

$(x - 2)(x + 1) = 0$

$x = 2, -1$

Sub. in (1) $x = 2 \quad \therefore y = 4$

$x = -1 \quad \therefore y = 1$

$\therefore$ Points of intersection (2, 4), (−1, 1).

b $y = x$ and $y = \frac{1}{x}$

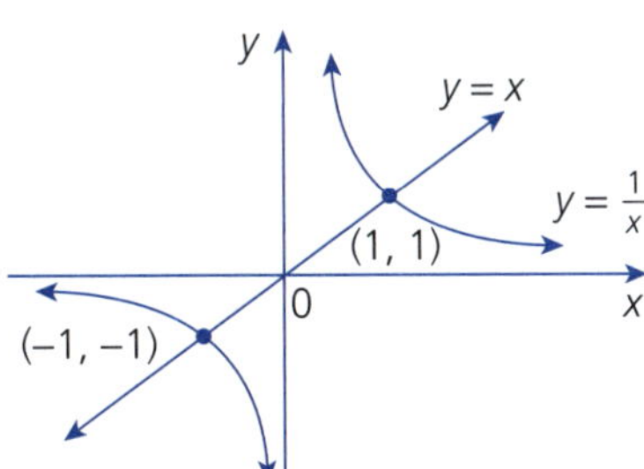

$y = x \quad (1)$

$y = \frac{1}{x} \quad (2)$

Let (1) = (2) $\quad x = \frac{1}{x}$

$\therefore x^2 = 1$

$\therefore x = \pm 1$

Sub. in (1) $x = 1, y = 1$

$x = -1, y = -1$

$\therefore$ Points of intersection are (1, 1), (−1, −1).

c $x^2 + y^2 = 4$ and $3x + 4y - 10 = 0$

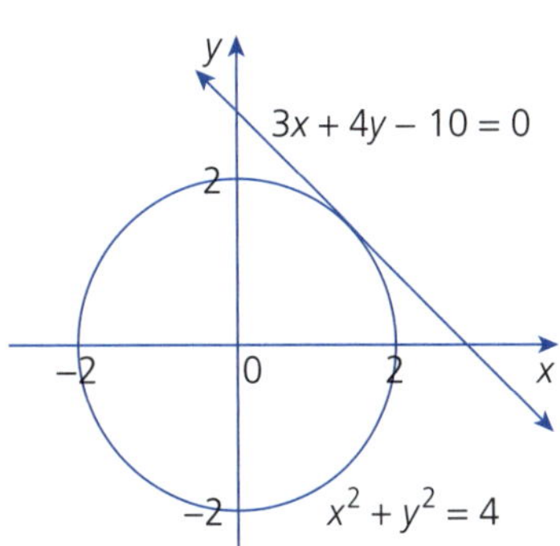

$x^2 + y^2 = 4 \quad (1)$

$3x + 4y - 10 = 0 \quad (2)$

$4y = -3x + 10$

$y = -\frac{3x - 10}{4} \quad (3)$

Substitute in (1)

$x^2 + \left(-\frac{3x - 10}{4}\right)^2 = 4$

$x^2 + \frac{9x^2 - 60x + 100}{16} = 4$

$16x^2 + 9x^2 - 60x + 100 = 64$

$25x^2 - 60x + 36 = 0$

$(5x - 6)^2 = 0$

$x = 1\frac{1}{5}$

Substitute in (3)

$y = -\frac{3\left(1\frac{1}{5}\right) - 10}{4}$

$= 1\frac{3}{5}$

$\therefore$ The point of intersection is $\left(1\frac{1}{5}, 1\frac{3}{5}\right)$.

Practise Practise

1 On separate number planes, graph the following:

a $y = x^2 - 2$ **b** $y = -2x^2$
c $y = 3 - x^2$ **d** $y = x^3 + 1$
e $y = \frac{2}{x}$ **f** $xy = 1$
g $y = -\frac{3}{x}$ **h** $y = 3^x$
i $y = \left(\frac{1}{2}\right)^x$ **j** $y = (x + 1)^2$
k $y = (2 - x)^2$

pp. 87–92

2 Determine whether the following parabolas are concave up or concave down when graphed:

a $y = x^2 - 5x + 2$ **b** $y = x - x^2$
c $y = 3 + x^2 - x$ **d** $y = 4x^2 - x + 1$

pp. 87–91

3 Without graphing, find the x-intercepts and y-intercept for the following parabolas:

a $y = x^2 + x - 2$ **b** $y = x^2 - x$
c $y = x^2 - 4$ **d** $y = 9 - x^2$

pp. 87–91

4 Find the equation of the axis of symmetry for each of the following:

a $y = 3x^2 - 6x + 2$ **b** $y = 4 - 5x - x^2$
c $y = 4x^2 - 8x$ **d** $y = 4 - x^2$

pp. 87–91

5 Find the maximum value of y if:

a $y = 4 - x - x^2$
b $y = -x^2 + 4x + 2$

pp. 87–91

6 Find the minimum value of y if:

a $y = x^2 + 6x + 12$
b $y = x^2 - 2x + 3$

pp. 87–91

7 Sketch the parabola with:

a maximum value = 3, axis of symmetry $x = 2$, and y-intercept = −1.
b minimum value = −4, axis of symmetry $x = 3$, and y-intercept = −2.
c minimum value = 0, axis of symmetry $x = -2$, and y-intercept = 3.

pp. 87–91

8 Find the vertex of:

a $y = x^2 - 5x - 6$ **b** $y = x^2 + 2x - 8$
c $y = 2x^2 - 3x + 1$ **d** $y = 5 - 2x - x^2$

pp. 87–91

9 Find the equation of:

a

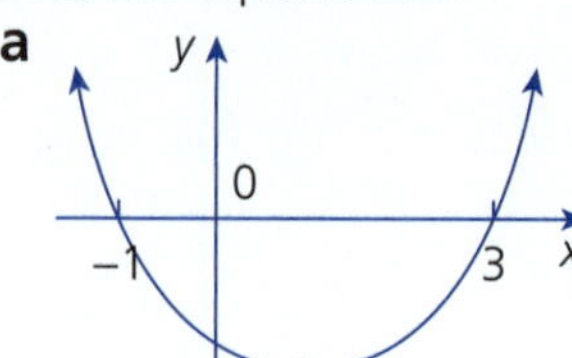

b

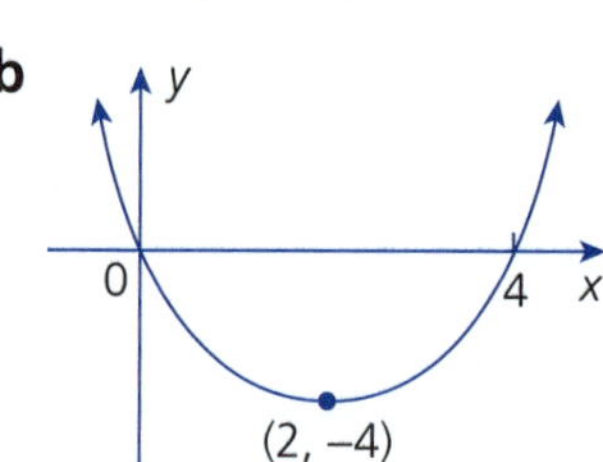

c

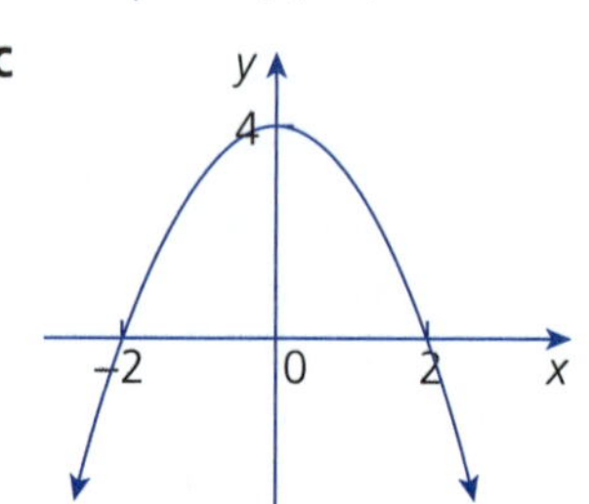

d

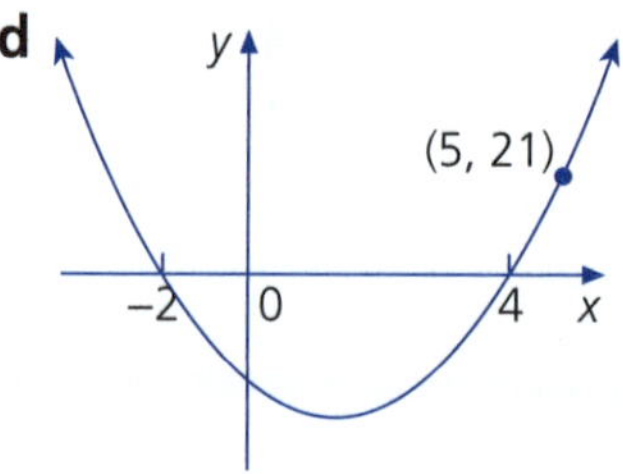

p. 91

10 On separate number planes graph:

a $y = x^3$ and $y = x^3 + 3$
b $y = 4 - x^3$ and $y = -2 - x^3$
c $y = x^3$ and $y = \frac{1}{2}x^3$
d $y = (x - 3)^3$ and $y = (x + 2)^3$

p. 91

11 Write down a possible equation for:

a

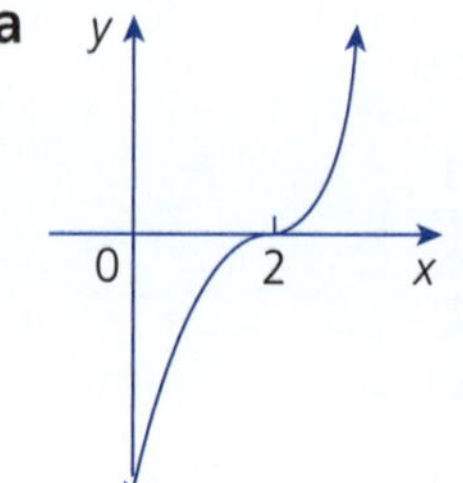

b

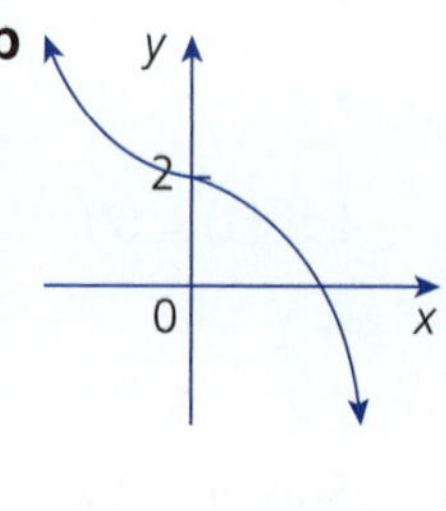

p. 91

12 On separate number planes graph:

a $y = \frac{1}{x}$ **b** $xy = -3$ p. 92

13 Match the following graphs with equations:

a $y = 2^x$ **b** $x^2 + y^2 = 4$
c $y = (x + 2)^2$ **d** $y = 4 - x^2$
e $y = -2x + 2$ **f** $y = x^3 + 2$
g $y = 2x - 1$ **h** $xy = 2$
i $y = (x - 2)^2$ **j** $y = -2 - x^3$ pp. 87–95

A

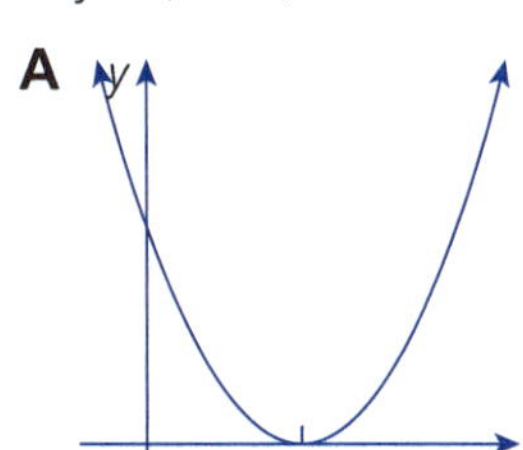

B

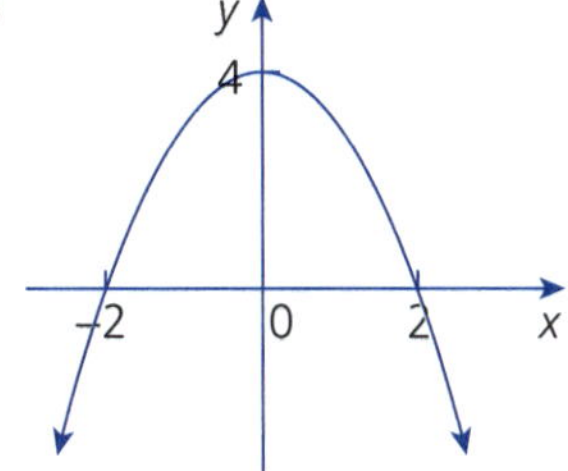

C

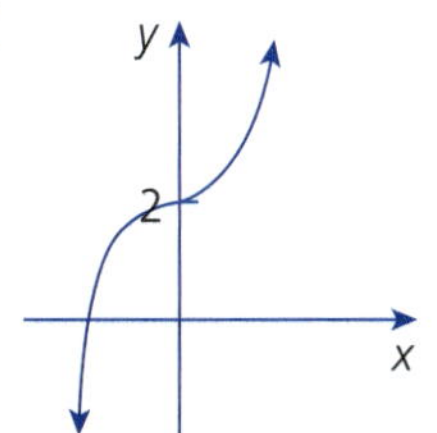

D

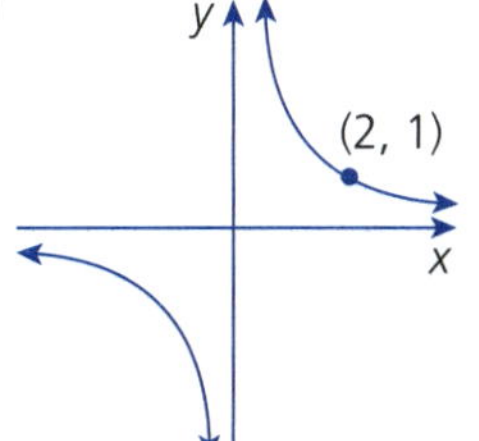

E

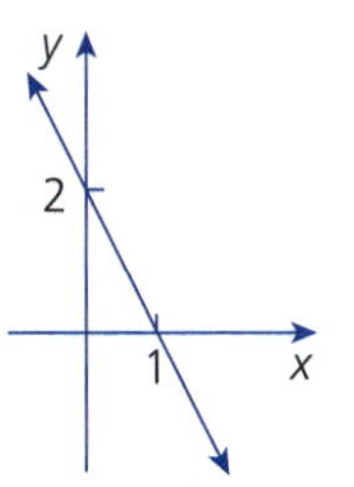

F

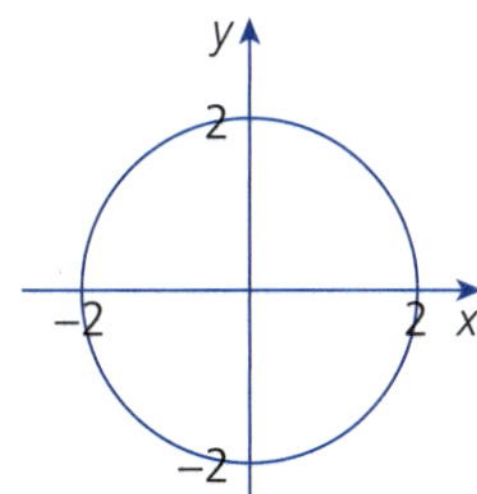

G

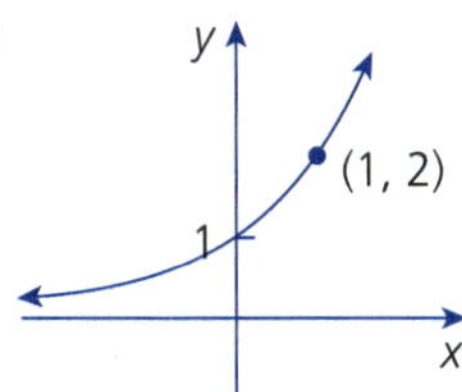

H

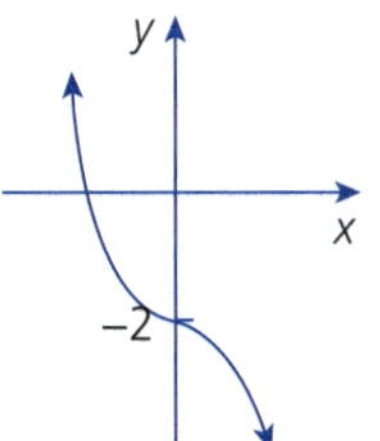

I

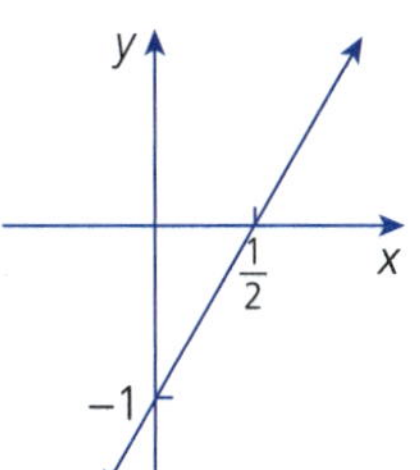

J

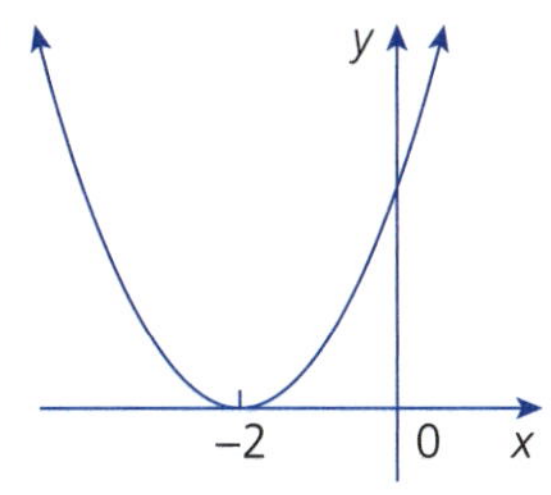

14 On separate number planes, graph:

a $y = x^7 - 2$
b $y = (x - 3)^4$
c $y = 2 - x^5$
d $y = x^{10} - 8$ p. 93

15 Sketch on separate number planes:

a $y = (x - 1)(x - 3)(x + 4)$
b $y = x(4 - x)(x + 1)$
c $y = 2x(x + 3)(x - 1)$
d $y = (4 - x)^2(x + 3)$ pp. 94–95

16 Find the centre and radius of the following circles:

a $x^2 + y^2 = 49$
b $x^2 + y^2 - 12 = 0$
c $x^2 + y^2 = 5$ p. 95

17 Find centre and radius of:

a $(x + 3)^2 + (y - 1)^2 = 16$
b $(x - 1)^2 + (y + 4)^2 = 48$
c $x^2 + (y + 1)^2 = 24$
d $x^2 + 8x + y^2 - 10y = 20$
e $x^2 - 6x + y^2 + 16y = 16$
f $x^2 - 3x + y^2 + 10y = 8$
g $2x^2 + 2y^2 + 4y = 36$ p. 96

18 Sketch the graph of:

a $(x - 1)^2 + (y + 3)^2 = 1$
b $x^2 + 4x + y^2 - 6y = 3$ p. 96

19 Find the point(s) of intersection of:

a $y = x^2$ and $y = 5x + 24$
b $y = x^2 + 2$ and $y = 3x$
c $y = x^2 + 2x + 10$ and $y = 2x^2 + 3x + 4$
d $y = \frac{4}{x}$ and $y = x + 3$
e $x^2 + y^2 = 8$ and $y = 2$
f $(x + 3)^2 + (y - 1)^2 = 8$ and $y = x$ p. 97

Go to p. 300 for **Quick Answers** or to pp. 324–328 for **Worked Solutions**

For a complete understanding of this topic, you must be able to:

✓	Graph parabolas with equations of the form $y = ax^2 + c$, $y = (x - k)^2$, $y = (x - m)(x - n)$		pp. 87–89
✓	Find the vertex of a parabola		pp. 89–91
✓	Find the axis of symmetry and minimum/maximum value of a parabola		pp. 89–91
✓	Determine the equation of a parabola		p. 91
✓	Graph a cubic, hyperbola and exponential		pp. 91–92
✓	Graph $y = x^n$ when n is even or odd		pp. 93–94
✓	Graph curves of the form $y = a(x - r)(x - s)(x - t)$		pp. 94–95
✓	Find the centre and radius of a circle of the form $x^2 + y^2 = r^2$		p. 95
✓	Find the centre and radius of a circle of the form $(x - h)^2 + (y - k)^2 = r^2$		p. 96
✓	Solve simultaneous equations to find points of intersection of curves.		p. 97

Now you are ready to do the tests!

Intermediate Test

Non-linear Relationships

(40 marks)

1 The equation of the parabola is:
A $y = x + 9$
B $y = x^2 + 9$
C $y = x^2 - 9$
D $y = 9 - x^2$

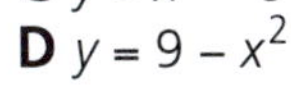

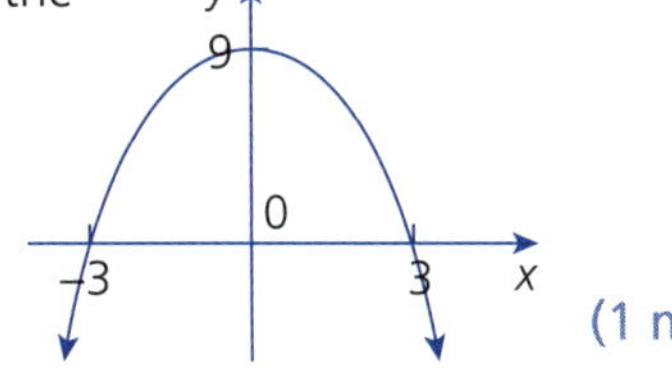

(1 mark)

2 The equation of the curve could be:
A $y = (x - 2)(x + 3)$
B $y = (x + 2)(x - 3)$
C $y = (x - 2)(x - 3)$
D $y = (x + 2)(x + 3)$

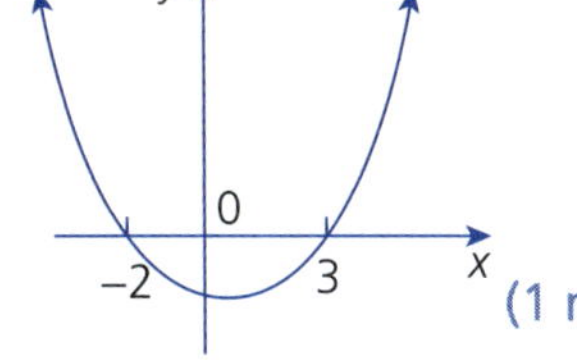

(1 mark)

3 On a number plane, graph $y = x(4 - x)$. (1 mark)

4 For the parabola $y = (x + 3)(x - 1)$:
a find the equation of its axis of symmetry (1 mark)
b find the minimum value (1 mark)
c hence, sketch the curve. (1 mark)

5 Find the axis of symmetry and maximum value of $y = 4 - 2x - x^2$. (2 marks)

6 Write down the equation of the parabola:

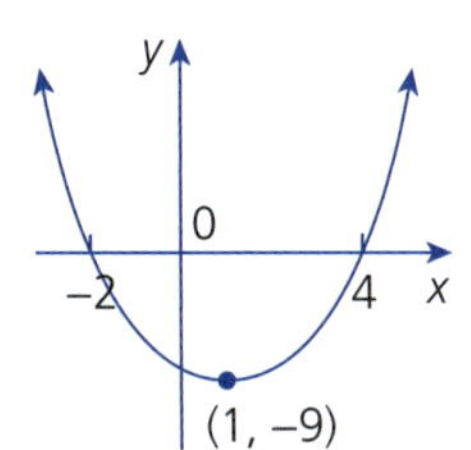

(2 marks)

7 On separate number planes, sketch:
a $y = x^3 - 2$ (1 mark)
b $y = \frac{3}{x}$ (1 mark)
c $y = 2^x$ (1 mark)

8 Write the equation of a circle with centre (0, 0) and radius 4 units. (1 mark)

9 Find the x and y-intercepts of $y = x^2 - x - 20$. (2 marks)

10 Find the equation of the axis of symmetry of $y = 25 - x^2$. (1 mark)

11 On the same number plane sketch:
a $y = x^2 - 2$ and $y = 3 - x^2$ (1 mark)
b $y = \frac{x^2}{2}$ and $y = 3x^2$ (1 mark)

12 Sketch:
a $y = (3 - x)(x + 2)$ (1 mark)
b $y = x(x + 2)$ (1 mark)
c $y = x^2 - 5x - 14$ (2 marks)

13 Write a possible equation for the following parabolas:

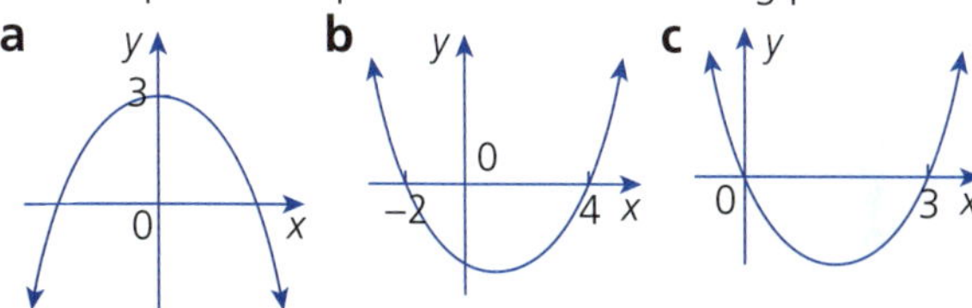

(3 marks)

14 On the same number plane, sketch:
a $y = x^3$ and $y = -x^3 + 1$ (2 marks)
b $y = \frac{2}{x}$ and $y = \frac{-2}{x}$ (1 mark)

15 Write the equation of the circle:
a with centre (5, −2) and radius 4 units (1 mark)
b with centre (0, 0) and radius $\sqrt{17}$ units. (1 mark)
c

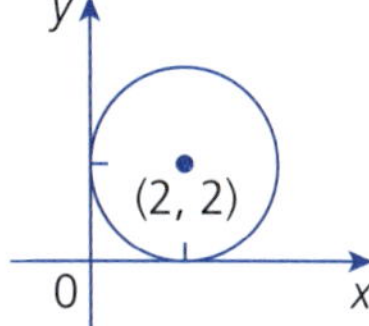

(2 marks)

16 On separate number planes, graph:
a $y = (x + 2)(x - 1)(x + 3)$ (2 marks)
b $y = 2x(x + 1)(x - 2)$ (2 marks)

17 Find the points of intersection of $y = x^2$ and $y = 2x + 3$. (3 marks)

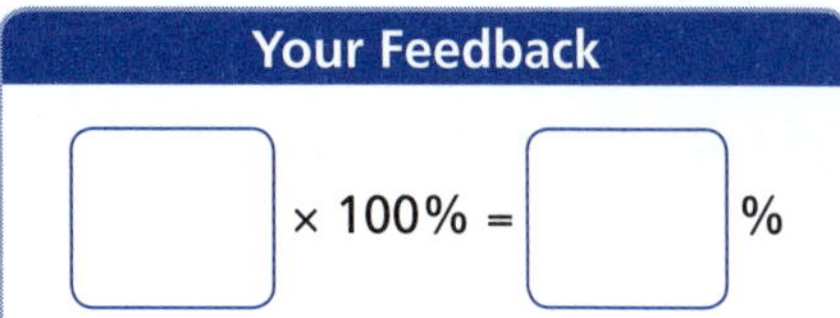

QA PAGE 305
WS PAGE 366

Advanced Test

Non-linear Relationships

(40 marks)

1 The equation of the curve could be:

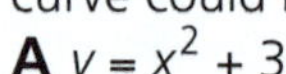

A $y = x^2 + 3$
B $y = (x + 3)^2$
C $y = (x - 3)^2$
D $y = x^2 - 3$

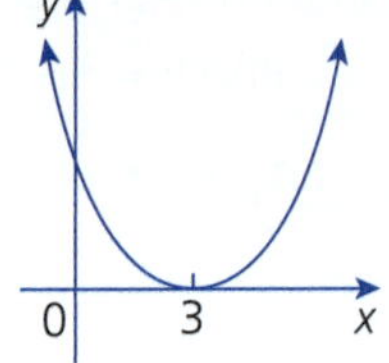

(1 mark)

2 $y = 2x^2$ and $y = x^2 + 2$ are graphed on the same number plane.

Which one of the following statements is true?

A Both parabolas pass through the origin.
B Both parabolas are concave up.
C Both parabolas have a y-intercept of 2.
D Both parabolas have the same minimum value.

(1 mark)

3 On a number plane, graph $y = x^2 - 4x - 5$. (2 marks)

4 For the parabola, $y = (x - 5)(2 - x)$, find the:

a equation of the axis of symmetry (1 mark)
b maximum value. (1 mark)

5 Find the axis of symmetry and minimum value of $y = 2x^2 - 4x + 5$. (2 marks)

6 Write down the equation of the parabola:

a

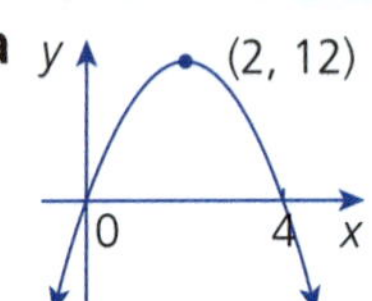

(2 marks)

b with x-intercepts of –4 and 6 and minimum value of –50. (2 marks)
c with x-intercepts of –2 and 8 and a y-intercept of –64. (2 marks)

7 A parabola passes through (1, 2) and has an axis of symmetry of $x = 3$. The equation of the parabola could be:

A $y = x^2 - 6x + 7$ **B** $y = x^2 - 6x$
C $y = x^2 - 3x + 4$ **D** $y = x^2 - 6x + 8$ (1 mark)

8 On a number plane, sketch $y = 4 - x^3$. (1 mark)

9 Find the radius of the circle with equation $2x^2 + 2y^2 = 96$. (1 mark)

10 Find the vertex of the parabolas with:

a $y = x(x + 4)$ (2 marks)
b $y = (x + 3)(x - 5)$ (2 marks)
c $y = 2x^2 - 4x + 1$ (2 marks)

11 Find the equation of the parabola:

a

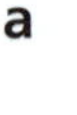

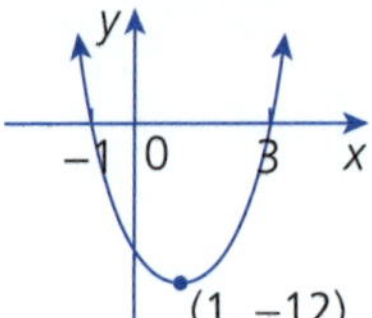

(2 marks)

b

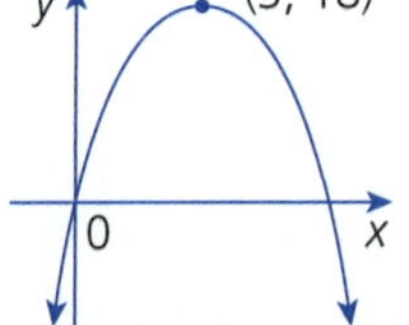

(2 marks)

c

y, –3, 0, 3, x, –36

(2 marks)

12 On separate number planes, sketch:

a $y = (x - 2)^3$ (1 mark)
b $y = \dfrac{1}{x + 3}$ (1 mark)

13 Find the coordinates of the centre and the radius of $x^2 - 4x + y^2 - 12y = 9$. (2 marks)

14 Graph $x^2 + y^2 + 6y = 0$. (2 marks)

15 On a number plane, graph $y = -(x + 2)^2(x - 1)$. (2 marks)

16 Find the point of intersection of $y = \dfrac{12}{x}$ and $y = 3x$. (2 marks)

17 Show that $y = 4x - 1$ and $y = x^2 + 3$ touch at one point. (1 mark)

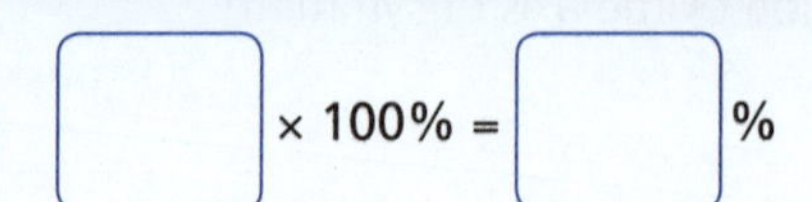

QA PAGE 305
WS PAGE 368

Chapter 7
Rates and Proportion

Converting between rates

A **rate** is the comparison of two quantities of different units.

Keywords

Rate
Direct proportion
Constant
Gradient
Positive
Negative
Increases
Decreases
Conversion
Inverse proportion

1 Convert 60 km/h to m/s.

S Change kilometres to metres and hours to seconds:

60 km = 60 000 m
1 h = 60 min = 3600 seconds

60 km/h = 60 000 m/3600 s

Divide by 3600:

$\therefore$ 60 km/h = $16\frac{2}{3}$ m/s

2 Convert 10 m/s to km/h.

S Change metres to kilometres and seconds to hours.

Multiply by 60, and then multiply by 60:

10 m/s = 36 000 m/h
36 000 m = 36 km
$\therefore$ 10 m/s = 36 km/h

3 Convert 8 km/L to L/100 km.

S Rewrite 8 km/L as 1 L/8 km

Divide by 8 then multiply by 100:
$1 \div 8 \times 100 = 12.5$
$\therefore$ 8 km/L = 12.5 L/100 km

Direct proportion

A **proportion** is a statement that two ratios are equal. For example, $a:b = c:d$ can be written as the proportion $\frac{a}{b} = \frac{c}{d}$.

Consider a pair of similar triangles:

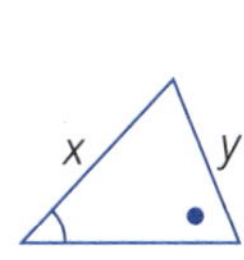

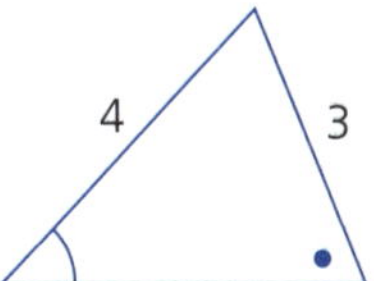

Using matching sides in proportion, $\frac{y}{3} = \frac{x}{4}$.

This can be rewritten as $y = \frac{3x}{4}$, which means that ***y* is directly proportional to *x***.

This can be expressed as $\boldsymbol{y \propto x}$, or $\boldsymbol{y = kx}$, where k is the **constant of proportionality (or the rate of change)**. (In the case above, $k = \frac{3}{4}$.)

As the **rate of change is constant**, the graph is a **straight line** passing through the origin with gradient k.

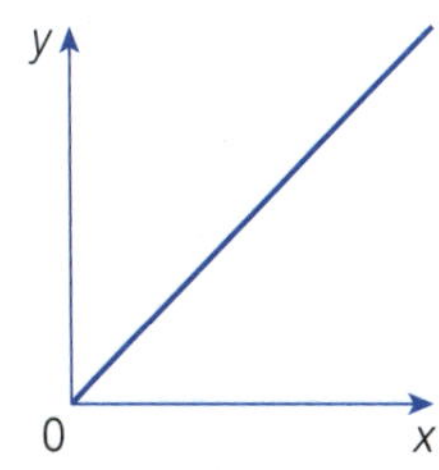

When a graph is drawn linking two variables that are directly proportional, the gradient will be **positive**.

Everyday examples of events that are in direct proportion are:

- The more hours worked, the more income is earned.
- As the side length of a square paddock decreases, the perimeter of the paddock decreases.

For Example

1 The table shows the height (h in cm) of a plant over time (t in weeks).

t	0	10	20	30
h	0	30	60	90

a Write a rule linking h and t.

S $\therefore h = 3t$

b Complete: As t increases, h __________.

S **As t increases, h increases.**

c What is the height of the plant after 50 weeks?

S **Subs. $t = 50$ in $h = 3t$:**

$h = 3 \times 50$

$= 150$

$\therefore$ Plant is 150 cm high.

e Draw a graph of h against t for values of $t \leq 100$.

S

f What is the gradient of the graph?

S $\frac{\text{Rise}}{\text{Run}} = \frac{300}{100} = 3$

2 A hose is used to fill an empty container with water. It takes 4 minutes to fill the container with 120 L of water.

a At what rate is the water filling the container, in L/min?

S **120 L in 4 minutes means 30 L/min.**

b Write a rule linking V (volume in litres) and t (time in minutes).

S $V = 30t$

c Complete: As t increases, V __________.

S **As t increases, V increases.**

d Draw a graph of volume against time for $t \leq 20$.

S

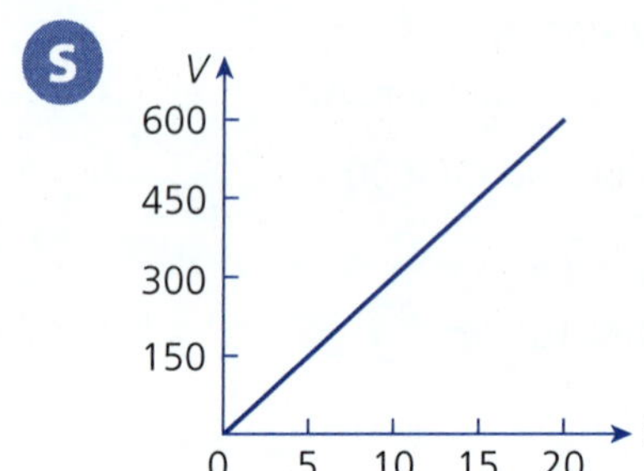

e Write down the gradient.

S **Gradient = 30**

f How much water is in the container after an hour?

S **Subs. $t = 60$ in $V = 30t$:**

$V = 30 \times 60$

$= 1800$

$\therefore$ 1800 litres

g How long would it take to fill a tank with a capacity of 2400 litres?

S **Subs. $V = 2400$ in $V = 30t$:**

$2400 = 30t$

$t = \frac{2400}{30}$

$= 80$

$\therefore$ 80 minutes

3 The graph shows the distance in kilometres (d) travelled by a motorist over a number of hours (h).

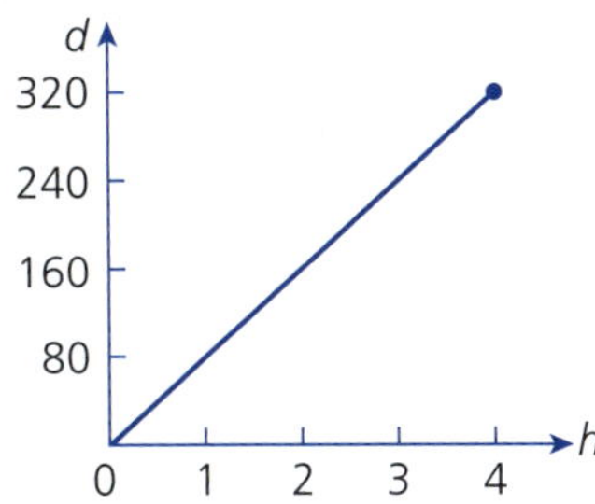

a What is the gradient of the graph?

S Gradient $= \frac{\text{Rise}}{\text{Run}}$

$= \frac{320}{4}$

$= 80$

b What is represented by the gradient?

S The gradient is the rate of change of the distance over time.

This is the speed of the motorist (80 km/h).

c Write a rule linking d with h.

S $d = 80h$

d How far will the motorist travel in $2\frac{1}{2}$ hours?

S Subs. $h = 2\frac{1}{2}$ in $d = 80h$:

$d = 80 \times 2\frac{1}{2}$

$= 200$

$\therefore$ Motorist travels 200 km.

Conversion graphs

A conversion graph is used to change one quantity to another when there is a change in the units.

For Example

1 Jannah converted speeds recorded in m/s into km/h and recorded them in the table below:

m/s	10	20	30	40	50
km/h	36	72	108	144	180

a Draw a graph to record the data.

S

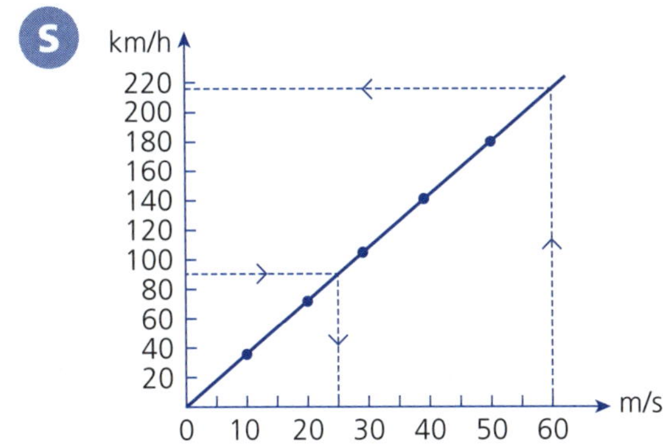

b Use the graph to convert:

i 60 m/s to km/h

S i 60 m/s = 216 km/h

ii 90 km to m/s

S ii 90 km/h = 25 m/s

2 Adam exchanged $100 Australian dollars for $80 US. He knows that $80 US is equal to 70 euros.

a Draw currency conversion graphs for $US against $A, and euros against $US.

S

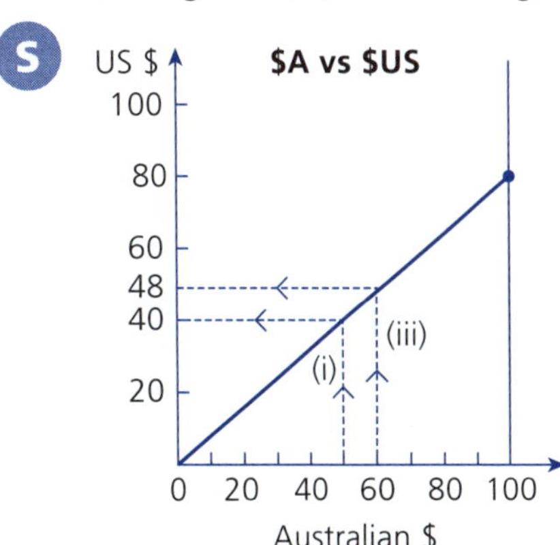

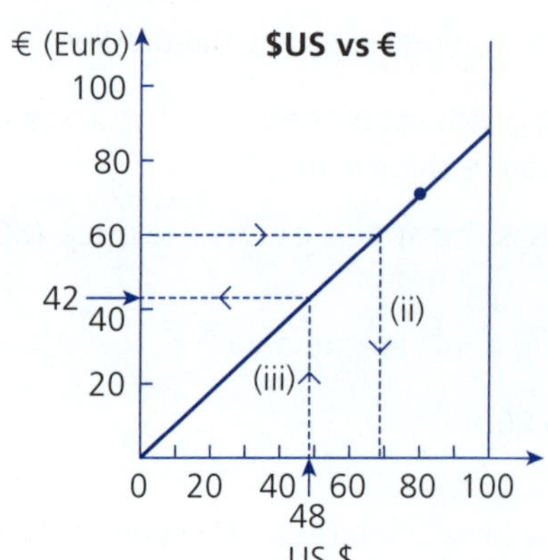

b Use the graphs to convert:

i $A50 to US dollars

S **$A50 = $US40**

ii 60 euros to US dollars

S **€60 = $US69 (nearest whole)**

iii $A60 to euros

S **$A60 = $US48 = €42**

Problem solving using direct proportion

If one quantity (y) is directly proportional to another quantity (x), then $y \propto x$, or $y = kx$, where k is the constant of proportionality.

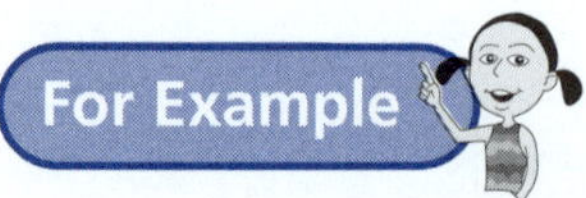

1 It is known that $y \propto x$ (y is proportional to x) and if $y = 24$ then $x = 8$.

a Write an equation that links y and x using the constant of proportionality k.

S $y = kx$

b Find the value of k.

S Subs. $y = 24$ and $x = 8$ in $y = kx$:
$24 = k \times 8$
$8k = 24$
$k = 3$

c Find the value of y when $x = 12$.

S Subs. $x = 12$ in $y = 3x$:
$y = 3 \times 12$
$= 36$

2 The cost $\$C$ of carpet is directly proportional to the length n of the carpet in metres. Twelve metres of carpet costs $1176.

a Write a formula for C in terms of n.

S C is directly proportional to n.
Subs. $C = 1176$ and $n = 12$ into $C = kn$:
$1176 = k(12)$
$12k = 1176$
$k = 98$
$\therefore C = 98n$

b Find the cost of 17 metres of carpet.

S Subs. $n = 17$ into $C = 98n$:
$C = 98 \times 17$
$= 1666$
$\therefore$ Carpet costs $1666.

c What length of carpet can be purchased for $1391.60?

S Subs. $C = 1391.6$ into $C = 98n$:
$1391.6 = 98n$
$$n = \frac{1391.6}{98}$$
$= 14.2$
$\therefore$ The carpet is 14.2 metres long.

3 The height of a certain termite mound is directly proportional to the square root of the number of termites. The height of this mound is 60 cm when the number of termites is 3600.

a By letting h = height of the mound in metres and n = number of termites, find an equation for h in terms of n.

S h is directly proportional to $\sqrt{n}$.

Subs. $h = 60$ and $n = 3600$ into $h = k\sqrt{n}$:
$60 = k\sqrt{3600}$
$60k = 60$
$k = 1$
$\therefore h = \sqrt{n}$

b What is the height of this mound, in centimetres, when there are 6400 termites?

S Subs. $n = 6400$ into $h = \sqrt{n}$:

$h = \sqrt{6400}$
$= 80$

$\therefore$ The mound is 80 cm high.

Inverse (indirect) proportion

If y is inversely (indirectly) proportional to x, we write $\boldsymbol{y \propto \frac{1}{x}}$, or $\boldsymbol{y = \frac{k}{x}}$, where k is the constant of proportionality.

When a graph is drawn linking two variables that are inversely proportional, the gradient will be **negative**.

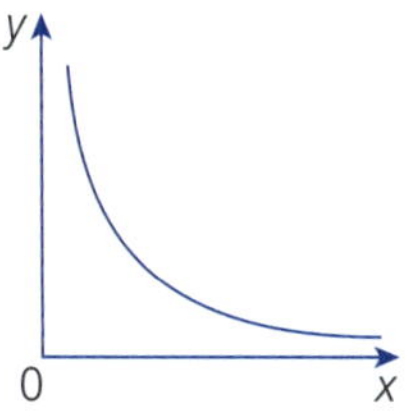

Everyday examples of events that are in inverse proportion are:

- The faster a car travels the less time for the trip.
- The fewer people mow a lawn, the longer the job will take.

For Example

1 The time (T) taken by workers (n) to paint a wall is detailed in the table below:

n	1	2	3	6	9	18
T	18	9	6	3	2	1

Use the table to draw a graph and explain why T is inversely proportional to n.

S

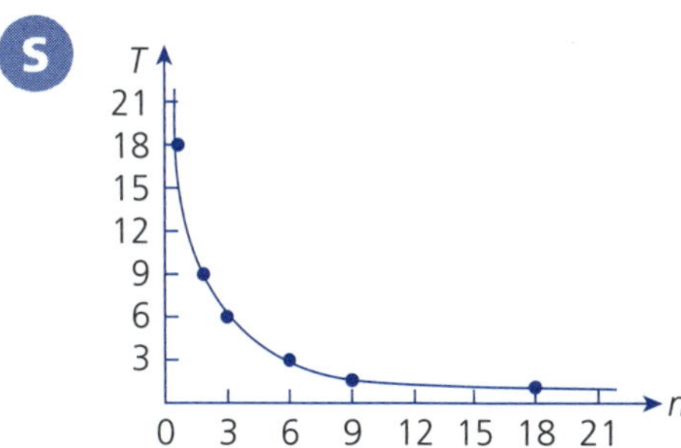

T is inversely proportional to n because as n increases, T decreases.

Problem solving using inverse proportion

If one quantity (y) is inversely proportional to another quantity (x), then $\boldsymbol{y \propto \frac{1}{x}}$, or $\boldsymbol{y = \frac{k}{x}}$, where k is the constant of proportionality.

For Example

1 It is known that $y \propto \frac{1}{x}$ (y is proportional to $\frac{1}{x}$) and if $y = 12$ then $x = 4$.

a Find the value of the constant of proportionality (k).

S Subs. $y = 12$ and $x = 4$ in $y = \frac{k}{x}$:

$12 = \frac{k}{4}$

$k = 4 \times 12$

$k = 48$

b Write an equation linking y and x.

S $y = \frac{48}{x}$

c Find the value of y when $x = 16$.

S Subs. $x = 16$ in $y = \frac{48}{x}$:

$y = \frac{48}{16}$

$= 3$

2 The time taken to complete a journey is inversely proportional to the speed of a car. A car takes 8 hours to complete a journey when travelling at 100 km/h.

a Write a formula for T in terms of s, if T = time for the journey in hours and s is the speed of the car in km/h.

S T is indirectly proportional to s.

Subs. $T = 100$ and $s = 8$ into $T = \frac{k}{s}$:

$100 = \frac{k}{8}$

$k = 100 \times 8$

$k = 800$

$\therefore T = \frac{800}{s}$

b How long would the same journey take if the car were travelling at 80 km/h?

S Subs. $s = 80$ into $T = \frac{800}{s}$:

$T = \frac{800}{80}$

$= 10$

$\therefore$ The journey takes 10 hours.

Matching graphs to physical phenomena

A line graph can represent physical phenomena. The following graphs show relationships between distance (d) and time (t).

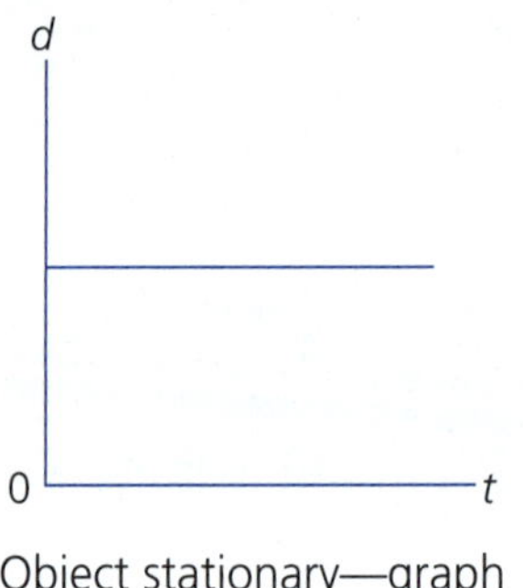

Object stationary—graph is horizontal.

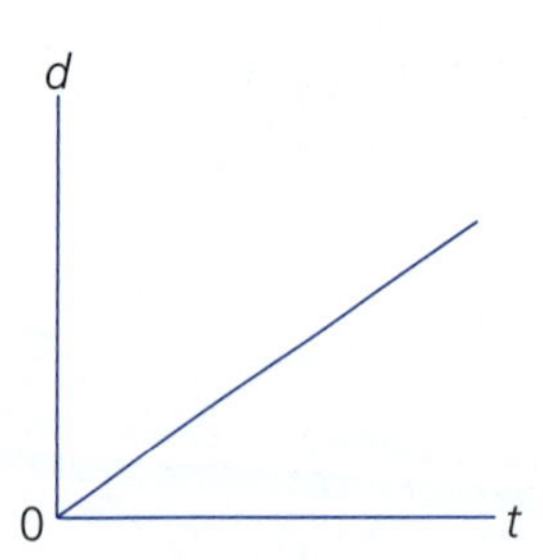

Object travelling at constant speed—graph is increasing at constant rate.

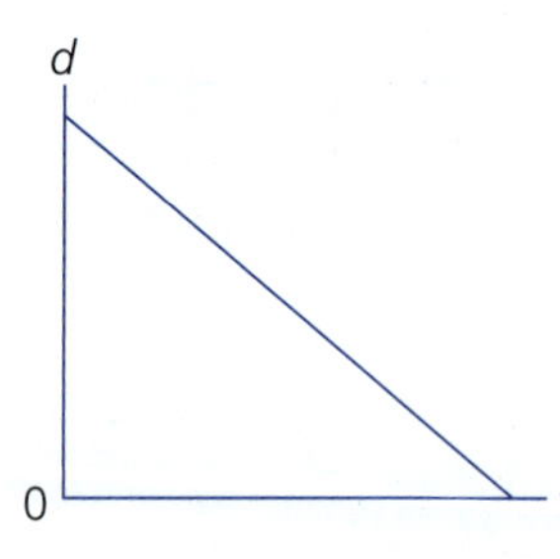

Object travelling at constant speed—graph is decreasing at constant rate.

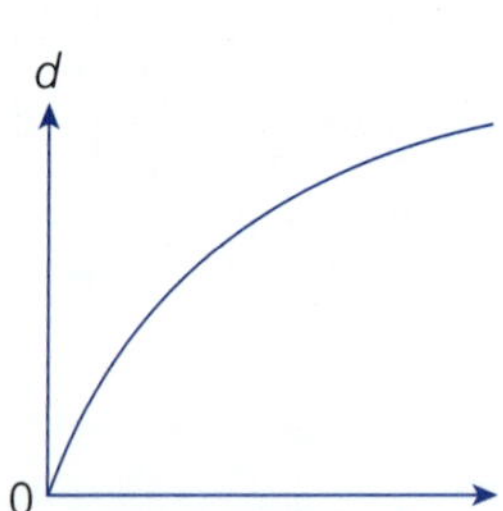

Object initially travels at high speed but slows down—graph is increasing at decreasing rate.

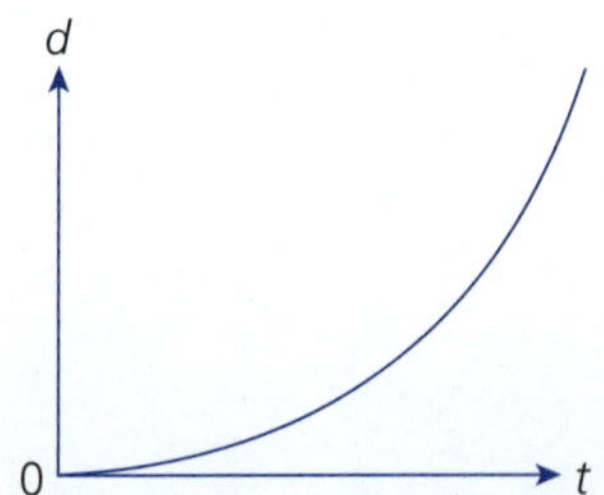

Object initially travels at slow speed but speeds up—graph is increasing at increasing rate.

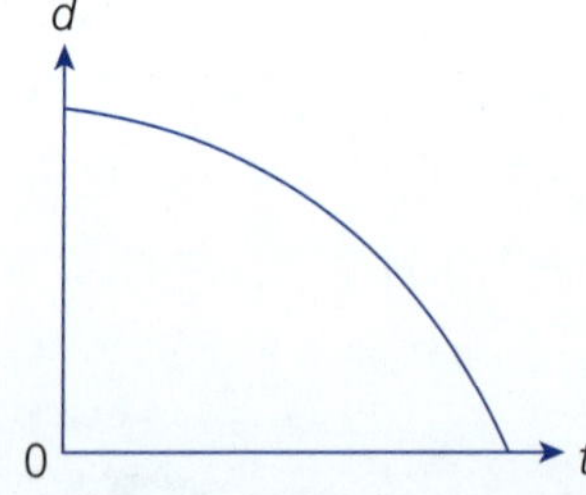

Object initially travels at slow speed but speeds up—graph is decreasing at an increasing rate.

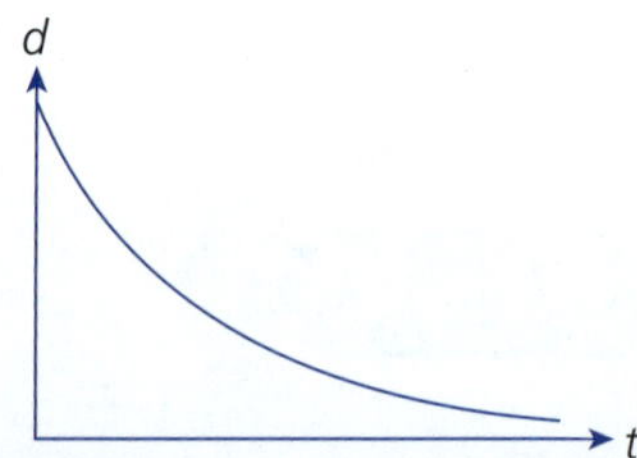

Object initially travels at high speed but slows down—graph is decreasing at a decreasing rate.

For Example

1 Water is poured into an empty tank at a steady rate and the rise of the water level is graphed. Which graph best represents the change in water level?

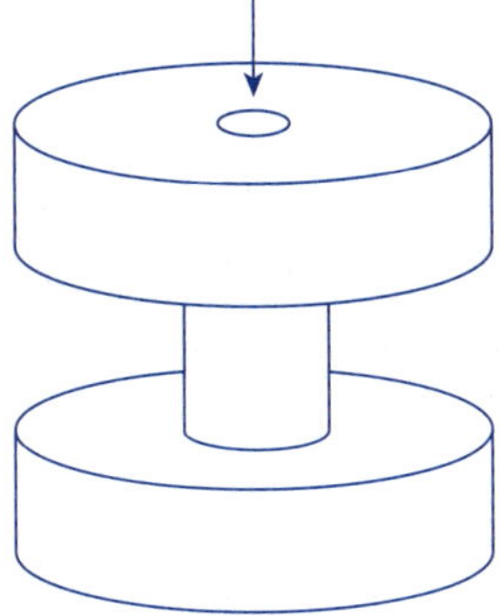

S

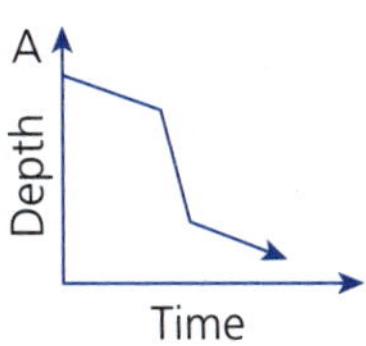

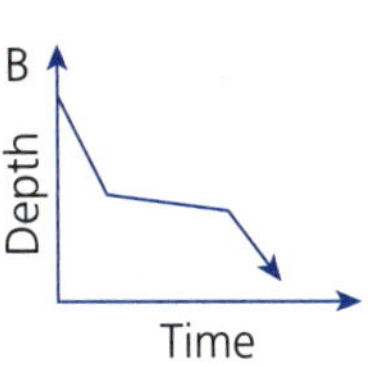

C

Depth

Time

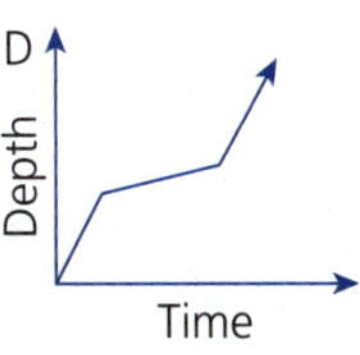

The water level would increase slowly, then quickly, and then slowly again, therefore the gradient is *gentle*, then *steep* and then *gentle*. The answer is C.

2 On a picnic Jack climbs a hill, spends some time looking at the view and then rolls down the hill back to where he started.

Draw a graph relating height (h) and time (t) to represent Jack's movement.

S

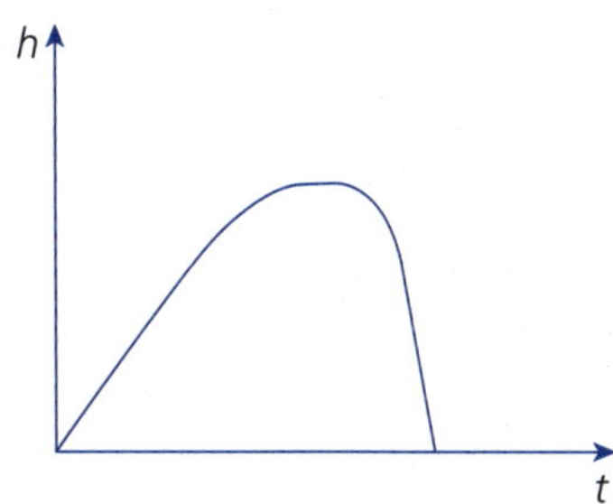

3 A container is empty and is filled with a constant rate of water. The graph shows the relationship between the depth of the water (d) over time (t).

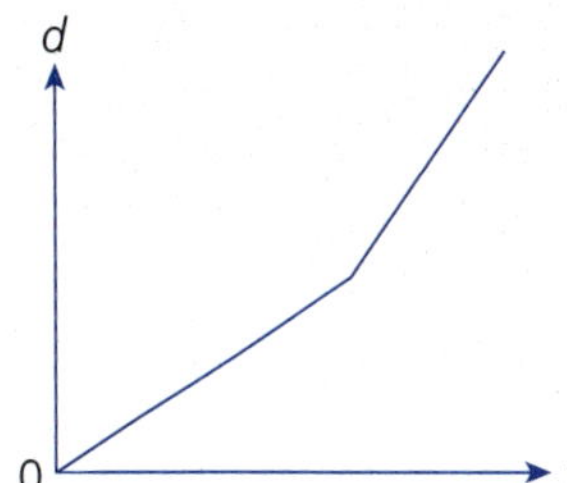

Draw a sketch of the container.

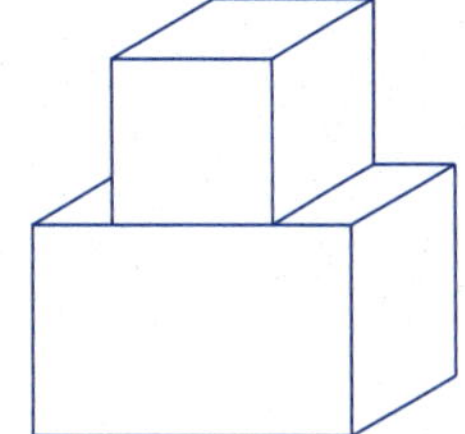

or

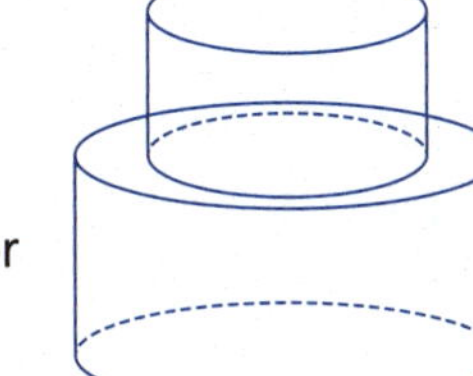

Practise Practise

1 Convert the following rates:

a 100 km/h to m/s
b 40 m/s to km/h
c 20 km/L to L/100 km
d 10 L/100 km to km/L
e 28 cm/s to km/h
f 80c/100 g to kg/$
g 160 c/L to L/$100
h 5 mL/min to h/L
i 2 L/h to s/mL.

p. 103

2 The table shows the cost of garden edging (C in dollars) over different lengths (d metres).

d	0	5	10	15	20
C	0	60	120	180	240

a Write a rule linking C and d.
b Complete: As d increases, C __________.
c What is the cost of 35 metres of edging?
d Find the length of edging if the cost is $504.
e Draw a graph of C against d for values of $d \leq 30$.
f What is the gradient of the graph?

pp. 103–105

3 The table below shows the wage (W) for hours worked (n) by Miriam:

n	0	8	16	24	32	40
W	0	132	264	396	528	660

a Write a rule linking W and n.
b Complete: As n increases, W __________.
c How much does Miriam earn when she works 30 hours?
d If Miriam earns $594, how many hours does she work?
e Draw a graph of W against n for values of $n \leq 40$.
f What is the gradient of the graph?

pp. 103–105

4 The table shows the distance a motorist travelled (D) for each litre of petrol (n):

n	0	5	10	15	20
D	0	45	90	135	180

a Write a rule linking D and n.
b How far does the motorist travel on 35 litres?
c The petrol tank has a capacity of 60 litres. What fraction of petrol remains after the motorist fills the tank and then travels 405 kilometres?

pp. 103–105

5 The table shows the conversion of temperatures between Celsius (°C) and Fahrenheit (°F):

°C	0	10	20	30	40
°F	32	50	68	86	104

a Draw the conversion graph.
b Use the graph to convert:
  i 38 °C to °F
  ii 59 °F to °C.

pp. 105–106

6 Daniel used a fitness book published in the USA to plan his training schedule. Using the conversion of 1 mile = 1.6 kilometres, he completed the table below:

Miles	5	10	15	30
Kilometres	8	16	24	48

a Draw the conversion graph.
b Use the graph to convert:
  i 20 miles to kilometres
  ii 40 kilometres to miles.

pp. 105–106

7 Mitch used the conversion of 1 kg = 2.2 pounds to make conversions between metric and imperial measures of mass.

a Complete the table below:

Kilograms	0	5	10	15	20
Pounds					

b Draw the conversion graph.
c Use the graph to convert:
  i 8 kilograms to pounds
  ii 150 kilograms to pounds.

pp. 105–106

8 Currency conversion graphs are drawn below to change Australian dollars ($A) to euros (€) and euros to British pounds (£).

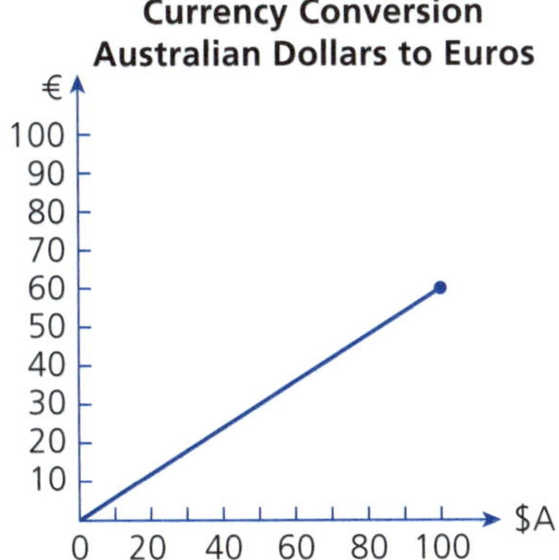

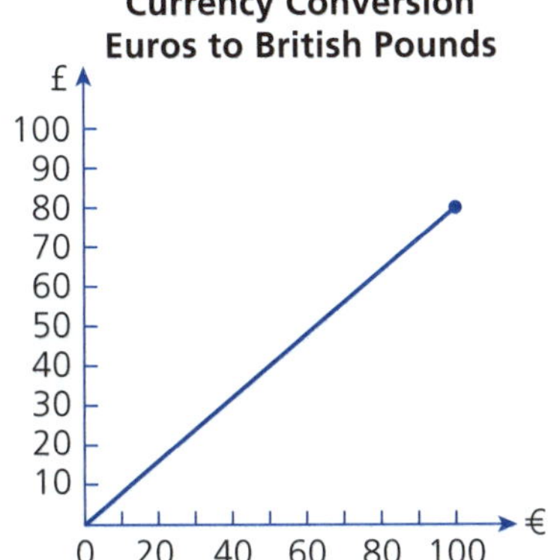

a Complete:
- **i** $A100 = € ________
- **ii** €100 = £ ________

b Use the graphs to convert:
- **i** $A60 to euros
- **ii** €42 to Australian dollars
- **iii** €60 to British pounds
- **iv** £20 to euros

c Make the following conversions:
- **i** $A5000 into euros
- **ii** €2400 into Australian dollars
- **iii** €750 into British pounds
- **iv** £100 into euros

d Use the conversion graphs to change:
- **i** $A100 into British pounds
- **ii** £60 into Australian dollars pp. 105–106

9 It is known that y is directly proportional to x and if $y = 12$ then $x = 4$.

a Write an equation that links y and x using the constant of proportionality k.
b Find the value of k.
c What is the value of y when $x = 20$?
d Find the value of x when $y = 18$. pp. 106–107

10 It is known that $P \propto t$ and when $P = 10$ then $t = 2$.

a Write an equation that links P and t using the constant of proportionality k.
b Find the value of k.
c If $t = 16$, find the value of P.
d When $P = 30$, what is the value of t.
pp. 106–107

11 Chloe's wage is directly proportional to the number of hours she works. When she works 26 hours she is paid $468. Let n = number of hours worked and W = wages in dollars.

a By finding the constant of proportionality, write an equation that links W and n.
b If Chloe worked for 27 hours, find her wage.
c If the wage is $648, how many hours did she work? pp. 106–107

12 The price of a Christmas pudding is directly proportional to its mass. The price of a pudding with a mass of 1.5 kg is $7.80. Let m = mass of pudding in kg and P = price of the pudding in $.

a By finding the constant of proportionality, write an equation that links P and m.
b What is the price of a pudding with a mass of 3.6 kg?
c Find the mass of a pudding if it costs $14.56.
pp. 106–107

13 In the table below, it is known that y is directly proportional to x:

x	1.3		4.5
y		10.44	16.2

a By finding the constant of proportionality, write an equation that links y and x.
b Complete the table. pp. 106–107

14 The horizontal force, F Newtons, needed to push a wooden block across a horizontal surface is directly proportional to its mass, m kg. It is known that it takes a force of 36 Newtons to move a block with mass 3 kg.

a By finding the constant of proportionality, write an equation that links F and m.
b How much force is required if the block has a mass of 6.2 kg?
c What is the mass of the block if the force applied was 43.2 Newtons? pp. 106–107

15 Given that y is directly proportional to $\sqrt{x}$ and if $y = 24$ then $x = 9$.

a Write an equation that links y and $\sqrt{x}$ using the constant of proportionality k.
b Find the value of k.
c What is the value of y when $x = 16$?
d Find the value of x when $y = 240$. pp. 106–107

16 Given that y is directly proportional to x^2 and if $y = 48$ then $x = 4$.

a Write an equation that links y and x^2 using the constant of proportionality k.

b Find the value of k.

c What is the value of y when $x = 6$?

d Find the value of x when $y = 588$ (if $x > 0$).

pp. 106–107

17 The table below shows the time taken (T hours) for a train to travel 600 kilometres at various average speeds (s km/h):

s	30	60	120	200	240	300
T	20	10	5	3	2.5	2

Use the table to draw a graph and explain why T is inversely proportional to s. pp. 107–108

18 It is known that y is inversely proportional to x and if $y = 3$ then $x = 8$.

a Write an equation that links y and x using the constant of proportionality k.

b Find the value of k.

c What is the value of y when $x = 16$?

d Find the value of x when $y = 80$. pp. 107–108

19 It is known that $M \propto \frac{1}{a}$ and when $M = 16$ then $a = 2$.

a Write an equation that links M and a using the constant of proportionality k.

b Find the value of k.

c If $a = 8$, find the value of M.

d When $M = 64$, what is the value of a.

pp. 107–108

20 The time (T minutes) taken to fill a large water tank is inversely proportional to the number (n) of water tankers used. Liam uses five tankers to fill the tank in 20 minutes.

a By finding the constant of proportionality, find an equation that links T and n.

b How long would it takes to fill the tank using two tankers?

c If the tank was filled in 25 minutes, how many tankers were used? pp. 107–108

21 In the table below, it is known that b is inversely proportional to a:

a	2		8
b		1	0.5

a By finding the constant of proportionality, find an equation that links y and x.

b Complete the table. pp. 107–108

22 The diagrams below show various bottles that are filled with water at a constant rate. Draw a separate graph for each bottle, showing the relationship between the height (h) of water and time (t).

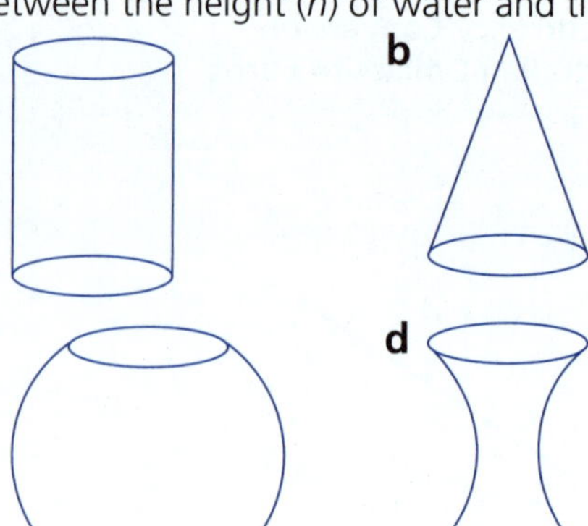

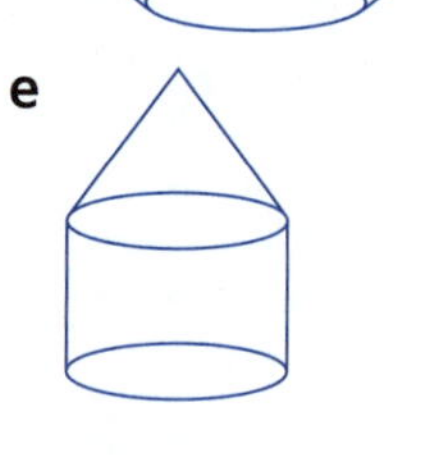

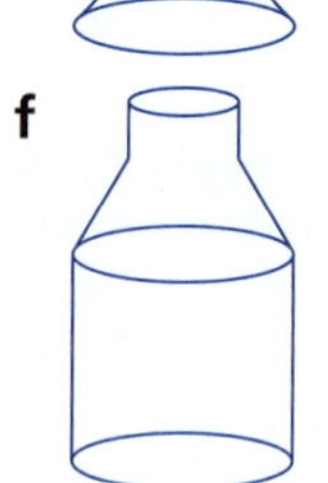

e

f

pp. 108–109

23 The graphs below show the change in height of water in a variety of bottles if water is poured in at a constant rate. Draw a sketch of each bottle.

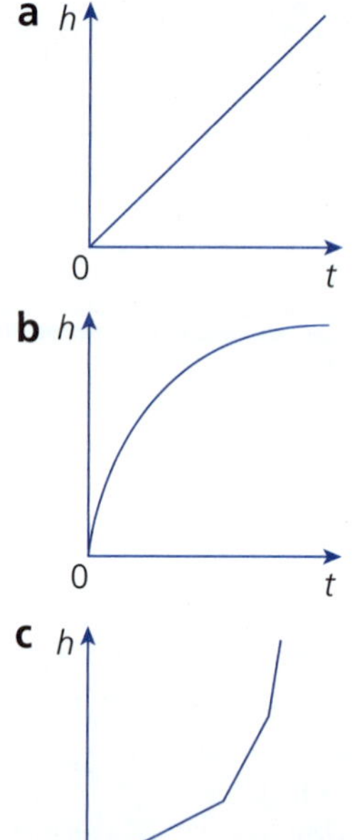

pp. 108–109

Go to p. 301 for **Quick Answers** or to pp. 328–331 for **Worked Solutions**

For a complete understanding of this topic, you must be able to:

✓	Convert between rates		p. 103
✓	Use a table of values to establish a rule linking two variables		pp. 103–105
✓	Draw and use conversion graphs		pp. 105–106
✓	Solve problems involving direct proportion		pp. 106–107
✓	Recognise direct and inverse proportion from graphs		p. 107
✓	Solve problems involving inverse proportion		pp. 107–108
✓	Match graphs to physical phenomena.		pp. 108–109

Now you are ready to do the tests!

Intermediate Test

Rates and Proportion

(25 marks)

1 Human hair grows at an average of 0.4 mm/day. Express this rate in cm/year. (1 mark)

2 Convert:

a 20 grams/m^2 to kg/hectare.

b 0.08 cents/cm^3 to \$/m^3. (2 marks)

3 Usain Bolt has run 100 metres in 9.58 seconds. A cheetah has run the same distance in 5.95 seconds. Find the difference between the two speeds, to the nearest km/h. (3 marks)

4 The graph below shows the relationship between distance travelled (D, in km) over time (h hours) when a car is driven:

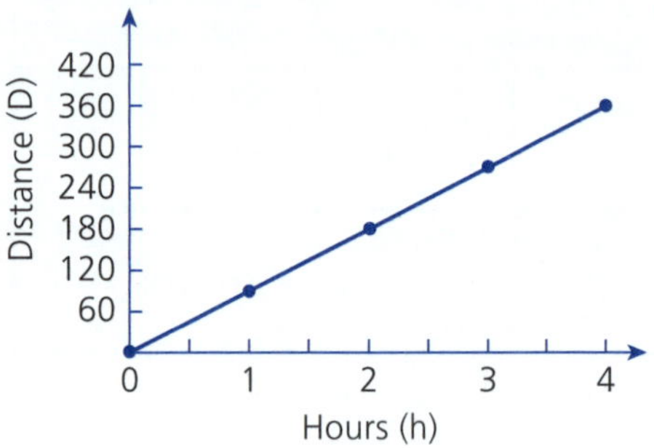

a Use the graph to complete the table:

h	0	1	2	3	4
D					

b Write the rule that links D and h.

c What is the speed of the car?

d How far will the car have travelled after 75 minutes?

e On the same graph, draw a second line to represent the distance travelled for a second car that averages 60 km/h.

f Find the distance between the two cars after travelling for 3.5 hours.

g How much longer will it take the second car to travel 180 km? (8 marks)

5 The cost (\$$C$) of building a single-storey house on a concrete slab is directly proportional to the area (A) of the floor space of the house in square metres. It costs \$235 000 to build a house with a floor space of 200 m^2.

a Find an equation for C in terms of A.

b Find the cost of a house with a floor space of 310 m^2. (3 marks)

6 The distance (d), in kilometres, to the horizon is directly proportional to the square root of the height (h), in metres, of the observer's eye above sea level.
A person whose eyes are 1.69 m above the ground can see 4.16 km out to sea.

a Find an equation for d in terms of h.

b How high above sea level, in metres, would a person's eyes need to be to see a boat that is 8 km out to sea?

c If the person stood on a headland so that their eyes were 18 metres above sea level, can they see an island 12 kilometres out to sea? Justify your answer with calculations. (4 marks)

7 Draw graphs representing the height (h) of water in the following containers, if water is poured into each container at a constant rate.

a (1 mark)

b (1 mark)

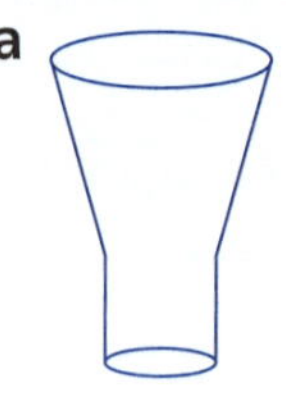

8 Draw separate sketches to show the relationship between y and x if:

a y is increasing at a constant rate (1 mark)

b y is decreasing at a decreasing rate. (1 mark)

Your Feedback

☐ × 100% = ☐ %

QA PAGE 306

WS PAGE 369

Advanced Test

Rates and Proportion

(25 marks)

1 125 mL of water is poured into a container. Twenty grams of salt is added to the water. What is the concentration of salt in the water in grams/L? (1 mark)

2 Given that y is directly proportional to x, what is the effect on y when x is doubled? (1 mark)

3 Given that y is directly proportional to x^2, what is the effect on y when x is doubled? (1 mark)

4 In the table, y is directly proportional to $\sqrt{x}$:

x	4	16	49	
y		8		24

a Find the constant of proportionality k.
b Complete the table. (5 marks)

5 Two trucks, A and B, leave the same location at the same time and travel the same road. The distance (D) in km and the time of travel (t) in hours of the trucks are recorded in the graph below:

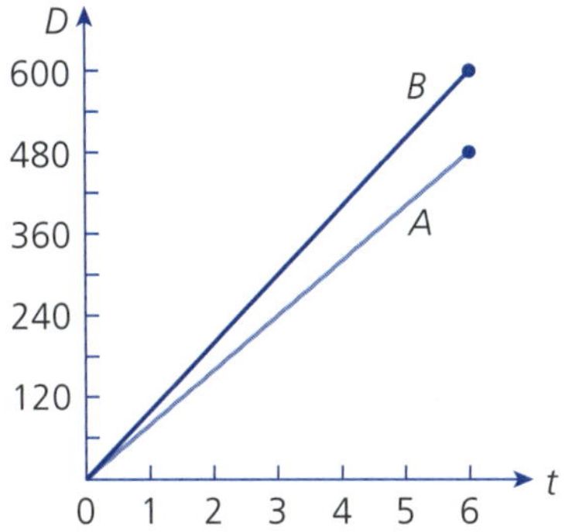

a As D and t are in direct proportion, write an equation for each truck linking D and t.
b How much further has B travelled than A after 4 hours?
c When are the two trucks 60 km apart? (6 marks)

6 Gina plans to buy some silver pendants. The cost (\$$C$) varies directly with the square of the length (n cm) of the pendant. A pendant of length 3 cm costs \$64.80.
a Write an equation linking C and n.
b What is the length of the pendant that costs \$45?
c What is the price of a pendant that is 4.3 cm in length? (4 marks)

7 Jason uses a microwave oven to heat lasagne. The time taken for heating is inversely proportional to the power setting (in watts). It takes 6 minutes at a power setting of 360 watts to heat the lasagne. How long would it take at a power setting of 500 watts? (2 marks)

8 Draw a sketch of y against x if:
a y is decreasing at an increasing rate (1 mark)
b y commences by decreasing at a constant rate, then increasing at a decreasing rate. (1 mark)

9 'The rate of unemployment is increasing at an increasing rate.' Draw a sketch to represent this statement. (1 mark)

10 Jack commenced his hike up a hill at a rapid pace but as he reached the summit he slowed down significantly. He rested when he got to the top, and then started walking back down the hill. However, just after he started his descent, he lost his footing and tumbled down the hill, getting faster and faster until he had reached the bottom. Draw a sketch using time (t) and height (h) of Jack throughout his journey. (2 marks)

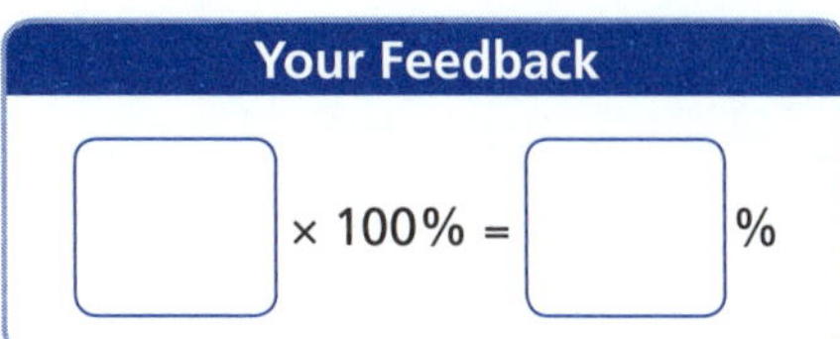

QA PAGE 306

WS PAGE 370

Chapter 8
Pythagoras and Trigonometry

Pythagoras' Theorem

In any right-angled triangle, the square of the **hypotenuse** is equal to the sum of the squares of the other two sides.

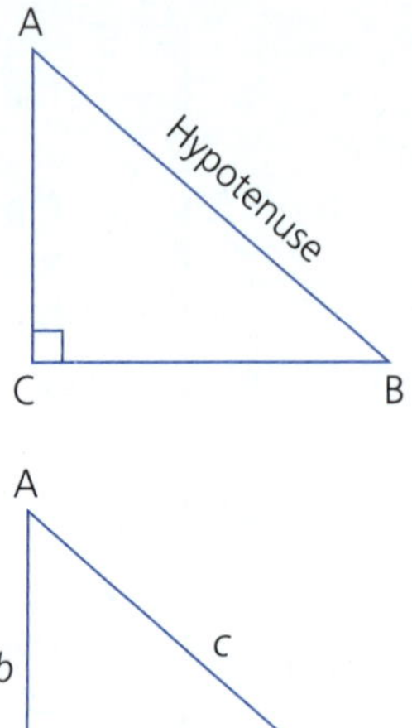

The hypotenuse is the longest side of the right-angled triangle—it is always the side opposite the right-angle.

In simpler terms:

$c^2 = a^2 + b^2$

where c is the length of the hypotenuse.

A, b, c, C, a, B

Keywords

Adjacent
Altitude
Angle of elevation
Area
Bearing
Cos θ
Cosine
Cosine Rule
Degree
Depression
Elevation
Exact value
Horizontal
Hypotenuse
Inclination
Minute
Opposite
Pythagoras
Pythagorean
Ratio
Reciprocal
Right-angled
Sin θ
Sine
Sine Rule
Subtend
Tan θ
Tangent
Theorem
Triad
Triangle
Trigonometry

Using Pythagoras' Theorem to find missing lengths

Given the lengths of any two sides of a right-angled triangle, it is possible to calculate the length of the third side.

For Example

1 Calculate the value of y in this triangle.

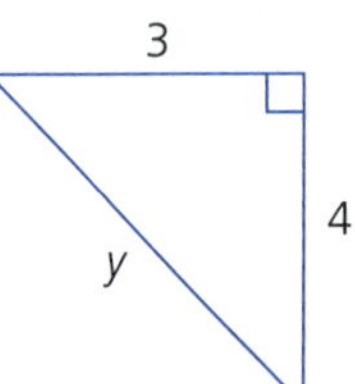

S

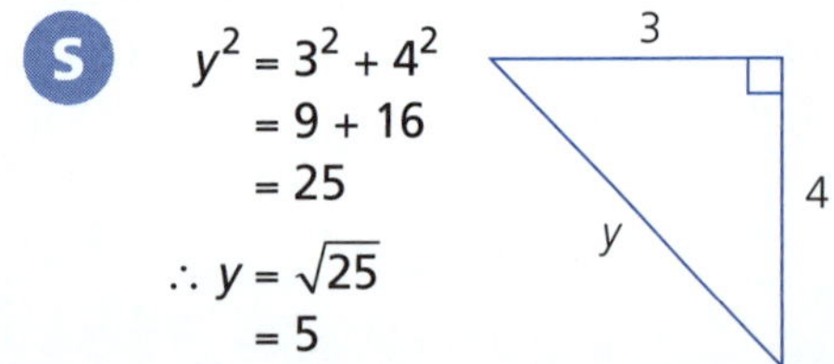

$$y^2 = 3^2 + 4^2$$
$$= 9 + 16$$
$$= 25$$
$$\therefore y = \sqrt{25}$$
$$= 5$$

2 Given that AB = 7 cm, AC = 4 cm, calculate the length of BC correct to one decimal place.

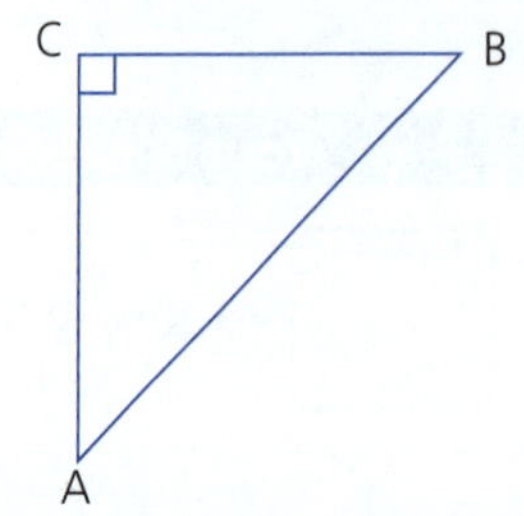

S

C, x, B, 4, 7, A

Let BC = x cm (AB is hypotenuse).

$$\therefore x^2 + 4^2 = 7^2$$
$$x^2 = 7^2 - 4^2$$
$$= 49 - 16$$
$$= 33$$
$$\therefore x = \sqrt{33}$$
$$= 5.744\,562\,6$$
$$= 5.7 \text{ (to 1 decimal place)}$$

Exact solution

That is, BC is 5.7 cm long.

3 In the $\triangle PQR$, PR = 5 cm, RQ = 4 cm and $\angle PRQ = 90°$. Calculate the length of PQ correct to three significant figures.

S Let PQ be y cm (PQ is hypotenuse):

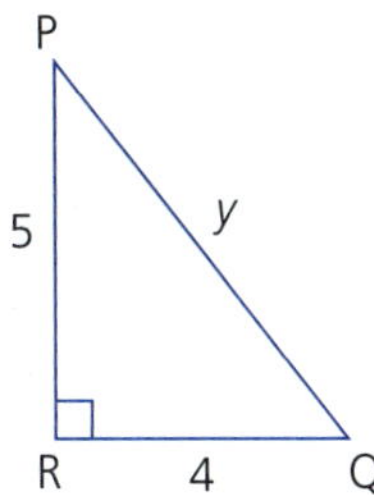

$\therefore y^2 = 5^2 + 4^2$
$= 25 + 16$
$= 41$

$\therefore y = \sqrt{41}$
$= 6.403\,124\,2$
$= 6.40$ (to 3 significant figures)

Exact solution

PQ is 6.40 cm long.

Proving a triangle is right-angled

For $\triangle ABC$ to be right-angled, we must be able to prove that $AB^2 = BC^2 + AC^2$.

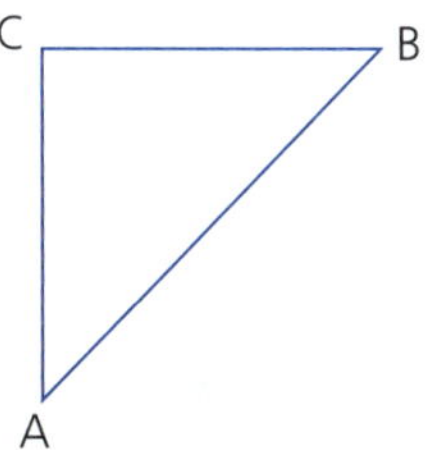

1 Show that the following triangle is right-angled.

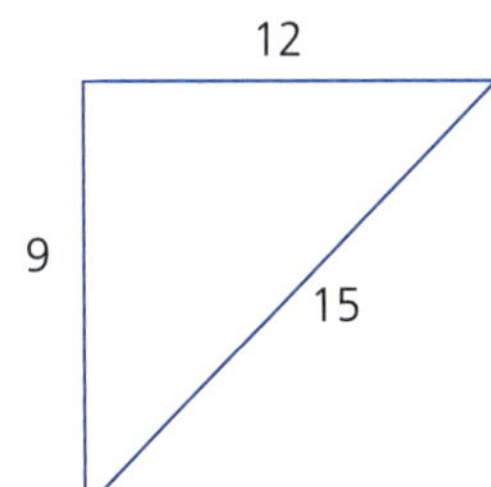

S Hypotenuse is the longest side, that is, 15 units.

We have to prove that:

$15^2 = 9^2 + 12^2$

Now LHS $= 15^2$
$= 225$

LHS means 'left-hand side' of expression.

RHS $= 9^2 + 12^2$
$= 81 + 144$
$= 225$
= LHS

RHS means 'right-hand side' of expression.

$\therefore 15^2 = 9^2 + 12^2$

Then $\triangle$ is right-angled.

2 Which of the following triangles are right-angled?

A $\triangle PQR$, where PQ = 6 cm, QR = 10 cm, PR = 11 cm

B $\triangle XYZ$, where XY = 8 cm, XZ = 15 cm, YZ = 20 cm

C $\triangle ABC$, where BC = 17 cm, AB = 8 cm, AC = 15 cm

D $\triangle DEF$, where EF = 16 cm, DE = 18 cm, DF = 12 cm

S Consider $\triangle ABC$.

We have to prove that:

$8^2 + 15^2 = 17^2$

Now LHS $= 8^2 + 15^2$
$= 64 + 225$
$= 289$

RHS $= 17^2$
$= 289$
= LHS

$\therefore 8^2 + 15^2 = 17^2$

$\therefore$ Right-angled $\therefore$ C

Pythagorean triads (or triples)

1 The numbers $\{a, b, c\}$ form a Pythagorean triad if:

a $c^2 = a^2 + b^2$

b a, b, and c are positive whole numbers. That is, {3, 4, 5} is a Pythagorean triad because $3^2 + 4^2 = 5^2$.

$$\begin{aligned} \text{LHS} &= 3^2 + 4^2 \\ &= 9 + 16 \\ &= 25 \end{aligned} \qquad \begin{aligned} \text{RHS} &= 5^2 \\ &= 25 \\ &= \text{LHS} \end{aligned}$$

2 But {1.2, 1.6, 2} is *not* a Pythagorean triad, even though $2^2 = 1.2^2 + 1.6^2$ because 1.2 and 1.6 are not whole numbers.

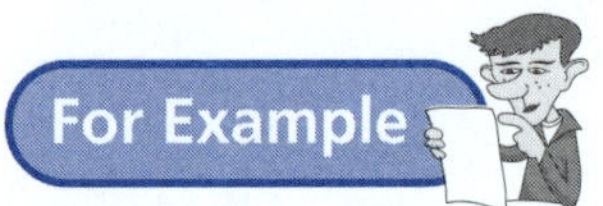

1 Find the value of t such that:

a $\{18, 24, t\}$ is a Pythagorean triad.

S
$$\begin{aligned} t^2 &= 18^2 + 24^2 \\ &= 900 \\ t &= \sqrt{900} \\ &= 30 \end{aligned}$$

b $\{t, 12, 13\}$ is a Pythagorean triad.

S
$$\begin{aligned} t^2 + 12^2 &= 13^2 \\ t^2 &= 13^2 - 12^2 \\ &= 169 - 144 \\ &= 25 \\ \therefore t &= \sqrt{25} \\ &= 5 \end{aligned}$$

Right-angle trigonometry

Sides and ratios

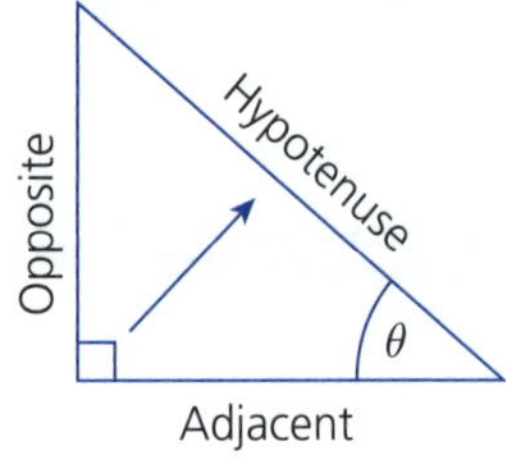

We name each of the sides. The hypotenuse is the side opposite the right-angle.

$$\sin\theta = \frac{\text{Opposite}}{\text{Hypotenuse}} \text{ (SOH)}$$

$$\cos\theta = \frac{\text{Adjacent}}{\text{Hypotenuse}} \text{ (CAH)}$$

$$\tan\theta = \frac{\text{Opposite}}{\text{Adjacent}} \text{ (TOA)}$$

← 'sin' is the abbreviation for sine, 'cos' for cosine and 'tan' for tangent.

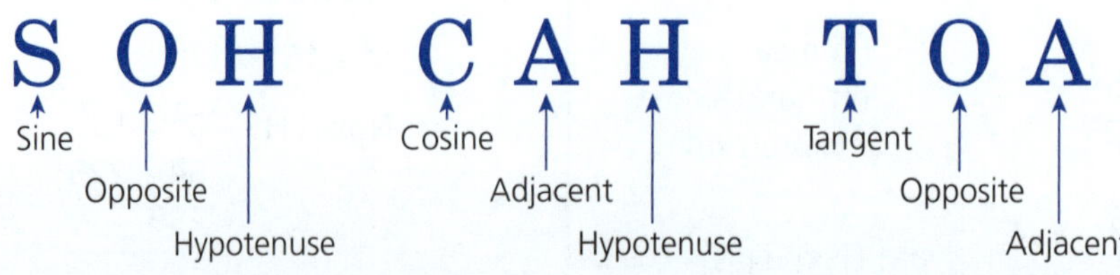

Some students use the following:

Old Hags All Hate Old Age OR Some Old Hags Can't Always Hide Their Old Age

For Example

1 Find expressions for $\sin\theta$, $\cos\theta$ and $\tan\theta$.

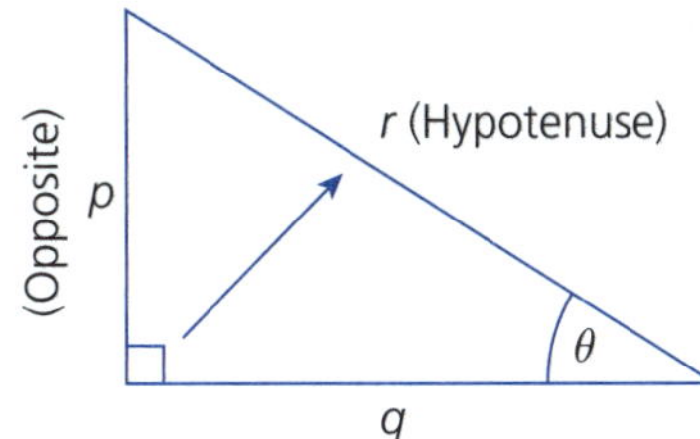

S

$$\sin\theta = \frac{\text{Opposite}}{\text{Hypotenuse}} = \frac{P}{r}$$

$$\cos\theta = \frac{\text{Adjacent}}{\text{Hypotenuse}} = \frac{q}{r}$$

$$\tan\theta = \frac{\text{Opposite}}{\text{Adjacent}} = \frac{p}{q}$$

2 Given that $\sin\theta = \frac{5}{13}$, find an expression for $\tan\theta$ as a simple fraction.

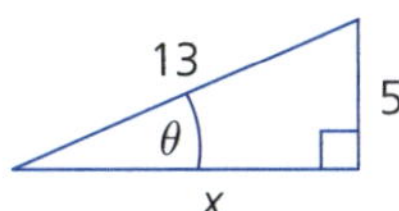

$$\sin\theta = \frac{5}{13} = \frac{\text{Opposite}}{\text{Hypotenuse}}$$

$\therefore$ opposite = 5, hypotenuse = 13

By Pythagoras' Theorem

$$x^2 + 5^2 = 13^2$$
$$\therefore x^2 = 169 - 25$$
$$= 144$$
$$\therefore x = \sqrt{144} = 12$$

$$\therefore \tan\theta = \frac{\text{Opposite}}{\text{Adjacent}} = \frac{5}{12}$$

3 Given that $\tan\phi = \frac{8}{15}$:

a Derive expressions for $\sin\phi$ and $\cos\phi$.

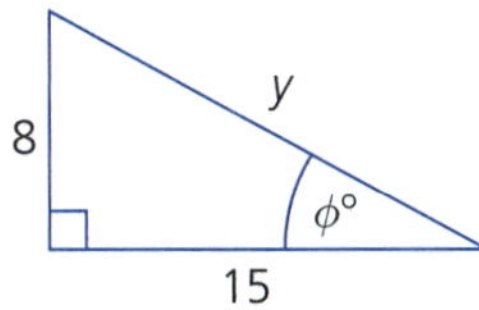

$$\tan\phi = \frac{8}{15}$$
$$= \frac{\text{Opposite}}{\text{Adjacent}}$$

Opposite = 8
Adjacent = 15

$$y^2 = 8^2 + 15^2$$
$$= 289$$
$$\therefore y = \sqrt{289}$$
$$= 17$$

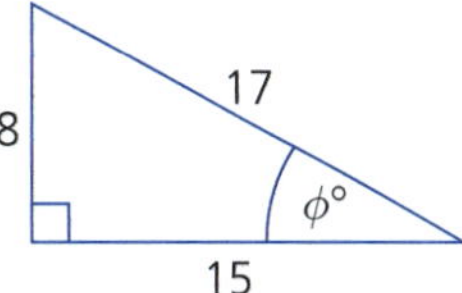

$$\sin\phi = \frac{\text{Opposite}}{\text{Hypotenuse}} = \frac{8}{17}$$

$$\cos\phi = \frac{\text{Adjacent}}{\text{Hypotenuse}} = \frac{15}{17}$$

b Show that $\tan\phi = \frac{\sin\phi}{\cos\phi}$.

S

$$\frac{\sin\phi}{\cos\phi} = \sin\phi \div \cos\phi = \frac{8}{17} \surd \frac{15}{17}$$

$$= \frac{8}{{}_1\cancel{17}} \times \frac{\cancel{17}^1}{15} = \frac{8}{15}$$

$$\tan\phi = \frac{8}{15}$$

$$\therefore \tan\phi = \frac{\sin\phi}{\cos\phi}$$

4 Find expressions for the following:

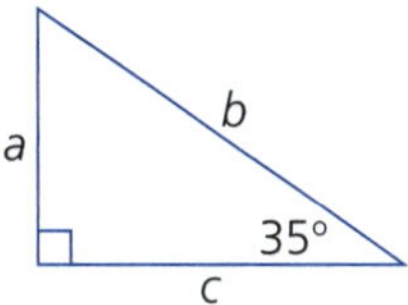

a $\sin 35°$, $\cos 35°$

S

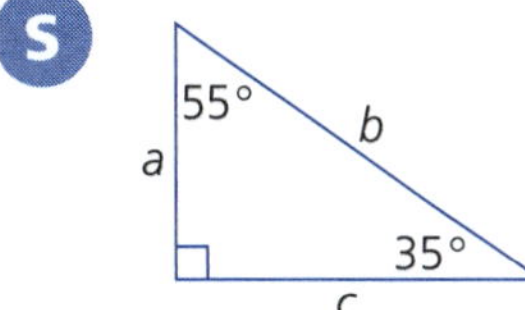

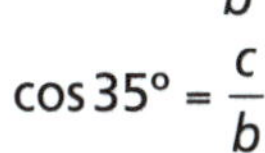

$$\sin 35° = \frac{a}{b}$$

$$\cos 35° = \frac{c}{b}$$

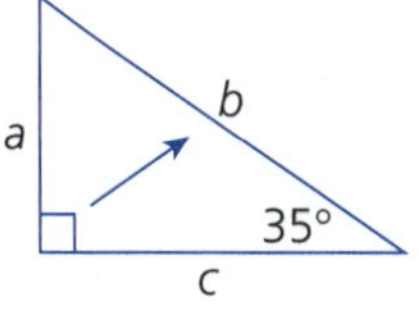

b $\sin 55°$, $\cos 55°$

S $\sin 55° = \dfrac{c}{b}$

$\cos 55° = \dfrac{a}{b}$

c Show that $\sin 35° = \cos 55°$

S $\sin 35° = \dfrac{a}{b}$, $\cos 55° = \dfrac{a}{b}$

$\therefore \sin 35° = \cos 55°$

> **In general:** $\tan\theta = \dfrac{\sin\theta}{\cos\theta}$, for all θ (except $\cos\theta = 0$)
>
> $\sin(90° - \theta) = \cos\theta$, for all θ $(0° < \theta < 90°)$
>
> $\cos(90° - \theta) = \sin\theta$, for all θ $(0° < \theta < 90°)$

These general results can be used to solve simple equations, or to simplify expressions.

For Example

1 If $\sin 60° = \cos x°$, find x.

S $\sin 60° = \cos 30°$ $\therefore x = 30$

2 If $\cos 10° = \sin x°$, find x.

S $\cos 10° = \sin 80°$ $\therefore x = 80$

3 Simplify the expressions:

a $\dfrac{\sin 28Υ}{\cos 62Υ}$

S $\dfrac{\sin 28Υ}{\cos 62Υ} = \dfrac{\sin 28Υ}{\sin 28Υ} = 1$

b $\dfrac{\sin 28Υ}{\sin 62Υ}$

S $\dfrac{\sin 28Υ}{\sin 62Υ} = \dfrac{\sin 28Υ}{\cos 28Υ} = \tan 28°$

> $\tan\theta = \dfrac{\sin\theta}{\cos\theta}$

Finding the sides of right-angled triangles

Real-life problems can be solved using trigonometry.

> Your calculator might use [° ′ ″] or [DMS] for Degree, Minutes and Seconds input.

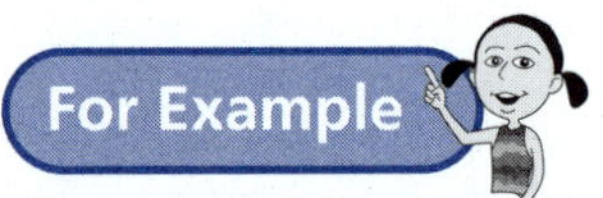

1 Find the side x.

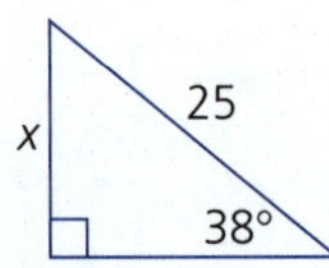

S $\begin{matrix}\text{Opp} \\ \text{Hyp}\end{matrix}\ \dfrac{x}{25} = \sin 38°$

$\therefore x = 25 \times \sin 38°$

$= 15.391\,537$

$= 15.4$ (to one decimal place)

2 Find the side y.

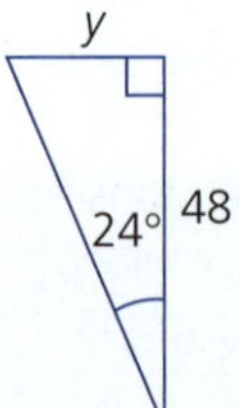

S $\begin{matrix}\text{Opp} \\ \text{Adj}\end{matrix}\ \dfrac{y}{48} = \tan 24°$

$y = 48 \times \tan 24°$

$= 21.370\,977$

$= 21.4$ (to one decimal place)

3 When the altitude of the sun is 63°14′, a tree casts a shadow 8.6 m long. Calculate the height of the tree correct to three significant figures.

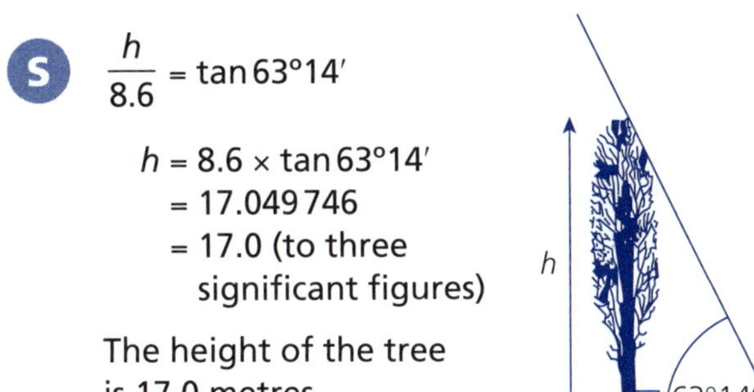

S

$$\frac{h}{8.6} = \tan 63°14'$$

$$h = 8.6 \times \tan 63°14'$$
$$= 17.049\,746$$
$$= 17.0 \text{ (to three significant figures)}$$

The height of the tree is 17.0 metres.

Note: the correct angle from the diagram is the angle between the sun's rays and ground level.

4 A 3-metre ladder leans against a brick wall and is inclined at 28°35′ to the wall. How far up the wall will the ladder reach? (Answer correct to one decimal place.)

S

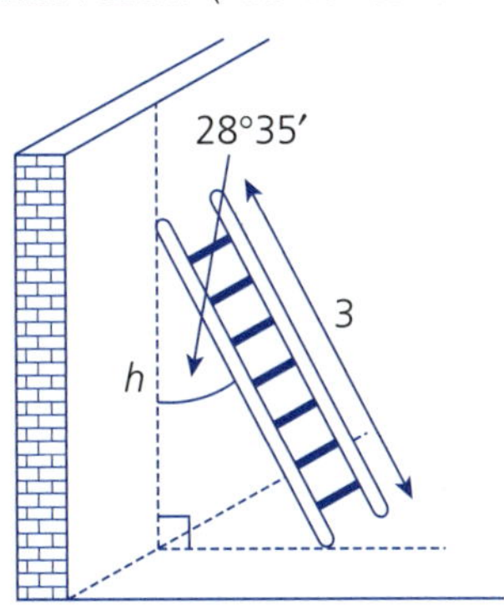

$$\frac{h}{3} = \cos 28°35'$$

$$h = 3 \times \cos 28°35'$$
$$= 2.634\,366$$
$$= 2.6 \text{ (one decimal place)}$$

The ladder reaches 2.6 metres up the wall.

5 Find the hypotenuse.

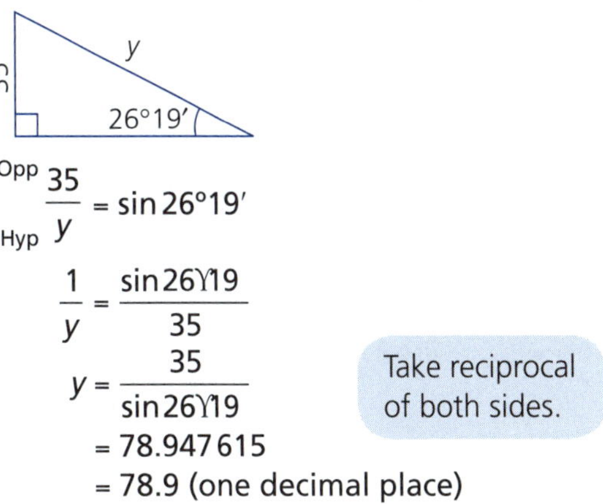

$$\frac{35}{y} = \sin 26°19' \quad \text{(Opp/Hyp)}$$

$$\frac{1}{y} = \frac{\sin 26°19'}{35}$$

$$y = \frac{35}{\sin 26°19'}$$
$$= 78.947\,615$$
$$= 78.9 \text{ (one decimal place)}$$

Take reciprocal of both sides.

6 A 25-metre high antenna is held in position by four wires fixed 7 metres from the top of the antenna. If the wires are attached so as to make an angle of 42° with the antenna, how much wire is used (to the nearest cm)?

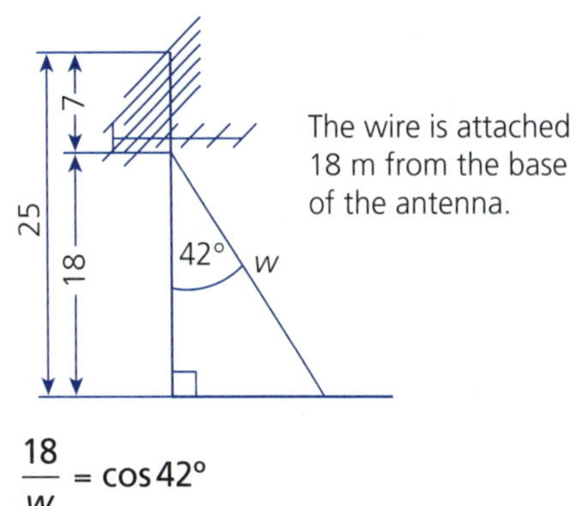

$$\frac{18}{w} = \cos 42°$$

$$\frac{1}{w} = \frac{\cos 42°}{18}$$

$$w = \frac{18}{\cos 42°}$$
$$= 24.221\,389$$

Do not round off until the end.

Length of wire needed
$= 4 \times 24.221\,389$
$= 96.885\,557$
$= 96.89$ (to the nearest cm)

The length of wire needed is 96.89 metres.

Finding the angles of right-angled triangles

An angle in a right-angled triangle can be found by using the calculator's 2nd function key or its SHIFT key (see your calculator's instruction book).

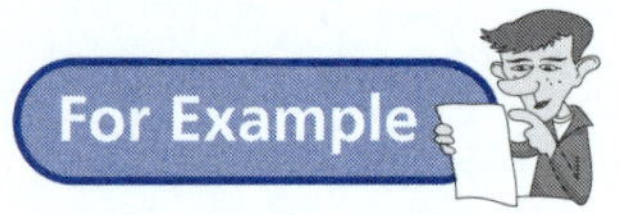

1 Find θ.

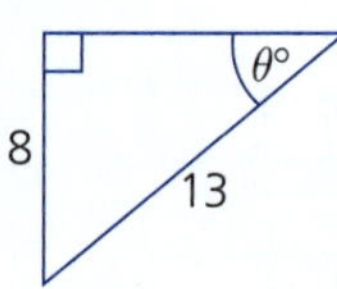

S $\sin\theta = \frac{8}{13}$

$\theta = 37°59'$

$\frac{\text{Opposite}}{\text{Hypotenuse}}$

2 Find x.

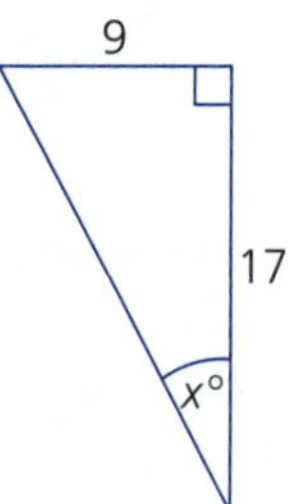

S $\tan x = \frac{9}{17}$

$\therefore x = 27°54'$

$\frac{\text{Opposite}}{\text{Adjacent}}$

3 Bill stands 34 m from a 60 m high building. Find the angle of elevation of the top of the building (to the nearest minute).

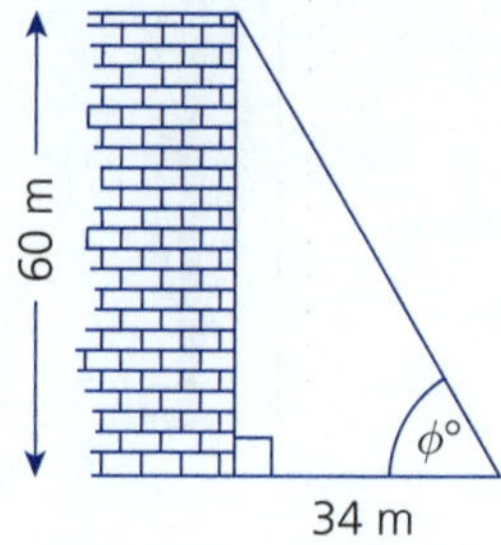

S Angle of elevation is $\phi°$.

$\tan\phi = \frac{60}{34}$

$\therefore \phi = 60°28'$

$\frac{\text{Opposite}}{\text{Adjacent}}$

The angle of elevation is 60°28′. (The term 'angle of elevation' is explained on the next pages.)

Two-triangle problems

When a problem involves drawing two triangles, the side or angle in one triangle needs to be found and used for calculations involving the second triangle.

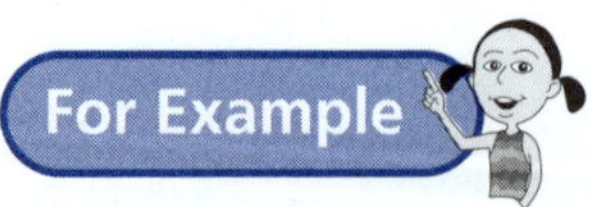

1 Mary and Bob are flying model planes so that Mary's plane is directly above Bob's. The planes, held by wires, are flying about the couple in a circle of 24 m radius. The angles that the wires make with the horizontal are 28° and 35° for Bob and Mary's planes respectively. Mary's plane is at C, Bob's at D.

Calculate the vertical distance between the two planes.

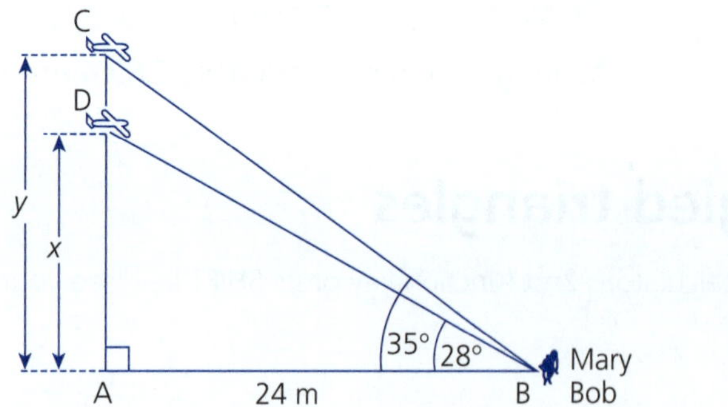

S The required distance is CD. This is the difference between AC and AD.

Let AD = x m and AC = y m.

In $\triangle$ABC, $\frac{y}{24} = \tan 35°$

$\therefore y = 24 \times \tan 35°$

In $\triangle$ABD, $\frac{x}{24} = \tan 28°$

$\therefore x = 24 \times \tan 28°$

Now CD $= y - x$

$= 24 \tan 35° - 24 \tan 28°$

$= 4.043\,9546$

$= 4$ (to the nearest metre)

Mary's plane is 4 metres above Bob's plane.

Some notes regarding angles

Angles of elevation and depression

Angle of elevation: angle between line of sight (looking upwards at something) and the horizontal.

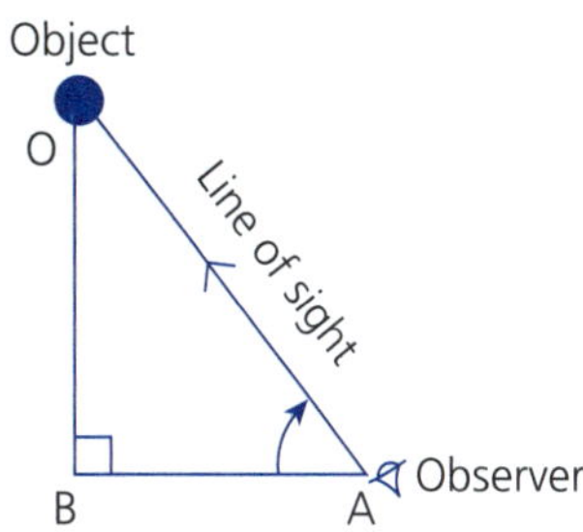

Angle of depression: angle between line of sight (looking downwards at something) and the horizontal.

Angle of elevation = Angle of depression

In the figure opposite, ∠HAO = ∠BOA (using parallel lines, alternate ∠s)

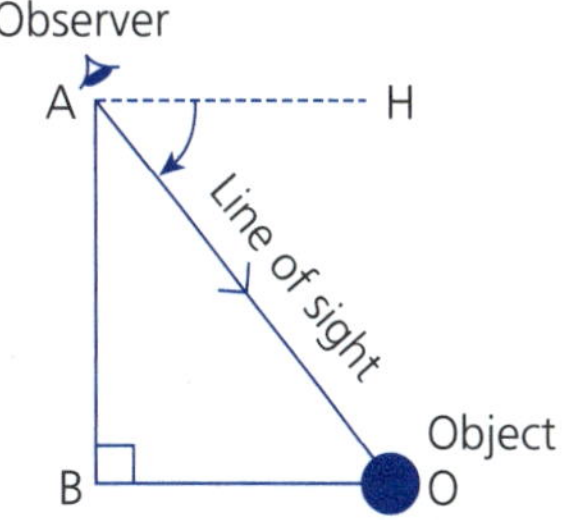

Angle of elevation of the sun

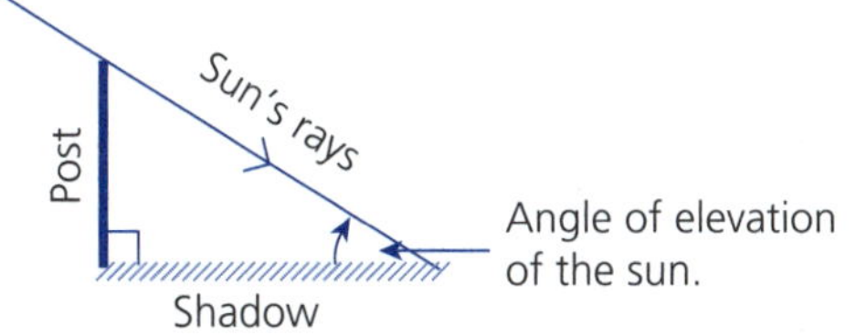

The angle concerned is that between the sun's rays and the horizontal.

Angles subtended by an interval

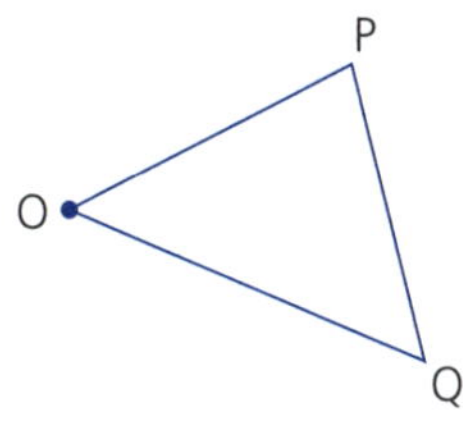

PQ subtends the angle POQ at the point O.

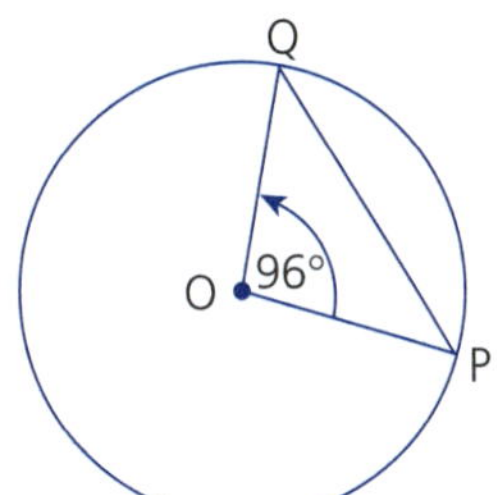

The chord PQ subtends an angle of 96° at the centre, O, of the circle. (This terminology is also used in geometry of the circle.)

Bearings

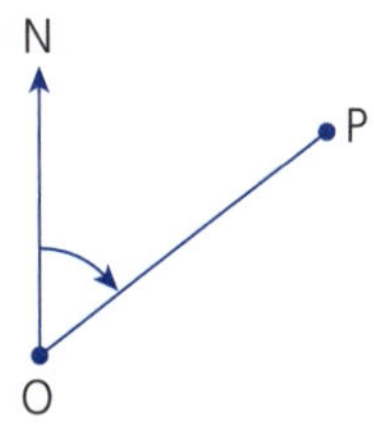

Bearings are measured clockwise from due north and the degrees in bearings are written as three-digit numbers.

The bearing of P from O is 047°. This means that ∠NOP = 47°.

Bearings of P from O.

1 140°

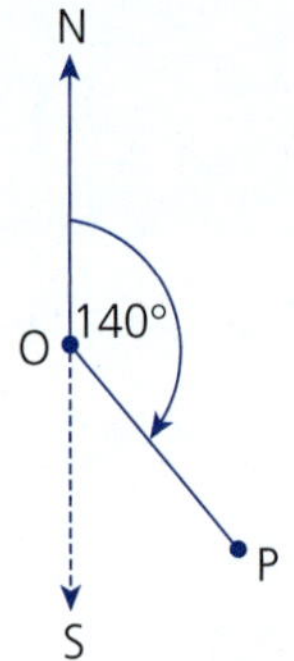

2 280°

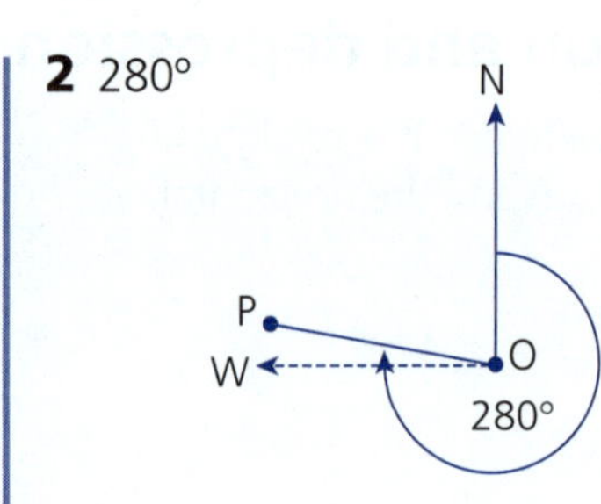

Special cases:

- NE has bearing 045°.
- Due E has bearing 090°.
- SE has bearing 135°.
- Due S has bearing 180°.
- The angle between NE and SE is 90°.
- The angle between SW and NW is 90°.
- The angle between W and NW is 45°.

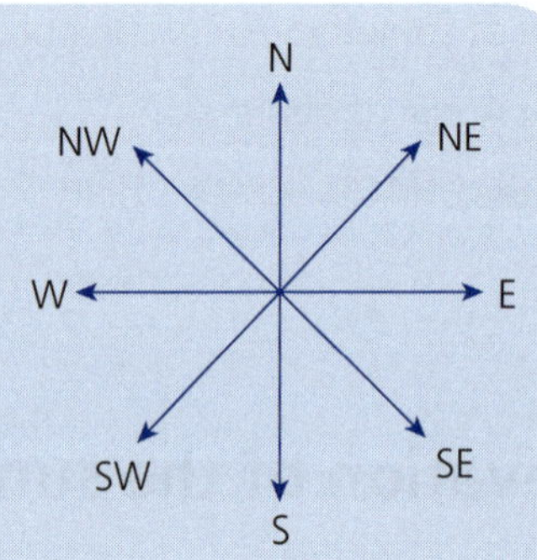

Examples involving trigonometry

A diagram is used to summarise the information given in the question and is then used to solve the problem.

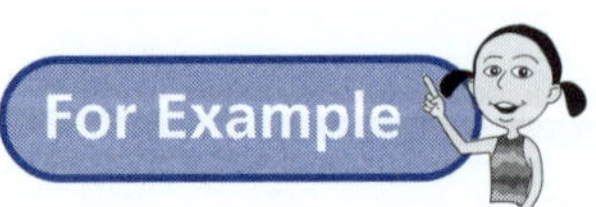

1 A ferry leaves Minmi Breakwater and steams 12 km on a bearing of 114° to a point S. Find the distance that the ferry is south of Minmi.

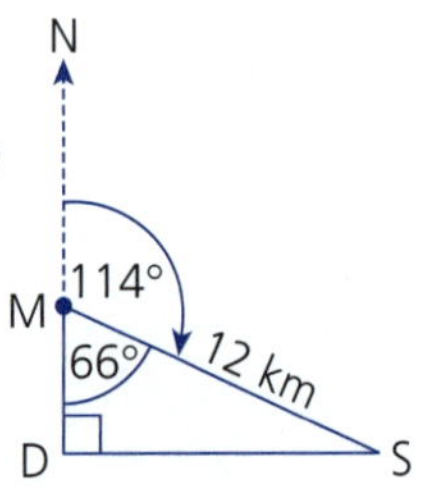

S This asks us to find the distance MD.

$\angle NMS = 114°$
$\angle DMS = 180° - 114°$
$= 66°$ (angle sum of straight lines)

In $\triangle DSM$, $\dfrac{MD}{12} = \cos 66°$

$\therefore MD = 12 \times \cos 66°$
$= 4.880\,8397$
$= 4.9$ (to one decimal place)

The ferry is 4.9 km south of Minmi.

2 Red Riding Hood walked 2.4 km due south from home (H) to an old tree (T) and then due west for 0.9 km. Calculate Red Riding Hood's bearing from home (to the nearest degree).

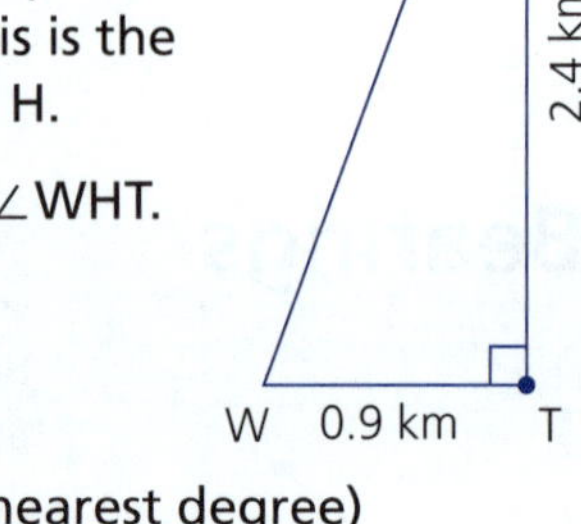

S The actual angle required is reflex $\angle NHW$. This is the bearing of W from H.

We must first find $\angle WHT$. Call it θ.

$\tan\theta = \dfrac{0.9}{2.4}$

$\theta = 21°$ (to the nearest degree)

$\therefore \angle NHW = 180° + 21°$
$= 201°$

The bearing of Red Riding Hood is 201°.

Three-dimensional problems

Trigonometry can be used with 3D figures to calculate lengths and angles. Sometimes it is used in conjunction with Pythagoras' Theorem.

In this problem we have the rectangular prism with sides of length 5 cm, 12 cm and 8 cm (as in the diagram). The problem is to find ∠AFD. This could be phrased in many ways but consider this one explained below.

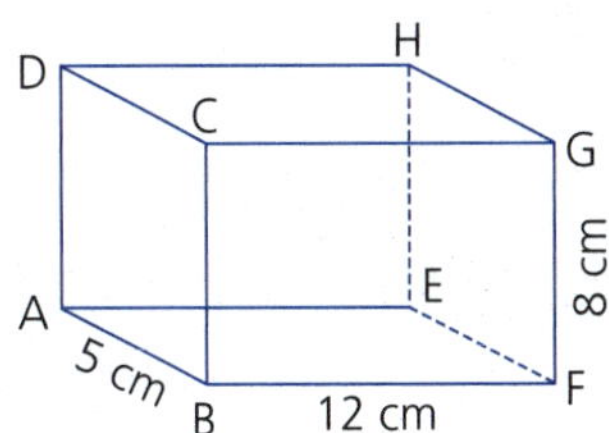

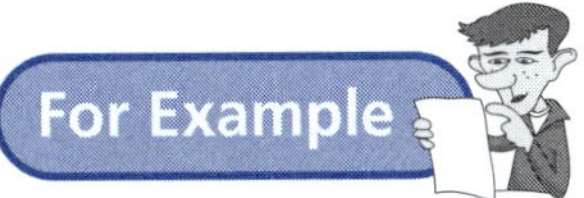

1 Louie the fly sits at F and wishes to fly up to see his girlfriend fly, Maisie, at D. At what angle to the base plane ABFE must Louie take off to fly the most direct route?

We can break the question into two parts. The first step is to find the length of the base diagonal AF.

∠ABF = 90°, as ABFE is a rectangle.
Using △ABF:

$$AF^2 = 5^2 + 12^2$$
$$= 169$$
$$AF = \sqrt{169}$$
$$= 13$$

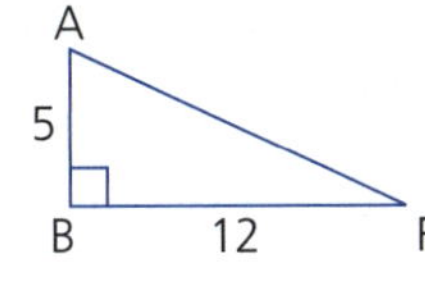

Step two is to use △AFD to calculate ∠AFD.

∠DAF = 90°, because AD is perpendicular to the base plane.

$$\tan\theta = \frac{8}{13}$$
$$\therefore \theta = 31°36'$$

That is, ∠AFD = 31°36′.

D, 8, A, 13, F, θ°

Louie must take off at an angle of 31°36′ to fly direct to Maisie.

Exact trigonometric ratios

You can use exact values for ratios involving angles of 30°, 45° and 60°, as follows.

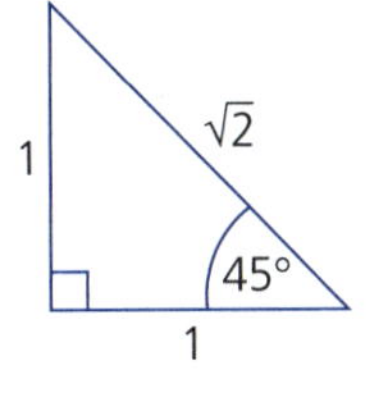
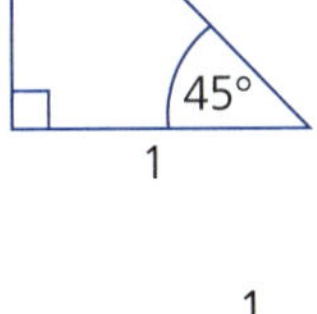

$\sin 45° = \frac{1}{\sqrt{2}}$

$\cos 45° = \frac{1}{\sqrt{2}}$

$\tan 45° = 1$

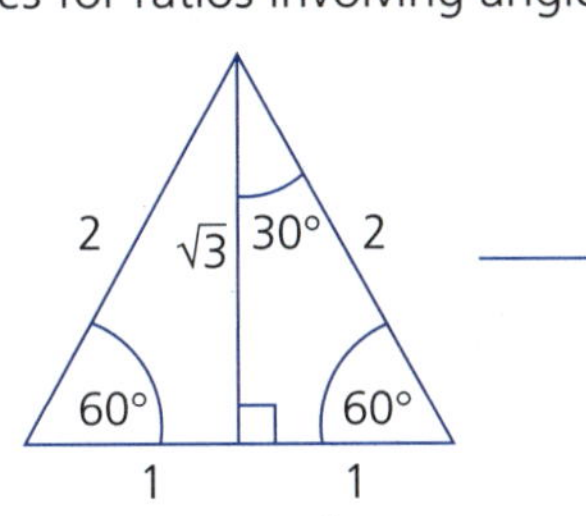

$\sin 30° = \frac{1}{2}$

$\cos 30° = \frac{\sqrt{3}}{2}$

$\tan 30° = \frac{1}{\sqrt{3}}$

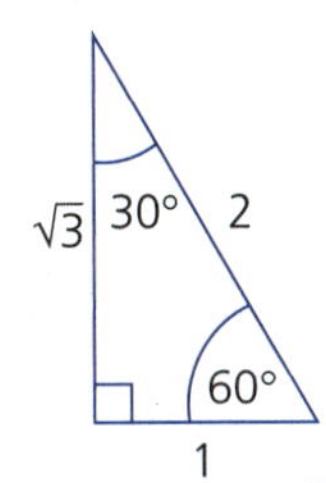

$\sin 60° = \frac{\sqrt{3}}{2}$

$\cos 60° = \frac{1}{2}$

$\tan 60° = \sqrt{3}$

Ratio	Angle		
	30°	**45°**	**60°**
sin	$\frac{1}{2}$	$\frac{1}{\sqrt{2}}$	$\frac{\sqrt{3}}{2}$
cos	$\frac{\sqrt{3}}{2}$	$\frac{1}{\sqrt{2}}$	$\frac{1}{2}$
tan	$\frac{1}{\sqrt{3}}$	1	$\sqrt{3}$

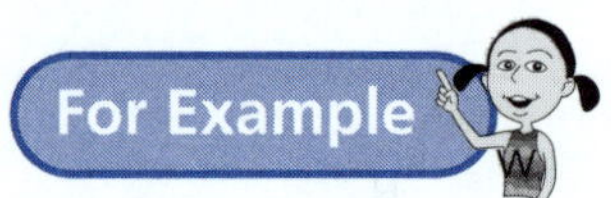

1 Find the exact value of y in the following diagrams:

a

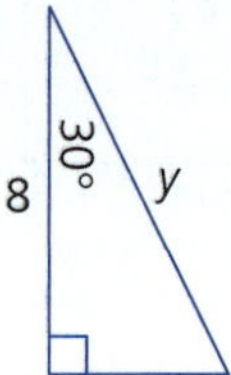

S

$$\frac{8}{y} = \cos 30°$$

$$\therefore \frac{8}{y} = \frac{\sqrt{3}}{2}$$

$$\therefore \sqrt{3}y = 16$$ Cross-multiply.

$$y = \frac{16}{\sqrt{3}}$$

b

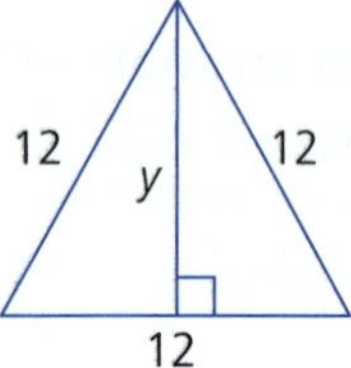

S

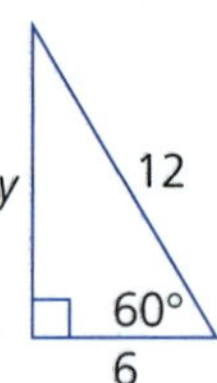

The triangle is equilateral. Each angle is 60°.

$$\therefore \frac{y}{12} = \sin 60°$$

$$\therefore y = 12 \sin 60°$$

$$= 12 \times \frac{\sqrt{3}}{2}$$

$$= 6\sqrt{3}$$

2 By first calculating the length x, find the area of the trapezium.

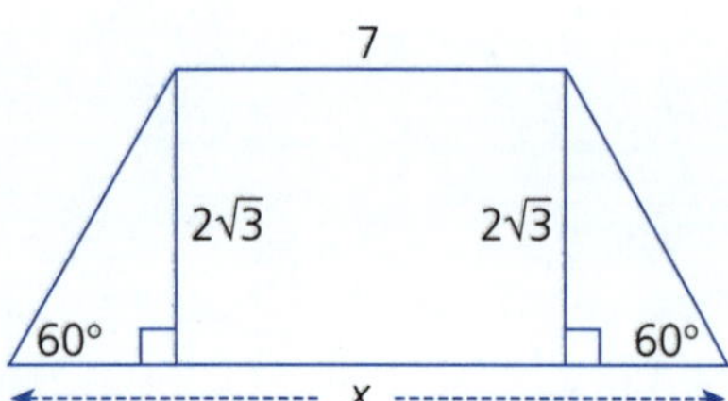

S

$$\frac{y}{2\sqrt{3}} = \tan 30°$$

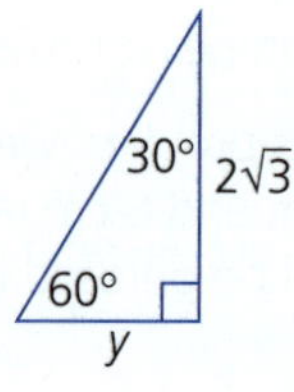

$$\therefore y = 2\sqrt{3}\tan 30)$$

$$= 2\sqrt{3} \times \frac{1}{\sqrt{3}}$$

$$= 2$$

Then $x = 2 + 7 + 2 = 11$

Area $= \frac{1}{2}h(a + b)$

$$= \frac{1}{2} \times 2\sqrt{3} \times (7 + 11)$$

$$= \sqrt{3}(18)$$

$$= 18\sqrt{3}$$

Area is $18\sqrt{3}$ units^2.

3 Find the perimeter of the triangle. Leave your answer in exact form.

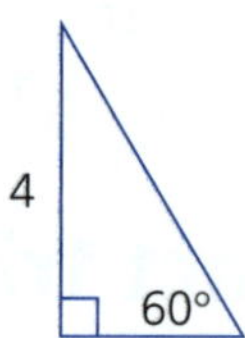

S

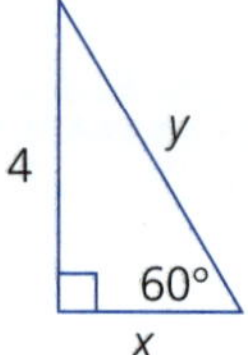

$$\frac{4}{x} = \cos 60°$$

$$\frac{4}{x} = \frac{1}{2}$$

$$x = 8$$

By Pythagoras:

$$y^2 = 4^2 + 8^2$$

$$= 16 + 64$$

$$= 80$$

$$y = \sqrt{80}$$

$$= 4\sqrt{5}$$

Perimeter $= 4 + 8 + 4\sqrt{5}$

$= 12 + 4\sqrt{5}$ $\quad \therefore (12 + 4\sqrt{5})$ units

Graphs of trigonometric functions

Trigonometric functions are graphed on the number plane using the values of x (or θ) stated in the question. The **period** is the 'time' taken for the graph to start repeating. This is measured in terms of the independent variable (the variable on the horizontal axis).

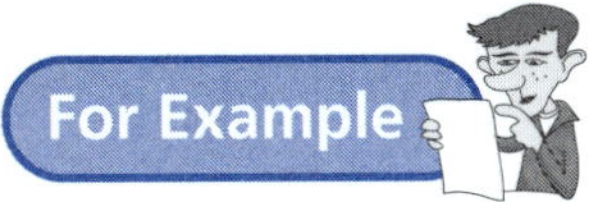

1 Graph $y = \sin x$, $y = \cos x$ and $y = \tan x$ for $0° \le x \le 360°$.

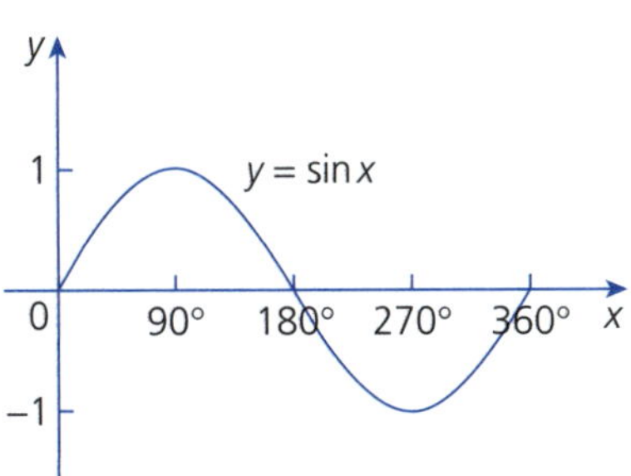

Period = 360°

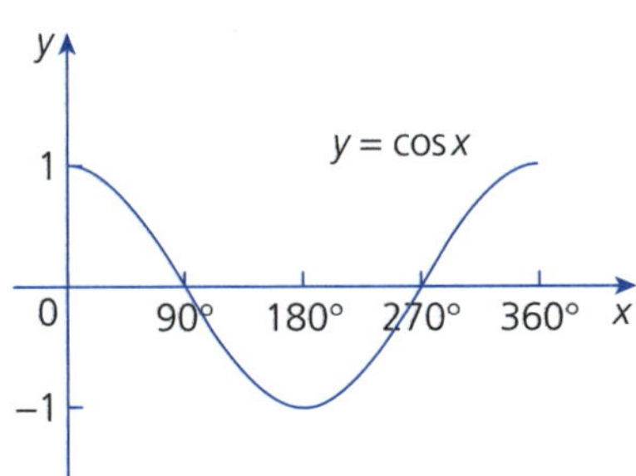

Period = 360°

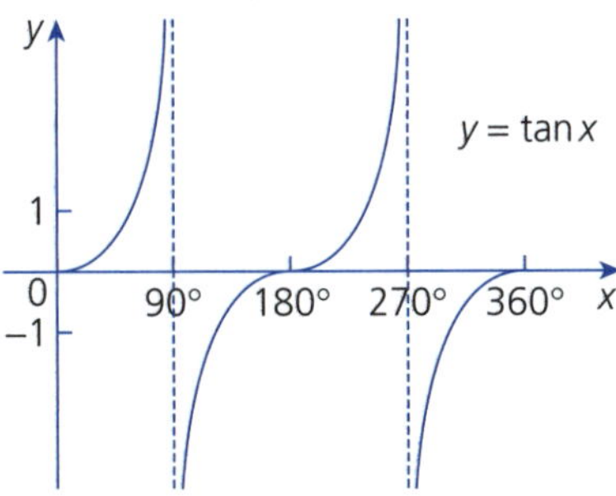

Period = 180°

2 Graph $y = 2 \sin x$, for $0° \le x \le 360°$.

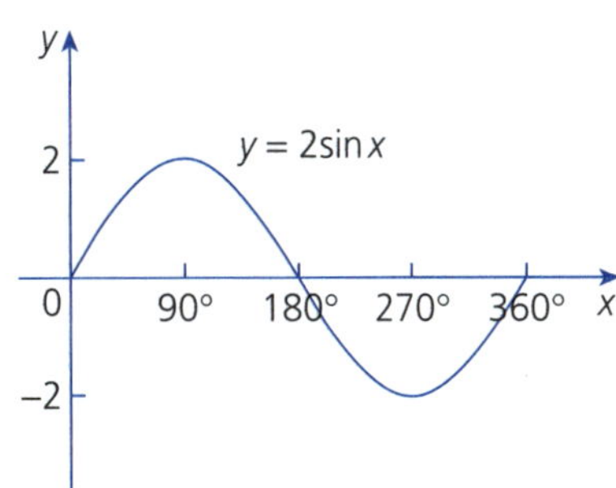

3 Graph $y = \cos x$, for $-180° \le x \le 180°$.

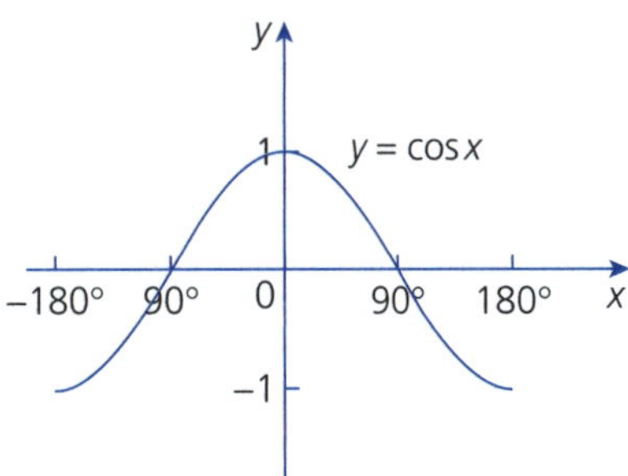

4 Graph $y = 3 \cos 2x$, for $0° \le x \le 360°$.

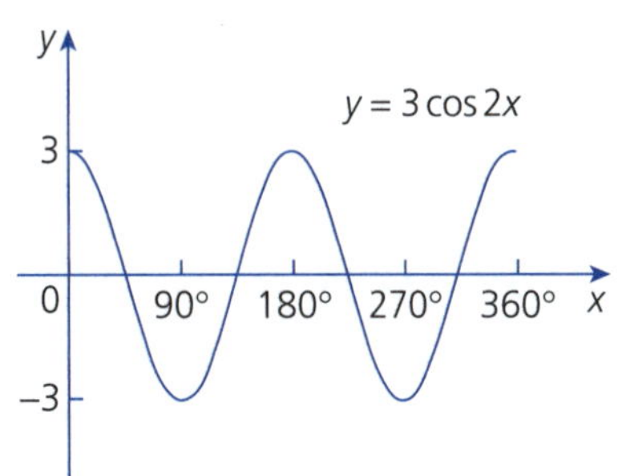

Trigonometric ratios for angles of any magnitude

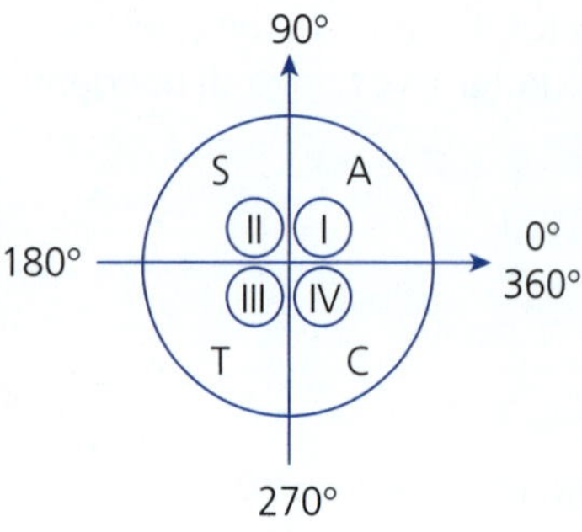

Trigonometric ratios can be found for angles that cannot exist in right-angled triangles.

Consider: **First quadrant**
Angles less than 90°.
All ratios positive.

Second quadrant
Angles between 90° and 180°.
Only the sin ratio is positive.

Third quadrant
Angles between 180° and 270°.
Only the tan ratio is positive.

Fourth quadrant
Angles between 270° and 360°.
Only the cos ratio is positive.

Remember:

S A / T C — All Stations To Central

$\sin(180° - A) = \sin A$
$\cos(180° - A) = -\cos A$
$\tan(180° - A) = -\tan A$

For Example

1 $\tan 140°$

S $\tan 140° = -\tan(180° - 140°)$
$= -\tan 40°$

Second quadrant.

2 $\cos 128°18'$

S $\cos 128°18' = -\cos(180° - 128°18')$
$= -\cos 51°42'$

Second quadrant.

3 Find the size of the obtuse angle θ, given that $\cos\theta = -0.025$.

S $\cos\theta = -0.025$
$\theta = 91°26'$

Triangles without right-angles

Trigonometry can be used in all triangles regardless of whether they contain a right-angle. There are two methods, and each requires a formula.

The Sine Rule (rule of opposites)

Use the Sine Rule when you know two angles and one side, or when you know two sides and an angle opposite one of the known sides.

$$\frac{a}{\sin A} = \frac{b}{\sin B} = \frac{c}{\sin C}$$

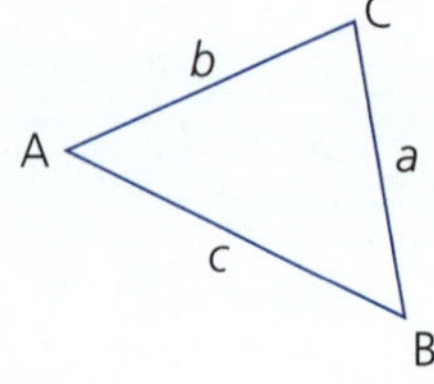

Note the opposites in the formula:

$$\frac{\text{side}}{\sin\,[\text{opposite angle}]} = \frac{\text{side}}{\sin\,[\text{opposite angle}]}$$

1 Find x, correct to three significant figures.

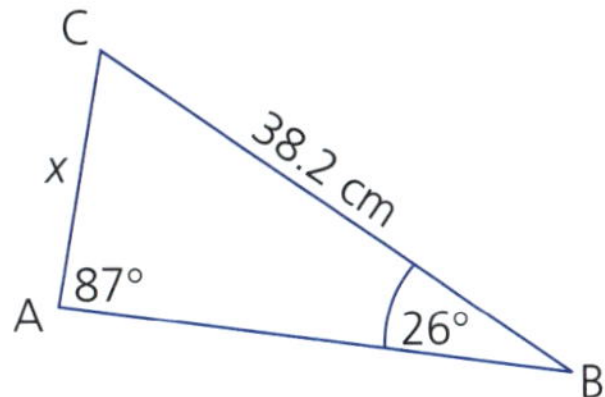

S

$$\frac{a^{\checkmark}}{\sin A^{\checkmark}} = \frac{b^{\checkmark}}{\sin B^{\checkmark}} = \frac{c}{\sin C}$$

We have $\hat{A} = 87°$, $\hat{B} = 26°$ and $a = 38.2$.

We want b.

$$\frac{38.2}{\sin 87\Upsilon} = \frac{x}{\sin 26\Upsilon}$$

Therefore $x = \frac{38.2 \sin 26\Upsilon}{\sin 87\Upsilon}$

$= 16.768\,759$

$= 16.8$ (to three significant figures)

2 Find θ to the nearest minute.

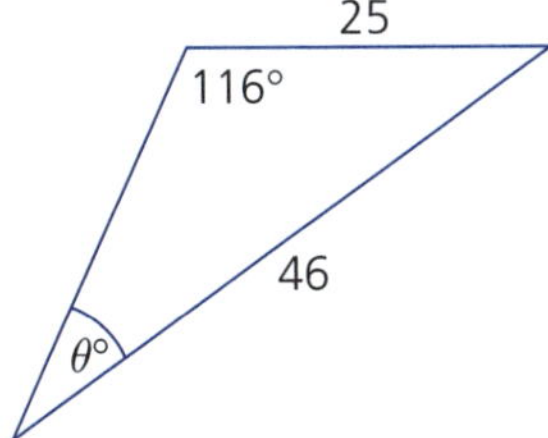

S

$$\frac{25}{\sin\theta} = \frac{46}{\sin 116\Upsilon} = \frac{\text{Side}}{\sin\text{(Opp)}}$$

$\therefore 46 \sin\theta = 25 \sin 116°$

$\sin\theta = \frac{25 \sin 116\Upsilon}{46}$

$\therefore \theta = 29°14'$

The ambiguous case

Note: the required angle in Example **2**, previously, could also be (180° – 29°14′), that is, 150°46′, but this would mean the angle sum of the triangle is greater than 180°. As both acute and obtuse angles have sine positive, an obtuse answer is possible. When a decision cannot be made between the acute result and the obtuse result, it is known as the ambiguous case.

3 Find θ to the nearest minute.

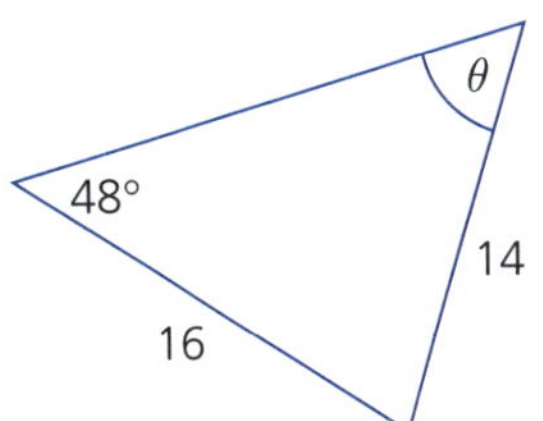

S

$$\frac{16}{\sin\theta} = \frac{14}{\sin 48\Upsilon}$$

$14 \sin\theta = 16 \sin 48°$

$\sin\theta = \frac{16 \sin 48\Upsilon}{14}$

$\theta = 58°8'$

But also, $\theta = (180° - 58°8')$

$= 121°52'$

Either answer is possible. The diagram below shows the two possible triangles with sides 14, 16 and angle 68°:

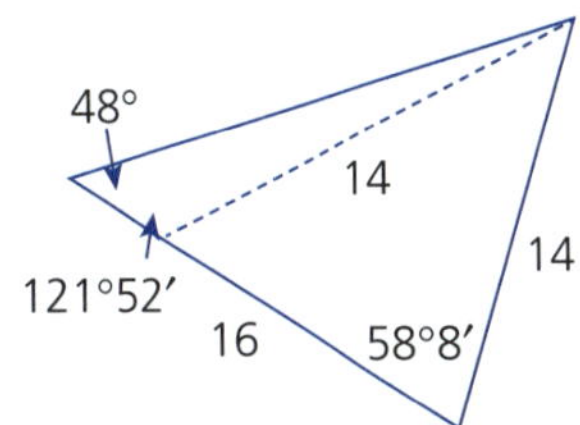

Check with the original information—the obtuse value may not be possible. Remember that the angle sum of the triangle is 180°.

The Cosine Rule—finding a side

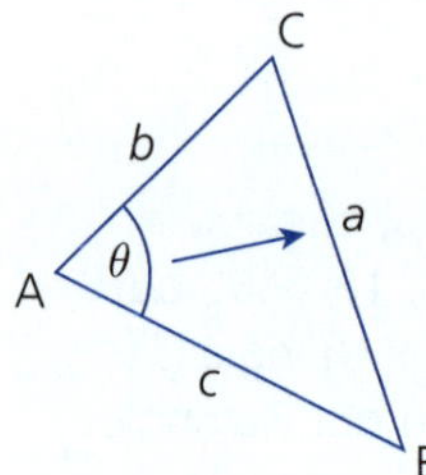

The Cosine Rule is used when the question involves two sides and the included angle.

$$a^2 = b^2 + c^2 - 2bc\cos A$$

Sides around angle A — Angle opposite the side being found

Similar formulae can be written for b^2 and c^2.

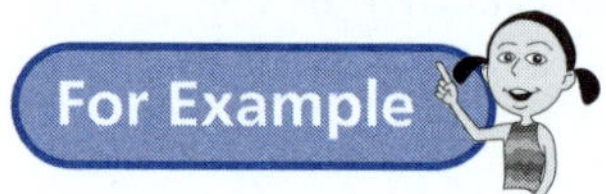

1 Find the length of BC to the nearest cm.

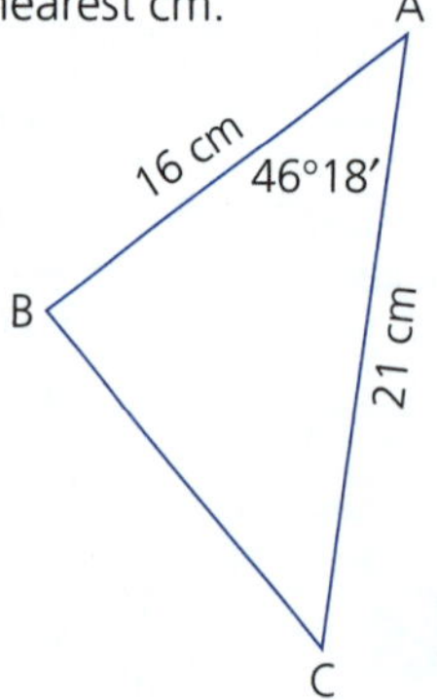

S

$$\begin{aligned} a^2 &= b^2 + c^2 - 2bc\cos A \\ &= 21^2 + 16^2 - 2 \times 21 \times 16 \times \cos 46°18' \\ &= 232.727\,02 \\ a &= \sqrt{232.727\,02} \\ &= 15.255\,393 \\ &\approx 15 \end{aligned}$$

BC is 15 cm.

The Cosine Rule—to find an angle

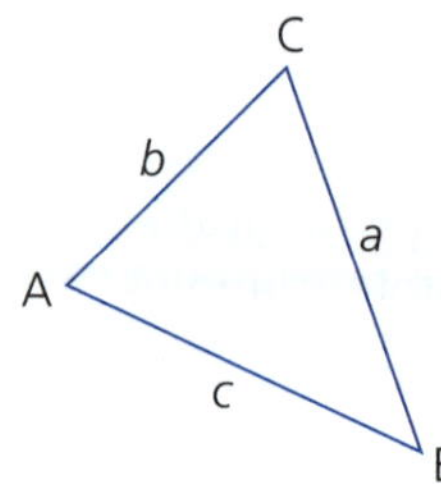

The question must contain values for all three sides to be able to use the angle form of the Cosine Rule.

$$\cos A = \frac{b^2 + c^2 - a^2}{2bc}$$

Similar forms can be written for $\cos B$ and $\cos C$. Note that the side subtracted is the one opposite the required angle.

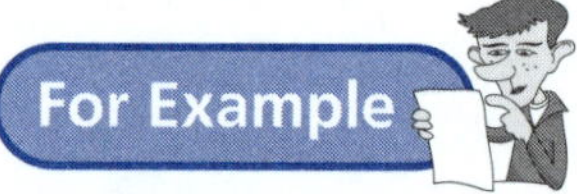

1 In the $\triangle ABC$, $a = 12$ cm, $b = 17$ cm and $c = 9$ cm. Find the size of the largest angle.

The largest angle is always opposite the longest side. Therefore the angle required is B.

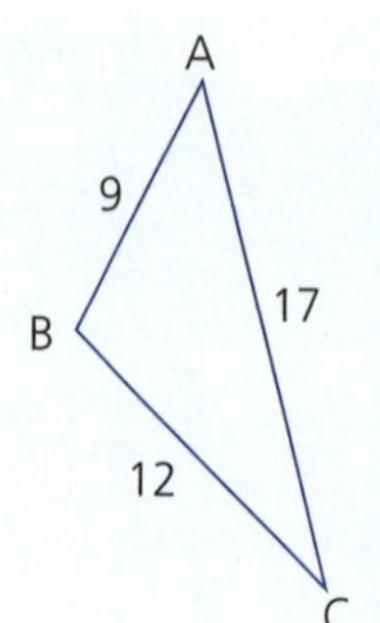

$$\begin{aligned} \cos B &= \frac{a^2 + c^2 - b^2}{2ac} \\ &= \frac{12^2 + 9^2 - 17^2}{2 \times 12 \times 9} \\ &= \frac{(12^2 + 9^2 - 17^2)}{(2 \times 12 \times 9)} \\ &= -0.296\,2963 \\ \therefore B &= 107°14' \end{aligned}$$

Use brackets around top and bottom.

Miscellaneous worked examples

You can now apply the formulae to real-world problems.

1 Two cannons X and Y, 100 metres apart, fire at a target Z. If $\angle ZXY = 38°$ and $\angle ZYX = 64°19'$, find the distance of X from the target Z (to one decimal place).

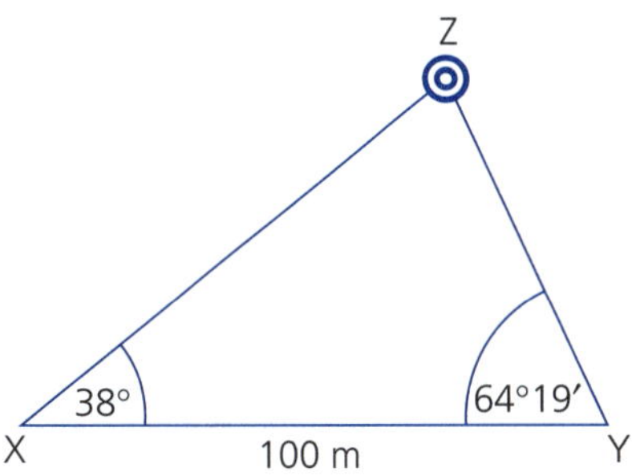

The question asks us to find the length of XZ. We have two angles and a side—use the Sine Rule. *But*, the Sine Rule needs the angles opposite the sides. We must calculate the size of $\hat{Z}$.

$$\hat{Z} = 180° - (64°19' + 38°) = 77°41'$$

Angle sum of a △.

Then

$\dfrac{\text{side}}{\sin\,[\text{opp. angle}]}$

$$\frac{XZ}{\sin 64°19'} = \frac{100}{\sin 77°41'}$$

$$\therefore XZ = \frac{100 \sin 64°19'}{\sin 77°41'} = 92.243\,428 = 92.2 \text{ (to one decimal place)}$$

The distance of X from the target is 92.2 m.

2 A ship (S) leaves port (P) and travels 50 km on a bearing of 058°, while a yacht (Y) also leaves (P) and travels 34 km on a bearing of 124°.

a Find the distance between the ship and the yacht.

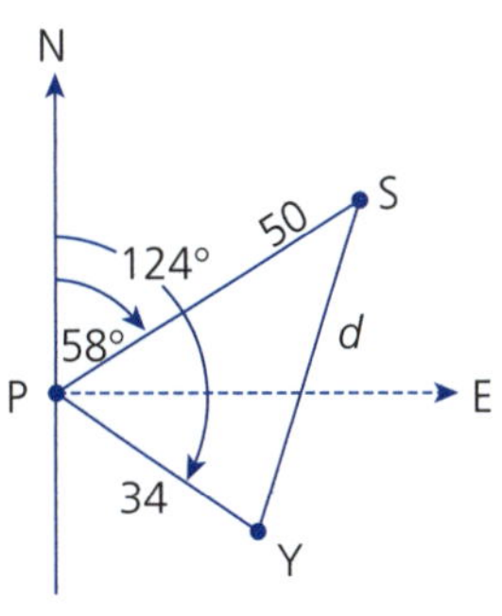

Let the distance SY be d km. (Use the Cosine Rule, as we know two sides and the included angle.)

$\angle SPY = 124° - 58° = 66°$

Then

$$d^2 = 50^2 + 34^2 - 2(50)(34)\cos 66° = 2273.0954$$

$$\therefore d = \sqrt{2273.0954} = 47.676\,99 = 47.7 \text{ (to one decimal place)}$$

The distance between the boats is 47.7 km.

b Find the bearing of the yacht as measured from the ship to the nearest degree.

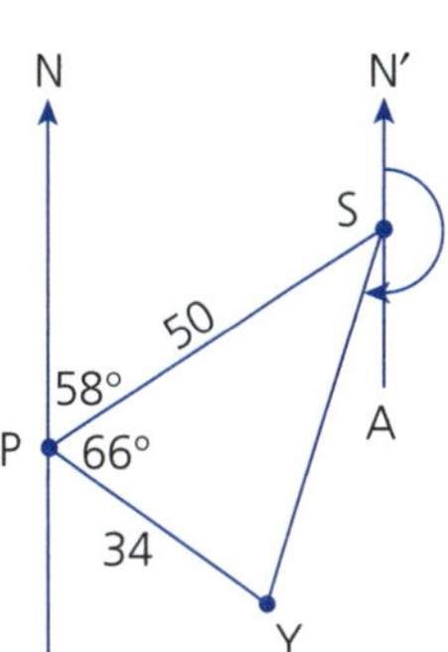

The bearing of Y from S is $\angle N'SY$. This can be calculated by subtracting $\angle N'SP$ and $\angle PSY$ from 360°.

Note $\angle N'SY$ is reflex; clockwise direction.

Now $\angle N'SP = 180° - 58° = 122°$

(co-interior angles, NP || N′A)

Use the Cosine Rule to calculate $\angle PSY$ as we have three sides of the triangle.

$$\cos\angle PSY = \frac{(50^2 + 47.7^2 - 34^2)}{(2 \times 50 \times 47.7)}$$

$$\angle PSY = 40°39'$$

$$\text{Bearing of Y} = 360° - (122° + 40°39') = 197°21' = 197° \text{ (nearest degree)}$$

The bearing of Y is 197°.

3 From a point, A, level with the base of a hill, the angle of elevation of the top of the hill, H, is measured as 15°. The angle of elevation from A to the top of the tower, T, built on the hill is 21°. If A is 1000 m in a direct line from the base (B) of the hill, calculate the length AH and use this information to calculate the height of the tower (TH) correct to one decimal place.

S

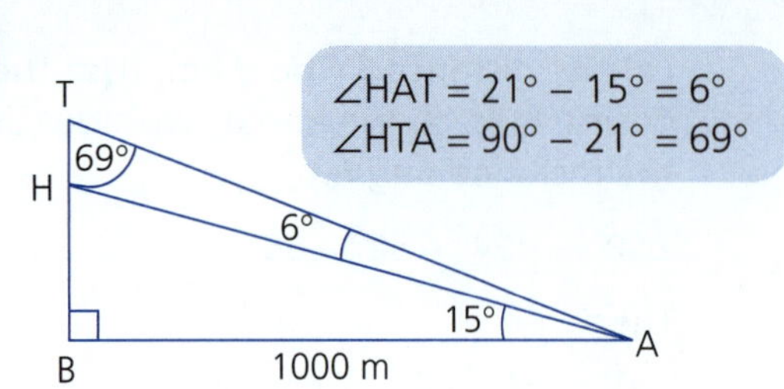

$\angle HAT = 21° - 15° = 6°$
$\angle HTA = 90° - 21° = 69°$

In $\triangle ABH$,

$$\frac{AB}{AH} = \cos 15°$$

$$\therefore AH = \frac{1000}{\cos 15\Upsilon}$$
$$= 1035.2762$$

This length is then used as a link to $\triangle AHT$.

In $\triangle AHT$,

$$\frac{TH}{\sin 6°} = \frac{AH}{\sin 69\Upsilon}$$

$$TH = \frac{AH \sin 6\Upsilon}{\sin 69\Upsilon}$$
$$= \frac{1035.2762 \times \sin 6°}{\sin 69°}$$
$$= 115.91484$$
$$= 115.9 \text{ (to one decimal place)}$$

The height of the tower is 115.9 metres.

The area of a triangle

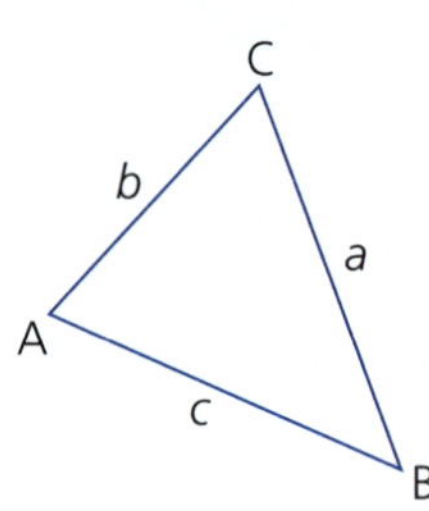

$$\text{Area} = \frac{1}{2}ab \sin C$$

Two sides and included angle.

A similar formula can be written for any combination of two sides and the angle between them.

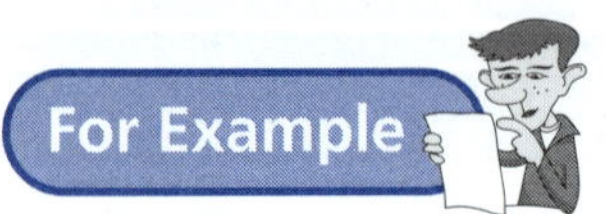

1 Find the area of the triangle ABC where $a = 14$ cm, $c = 15$ cm and $\angle B = 48°$.

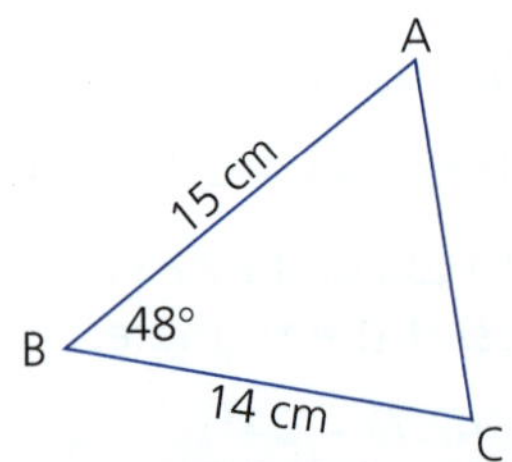

S

$$\text{Area} = \frac{1}{2}ac \sin B$$

Equivalent formula using Angle B.

$$= \frac{1}{2} \times 14 \times 15 \times \sin 48°$$
$$= 78.030207$$
$$= 78 \text{ (to the nearest whole number)}$$

The area is 78 cm^2.

Trigonometry and the number plane

The angle between any straight line and the positive (→) direction of the x-axis, $\alpha°$ in these diagrams, can be calculated from the gradient of the line.

$\tan \alpha° = m$

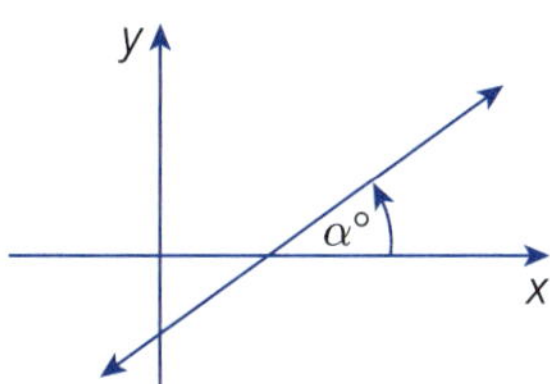

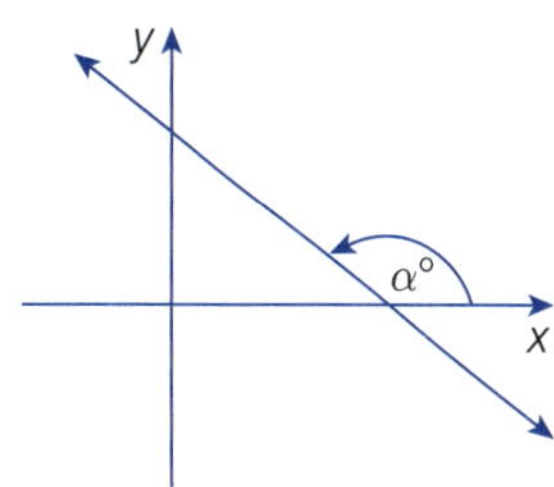

1 Find the angle between $y = 2x - 5$ and the positive direction of the x-axis.

S

$$m = 2 \text{ (compare with } y = mx + b)$$
$$\therefore \tan \alpha° = 2$$
$$\therefore \alpha° = 63°26'$$

2 Find the angle between $2x + 3y = 5$ and the positive direction of the x-axis.

S

$$2x + 3y = 5$$
$$3y = -2x + 5$$
$$y = \frac{-2}{3}x + \frac{5}{3}$$

$$\therefore \tan \alpha° = \frac{-2}{3}$$

(compare with $y = mx + b$)

Now, $\tan \alpha° = \frac{2}{3}$

$\therefore \alpha° = 33°41'$

But $\tan \alpha° < 0$, $\therefore$ obtuse angle

As $180° - 33°41' = 146°19'$

Then $\alpha° = 146°19'$

3 Find the equation of the line passing through the point (3, –2) and making an angle of 135° with the positive direction of the x-axis.

S

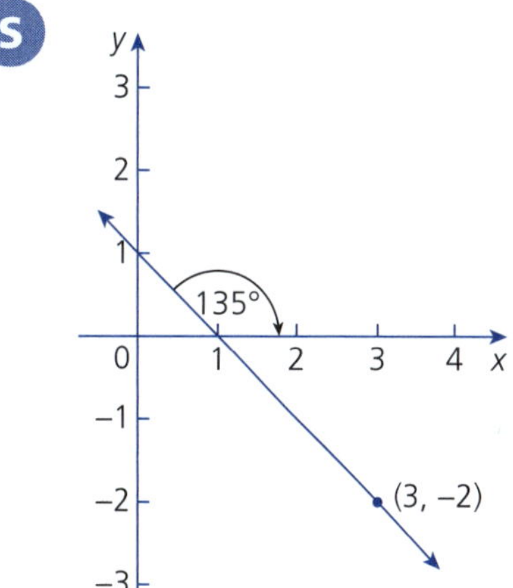

As $m = \tan \alpha$

$\therefore m = \tan 135°$

$m = -1$

$\therefore$ Using $y - y_1 = m(x - x_1)$

$$y + 2 = -1(x - 3)$$
$$y + 2 = -x + 3$$
$$x + y - 1 = 0$$

$\therefore$ The equation of the line is $x + y - 1 = 0$.

Practise Practise

1 Calculate the exact length of the marked side:

a

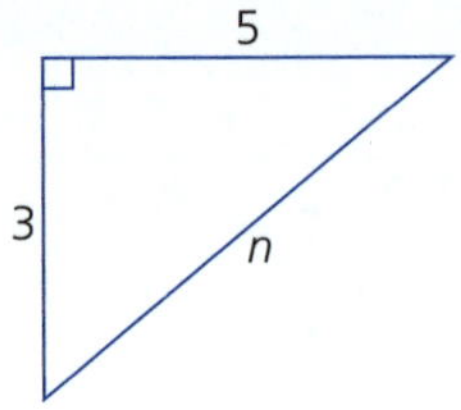

b

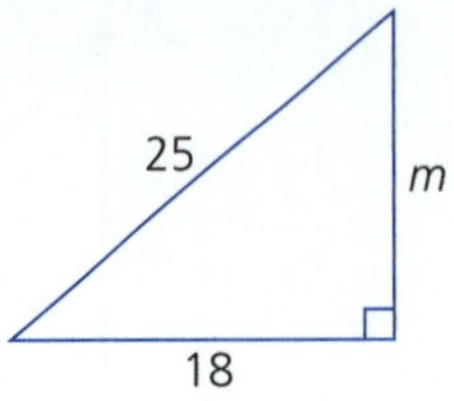

c

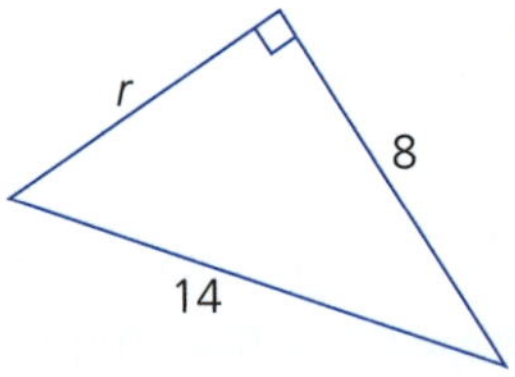

d

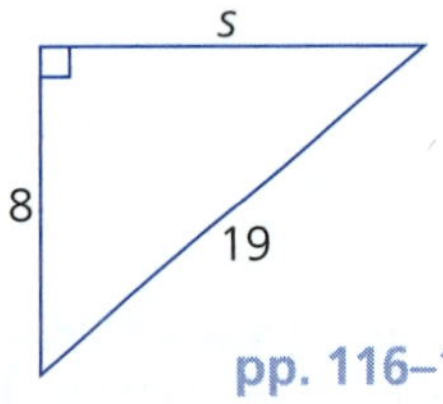

pp. 116–117

2 A rectangle is 16 cm long and 20 cm wide. Calculate the length of the diagonal of the rectangle. pp. 116–117

3 Calculate the length of a diagonal of a square with sides of 3 cm. Answer correct to one decimal place. pp. 116–117

4 A metal brace 35 cm long is screwed into place so that one end is 21 cm from a wall. How far up the wall does the brace reach?

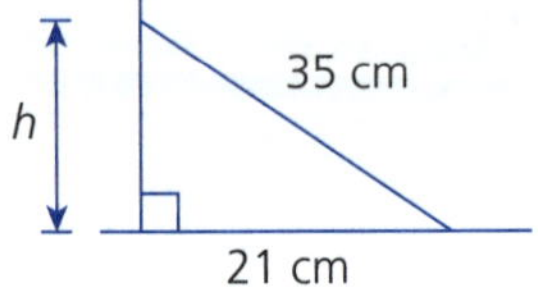

pp. 116–117

5 An 8-metre ladder reaches 7.5 metres up a vertical brick wall. How far is the foot of the ladder from the wall? Answer correct to one decimal place.

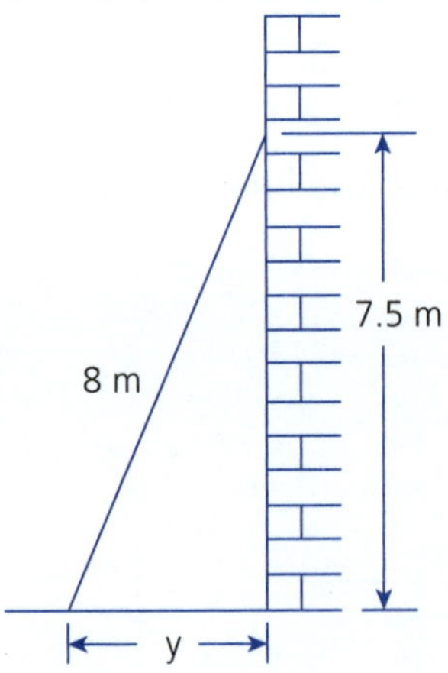

pp. 116–117

6 A pollution tower 45 metres high is supported by wire 40 metres long. Each wire is attached at ground level 21 metres from the base of the tower. How far from the top of the tower are the wires attached? Answer correct to four significant figures.

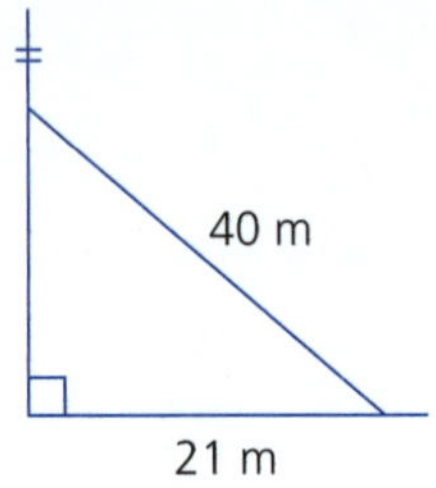

pp. 116–117

7 The square MNOP has diagonal of length 18 cm. Calculate the length of a side of the square. Answer correct to one decimal place.

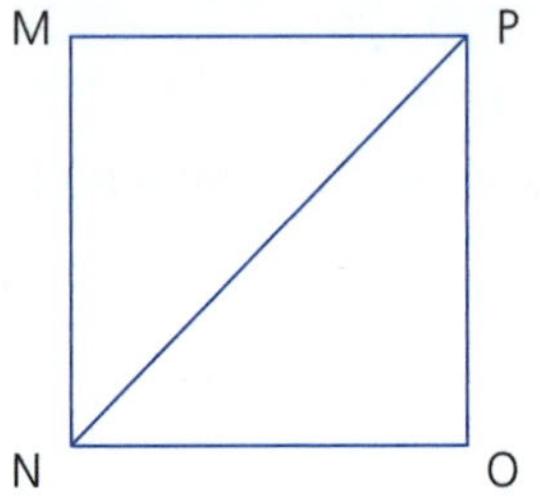

pp. 116–117

8 Determine whether the following triangles are right-angled:

a AB = 8 cm, BC = 15 cm, AC = 17 cm
b PQ = 12 cm, QR = 15 cm, PR = 20 cm
c XY = 401 cm, XZ = 40 cm, YZ = 399 cm p. 117

9 Complete the Pythagorean triads:

a $\{7, 24, a\}$ **b** $\{9, 12, y\}$
c $\{18, n, 30\}$ **d** $\{9, t, 41\}$
e $\{13, N, 85\}$ **f** $\{21, 72, p\}$ p. 118

10 **a** If $\sin 70° = \cos x°$, find x.
b If $\cos 5° = \sin y°$, find y. pp. 118–120

11 Simplify $\frac{\sin 10\Upsilon}{\cos 80\Upsilon}$. pp. 118–120

12 Find y from the following figures (correct to one decimal place):

a

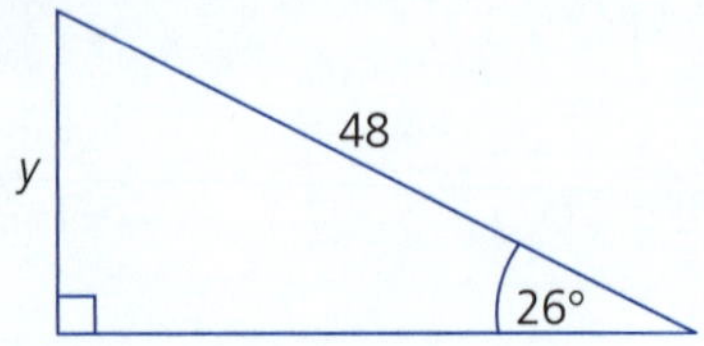

b

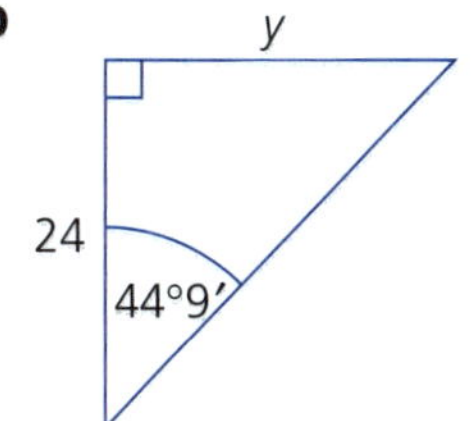

c

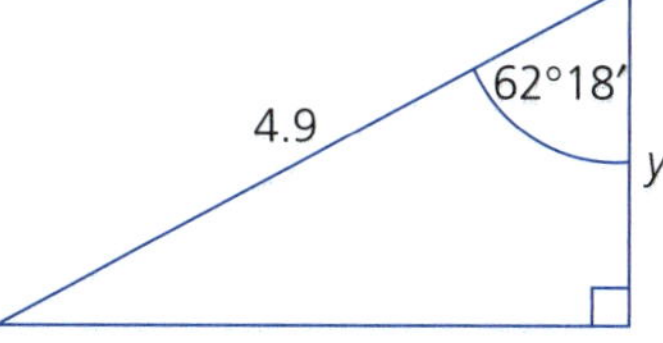

d

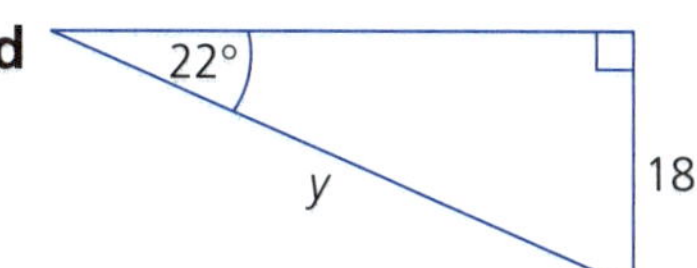

e

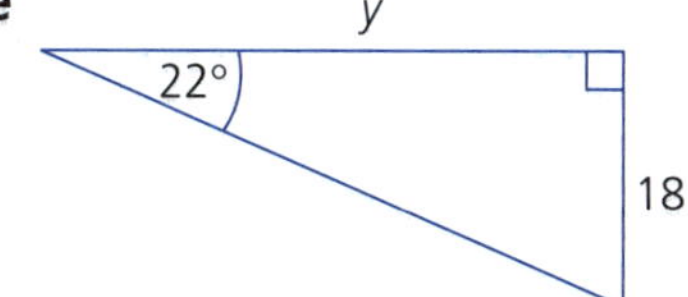

f

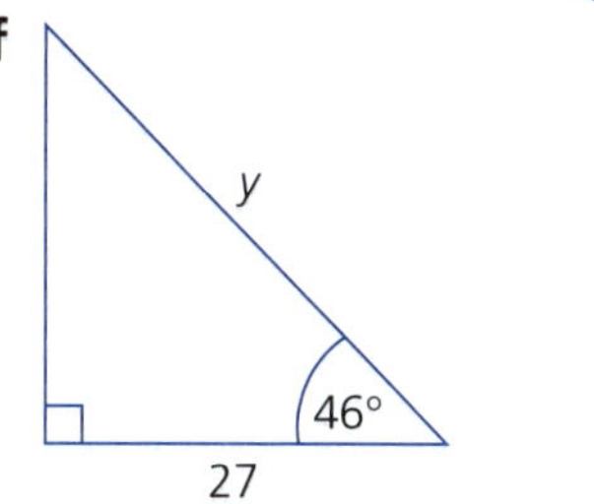

g

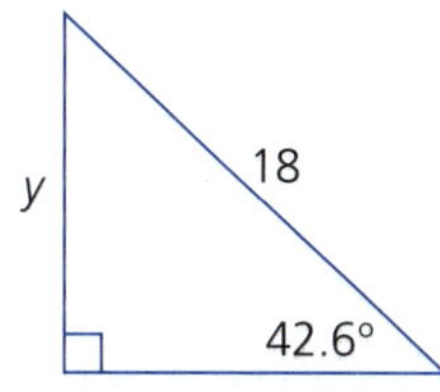

h

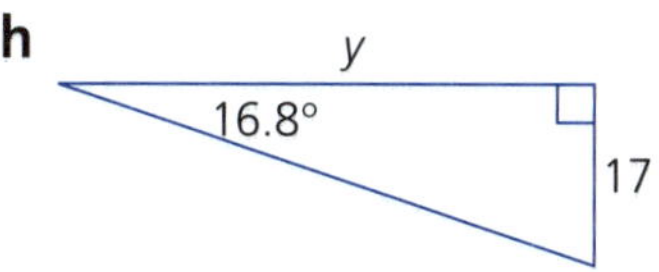

pp. 120–121

13 Find the size of the marked angle (answer correct to the nearest minute):

a

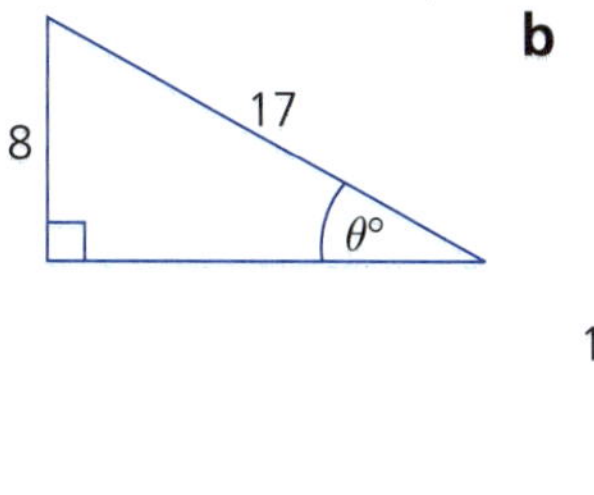

b

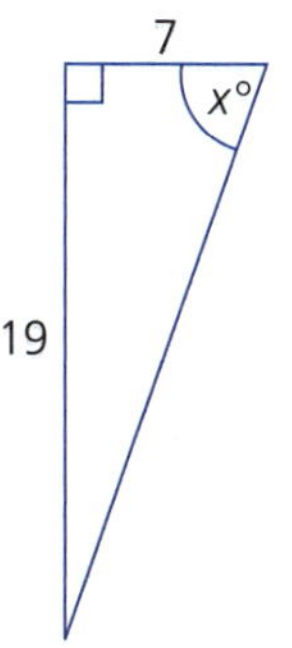

c

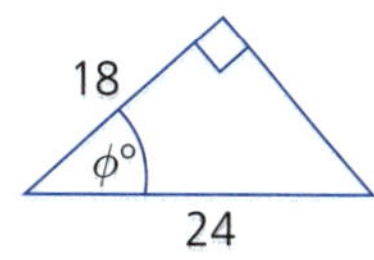

d

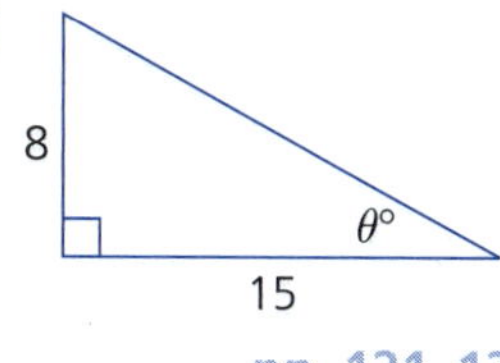

pp. 121–122

14 **a** If $\tan\theta = \frac{4}{7}$, find θ to the nearest degree.

b If $\cos\theta = \frac{4}{7}$, find θ to the nearest degree.

c If $\sin f = \frac{4}{7}$, find f correct to the nearest tenth of a degree. pp. 121–122

15 Given $\cos\theta = \frac{8}{10}$, derive expressions for $\sin\theta$ and $\tan\theta$.

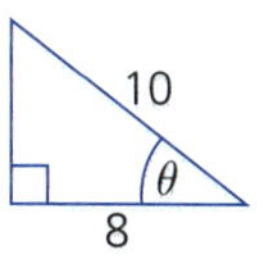

pp. 121–122

16 Given that $\tan\theta = \frac{3}{4}$, derive expressions for $\sin\theta$ and $\cos\theta$. pp. 121–122

17 Captain Hook notes from the bow of his ship that the angle of depression of a whale is 18°. If the bow of the ship is 3.2 m above the water level, calculate the distance of the whale from the ship (answer to one decimal place).

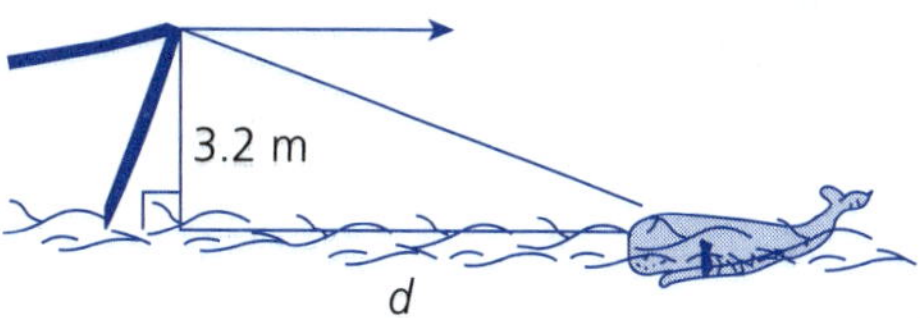

pp. 120–122

18 Billy Blogs measures the angle of elevation of a 35 m building from a position 100 m from the base of the building. Find the angle of elevation, correct to the nearest minute.

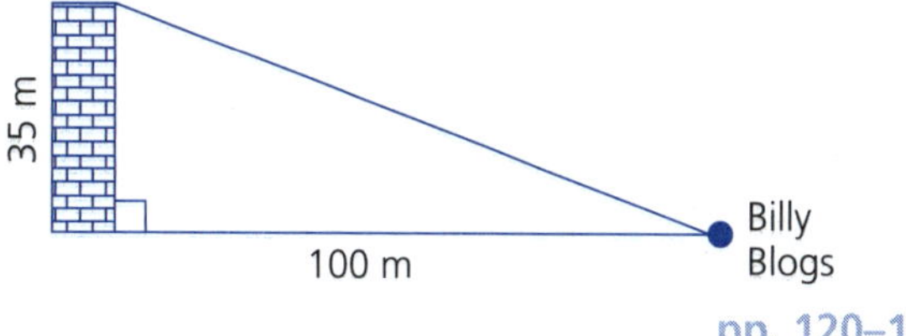

pp. 120–123

19

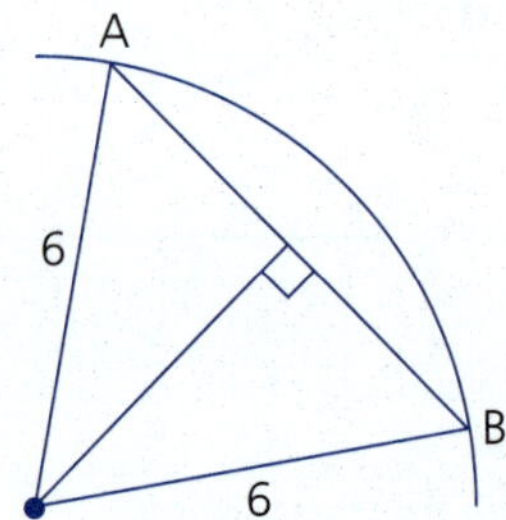

A chord of a circle of radius 6 m, subtends an angle of 84° at the centre. Find the length of the chord correct to one decimal place. pp. 120–122

20 Two sails on a yacht are cut so that AD = 2.8 m, ∠DAB = 48°, BE = 0.9 m, and DC = 3.7 m. Calculate the length BD from △ADB and hence calculate ∠DCE to the nearest minute.

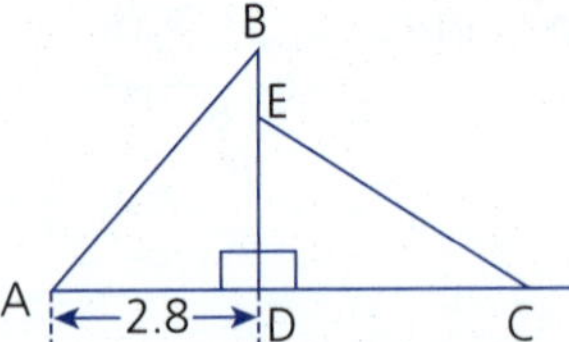

pp. 120–122

21 The rugby league goalposts at a stadium are 3.5 m high. Peter observed that the shadow cast by the posts was 5 m long. Calculate the angle of elevation of the sun (to the nearest degree).

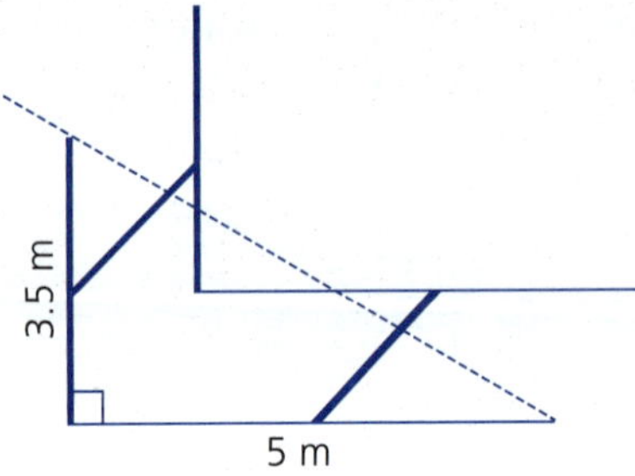

pp. 120–123

22 Marko and Greg set up garbage cans 4 m apart for soccer posts. If Marko is directly in line with one post, for what range of angles will the soccer ball pass between the posts if he is:

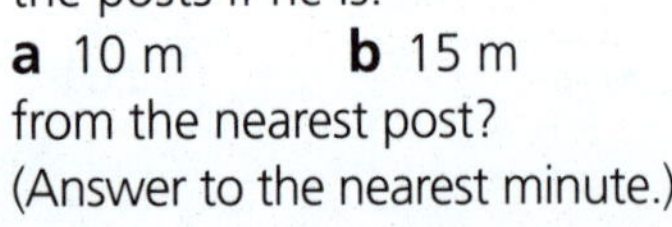

a 10 m **b** 15 m

from the nearest post?

(Answer to the nearest minute.)

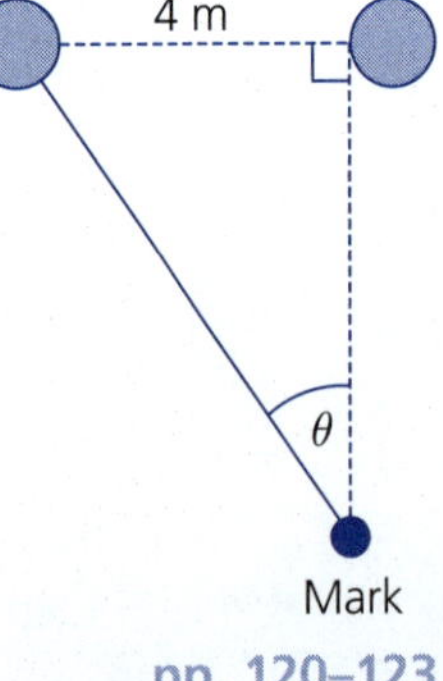

pp. 120–123

23 A large black bear ambles 8 km south and then 14 km east. She then calculates the bearing of her starting point from her new position. Find this bearing to the nearest minute.

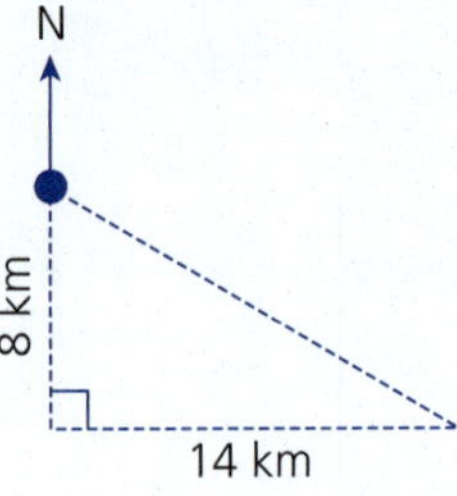

pp. 123–124

24 A metal support PQ holds up a brick wall AB. If the foot of the support is 5.2 m from the wall and reaches 7.8 m up the wall, calculate the:

a length of the support

b angle the support makes with the ground.

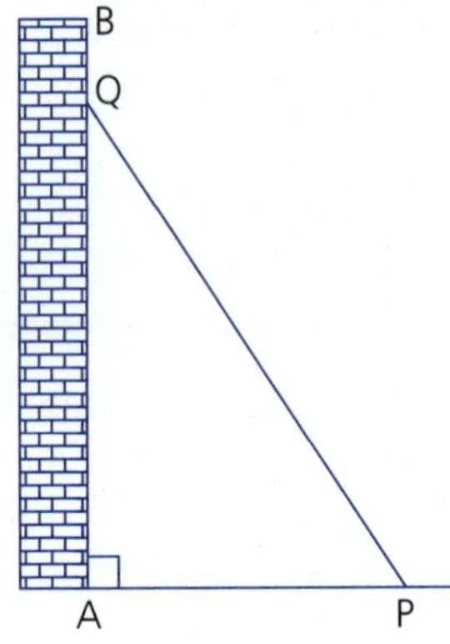

pp. 120–123

25 If $\cos\phi = \frac{1}{2}$, find the *exact* values of $\sin\phi$ and $\tan\phi$. From any of these ratios, find ϕ. p. 125

26 If $\tan\alpha = 1$, find *exact* values for $\sin\alpha$ and $\cos\alpha$. From one of these ratios, calculate the value of α. p. 125

27 On separate number planes, draw the graphs of:

a $y = \sin\theta$ for $0° \le \theta \le 360°$

b $y = \cos\theta$ for $-180° \le \theta \le 180°$

c $y = \tan\theta$ for $0° \le \theta \le 180°$ p. 127

28 Sketch the following and write down the period:

a $y = 3\sin\theta$ for $0° \le \theta \le 360°$

b $y = \sin 2\theta$ for $0° \le \theta \le 360°$

c $y = \sin\frac{\theta}{2}$ for $0° \le \theta \le 360°$ p. 127

29

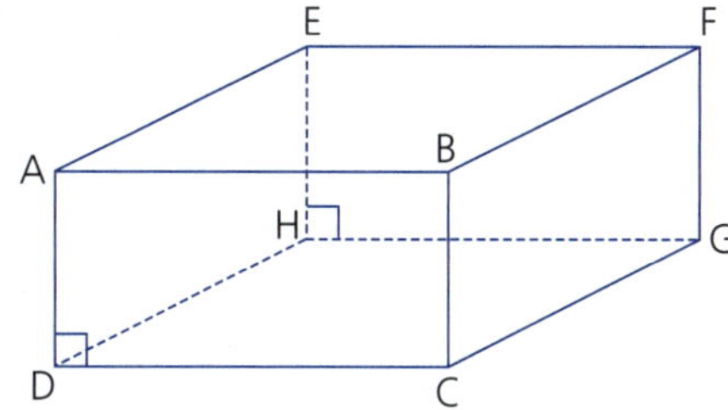

The diagram shows a rectangular prism, where AB = 8 cm, BF = 6 cm, BC = 5 cm. Find the angle of elevation of F from D. p. 125

30 Find the size of the obtuse angle θ, given that:

a $\cos\theta = -0.3165$

b $\tan\theta = -1.2851$

c $\sin\theta = 0.3896$ p. 128

31 Find the length f in the following diagrams (to one decimal place):

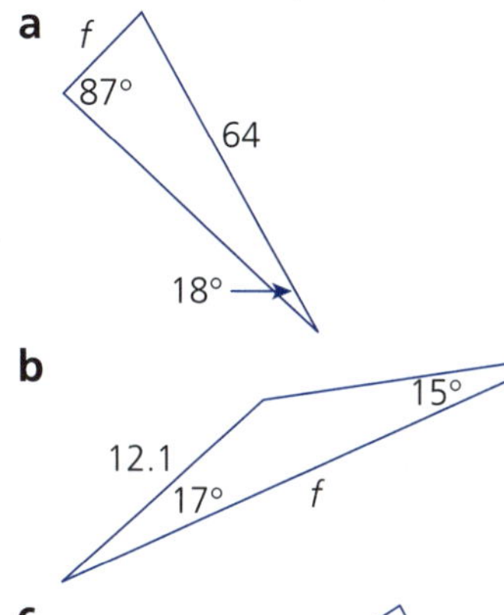

c

86°27′

f

32°16′

40.5

pp. 128–129

32 A vertical post 12 m high stands on a uniform slope of 12°18′. Find the length of the shadow of the pole when the angle of elevation of the sun is 54°34′ and the shadow is being cast directly down the hill.

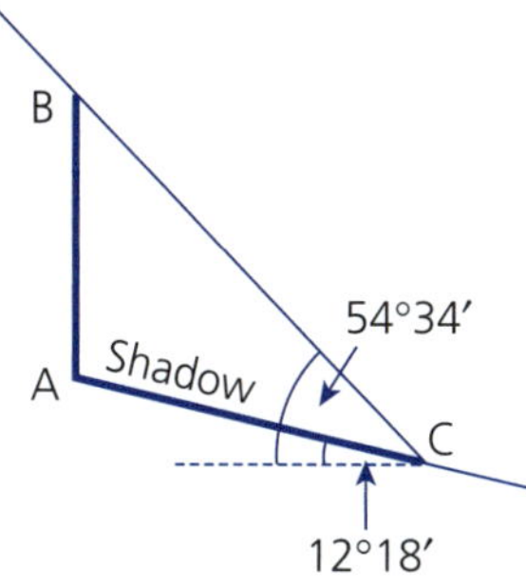

pp. 128–129

Hint: use the slope of the rise to find ∠BAC.

33

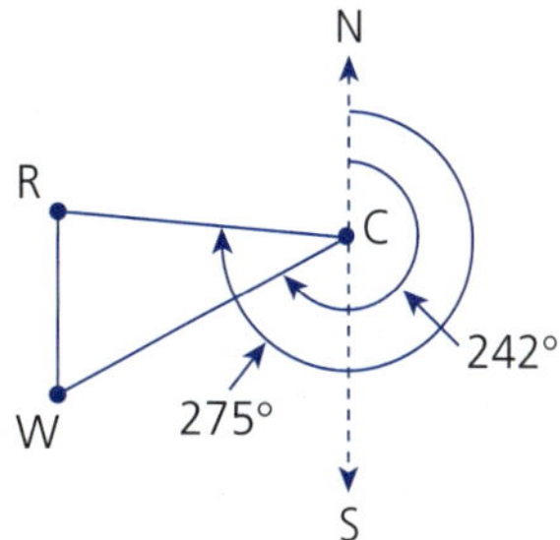

A cabin cruiser in difficulty observes that two ports, Redhead (R) and Wallsend (W), have bearings 275° and 242° respectively. Redhead is 25 km due north of Wallsend. How far must the ship travel to the nearest port? pp. 128–129

34 **a** Find the length t in the following diagrams (answer to one decimal place):

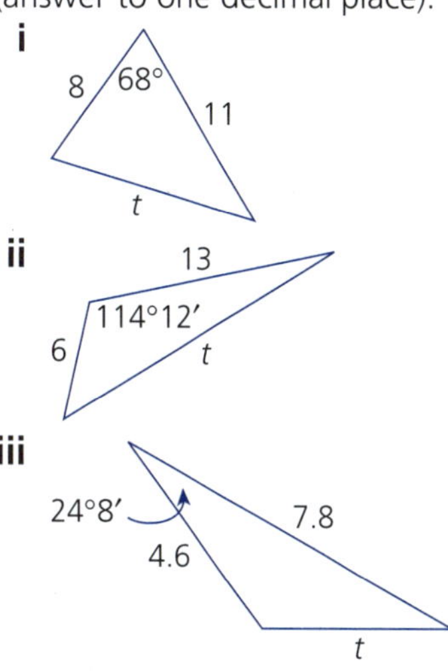

b Find the area of each triangle, correct to one decimal place. p. 130

35 Calculate the size of the following marked angles to the nearest minute:

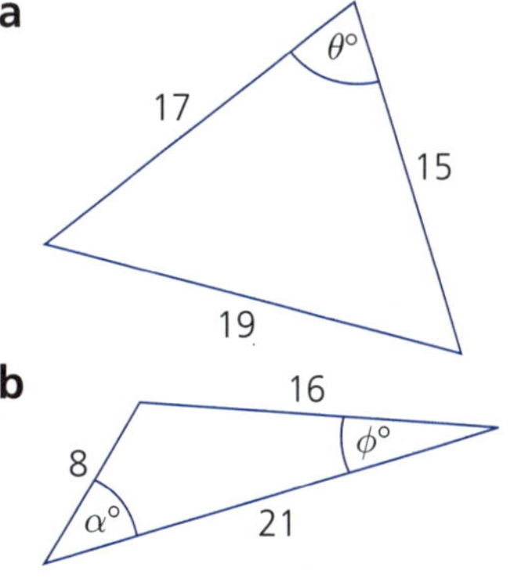

p. 130

36 Maria cycles 20 km due south from O and then 13 km on a bearing of 035°. How far is she from her starting point?

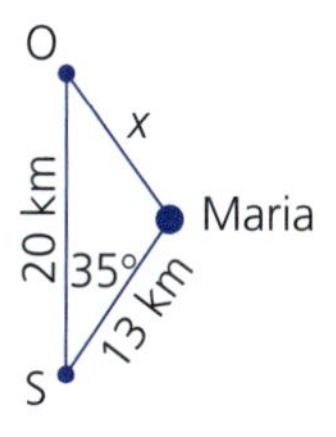

pp. 131–132

37 Find the size of the largest and smallest angles in a triangle with sides 18 cm, 23 cm and 8 cm. Hence find the area of this triangle (to the nearest 0.1 cm^2). pp. 131–132

38 Find the angles of a parallelogram and the length of the other diagonal given that it has adjacent sides 30 cm and 40 cm with one diagonal 35 cm. pp. 131–132

39

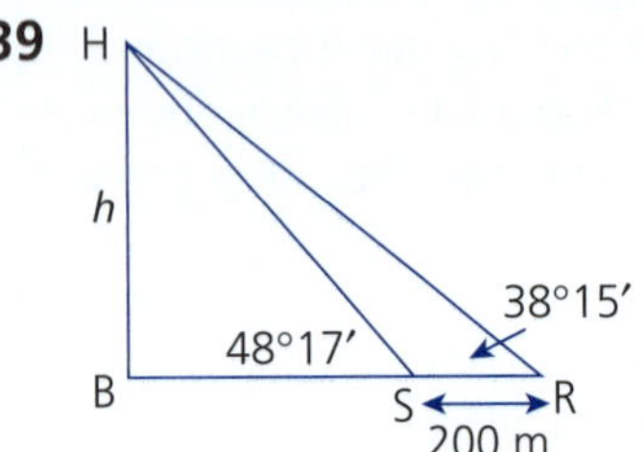

From two points R and S on level ground, 200 m apart and in a direct line with the base of a hill, Pierre observes the angles of elevation of the top of the hill to be 38°15′ and 48°17′ respectively. Find the height h of the hill. pp. 131–132

40

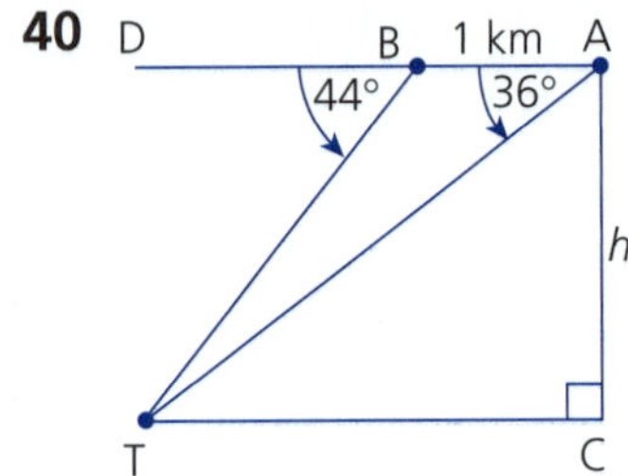

Biggles flying at a constant height on a bombing run observes the angle of depression of his target (T) to be 36°. After flying a further 1 km directly towards the target and at the same altitude, the angle of depression is noted as 44°. Find the altitude h of the plane in km, correct to two significant figures. pp. 122–132

41 From the top of a tall building 160 m high, the angles of depression of the top and the base of another building are 30° and 56° respectively. Find the height h of the second building. pp. 123–132

42

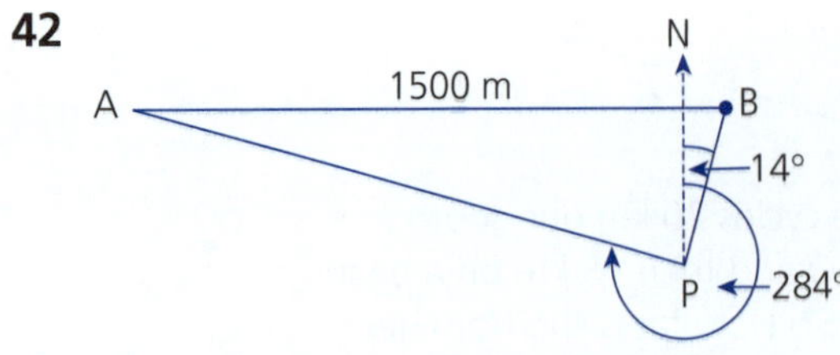

Minmi Breakwater runs east–west for 1500 m. Observers from a prawn trawler note that the bearing of one end of the breakwater is 014°, while that of the other is 284°. Calculate the distance from the prawn trawler to both ends of the breakwater. pp. 123–132

43 Two straight roads PQ and PR are inclined to each other at 58°30′. Two bike riders start simultaneously from P and travel along the roads at 18 and 24 km/h respectively. How long is it before they are 80 km apart? (Answer to the nearest minute.) pp. 123–132

44 A balloon is observed simultaneously from two points, X and Y, on opposite sides of the balloon. From X, the angle of elevation of the balloon is 62°18′, while from Y, it is 68°12′. If X and Y are 1200 m apart, find the height of the balloon.

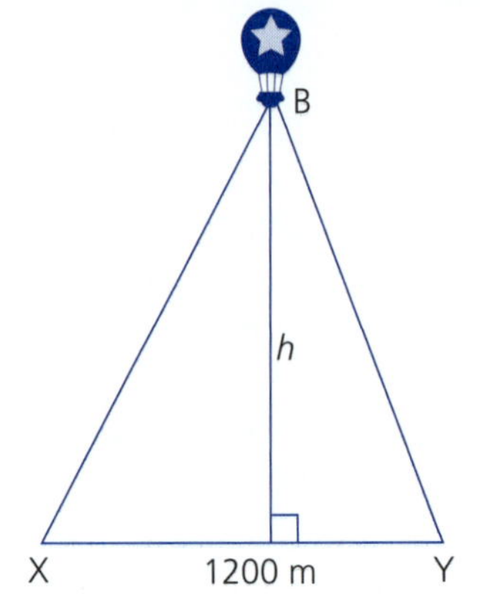

pp. 122–132

45

T
S
A
Monika
B
Mabel

Monika and Mabel observe the Eyefull Tower. Monica is due south and Mabel due east of the tower. They note that the angles of elevation of the tower are 48° and 36° respectively. If the tower is 120 m high, calculate the distance between Monika and Mabel (to the nearest metre). pp. 122–132

46 Find the size of the angle between the given straight line and the positive direction of the x-axis:

a $y = \frac{-1}{2}x + 1$

b $3y = x + 4$

c $3x - 2y = 5$ p. 133

Go to p. 301 for **Quick Answers** or to pp. 331–336 for **Worked Solutions**

Your ✓ Checklist

For a complete understanding of this topic, you must be able to:

✓	Use Pythagoras' Theorem to calculate the length of a side of any right-angled triangle		pp. 116–117
✓	Use Pythagoras' Theorem to prove that a triangle is right-angled		p. 117
✓	Use Pythagoras' Theorem to show that a set of three numbers is a Pythagorean triad		p. 118
✓	Find sides and angles in right-angled triangles		pp. 118–122
✓	Apply right-angle trigonometry to word problems (diagram or partial diagram sometimes provided)		pp. 120–123
✓	Apply trigonometry to bearing questions		pp. 123–124
✓	Use trigonometry to solve three-dimensional problems		p. 125
✓	Use exact values for ratios involving 30°, 45° and 60°		pp. 125–126
✓	Draw graphs of trigonometric functions		p. 127
✓	Work with ratios for obtuse angles		p. 128
✓	Apply Sine Rule and Cosine Rule to find sides and angles in triangles without right-angles.		pp. 128–131
✓	Find the area of non-right-angled triangles		p. 131
✓	Use $m = \tan\alpha$ to find the angle between a line and the positive direction of the x-axis.		p. 132

Now you are ready to do the tests!

Intermediate Test

Pythagoras and Trigonometry

(35 marks)

1 Find the value of the pronumeral, correct to two decimal places:

a (2 marks)

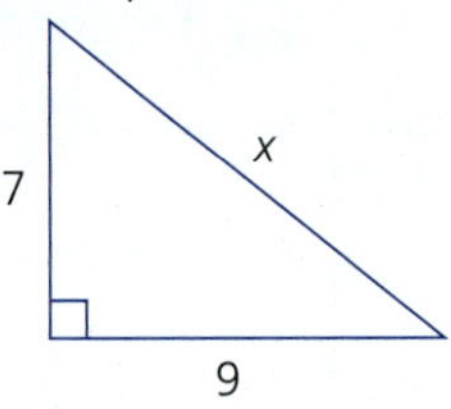

b (2 marks)

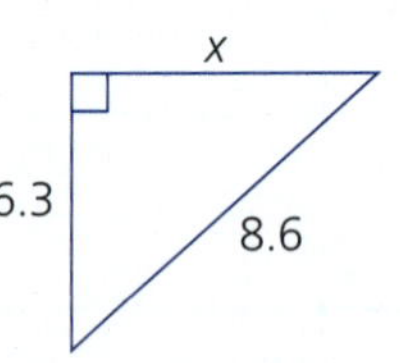

2 Determine whether the triangle below is right-angled: (2 marks)

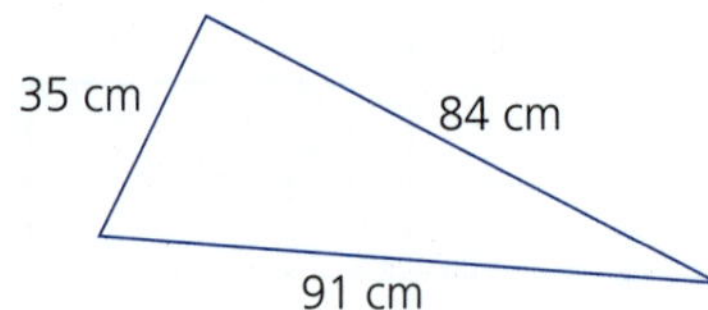

3

5
θ
12

Write down a ratio for $\tan\theta$ and $\sin\theta$. (3 marks)

4 If $\sin 70° = \cos x°$, then $x = ?$ (1 mark)

5 Find y, if $\cos(2y + 1)° = \sin 65°$. (2 marks)

6 Find the value of m, correct to one decimal place:

a (2 marks)

m
42°14′
18

b (2 marks)

27
m
39°15′

c (2 marks)

56°
m
16.7

7 Calculate the size of the marked angle, correct to the nearest minute:

a (2 marks)

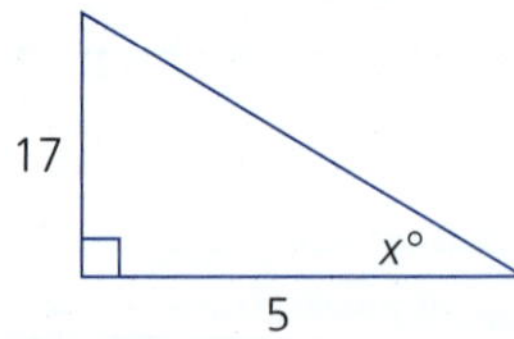

b (2 marks)

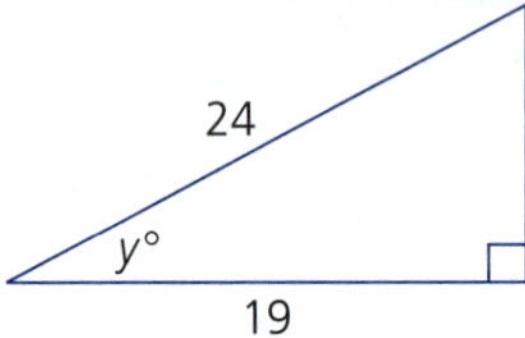

8 A 5.2-metre ladder rests against a wall, making an angle of 18°15′ with the wall. How far up the wall will the ladder reach? (2 marks)

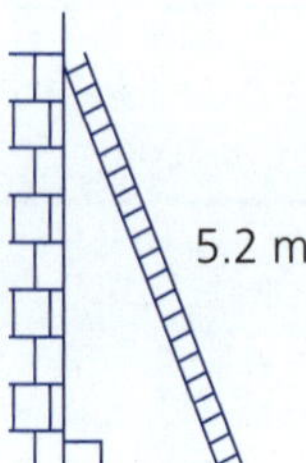

9 Find the length of the marked side in each diagram:

a

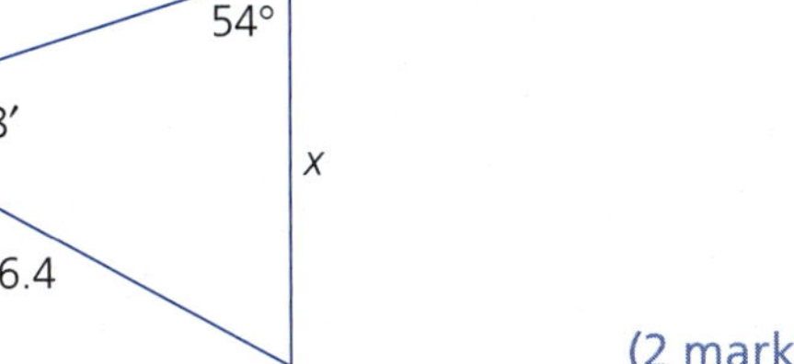

(2 marks)

b

8.7
y
87°18′
9.5

(2 marks)

10 Calculate the area of the triangle XYZ where XY = 17 cm, ZX = 15 cm, ∠ZXY = 16°40′.

(2 marks)

11 Find the angle made by the line $y = \frac{3}{4}x - 5$ and the positive direction of the x-axis. (2 marks)

12

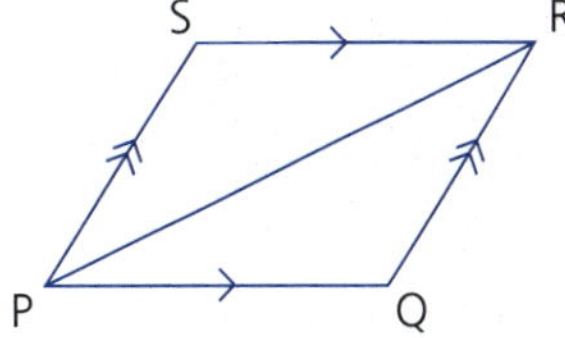

PQRS is a parallelogram with PQ = 14 cm, QR = 9.2 cm. The longer diagonal is 21.5 cm. Find the size of the angles of the parallelogram.

(3 marks)

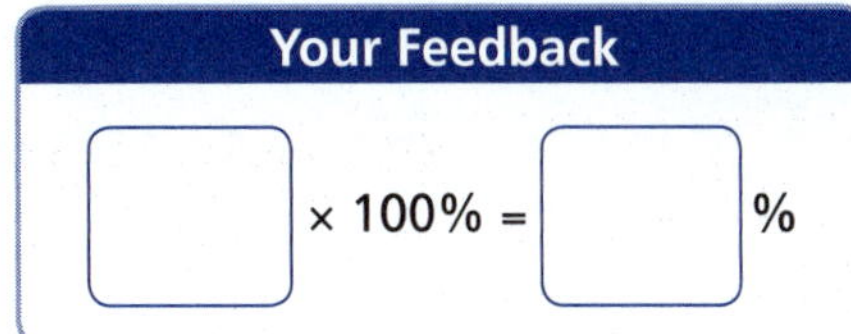

QA PAGE 306

WS PAGE 370

Advanced Test

Pythagoras and Trigonometry

(40 marks)

1 A sail is to be cut in the shape of a right-angled triangle. If the longest side is 25 metres and the shortest is 7 m, calculate the length of the third side. (2 marks)

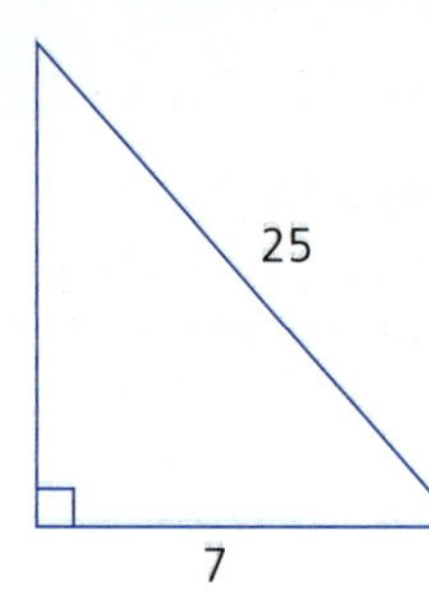

2 ABC is a triangle right-angled at B. A semicircle is drawn with AC as diameter. If AB = 12 cm and BC = 16 cm, calculate the area of the semicircle. (2 marks)

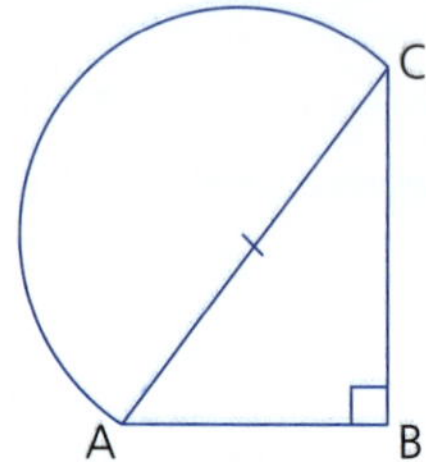

3 Draw the graph of $y = \sin(\theta + 90°)$, where $0° \le \theta \le 360°$. (2 marks)

4 Given that $\cos\phi = \frac{8}{17}$, find the values of $\sin\phi$ and $\tan\phi$ as fractions in their simplest form. (3 marks)

5 Write down exact values for:

a $\cos 45°$

b $\tan 150°$

c $(\sin 120°)^2 + (\cos 120°)^2$ (4 marks)

6

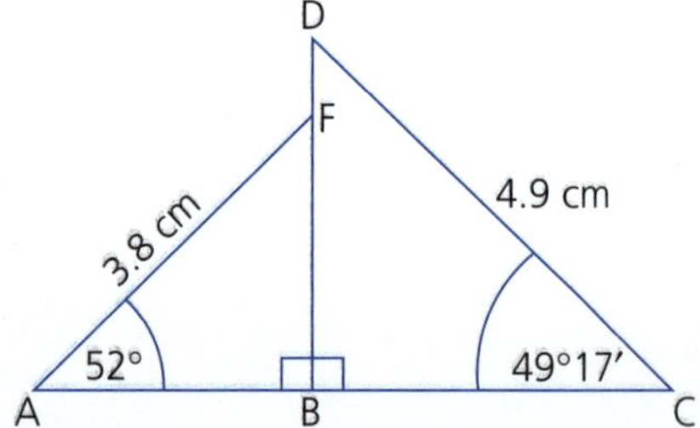

Calculate the length, correct to one decimal place, of:

a BD

b BF

c FD (5 marks)

7 If $0° \le \theta \le 90°$ and:

a $\cos\theta = \frac{5}{12}$, find θ to the nearest degree. (1 mark)

b $\tan m° = 1\frac{1}{4}$, find $m°$ to the nearest minute. (1 mark)

8

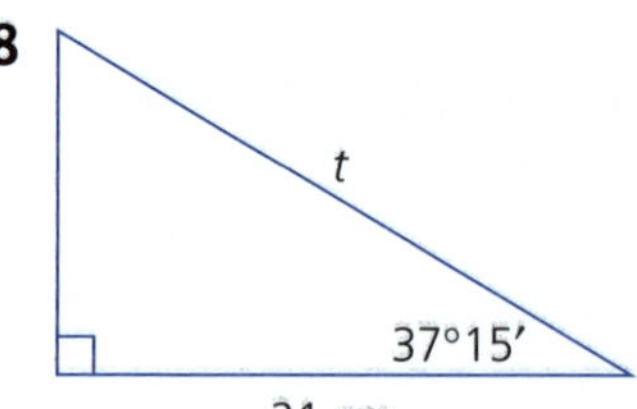

Calculate t correct to one decimal place. (2 marks)

9

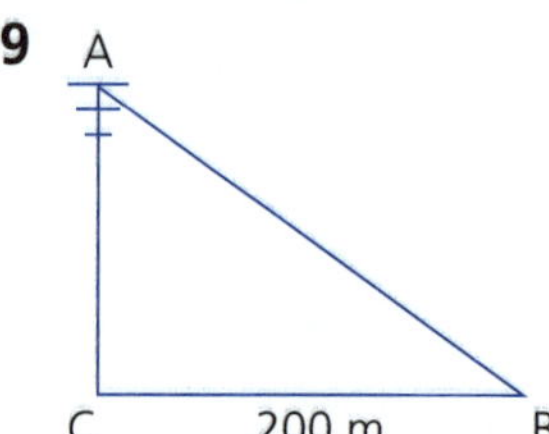

The angle of elevation of the top of an aerial measured from a point B, 200 m from the base of the aerial is 62°27′. Calculate the height of the aerial. (2 marks)

10 Find the length of the marked side in each diagram.

a 16.9 cm, 27°15′, 112°, y (2 marks)

b 5.9 m, 110°27′, 7.8 m, x (2 marks)

11 Calculate the size of the largest angle in a triangle with sides of length 17 cm, 12 cm and 8 cm. (2 marks)

12 Calculate the area of the parallelogram: (2 marks)

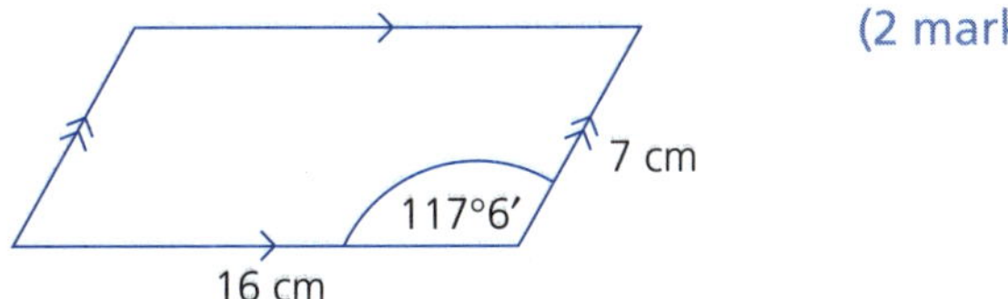

13 In the diagram are drawn the lines with equations, $y = 4 - 2x$ and $4x - 3y + 12 = 0$:

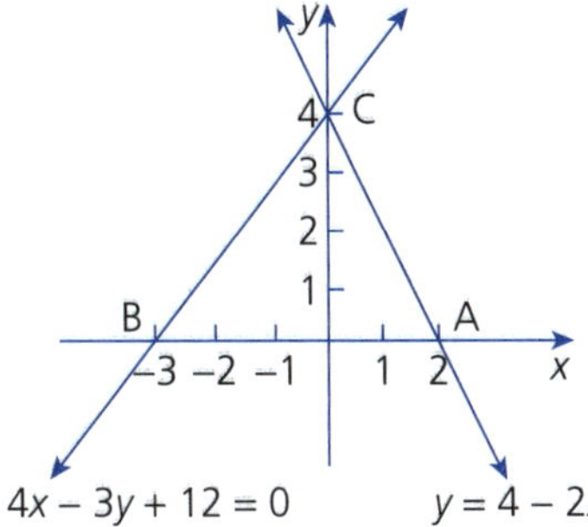

If the lines cut the x-axis at A and B and meet on the y-axis at C, calculate all of the angles of the triangle ACB. (3 marks)

14 A plane left X and flew for 4 hours at an average speed of 400 km/h, on a bearing of 136°, to Y. It then changed to a bearing of 250° and travelled for 1.5 hours at an average speed of 350 km/h to Z.

a Draw a diagram showing the plane's flight path. (1 mark)

b How far is the plane from its starting point X, to the nearest kilometre? (2 marks)

c Find the bearing of Z from X. (2 marks)

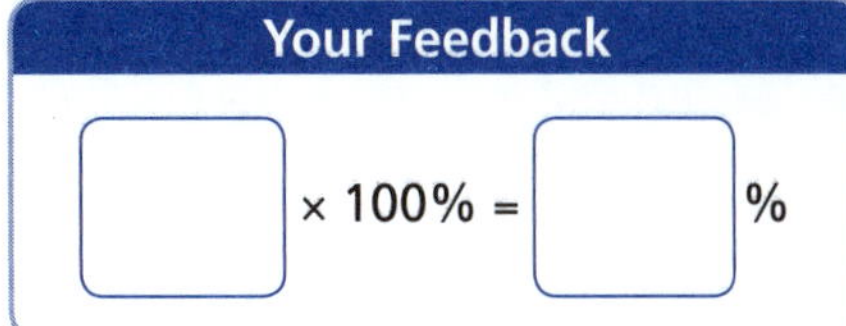

QA PAGE 306

WS PAGE 371

Chapter 9
Measurement

Area of special quadrilaterals

Students should already be familiar with the formulae used to find the area of special quadrilaterals:

Name	Figure	Formula
Parallelogram		$A = bh$
Trapezium		$A = \frac{1}{2}h(a + b)$
Rhombus		$A = \frac{1}{2}$ (product of diagonals) $A = \frac{1}{2}xy$

Keywords

Closed cylinder
Cone
Cylinder
Curved surface area
Hemisphere
Open cylinder
Pyramid
Sphere

For Example

1 Find the area of the following quadrilaterals:

a

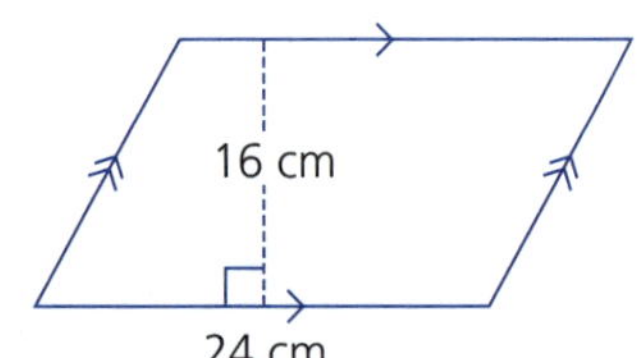

S $A = bh$
$= 24 \times 16$
$= 384$

$\therefore$ 384 cm^2

b

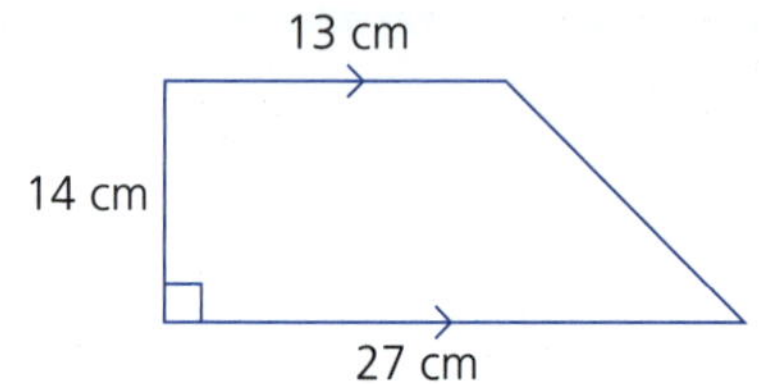

S $A = \frac{1}{2}h(a + b)$
$= \frac{1}{2} \times 14(13 + 27)$
$= 7(40)$
$= 280$

$\therefore$ 280 cm^2

c

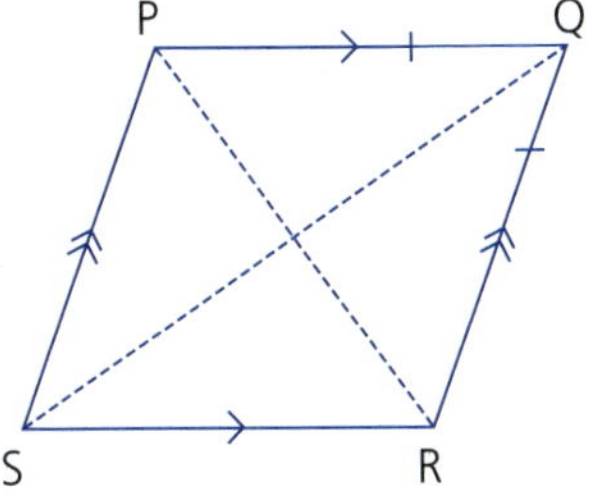

PR = 6 cm, QS = 10 cm

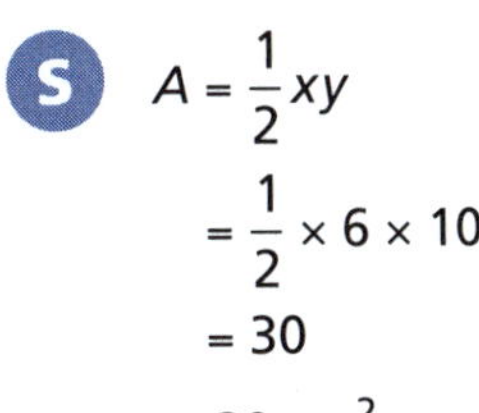

S $A = \frac{1}{2}xy$

$= \frac{1}{2} \times 6 \times 10$

$= 30$

$\therefore$ 30 cm^2

d

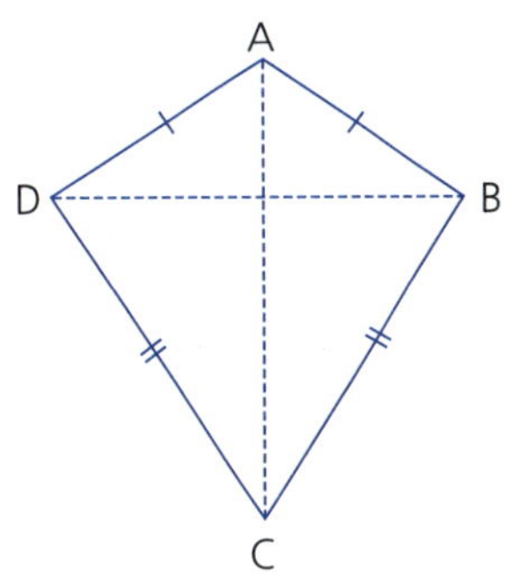

AC = 8 cm, DB = 5 cm

S $A = \frac{1}{2}xy$

$= \frac{1}{2} \times 8 \times 5$

$= 20$

$\therefore$ 20 cm^2

Volume of a cylinder, right-pyramid, cone and sphere

Name	Figure	Formula
Cylinder		$V = \pi r^2 h$
Pyramid		$V = \frac{1}{3}Ah$ $= \frac{1}{3}$ Area of base × Height
Cone		$V = \frac{1}{3}Ah$ $= \frac{1}{3}\neq r^2 h$ $= \frac{1}{3}$ Area of circular base × Height
Sphere		$V = \frac{4}{3}\neq r^3$

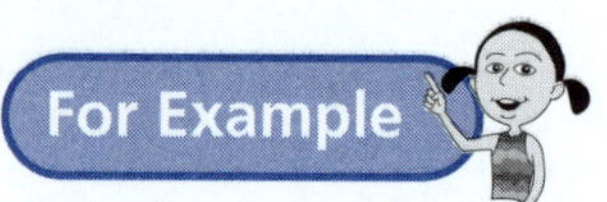

1 Find the volume of the following pyramids, cones and spheres:

a

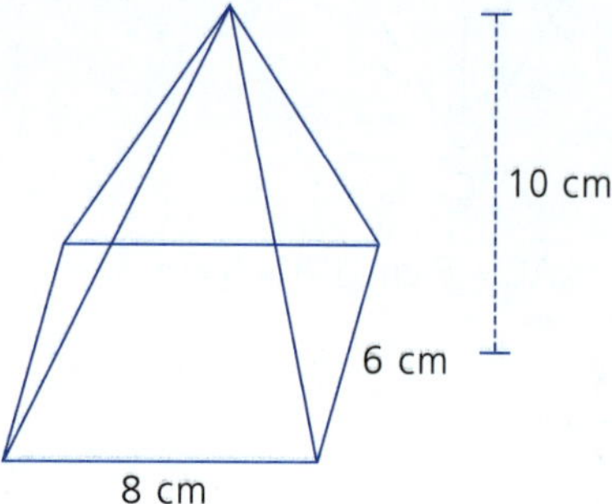

S $V = \frac{1}{3}Ah$

$= \frac{1}{3} \times 8 \times 6 \times 10$

$= 160$

$\therefore$ The volume is 160 cm^3.

b

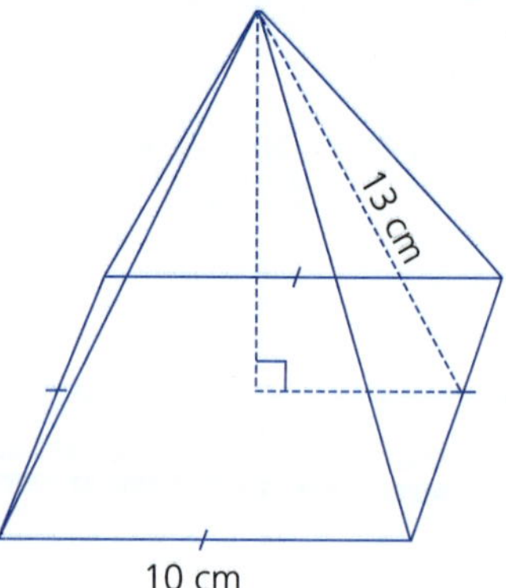

S

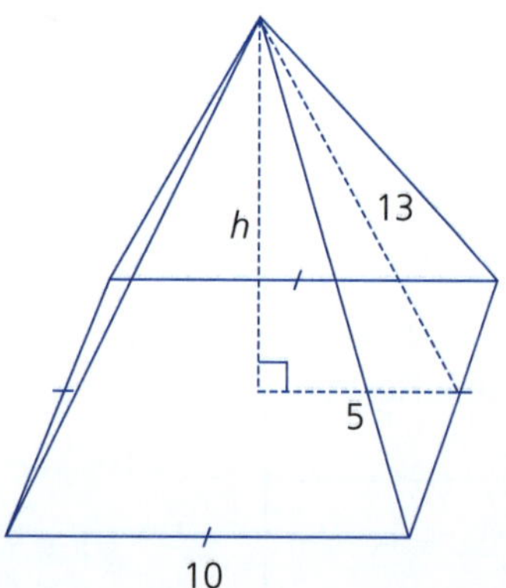

Let h = height.

$h^2 = 13^2 - 5^2$

$= 169 - 25$

$= 144$

$h = \sqrt{144}$

$= 12$

$\therefore V = \frac{1}{3}Ah$

$= \frac{1}{3} \times 10 \times 10 \times 12$

$= 400$

$\therefore$ The volume is 400 cm^3.

c

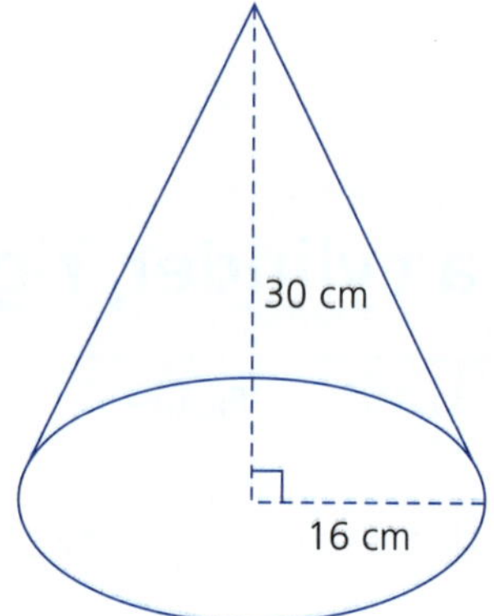

S $V = \frac{1}{3}\pi r^2 h$

$= \frac{1}{3} \times \pi \times 16^2 \times 30$

$= 8042.477\,193$

$= 8042.48$ (to 2 decimal places)

$\therefore$ The volume is 8042.48 cm^3.

d

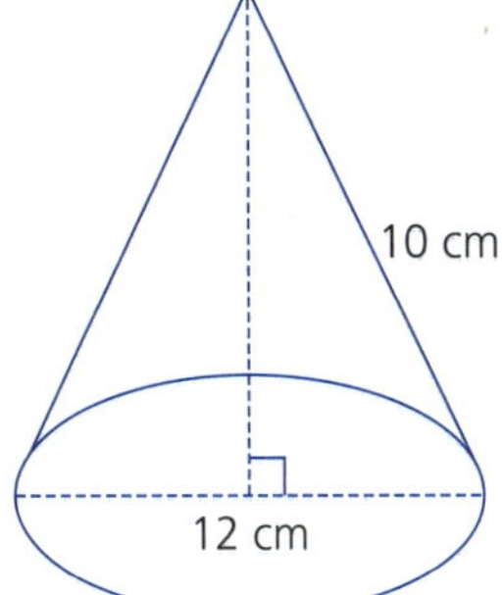

S

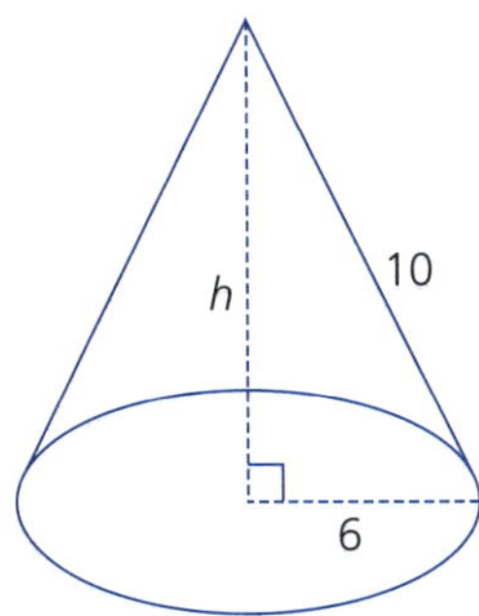

Diameter = 12 cm
$\therefore$ Radius = 6 cm

Let h = height

$\therefore h^2 = 10^2 - 6^2$
$= 100 - 36$
$= 64$
$h = \sqrt{64}$
$= 8$

$V = \frac{1}{3} \neq r^2 h$
$= \frac{1}{3} \times \pi \times 6^2 \times 8$
$= 301.592\,8947$
$= 301.59$ (to 2 decimal places)

$\therefore$ The volume is 301.59 cm^3.

e

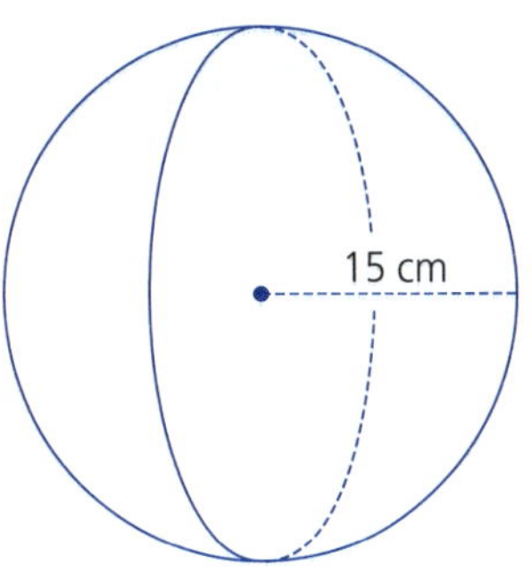

S $V = \frac{4}{3} \neq r^3$
$= \frac{4}{3} \times \pi \times 15^3$
$= 14\,137.166\,94$
$= 14\,137.17$ (to 2 decimal places)

$\therefore$ The volume is 14 137.17 cm^3.

f

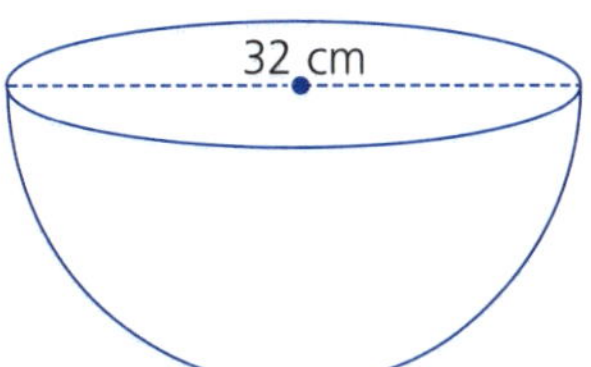

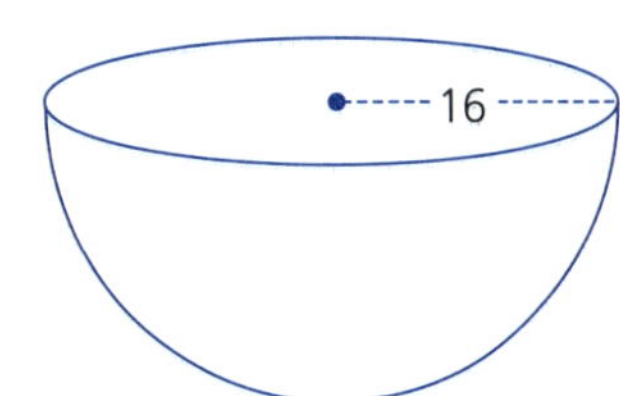

Shape is a hemisphere.
Diameter = 32 cm
$\therefore$ Radius = 16 cm

$V = \frac{1}{2} \times \frac{4}{3} \pi r^3$
$= \frac{1}{2} \times \frac{4}{3} \times \pi \times 16^3$
$= 8578.642\,339$
$= 8578.64$ (to 2 decimal places)

$\therefore$ The volume is 8578.64 cm^3.

Surface area of right-cylinder, pyramid, right-cone and sphere

The surface area of a solid is the sum of all of the areas of its faces.

Name	Figure	Formula
Cylinder	h, r	Total surface area = $2\pi r^2 + 2\pi rh$ Curved surface area = $2\pi rh$
Pyramid		Surface area (SA) = Sum of the areas of all faces
Cone	l, r	Total surface area = $\pi r^2 + \pi rl$ Curved surface area = πrl
Sphere	r	Surface area = $4\pi r^2$

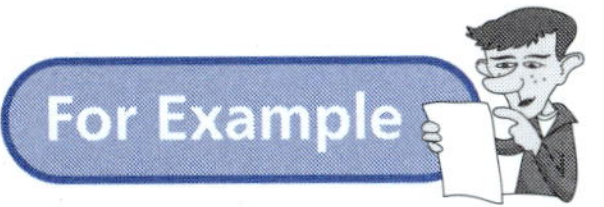

1 Find the surface area of the following shapes:

a

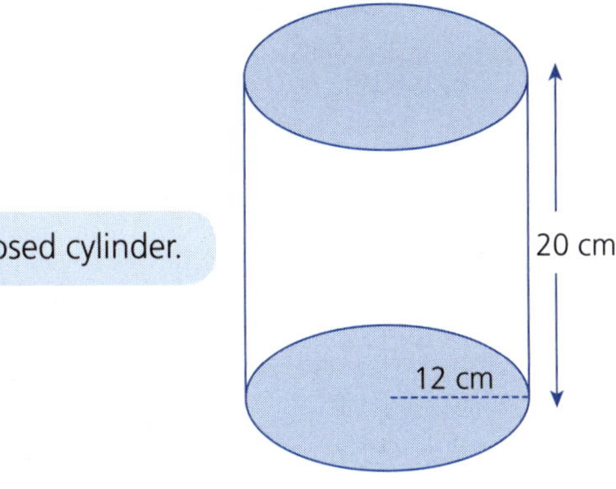

S $SA = 2\pi r^2 + 2\pi rh$
$= 2 \times \pi \times 12^2 + 2 \times \pi \times 12 \times 20$
$= 2412.743\,158$
$= 2412.74$ (to 2 decimal places)

$\therefore$ The surface area is $2412.74\ \text{cm}^2$.

b

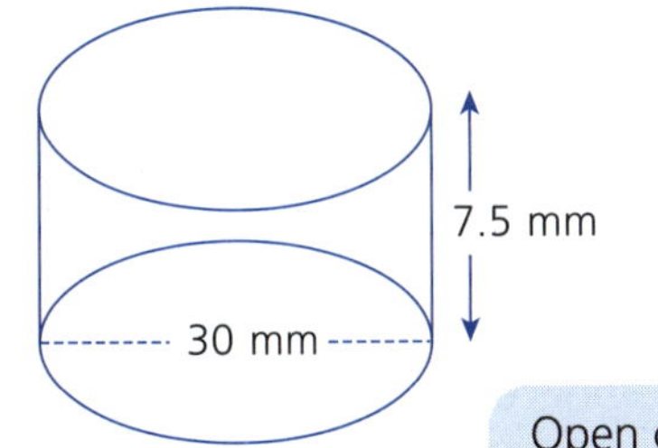

S Diameter = 30 mm
$\therefore$ Radius = 15 mm

$SA = 2\pi rh$
$= 2 \times \pi \times 15 \times 7.5$
$= 706.858\,3471$
$= 706.86$ (to 2 decimal places)

$\therefore$ The surface area is $706.86\ \text{mm}^2$.

c

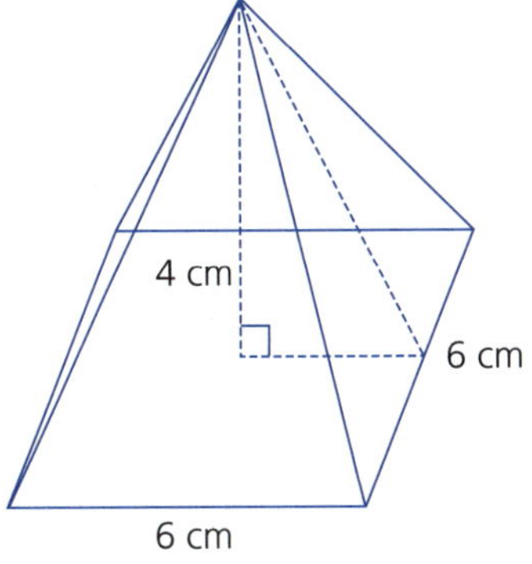

S As base = 6 cm
then BC = 3 cm

$\therefore x^2 = 3^2 + 4^2$
$= 9 + 16$
$= 25$
$x = \sqrt{25}$
$= 5$

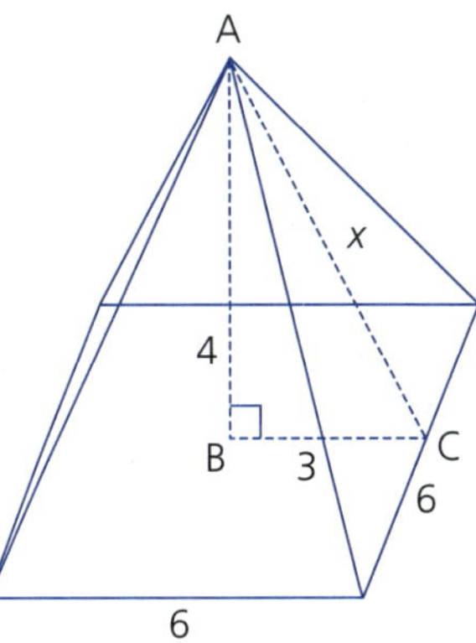

$SA = 6 \times 6 + 4 \times \frac{1}{2} \times 6 \times 5$
$= 36 + 60$
$= 96$

$\therefore$ The surface area is $96\ \text{cm}^2$.

d

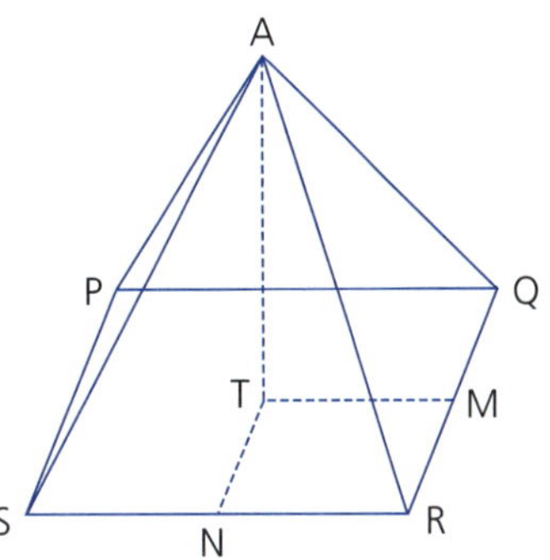

PQ = 18 cm, QR = 10 cm, AT = 12 cm

S In $\triangle$ATM, MT = 9 cm, AT = 12 cm
$\therefore AM^2 = 9^2 + 12^2$
$= 81 + 144$
$= 225$
$AM = \sqrt{225}$
$= 15$

In $\triangle$ATN, NT = 5 cm,
AT = 12 cm

$\therefore AN^2 = 5^2 + 12^2$
$= 25 + 144$
$= 169$
$AN = \sqrt{169}$
$= 13$

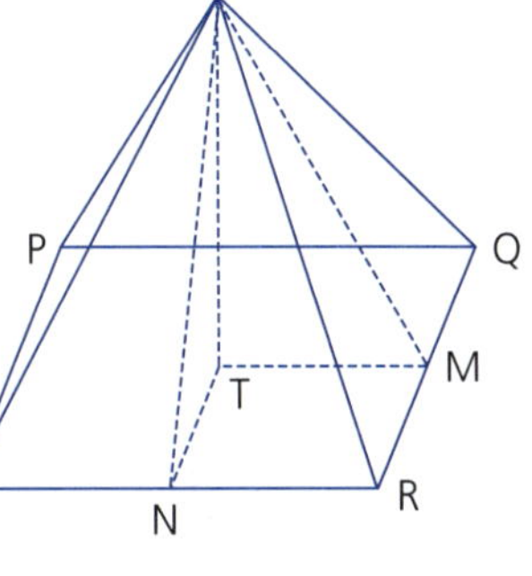

SA = Rectangle (18×10)
$+ 2\triangle s\ \frac{1}{2} \times 10 \times 15$
$+ 2\triangle s\ \frac{1}{2} \times 18 \times 13$
$= 18 \times 10 + 2 \times \frac{1}{2} \times 10 \times 15 + 2 \times \frac{1}{2} \times 18 \times 13$
$= 564$

$\therefore$ The surface area is $564\ \text{cm}^2$.

e

Closed cone.

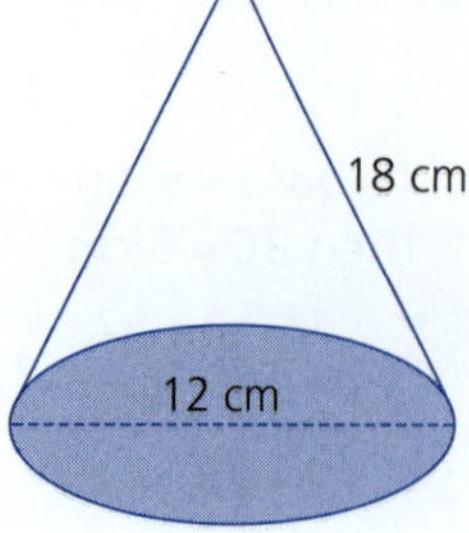

S

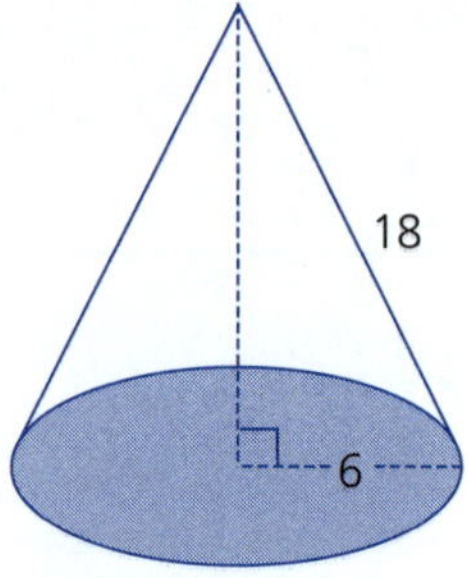

Diameter = 12 cm
$\therefore$ Radius = 6 cm

$$\begin{aligned} SA &= \pi r^2 + \pi r l \\ &= \pi \times 6^2 + \pi \times 6 \times 18 \\ &= 452.389\,3421 \\ &= 452.39 \text{ (to 2 decimal places)} \end{aligned}$$

$\therefore$ The surface area is 452.39 cm^2.

f

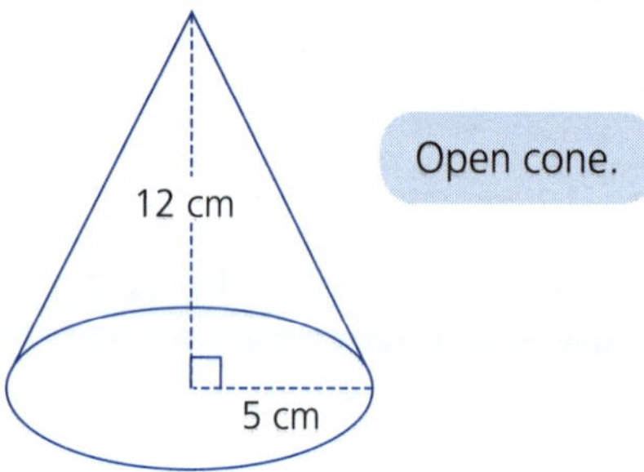

S

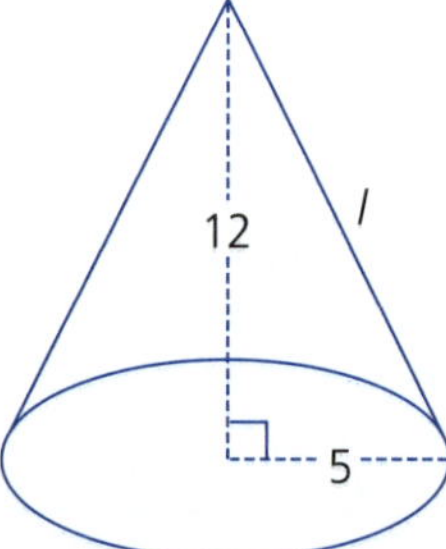

$$\begin{aligned} l &= 12^2 + 5^2 \\ &= 144 + 25 \\ &= 169 \\ l &= \sqrt{169} \\ &= 13 \end{aligned}$$

$$\begin{aligned} SA &= \pi r l \quad \text{[curved surface area]} \\ &= \pi \times 5 \times 13 \\ &= 65\pi \end{aligned}$$

$\therefore$ The surface area is 65π cm^2.

g

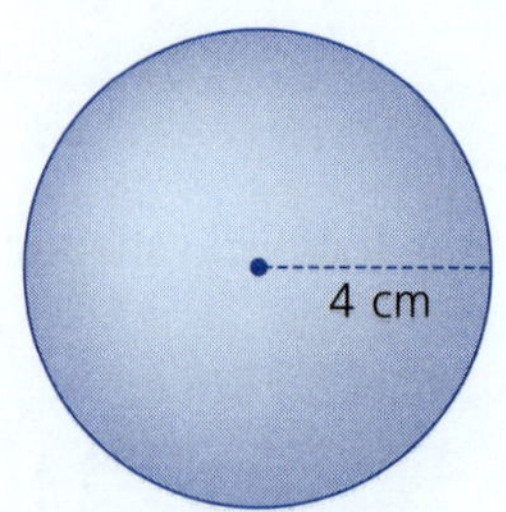

S

$$\begin{aligned} SA &= 4\pi r^2 \\ &= 4 \times \pi \times 4^2 \\ &= 64\pi \end{aligned}$$

$\therefore$ The surface area is 64π cm^2.

h

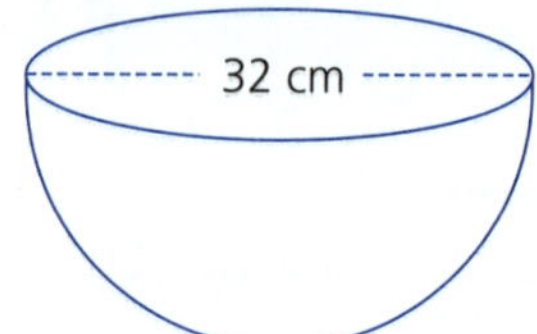

S Diameter = 32 cm
$\therefore$ Radius = 16 cm

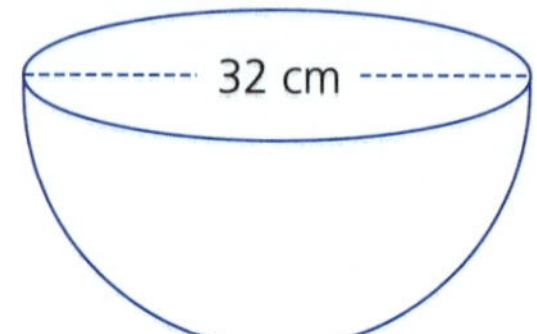

$$\begin{aligned} SA &= \frac{1}{2}\text{ sphere} + \text{circle} \\ &= \frac{1}{2} \times 4\pi r^2 + \pi r^2 \\ &= 2\pi r^2 + \pi r^2 \\ &= 3\pi r^2 \\ &= 3 \times \pi \times 16^2 \\ &= 2412.743\,158 \\ &= 2412.74 \text{ (to 2 decimal places)} \end{aligned}$$

$\therefore$ The surface area is 2412.74 cm^2.

Practise Practise

1 Find the area of the following quadrilaterals:

a

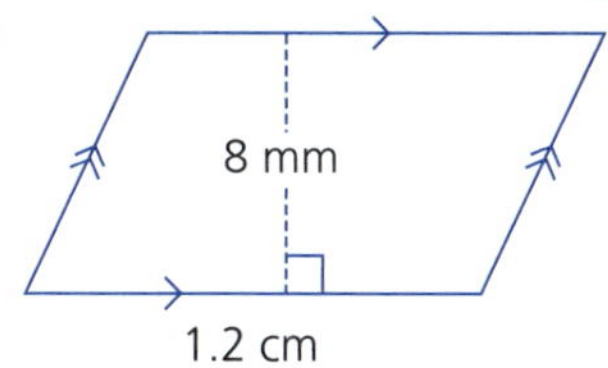

b

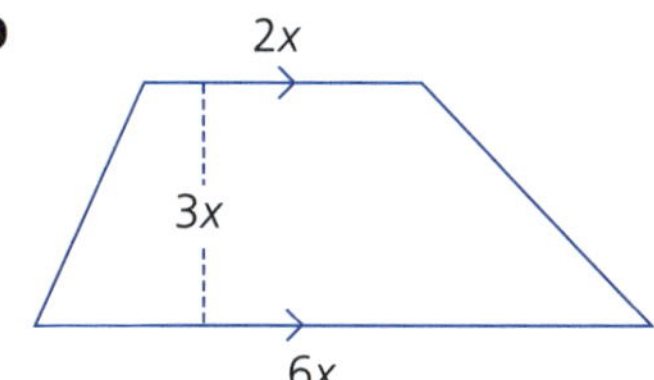

c

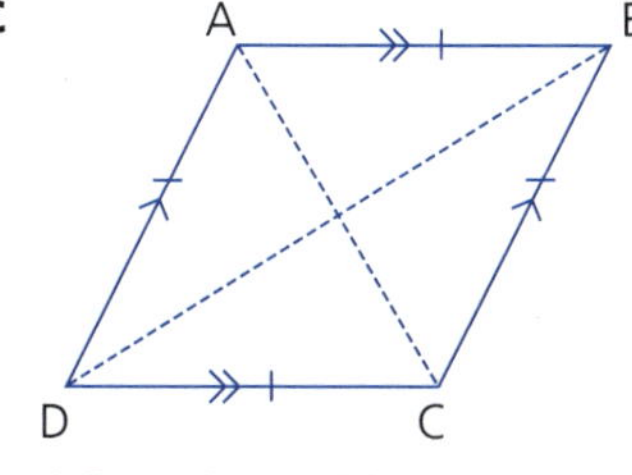

AC = 12 cm, BD = 17 cm

d

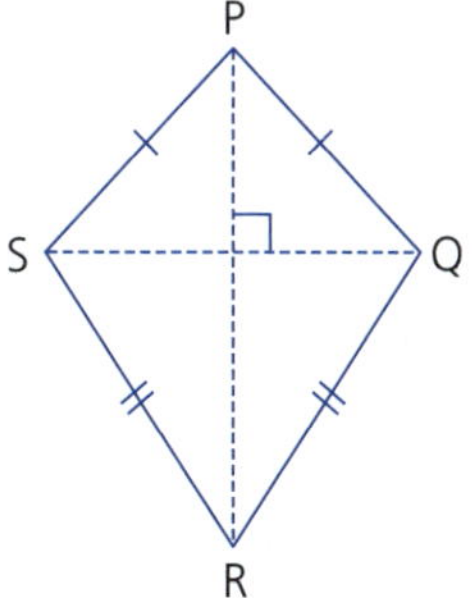

PR = $3a$, SQ = $2a$

pp. 144–145

2 Find the area and perimeter of the following shapes:

a

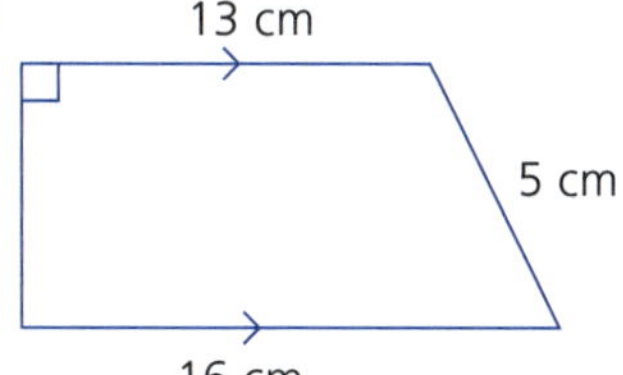

b

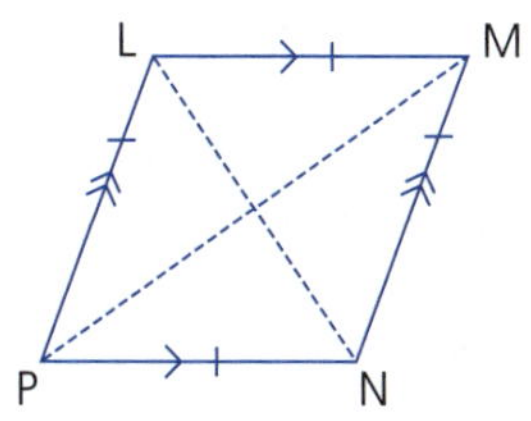

LN = 6 cm, PM = 8 cm

c

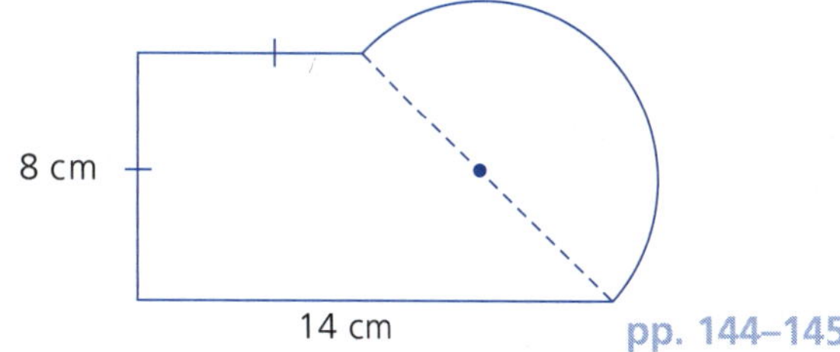

pp. 144–145

3 Find the volume of the following shapes, correct to two decimal places:

a

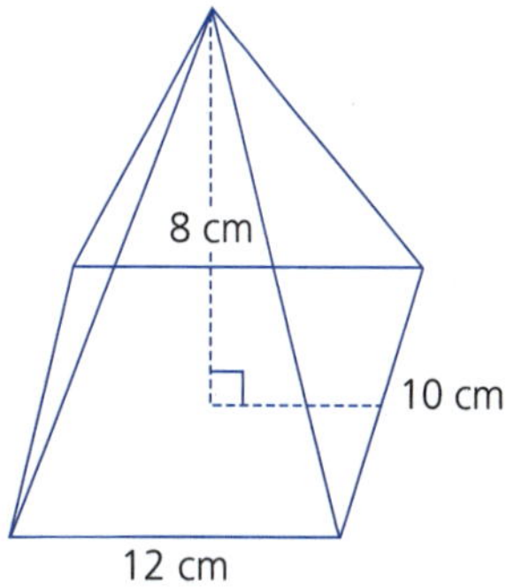

b

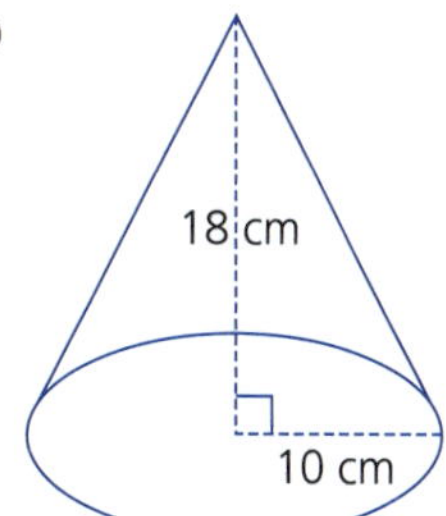

c

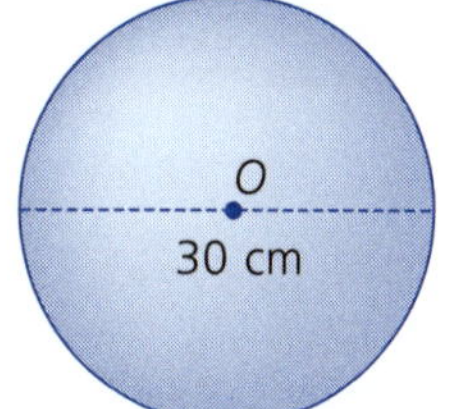

d

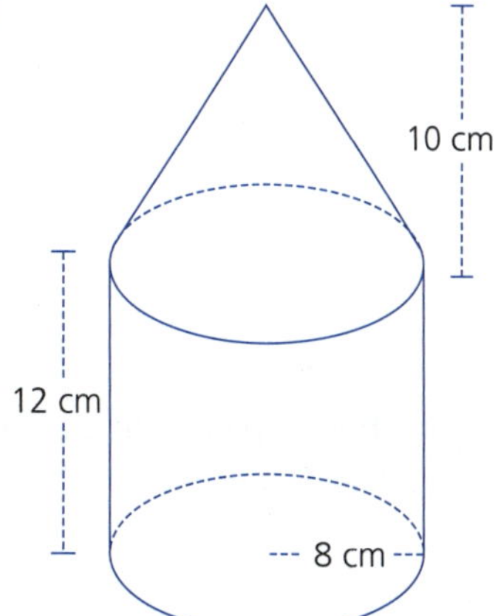

pp. 145–147

4 Calculate the capacity, in litres, of a basketball with a diameter of 25 cm. pp. 145–147

5 Find the volume of the solids, correct to three significant figures:

a

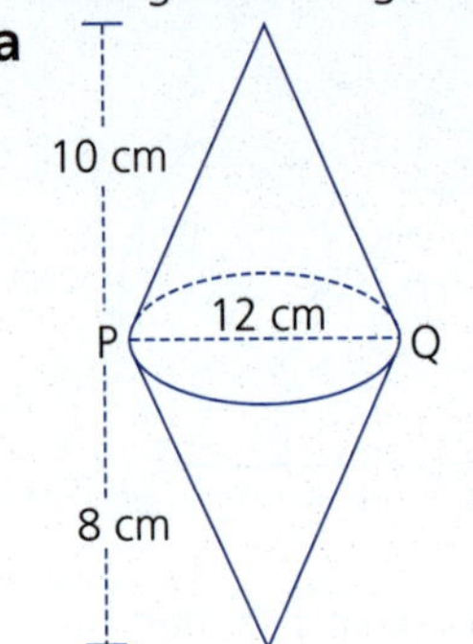

b

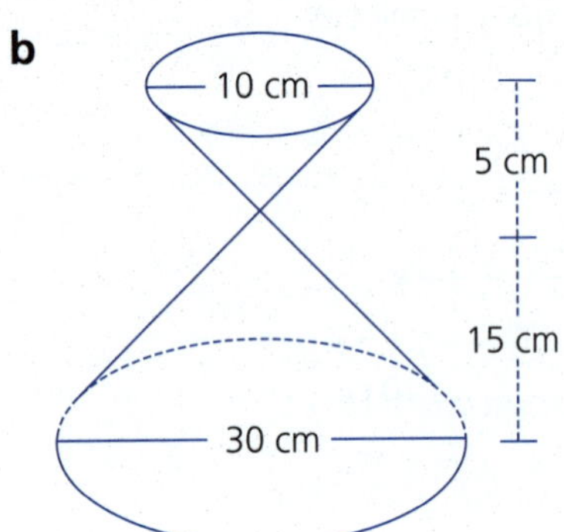

c

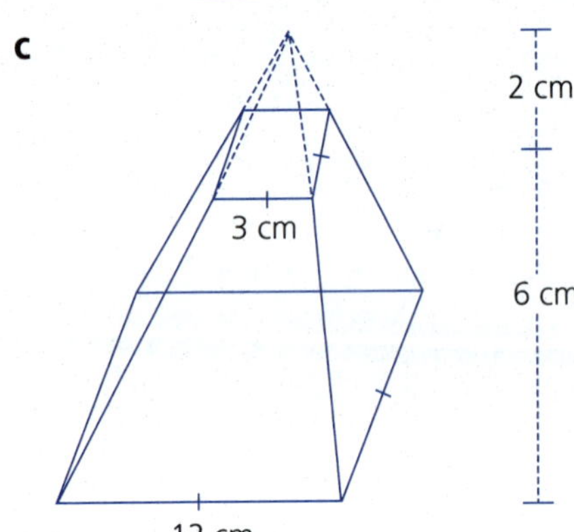

d

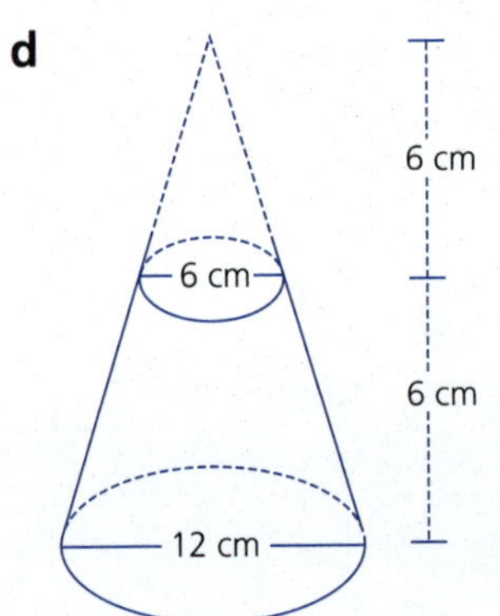

pp. 145–147

6 Find the volume of a hemisphere with diameter 24 cm. Leave your answer in terms of π. pp. 145–147

7 The central section of a gift box of chocolates is a cylinder with cones on both ends. Calculate the volume of the chocolate box.

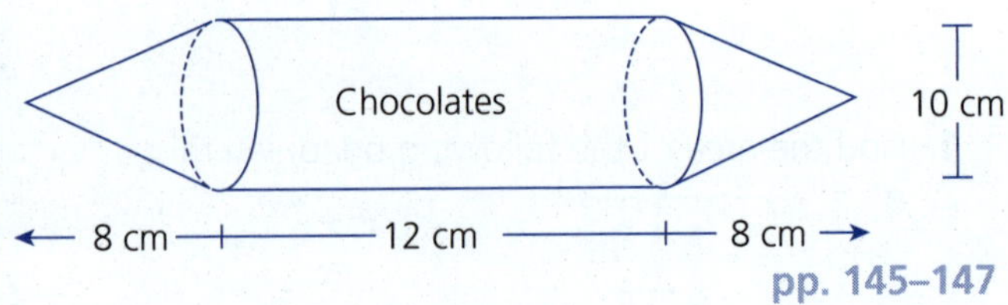

pp. 145–147

8 The diagram shows three solid spheres that fit tightly into a cylinder.

a Find the volume of each solid sphere, correct to three decimal places.

b Find the volume of space remaining in the cylinder, correct to three decimal places.

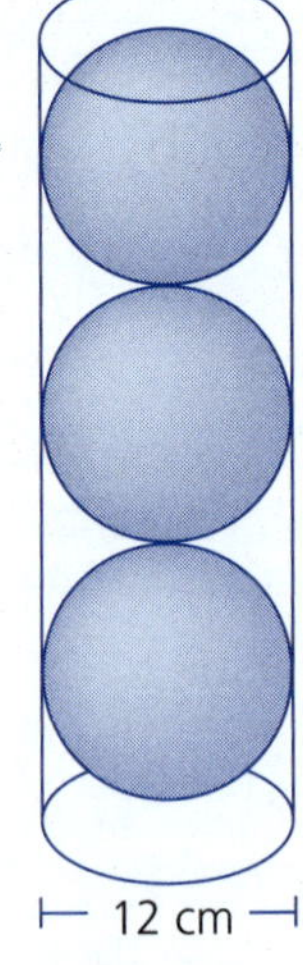

pp. 145–147

9 Louise makes spherical rum balls, each with a volume of 8 cm^3. What is the radius of each ball, leaving your answer in terms of π? pp. 145–147

10 The radius of a cone is 8 cm and its volume is 120 cm^3. Correct to one decimal, what is the height of the cone? pp. 145–147

11 A square pyramid has a volume of 270 cm^3 and each side of its base is 18 cm. Find the height of the pyramid. pp. 145–147

12

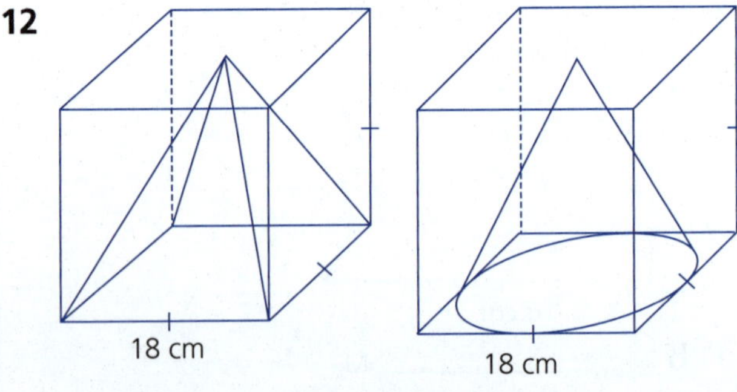

The diagrams show a solid pyramid and a solid cone inside identical cubes. Find the:

a volume of each cube

b volume of the pyramid

c volume of the cone

d percentage of each cube filled. pp. 145–147

13 Find the surface area of a closed cylinder with radius 4 cm and height 2 cm. Leave your answer in terms of π. pp. 148–150

14 Find the surface area of the following shapes, correct to two decimal places if necessary:

a

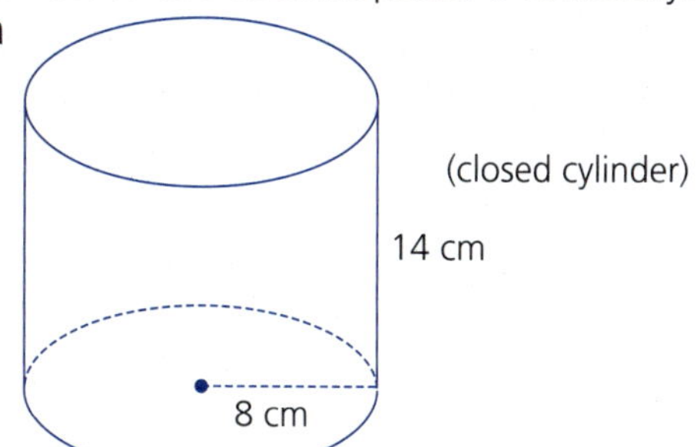

b

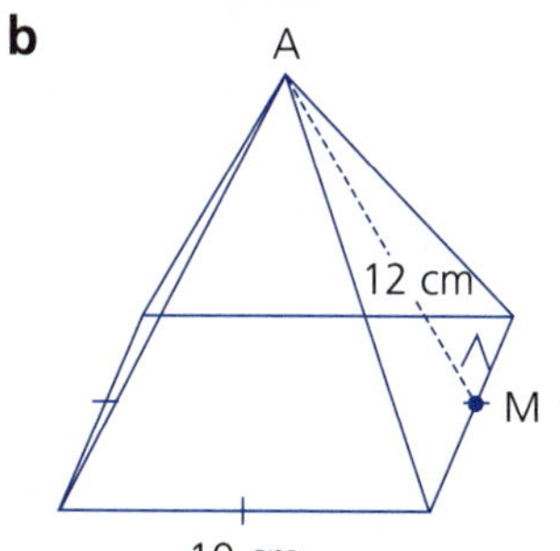

c

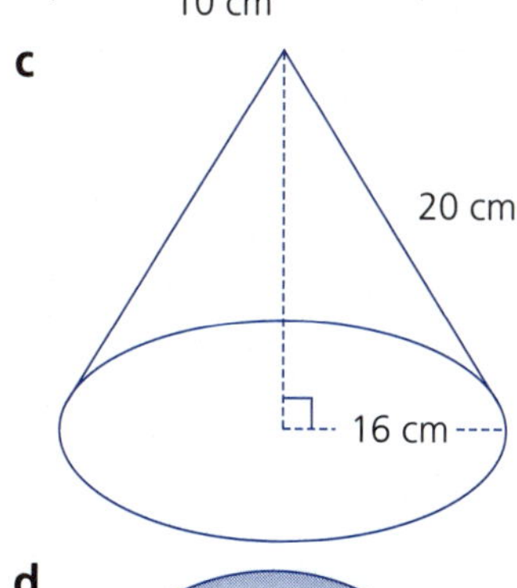

d

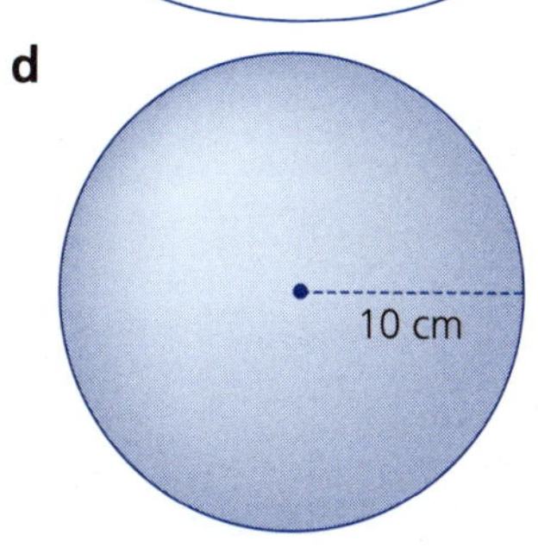

pp. 148–150

15 Find the surface area:

a

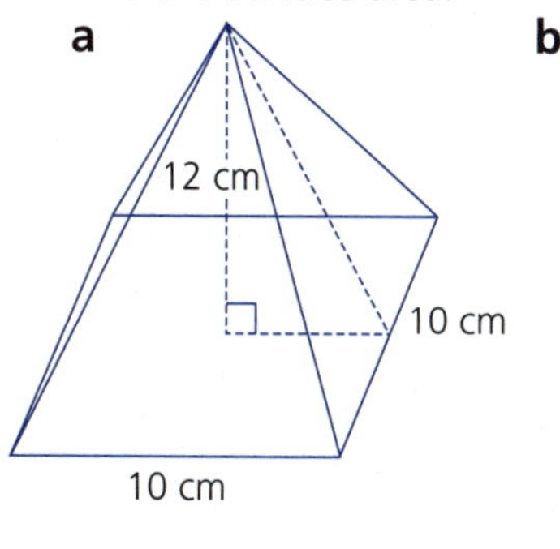

b

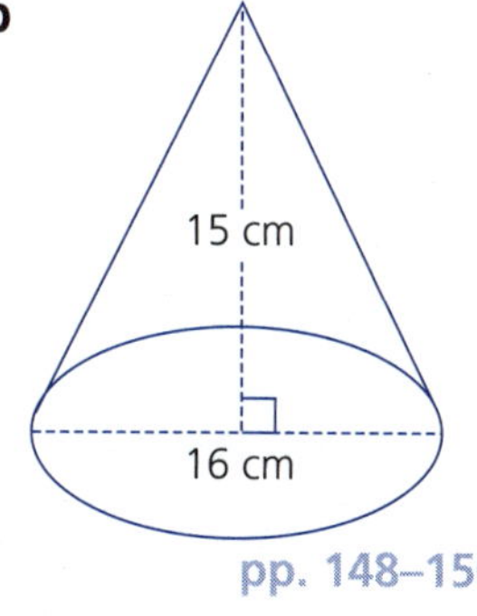

pp. 148–150

16 Find the curved surface area of these shapes, keeping your answer in terms of π:

a **b**

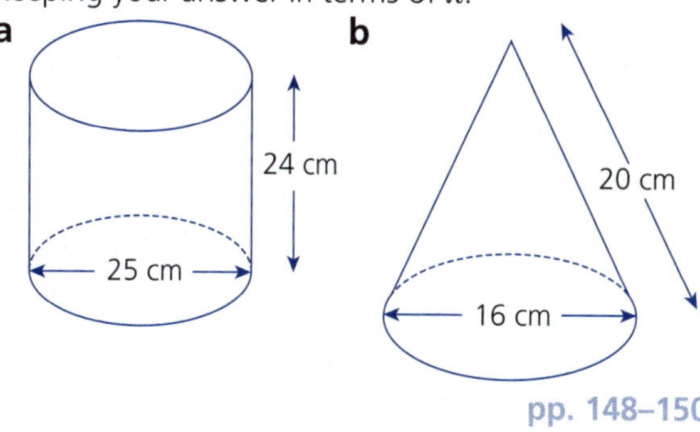

pp. 148–150

17 Find the surface area of a hemisphere with radius 8 cm. Leave your answer in terms of π.

pp. 148–150

18 The diagram shows a can of tomatoes. The radius of the can is 4 cm and the height 10 cm.

a Find the volume of the can, correct to three decimal places.

b Find the capacity of the can, to the nearest mL.

c Find the area of sheet metal used to make the can.

d Find the area of the label.

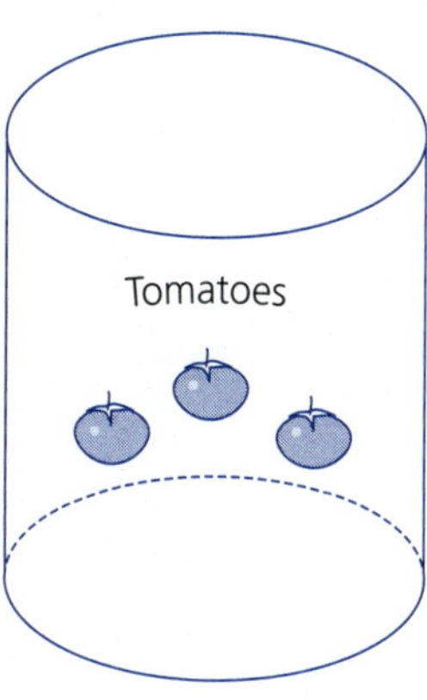

pp. 145–150

19

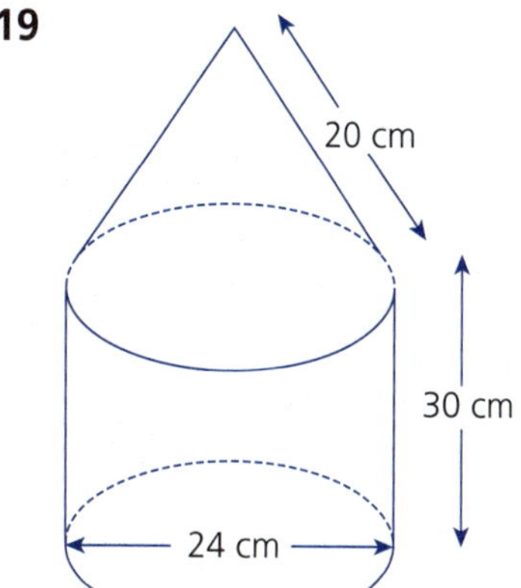

Find the surface area of the solid, correct to two decimal places. pp. 148–150

Go to p. 301 for **Quick Answers** or to pp. 336–338 for **Worked Solutions**

For a complete understanding of this topic, you must be able to:

✓	Apply formulae for areas of special quadrilaterals		pp. 144–145
✓	Calculate the volume of right-pyramids, cones, cylinders and spheres		pp. 145–147
✓	Calculate the surface area of right-cylinders, pyramids, right-cones and spheres.		pp. 148–150

Now you are ready to do the tests!

Intermediate Test

Measurement

(30 marks)

1 Find the volume of the following shapes:

a (1 mark)

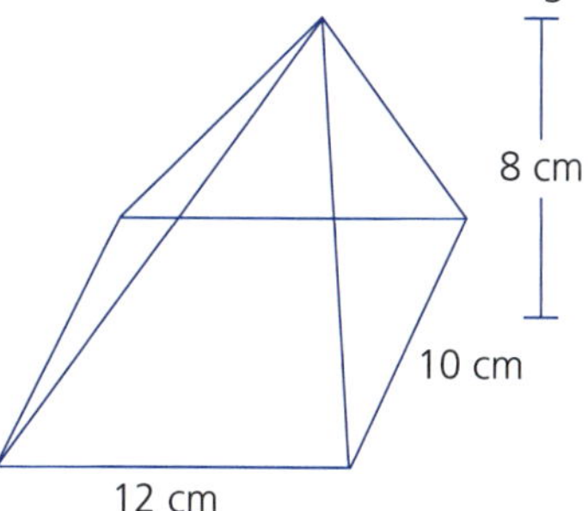

b

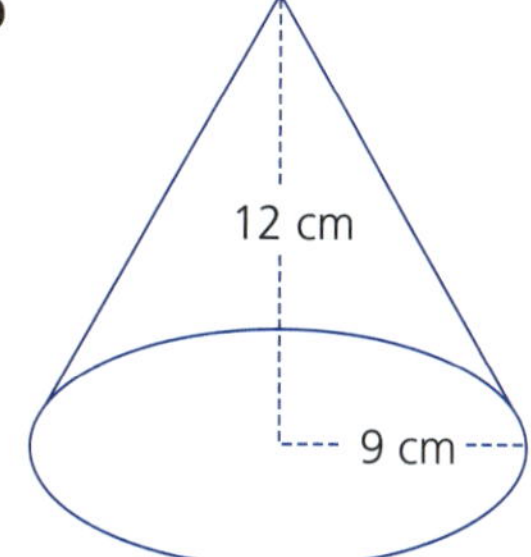

Leave your answer correct to two decimal places. (2 marks)

c

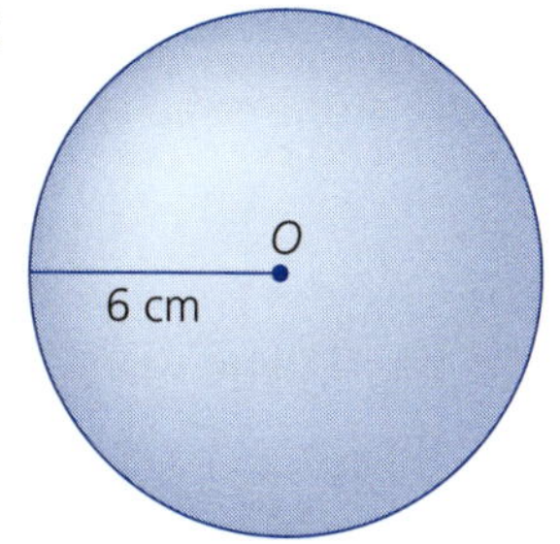

Leave your answer in terms of π. (2 marks)

d

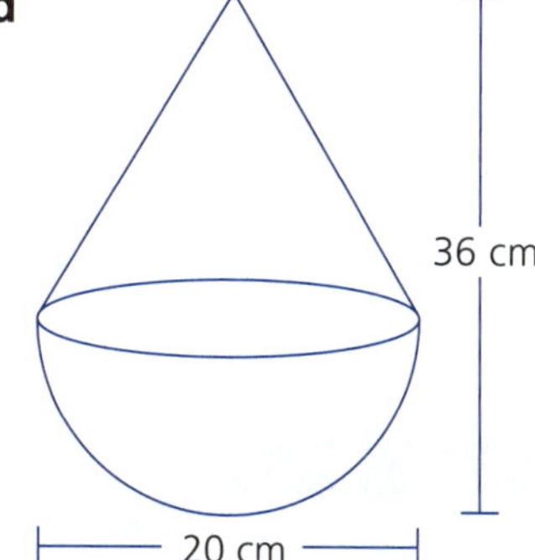

Give your answer correct to two decimal places. (3 marks)

2 Find the surface area of these shapes:

a

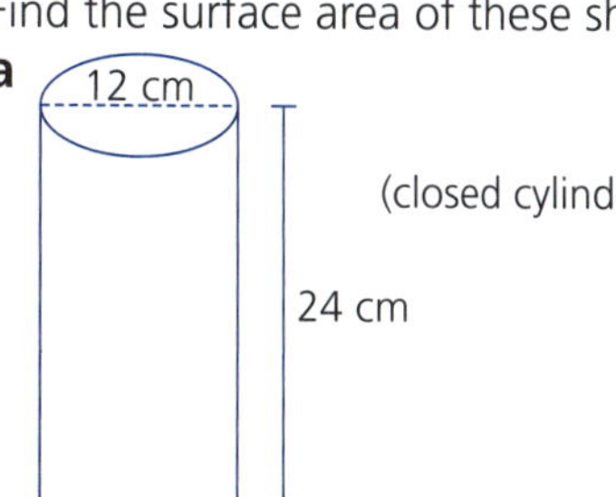

Give your answer correct to two decimal places. (2 marks)

b

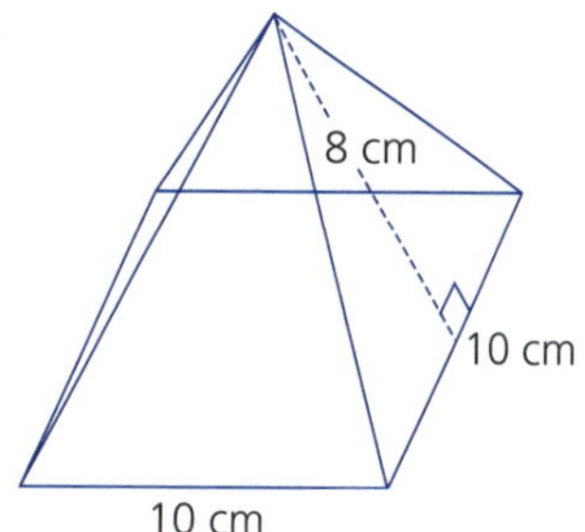

(2 marks)

c

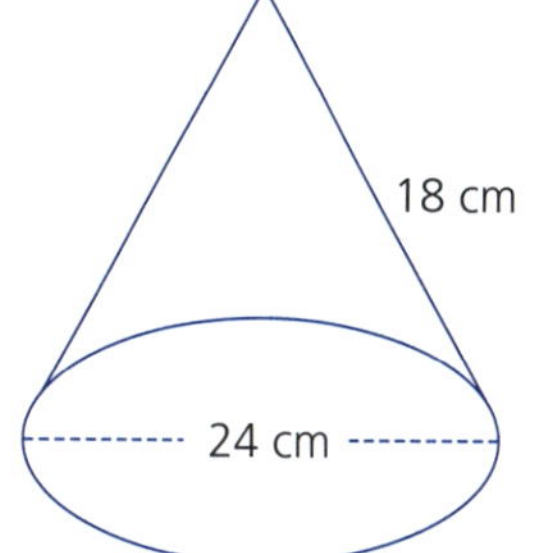

Give your answer correct to two decimal places. (2 marks)

d

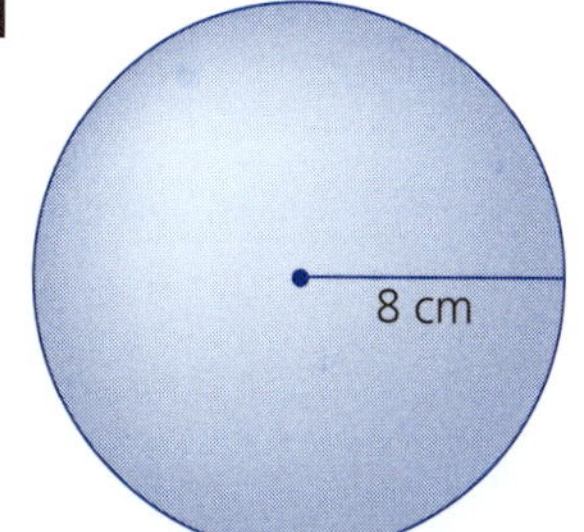

Give your answer in terms of π. (2 marks)

3 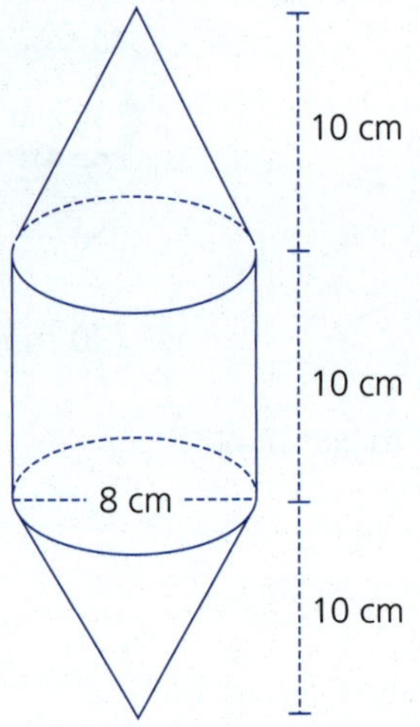

a Find the volume of the solid, correct to three significant figures. (2 marks)

b What fraction of the entire solid is the volume of the top cone? (1 mark)

4 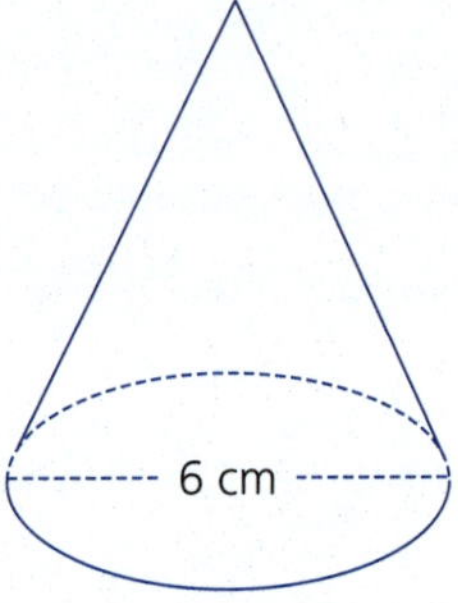

The perpendicular height is twice the diameter of the cone. Find the curved surface area of the cone, correct to two decimal places. (2 marks)

5 Find the curved surface area of a cylinder with radius 4 cm and volume 80π cm^3. Give your answer in terms of π. (2 marks)

6 An enclosed cone has a perpendicular height of 3 cm and a volume of 16π cm^3.

a What is the radius of the cone? (2 marks)

b Find the surface area of the cone, in terms of π. (3 marks)

7 A cone and a sphere each have a radius of 15 cm. If the solids have the same volume, find the height of the cone. (2 marks)

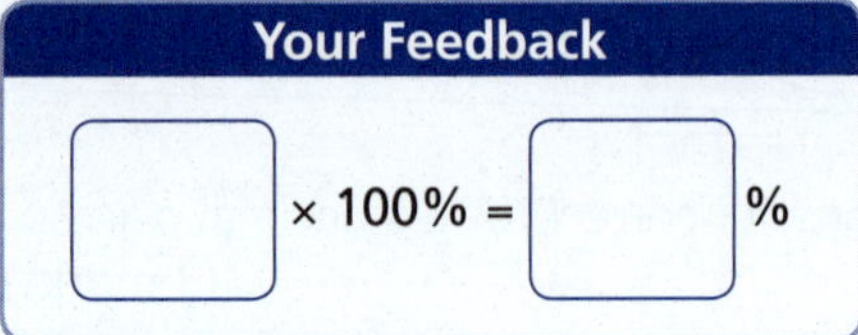

QA PAGE 306

WS PAGE 372

Advanced Test

Measurement

(30 marks)

1 Find the volume of the following:

a (2 marks)

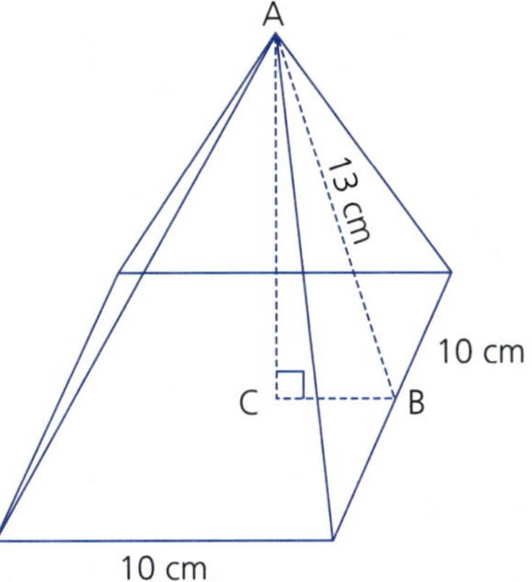

b

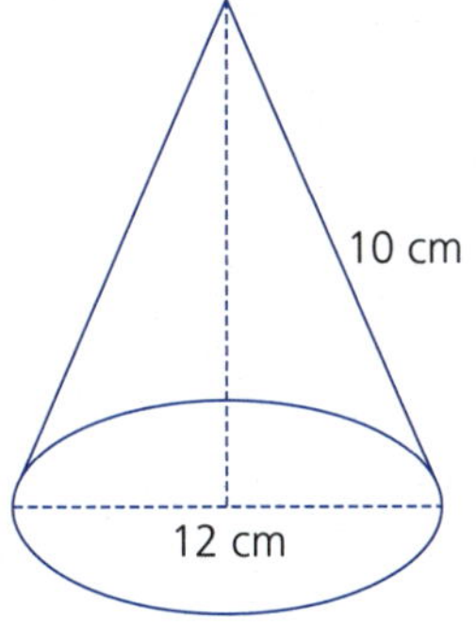

Leave your answer in terms of π. (2 marks)

c

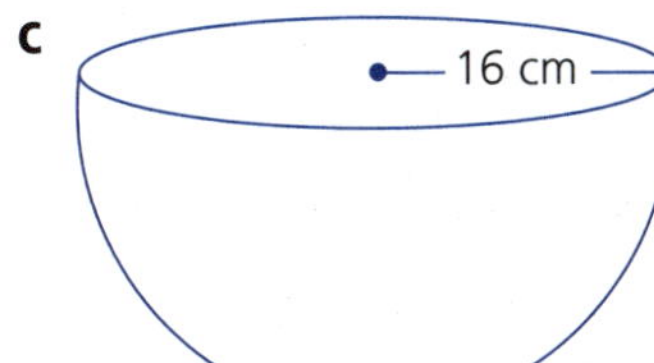

Leave your answer correct to two decimal places.

(2 marks)

2 The diagram shows two solid spheres inside a cylinder.

Find, correct to two decimal places, the volume of:

a each sphere. (1 mark)

b the air inside the cylinder.

(2 marks)

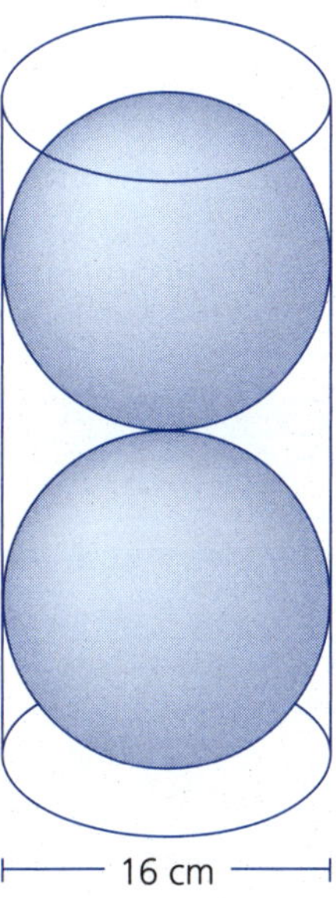

3 Find the surface area of an open cylinder with a diameter of 16 cm and a height of 10 cm. Leave your answer in terms of π. (2 marks)

4 Find the surface area of the following solids:

a

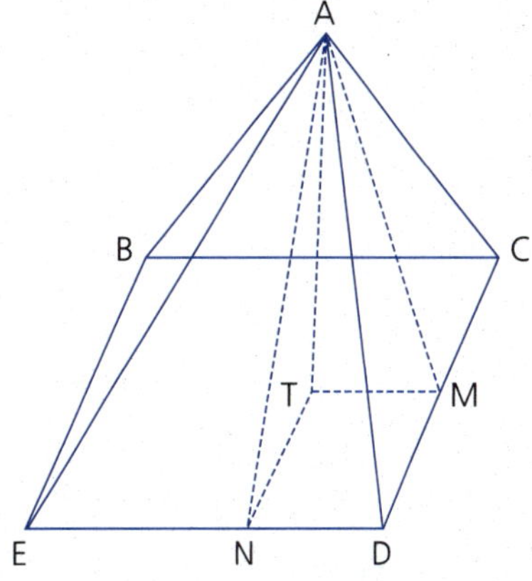

DE = 16 cm, CD = 10 cm, AT = 6 cm

(3 marks)

b

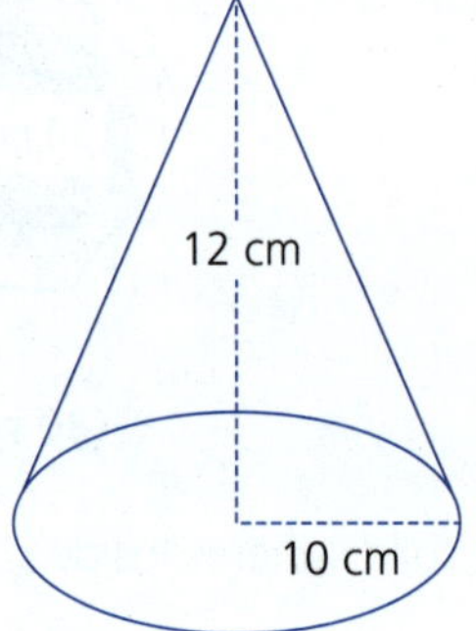

Leave your answer correct to two decimal places. (2 marks)

c

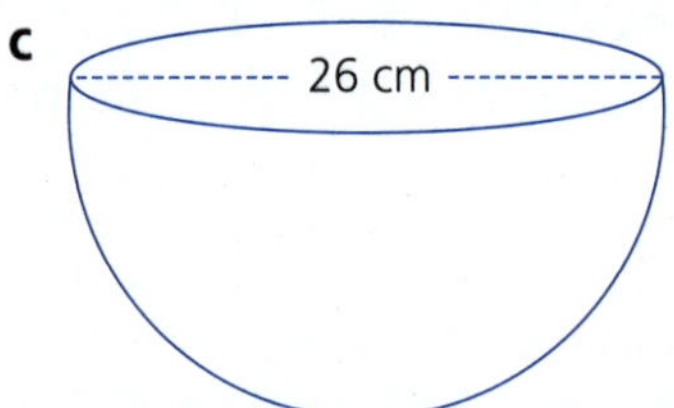

Leave your answer correct to two decimal places. (2 marks)

5

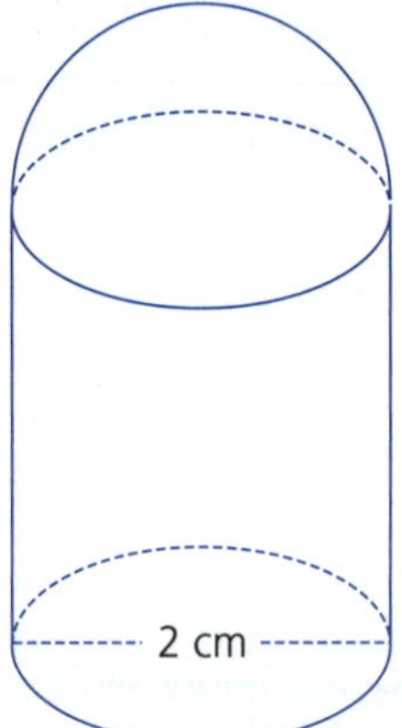

A hemisphere is surmounted on a cylinder.
The cylinder is the same height as its diameter.

a What fraction of the total volume is the volume of the cylinder? (2 marks)

b What fraction of the total surface area is the surface area of the hemisphere? (2 marks)

6 A sphere's radius has doubled. By how much has the surface area increased? (2 marks)

7 A basketball fits exactly into a cube, which has side lengths equal to the diameter of the basketball. What percentage of the cube is taken up by the basketball? (4 marks)

8

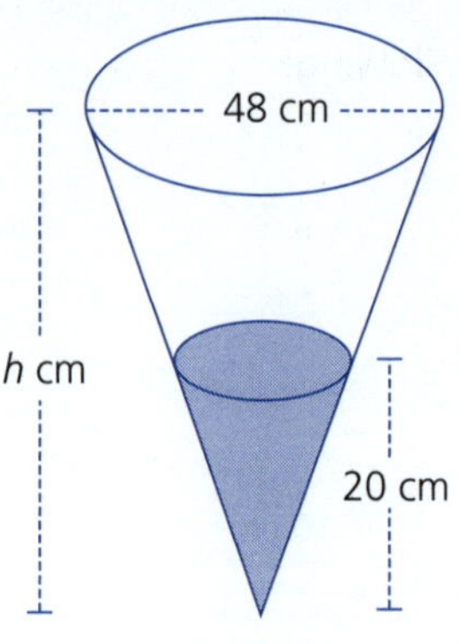

A cone has a diameter of 48 cm. The cone is partly filled with 240π mL to a depth of 20 cm.

What is the height (h) of the cone? (2 marks)

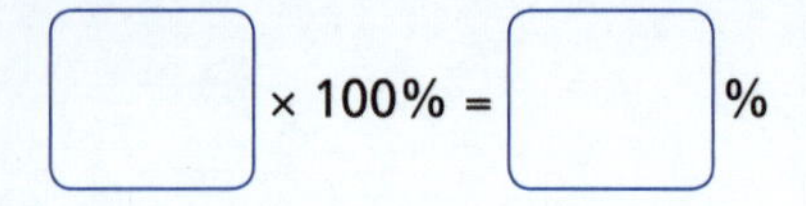

QA PAGE 306

WS PAGE 373

Chapter 10
Geometry

Angles

The basic angle properties are as follows:

- **Complementary angles have a sum of 90°.**

  $\angle ABC = 90°$ $x + y = 90$

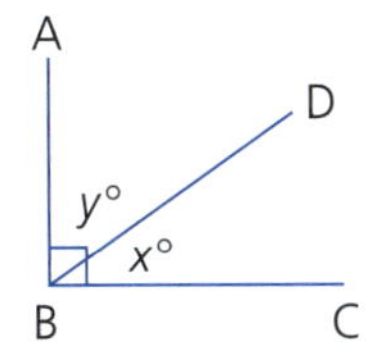

- **Supplementary angles have a sum of 180°.**

  AC is a straight line.
  $x + y = 180$

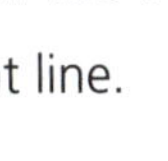

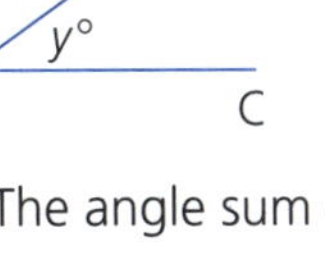

Note that this could also be phrased: 'The angle sum of a straight line is 180°.'

- **Vertically opposite angles are equal.**

  AC and DB are straight lines. $x = z$
  $y = a$

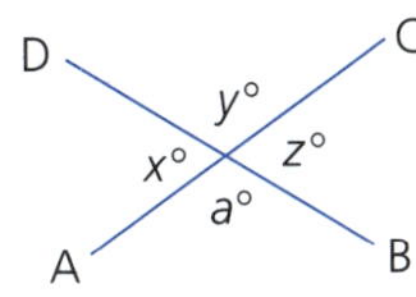

- **The angle sum of a revolution is 360°.**

  $p + q + r + s = 360$

  This property is often referred to as 'angles at a point'.

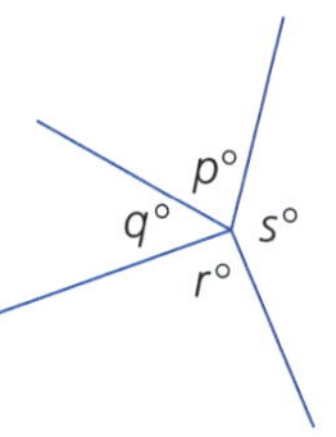

Note also:
$y + z = 180$
$x + a = 180$
$x + y = 180$
$a + z = 180$
(angle sum of a straight line)

Keywords

Alternate	Isosceles	Square
Co-interior	Parallel	Supplementary
Complementary	Parallelogram	Transversal
Congruent	Polygon, regular	Trapezium
Corresponding	Proportion	Triangle
Deductive	Quadrilateral	Vertically opposite
Equilateral	Ratio	
Exterior	Rectangle	
Included	Rhombus	
Interior	Similar	

For Example

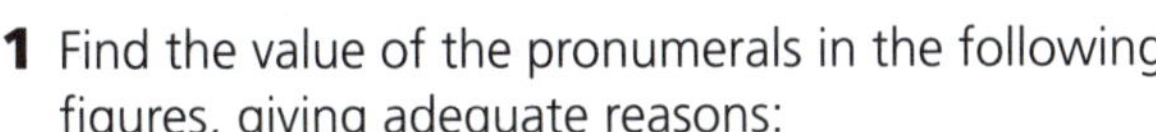

1 Find the value of the pronumerals in the following figures, giving adequate reasons:

a

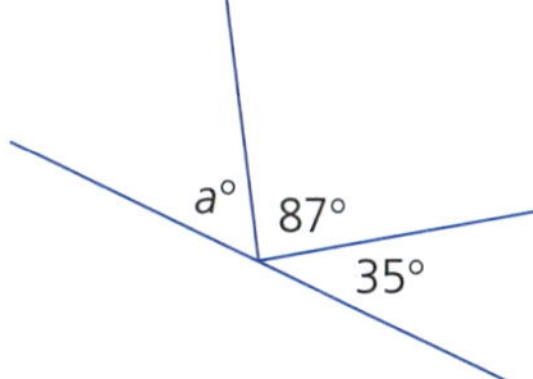

S $a + 87 + 35 = 180$ ($\angle$ sum of straight line)
$\therefore a = 180 - 122$
$= 58$

b

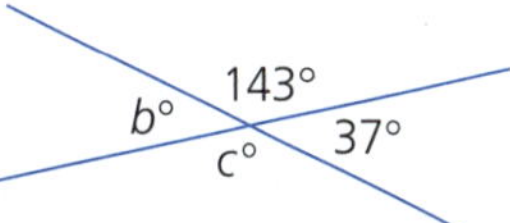

S $b = 37$ (vertically opposite angles)
$c = 143$ (vertically opposite angles)

c

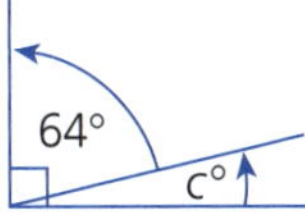

S $c + 64 = 90$ (complementary angles)
$\therefore c = 90 - 64$
$= 26$

d

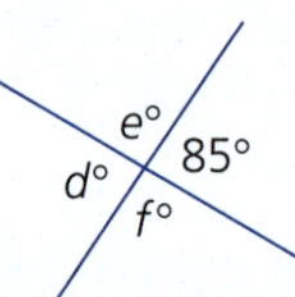

S $d = 85$ (vert. opp. angles)
$e + 85 = 180$ ($\angle$ sum of straight line)
$\therefore e = 180 - 85$
$= 95$
Also $f = 95$ (vert. opposite angles)

e

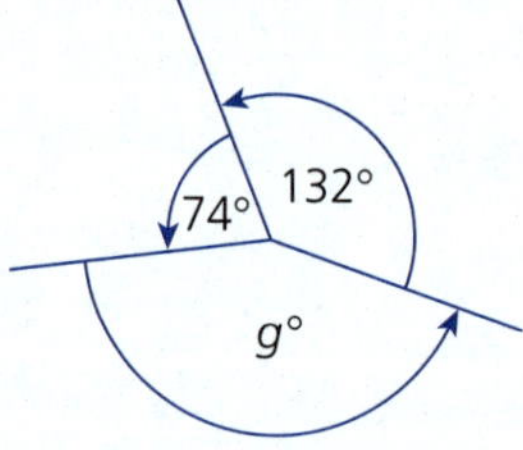

S $g + 74 + 132 = 360$ (angles at a point)
$\therefore g = 360 - 206$
$= 154$

f

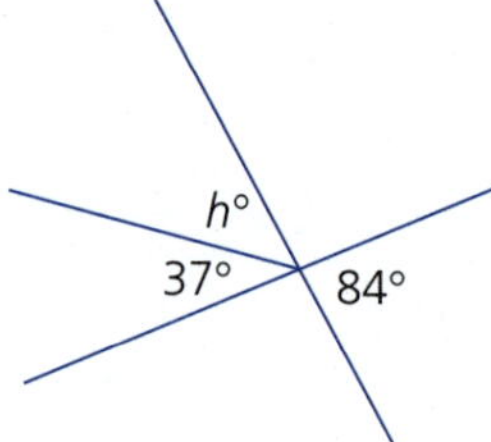

S $h + 37 = 84$ (vert. opp. angles)
$\therefore h = 84 - 37$
$= 47$

g

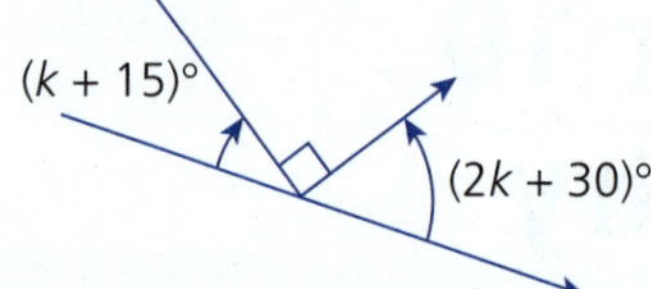

S $(k + 15) + 90 + (2k + 30) = 180$
(angle sum of a straight line)

$\therefore 3k + 135 = 180$
$3k = 45$
$\therefore k = 15$

If the question had asked for the size of each unknown angle, then:

$k + 15 = 15 + 15 = 30$

$2k + 30 = 2 \times 15 + 30 = 60$

$\therefore$ The other angles are 30° and 60°. (Check that they sum to 180°.)

h

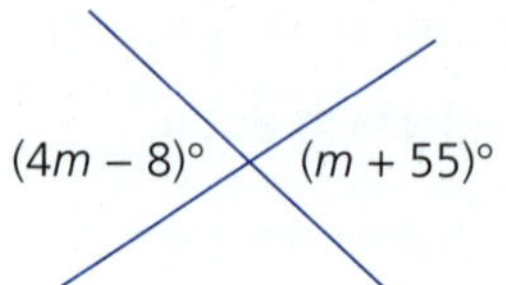

S $4m - 8 = m + 55$ (vert. opp. angles)
$\therefore 3m = 63$
$m = 21$

If you need to find the size of each angle:

$4m - 8 = 4 \times 21 - 8 = 76$

$\therefore$ Each angle is 76°.

Parallel lines

Parallel lines cut by a transversal

When a system of parallel lines is cut by a transversal:

a alternate angles so formed are equal

b corresponding angles so formed are equal

c co-interior angles so formed are supplementary.

These angles are illustrated on the following page.

These angles types appear in pairs and can be remembered by associating them with the relevant letter:

- Alternate—Z
- Corresponding—F
- Co-interior—C (actually [)

Alternate angles (Z)

Co-interior angles ([)

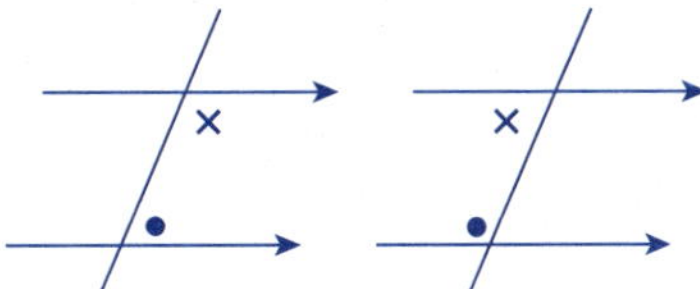

Corresponding angles (F)

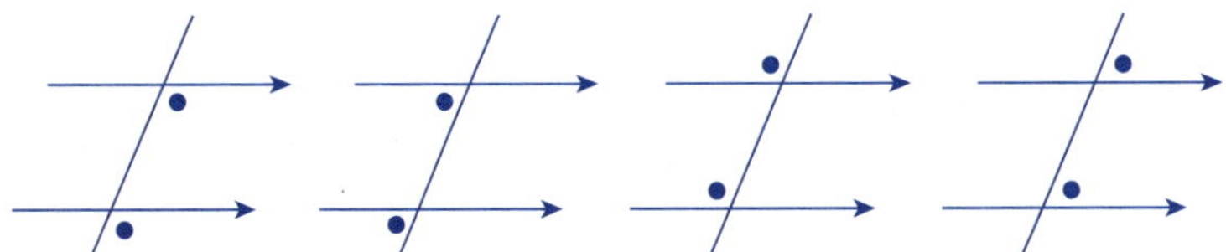

Note: these angle properties are true only when the lines are **parallel**. This means that alternate angles are only equal when they are between parallel lines. Similarly, corresponding angles are only equal and co-interior angles are only supplementary when the lines are parallel.

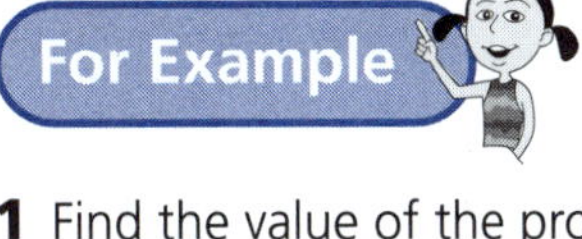

1 Find the value of the pronumeral in the following figures, giving adequate geometric reasons:

a

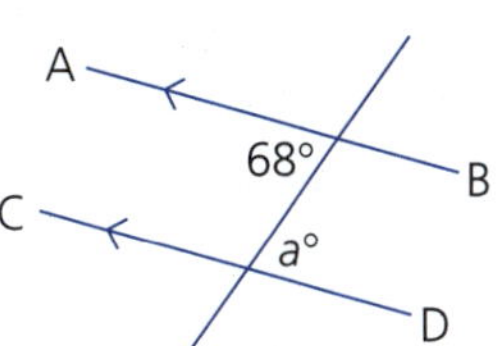

S $a = 68$ (alternate angles, AB || CD)

b

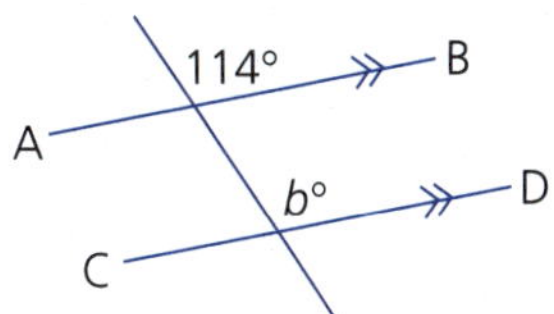

S $b = 114$ (corresponding angles, AB || CD)

c

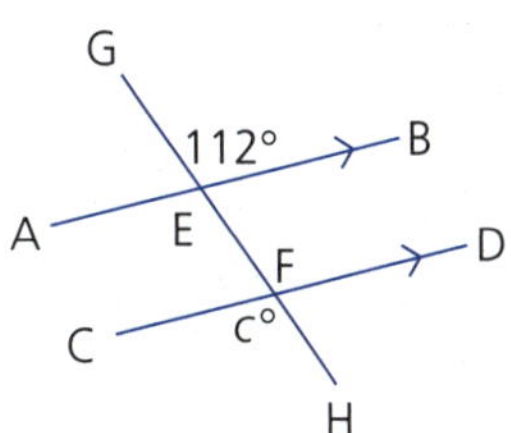

S ∠AEF = 112° (vert. opp. angles)
∴ $c = 112$ (corresponding angles, AB || CD)

d

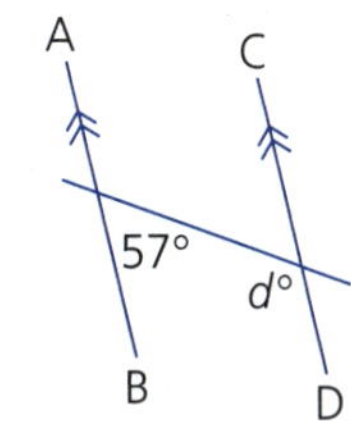

S $d = 180 - 57$ (co-interior angles, AB || CD)
$= 123$

e

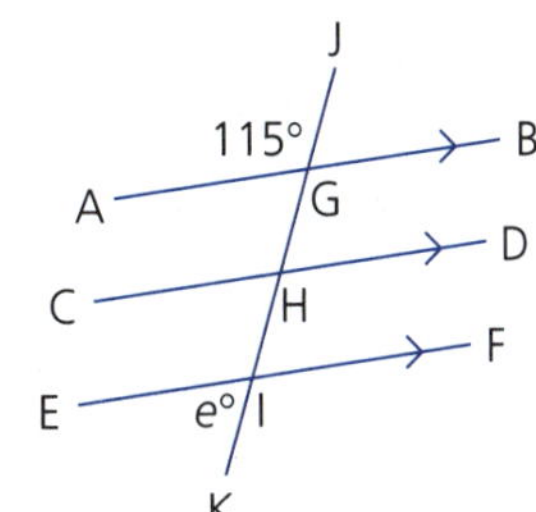

S ∠CHG = 115° (corresponding angles, AB || CD)
∠EIH = 115° (corresponding angles, CD || EF)
∴ $e = 180 - 115$ (angle sum of straight line)
$e = 65$

f

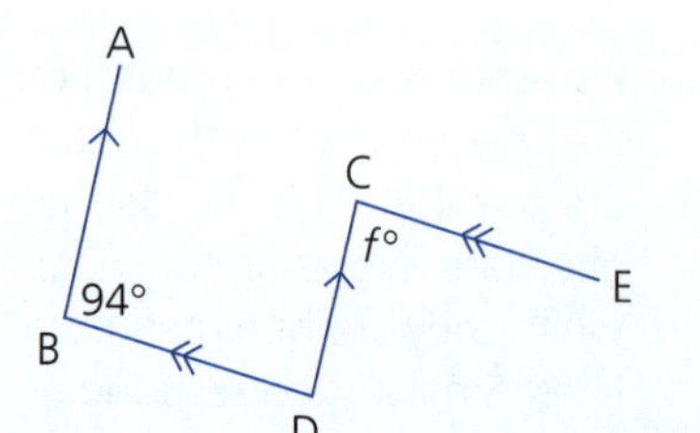

S $\angle BDC = 180° - 94°$ (co-interior angles, AB ∥ CD)
$= 86°$
$\therefore f = 86$ (alternate angles, BD ∥ CE)

Note: always draw a clear diagram. When more than one step is required, fill in the angles as you move around the diagram towards the required angle.

Consider Example **1e** again. The stages of your solution would be:

1 Question stage

2 $\angle CHG = 115°$,
filled in on diagram.

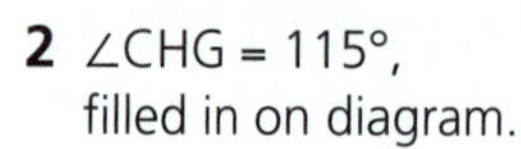

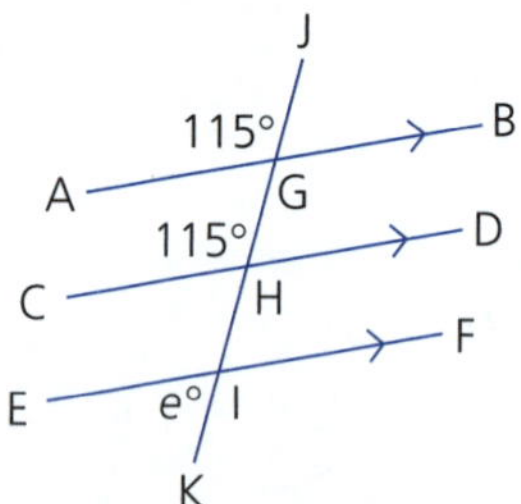

3 $\angle EIH = 115°$,
filled in on diagram.

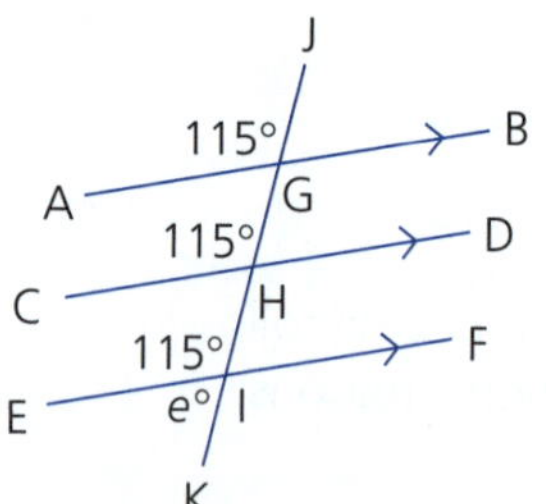

4 $e = 65$,
filled in on diagram.

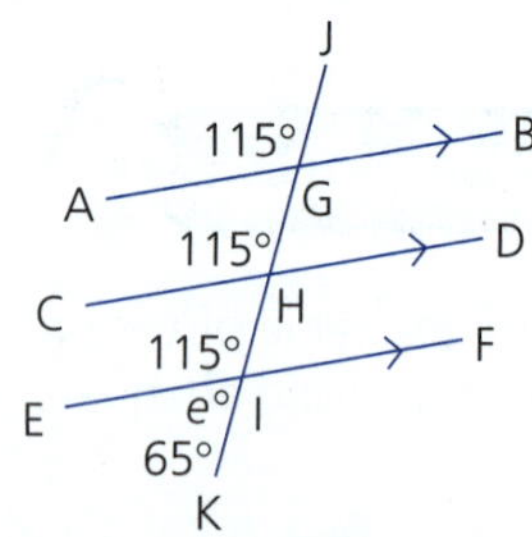

This sequence is drawn to illustrate the procedure used on a diagram in moving from a known angle to an unknown. (It would, of course, all happen on the one diagram.)

Also remember that in many figures there is more than one way of proceeding around a diagram.

Tests for parallel lines

Two lines are parallel if:

- a pair of alternate angles are equal; or
- a pair of corresponding angles are equal; or
- a pair of co-interior angles are supplementary; or
- they are both parallel to a third line.

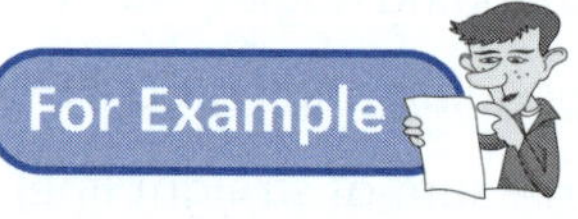

1 Determine if lines are parallel, giving reasons.

a

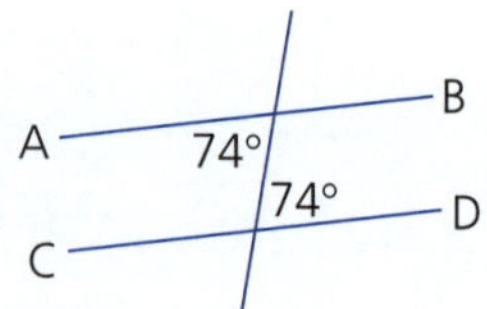

S AB ∥ CD because a pair of alternate angles are equal.

b

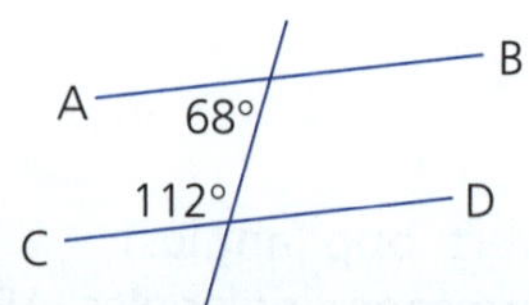

S AB ∥ CD because a pair of co-interior angles are supplementary.

c

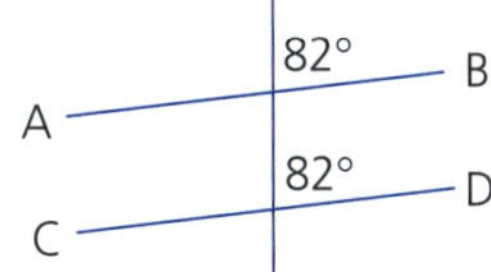

S AB ∥ CD because a pair of corresponding angles are equal.

d

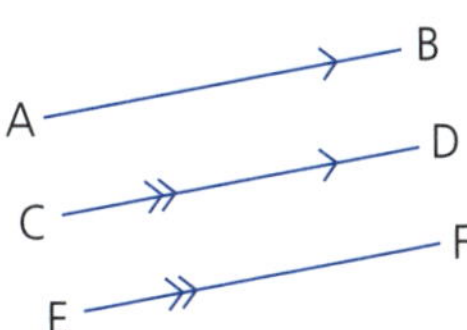

S Given AB ∥ CD, and CD ∥ EF, then AB ∥ EF, because both AB and EF are parallel to CD.

Triangles

Some definitions: a **scalene** triangle is a triangle in which no two sides are equal in length. An **isosceles** triangle is a triangle in which there are two sides equal in length. An **equilateral** triangle is a triangle in which all sides are equal in length.

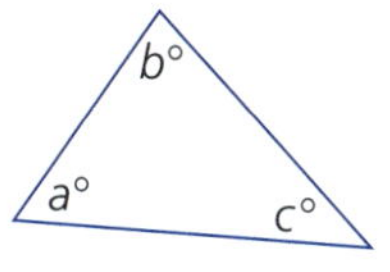

The **angle sum of a triangle is 180°**, that is, $a + b + c = 180$.

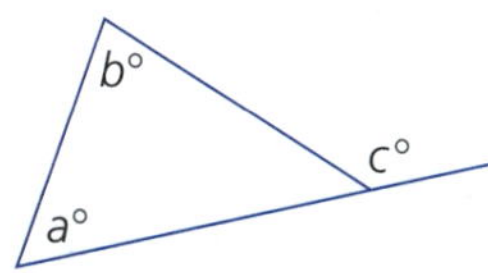

The **exterior angle of a triangle equals the sum of the two interior opposite angles**, that is, $a + b = c$.

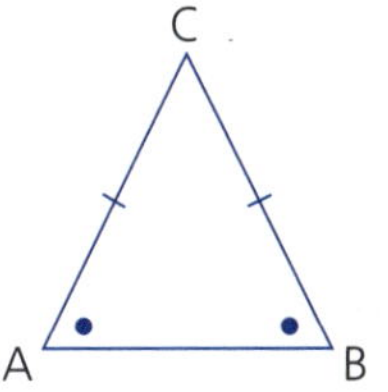

The **base angles of an isosceles triangle are equal**.
Given that AC = BC, then ∠A = ∠B.

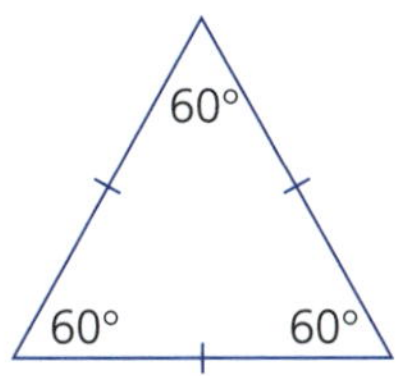

All angles of an equilateral triangle are equal to 60°.

For Example

1 Find the values of the pronumerals in the following figures:

a

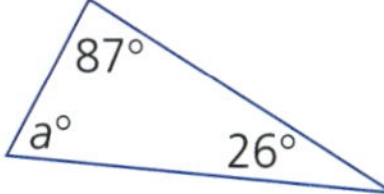

S $a + 26 + 87 = 180$ (angle sum of a △)
$\therefore a = 180 - 113$
$= 67$

b

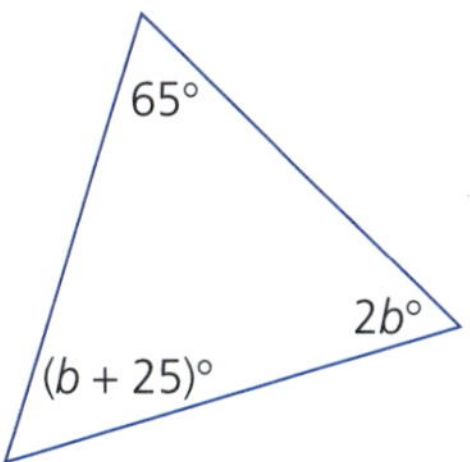

S $(b + 25) + 2b + 65 = 180$ (angle sum of a △)
$\therefore 3b + 90 = 180$
$3b = 90$
$b = 30$

c

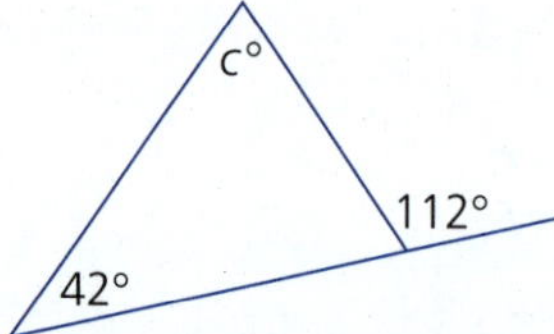

S $c + 42 = 112$ (exterior angle of a triangle)
$\therefore c = 112 - 42$
$= 70$

d

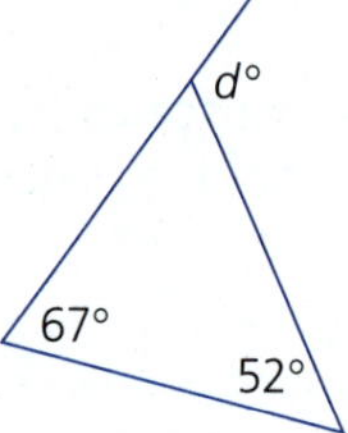

S $d = 67 + 52$ (exterior angle of a △)
$\therefore d = 119$

e

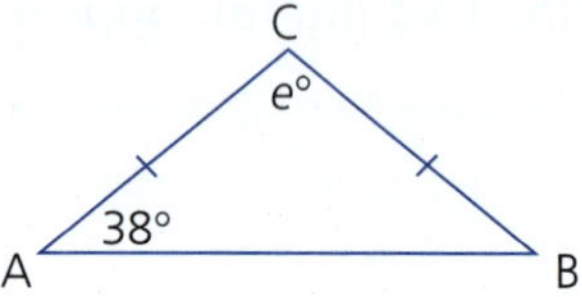

S ∠ABC = 38° (base angles of isosc. △)
$e + 38 + 38 = 180$ (angle sum of a △)
$\therefore e = 180 - 76$
$= 104$

f

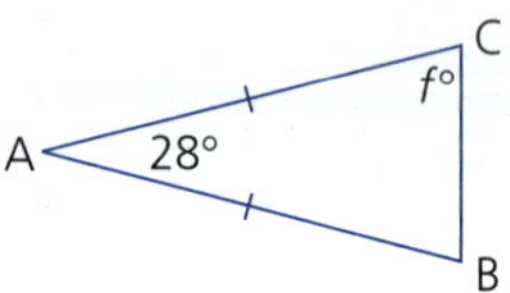

S ∠ABC = f° (base angles of isosc. △)
$\therefore f + f + 28 = 180$ (angle sum of a △)
$2f = 180 - 28$
$2f = 152$
$f = 76$

g

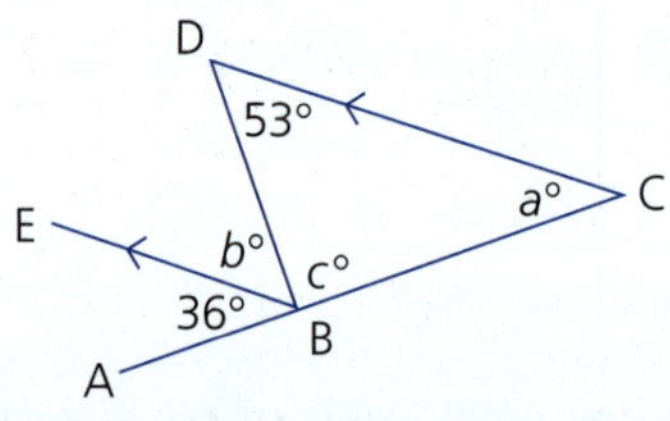

S $a = 36$ (corresponding angles, EB ∥ DC)
$b = 53$ (alternate angles, EB ∥ DC)

$\therefore 36 + 53 + c = 180$ (angle sum of a △)
$c + 89 = 180$
$c = 91$

h

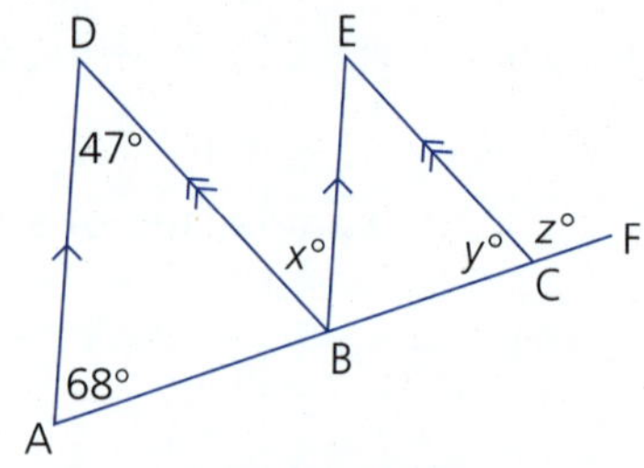

S $x = 47$ (alternate angles, AD ∥ BE)
∠EBC = 68° (corresponding angles, AD ∥ BE)
∠BEC = 47° (alternate angles, DB ∥ EC)

$\therefore y + 68 + 47 = 180$ (angle sum of a △)
$y = 180 - 115$
$= 65$

Also, $z = 68 + 47$ (exterior angle of a △)
$= 115$

Quadrilaterals

The angle sum of a quadrilateral

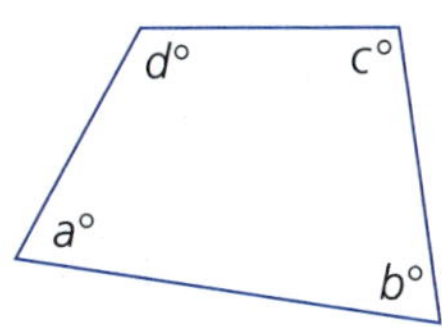

The angle sum of a quadrilateral is 360°, that is,
$a + b + c + d = 360$.

1

84°
112°
86°
$x°$

S $x + 86 + 112 + 84 = 360$
(angle sum of a quadrilateral)

$$\therefore x + 282 = 360$$
$$x = 360 - 282$$
$$x = 78$$

2

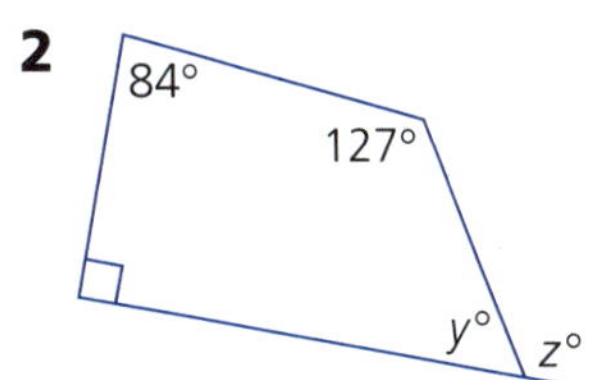

S $y + 90 + 84 + 127 = 360$
(angle sum of a quadrilateral)

$$\therefore y + 301 = 360$$
$$y = 59$$

Also, $z + 59 = 180$
(angle sum of a straight line)

$$\therefore z = 180 - 59$$
$$z = 121$$

Special quadrilaterals

A trapezium is a quadrilateral in which one pair of opposite sides is parallel.

A parallelogram is a quadrilateral with both pairs of opposite sides parallel. Its special properties are that its:
- opposite sides are equal
- opposite angles are equal
- diagonals bisect each other.

A rectangle is a parallelogram with one angle a right-angle. Its special properties are that:
- all the properties of the parallelogram
- all angles are right-angles
- the diagonals are equal.

A square is a rectangle with one pair of adjacent sides equal. Its special properties are that:
- all the properties of the rectangle
- all sides are equal
- the diagonals bisect each other at right-angles
- the diagonals bisect the angles of the square.

A rhombus is a parallelogram with one pair of adjacent sides equal. Its special properties are that:
- all the properties of a parallelogram
- the diagonals bisect each other at right-angles
- the diagonals bisect the angles of the rhombus.

The **square** can also be defined in terms of the **rhombus**. A square is a rhombus with one angle a right-angle. A square is a quadrilateral, which is both a rhombus and a rectangle.

The quadrilateral family tree

Each quadrilateral in the tree has all the properties of those above it, plus its own distinct properties.

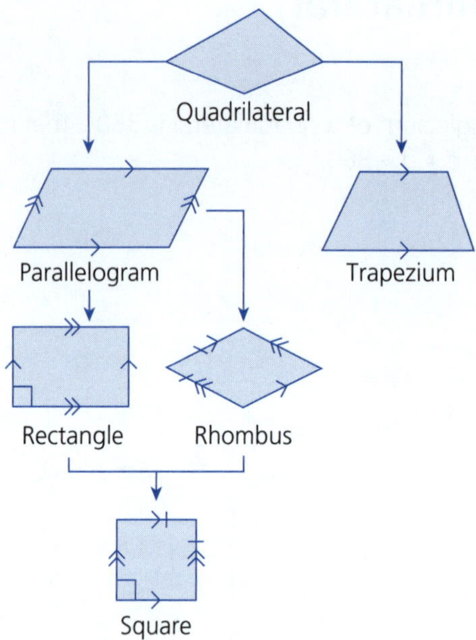

Angle sum of a polygon

- Polygons have an angle sum given by $S = 180(n - 2)$ degrees, where n = number of sides.
- The sum of all external angles of a polygon is 360°.

1 Find the angle sum of an octagon.

S An octagon has eight sides.

Subs. $n = 8$ in $S = 180(n - 2)$
$= 180(8 - 2)$
$= 1080$

∴ Angle sum of 1080°.

2 What is the size of each interior angle in a regular decagon?

S A decagon has 10 sides.

Subs. $n = 10$ in $S = 180(n - 2)$
$= 180(10 - 2)$
$= 1440$

∴ Angle sum of 1440°.

Size of each angle $= 1440 \div 10$
$= 144$

∴ Each angle is 144°.

3 Find the size of each exterior angle of a regular hexagon?

S A hexagon has six sides.

Size of each exterior angle $= 360 \div 6$
$= 60$

∴ Each exterior angle is 60°.

4 Each exterior angle of a regular polygon is 4°. What is the angle sum of the polygon?

S Number of exterior angles $= 360 \div 4$
$= 90$

∴ The polygon has 90 sides.

Each interior angle = 176°

Angle sum of polygon $= 176 \times 90$
$= 15\,840$

∴ The angle sum is 15 840°.

Congruent triangles

Congruent triangles are equal in all respects.

Using tests for congruent triangles

Tests for congruent triangles can be applied to prove assumptions. The tests for congruent triangles are:

a Two triangles are congruent if the three sides of one triangle are respectively equal to the three sides of the other. (**SSS**)

b Two triangles are congruent if two sides of one and the included angle are respectively equal to two sides and the included angle of the other. (**SAS**)

c Two triangles are congruent if two angles and one side of one triangle are respectively equal to two angles and the corresponding side of the other. (**ASA**)

Note 1: the included angles are between the equal sides:

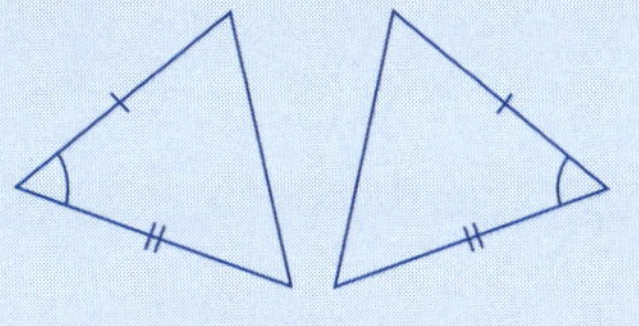

Note 2: the corresponding sides are opposite the **same** equal angle in each triangle, as follows:

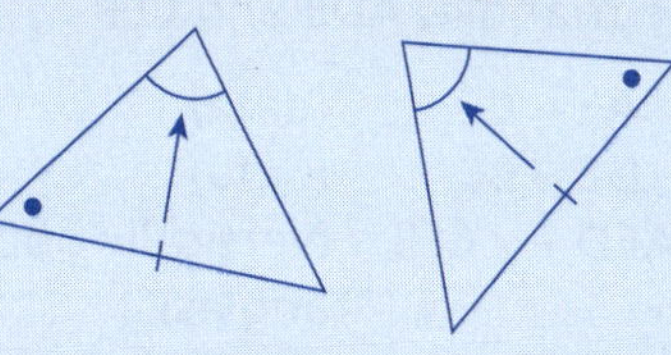

d Two right-angled triangles are congruent if the hypotenuse and one side of one triangle are respectively equal to the hypotenuse and one side of the other. (**RHS**)

Important
- Mark the data on the diagram and then mark any other angles and sides that you can show are equal.
- A congruence proof consists of **three** (3) equality statements.
- All congruence assumptions may be applied to right-angled triangles—RHS is just an additional assumption for right-angled triangles.

1

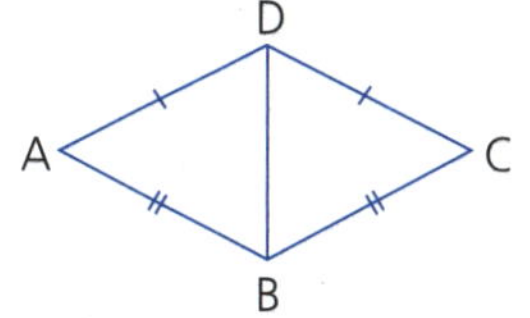

Given that AD = DC and AB = BC, prove that △ABD ≡ △CBD.

Note: ≡ means 'is congruent to'.

S Proof:

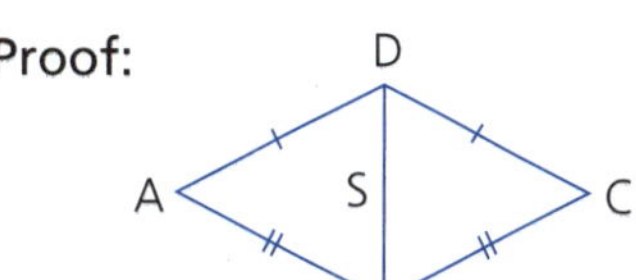

In the triangles ABD and CBD

AB = CB	(data)
AD = CD	(data)
BD = BD	(common side)
∴△ABD ≡ △CBD	(SSS)

2 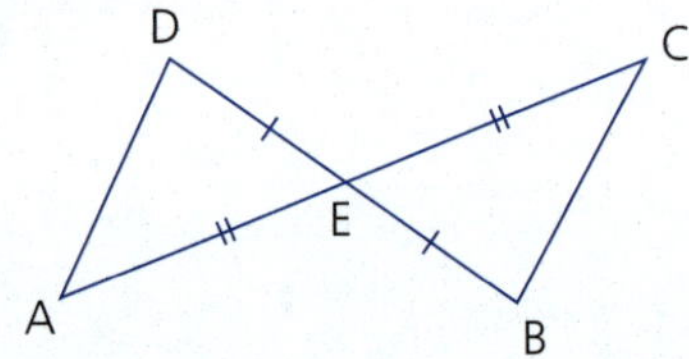

Given AE = EC and DE = EB, prove that $\triangle AED \equiv \triangle CEB$.

 Proof:

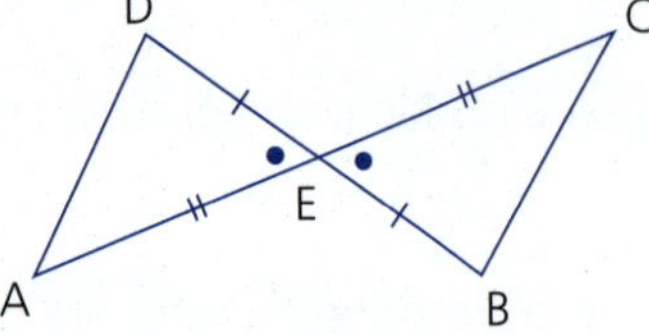

In the triangles ADE and CEB

AE = CE	(data)
DE = BE	(data)
$\angle AED = \angle CEB$	(vertically opposite angles)
$\therefore \triangle AED \equiv \triangle CEB$	(SAS)

3 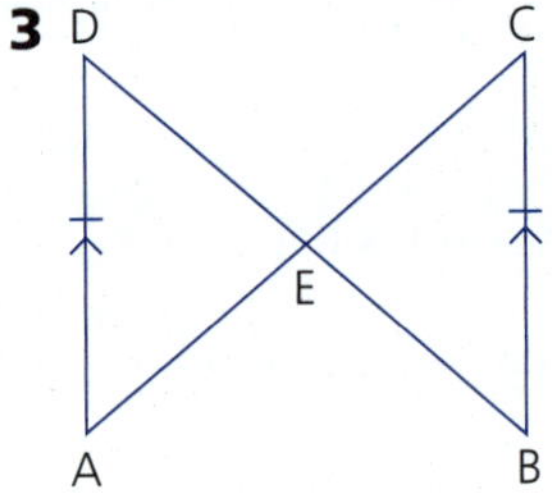

Given AD = BC and AD $\parallel$ BC, prove that $\triangle AED \equiv \triangle CEB$.

S Proof:

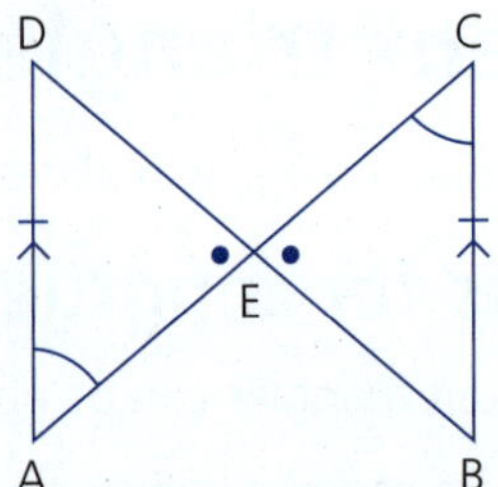

In the triangles AED and CEB

AD = CB	(data)
$\angle DAE = \angle BCE$	(alternate angles, AD $\parallel$ BC)
$\angle AED = \angle CEB$	(vertically opposite angles)
$\therefore \triangle AED \equiv \triangle CEB$	(ASA)

4 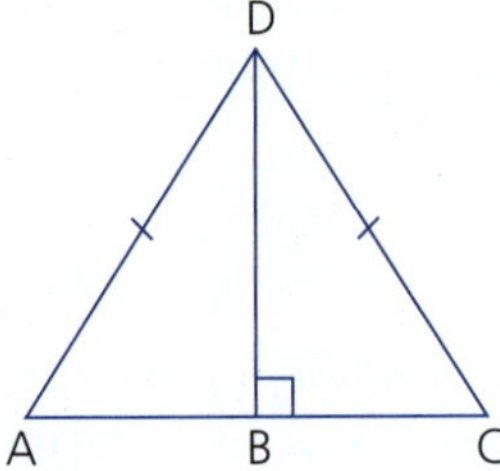

Given BD $\perp$ AC and AD = DC, prove that $\triangle ABD \equiv \triangle CBD$.

S Proof:

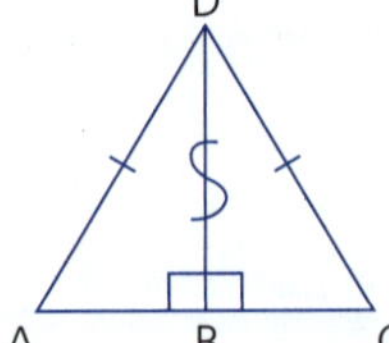

In the triangles ABD and CBD

AD = CD	(data)
BD = BD	(common side)
$\angle ABD = \angle CBD = 90°$	(BD $\perp$ AC)
$\therefore \triangle ABD \equiv \triangle CBD$	(RHS)

Beyond the congruence assumptions

Many important properties can be proved using congruent triangles. Once the triangles have been shown to be congruent, all other corresponding sides and angles of the triangles are then known to be equal.

1 ABCD is a parallelogram with diagonals AC and DB meeting at E.

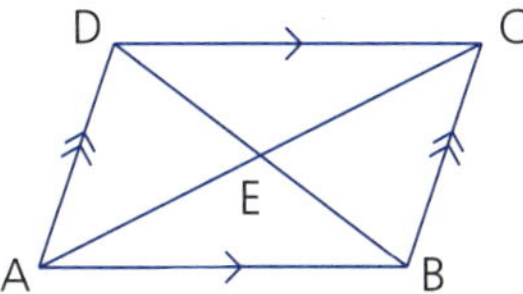

a Prove that $\triangle AEB \equiv \triangle CED$.

S In triangles AEB and CED

AB = CD (opposite sides of a parallelogram)
$\angle EAB = \angle ECD$ (alternate angles, AB ∥ DC)
$\angle ABE = \angle EDC$ (alternate angles, AB ∥ DC)
$\therefore \triangle AEB \equiv \triangle CED$ (ASA)

b Prove that AE = EC and EB = DE (that is, that the diagonals bisect each other).

S Proof:

As triangles AEB and CED are congruent,

AE = CE (corresponding sides of congruent triangles)
and EB = ED (corresponding sides of congruent triangles)

Deductive geometry

Each step should be justified by stating a reason.

1

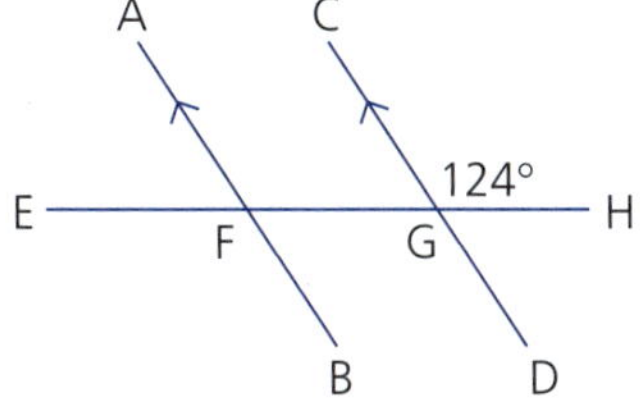

Given AB ∥ CD and $\angle CGH = 124°$, find $\angle AFE$, giving reasons.

$\angle AFG = 124°$ (corresponding angles, AB ∥ CD)
$\therefore \angle AFE = 180° - 124°$ (angle sum of a straight line)
$\angle AFE = 56°$

2

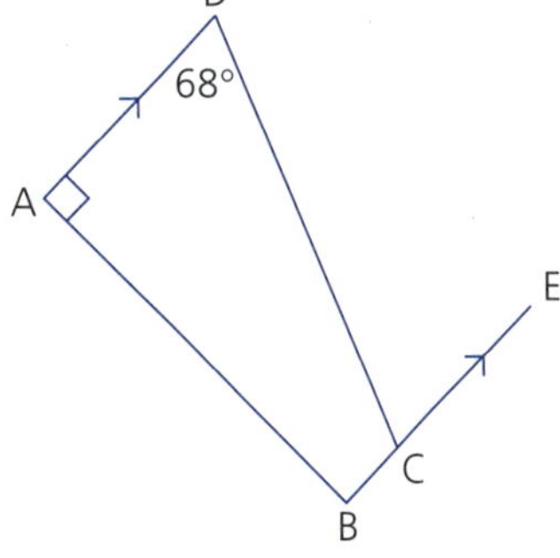

Given $\angle BAD = 90°$, AD ∥ BE and $\angle ADC = 68°$, find $\angle DCE$ giving reasons.

Remember to solve the problem on your diagram first.

S $\angle DCE = 68°$ (alternate angles AD ∥ BE)

3

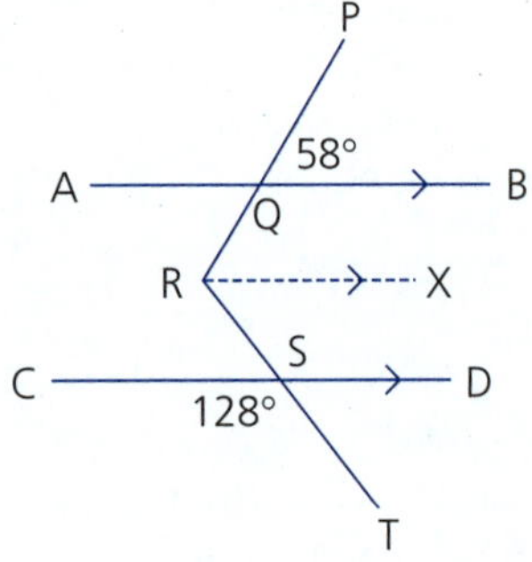

Given $\angle PQB = 58°$, $\angle CST = 128°$, and $AB \parallel CD$, find $\angle QRS$.

S Construction: draw RX parallel to AB and CD.

$\angle CSR = 180° - 128°$
$= 52°$ (angle sum of a straight line)
$\angle SRX = 52°$ (alternate angles, $CD \parallel RX$)
$\angle QRX = 58°$ (corresponding angles, $AB \parallel RX$)

Now
$\angle QRS = \angle QRX + \angle SRX$
$= 58° + 52°$
$= 110°$

Now look at the steps on the diagram:

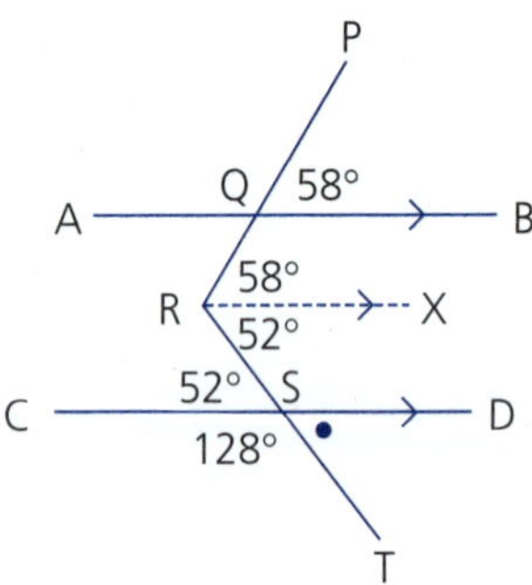

By working around the diagram, from known to unknown, the logical steps for your proof become obvious.

4

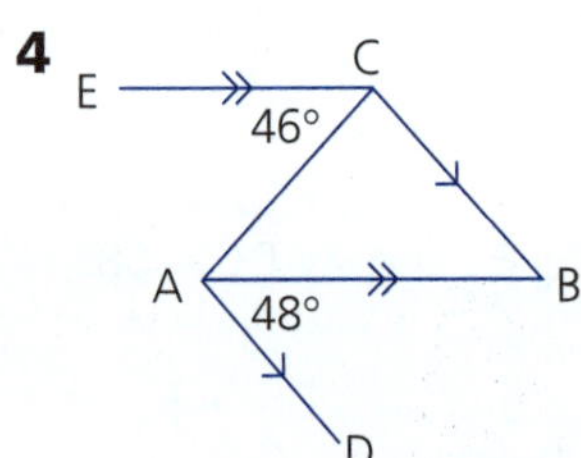

Given $AD \parallel CB$, $EC \parallel AB$, $\angle ECA = 46°$ and $\angle DAB = 48°$, find $\angle ACB$, giving reasons.

S

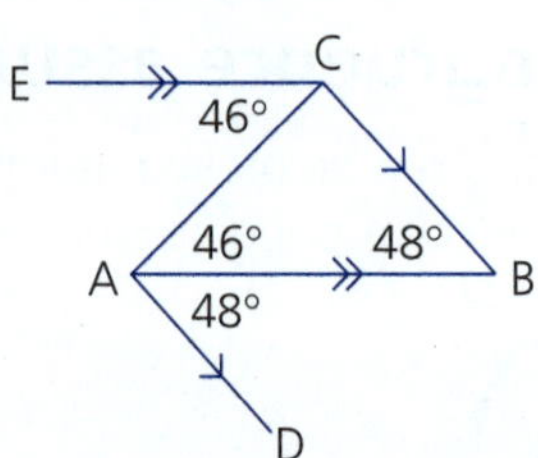

$\angle ABC = 48°$ (alternate angles, $AD \parallel CB$)
$\angle CAB = 46°$ (alternate angles, $EC \parallel AB$)
$\therefore \angle ACB = 180° - 48° - 46°$ (angle sum of a triangle)
$\angle ACB = 86°$

5

Given $AC = BC$ and $\angle ACB = 48°$, find $\angle CAB$, giving reasons.

S Let $\angle CAB = x°$
$\therefore \angle ABC = x°$ (base angles of an isosceles triangle)
$\therefore 2x + 48 = 180$ (angle sum of a triangle)
$\therefore 2x = 132$
$x = 66$
$\therefore \angle CAB = 66°$

6

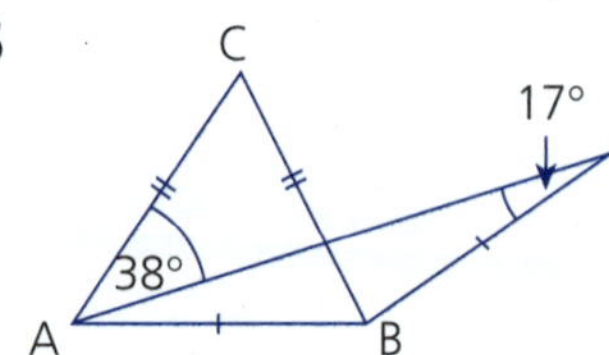

Given $AC = CB$, $AB = BD$, $\angle ADB = 17°$ and $\angle CAD = 38°$, find $\angle ACB$, giving reasons.

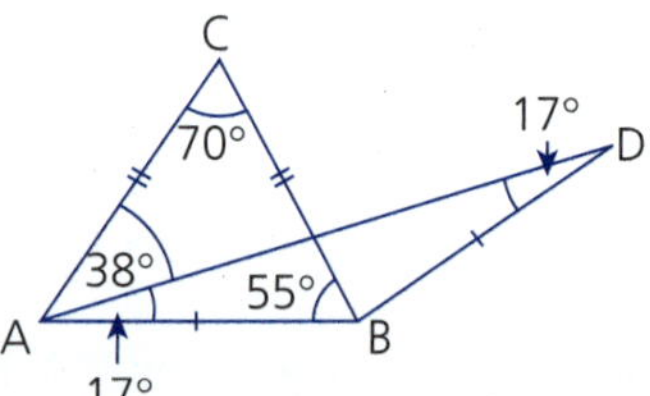

$\angle DAB = 17°$ (base angles of isosceles $\triangle ABD$)
$\therefore \angle CAB = 17° + 38°$ ($\angle CAD + \angle DAB$)
$= 55°$
$\therefore \angle ABC = 55°$ (base angles of isosceles $\triangle ABC$)

Hence
$\angle ACB = 180° - 55° - 55°$ (angle sum of a triangle)
$\angle ACB = 70°$

7 AB is a diameter of a circle with centre O.
$\angle ACB$ is subtended by AB at C.
Prove that $\angle ACB = 90°$.

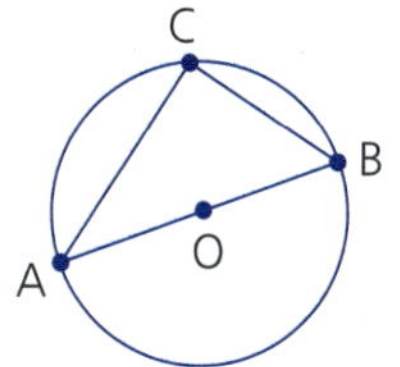

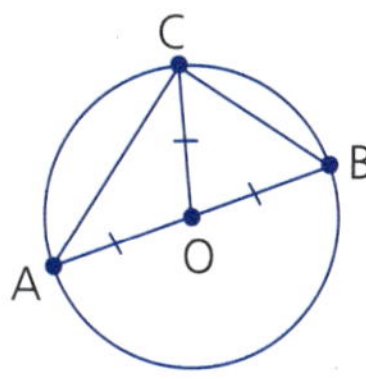

Join OC. Let $\angle CAO = x°$, $\angle CBO = y°$

In $\triangle AOC$,

$AO = CO$ (equal radii)

$\therefore \angle ACO = x°$ ($\angle$s opposite equal sides)

Similar triangles

Definition

Two triangles are similar if:

a two angles of one are equal to two angles of the other (equiangular), or

b the lengths of their corresponding sides are in proportion, or

c the lengths of two sides are in proportion and the included angles are equal.

The following figures illustrate the three tests:

a

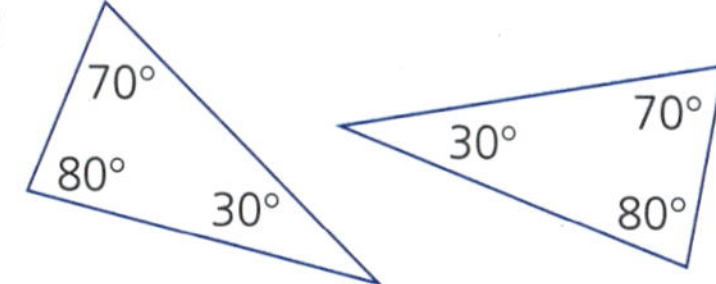

Equiangular

In $\triangle COB$,

$CO = BO$ (equal radii)

$\therefore \angle OCB = y°$ ($\angle$s opposite equal sides)

Then $\angle ACB = x° + y°$

In $\triangle ABC$,

$x° + y° + x° + y° = 180°$ ($\angle$ sum of $\triangle$)

$\therefore 2x + 2y = 180$

$\therefore x + y = 90$

$\therefore \angle ACB = 90°$

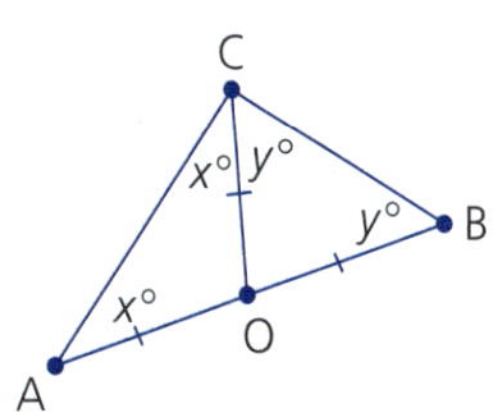

b

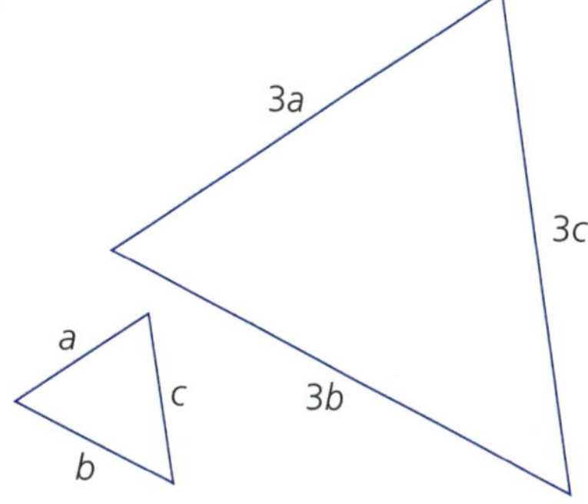

$$\frac{a}{3a} = \frac{b}{3b} = \frac{c}{3c} = \frac{1}{3}$$

Three sides in proportion

c

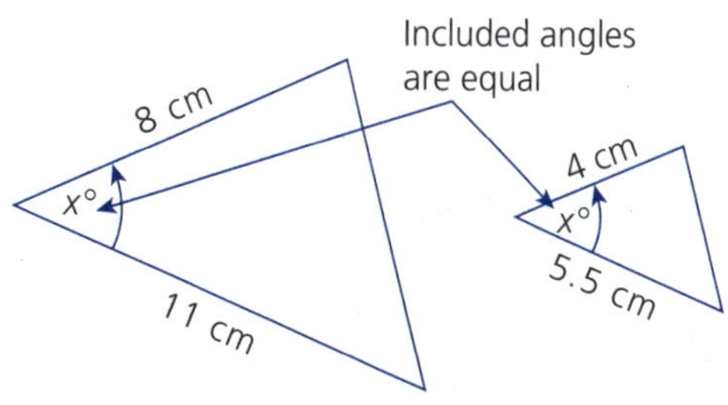

$$\frac{8}{4} = \frac{11}{5.5} = \frac{2}{1}$$

Two sides in proportion, and included angle equal

Special case

Two right-angled triangles are similar if the hypotenuse and one other side of one right-angled triangle are proportional to the hypotenuse and another side of the second right-angled triangle.

Note: this is a special instance of case **c** on the opposite page.

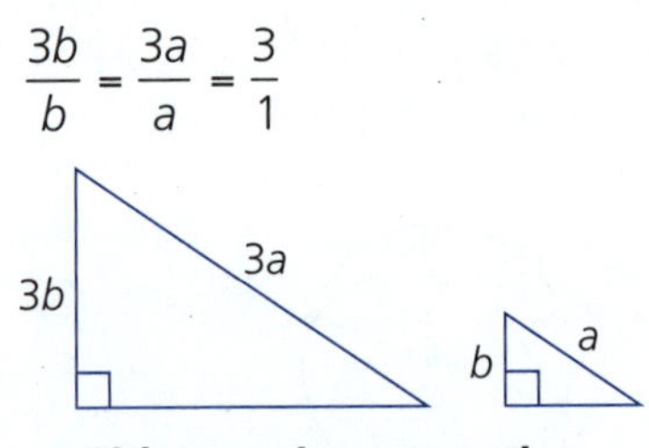

Sides are in proportion

Proving two triangles similar

A formal approach is used to establish similar triangles, giving reasons for each step.

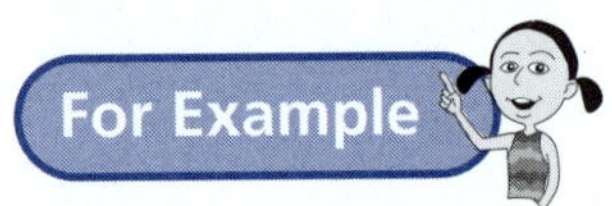

1

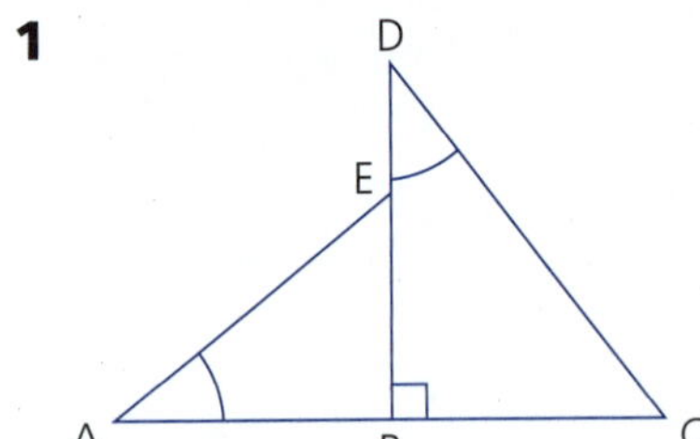

Given that DB ⊥ AC and that ∠BAE = ∠BDC, prove that the triangles ABE and DBC are similar.

Proof:

In the △s ABE and DBC

∠BAE = ∠BDC	(data)
∠ABE = ∠DBC	(both right-angles, DB ⊥ AC)
∴ △ABE ||| △DBC	(equiangular)

||| means 'is similar to'.

Also remember that if two corresponding angles are equal then the third angle must be equal.

2

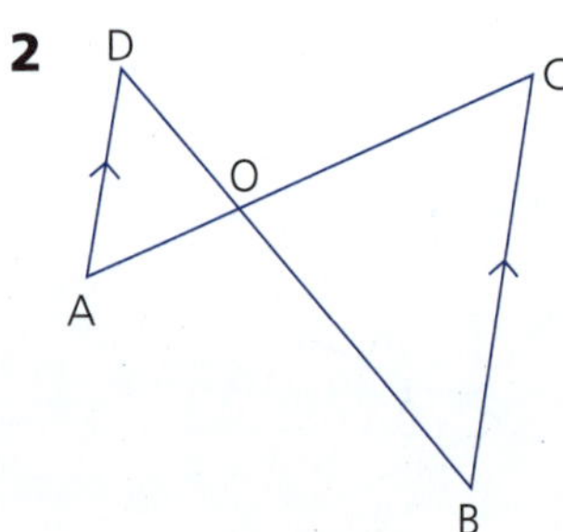

Given AD || BC, prove that △s AOD and COB are similar. Hence write down the ratios of the corresponding sides.

S Proof:

In △s AOD and COB

∠AOD = ∠COB	(vertically opposite)
∠DAO = ∠BCO	(alternate angles, AD || BC)
∴ △AOD ||| △COB	(equiangular)

Note: alternate angles is not a sufficient reason for ∠DAO = ∠BCO; the alternate angles are only equal because AD || BC. This must also be mentioned.

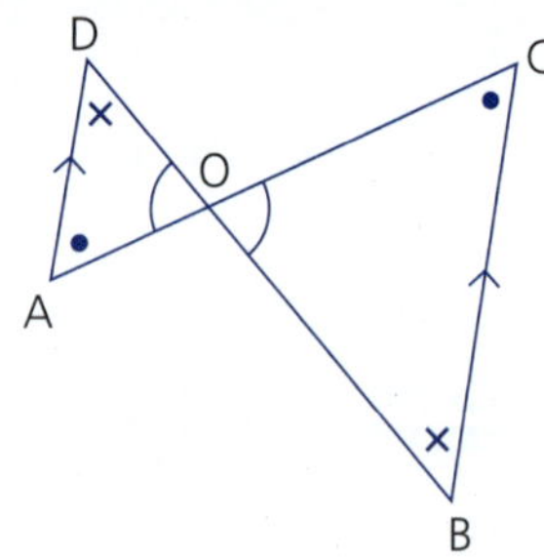

The corresponding sides will be opposite the same equal angles in both triangles. Mark the equal angles in the diagram in a distinctive way (include the third angle in each triangle).

Now make up the ratios of the opposite sides.

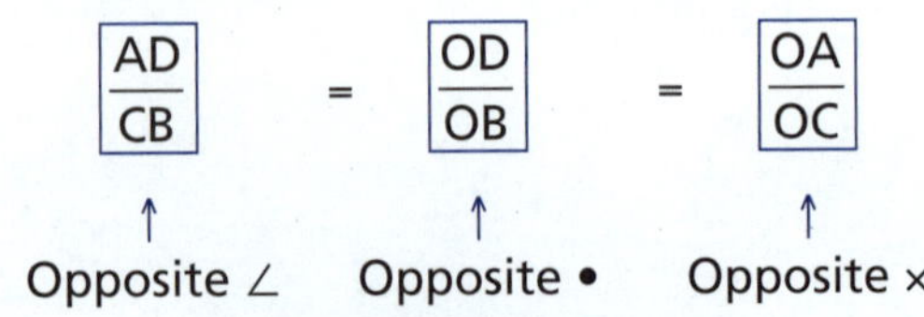

$\frac{\text{All sides from } \triangle AOD}{\text{All sides from } \triangle COB}$

The ratios are equal because the lengths of the corresponding sides of similar triangles are in proportion. This property is useful to calculate unknown sides of similar triangles.

Referring to the same diagram again:
If AD = 12 cm, BC = 18 cm and AO = 8 cm, calculate the length of OC.

We have:

$$\frac{AD^{\checkmark}}{CB^{\checkmark}} = \cancel{\frac{OD}{OB}} = \frac{OA^{\checkmark}}{OC^{?}}$$

AD = 12
BC = 18
AO = 8
OC = ?

We do not need the middle ratio as we have no information about OD or OB. Substituting into the proportion:

$$\frac{12}{18} = \frac{8}{OC}$$

Cross-multiply

$$12 \times OC = 8 \times 18$$

$$OC = \frac{8 \times 18}{12} = 12$$

OC is 12 cm long.

3

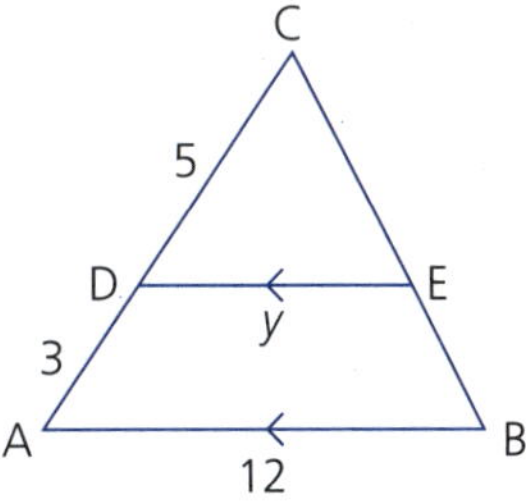

a Given that DE ∥ AB, prove that the △s ABC and DEC are similar.

S In the △s ABC and DEC

∠ACB = ∠DCE	(common angle)
∠BAC = ∠EDC	(corresponding angles, AB ∥ DE)
∴ △ABC ||| △DEC	(equiangular)

b If AD = 3, DC = 5 and AB = 12, find the value of y.

S The ratios of the corresponding sides of congruent triangles are equal.

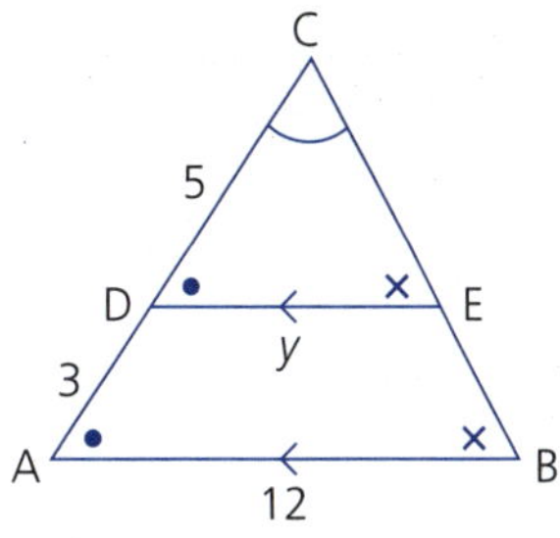

$$\frac{DE^{?}}{AB^{\checkmark}} = \cancel{\frac{CE}{CB}} = \frac{CD^{\checkmark}}{CA^{\checkmark}}$$

CA = 5 + 3 = 8

$$\frac{y}{12} = \frac{5}{8}$$

$$8y = 60$$

$$y = \frac{60}{8} = 7.5$$

The value of y is 7.5.

Mark equal angles in the same manner. Read off the opposite sides. Make sure the sides of one triangle are all on top of the ratio, that is: $\frac{\text{sides of } \triangle CDE}{\text{sides of } \triangle CAB}$

c If EB = 1.8, let CE be x and calculate x.

S

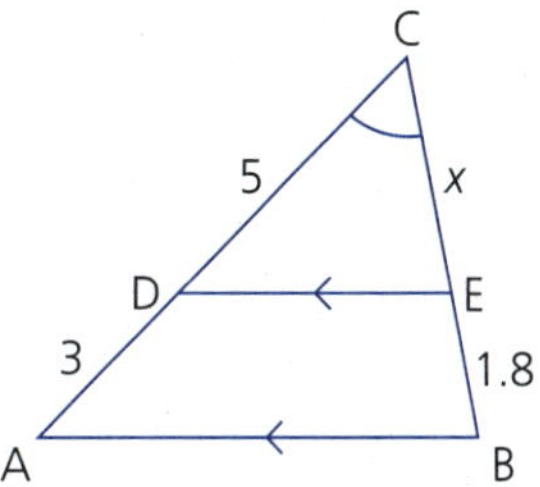

$$\cancel{\frac{DE}{AB}} = \frac{CE^{?}}{CB^{\checkmark}} = \frac{CD^{\checkmark}}{CA^{\checkmark}}$$

From diagram

CE = x CD = 5
CB = x + 1.8 CA = 8

$$\frac{x}{x + 1.8} = \frac{5}{8}$$

$$8x = 5(x + 1.8)$$
$$8x = 5x + 9$$
$$3x = 9$$
$$x = 3$$

The value of x is 3.

Additional triangle properties (proven using similar triangles)

The interval joining the midpoints of two sides of any triangle is parallel to the third side and is half of its length.

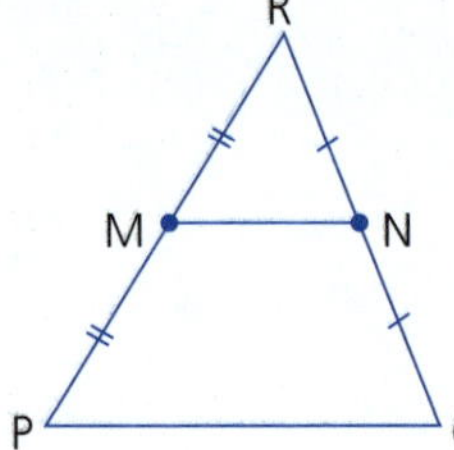

If RM = MP and RN = NQ then MN || PQ and MN = $\frac{1}{2}$PQ.

The line through the midpoint of one side of a triangle and parallel to another side must bisect the third side.

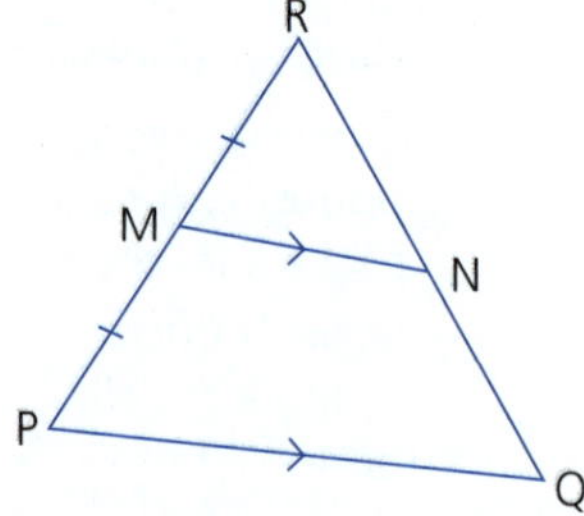

If RM = MP and MN || PQ then RN = NQ.

Areas of similar figures

Similar figures have the same shape, with the lengths of corresponding dimensions in the same ratio. Similar figures have:

- perimeters proportional to the lengths of corresponding sides
- areas proportional to the squares of the lengths of corresponding sides.

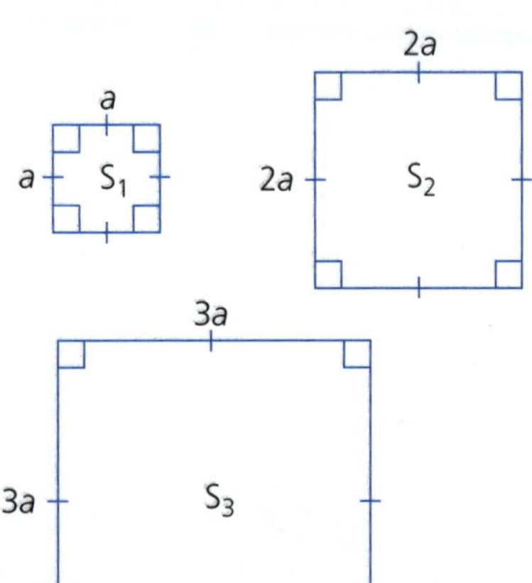

Area $S_1 = a \times a = a^2$ units^2

Area $S_2 = 2a \times 2a = 4a^2$ units^2

Area $S_3 = 3a \times 3a = 9a^2$ units^2

Consider S1 and S2 : Ratio of sides $= \frac{a}{2a} = \frac{1}{2}$

Ratio of areas $= \frac{a^2}{4a^2} = \frac{1}{4}$

Consider S2 and S3 : Ratio of sides $= \frac{2a}{3a} = \frac{2}{3}$

Ratio of areas $= \frac{4a^2}{9a^2} = \frac{4}{9}$

Note: the ratio of the areas is the square of the ratio of the sides. The areas of similar shapes are proportional to the squares of the corresponding sides. Thus, if the ratio of corresponding sides is $r : s$, then the ratio of the areas of the figures will be $r^2 : s^2$.

1

2 cm

5 cm

Find the ratio of the areas of circles with radii 2 cm and 5 cm.

S Ratio of areas $= \dfrac{2^2}{5^2} = \dfrac{4}{25}$

2

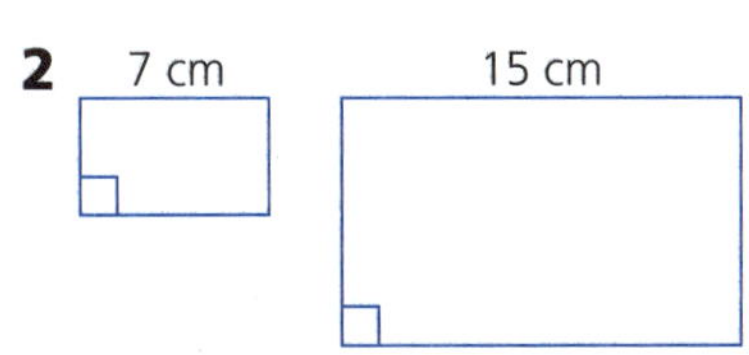

Similar rectangles have lengths 7 cm and 15 cm. Find the ratios of their areas.

S Ratio of areas $= \dfrac{7^2}{15^2} = \dfrac{49}{225}$

3 A scale drawing of a lake is made, such that the actual width of the lake, 180 m, becomes 36 cm on the plan. If the area of the lake on the plan is 200 cm^2, calculate the actual area of the lake.

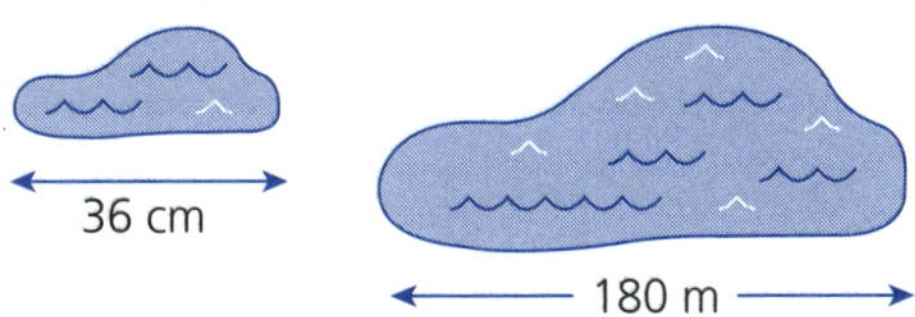

S Ratio of diameters $= 36$ cm : 180 m
$= 36 : 18000$
$= 1 : 500$ [÷ by 36]

Ratio of areas $= 1^2 : 500^2$
$= 1 : 250000$

Let the area of the lake be A cm^2.

Then $\dfrac{200}{A} = \dfrac{1}{250000}$

$A = 50000000$

The area is 50 000 000 cm^2, that is, 5000 m^2.

1 m = 100 cm
1 m^2 = 100^2 cm^2
∴ 1 m^2 = 10 000 cm^2

4 It has been suggested that the dimensions of a rugby in-goal area be increased to $1\frac{1}{4}$ times the present size. If the area of the present in-goal area is 800 m^2, find the area of the proposed in-goal area.

800 m^2 ? m^2

S Ratio of sides $= \dfrac{5}{4}$

Ratio of areas $= \dfrac{25}{16}$

Let the new area be A m^2.

$$\frac{A}{800} = \frac{25}{16}$$

That is, $16A = 25 \times 800$

$$A = \frac{25 \times 800}{16}$$
$$= 1250$$

The new in-goal area would be 1250 m^2.

Volumes of similar solids

Similar solids have the same shape, with the lengths of corresponding dimensions in the same ratio. Similar solids have:

- surface areas proportional to the squares of the lengths of corresponding sides
- volumes proportional to the cubes of the lengths of corresponding sides.

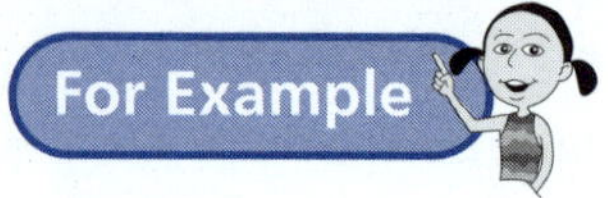

1

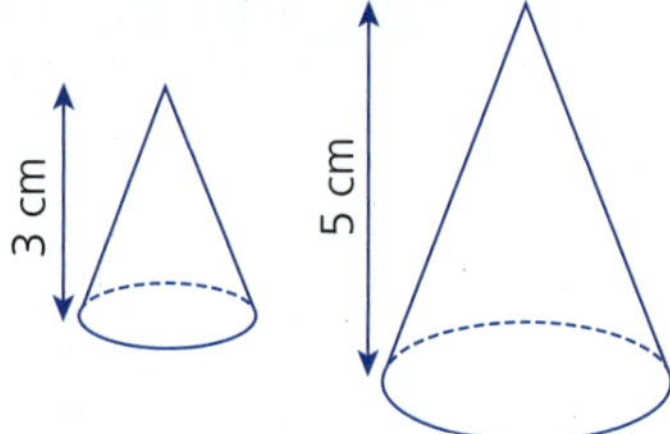

For these similar cones, find the ratio of their:

a surface areas

b volumes.

S Ratio of heights $= \frac{3}{5}$

Ratio of surface areas $= \frac{3^2}{5^2} = \frac{9}{25}$

Ratio of volumes $= \frac{3^3}{5^3} = \frac{27}{125}$

2 If the volume of the larger cone in Example **1** is 500 cm^3, calculate the volume of the smaller cone.

S Let the volume be V cm^3.

$$\frac{V}{500} = \frac{27}{125}$$
$$125V = 27 \times 500$$
$$V = \frac{27 \times 500}{125}$$
$$= 108$$

The volume of the smaller cone is 108 cm^3.

3 A scale model of a yacht displaces 800 cm^3 of water. If the scale model has a length of 10 cm, compared with the actual length of 20 m, find the volume of water displaced by the real yacht.

S Ratio of lengths = 10 cm : 20 m
= 10 : 2000
= 1 : 200

Ratio of volumes $= 1^3 : 200^3$

Let the volume of the actual yacht be V.

Then

$$\frac{800}{V} = \frac{1}{200^3}$$

Plan / Actual

$$V = 800 \times 200^3 \text{ cm}^3$$
$$= \frac{800 \times 8\,000\,000}{1\,000\,000} \text{ m}^3$$
$$= 6400 \text{ m}^3$$

$1 \text{ m}^3 = 100 \times 100 \times 100 \text{ cm}^3$
$= 1\,000\,000 \text{ cm}^3$

The actual displacement of water is 6400 m^3.

4 A $\frac{1}{50}$ scale model of the Boganbrey Town Hall requires 80 mL of paint to cover its outside walls. Calculate the quantity of paint needed for the outside walls of the actual Town Hall.

S Let the required area to be covered be A.

Ratio of lengths = 1 : 50
Ratio of surface areas $= 1^2 : 50^2$
= 1 : 2500

Area of scale mode / Area of actual hal

Then $\frac{0.08}{A} = \frac{1}{2500}$ 80 mL = 0.08 L

$$A = 0.08 \times 2500$$
$$= 200$$

The Town Hall surface will require 200 L of paint.

Practise Practise

In Questions 1–31, find the values of the pronumerals, giving reasons for your answers.

1

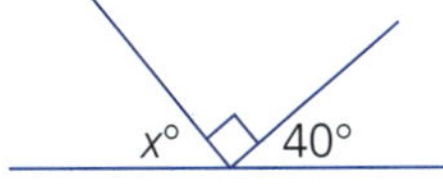

pp. 159–160

2

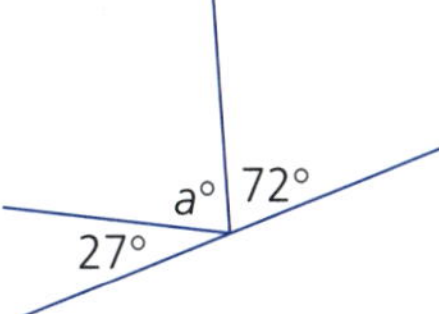

pp. 159–160

3

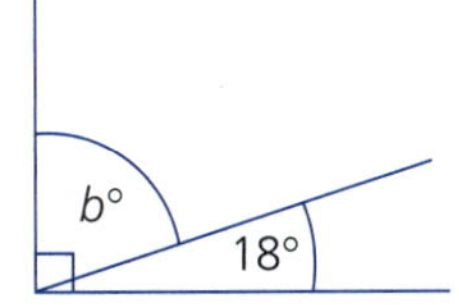

pp. 159–160

4

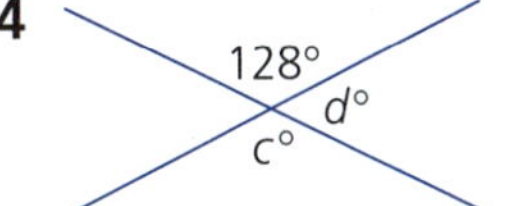

pp. 159–160

5

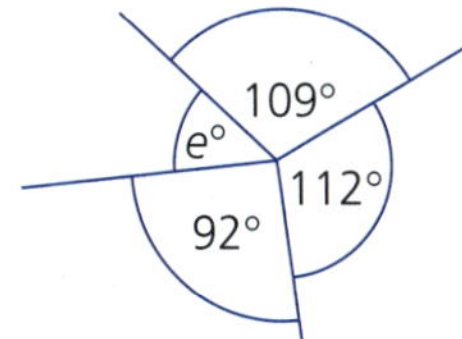

pp. 159–160

6

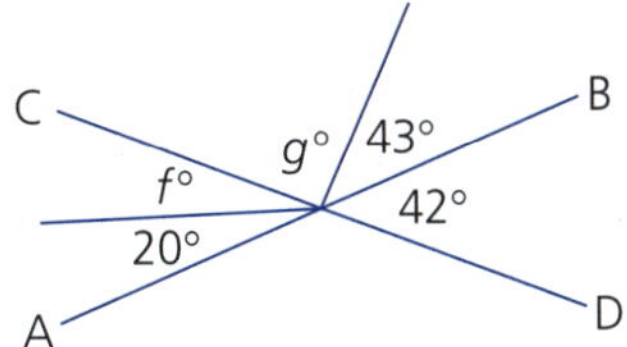

AB and CD are straight lines.

pp. 159–160

7

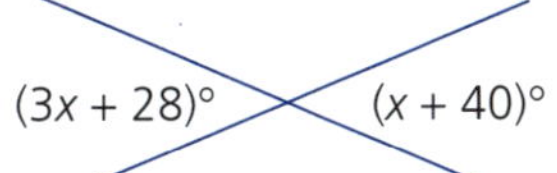

Also find the size of each marked angle.

pp. 159–160

8

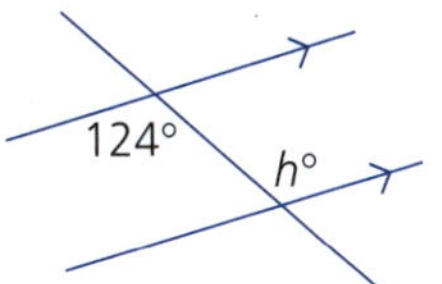

pp. 160–163

9

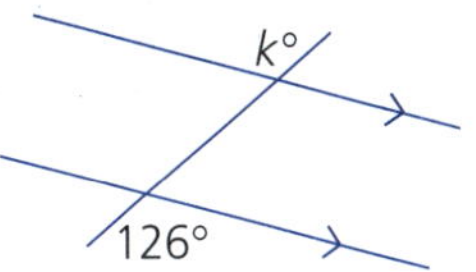

pp. 160–163

10

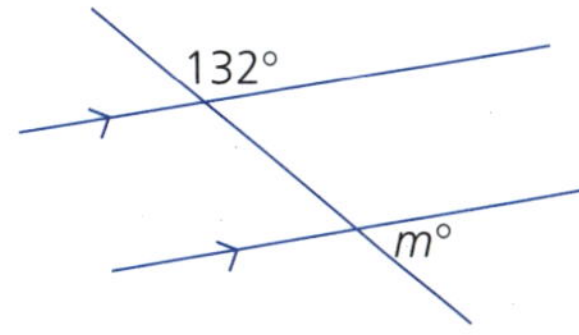

pp. 160–163

11

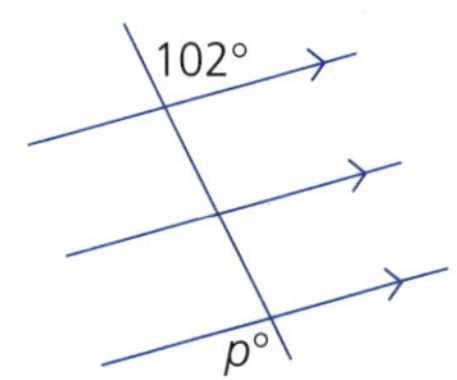

pp. 160–163

12

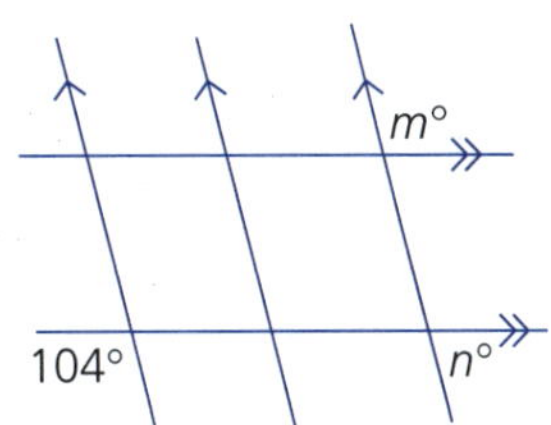

pp. 160–163

13

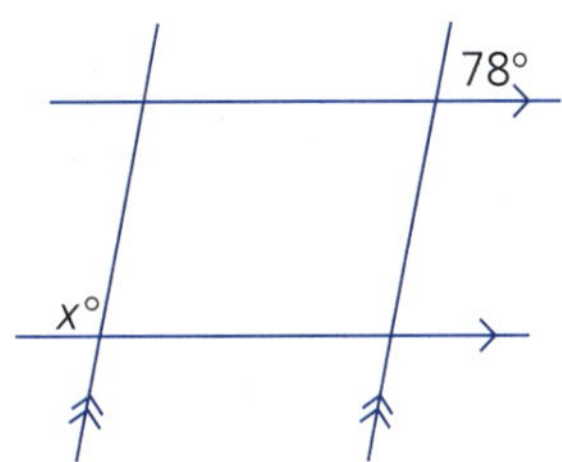

pp. 160–163

14

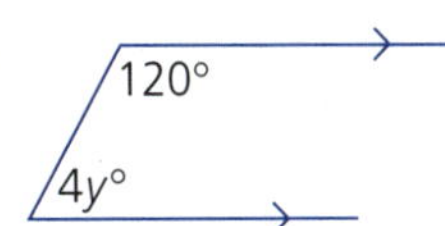

pp. 160–163

15

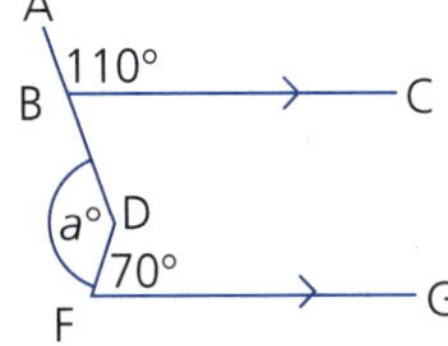

pp. 160–163

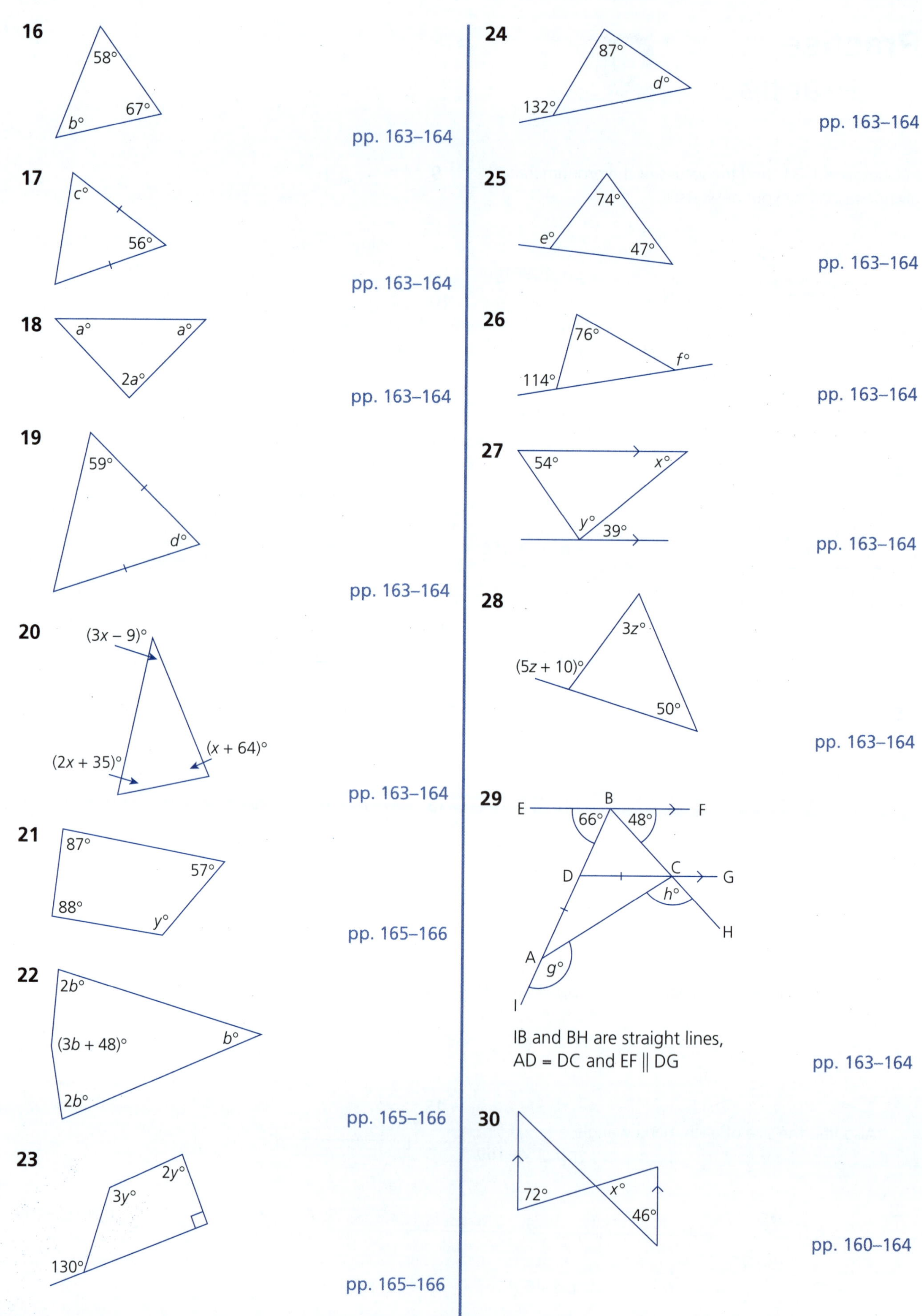
16
58°
67°
b°
pp. 163–164
17
c°
56°
pp. 163–164
18
a°
a°
2a°
pp. 163–164
19
59°
d°
pp. 163–164
20
(3x – 9)°
(2x + 35)°
(x + 64)°
pp. 163–164
21
87°
57°
88°
y°
pp. 165–166
22
2b°
(3b + 48)°
b°
2b°
pp. 165–166
23
2y°
3y°
130°
pp. 165–166
24
87°
d°
132°
pp. 163–164
25
74°
e°
47°
pp. 163–164
26
76°
f°
114°
pp. 163–164
27
54°
x°
y°
39°
pp. 163–164
28
3z°
(5z + 10)°
50°
pp. 163–164
29
E
B
F
66°
48°
D
C
G
h°
H
A
g°
I
IB and BH are straight lines,
AD = DC and EF || DG
pp. 163–164
30
72°
x°
46°
pp. 160–164

31 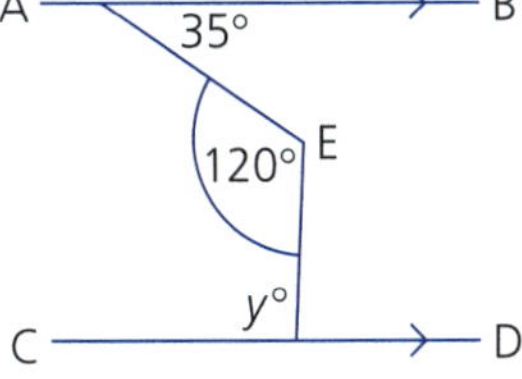

pp. 160–164

32 Find the angle sum of a polygon with:

a 5 sides
b 12 sides
c 20 sides. p. 166

33 What is the size of each interior angle of a regular polygon with:

a 8 sides?
b 12 sides?
c 32 sides? p. 166

34 What is the size of each exterior angle of a regular polygon with:

a 6 sides?
b 10 sides?
c 20 sides? p. 166

35 A regular polygon has each exterior angle $n°$. How many sides has the polygon? p. 166

In Questions 36 to 46, prove that these triangles are congruent, giving adequate reasons.

36

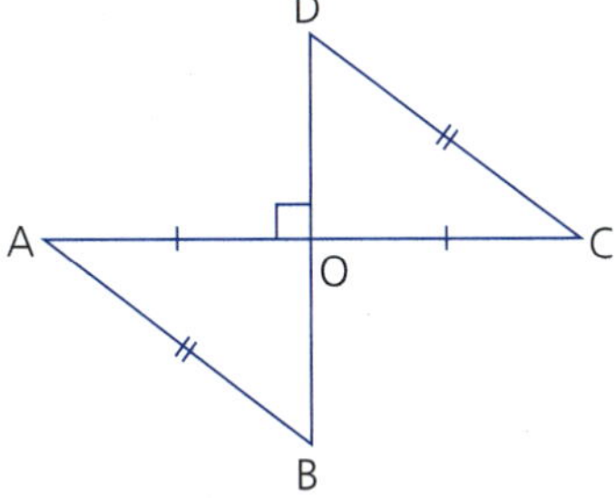

Given AO = OC, AB = CD, AC ⊥ DB, prove △AOB ≡ △COD. pp. 167–168

37

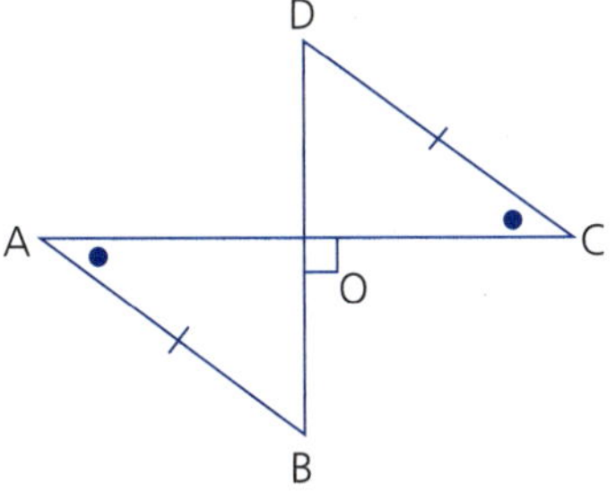

Given AB = DC, ∠OAB = ∠OCD, AC ⊥ DB, prove △AOB ≡ △COD. pp. 167–168

38

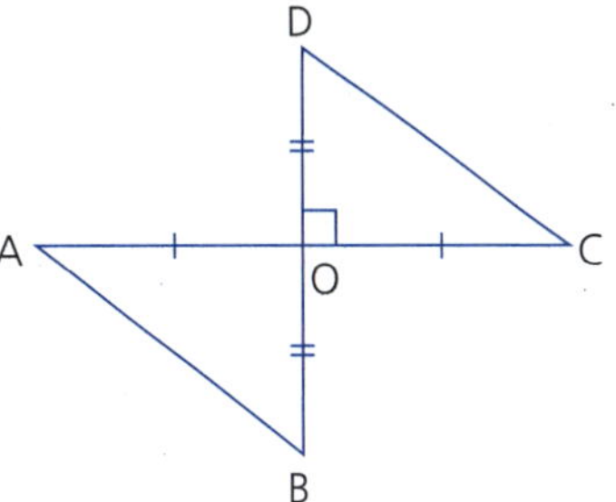

Given AO = OC, BO = OD, AC ⊥ DB, prove △AOB ≡ △COD. pp. 167–168

39

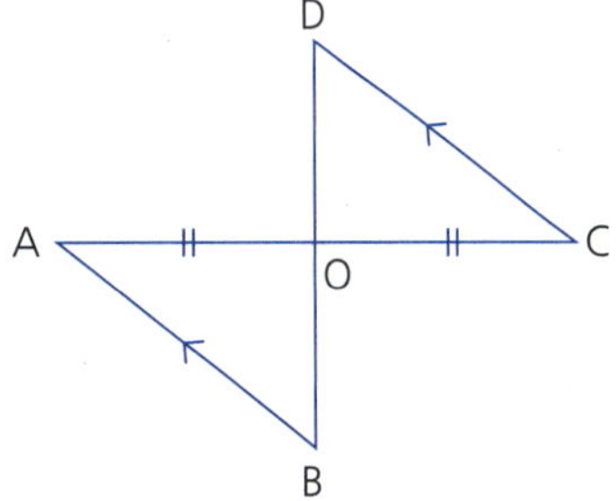

Given AB || DC, AO = OC, prove △AOB ≡ △COD. pp. 167–168

40

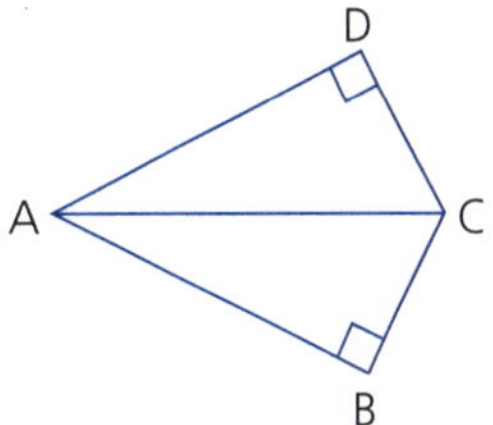

Given that CA bisects ∠DAB, CB ⊥ AB and CD ⊥ AD, prove that △ABC ≡ △ADC. pp. 167–168

41

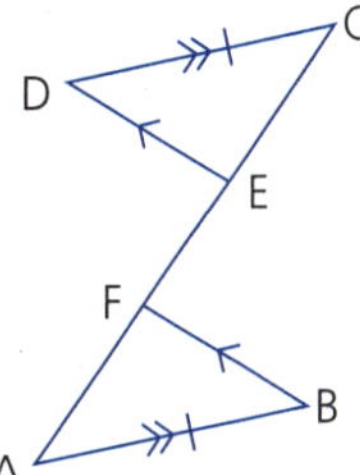

Given CD || AB, DE || FB, CD = AB, and AC is a straight line, prove that △ABF ≡ △CDE and hence that:

a AF = CE
b AE = CF. pp. 169–171

42

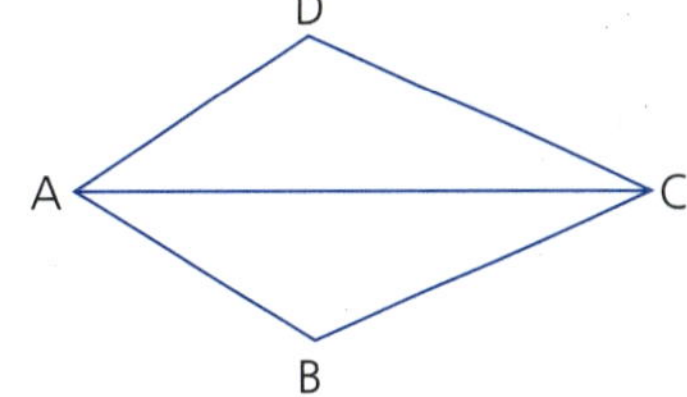

Given AD = AB, DC = BC, prove that $\triangle ADC \equiv \triangle ABC$ and hence that $\angle DAC = \angle BAC$. pp. 169–171

43

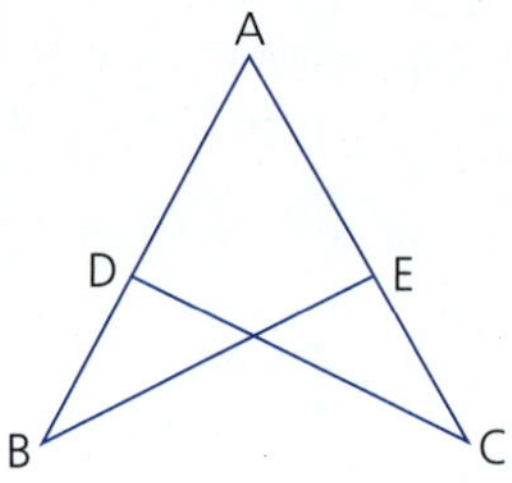

Given AD = AE, BD = EC, prove that $\triangle ABE \equiv \triangle ACD$ and hence that $\angle ABE = \angle ACD$. pp. 169–171

44

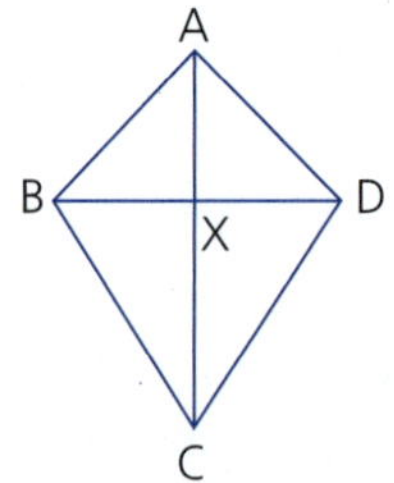

ABCD is a quadrilateral with AB = AD and BC = DC. Prove that:

a $\triangle ABC \equiv \triangle ADC$

b $\triangle ABX \equiv \triangle ADX$

c AC bisects BD at right-angles. pp. 169–171

45

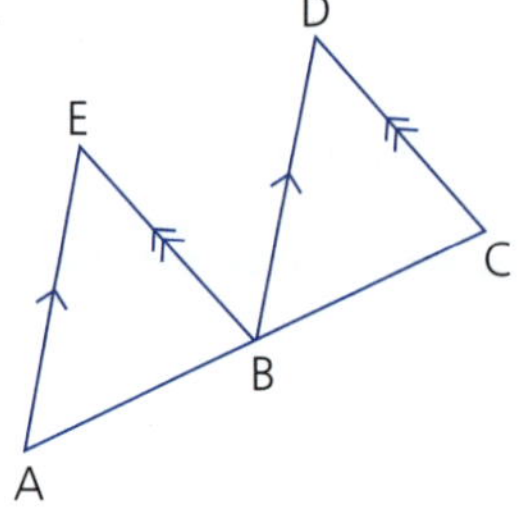

Given B is the midpoint of AC, and that AE || BD, EB || DC, prove that $\triangle$s ABE and BCD are congruent. pp. 169–171

46

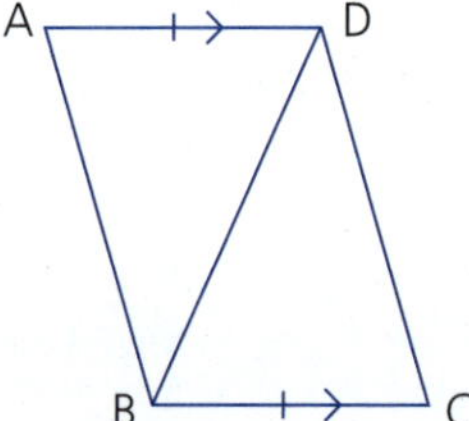

Given AD = BC, AD || BC, prove that $\triangle$s ABD and CDB are congruent. Hence prove that:

a AB = DC

b $\angle ABD = \angle BDC$

c AB || DC. pp. 169–171

For Questions 47–64, give full reasons in each answer.

47

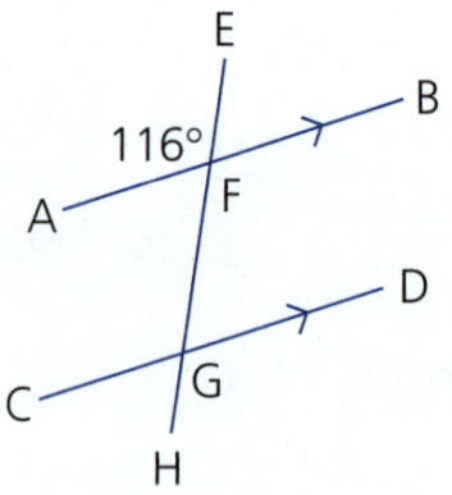

Given AB || CD and $\angle AFE = 116°$, find $\angle CGH$. pp. 169–171

48

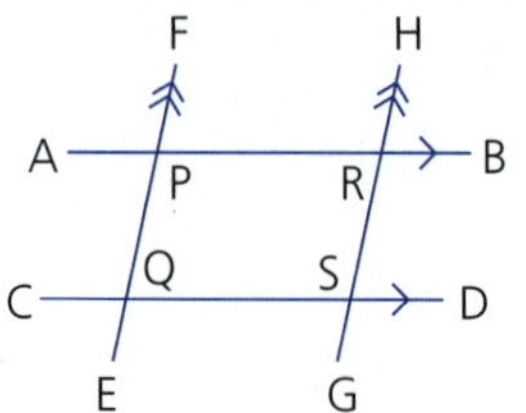

Given AB || CD, EF || GH and $\angle CQE = 78°$, find $\angle BRS$. pp. 169–171

49

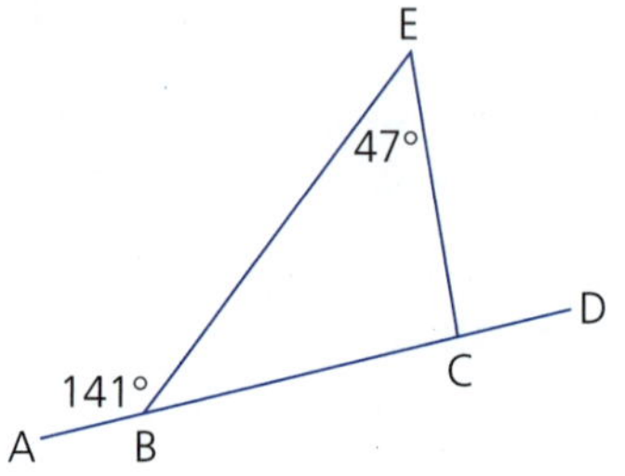

AD is a straight line, $\angle ABE = 141°$ and $\angle BEC = 47°$. Find $\angle ECD$. pp. 169–171

50

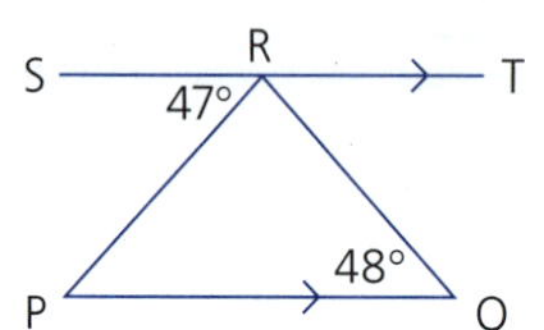

Given PQ || ST, $\angle SRP = 47°$ and $\angle PQR = 48°$, find $\angle PRQ$. pp. 169–171

51

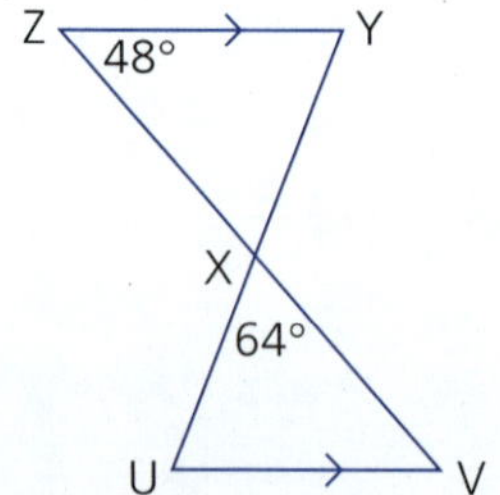

Given UV || ZY, $\angle YZX = 48°$ and $\angle UXV = 64°$, find $\angle XUV$. pp. 169–171

52

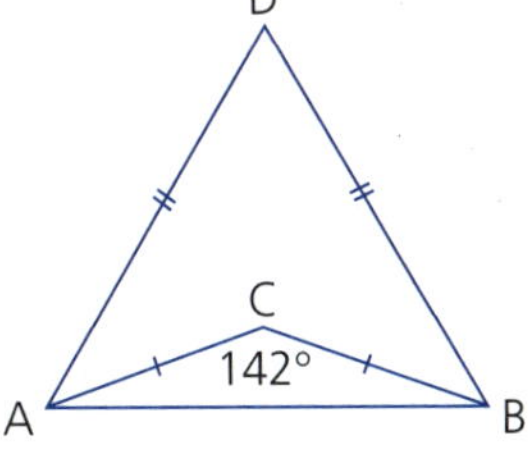

Given AC = BC, AD = BD, ∠ACB = 142° and that AC bisects ∠DAB, find ∠ADB.

pp. 169–171

53

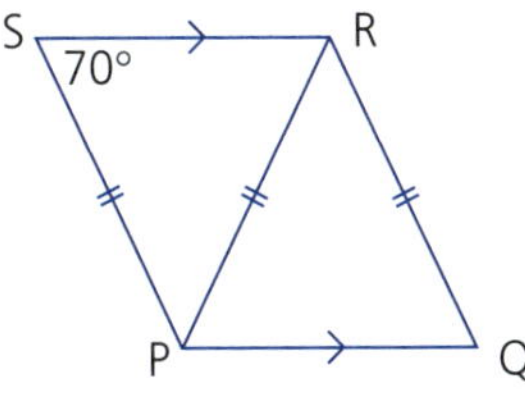

Given PQ || SR and SP = PR = RQ, find ∠PQR.

pp. 169–171

54

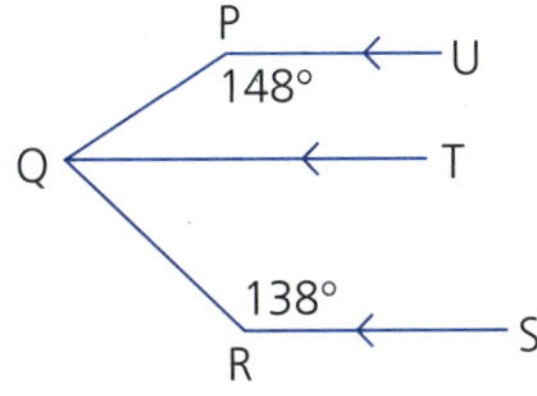

Given PU || QT || RS, ∠QPU = 148° and ∠QRS = 138°, find reflex ∠PQR.

pp. 169–171

55

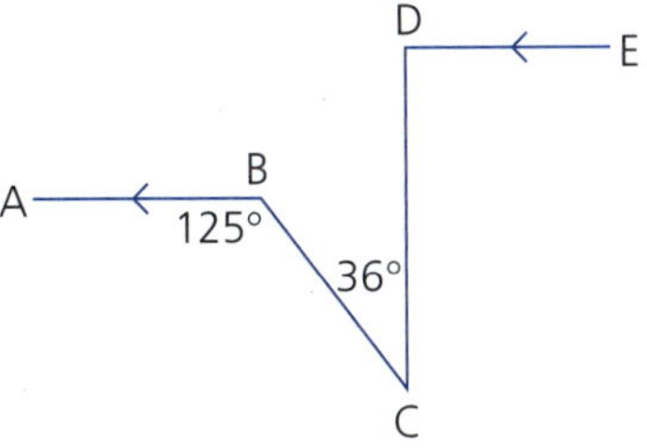

Given AB || DE, ∠ABC = 125° and ∠BCD = 36°, find ∠CDE.

pp. 169–171

56

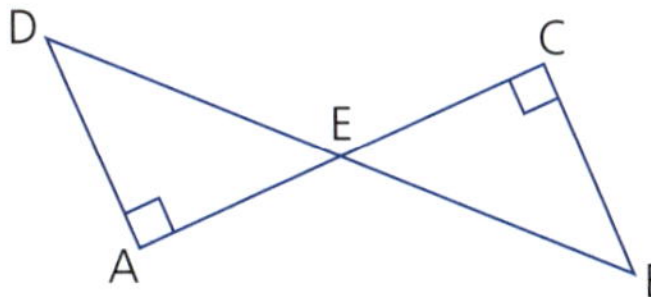

E is the midpoint of BD, DA ⊥ AC and BC ⊥ AC. Prove that DB bisects AC.

pp. 169–171

57 Prove that the diagonals of any rhombus bisect each other at right-angles.

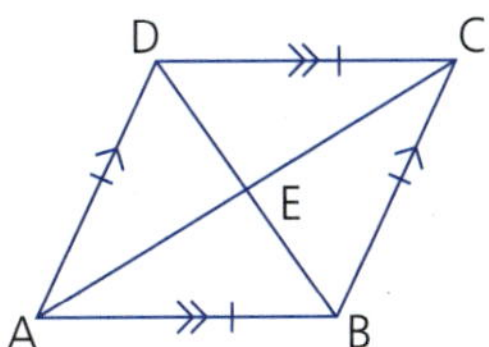

Hint: prove that:
a △DEC ≡ △BEA
b △DEC ≡ △BEC.

pp. 169–171

58 Prove that, if all sides of a quadrilateral are equal then it is a rhombus.

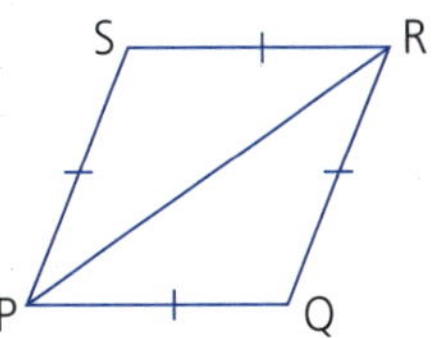

Given PQ = QR = RS = SP.
Draw diagonal PR.

pp. 169–171

Hint: prove that △PSR ≡ △RQP.

59

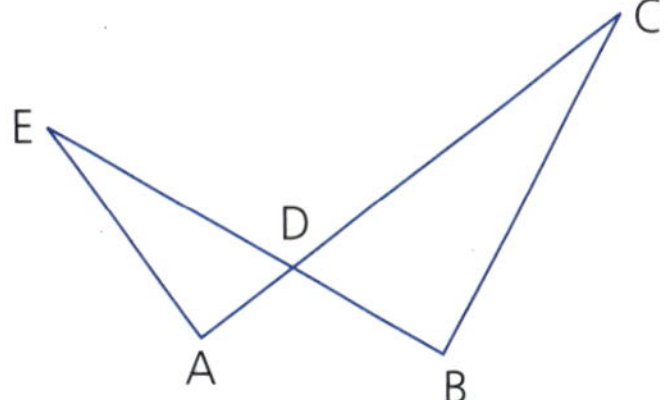

Given that ∠AED = ∠BCD, prove that the triangles ADE and BDC are similar.

pp. 171–174

60

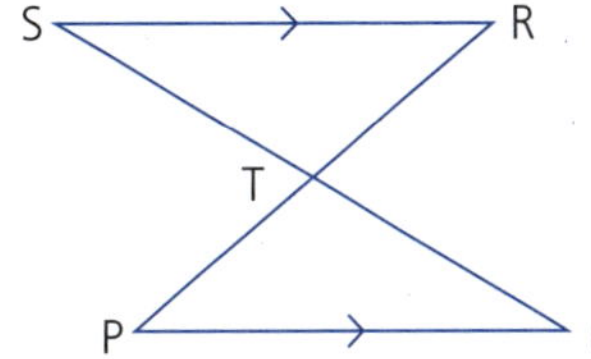

a Given PQ || SR, prove that △PTQ ||| △RTS.
b Complete the ratios:

$$\frac{PQ}{\square} = \frac{QT}{\square} = \frac{TP}{\square}$$

pp. 171–174

61

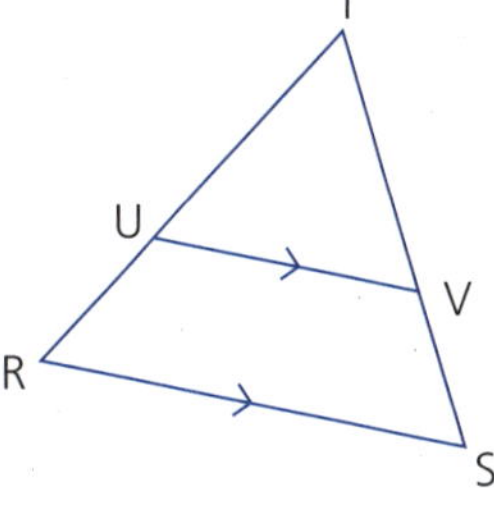

a Given UV || RS, prove that the triangles UVT and RST are similar.
b Write down the equality linking the ratios of the corresponding sides.

pp. 171–174

62

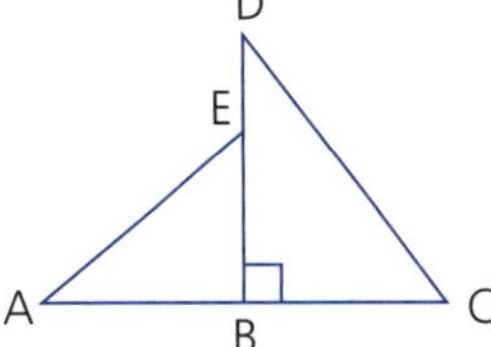

Given DB ⊥ AC, and ∠AEB = ∠BCD, prove that △ABE ||| △DBC.

pp. 171–174

63

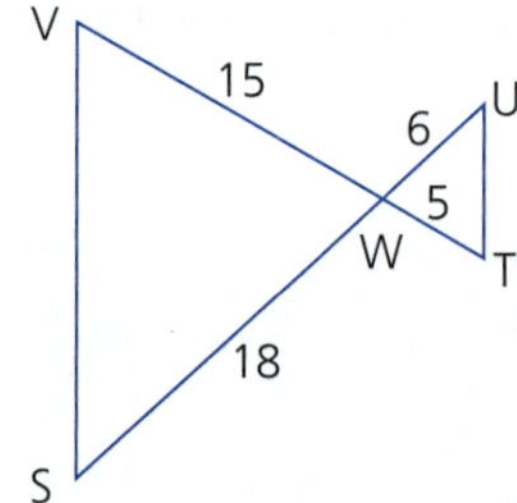

If VW = 15, SW = 18, UW = 6, WT = 5, prove that the triangles are similar. pp. 171–174

64

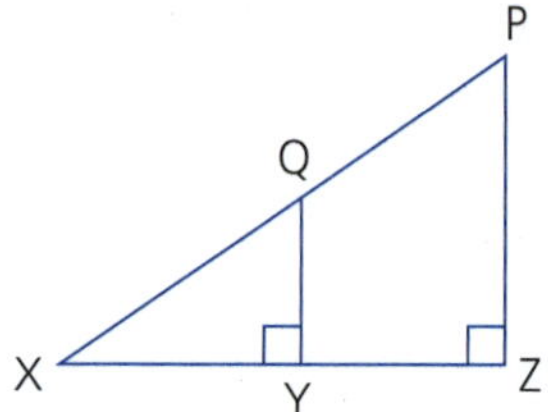

a If ∠QYX and ∠PZY are both 90°, prove that △XYQ ||| △XZP.

b Given that QY = 2.4 m and that XY = 3.6 m and YZ = 4.8 m, find the length PZ
. pp. 171–174

65 Two semicircles have radii of 3 cm and 6 cm. Find the ratio of their:

a perimeters

b areas. pp. 174–175

66 Two regular octagons have sides of length 3 cm and 5 cm. If the area of the larger octagon is 45 cm^2, calculate the area of the smaller one. pp. 174–175

67 A giant Deadwood tree has a base circumference of 40 m. Scientists have theorised that this circumference reduces in a fixed ratio for every 10 m of height. The circumference 10 m from the base is measured to be 32 metres.

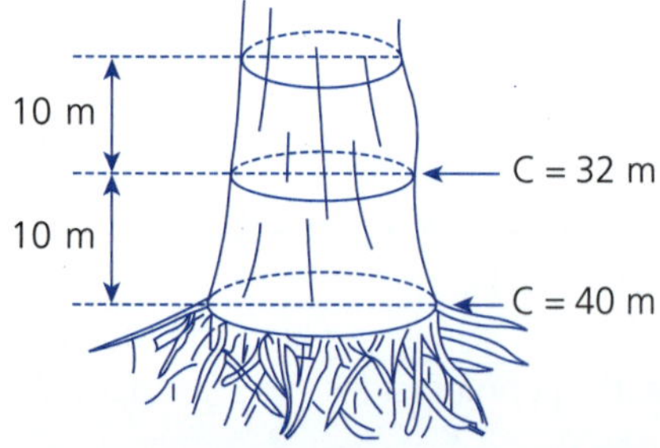

a Find the reduction ratio.

b Assuming, as have the scientists, that this ratio is fixed, calculate the circumference 20 m from the base of the tree.

c Calculate the area of cross-section of the tree, both at the 10 m height and at the 20 m height, given that the area of the base is estimated to be 750 m^2. pp. 174–176

68

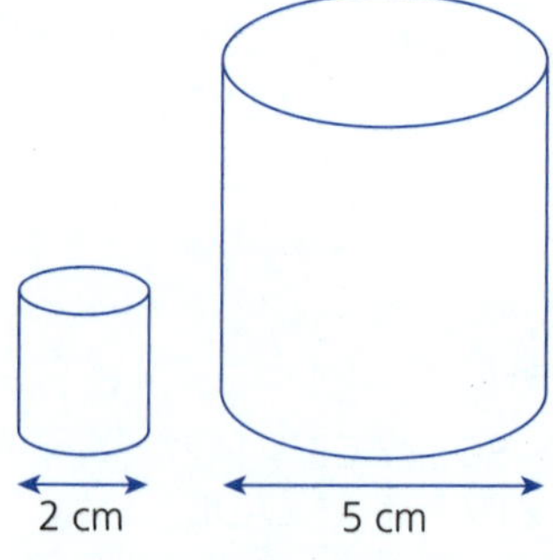

Similar cylinders have base diameters of 2 cm and 5 cm.

a Find the ratio of their:

  i surface areas

  ii volumes.

b If the larger cylinder has a volume of 60 cm^3, find the volume of the smaller one.

c If the smaller cylinder has a surface area of 18 cm^2, calculate the surface area of the larger one. pp. 174–176

69 Two similar cubes with volumes 32 cm^3 and 108 cm^3 contain spherical glass balls that fit snugly in each one.

a Calculate the ratio of the diameters of the glass balls.

b Find the ratio of the surface areas of the glass balls. pp. 174–176

70

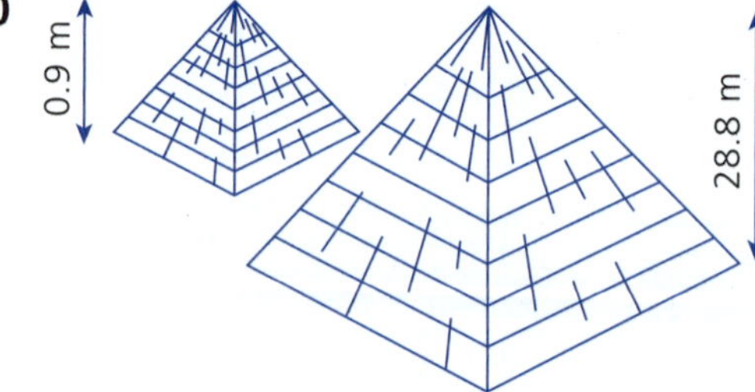

A scale model of the Great Pyramid of Kotara is to be constructed. The height of the Great Pyramid is 28.8 m, while the height of the scale model is 0.9 m.

a Calculate the scale used.

b Calculate the ratio of the surface areas of the pyramids.

c If 48 L of paint is used to coat the Great Pyramid, calculate the quantity required to coat the scale model.

d If the quantity of air contained within the Great Pyramid is 8192 kL, calculate the amount contained within the scale model. pp. 174–176

Go to p. 301 for **Quick Answers** or to pp. 338–343 for **Worked Solutions**

Your ✓ Checklist

For a complete understanding of this topic, you must be able to:

✓	Use basic angle properties to answer geometric questions		pp. 159–163
✓	Use angle properties of triangles to answer geometric questions		pp. 163–164
✓	Use properties of quadrilaterals to answer geometric questions		pp. 165–166
✓	Find the angle sum of an *n*-sided polygon		p. 166
✓	Prove two triangles congruent		pp. 167–168
✓	Apply geometric knowledge to deductive geometry		pp. 169–171
✓	Prove two triangles similar		pp. 171–173
✓	Find areas of similar figures		pp. 174–175
✓	Find volumes of similar solids.		p. 176

Now you are ready to do the tests!

Intermediate Test

Geometry

(30 marks)

1 Find the value of the letter in the following diagrams:

a (1 mark)

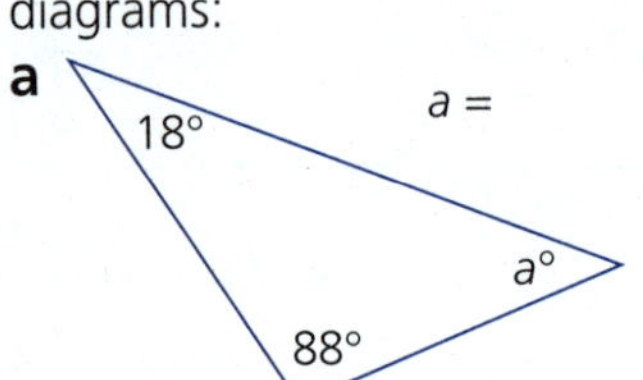

a =

b (1 mark)

69° 57° b°

b =

c (1 mark)

c° 47°

c =

d (1 mark)

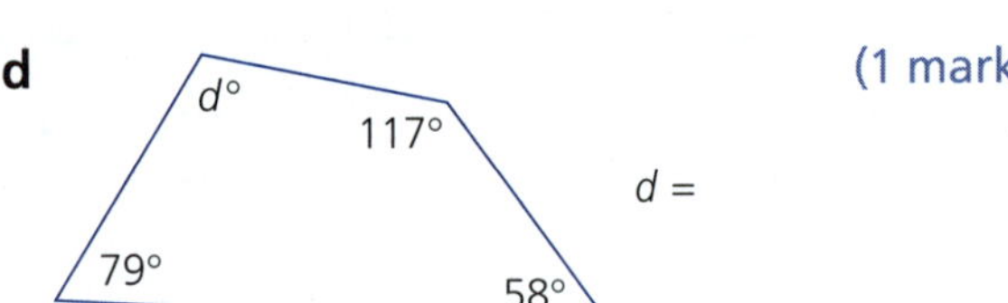

d =

e (1 mark)

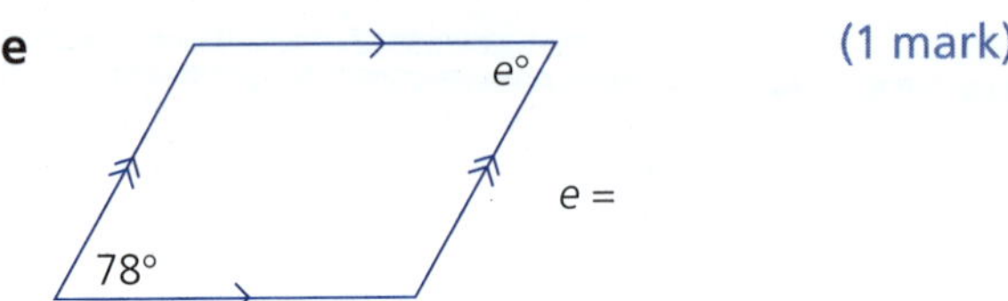

e =

f (1 mark)

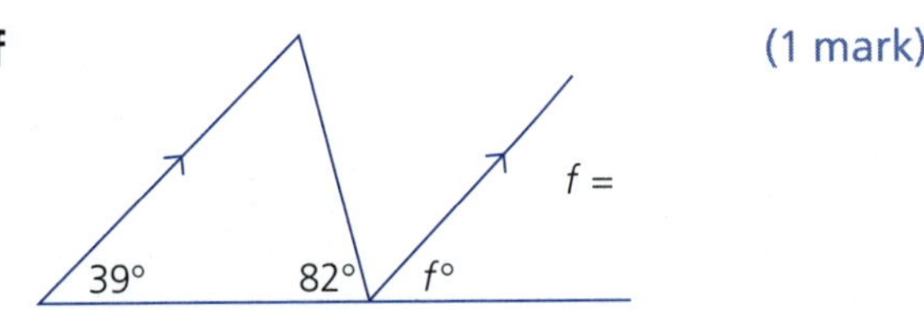

f =

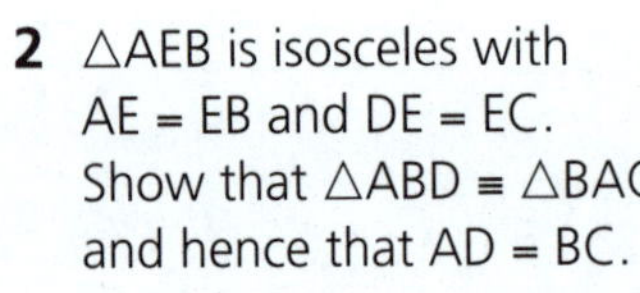

2 △AEB is isosceles with AE = EB and DE = EC. Show that △ABD ≡ △BAC and hence that AD = BC.

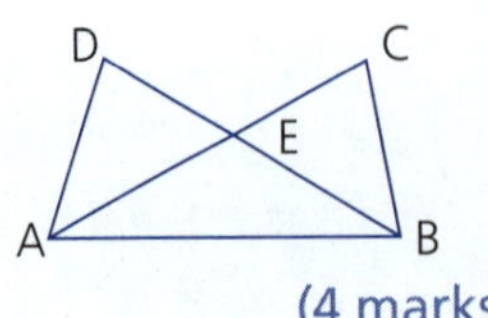

(4 marks)

3 ABCD is a parallelogram. By use of congruent triangles prove that the opposite angles of the parallelogram are equal.

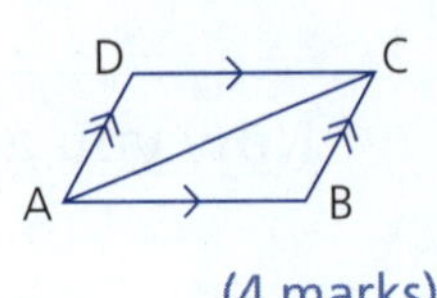

(4 marks)

4 In the △PQR, RQ || TS. Show that the triangles PQR and PST are similar.

Hence, or otherwise show that

$$QR = \frac{PQ \times ST}{PS}.$$

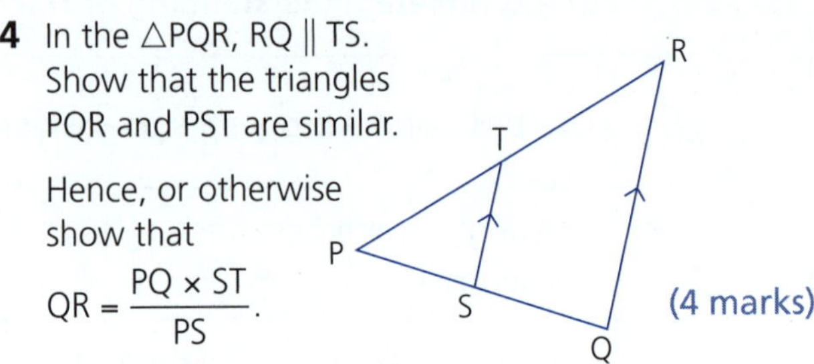

(4 marks)

5

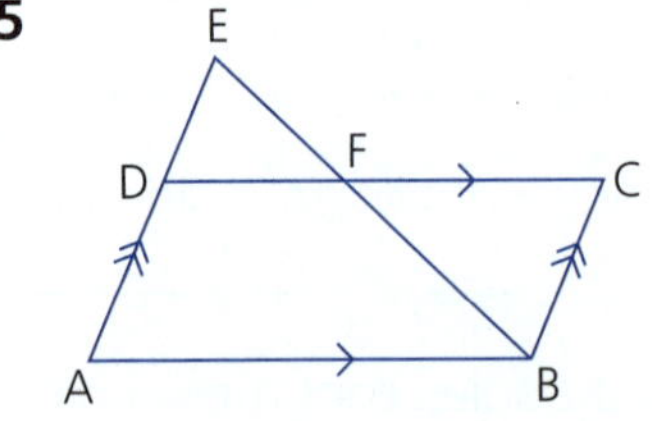

ABCD is a parallelogram. AD is produced to E and E is joined to B.

a Show that ∠DFB = ∠DAB + ∠FBC. (3 marks)

b Prove that △DFE ||| △CFB. (2 marks)

c If DE = 4 cm, DF = 6 cm, FC = 8 cm, calculate the length of BC. (2 marks)

6 Find the size of each angle in a regular decagon. (1 mark)

7

3 cm   4 cm

Two similar cones have diameters of 3 cm and 4 cm.

a What is the ratio of their:

i perpendicular heights? (1 mark)

ii surface areas? (1 mark)

iii volumes? (1 mark)

b If the larger cone has a volume of 640 cm^3, what is the volume of the smaller cone? (1 mark)

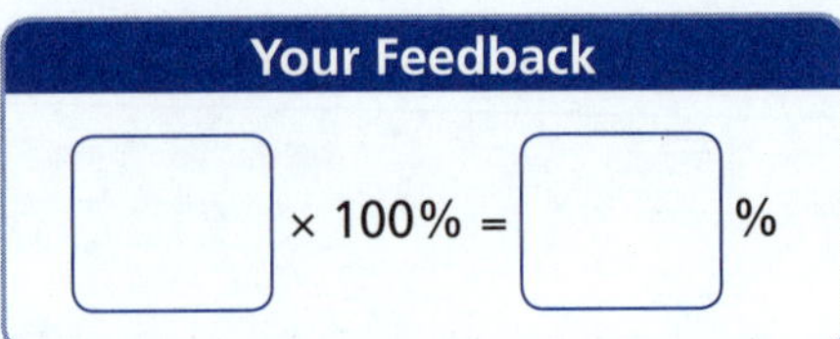

QA PAGE 306

WS PAGE 374

Advanced Test

Geometry

(30 marks)

1 Find the value of the letter in the followng diagrams:

a

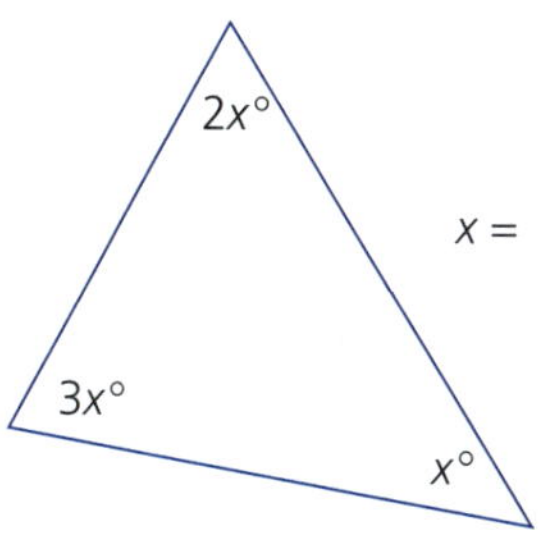

$x =$

(1 mark)

b

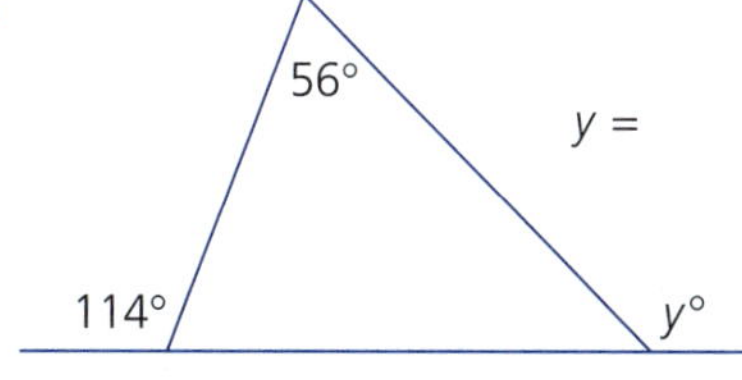

$y =$

(1 mark)

c

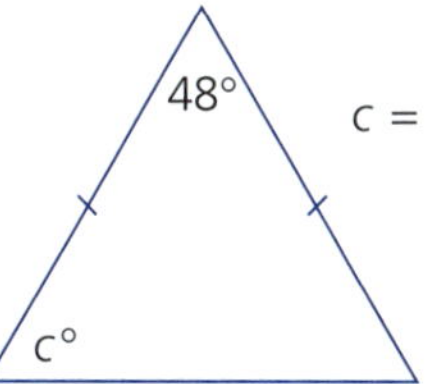

$c =$

(1 mark)

d

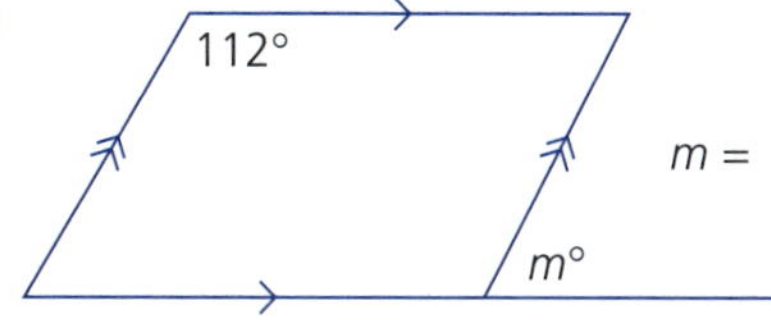

$m =$

(1 mark)

e

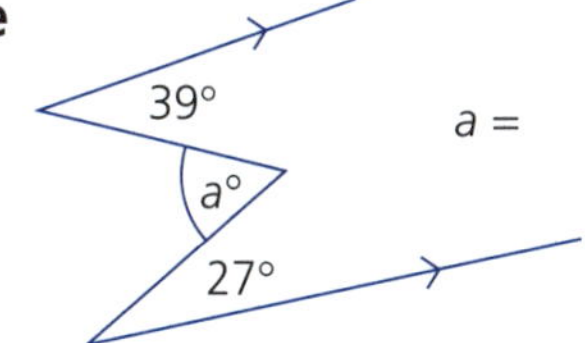

$a =$

(1 mark)

2

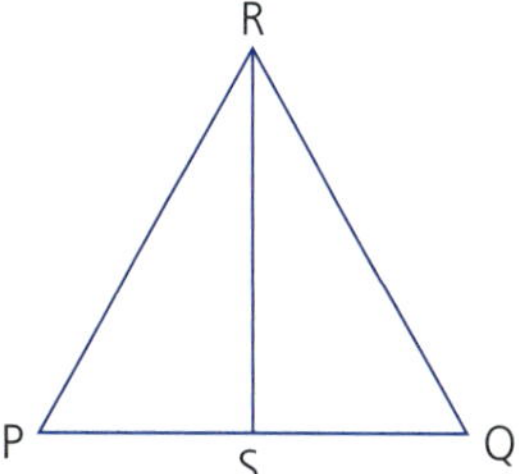

The △PQR is isosceles with PR = QR.
RS is perpendicular to PQ.

a Draw this diagram on paper and mark in the given information. (1 mark)

b Prove that △PSR ≡ △QSR. (2 marks)

c Hence, show that PS = SQ, and ∠PRS = ∠QRS. (2 marks)

3

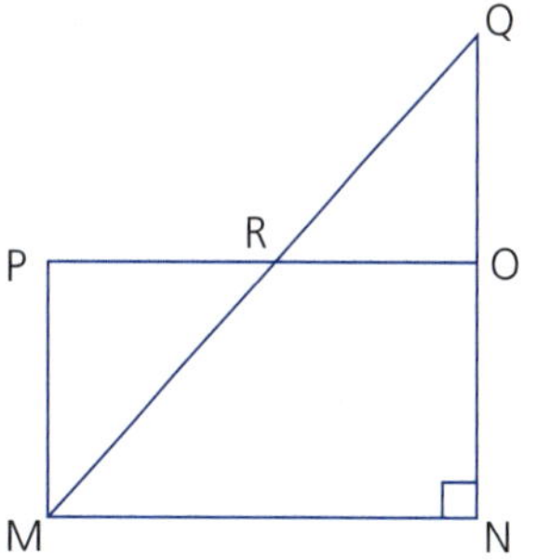

MNOP is a rectangle. NO is produced to Q such that NO = OQ. Q is joined to M cutting PO at R.

a Draw this diagram on paper and mark in all important information. (1 mark)

b Prove that △PMR ≡ △OQR. (2 marks)

c Show that MQ bisects PO. (1 mark)

4

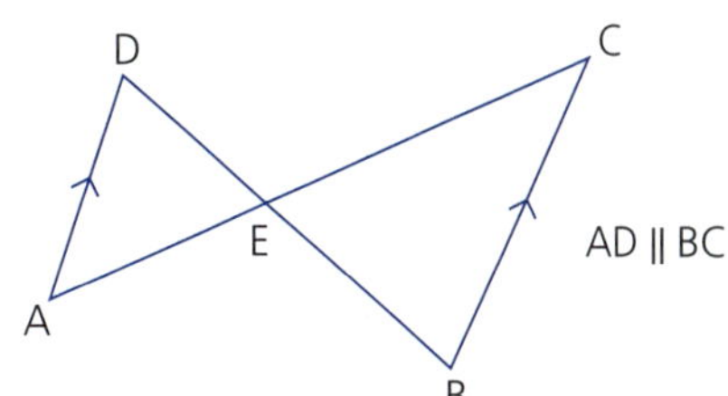

a Prove that the triangles AED and CEB are similar. (2 marks)

b Prove that $\frac{DE}{EB} = \frac{AE}{EC}$. (1 mark)

5

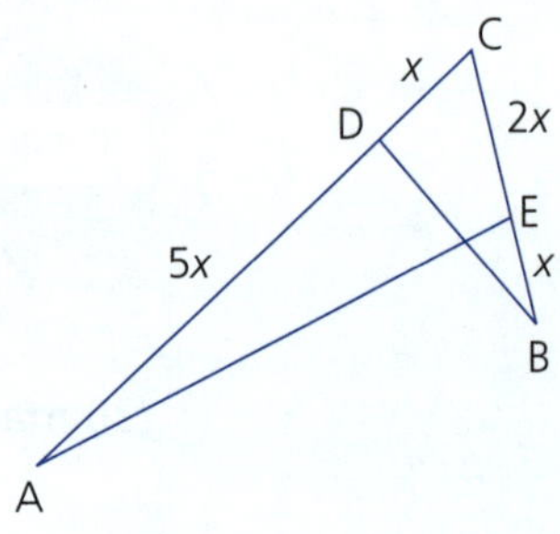

In the $\triangle$s AEC and DBC, BE = x cm, CE = $2x$ cm, AD = $5x$ cm and DC = x cm.

a Show that $\triangle$AEC ||| $\triangle$BDC. (3 marks)

b If BD = y cm, find the length of AE in terms of y. (1 mark)

6

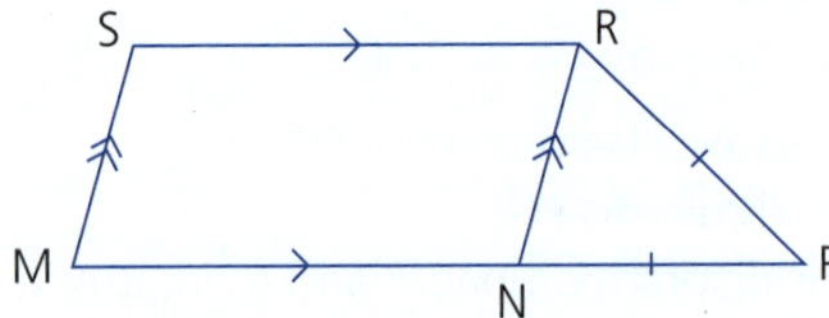

MNRS is a parallelogram. MN is produced to P such that NP = PR.

Show that:

a $\angle$SMN = $\angle$NRP

b $\angle$NPR = $\angle$MNR − $\angle$SMN. (4 marks)

7 Each exterior angle of a regular polygon is 18°. Find the angle sum of the polygon. (1 mark)

8

The diagram above shows two similar cylinders where the capacity of the larger cylinder is eight times the capacity of the smaller cylinder.

The diameter of the smaller cylinder is 3 cm.

a Find the radius of the larger cylinder. (2 marks)

b The surface area of the smaller cylinder is 40 cm^2. What is the surface area of the larger cylinder? (2 marks)

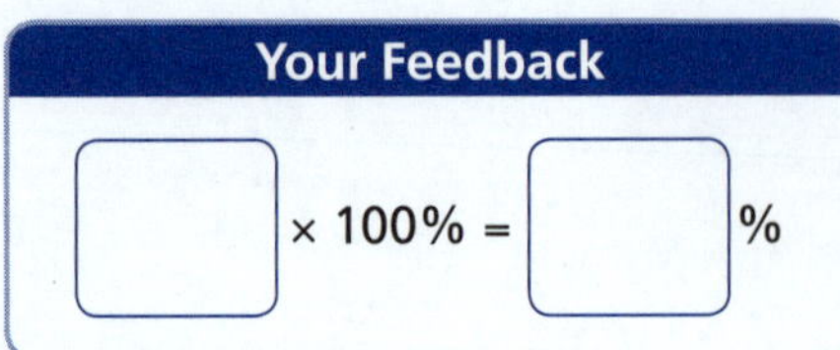

QA PAGE 307

WS PAGE 375

Chapter 11
Statistics

Single variable (univariate) data analysis

Some definitions and formulae

Range = Highest score – Lowest score

Mode = Most common score or score with the highest frequency

Median = Middle score when scores are arranged in *ascending* order

Mean = Average = $\dfrac{\text{Total of the scores}}{\text{Number of scores}}$

$= \dfrac{\Sigma xf}{\Sigma f}$

A note on notation

x = score

f = frequency

c.f. = cumulative frequency

xf = (score) × (frequency)

$\sum$ = sum

$\sum f$ = sum of frequency column

$\sum xf$ = sum of xf column

σ = standard deviation

$\bar{x}$ = mean

Keywords

Box plot	Median
Bi-modal	Mode
Bivariate	Normal
Cumulative	Outcome
Dependent	Prediction
Distribution	Range
Dot plot	Relative
Frequency	Relationship
Frequency polygon	Scatter plot
Histogram	Skewed
Independent	Standard deviation
Interquartile range	Stem-and-leaf
Lower quartile	Trend
Mean	Univariate
	Upper quartile

For Example

1 The results for 25 Year 10 students in a weekly spelling test are as follows:

7	8	9	(10)	4	7	6
4	5	(3)	8	5	6	7
9	9	8	7	6	5	7
8	6	4	9			

a Construct a frequency distribution table to answer the following questions.

S Frequency distribution table:

Score (x)	Tally	Frequency (f)	Cumulative frequency (c.f.)	Score × frequency (xf)
3	|	1	1	3 ⇐ (3 × 1)
4	|||	3	4	12 ⇐ (4 × 3)
5	|||	3	7	15 ⇐ (5 × 3)
6	||||	4	11	24
7	~~||||~~	5	16	35
8	||||	4	20	32
9	||||	4	24	36
10	|	1	25	10
	Σf	25	Σxf	167

b How many students scored:

- **i** 8 marks?
- **ii** 8 or less marks?
- **iii** less than 8 marks?
- **iv** more than 8 marks?

S

- **i** 4 (from frequency column)
- **ii** 20 (from c.f. column)
- **iii** 16 (same as 7 or less)
- **iv** 25 – 20 = 5 (total minus 8 or less)

c What percentage of students scored:

- **i** 8 marks?
- **ii** 8 or less marks?
- **iii** less than 8 marks?
- **iv** more than 8 marks?

S

i Percentage $= \frac{4}{25} \times 100\% = 16\%$

ii Percentage $= \frac{20}{25} \times 100\% = 80\%$

iii Percentage $= \frac{16}{25} \times 100\% = 64\%$

iv Percentage $= \frac{5}{25} \times 100\% = 20\%$

d Find the range and the mode.

S Range = Highest score – Lowest score
= 10 – 3 = 7

Mode = Score with highest frequency
= 7 (frequency is 5)

e Calculate the mean.

S Mean $= \frac{\Sigma xf}{\Sigma f}$
$= \frac{167}{25}$
$= 6.68$

f Find the median score.

S To find the median, we need the middle score. This is the 13th score. From the c.f. column, the 13th score is a 7. (The 12th, 13th, 14th, 15th and 16th scores are all 7.)

When we have 25 scores the 13th is the middle score, that is, 12 before it and 12 after it.

g Construct a:

- **i** frequency histogram
- **ii** cumulative frequency histogram.

S **i Histogram**

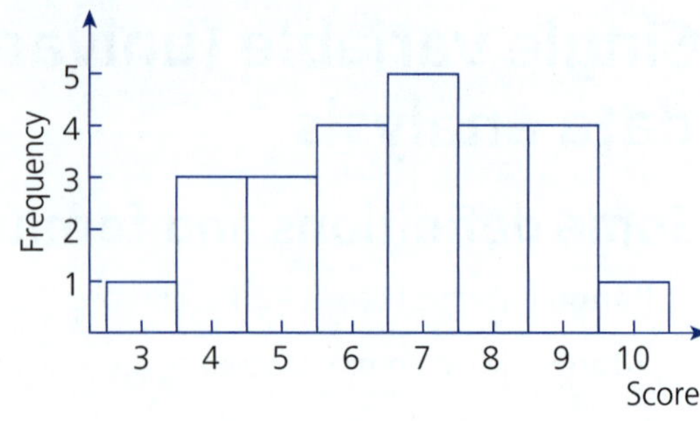

ii Cumulative frequency histogram

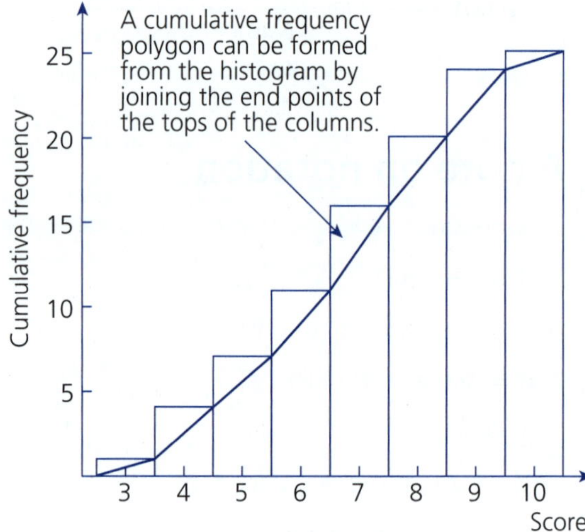

Note that if a frequency polygon had been required, the graph would be as follows:

Frequency polygon

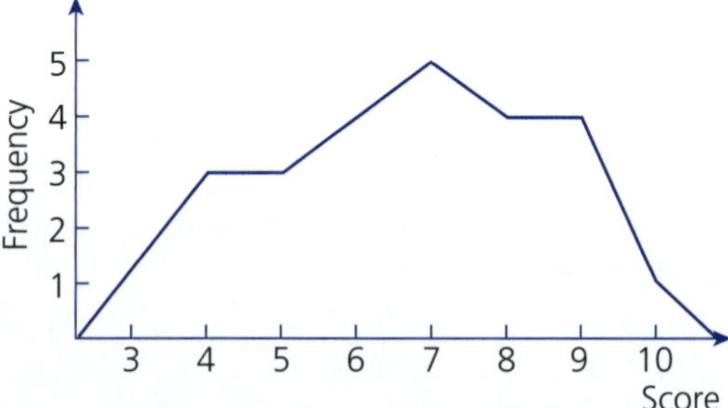

This can also be achieved by joining the midpoints of the tops of the columns in the histogram.

Notes

- When there are only a few scores, a table is not necessary.
- Remember that your calculator will calculate means.

2 Gregory, over his first six innings, managed to score 18, 32, 12, 12, 20 and 12.

a Find the range of these scores.

S Range = Highest score − Lowest score
= 32 − 12
= 20

b Write down the modal score.

S Mode = Most common score
= 12

c Calculate the mean and the median.

S $\text{Mean} = \dfrac{\text{Total of scores}}{\text{Number of scores}}$

$= \dfrac{18 + 32 + 12 + 12 + 20 + 12}{6}$

$= \dfrac{106}{6}$

= 17.7 (to 1 decimal place)

Median = middle score when scores are in ascending order

The scores in ascending order are:
12, 12, 12, 18, 20, 32

Middle scores

$\text{Median} = \dfrac{12 + 18}{2}$
= 15

When there is an even number of scores, the median will be the average of the two middle scores, that is, half-way between the two middle scores.

d If in his next three innings he scores 17, 15 and 20, find the new range, mode, mean and median.

S The scores are now 18, 32, 12, 12, 20, 12, 17, 15 and 20.

Range = 32 − 12 = 20 — unchanged

Mode = 12 — unchanged

Mean

$= \dfrac{18 + 32 + 12 + 12 + 20 + 12 + 17 + 15 + 20}{9}$

$= \dfrac{158}{9}$ = 17.6 (to 1 decimal place)

The scores in ascending order are:

12, 12, 12, 15, 17, 18, 20, 20, 32.

Middle score

When there is an odd number of scores there is a single middle score.

The mean has dropped marginally while the median has risen.

e How many runs must he score in his tenth innings to have an overall average of 18?

S For Gregory to average 18 over ten innings, his total for ten innings would be (10 × 18) runs, that is, 180 runs. Over nine innings, he has scored 158 runs, thus he would have to score (180 − 158) runs, that is, 22 runs in his tenth innings.

3 Naomi averaged 17 points per game over the first five games of the basketball season. Over the following seven games her average was 23 points per game. Calculate Naomi's average for the season.

S Over the first five games Naomi scored (5 × 17) = 85 points.

5 games with 17 points per game

Over the next seven games Naomi scored (7 × 23) = 161 points.

7 games with 23 points per game

Overall score = 85 + 161
= 246 points

$\text{Season average} = \dfrac{246}{12}$
= 20.5

$\dfrac{\text{Total points}}{\text{Number of games}}$

Naomi's season average was 20.5 points.

4 Find the mode of the scores 3, 6, 5, 6, 2, 1, 5.

S The scores have two modes: 5 and 6.
This means the set of scores is **bi-modal**.

Dot plots

A dot plot is an alternative method of displaying information. It is simpler to draw and simpler to read than a histogram.

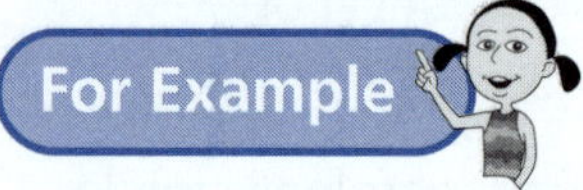

1 The type of cars passing along Macquarie Street over a 10-minute period was noted and the results recorded in the following table:

Brand	Frequency
Ford	5
Holden	7
Toyota	8
Hyundai	4
Other	6
Total	30

S A dot diagram was constructed from the information in the table.

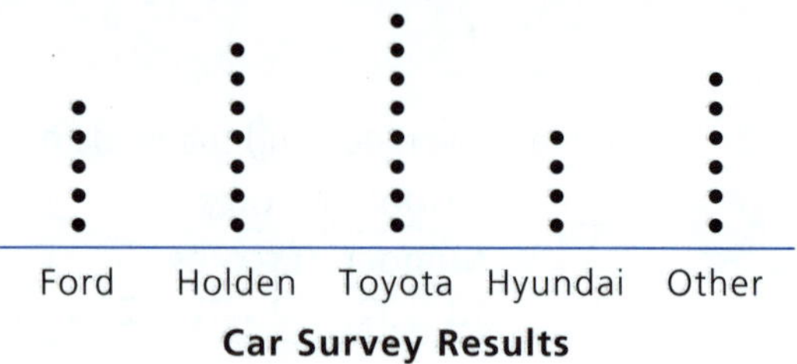

Car Survey Results

A dot diagram can also be used in the same way as a tally to organise raw data.

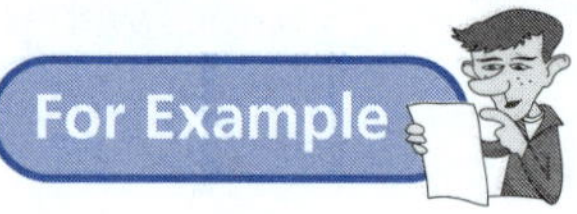

2 Consider the results of 25 Year 10 students in a weekly spelling test:

7	8	9	10	4
7	6	4	5	3
8	5	6	7	9
9	8	7	6	5
7	8	6	4	9

Record the data in a dot plot and a frequency table.

S

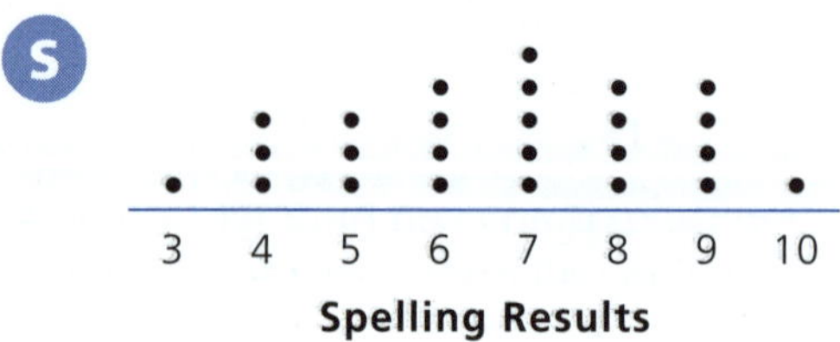

Spelling Results

This information is then transferred to a frequency distribution table:

Score	Frequency
3	1
4	3
5	3
6	4
7	5
8	4
9	4
10	1
	25

Stem-and-leaf plots

Stem-and-leaf plots are another method of displaying numerical data. The leaf is the *final* digit of a number, where the stem is the *first* digit or digits. The leaf is always a single digit. The stem can contain any number of digits. For example, if 68, 647 and 24 317 are organised into stem-and-leaf then:

68	6 stem	8 leaf
647	64 stem	7 leaf
24 317	2431 stem	7 leaf

1 The results in a mathematics test out of 50 were:

19	48	36	40	31	22
18	27	18	20	18	36
49	50	45	13	9	17
22	31	39	26	28	30
44	8	23	19	46	33

Draw a stem-and-leaf plot for this data.

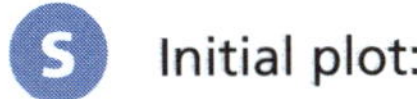

S Initial plot:

Stem is tens digit ⇓ Leaf is units digit ⇓

Stem	Leaf
0	9 8
1	9 8 8 8 3 7 9
2	2 7 0 2 6 8 3
3	6 1 6 1 9 0 3
4	8 0 9 5 4 6
5	0

← This row represents the scores 48, 40, 49, 45, 44, 46

Note: in the initial plot the numbers are entered as they occur.

Refined plot:

Stem	Leaf
0	8 9
1	3 7 8 8 8 9 9
2	0 2 2 3 6 7 8
3	0 1 1 3 6 6 9
4	0 4 5 6 8 9
5	0

The digits in the leaf are here arranged in increasing order in order to enable greater understanding of the scores.

A stem-and-leaf plot allows the smallest and largest numbers to be seen, as well as the size of the distribution and any clustering of data. It gives an indication of the groupings to be used in a grouped table and makes the tallying easier.

Outlying scores (outliers) stand out.

Back-to-back stem-and-leaf plots

Two sets of data can be compared easily when arranged as a back-to-back stem-and-leaf plot.

1 The results of a history test for 9A and 9B were recorded:

9A		9B
	4	1
2	5	0 3 4
9 7 4 1 1 0	6	3 6 6 7 8 8 9
9 7 6 6 6 5 3	7	0 0 2 4 4 4 8
9 8 6 5 5 0	8	0 2 4 5 5 6
6 5 4 2 0	9	2

Comment on the results of the students in the classes regarding the:

a number of students

S Each class has 25 students.

b outliers.

S The outlier in 9B scored 41.

An outlier is a score considerably higher or lower than other scores.

Using the stem-and-leaf plot to find the range, mode and median

A stem-and-leaf plot can be used to find statistical measures.

1 From the stem-and-leaf plot (below) showing the results of 30 students in a maths test, find the range, mode and median:

Stem	Leaf
0	8 9
1	3 7 8 8 8 9 9
2	0 2 2 3 6 7 8
3	0 1 1 3 6 6 9
4	0 4 5 6 8 9
5	0

Range = Highest score − Lowest score
= 50 − 8 = 42

Mode = Score(s) with highest frequency
= 18

Look for the number occuring the most in each stem.

Median = middle score when scores are arranged in order.

The stem-and-leaf plot has organised the scores in the correct order.

Begin by crossing a number from the start and from the end one at a time until you reach the middle number (odd number of scores) or middle numbers (even number of scores).

Start here …
⇓

Stem	Leaf
0	8 9
1	3 7 8 8 8 9 9
2	0 2 2 3 6 (7 8)
3	0 1 1 3 6 6 9
4	0 4 5 6 8 9
5	0

⇑
then here.

In this case

Median = Average of 27 and 28

$$= \frac{27 + 28}{2}$$

$$= 27.5$$

Then range = 42
mode = 18
median = 27.5

2 Twenty boys and 16 girls from Year 9 had their height measured (in centimetres).

Boys: 180, 186, 164, 166, 158, 170, 172, 175, 182, 154, 151, 158, 167, 169, 170, 174, 179, 177, 181, 171

Girls: 147, 142, 165, 168, 174, 142, 155, 159, 160, 179, 181, 176, 166, 159, 154, 162

a Display the two sets of data back-to-back on a stem-and-leaf plot.

S Initial plot:

Leaf (G)	Stem	Leaf (B)
2 2 7	14	
4 9 9 5	15	8 4 1 8
2 6 0 8 5	16	4 6 7 9
6 9 4	17	0 2 5 0 4 9 7 1
1	18	0 6 2 1

Final plot:

Leaf (G)	Stem	Leaf (B)
7 2 2	14	
9 9 5 4	15	1 4 8 8
8 6 5 2 0	16	4 6 7 9
9 6 4	17	0 0 1 2 4 5 7 9
1	18	0 1 2 6

b Calculate for each plot the range and median.

S Boys

15	1 4 8 8
16	4 6 7 9
17	0 0 1 2 4 5 7 9
18	0 1 2 6

Range = 186 − 151 = 35 cm

$$\text{Median} = \frac{170 + 171}{2} = 170.5 \text{ cm}$$

Girls

7 2 2	14
9 9 5 4	15
8 6 5 2 0	16
9 6 4	17
1	18

Range = 181 − 142 = 39 cm

$$\text{Median} = \frac{160 + 162}{2} = 161 \text{ cm}$$

c Comment on the two sets of data.

S

- There is a greater spread of heights for the girls.
- The median is higher for boys.
- Clustering for girls is in the 160s; for boys in 170s.
- From this data it is appears that boys in Year 9 are taller than girls on average, but there is a greater variation among girls.

Interquartile range

When scores are arranged in ascending order:

- The **median** is the middle score of the distribution.
- The **upper quartile** is the middle score between the median and the highest score.
- The **lower quartile** is the middle score between the median and the lowest score.
- The **interquartile range** is equal to the upper quartile minus the lower quartile.

Note: 50% of the scores are represented by the interquartile range.

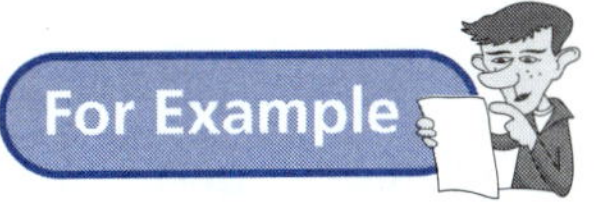

1 Consider the following scores for **a** and **b**:

a 3, 5, 4, 6, 4, 4, 8, 7, 6, 8
Find the:

i median

ii upper quartile

iii lower quartile

iv interquartile range.

S Arrange in ascending order:

⇓
3, 4, 4, 4, 5, 6, 6, 7, 8, 8
⇑

i Median = Average of 5th and 6th scores

$$= \frac{5 + 6}{2}$$
$$= 5.5$$

Even number of scores ∴ two middle numbers.

ii For the upper quartile, use only the scores above the median, that is:

6, 6, 7, 8, 8

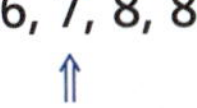

Upper quartile
= Middle number of scores above median
= 7

Odd number of scores ∴ one middle number.

iii For the lower quartile, use only the scores below the median:

3, 4, 4, 4, 5

⇑

Lower quartile

= Middle number of scores below median

= 4

iv Interquartile range

= Upper quartile – Lower quartile

= 7 – 4

= 3

b 18, 20, 16, 16, 18, 14, 18, 21, 19

Find the:

i median

ii upper quartile

iii lower quartile

iv interquartile range.

S Arrange in ascending order

⇓

14, 16, 16, 18, 18, 18, 19, 20, 21

⇑

i Median = middle score = 18

One middle score

ii For the upper quartile, use only the scores above the median:

18, 19, 20, 21

⇑

Upper quartile

= Average of 2nd and 3rd scores

$= \frac{19 + 20}{2} = 19.5$

iii For the lower quartile, use only the scores below the median:

14, 16, 16, 18

⇑

Lower quartile

= Average of 2nd and 3rd scores

= 16

iv Interquartile range

= Upper quartile – Lower quartile

= 19.5 – 16

= 3.5

Interquartile range from stem-and-leaf plots

A stem-and-leaf plot can be used to find both the median and the quartiles.

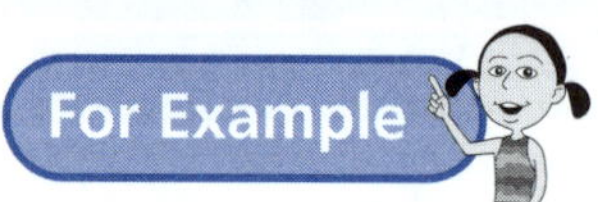

1 Consider the stem-and-leaf plot from before. The results of 30 students in a mathematics test out of 50 have been tabulated. Find the interquartile range.

Stem	Leaf
0	8 9
1	3 7 8 8 8 9 9
2	0 2 2 3 6 (7 8)
3	0 1 1 3 6 6 9
4	0 4 5 6 8 9
5	0

← There are 30 scores so that the median is the average of the 15th and 16th scores.

That is, median $= \frac{27 + 28}{2}$

$= 27.5$

For the upper quartile consider only numbers above the median (28, 30, ...). There are 15 of these so the upper quartile will be the 8th number above the median.

Cross off 7 numbers

⇓

2	8
3	0 1 1 3 6 6 (9)
4	0 4 5 6 8 9
5	0

15 numbers

Upper quartile = 39

For the lower quartile consider only numbers below the median (8, 9, ..., 27). There are 15 of these so the lower quartile will be the 8th number above the smallest score. Cross off the first seven numbers as follows:

Stem	Leaf
0	8 9 (crossed off)
1	3 7 8 8 8 (crossed off) (9) 9
2	0 2 2 3 6 7

15 numbers

The lower quartile = 19
Interquartile range = 39 − 19 = 20

Box plots

A box plot uses five selected pieces of information—lowest score, highest score, median, and upper and lower quartiles—to give an alternative display for a set of scores.

These five numbers are also called the **five point summary**.

1 Consider a set of data with lowest score 8, highest score 50, median 27.5, upper quartile 39 and lower quartile 19. Draw a box plot for the data.

S

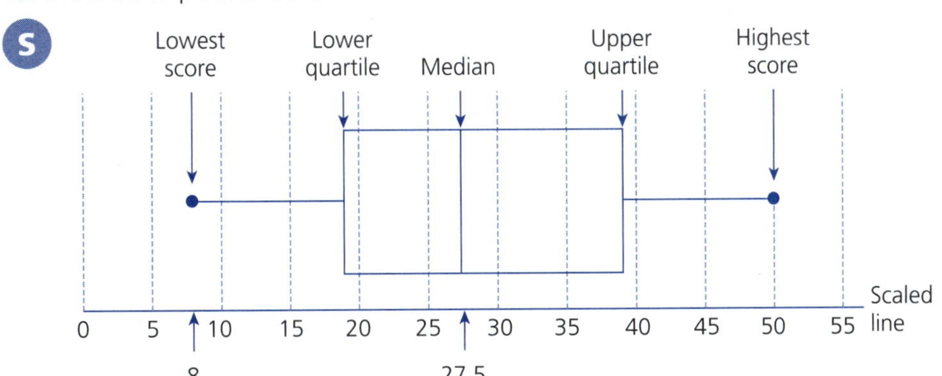

This is sometimes called a box-and-whiskers plot. The lines at each end are the whiskers, while the centre rectangle is the box. The box contains 50% of the scores.

A box plot can be drawn to illustrate any numerical data regardless of how the quartiles and the median are calculated.

1 Use the information from Example **1a** on pages 193–194 to write a five point summary and a box plot for the data.

S Lowest score = 3
Lower quartile = 4
Median = 5.5
Upper quartile = 7
Highest score = 8

Box plot:

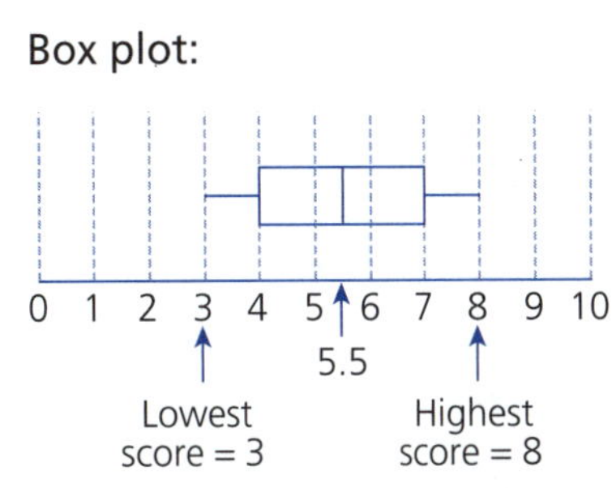

2 A stem-and leaf plot has been constructed to represent the scores of 25 students in a spelling test out of 30. Construct a box plot from the information.

Stem	Leaf
0	7 7 8 9 9
1	0 1 1 4 4 6 6 9
2	1 1 2 2 5 8 9 9 9
3	0 0 0

← Median (the circled 9 in row 1)

Middle of lower scores. (marked 0 1 in row 1)

Middle of upper scores. (marked 8 9 in row 2)

S There are 25 scores,
∴ the median is the 13th score.

Median = 19

For lower quartile consider only lower 12 scores. The two middle scores (6th and 7th) are 10 and 11:

$$\text{Lower quartile} = \frac{10 + 11}{2} = 10.5$$

For upper quartile consider only upper 12 scores. The two middle scores are 28 and 29:

$$\text{Upper quartile} = \frac{28 + 29}{2} = 28.5$$

Summary:

Lowest score = 7
Highest score = 30
Lower quartile = 10.5
Median = 19
Upper quartile = 28.5

This is called the 'five point summary'.

Box plot:

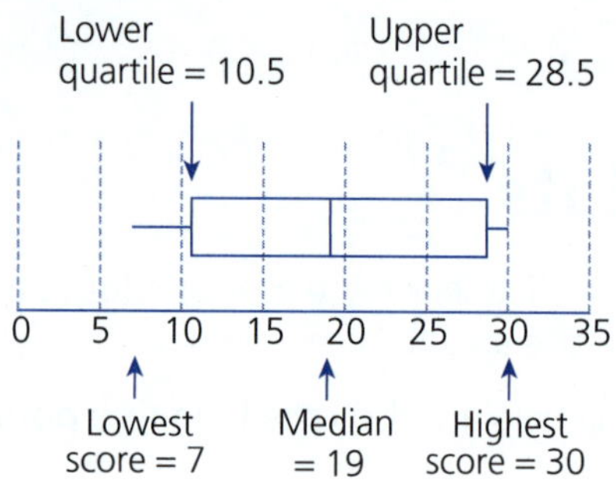

Comparing data sets

Comparing data sets using box plots

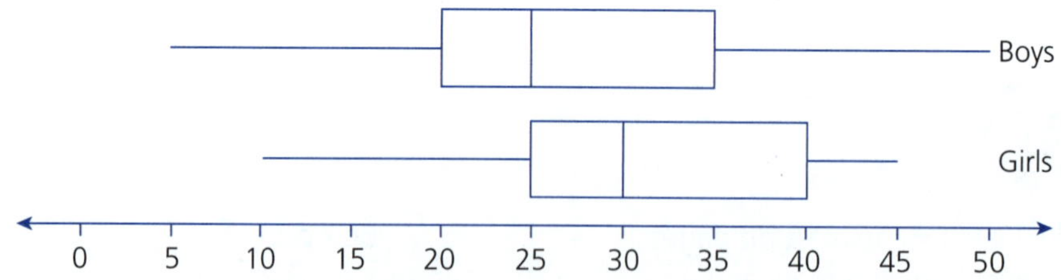

Box plots can be used to compare different data sets:

- Both highest and lowest scores were boys (50, 5).
- Both have the same interquartile range (15). Clustering of scores is thus the same for the middle 50%, but boys have a greater variation in overall scores.
- 75% of girls scored 25 or more; 50% of boys scored 25 or more.
- Girls have, as a whole, performed better than the boys.

Comparing data sets using back-to-back stem-and-leaf plots

Test marks in Class 10T out of 70:

Boys		Girls
7	1	3
5	2	4
6 3	3	2 2 5
6 1 1	4	0 1 5
8 7 5 2	5	3 3 3 9
9 4 1	6	2 4 8
0	7	

From the stem-and-leaf plot:

Median boys = 52
Median girls = 45

Lower quartile boys = 36
Lower quartile girls = 32

Upper quartile boys = 61
Upper quartile girls = 59

Interquartile range boys = 61 – 36 = 25

Interquartile range girls = 59 – 32 = 27

Although the boy's median is significantly higher, the interquartile ranges are similar.

Skew

When data sets are compared, the distribution of the data can be considered.

This histogram is **symmetrical** and the mode, mean and median are all equal. The majority of scores are **clustered** around the middle and this is known as a **normal distribution**.

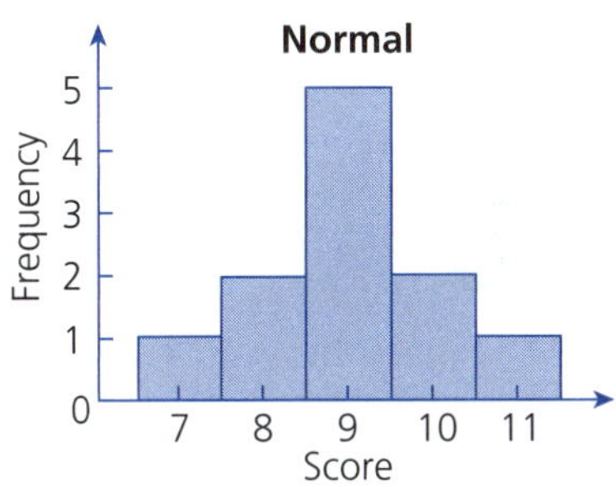

This histogram is **symmetrical** but there are two modes at either end of the distribution. The distribution is **bi-modal**.

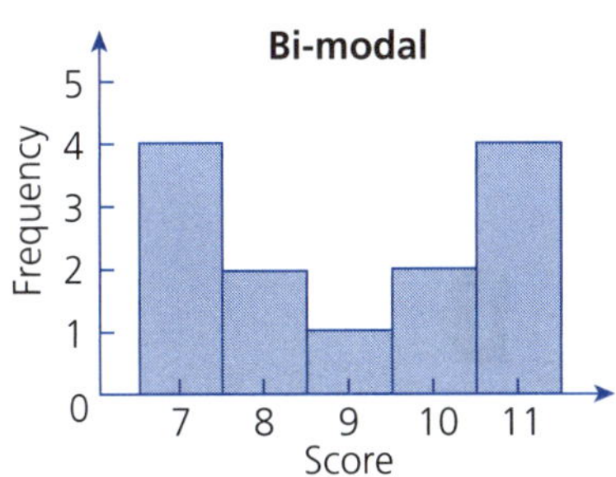

The histograms below are **not symmetrical** and the scores are gathered at the lower end, or the upper end, of the distribution. The distributions are said to be **skewed**.

This is an example of a **positively skewed** distribution.
(Note that the tail of the graph is towards the positive direction.)

Positively skewed

This is an example of a **negatively skewed** distribution.
(Note that the tail of the graph is towards the negative direction.)

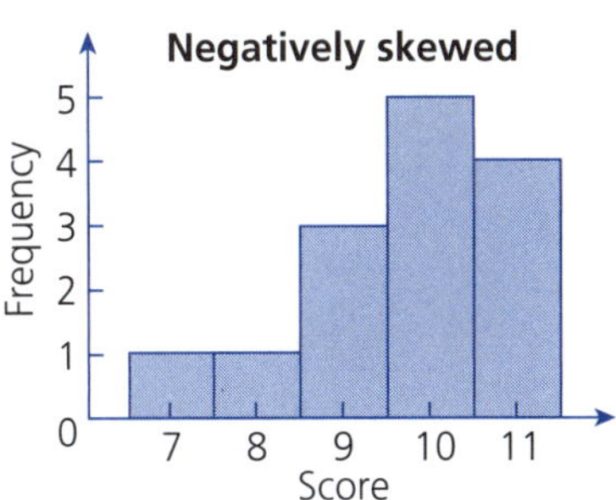

For Example

1

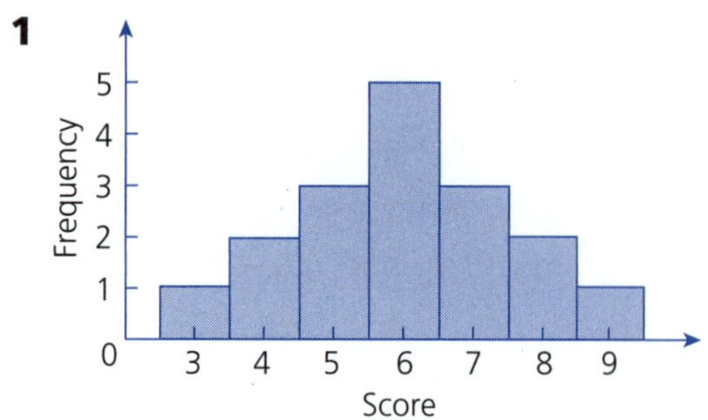

Refer to the above histogram to complete the following tasks.

a Draw a frequency table.

S

Score (x)	Freq (f)	$f \times x$
3	1	3
4	2	8
5	3	15
6	5	30
7	3	21
8	2	16
9	1	9
Total	17	102

b Find the mean, mode and median.

S Mean $= \frac{102}{17}$

$= 6$

Mode = 6

Median: as there are 17 scores, then the 9th score is the middle.

∴ Median is 6.

c Determine whether the data is normally distributed.

S Yes, normal distribution.

2 The diagrams below show the frequency polygons of two distributions. Comment on the skew of each data set.

a

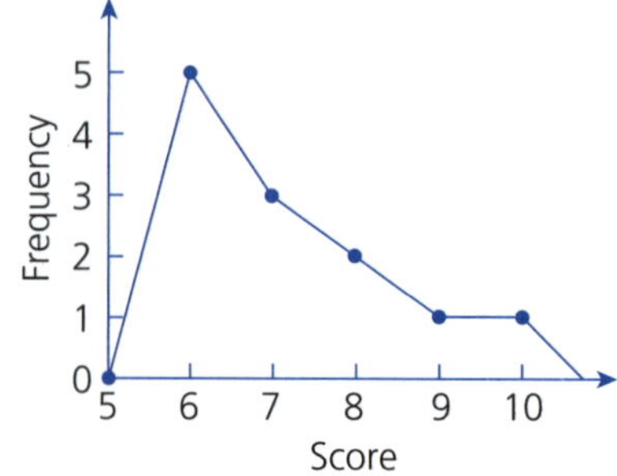

S Positively skewed.

b

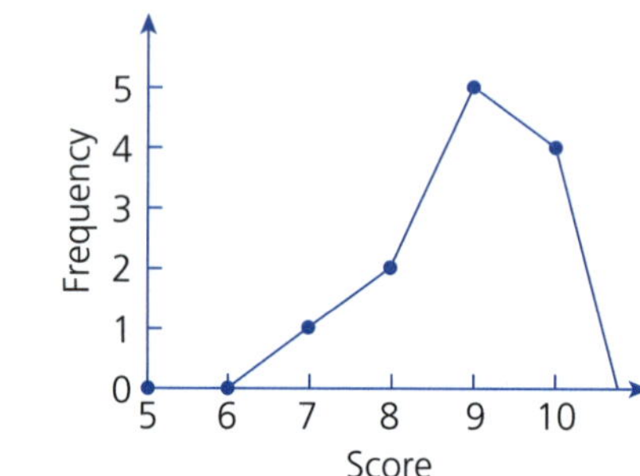

S Negatively skewed.

3 A teacher gave two tests to a class. As the class started a new topic they sat a pre-test, and then a post-test when the topic was completed. The back-to-back stem-and-leaf plot shows the results for both tests:

Pre-test		Post-test
9 9 9 8 7 5 5 3	0	2
9 8 8 6 3 1	1	4 5
8 5 4	2	0 1 8 8
1	3	2 4 5 7 8 8
	4	6 7 7 8 9

Comment on the skew of the distributions.

S The pre-test was positively skewed and the post-test was negatively skewed.

4 Draw a box plot for each and comment on the skew of each distribution:

a 2, 4, 4, 6, 6, 6, 8, 8, 10

S Lowest score = 2
Lower quartile = 4
Median = 6
Upper quartile = 8
Highest score = 10

Normal distribution.

b 2, 4, 6, 7, 7, 8, 8, 8, 9, 9, 9, 9, 10

S Lowest score = 2
Lower quartile = 6.5
Median = 8
Upper quartile = 9
Highest score = 10

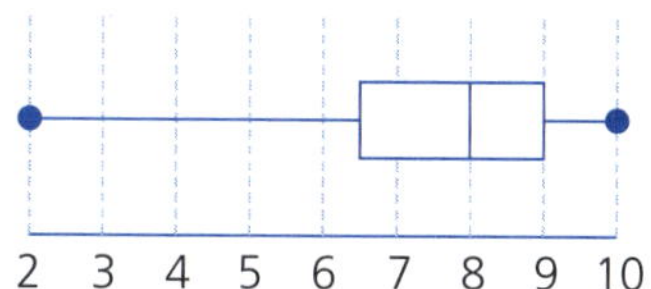

Negatively skewed.

5 A class was given a pre-test and a post-test on statistics and the results recorded in the box plots below:

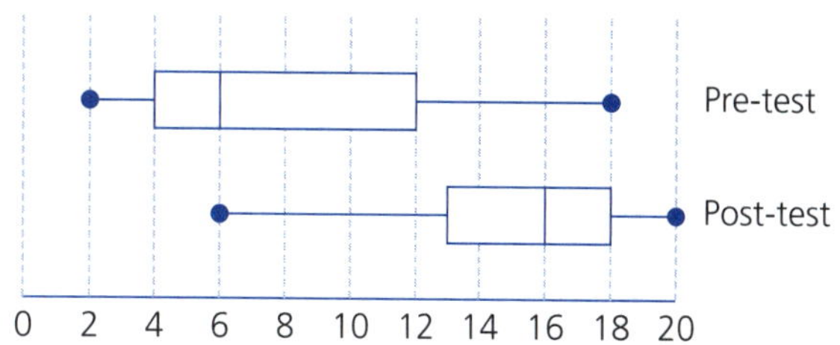

Compare the results of the two tests, making reference to:

a spread

S The pre-test results had a greater spread with IQR of 8 and range of 16 compared with the post-test with an IQR of 5 and range of 14.

b skew.

S The pre-test results were positively skewed and the post-test results were negatively skewed.

Standard deviation

The range and the interquartile range are measures of spread that involve the difference between two measures. A better indication of spread is the **standard deviation**, which measures the degree to which each score (x) deviates from the mean ($\bar{x}$). This means we find ($x - \bar{x}$) for each score. To ensure a positive result, these difference are squared and the average calculated.

For n scores with mean $\bar{x}$, the standard deviation is $\sigma = \sqrt{\frac{\Sigma(x - \bar{x})^2}{n}}$

1 For the scores 1, 4, 5, 6, 7, 7, 8, 10:

a find the mean

S Mean $= \frac{1 + 4 + 5 + 6 + 7 + 7 + 8 + 10}{8}$

$= 6$

$\therefore$ The mean is 6.

b draw a table to find the standard deviation.

S

x	$x - \bar{x}$	$(x - \bar{x})^2$
1	−5	25
4	−2	4
5	−1	1
6	0	0
7	1	1
7	1	1
8	2	4
10	4	16
	Total	52

$$\sigma = \sqrt{\frac{\Sigma(x - \bar{x})^2}{n}}$$

$$= \sqrt{\frac{52}{8}}$$

$= 2.549\,509\,757\ldots$

$= 2.55$ (2 dec. pl.)

$\therefore$ The standard deviation is 2.55.

Check using the calculator—see the calculator manual.

2 Calculate the mean and standard deviation, correct to two decimal places:

a 18, 21, 16, 15, 23, 20, 17

S $\bar{x} = 18.57$ $\sigma = 2.66$

b 43, 89, 105, 11, 56, 87, 198

S $\bar{x} = 84.14$ $\sigma = 55.12$

3 The ages of 10 students are listed below:

15, 14, 16, 15, 15, 18, 14, 16, 17, 20

Compare the mean and standard deviation at present to the mean and standard deviation in 10 years.

S Existing: $\bar{x} = 16$ $\sigma = 1.79$ (2 dec. pl.)

In 10 years' time, scores are 25, 24, 26, 25, 25, 28, 24, 26, 27, 30.

New: $\bar{x} = 26$ $\sigma = 1.79$ (2 dec. pl.)

$\therefore$ The mean increases by 10 but the standard deviation remains the same.

Bivariate data analysis

Introduction

In this chapter we have concentrated on single variable data such as marks in a test, or height of students. Now we consider **pairs** of variables such as height and weight and any **relationship** that may exist between the two variables. One variable is the **independent** variable (e.g. time) and the other is the **dependent** variable. When data is written in a table the independent variable is usually the top row, or the left-hand column. When graphed, the independent variable is placed on the horizontal axis.

The simplest way to graph bivariate data is to use a **scatter plot**. The relationship between the two variables can be described as:

- Positive or negative

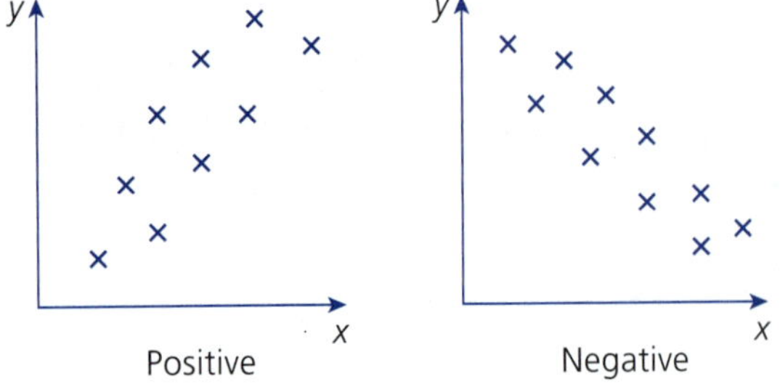

- Strong or weak

The scatter plot can be used to identify **trends** and make **predictions**.

1 State the independent and dependent variables in each of the following pairs of data:

a Age and height of a primary school student

S Independent: age; dependent: height

b Speed of a car and the distance travelled

S Independent: speed; dependent: distance

c Wage earned and number of hours worked.

S Independent: hours worked; dependent: wage

2 State the type of relationship between x and y for each of the following scatter plots:

a

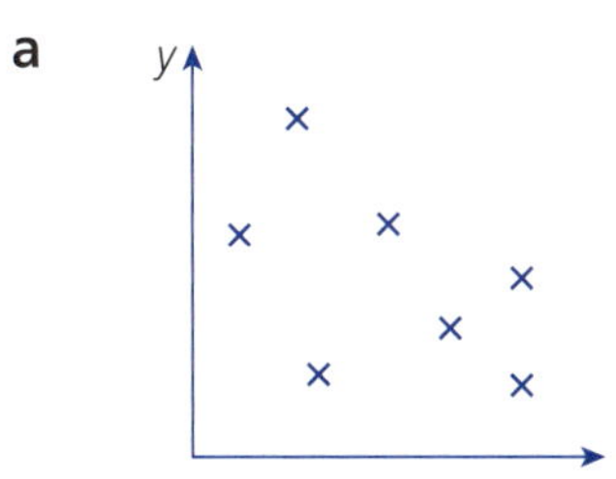

S Weak negative relationship.

b

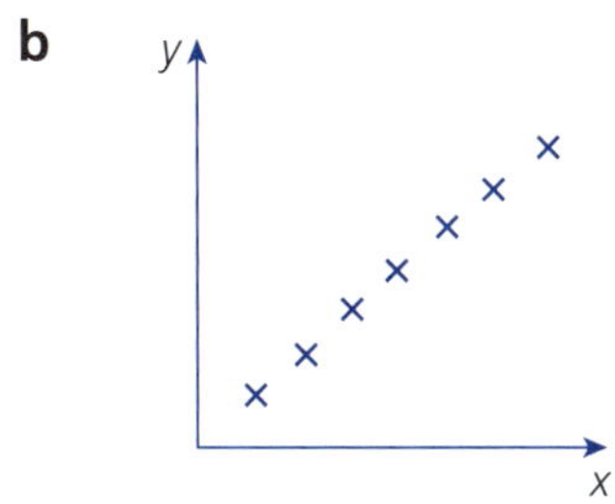

S Strong positive relationship.

3 A survey was conducted to find the time (t, in weeks) people had spent on a particular diet and the corresponding mass loss (m, in kg). The results of the survey are shown in the table:

t	6	2	8	4	5	2	4	7	2	3
m	5	1	6	3	6	4	2	6	2	3

a Draw a scatter plot.

S

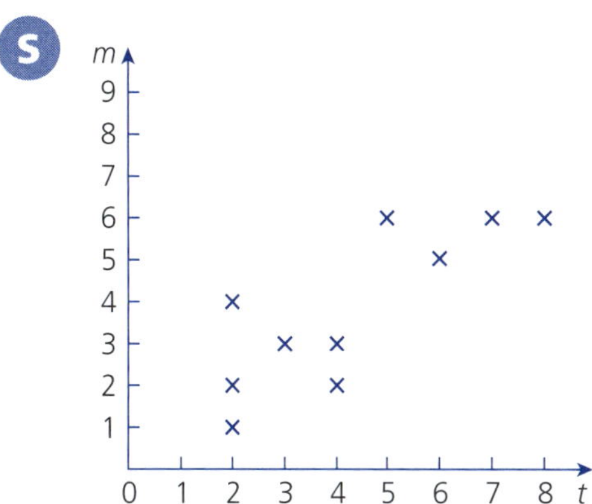

b Describe the relationship between t and m.

S Weak positive relationship.

c What is the trend suggested by the scatter plot?

S The longer a person stays on the diet the more mass will be lost.

Constructing a line of best fit

When there is a high level of correlation between two variables, the scatter plot can be used to draw a line of best fit. The equation of the line can be found and is useful to make predictions about the data. The 'Least squares' method is often used to draw the line, but in this study guide the less formal approach of 'Line of best fit by eye' is used.

For Example

1 The scatter plot shows the linear relationship between Mathematics results (x) and English results (y). Aaron has drawn a line of best fit that passes through two points on the scatter plot.

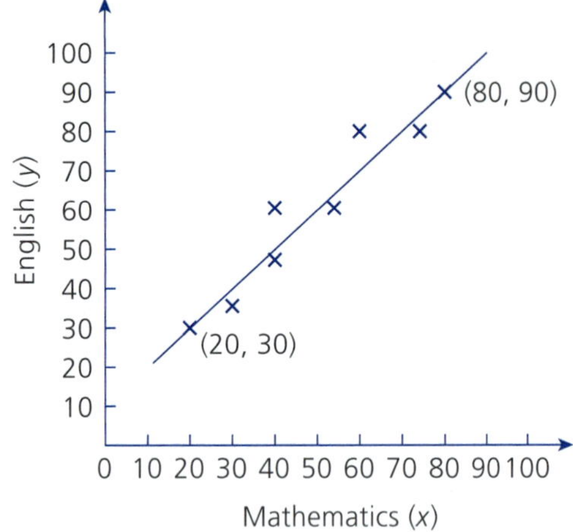

a Find the equation of the line of best fit.

S The line is of the form $y = mx + b$ and passes through (20, 30) and (80, 90).

$$\text{Gradient } m = \frac{90 - 30}{80 - 20} = 1$$

$\therefore$ Equation of the form $y = x + b$.

Now, subs. (20, 30) in $y = x + b$

$$30 = 20 + b$$
$$b = 10$$

$\therefore$ The line is $y = x + 10$.

b Use the equation to estimate the English result for a student who scored 75 in the Mathematics test.

S Subs. $x = 75$ in $y = x + 10$
$$= 75 + 10$$
$$= 85$$

$\therefore$ The student scored 85 in the English test.

c Use the equation to estimate the Mathematics result for a student who scored 48 in the English test.

S Subs. $y = 48$ in $y = x + 10$
$$48 = x + 10$$
$$x = 48 - 10$$
$$= 38$$

$\therefore$ The student scored 38 in the Mathematics test.

2 The scatter plot shows the linear relationship between the age of cars (x) in years and the price (y) in dollars of vehicles at a second-hand car yard.

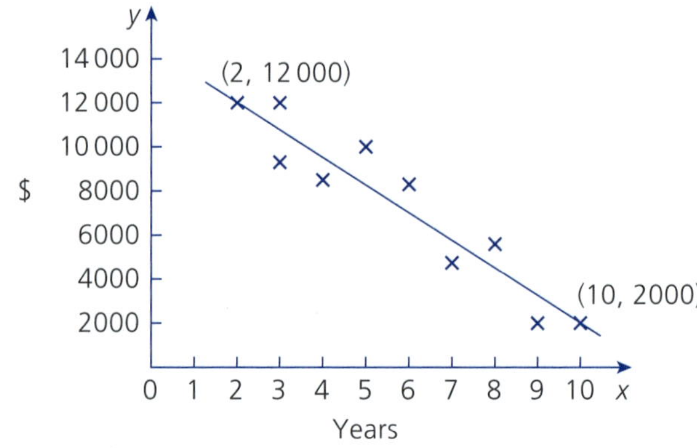

a Find the equation of the line of best fit.

S The line passes through (2, 12 000) and (10, 2000).

$$\text{Gradient } m = \frac{12000 - 2000}{2 - 10} = \frac{10000}{-8} = -1250$$

$\therefore$ Equation of the form $y = -1250x + c$.

Now, subs. (2, 12 000) in $y = -1250x + c$
$$12\,000 = -1250(2) + c$$
$$c = 14\,500$$

$\therefore$ The line is $y = 14\,500 - 1250x$.

b Use the equation to estimate the cost of a 6-year-old car.

S Subs. $x = 6$ in $y = 14\,500 - 1250x$
$$= 14\,500 - 1250(6)$$
$$= 7000$$

$\therefore$ The cost would be $7000.

Practise Practise

1 Gai has been playing soccer for six years and has recorded her goal-scoring record. Over these years Gai scored 8, 7, 4, 8, 6 and 3 goals.

a **i** Calculate Gai's mean number of goals.
ii Find the range and the mode.
iii Find the median.

b In her seventh year of soccer Gai struck brilliant form to score 13 goals. How will this affect Gai's mean and median?

c After eight years, Gai's mean number of goals slumped to 6.5. How many goals did she score in her eighth year?

pp. 187–189

2 For the following sets of scores, find the mean and median (to one decimal place if necessary).

a 7, 4, 7, 3, 5, 6, 4, 8, 4
b 2, 4, 6, 3, 5, 7, 9, 11, 7, 4

pp. 187–189

3 Zoltan needs to average 75 in his yearly examinations to be allowed to drive the family car. His average for his first four exams is 70. What mark must he score in his fifth and final exam to average 75?

pp. 187–189

4 Twenty-five students were surveyed. They were asked the question: 'What colour is your favourite cap?'

The results of the survey were: green 8, red 7, blue 3, black 2, yellow 1, other colour 4. Draw a dot plot to illustrate this information.

pp. 187–189

5 The shoe sizes of 22 students from a Year 10 class were collected:

$6\frac{1}{2}$	5	8	$7\frac{1}{2}$	7	5	5	$5\frac{1}{2}$
7	6	$4\frac{1}{2}$	4	7	6	$5\frac{1}{2}$	6
5	4	5	7	7	$6\frac{1}{2}$		

By first making a dot plot, draw up a frequency distribution table to illustrate this information.

p. 190

6 The height of 20 Year 9 boys was measured and recorded. The heights in cm were:

148	150	151	155	149	154	163
159	152	162	165	167	147	158
168	154	149	150	160	152	

Construct a stem-and-leaf plot to illustrate this information.

p. 191

7 The results of students in an English test out of 70 were as follows:

26	47	39	64	48	50	52	33	61	55
47	62	46	52	60	39	28	54	39	62
49	53	57	44	64	43	40	33	47	50

a Construct a stem-and-leaf plot to illustrate this information.

b Write down the range, mode and find the median.

pp. 191–193

8 For the following sets of scores find the median:

a 7, 9, 8, 6, 6, 5, 4, 8
b 9, 8, 9, 6, 7, 6, 6, 5, 9, 8, 7
c 18, 14, 17, 12, 19, 16, 16, 14, 18, 15
d 11, 7, 9, 12, 12, 7, 8, 11, 12, 8, 7, 14, 13, 10, 9

pp. 187–193

9 From the following stem-and-leaf plots find the median:

a

Stem	Leaf
1	0 7 8
2	1 2 4 6 9
3	7 7 7 8 8 9
4	0 0 6 7 8 8 9
5	1 4

b

Stem	Leaf
10	5 8
11	3 3 6 6 9
12	4 4 7
13	5 5 6 8 8
14	7

pp. 192–193

10 For the following sets of scores, find the median, the upper and lower quartiles and the interquartile range and draw a box plot:

a 7, 9, 8, 6, 6, 5, 4, 8
b 9, 8, 9, 6, 7, 6, 6, 5, 9, 8, 7
c 18, 14, 17, 12, 19, 16, 16, 14, 18, 15
d 11, 7, 9, 12, 12, 7, 8, 11, 12, 8, 7, 14, 13,10, 9

pp. 193–196

11 From the following box plots:

a write down the median

b calculate the interquartile range.

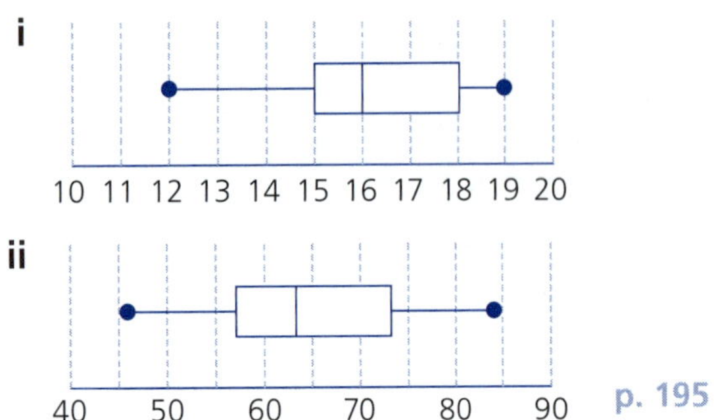

p. 195

12 From the following stem-and-leaf plots find the median, the upper and lower quartiles and construct a box plot:

a

Stem	Leaf
1	0 7 8
2	1 2 4 6 9
3	7 7 7 8 8 9
4	0 0 6 7 8 8 9
5	1 4

b

Stem	Leaf
10	5 8
11	3 3 6 6 9
12	4 4 7
13	5 5 6 8 8
14	7

pp. 192–196

13 The heights of 30 Year 9 boys were measured and tabulated. The results were as follows:

161	149	150	161	167	169
172	180	182	150	174	167
168	170	175	182	181	168
154	166	162	160	159	169
166	158	164	174	172	170

a Construct a stem-and-leaf plot from this information.

b From the stem-and-leaf plot find the median, the upper and lower quartiles and the interquartile range.

c Construct a box plot to illustrate this information.

pp. 192–196

14 Here is a list of seven scores: 14, 10, 18, 15, 9, 12, 6

a Find the mean.

b Complete the table to find the standard deviation, correct to two decimal places:

x	$x - \bar{x}$	$(x - \bar{x})^2$
	Total	

pp. 199–200

15 Calculate the mean and standard deviation of the following scores. Leave your answer correct to three significant figures:

a 19, 27, 21, 18, 15, 17

b 103, 101, 109, 97, 110, 100, 106

c 1.6, 1.71, 1.58, 1.66, 1.63

pp. 199–200

16 Compare the following sets of data by commenting on their means and standard deviations:

a Set A: 16, 18, 12, 19, 22, 15
Set B: 23, 11, 19, 21, 13, 15

b Set A: 2, 4, 6, 8, 10
Set B: 17, 19, 21, 23, 25

pp. 199–200

17 The mean of a set of scores is 12 and the standard deviation is 3. How will the mean and standard deviation be affected if:

a each score is increased by 2?

b each score is doubled?

c another score of 12 is included?

pp. 199–200

18 For the scores 23, 19, 25, 19, 23, 25, 28, find the range, interquartile range and the standard deviation.

pp. 192–200

19 Comment on the skew of the following data sets:

a

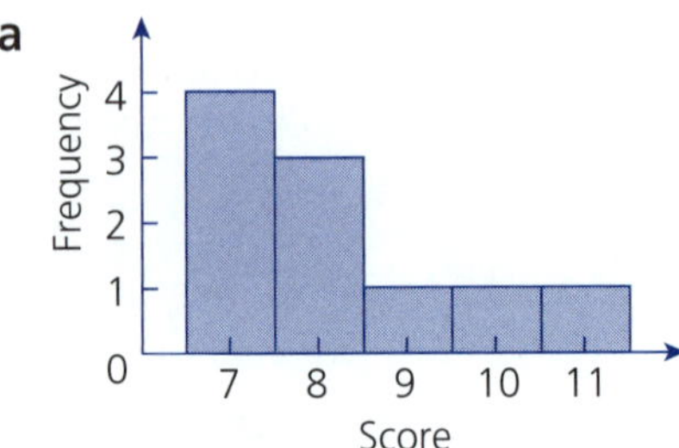

b

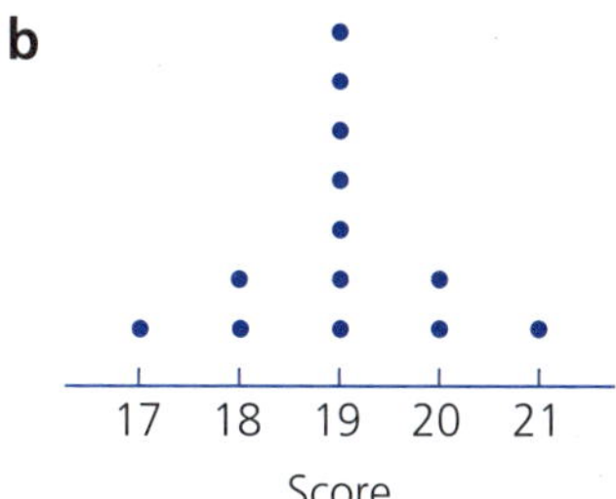

c

Stem	Leaf
1	4 7 7 8 9
2	3 5 6
3	1
4	4

d

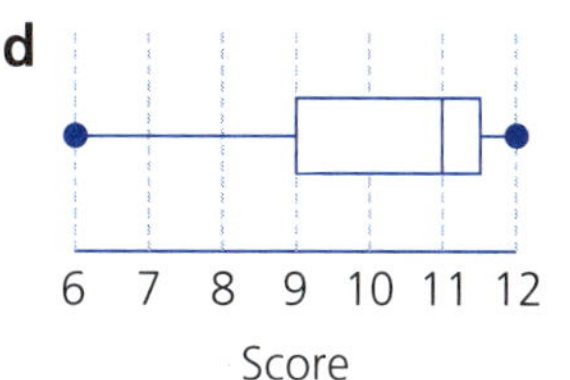

e

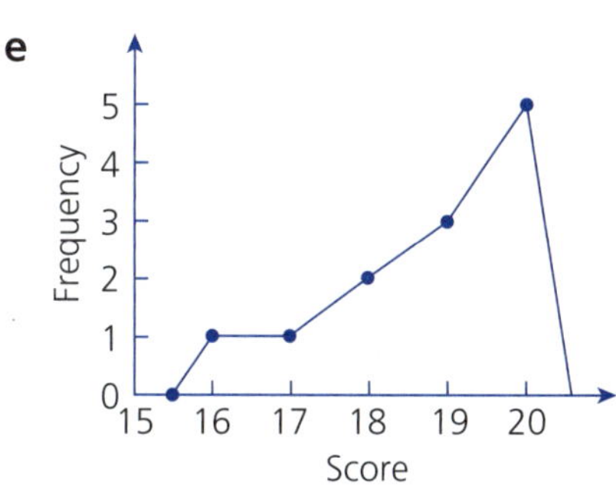

f

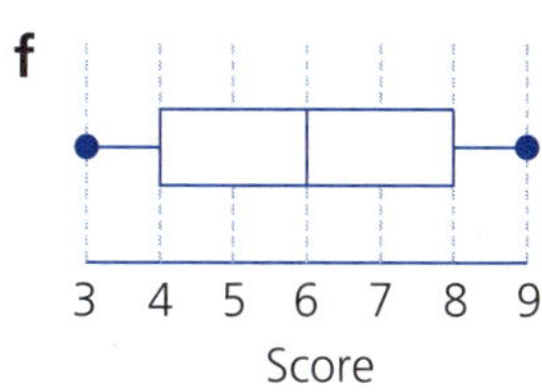

p. 197

20

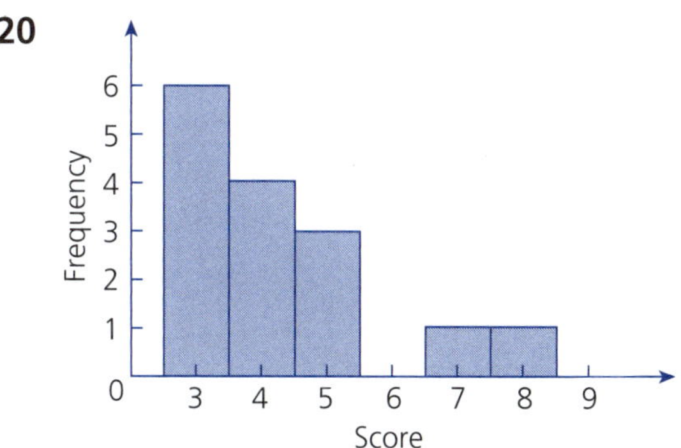

Use the histogram to:

a draw a frequency table
b find the mean, mode and median
c comment on the skew of the data.

pp. 187–199

21 The diagrams below show the frequency polygons of two distributions. Comment on the skew of each data set:

a

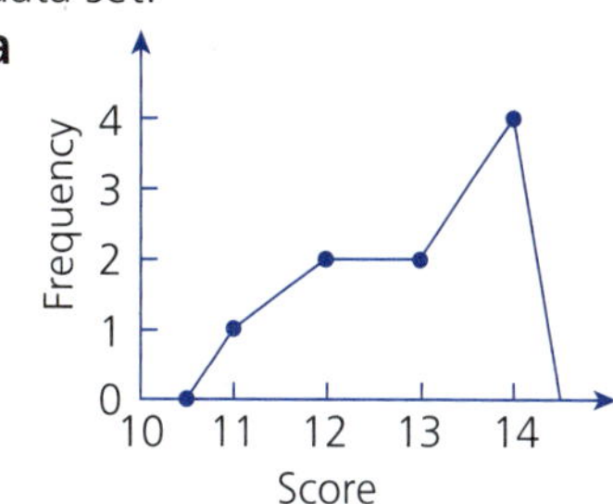

b

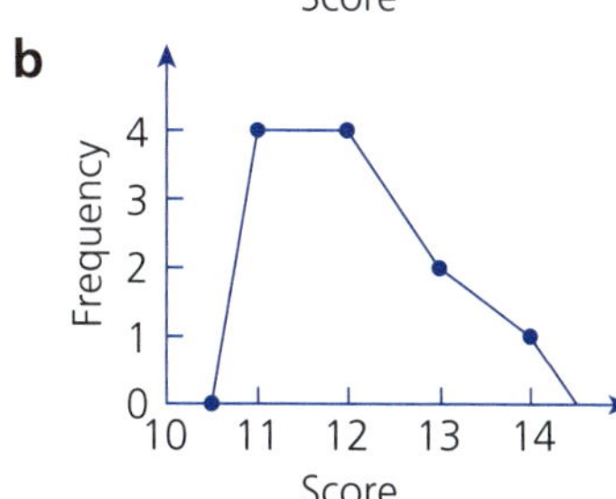

pp. 197–199

22 A group of students chose ten-pin bowling for sport. The dot plot shows the shoe sizes that were used by the group:

Shoe sizes

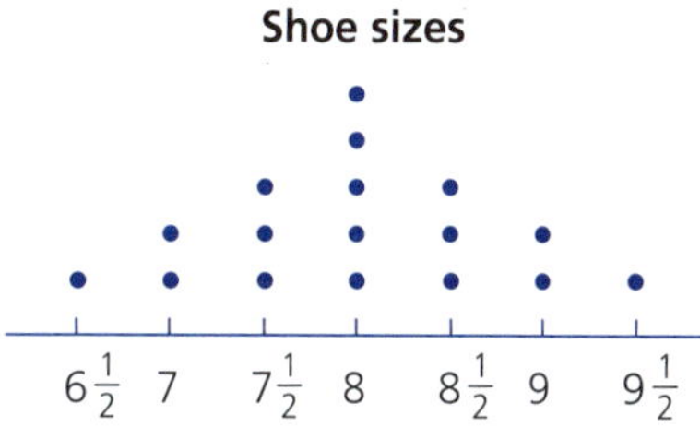

a Comment on the skew.
b Find the median, mode and mean of the data.

pp. 197–199

23 The results of two maths tests are recorded below in the back-to-back stem-and-leaf plot.

Comment on the skew of both tests:

Results of tests		
Trigonometry		**Equations**
9 9 6	0	
8 6 5 4 4 2 0	1	8 9
7 4 3	2	3 5 7 8 9
5 4	3	2 4 6 7 7 8 9 9
0	4	0

pp. 197–199

24 For the following data sets, draw a box plot and comment on the skew of each distribution:

a 9, 13, 15, 16, 16, 17, 17, 17, 18
b 8, 9, 10, 10, 10, 11, 18
c 10, 12, 12, 14, 14, 14, 16, 16, 18

pp. 195–199

25 Determine the independent variable in each of the following pairs of data:

a rainfall (r in mm) and crop yield (A in kg)
b height of plant (h in cm) and age (n in weeks)
c car's engine size (s in L) and fuel economy (E in L/100 km).

pp. 200–201

26 The graphs show a relationship between x and y. Comment on the type of relationship, using terms such as weak/strong and positive/negative:

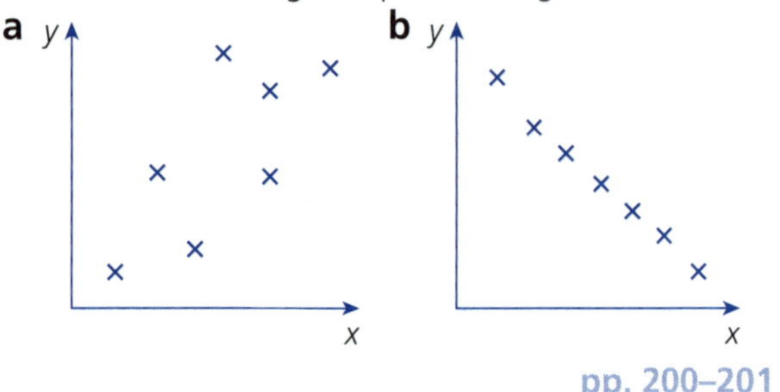

pp. 200–201

27 Eight students measured their heights (h) and arm spans (s) and the results are recorded in the table below. All lengths are in cm:

h	161	165	158	157	159	162	160	158
s	163	167	159	159	162	165	162	160

a Draw a scatter plot.
b Describe the relationship between h and s.
c Use the scatter plot to predict the arm span of a student with a height of 164 cm.
d What is the trend suggested by the scatter plot?

pp. 200–201

28 The same eight students then compared their heights (h) with the distance from the top of their heads to the middle of their chests (d). The results are recorded in the table below. All lengths are in cm:

h	161	165	158	157	159	162	160	158
d	41	42	40	39	40	41	40	39

a Draw a scatter plot.
b Use the scatter plot to predict the arm span of a student with a height of 164 cm.

pp. 200–201

29 Consider the data set:

x	4	11	9	5	2	7	1	8	4	6
y	7	0	2	5	8	3	9	3	8	4

a Draw a scatter plot for the data.
b Describe the relationship between x and y as positive or negative.
c Describe the relationship between x and y as strong or weak.

pp. 200–201

30 A teacher compared the students' results in a topic test with their final result in the course. The results are displayed in the scatter plot:

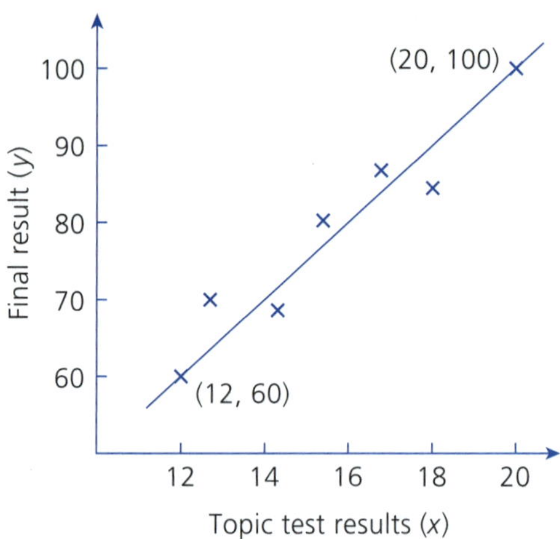

a Find the equation of the line of best fit.
b Use the equation to estimate the topic test result for a student who scored a final result of 65.

p. 202

31 The results of a survey linking student height and mass are recorded in the scatter plot below:

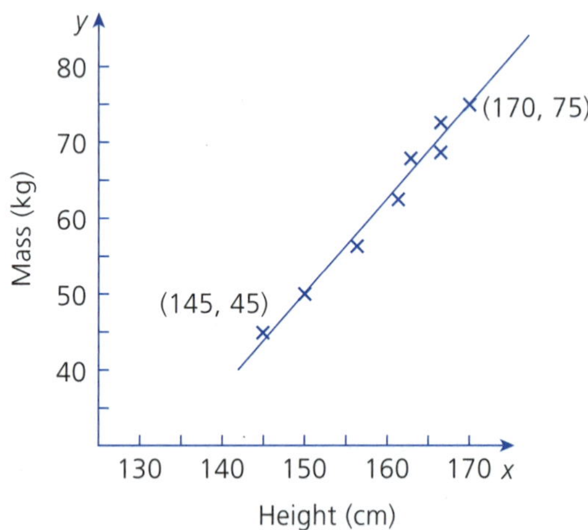

a Find the equation of the line of best fit.
b Use the equation to estimate the:
i mass of a student with a height of 160 cm
ii height of a student with a mass of 81 kg.

p. 202

Go to p. 302 for **Quick Answers** or to pp. 343–346 for **Worked Solutions**

For a complete understanding of this topic, you must be able to:

✓	Construct dot plots and stem-and-leaf plots and use them to find range, mode and median		pp. 187–193
✓	Calculate upper and lower quartiles, interquartile range and construct box plots		pp. 193–196
✓	Compare data sets using stem-and-leaf and box plots		pp. 196–197
✓	Recognise normal and skewed distributions		pp. 197–199
✓	Calculate the standard deviation of a set of numbers		pp. 199–200
✓	Recognise the independent and dependent variables in bivariate data displays		p. 200
✓	Display bivariate data in a scatter plot		pp. 200–201
✓	Recognise relationships of bivariate data as positive/negative and strong/weak		pp. 200–201
✓	Determine the equation of a line of best fit drawn on a scatter plot		p. 202
✓	Use the line of best fit to make predictions.		p. 202

Now you are ready to do the tests!

Intermediate Test

Statistics

(35 marks)

1 Which of the distributions shown below is bi-modal? (1 mark)

A 2, 2, 3, 3, 3, 4, 5 **B** 14, 16, 16
C 1, 2, 3, 3, 4, 4 **D** 0, 0, 1, 2

2 Grace's netball team scored 26, 32, 28, 39 and 30 points in their first five games of the season.

a Find the mean number of points scored in the first five games. (1 mark)

b How many points does her team need to score in the next game to maintain the same mean score? (1 mark)

c If Grace's team scores 37 points in the sixth game, what is their overall mean score for the six games? (1 mark)

d After the seventh game the mean score was calculated and it had dropped by two points. If the opposing team scored 28 points, by how many points did Grace's team lose? (2 marks)

3 Each day for a week Thomas ate two mandarins at lunchtime and counted the number of seeds he found in each mandarin. The number of seeds were:

4 6 3 5 4 2 6
4 5 3 5 3 1 5

a Calculate the mean, mode and range. (3 marks)

b Find the median and the interquartile range. (2 marks)

c Draw a box plot. (2 marks)

4 Find the mean and standard deviation of the following scores:

a 34, 23, 37, 29, 31, 35 (2 marks)

b 0, 0, 4, 2, 6, 7, 3, 5, 1, 0 (2 marks)

5 A box plot is drawn to show the distribution of marks for a class of 28 students. The test was marked out of 25:

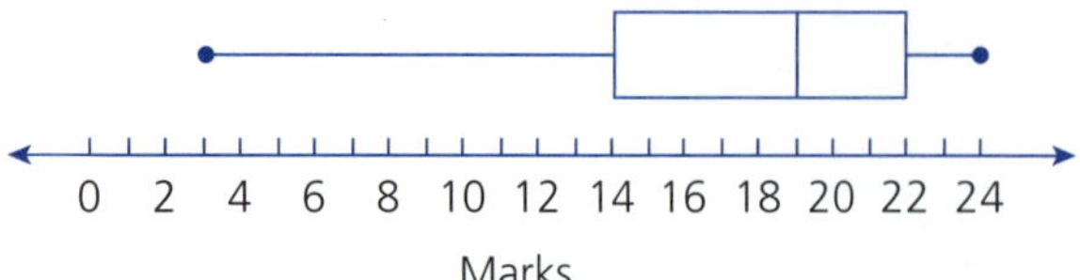

a Find the highest and lowest marks for the test. (2 marks)

b Approximately what percentage of students scored:

i 19 or less? (1 mark)

ii 14 or more? (1 mark)

iii 88% or more? (1 mark)

c How many students scored 14 or less? (1 mark)

d If you scored 72%, would you be placed in the bottom half or top half of the class? (1 mark)

e Comment on the skew of the distribution. (1 mark)

6 The Wildfires and the Diamonds play in the same basketball competition and the number of points scored in each game of the season is recorded in the back-to-back stem-and-leaf plot.

Wildfires		Diamonds
Leaf	Stem	Leaf
9 8	0	
9 9 7 6 4 2	1	9
4 2 0	2	1
2 1	3	2 6 6 7
1	4	0 2 4 7 8 9
	5	6 7

a Comment on the team's:

i range of scores (1 mark)

ii mode (1 mark)

iii median. (1 mark)

b Comment on the skew of each team's points. (1 mark)

7 Consider the bivariate data set:

x	2	5	3	6	5	2	1	6	4	4
y	4	9	7	10	10	5	3	11	8	7

a Draw a scatter plot for the data. (2 marks)

b Which is the dependent variable? (1 mark)

c Is the relationship positive or negative? (1 mark)

d Is the relationship weak or strong? (1 mark)

8 Hillary used a scatter plot to determine that the equation of the line of best fit was $y = 2x + 20$. What is the value of y when $x = 50$? (1 mark)

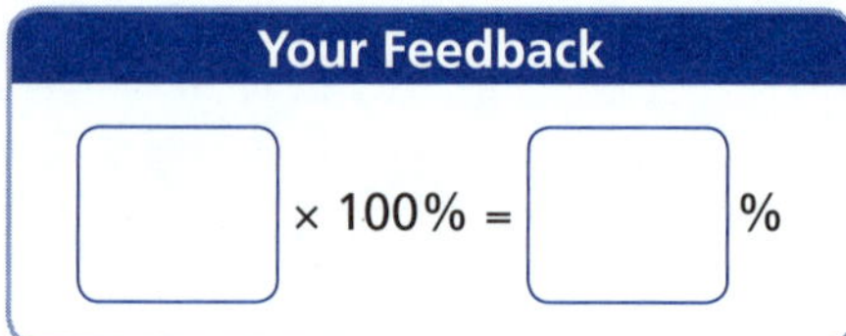

QA PAGE 307

WS PAGE 376

Advanced Test

Statistics

(35 marks)

1 Match the terms 'negatively skewed', 'positively skewed', 'bi-modal' and 'normally distributed' to each of the following histograms: (4 marks)

a

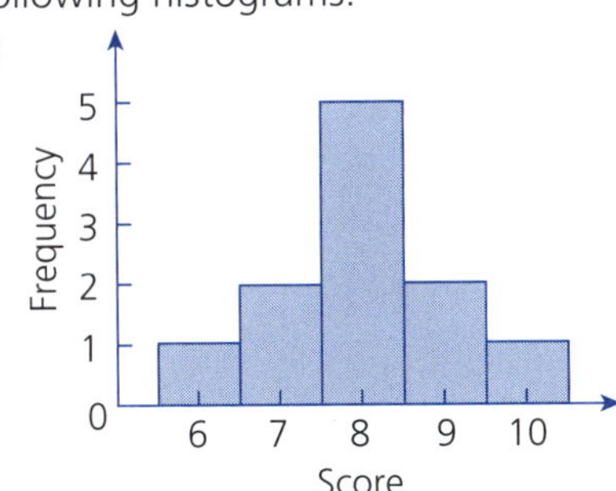

b

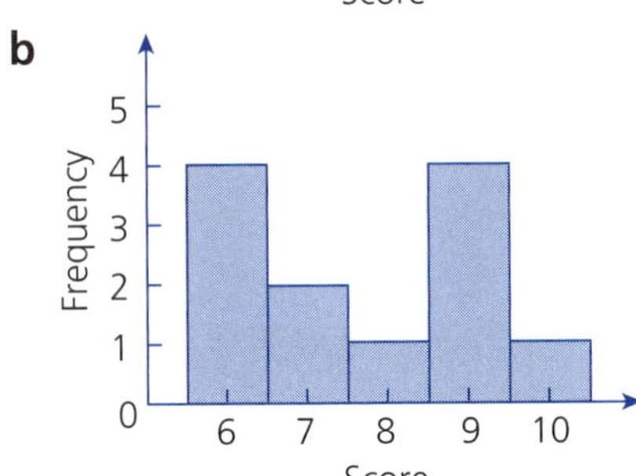

c

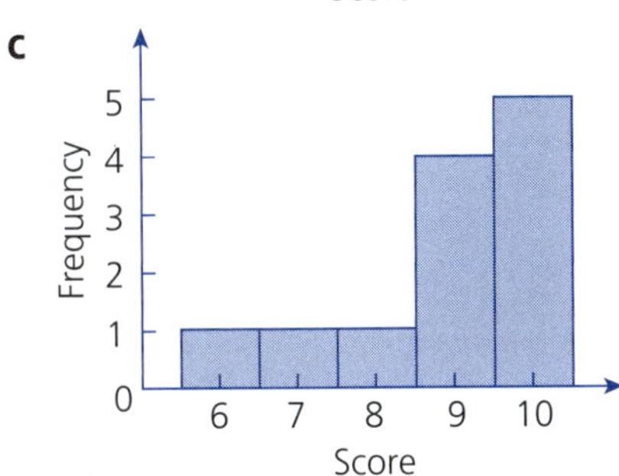

d

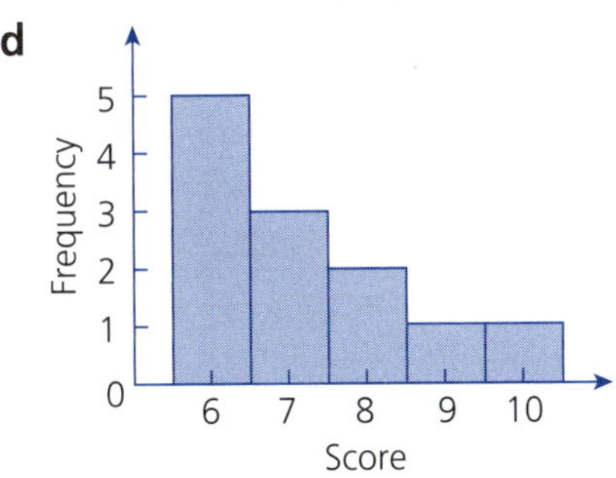

2 A set of scores is displayed in a stem-and-leaf plot:

Stem	Leaf
0	7
1	0 0 2 5 8
2	6 7
3	1 6
4	2

a Find the mode and range. (2 marks)
b Find the median. (1 mark)
c Find the interquartile range. (2 marks)
d Draw a box plot. (2 marks)

3 The table shows the results of a spelling test for a particular class of Year 10 students:

Score	Frequency
6	1
7	3
8	7
9	12
10	7

a The data set is skewed. Is it positive or negative? (1 mark)
b Find the standard deviation of the scores. (1 mark)
c The mean result in all Year 10 classes was 8.2. How does this class compare? (2 marks)

4 The prices of books in two shops were recorded and displayed in the box plot below:

Prices of Books

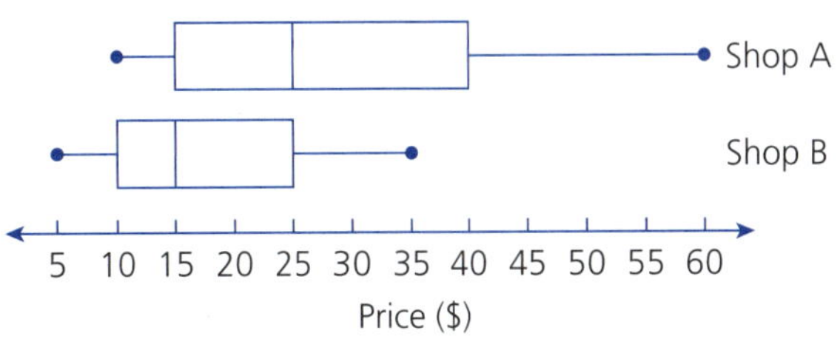

Shop A

Shop B

5 10 15 20 25 30 35 40 45 50 55 60

Price ($)

a What was the price of the cheapest book in Shop A? (1 mark)
b What percentage of books in Shop B were priced at $25 or less? (1 mark)

c If Shop A had 400 books for sale, how many were priced at \$40 or more? (1 mark)

d Compare the two shops' interquartile ranges. (2 marks)

5 A set of scores has a mean of 24 and a standard deviation of 3.5. Find the new mean and standard deviation of the scores if 6 is added to each score. (2 marks)

6 The number of people (n) at a beach at midday compared with the temperature (T) over a 10-day period is recorded in the table below:

T	27	29	31	20	23	26	31	36	41	38
n	86	102	105	11	21	46	68	121	85	92

a Draw a scatter plot. (2 marks)

b Is the relationship positive or negative? (1 mark)

c Comment on the strength of the relationship and give a reason. (2 marks)

7 Find the interquartile range for a data set if 75% of the scores are 7.4 or above and 25% of the scores are 10.1 or above. (1 mark)

8 Ten students sat a history test and a science test and the results are presented on the scatter plot below:

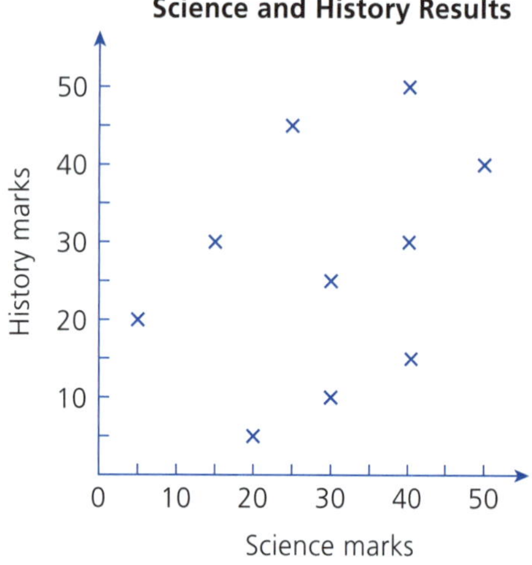

a Is the relationship between the pair of results weak or strong? (1 mark)

b What was the highest science mark? (1 mark)

c The results from the scatter plot are recorded in the table below:

Science	25	30	40	40	50	30	b	5	20	15
History	a	10	15	50	40	25	30	20	5	30

Find the missing values of a and b. (2 marks)

9 The relationship between the temperature (x) and the number of people (y) at a local swimming pool was recorded and the results shown in the scatter plot below:

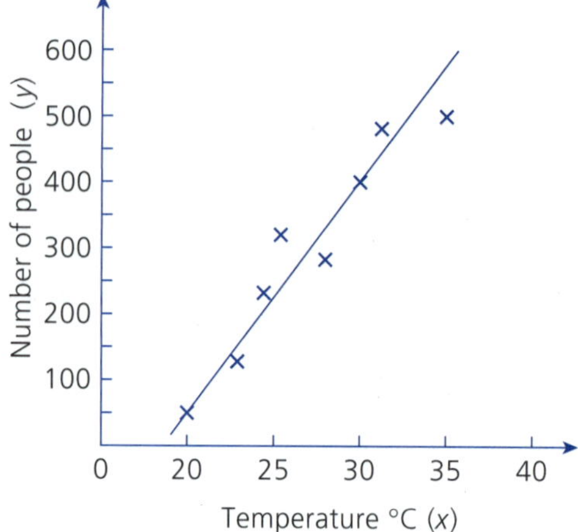

a Use the scatter plot to find an equation linking the temperature and the number of pool visitors. (2 marks)

b Use the equation to estimate the number of visitors if the temperature was 37°. (1 mark)

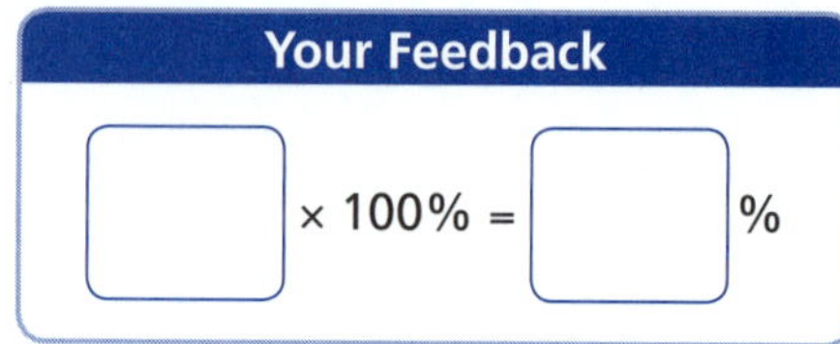

QA PAGE 307

WS PAGE 376

Chapter 12
Probability

Review

A coin is tossed once. There is one chance in two that it will land with a head facing up. We say that the *probability* of throwing a head is one out of two, or:

$P(\text{Head}) = \frac{1}{2}$.

As each outcome is equally likely, the probability of throwing a tail is also $\frac{1}{2}$, that is,

$P(\text{Tail}) = \frac{1}{2}$.

Keywords

Complementary event
Conditional
Favourable
Multistage event
Outcome
Tree diagram

When throwing a die, there are six faces and each is equally likely to turn up. It cannot be predicted which number will be face up. The probability of throwing a six is one chance out of six, that is, $P(6) = \frac{1}{6}$.

Definition of probability

The probability of an event occurring is defined as follows:

$$\text{Probability (event)} = \frac{\text{Number of favourable outcomes}}{\text{Number of possible outcomes}}$$ provided each outcome is equally likely to occur.

Consider the die. The number of favourable outcomes was one (throwing a six), while the number of possible outcomes was six (each face):

$$\therefore P(6) = \frac{\text{Number of favourable outcomes}}{\text{Number of possible outcomes}} = \frac{1}{6}$$

For Example

1 A fair die is thrown. Find the probability of rolling:

a a 3

S $P(3) = \frac{1}{6}$

b a 5

S $P(5) = \frac{1}{6}$

c an odd number

S There are three odd numbers (1, 3, 5):

$\therefore P(\text{odd}) = \frac{3}{6} = \frac{1}{2}$

d an even number

S $P(\text{even}) = \frac{3}{6} = \frac{1}{2}$

e a number larger than 4.

S There are two numbers larger than 4: 5 and 6.

$P(>4) = \frac{2}{6} = \frac{1}{3}$

Range of probability

A certainty has a probability of 1. An impossibility has a probability of 0.

The probability of any specific event occurring has a value between 0 and 1. A value outside this range is meaningless. Thus, if there are two possibilities, the probability of the event (say A) not occurring is $1 - P(A)$. They are **complementary events**.

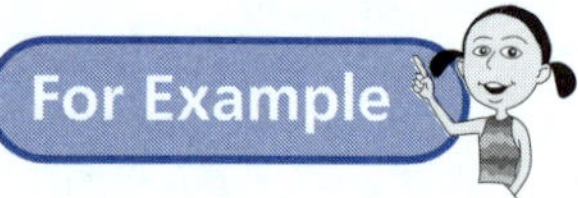

1 Michelle's school bag contains two black socks and three grey socks. If one sock is drawn out at random, what is the probability that the sock is:

a black

S $P(\text{black}) = \dfrac{\text{Number of favourable outcomes}}{\text{Number of possible outcomes}}$

$= \frac{2}{5}$

Total possible outcomes = total number of socks.

b grey.

S $P(\text{grey}) = 1 - P(\text{black})$

$= 1 - \frac{2}{5}$

$= \frac{3}{5}$

Relative frequency and probability

The relative frequency column can be used to find experimental probability.

Nicholas throws an unbiased die 180 times. The results are below.

The probability of throwing a one is theoretically $\frac{1}{6}$, or 0.167 (to three decimal places). Nicholas has thrown 28 ones out of 180 throws.

Score	Frequency	Relative frequency
1	28	0.156
2	30	0.167
3	31	0.172
4	29	0.161
5	32	0.178
6	30	0.167
Σ	180	1.000

Experimental probability of throwing a 1

$= \frac{28}{180} = 0.156$ (to three decimal places)

← Remember relative frequency $= \dfrac{f}{\text{Total } f}$

Total of relative frequency is 1, even though rounding here means a total of 1.001.

Venn Diagrams

Venn Diagrams provide a means of representing outcomes visually. Usually consisting of circles and a rectangle, the overlapping sections identify common outcomes.

For Example

1 A survey found the number of students who enjoyed Science and History and the results detailed on the Venn Diagram:

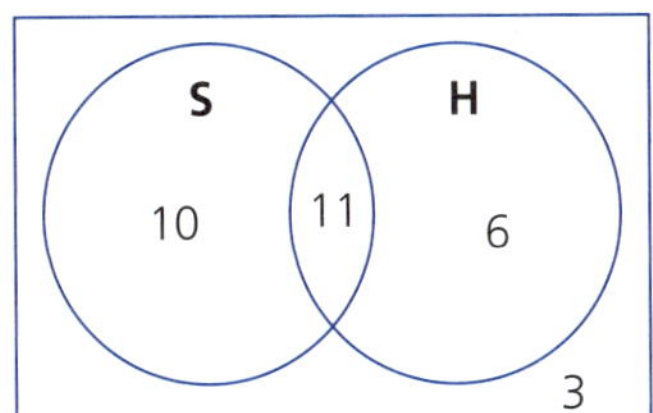

S: Science
H: History

a How many students were surveyed?

S 30

b If a student was chosen at random, what is the probability that the student:

i enjoys both subjects?

S $\therefore P(\text{both}) = \frac{11}{30}$

ii enjoys History but not Science?

S $\therefore P(\text{History, not Science}) = \frac{6}{30} = \frac{1}{5}$

iii does not enjoy either subject?

S $\therefore P(\text{neither}) = \frac{3}{30} = \frac{1}{10}$

2 On a particular day, the training patterns of a squad of female triathletes were recorded on the Venn diagram.

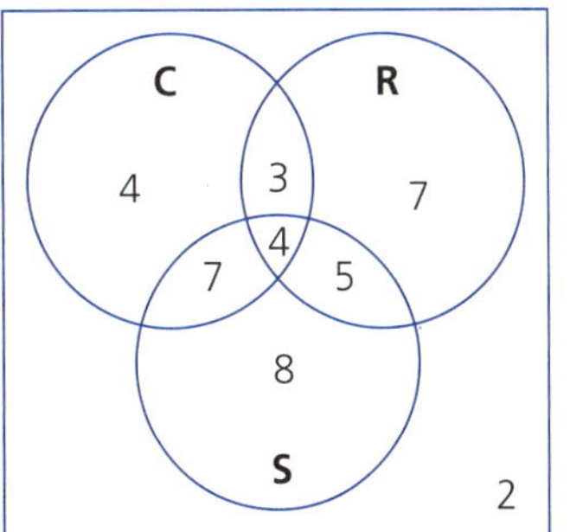

C: Cycling
R: Running
S: Swimming

a If a triathlete was chosen at random, what is the probability that she

i swam and cycled

S $\therefore P(\text{swam and cycled}) = \frac{11}{40}$

ii ran, cycled and swam

S $\therefore P(\text{all three}) = \frac{4}{40} = \frac{1}{10}$

b If a triathlete who had cycled that day was chosen at random, find the probability that she

i ran

S $\therefore P(\text{cyclist also ran}) = \frac{7}{18}$

ii did not swim.

S $\therefore P(\text{cyclist did not swim}) = \frac{7}{18}$

Multiple events using tables

Multiple events are sometimes called multistage events and involve more than one event, such as flipping a coin twice, or choosing two balls from a bag. If replacement is allowed then the outcomes from each selection can be repeated. If replacement is not allowed then the outcomes cannot be repeated.

For Example

1 A bag contains three balls numbered 1, 2 and 3. The first ball chosen is replaced before the second ball is selected.

a Complete a table listing the possible outcomes.

S

		1st choice		
		1	2	3
2nd choice	1	(1, 1)	(2, 1)	(3, 1)
	2	(1, 2)	(2, 2)	(3, 2)
	3	(1, 3)	(2, 3)	(3, 3)

b What is the probability of:

i selecting a 3 followed by a 1?

S $\therefore P(3, 1) = \frac{1}{9}$

ii selecting a double?

S $\therefore P[(1,1) \text{ or } (2, 2) \text{ or } (3, 3)] = \frac{3}{9} = \frac{1}{3}$

iii selecting a total of 5?

S $\therefore P(\text{total of } 5) = \frac{2}{9}$

iv selecting a total less than 4?

S $\therefore P(\text{less than } 4) = \frac{3}{9} = \frac{1}{3}$

2 Another bag contains three balls numbered 4, 5 and 6. The first ball chosen is **not** replaced before the second ball is selected.

a Complete a table listing the possible outcomes.

S

		1st choice		
		4	5	6
2nd choice	4	×	(5, 4)	(6, 4)
	5	(4, 5)	×	(6, 5)
	6	(4, 6)	(5, 6)	×

b What is the probability of:

i selecting a total of 9?

S $\therefore P(\text{total of } 9) = \frac{2}{6} = \frac{1}{3}$

ii selecting a total more than 10?

S $\therefore P(\text{more than } 10) = \frac{2}{6} = \frac{1}{3}$

3 Mitchell throws two dice and the scores are added. Draw up a table to show all possibilities.

S **From the table, count the total number of possible outcomes—36. So, all answers will be out of 36.**

First die (columns); Second die (rows)

+	1	2	3	4	5	6
1	2	3	4	5	6	7
2	3	4	5	6	7	8
3	4	5	6	7	8	9
4	5	6	7	8	9	10
5	6	7	8	9	10	11
6	7	8	9	10	11	12

Find the probability that the sum of numbers face up will be:

a 7

S $P(7) = \frac{6}{36} = \frac{1}{6}$ Highlighted in table.

b 8

S $P(8) = \frac{5}{36}$

c 12

S $P(12) = \frac{1}{36}$

d greater than 7

S $P(>7) = \frac{15}{36} = \frac{5}{12}$

e 7 or more

S $P(\geq 7) = \frac{21}{36} = \frac{7}{12}$

f 1

S $P(1) = 0$ impossible with 2 dice

g even or less than 7

S $P(\text{even or less than 7})$

$= \frac{24}{36}$

$= \frac{2}{3}$

h even and less than 7.

S $P(\text{even and less than 7})$

$= \frac{9}{36}$

$= \frac{1}{4}$

4 Scott throws two dice and then calculates the product of the two numbers, face up.

S The first step is to draw up the table.

×	1	2	3	4	5	6
1	1	2	3	4	5	6
2	2	4	6	8	10	12
3	3	6	9	12	15	18
4	4	8	12	16	20	24
5	5	10	15	20	25	30
6	6	12	18	24	30	36

Now count the total number of possible outcomes—again it is 36. This then becomes the denominator in all the probabilities.

Find the probability that the product of numbers face up will be:

a 6

S $P(6) = \frac{4}{36} = \frac{1}{9}$ Highlighted in table.

b 36

S $P(36) = \frac{1}{36}$

c less than 24

S $P(<24) = \frac{30}{36} = \frac{5}{6}$

d even

S $P(\text{even}) = \frac{27}{36} = \frac{3}{4}$

e even and less than 24

S $P(\text{even and less than 24}) = \frac{22}{36} = \frac{11}{18}$

f even or less than 24

S $P(\text{even or less than 24}) = \frac{35}{36}$

g is not 36.

S $P(\text{not 36}) = 1 - P(36) = 1 - \frac{1}{36} = \frac{35}{36}$

5 The table represents the dominant hand of boys and girls in a class of 30 students:

Predisposition	**Boy**	**Girl**	Totals
Left-handed	4	2	6
Right-handed	10	14	24
Totals	14	16	30

a If a student is chosen at random, what is the probability that the person is a right-handed boy?

S Ten right-handed boys out of 30 students:

$\therefore P(\text{right-handed boy}) = \frac{10}{30}$

$= \frac{1}{3}$

b If a girl is chosen at random what is the probability that she is left-handed?

S Two left-handed girls out of 16 students:

$\therefore P(\text{left-handed girl}) = \frac{2}{16}$

$= \frac{1}{8}$

Multiple events using tree diagrams (or probability trees)

A tree diagram is a branching diagram that is useful to list the possible outcomes.

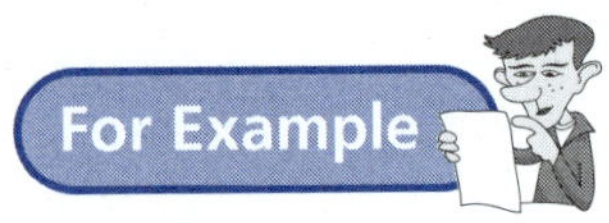

1 Lee tosses two coins.

S Draw the tree diagram.
Total number of outcomes = 4.

Diagram: First coin	Diagram: Second coin	Outcomes
H	H	H H
	T	H T
T	H	T H
	T	T T

Four outcomes

Find the probability that Lee throws:

a 2 heads

S $P(\text{H H}) = \frac{1}{4}$

b 2 tails

S $P(\text{T T}) = \frac{1}{4}$

c a head and a tail.

S $P(\text{H and T}) = \frac{2}{4} = \frac{1}{2}$

Could be H, T or T, H. The order is not specified.

2 Susie has three red and two blue socks in her drawer. She picks out two socks at random (the lights are out).

S Total number of outcomes = 20

Diagram: First sock	Diagram: Second sock	Outcomes
R	R	RR –
	R	RR –
	B	RB
	B	RB
R	R	RR –
	R	RR –
	B	RB
	B	RB
R	R	RR –
	R	RR –
	B	RB
	B	RB
B	R	BR
	R	BR
	R	BR
	B	BB
B	R	BR
	R	BR
	R	BR
	B	BB

Find the probability that Susie has chosen:

a 2 red

S $P(\text{RR}) = \frac{6}{20} = \frac{3}{10}$

Indicated in table by –

b 2 blue

S $P(\text{BB}) = \frac{2}{20} = \frac{1}{10}$

c one of each colour

S $P(\text{one of each}) = \frac{12}{20} = \frac{3}{5}$

d a pair of the same colour.

S $P(\text{pair}) = \frac{8}{20} = \frac{2}{5}$ [RR or BB]

Note: $P(\text{not a pair}) + P(\text{pair}) = 1$.
If the probabilities for all possible outcomes are totalled, the result will always be 1.

3 The letters A, B, C are written on identical cards and placed in a bag. Three cards are drawn from the bag without replacement to make a three-letter word. Use a tree diagram to list the possible outcomes and find the probability that the word is BAC.

S

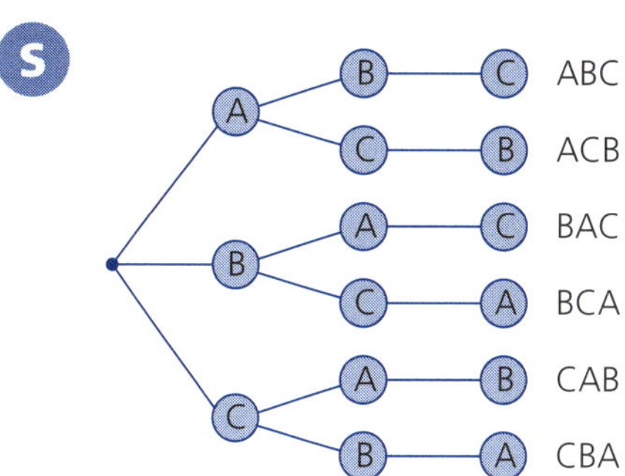

From the tree diagram, $P(\text{BAC}) = \frac{1}{6}$.

The tree diagram can be simplified by writing probabilities corresponding to a branch. The probability of each outcome is found by multiplying the probabilities on each branch.

1 René's top drawer contains three grey headbands and two red ones. Two headbands are taken from the drawer, one after the other, and the colours noted. Find the probability that the headbands are either both red, or both grey, or different colours.

Draw up a tree as before but mark the probabilities for each possible outcome along the branch.

Diagram (First pick, Second pick)	Outcome	Probability (Multiply along each branch)
$\frac{2}{5}$ R, then $\frac{1}{4}$ R	R R	$\Pr(\text{RR}) = \frac{2}{5} \times \frac{1}{4} = \frac{1}{10}$
$\frac{2}{5}$ R, then $\frac{3}{4}$ G	R G	$\Pr(\text{RG}) = \frac{2}{5} \times \frac{3}{4} = \frac{3}{10}$
$\frac{3}{5}$ G, then $\frac{2}{4}$ R	G R	$\Pr(\text{GR}) = \frac{3}{5} \times \frac{2}{4} = \frac{3}{10}$
$\frac{3}{5}$ G, then $\frac{2}{4}$ G	G G	$\Pr(\text{GG}) = \frac{3}{5} \times \frac{2}{4} = \frac{3}{10}$

Grey selected first, so still 2 reds out of 4.

Grey selected first, so only 2 left out of 4.

Prob. (grey first) = $\frac{3}{5}$

Prob. (red first) = $\frac{2}{5}$

Headbands may be:

a both red

S $P(\text{RR}) = \frac{2}{5} \times \frac{1}{4} = \frac{1}{10}$

b both grey

S $P(\text{GG}) = \frac{3}{5} \times \frac{2}{4} = \frac{3}{10}$

c different colours.

S $P(\text{different}) = P(\text{RG}) + P(\text{GR})$

$= \frac{3}{10} + \frac{3}{10}$ (From diagram.)

$= \frac{3}{5}$

3 grey, 2 red = 5 altogether

Note: to find the probability of two or more branches combined we add the probabilities of the branches.

2 The names of eight boys and four girls are placed in one hat, while the names of six boys and four girls are placed in a second hat. A name is drawn from each hat. Find the probability that the names drawn are either both girls, or both boys, or one boy, or at least one girl.

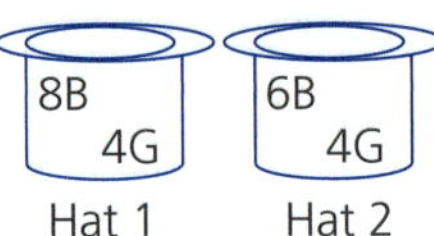

Diagram	Outcomes	Probability
First hat / Second hat		
$\frac{8}{12}$ B — $\frac{6}{10}$ B	BB	$Pr(BB) = \frac{8}{12} \times \frac{6}{10} = \frac{2}{5}$
$\frac{8}{12}$ B — $\frac{4}{10}$ G	BG	$Pr(BG) = \frac{8}{12} \times \frac{4}{10} = \frac{4}{15}$
$\frac{4}{12}$ G — $\frac{6}{10}$ B	GB	$Pr(GB) = \frac{4}{12} \times \frac{6}{10} = \frac{1}{5}$
$\frac{4}{12}$ G — $\frac{4}{10}$ G	GG	$Pr(GG) = \frac{4}{12} \times \frac{4}{10} = \frac{2}{15}$

The names may be:

a both girls

S $P(GG) = \frac{2}{15}$ From diagram.

b both boys

S $P(BB) = \frac{2}{5}$ From diagram.

c one boy

S $P(\text{one boy}) = P(BG) + P(GB)$
$= \frac{4}{15} + \frac{1}{5}$
$= \frac{7}{15}$

d at least one girl.

At least one girl means 1 G, 2 G, but not 2 boys, that is, everything except 2 boys.

↓

S $P(\text{at least one girl}) = 1 - P(BB)$
$= 1 - \frac{2}{5}$
$= \frac{3}{5}$

3 A box contains five red balls and three blue balls. Brae chooses three balls are chosen at random, without replacement.

a Draw a tree diagram.

S

$\frac{5}{8}$ R — $\frac{4}{7}$ R — $\frac{3}{6}$ R RRR
$\frac{5}{8}$ R — $\frac{4}{7}$ R — $\frac{3}{6}$ B RRB
$\frac{5}{8}$ R — $\frac{3}{7}$ B — $\frac{4}{6}$ R RBR
$\frac{5}{8}$ R — $\frac{3}{7}$ B — $\frac{2}{6}$ B RBB
$\frac{3}{8}$ B — $\frac{5}{7}$ R — $\frac{4}{6}$ R BRR
$\frac{3}{8}$ B — $\frac{5}{7}$ R — $\frac{2}{6}$ B BRB
$\frac{3}{8}$ B — $\frac{2}{7}$ B — $\frac{5}{6}$ R BBR
$\frac{3}{8}$ B — $\frac{2}{7}$ B — $\frac{1}{6}$ B BBB

b What is the probability that Brae chose:

i three red balls?

S $P(RRR) = \frac{5}{8} \times \frac{4}{7} \times \frac{3}{6}$
$= \frac{5}{28}$

ii balls of the same colour?

S $P(RRR \text{ or } BBB) = \frac{5}{8} \times \frac{4}{7} \times \frac{3}{6} + \frac{3}{8} \times \frac{2}{7} \times \frac{1}{6}$
$= \frac{11}{56}$

iii two blue balls and a red ball?

S $P(RBB \text{ or } BRB \text{ or } BBR)$
$= \frac{5}{8} \times \frac{3}{7} \times \frac{2}{6} + \frac{3}{8} \times \frac{5}{7} \times \frac{2}{6} + \frac{3}{8} \times \frac{2}{7} \times \frac{5}{6}$
$= \frac{15}{56}$

iv at least 1 blue ball?

S $P(\text{at least 1 blue ball})$
$= 1 - P(\text{no blue balls})$
$= 1 - P(\text{three red balls})$
$= 1 - \frac{5}{28}$ (from **i**)
$= \frac{23}{28}$

Conditional probability

Conditional probability is when the chance of something occurring depends on (is conditional on) another event happening first. The probability that B occurs, given that A occurs, can be written as $P(B \mid A)$.

1 A die is rolled. Given that the number is odd, what is the probability that it is less than 4?

S **Odd numbers are 1, 3, 5**

$\therefore P(\text{less than } 4) = \frac{2}{3}$

2 A survey of 40 students found the number of students who use mobile phones and tablets for study. The results are recorded in the Venn Diagram.

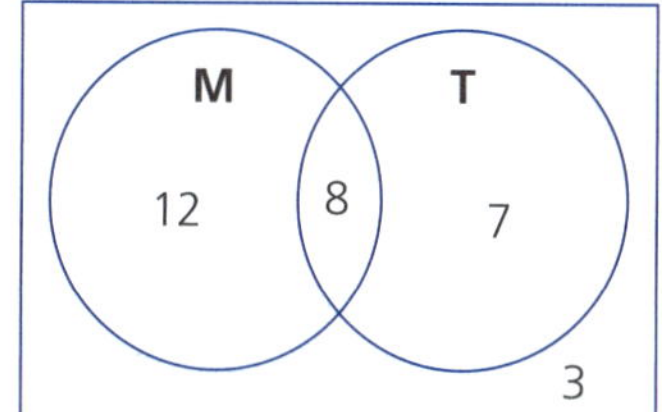

M: Mobile phone
T: Tablet

A student is chosen at random. What is the probability that they use a

a mobile phone, given that they use a tablet?

S **8 + 7 = 15 students use a tablet; 8 of these students also use a mobile phone:**

$\therefore P(\text{mobile} \mid \text{tablet}) = \frac{8}{15}$

b tablet, given that they do not use a mobile phone?

S **7 + 3 = 10 students do not use a mobile phone; 7 of these students use a tablet:**

$\therefore P(\text{tablet} \mid \text{not mobile}) = \frac{7}{10}$

3 In the music class of 24 students, 16 play the piano and 12 play the guitar. There are three students who neither play the piano nor the guitar.

a Draw a Venn Diagram to summarise this information.

S

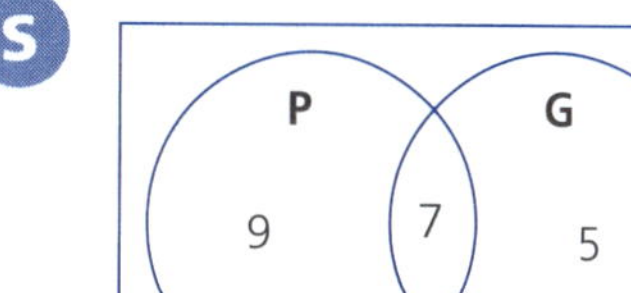

P: Piano
G: Guitar

b What is the probability that a student chosen at random plays the piano, given that they do not play the guitar?

S **9 + 3 = 12 students do not play guitar; 9 of these students play the piano:**

$\therefore P(\text{piano} \mid \text{not guitar}) = \frac{9}{12} = \frac{3}{4}$

More formally, $P(B \mid A) = \dfrac{P(\text{both } B \text{ and } A)}{P(A)}$.

1 A survey was conducted to find the number of people who regularly exercise and the results are shown below.

	Females	Males	Total
Do exercise	12	10	22
Do not exercise	14	14	28
Total	26	24	50

A person was chosen at random. What is the probability that they regularly exercise, given that they are female?

S

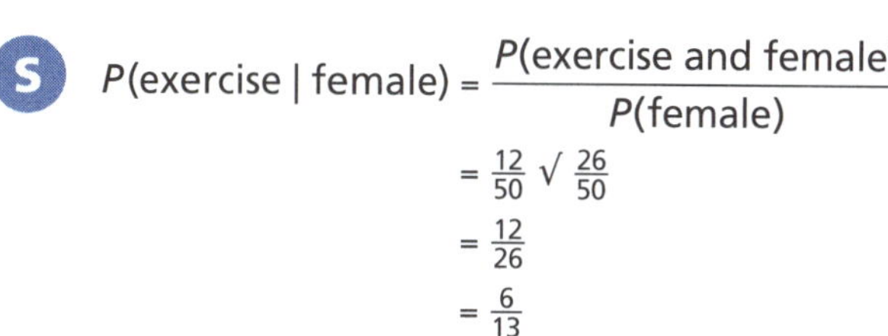

$$P(\text{exercise} \mid \text{female}) = \frac{P(\text{exercise and female})}{P(\text{female})}$$
$$= \frac{12}{50} \surd \frac{26}{50}$$
$$= \frac{12}{26}$$
$$= \frac{6}{13}$$

Practise Practise

1 A pack has 52 cards divided into four suits: diamonds, hearts, clubs and spades. Each suit contains a 2, 3, 4, 5, … , 10, J(ack), Q(ueen), K(ing), A(ce). If one card is drawn at random from the pack, find the probability that it is:
a an A
b a K
c a heart
d a spade
e a black suit
f a picture card (J, Q, K)
g not a picture card
h a *number* less than 7
i the six of hearts
j red card or a 6
k a red six. pp. 211–212

2 The probability of rain at the weekend is assessed as $\frac{7}{10}$. What is the probability that it will not rain at the weekend? pp. 211–212

3 A jar contains seven red, five black and three white jelly babies. If one jelly baby is drawn at random from the jar, find the probability that it is:
a red
b not red
c black
d red or white
e green. pp. 211–212

4 In a supermarket promotion, 10 gold tokens numbered 1 to 10 are hidden. Melville found one of the tokens. Find the probability that Melville found the token with number:
a 7
b even
c less than 7
d more than 7. pp. 211–212

5 Robyn keeps loose change in her car ashtray. Last Saturday she noted that there were four 50-cent coins and six 20-cent coins. She selected one coin from the ashtray to pay for a 45c iceblock. What is the probability that the coin selected is sufficient? pp. 211–212

6 The table summarises the birth place of a group of students:

Continent	Number
Africa	14
America	18
Asia	36
Australia	112
Europe	20

If a student is chosen at random what is the probability that they will be:
a born in Asia?
b born in Europe or Africa?
c not born in Australia? pp. 211–212

7 The graph below represents the results of a survey of the people living in the households in Mooney Street?

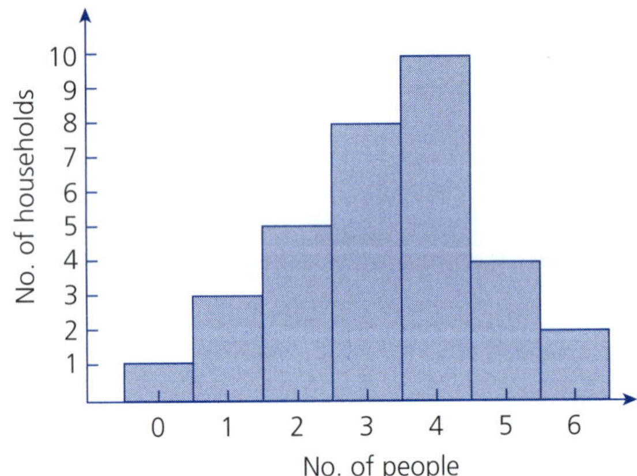

a How many households in Mooney Street?
b If a household was chosen at random, what is the probability that it contains:
i three people?
ii less than two people?
iii at least four people? pp. 211–212

8

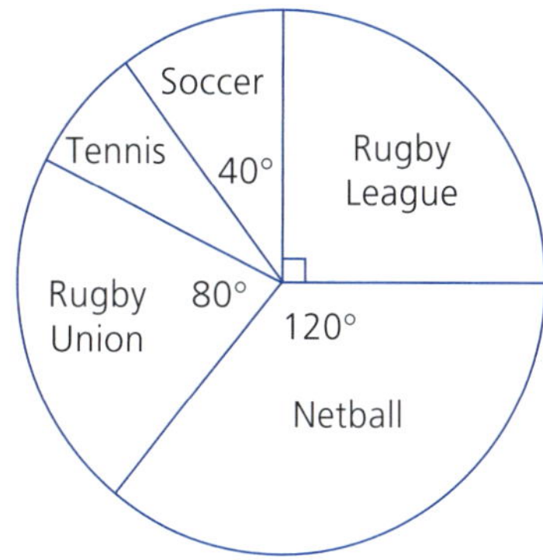

The sector graph represents the favourite sports of a number of students. If a student is chosen at random, what is the probability that the student favours:

a Netball?
b Tennis?
c Soccer or Rugby League? pp. 211–212

9 A bag contains red balls, green balls and blue balls. A ball is chosen from the bag, its colour noted and then returned to the bag. This occurs 100 times. The results are recorded in the table below:

Colour	Frequency	Relative frequency
Red	48	
Green	19	
Blue	33	
Total	100	

Complete the relative frequency column. p. 212

10 The diagram below represents the students who have travelled to Queensland and Victoria in the past 5 years:

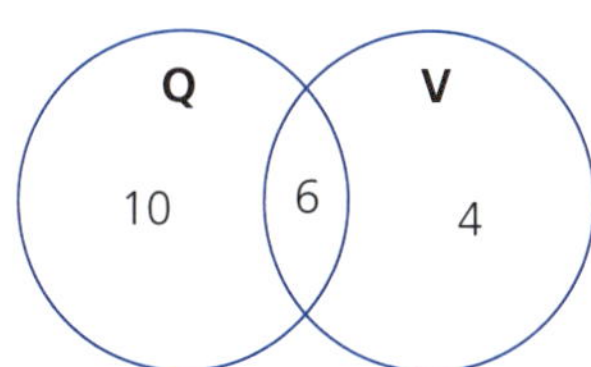

a How many students are represented in the diagram?
b If a student is chosen at random, find the probability that this student visited:
  i Queensland
  ii Victoria but not Queensland
  iii Victoria and Queensland. p. 213

11

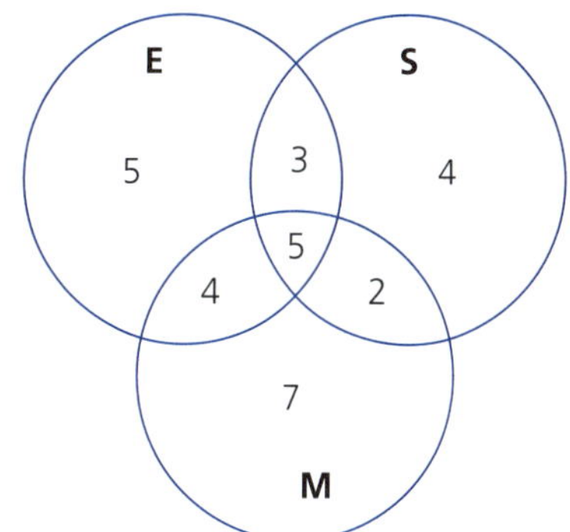

The diagram above represents the students who enjoy English (E), Science (S) and Mathematics (M).

a How many students are represented on the diagram?
b If a student is chosen at random, what is the probability that the student likes:
  i Mathematics?
  ii English and Mathematics?
  iii all three subjects?
  iv English or Science but not Mathematics?

p. 213

12 A coin is tossed and a die is thrown.

a Use a table to list all the possible outcomes.
b What is the probability of getting a:
  i head and a 4?
  ii tail and an odd number? pp. 214–215

13 Two dice are thrown (faces numbered 1 to 6). The sum of the two uppermost faces is noted. Find the probability that the sum is:

a 9
b 5
c odd
d greater than 9
e 9 or less
f not 9
g odd or less than 5
h odd and less than 5. pp. 214–215

14 If, instead of adding the numbers on the dice, the product is calculated, find the probability that this product is:

a 24
b odd
c less than 12
d 12 or less
e divisible by 3
f odd and divisible by 3
g odd or divisible by 3. pp. 214–215

15 Two dice (faces numbered 1 to 6) are thrown and the highest number from the two faces is noted. For example, if a 3 and 5 are thrown, the 5 only is noted. Draw up a table to illustrate all possible outcomes. Find the probability that the number noted is:

a 6
b 1
c not 6
d not 6 or 1
e even
f a prime
g an even prime
h a prime or even
i divisible by 3
j greater than 6. pp. 214–215

16 A sample of students was conducted and the results shown in the following table:

	Born in Australia	**Born overseas**	Totals
Female	74	46	
Male	56	24	
Totals			

a Complete the table.
b If a student is chosen at random what is the probability that the student is a:
  i male born in Australia?
  ii female born overseas?
c If a male is chosen at random, find the probability that he was born:
  i overseas
  ii in Australia. pp. 214–215

17 Mia has two tickets in a raffle that has two prizes. Draw a tree diagram to list the possible outcomes using W (wins) and L (loses). pp. 216–218

18 There are three children in a family. By drawing a tree diagram, list all possible outcomes using M (male) and F (female). pp. 216–218

19 A die is thrown twice. Complete the tree diagram listing to possible ways of tossing a 6 (S) and not tossing a 6 (N).

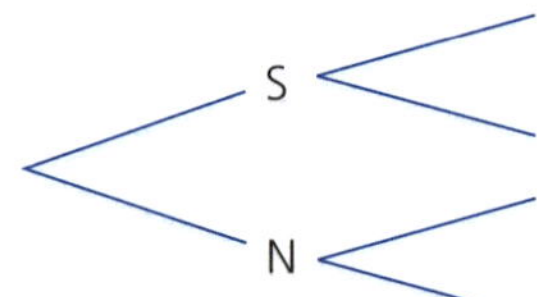

pp. 216–218

20 Three coins are tossed and the uppermost faces noted. Draw a tree diagram to illustrate the possible outcomes. Find the probability that the toss results in:

a three heads
b two heads
c one head
d no heads
e at least one head. pp. 216–218

21 A bag of Smarties contains three red and two blue ones. Two are drawn at random from the bag, the colour noted and then eaten. Find the probability that they are:

a both red
b both blue
c different colours
d the same colour. pp. 216–218

22 A biscuit barrel contains five choc-chip and four walnut biscuits. Two biscuits are drawn in succession and consumed. Find the probability that of the two biscuits:

a both are choc-chip
b one is a choc-chip
c only the first is a choc-chip
d neither is a choc-chip
e at least one is a choc-chip. pp. 216–218

23 One hundred tickets are sold in a Chook Raffle. Gregory John purchases two tickets, hoping to win both first and second prizes. Find the probability that he wins:

a first prize
b both prizes
c only second prize
d a prize
e no prize
f at least one prize. pp. 216–218

24 A die is rolled. Given that the number is even, what is the probability that it is:

a a 4
b less than 5? pp. 216–218

25 The letters of the alphabet are written on 26 identical cards and placed in a box. One card is chosen at random.

a What is the probability that the card is a vowel?
b Given that the card is a vowel, what is the probability that it is an E?
c Given that the card is a consonant, what is the probability that it is a P?

pp. 216–218

26 The gender and age of bus travellers were recorded and the results shown in the table below:

Survey of Bus Travellers

		Age		
		12 and under	**13 to 17**	**18 and over**
Gender	**Female**	5	12	10
	Male	8	11	4

A person is chosen at random.

a What is the probability that the person is female and age 13 to 17?

b Given that the person is female, what is the probability that she is aged 18 and over?

c Given that the person is aged 12 and under, what is the probability that the person is male?

p. 219

27 Data on the cholesterol level of a group of people was collected and summarised in the table below:

	Low	**Normal**	**High**	Totals
Men	16		26	
Women		22	14	60
Totals		50		

a Complete the table.

b If a person is chosen at random, what is the probability that they are:

- **i** male with a normal cholesterol level?
- **ii** female with a low cholesterol level?

c If a women is chosen at random, what is the probability that she:

- **i** has a normal cholesterol level?
- **ii** does not have a low cholesterol level?

d If a person with a high cholesterol level is chosen what is the probability that the person is female?

p. 219

28 A card is chosen from a deck of cards.

a If a red card is chosen, what is the probability that is a diamond?

b If a 7 is chosen, what is the probability that it is black?

c If a red queen is chosen, what is the probability that it is the queen of hearts?

p. 219

29 The Venn Diagram shows what was consumed by people at a morning tea, using T = tea, C = coffee and P = pastry.

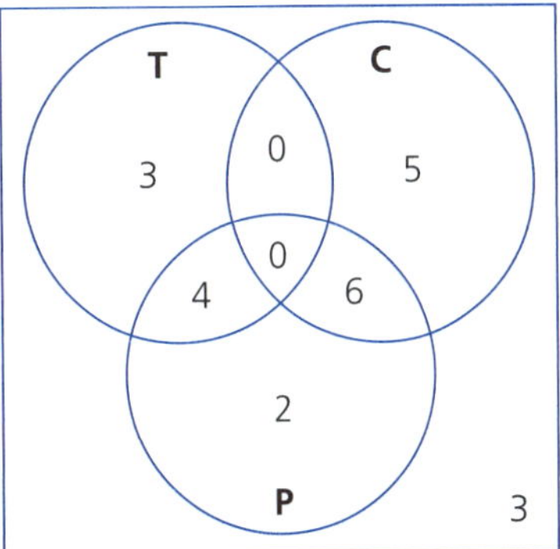

What is the probability that a randomly chosen person had a:

a coffee?

b tea and a pastry?

c pastry and a cup of coffee?

p. 219

30 Of a group of 30 supporters at a football game, 17 were wearing a team jersey and 12 were wearing a team cap. Nine fans were wearing both.

a Draw a Venn Diagram.

b What fraction of those wearing a jersey were also wearing a cap?

c What fraction of those who were not wearing a cap, were wearing a jersey?

p. 219

Go to p. 302 for **Quick Answers** or to pp. 346–348 for **Worked Solutions**

For a complete understanding of this topic, you must be able to:

✓	Recall basic probability concepts		pp. 211–212
✓	Calculate probabilities of events contained in Venn Diagrams		p. 213
✓	Calculate probabilities of compound events contained in tables		pp. 214–215
✓	Find the probability of compound events using tree diagrams		pp. 216–218
✓	Calculate the conditional probability of an event.		p. 219

Now you are ready to do the tests!

Intermediate Test

Probability

(30 marks)

1 A coin is tossed twice. What is the probability that:

a two heads were tossed (1 mark)

b at least one head is tossed. (1 mark)

2 A bag contains three red and five black balls. Two balls are selected at random, the first replaced before the second is drawn.

a Complete the probability tree. (1 mark)

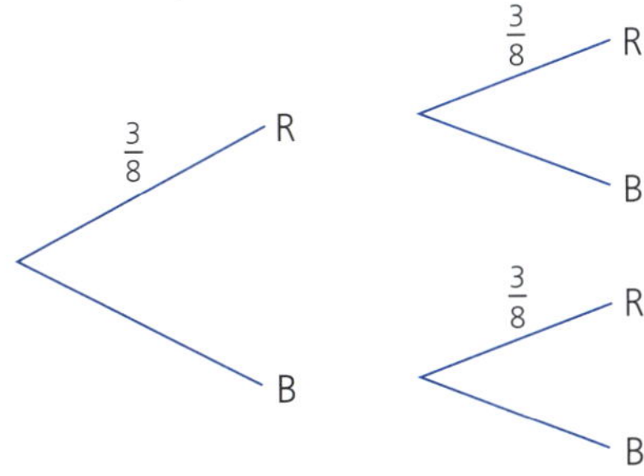

b What is the probability of choosing:

i two black balls? (1 mark)

ii balls of different colours? (1 mark)

3 A box has four balls marked 1, 2, 3, 4. A ball is selected at random from the box and the pointer, shown below, is rotated on the circular disc.

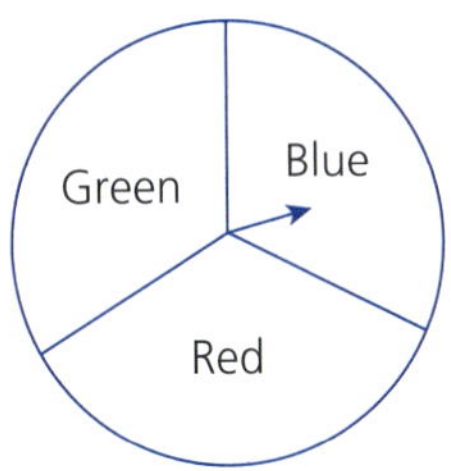

What is the probability of:

a a 3 and a green? (1 mark)

b an even number and not blue? (1 mark)

4 A survey was conducted to find the smoking habits of 200 people at a suburban shopping centre. The summary of the survey results are:

	Males	Females	Total
Smoker	27	36	63
Non-smoker	73	64	137
Total	100	100	200

a A male is chosen at random. What is the probability that he is a smoker? (1 mark)

b A female is selected at random. What is the probability that she does not smoke? (1 mark)

c A person is chosen at random. What is the probability that this person is a non-smoker? (1 mark)

5 City commuters were surveyed to find whether they had used a bus (B) or a train (T) in the previous week. The results are summarised below:

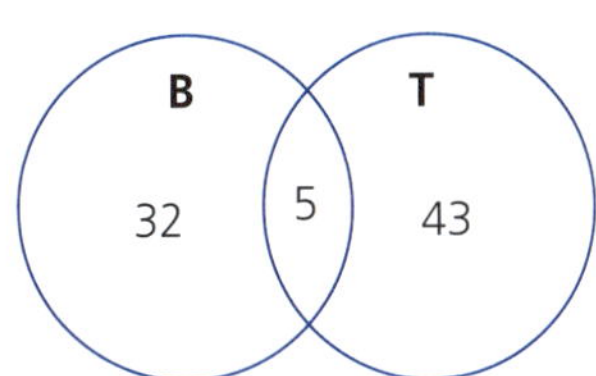

A commuter is chosen at random. What is the probability that this person travelled by:

a bus? (1 mark)

b train only? (1 mark)

6 A coin if flipped and a die is rolled. What is the probability of getting a:

a head and a 5? (1 mark)

b tail and an odd number? (1 mark)

c head and a number less than 3? (1 mark)

7 A bag contains three cards with the letters A, B and C. A card is chosen at random from the bag, and then returned to the bag before a second card is chosen.

a Draw a tree diagram to list all possible outcomes. (2 marks)

b What is the probability of choosing:

i BB? (1 mark)

ii BA? (1 mark)

iii at least one B? (1 mark)

iv AA or BB or CC? (1 mark)

v different letters? (1 mark)

8 A box contains three white discs and four blue discs.

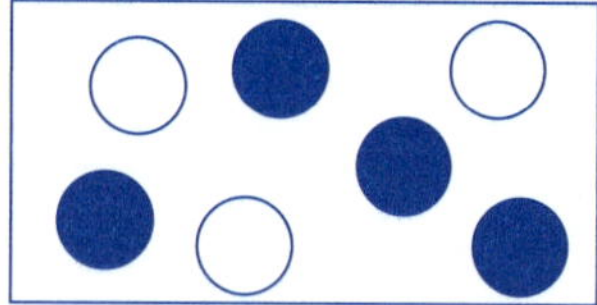

Two discs are chosen from the box. The first not replaced before the second is chosen.

a Complete the tree diagram: (2 marks)

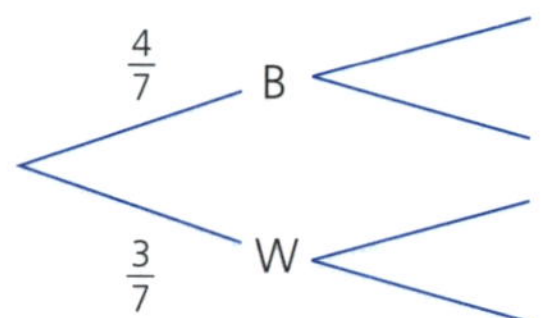

b What is the probability of selecting:

i BW? (1 mark)

ii WB? (1 mark)

iii WW? (1 mark)

9 Cards with the numbers from 1 to 20 are placed in a bag. A card is chosen at random from the bag.

a Find the probability that the number is less than 6. (1 mark)

b If the number is more than 10, what is the probability that it is even? (1 mark)

c The number is a multiple of 4. What is the probability that it is a factor of 24? (1 mark)

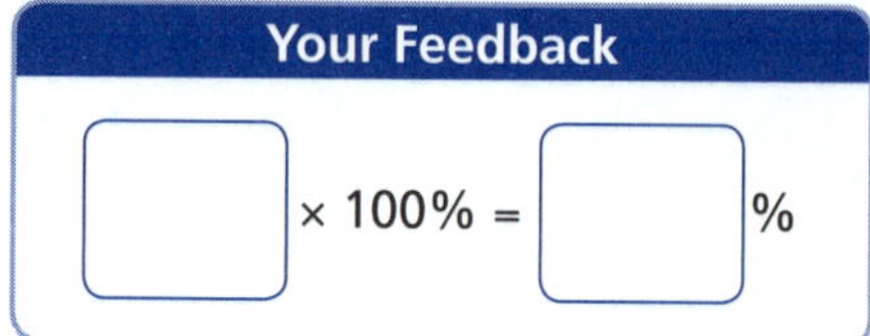

QA PAGE 307

WS PAGE 377

Advanced Test

Probability

(30 marks)

1 A bag contains four green and six yellow balls. Two balls are selected at random, the first not replaced before the second is drawn.

a Draw a probability tree. (1 mark)

b What is the probability of choosing:

i different coloured balls? (1 mark)

ii at least a yellow ball? (1 mark)

2 The table shows the results of a survey of voters at a recent state election:

	Males	Females
Liberal–NP	35	45
ALP	45	40
Other	20	15

a A woman is chosen at random. What is the probability that she voted Liberal–NP? (1 mark)

b A person is chosen at random. What is the probability that this person voted ALP? (1 mark)

3 A survey of 100 Year 10 students was conducted to find the subjects that they enjoyed. (Mathematics was not an option, as everyone enjoyed it.) The three subjects were English (E), Science (S) and PDHPE (P). The results of the survey are:

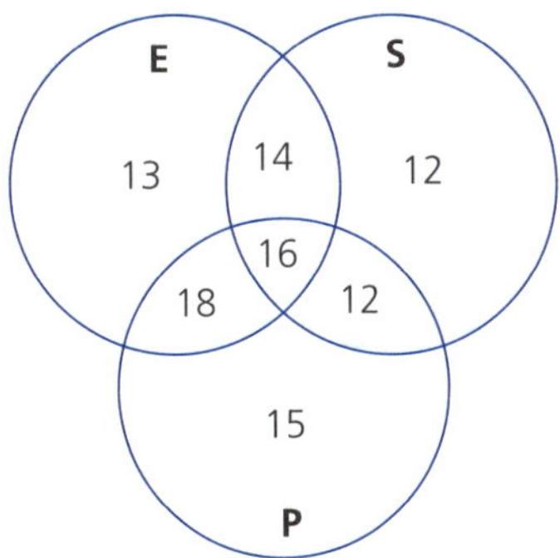

a A student is chosen at random. What is the probability that the student:

i likes PDHPE? (1 mark)

ii likes English, but not Science? (1 mark)

b A student who likes Science was chosen at random. What is the probability that this person:

i likes English? (1 mark)

ii likes all three subjects? (1 mark)

4 A survey of 100 students was conducted at a local school. Forty-five per cent of the students play team sport on a weekend, 35% play team sport during the week and 40% play do not play a team sport.

a Use a Venn Diagram to display the information. (1 mark)

b If a student is chosen at random, what is the probability that they play during the week as well as the weekend? (1 mark)

c What is the probability that a person who does not play on the weekend, plays during the week? (1 mark)

5 A group of cricket bowlers were surveyed and the results shown in the table below:

		Bowl	
		Left-handed	Right-handed
Bat	Left-handed	6	4
	Right-handed	5	15

a If a player was chosen at random, what is the probability that they bat and bowl left-handed? (1 mark)

b If a left-handed batman was chosen at random, what is the probability that they bowl right-handed? (1 mark)

6 A coin is tossed three times.

a Draw a tree diagram to list the possible outcomes. (2 marks)

b Find the probability of tossing:

i three tails (1 mark)

ii a head and two tails (1 mark)

iii at least one head. (1 mark)

7 Bag A contains three red and four blue balls. Bag B contains five red and six blue balls. A ball is chosen from each bag.

a Complete the tree diagram: (1 mark)

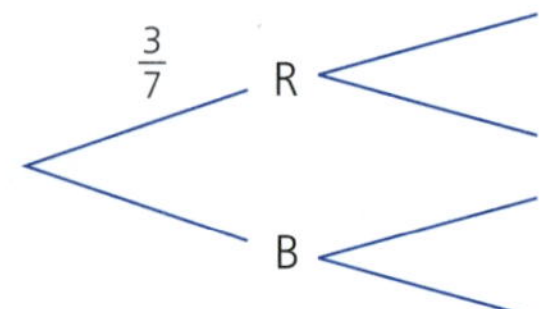

b Find the probability of choosing:

i two reds (1 mark)

ii the same colour (1 mark)

iii different colours (1 mark)

iv at least one blue. (1 mark)

8 William has 0.7 chance of passing his Science test, a 0.8 chance of passing his English test and a 0.9 chance of passing his Mathematics test.

a Complete the tree diagram: (3 marks)

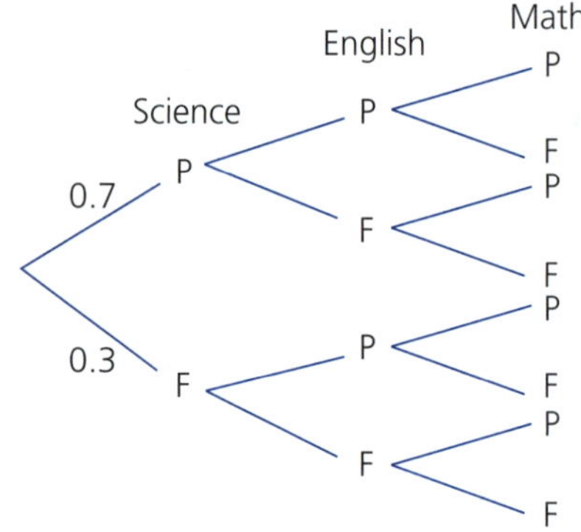

b Find the probability that William:

i passed all three tests (1 mark)

ii passed Maths but failed English and Science (1 mark)

iii failed only one test (1 mark)

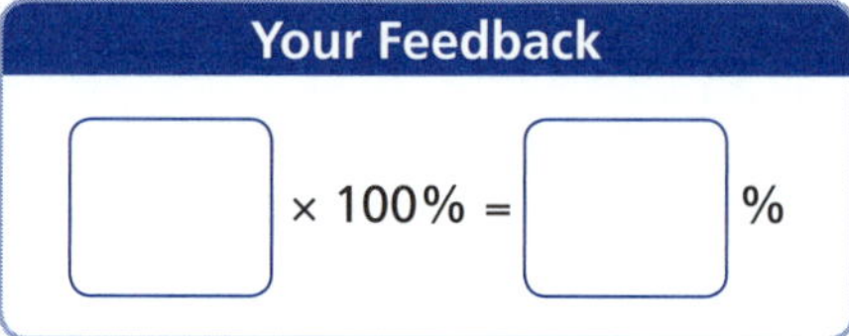

QA PAGE 307

WS PAGE 377

Chapter 13
Polynomials

Definition of a polynomial

Keywords

Coefficient, Constant, Degree, Factor, Intersection, Monic, Polynomial, Remainder, Roots, Zeros

$P(x)$ is a polynomial expression in x if:

$P(x) = a_n x^n + a_{n-1}x^{n-1} + \ldots + a_2x^2 + a_1x + a_0$,
where $a_n, a_{n-1}, \ldots, a_0$ are real numbers
n is an integer
$n \geq 0$.

Note: n cannot be a fraction or negative.

Therefore, expressions such as $\frac{1}{5}x^4$, 3, $x^3 - x$, are all polynomials in x, but 2^x, $\sqrt{x}$, $\frac{1}{x}$, x^{-2} are not polynomials in x.

From the definition of $P(x)$ above:

n = degree of the polynomial

a_n = leading coefficient

a_nx^n = leading term

a_0 = constant term.

Note: if the leading coefficient is 1, then $P(x)$ is a *monic* polynomial, for example, $x^3 - 4x + 1$.

For Example

1 For the polynomial $P(x) = 3x^4 - 7x - 2$, find the:

a degree of $P(x)$

S 4

b leading coefficient

S 3

c leading term

S $3x^4$

d constant term.

S -2

2 Find the degree of the following polynomials:

a $4 - 2x^3 + 6x$

S 3

b $(4x - 7)(3x^3 - 4)$

S 4 [found by multiplying $4x$ and $3x^3$, that is, $12x^4$]

Note: we could have expanded completely to get the same result.

3 Determine whether the following are polynomials in x:

a $\frac{x^2 - x}{x}$

S $\frac{x^2 - x}{x} = \frac{{}^1\cancel{x}(x-1)}{\cancel{x}_1}$
$= x - 1,$

which is a polynomial in x,

$\therefore \frac{x^2 - x}{x}$ is a polynomial.

b 3^{-1}

S $3^{-1} = \frac{1}{3} = \frac{1}{3}x^0$, which is a polynomial in x.

c $\frac{x-1}{x}$

S $\frac{x-1}{x} = 1 - \frac{1}{x}$
$= 1 - x^{-1}$, which is not a polynomial.

Operations with polynomials

Addition of polynomials

Remember only like terms can be added.

1 If $P(x) = 3x^2 - 4x + 2$ and $Q(x) = 5x^2 - 3x - 6$, find $P(x) + Q(x)$.

S

$$\begin{aligned} P(x) + Q(x) &= 3x^2 - 4x + 2 \\ &\quad + (5x^2 - 3x - 6) \\ &= 3x^2 - 4x + 2 \\ &\quad + 5x^2 - 3x - 6 \\ &= 8x^2 - 7x - 4 \end{aligned}$$

Adding like terms

Or

$$\begin{array}{r} 3x^2 - 4x + 2 \\ +\ \ 5x^2 - 3x - 6 \\ \hline 8x^2 - 7x - 4 \\ \hline \end{array}$$

We can work left to right across this addition.

$\therefore P(x) + Q(x) = 8x^2 - 7x - 4.$

2 If $P(x) = x^4 - 5x^2 + 3x + 1$, and $Q(x) = 4x^3 + 2x^2 - 5x - 3$, find $P(x) + Q(x)$.

S

$$\begin{array}{r} x^4 + 0x^3 - 5x^2 + 3x + 1 \\ +\quad\quad 4x^3 + 2x^2 - 5x - 3 \\ \hline x^4 + 4x^3 - 3x^2 - 2x - 2 \\ \hline \end{array}$$

We could have left a gap instead of $0x^3$, but less mistakes will be made this way.

$\therefore P(x) + Q(x) = x^4 + 4x^3 - 3x^2 - 2x - 2.$

Subtraction of polynomials

As with addition, only like terms can be subtracted.

For Example

1 Find the difference between $P(x)$ and $Q(x)$ if $P(x) = 5x^3 - 3x + 2$ and $Q(x) = 2x^2 - 5x + 1$.

S

$$\begin{aligned} P(x) - Q(x) &= 5x^3 - 3x + 2 - (2x^2 - 5x + 1) \\ &= 5x^3 - 3x + 2 - 2x^2 + 5x - 1 \\ &= 5x^3 - 2x^2 + 2x + 1 \end{aligned}$$

Remember, a negative in front of the bracket means −1 multiplies contents, that is, it changes signs.

or

$$\begin{array}{r} 5x^3 + 0x^2 - 3x + 2 \\ -\quad\quad 2x^2 - 5x + 1 \\ \hline 5x^3 - 2x^2 + 2x + 1 \\ \hline \end{array}$$

$\therefore P(x) - Q(x) = 5x^3 - 2x^2 + 2x + 1.$

2 If $P(x) = 5x^4 - 3x^2 - 5x$ and $Q(x) = 3x^3 - x^2 + 2$, find $P(x) - Q(x)$.

S

$$\begin{array}{r} 5x^4 + 0x^3 - 3x^2 - 5x + 0 \\ - \quad 3x^3 - x^2 + 0x + 2 \\ \hline 5x^4 - 3x^3 - 2x^2 - 5x - 2 \\ \hline \end{array}$$

$\therefore P(x) - Q(x) = 5x^4 - 3x^3 - 2x^2 - 5x - 2.$

Multiplication of polynomials

We have done this many times using quadratics (degree 2), but we now extend the procedure to involve cubics (degree 3), quartics (degree 4), and so on.

1 Find $P(x) \cdot Q(x)$ and determine its degree if $P(x) = x^4 - 3x$ and $Q(x) = x - 7$.

S

Remember $P(x) \cdot Q(x)$ means $P(x) \times Q(x)$.

$$\begin{aligned} P(x) \cdot Q(x) &= (x^4 - 3x)(x - 7) \\ &= x^4(x - 7) - 3x(x - 7) \\ &= x^5 - 7x^4 - 3x^2 + 21x \end{aligned}$$

Therefore, the degree = 5.

2 Find the constant term of $P(x) \cdot Q(x)$ if $P(x) = 4x^2 - 3 - x$ and $Q(x) = 5x - 7$.

S

$$\begin{aligned} P(x) \cdot Q(x) &= (4x^2 - 3 - x)(5x - 7) \\ &= 4x^2(5x - 7) - 3(5x - 7) - x(5x - 7) \\ &= 20x^3 - 28x^2 - 15x + 21 - 5x^2 + 7x \\ &= 20x^3 - 33x^2 - 8x + 21 \end{aligned}$$

Therefore, the constant term is 21.

We could have realised that the constant term was to be the product of −3 and −7.

Division of polynomials

When we divide polynomials, the procedure is similar to the long division method using rational numbers, for example: $5764 \div 43$ or $43\overline{)5764}$

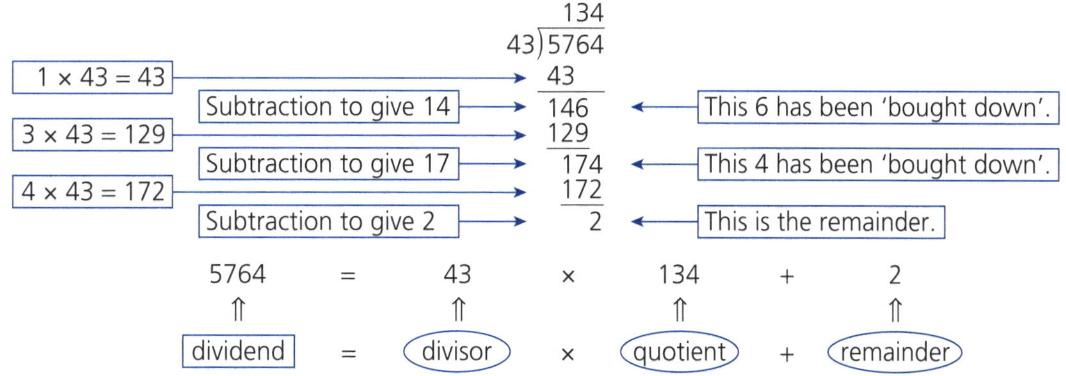

Let's now try with some polynomials in x.

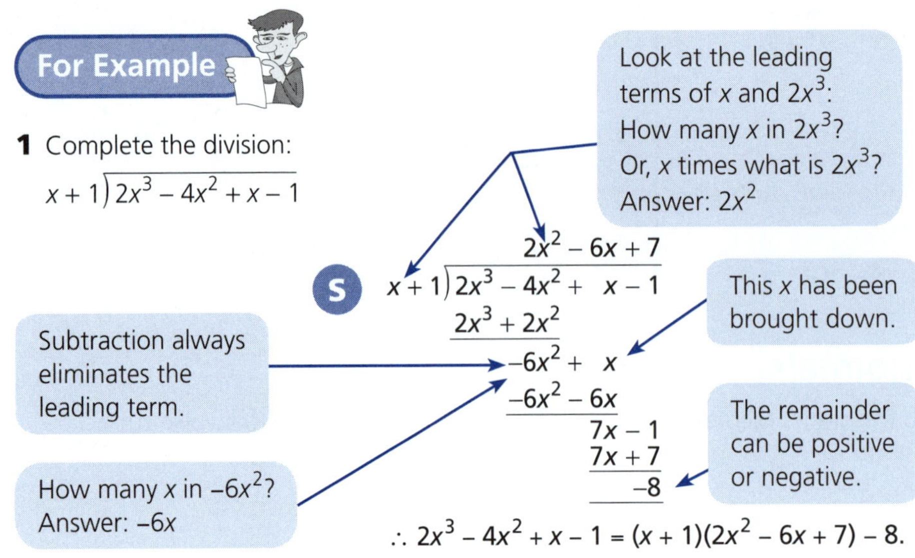

2 $(x^3+6x^2-2)\div(x+3)$

S

$$\begin{array}{r} x^2+3x-9 \\ x+3\overline{)\,x^3+6x^2+0x-2} \\ \underline{x^3+3x^2} \\ 3x^2+0x \\ \underline{3x^2+9x} \\ -9x-2 \\ \underline{-9x-27} \\ 25 \end{array}$$

In the absence of a term we can put in a 0 or we can leave a gap.

$\therefore x^3+6x^2-2=(x+3)(x^2+3x-9)+25.$

3 $(-x^3+2x^2+x-1)\div(x+2)$

S

$$\begin{array}{r} -x^2+4x-7 \\ x+2\overline{)\,-x^3+2x^2+x-1} \\ \underline{-x^3-2x^2} \\ 4x^2+x \\ \underline{4x^2+8x} \\ -7x-1 \\ \underline{-7x-14} \\ 13 \end{array}$$

$\therefore -x^3+2x^2+x-1=(x+2)(-x^2+4x-7)+13.$

4 $(2x^3-x^2-x+1)\div(x^2-1)$

S

$$\begin{array}{r} 2x-1 \\ x^2-1\overline{)\,2x^3-x^2-x+1} \\ \underline{2x^3-2x} \\ -x^2+x+1 \\ \underline{-x^2+1} \\ x \end{array}$$

Here we can leave gaps.

Note: the remainder is not always a constant.

$\therefore 2x^3-x^2-x+1=(x^2-1)(2x-1)+x$

5 $(x^3+x^2-x-1)\div(x^2+2x+1)$

S

$$\begin{array}{r} x-1 \\ x^2+2x+1\overline{)\,x^3+x^2-x-1} \\ \underline{x^3+2x^2+x} \\ -x^2-2x-1 \\ \underline{x^2-2x-1} \\ 0 \end{array}$$

$\therefore x^3+x^2-x-1=(x-1)(x^2+2x+1)$
$\qquad\qquad\qquad\quad=(x-1)(x+1)^2$

The remainder theorem

The remainder theorem can be used when the question is solely about remainders. It states:

If a given polynomial $P(x)$ is divided by $(x-a)$, then the remainder is $P(a)$.

1 Find the remainder when $2x^3 - 4x^2 + x + 7$ is divided by $(x - 2)$.

Let $P(x) = 2x^3 - 4x^2 + x + 7$

Now, as the divisor is $(x - 2)$, the remainder $= P(2)$

$$\begin{aligned} \therefore P(2) &= 2(2)^3 - 4(2)^2 + (2) + 7 \\ &= 16 - 16 + 2 + 7 \\ &= 9 \end{aligned}$$

This is a lot shorter than the division method used previously.

$\therefore$ The remainder is 9.

2 If $x^3 - x^2 + mx + 4$ has a remainder of 4 when divided by $(x + 2)$, find the value of m.

Let $P(x) = x^3 - x^2 + mx + 4$

Now, as the divisor is $(x + 2)$, that is, $[x - (-2)]$, the remainder $= P(-2) = 4$

$\therefore P(-2) = (-2)^3 - (-2)^2 + m(-2) + 4 = 4$

$$\begin{aligned} -8 - 4 - 2m + 4 &= 4 \\ -2m - 8 &= 4 \\ -2m &= 4 + 8 \\ -2m &= 12 \\ m &= -6 \end{aligned}$$

The factor theorem

$(x - a)$ divides $P(x)$, if and only if, $P(a) = 0$.

Thus $(x - a)$ is a factor of $P(x)$, if and only if, $P(a) = 0$ (as there is no remainder following the division).

1 Show that $(x + 2)$ is a factor of $P(x) = x^3 + x^2 - 4x - 4$. Express $P(x)$ as a product of its factors and hence solve $x^3 + x^2 - 4x - 4 = 0$.

For $(x + 2)$ to be a factor, $P(-2)$ must equal zero:

$$\begin{aligned} \therefore P(-2) &= (-2)^3 + (-2)^2 - 4(-2) - 4 \\ &= -8 + 4 + 8 - 4 \\ &= 0 \end{aligned}$$

$\therefore (x + 2)$ is a factor.

Now, to find other factors, we need to divide!

$$\begin{array}{r|l} & x^2 - x - 2 \\ \hline x + 2 & x^3 + x^2 - 4x - 4 \\ & \underline{x^3 + 2x^2} \\ & -x^2 - 4x \\ & \underline{-x^2 - 2x} \\ & -2x - 4 \\ & \underline{-2x - 4} \\ & 0 \end{array}$$

We knew that the remainder was zero.

$$\begin{aligned} \therefore x^3 + x^2 - 4x - 4 &= (x + 2)(x^2 - x - 2) \\ &= (x + 2)(x - 2)(x + 1) \end{aligned}$$

Factorising $(x^2 - x - 2)$ gives $(x - 2)(x + 1)$.

Now, as $x^3 + x^2 - 4x - 4 = 0$

$\therefore (x + 2)(x - 2)(x + 1) = 0$

$\therefore x = -2, 2, -1$

Remember that in quadratics, if $(x - a)(x - b) = 0$, then either $x - a = 0$ or $x - b = 0$, $\therefore x = a$ or b.

2 If $x^3 + ax + b$ is divisible by both $x + 2$ and $x - 3$, find the values of a and b.

Let $P(x) = x^3 + ax + b$.

- Divisible by $x + 2$:

$$\begin{aligned} \therefore P(-2) &= 0 \\ \therefore (-2)^3 + a(-2) + b &= 0 \\ -8 - 2a + b &= 0 \\ -2a + b &= 8 \\ \therefore 2a - b &= -8 \qquad (1) \end{aligned}$$

- Divisible by $x - 3$:

$$\therefore P(3) = 0$$
$$\therefore 3^3 + a(3) + b = 0$$
$$27 + 3a + b = 0$$
$$\therefore 3a + b = -27 \quad (2)$$

- Solving (1) and (2) simultaneously:

$$2a - b = -8$$
$$3a + b = -27$$
(1) + (2): $$5a = -35$$
$$a = -7$$

Substitute $a = -7$ in (1):

$$\therefore 2(-7) - b = -8$$
$$-14 - b = -8$$
$$-b = -8 + 14$$
$$-b = 6$$
$$b = -6$$
$$\therefore a = -7, b = -6$$

3 Solve $x^3 + 2x^2 - 5x - 6 = 0$.

S We look for a factor, and the clue is the constant term of –6. Therefore try the factors of –6, that is, ±1, ±2, ±3, ±6.

Let $P(x) = x^3 + 2x^2 - 5x - 6$.

Start with the simplest, i.e. $P(1)$, then the next simplest, etc.

$$\therefore P(1) = 1^3 + 2(1)^2 - 5(1) - 6$$
$$= 1 + 2 - 5 - 6$$
$$= -8 \neq 0$$

Try, $P(-1) = (-1)^3 + 2(-1)^2 - 5(-1) - 6$
$$= -1 + 2 + 5 - 6$$
$$= 0$$

$\therefore P(-1) = 0$

$\therefore x + 1$ is a factor.

Now we divide $P(x)$ by $(x + 1)$.

$$\begin{array}{r|l} & x^2 + x - 6 \\ x+1 & x^3 + 2x^2 - 5x - 6 \\ & x^3 + x^2 \\ \hline & \quad x^2 - 5x \\ & \quad x^2 + x \\ \hline & \qquad -6x - 6 \\ & \qquad -6x - 6 \\ \hline & \qquad\qquad 0 \end{array}$$

$$\therefore x^3 + 2x^2 - 5x - 6 = (x + 1)(x^2 + x - 6)$$
$$= (x + 1)(x + 3)(x - 2)$$

Now, as $x^3 + 2x^2 - 5x - 6 = 0$,
then $(x + 1)(x + 3)(x - 2) = 0$,
$\therefore x = -1, -3, 2$

Roots and zeros

In Example **3** above, we solved the equation and the solution was $x = -1, -3, 2$. We say that the **roots** of $x^3 + 2x^2 - 5x - 6 = 0$ are $x = -1, -3, 2$.

Also, these values –1, –3 and 2 are **zeros** of the polynomial $x^3 + 2x^2 - 5x - 6 = 0$. **That is, if a is a zero of $P(x) = 0$, then $(x - a)$ is a factor of $P(x)$ and $P(a) = 0$.**

Graphs of polynomial functions

When we have to sketch a polynomial function, the zeros tell us where the function cuts the x-axis, so we factorise to find the zeros.

1 Sketch $y = (x + 2)(x - 3)(4 - x)$.

Put $y = 0$.

We put $y = 0$ because this is the x-axis.

- By solving $y = 0$,
  $(x + 2)(x - 3)(4 - x) = 0$
  $\therefore x = -2, 3, 4$

  Therefore there are zeros at $-2, 3, 4$. The curve cuts the x-axis at these points.

- Now choose a point other than these zeros and determine the sketch.

Note: $y(5)$ means substitute $x = 5$ in y.

For example: $y(5) = (5 + 2)(5 - 3)(4 - 5)$ is negative.

This means that at $x = 5$, the curve is below the x-axis and so for all $x > 5$ it must be below the axis—it cuts the x-axis only at $x = 4, 3, -2$.

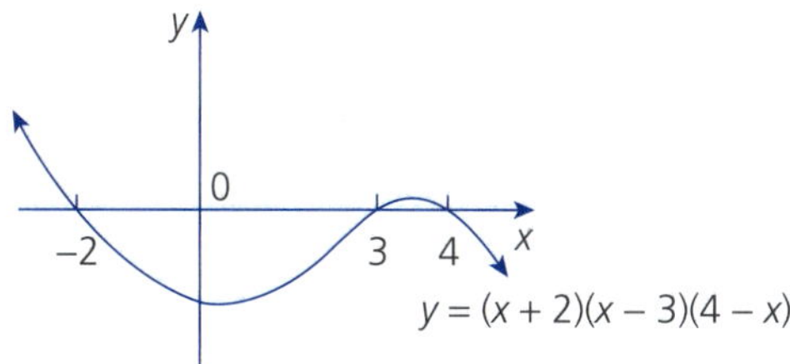

2 Sketch $y = (x + 2)^2(3 - x)$.

By solving $(x + 2)^2(3 - x) = 0$ we get a double root at $x = -2$ and another root at $x = 3$.

- At the double root the curve just touches the x-axis.
- Substitute point $x = 4$.

Then $y(4) = (4 + 2)^2(3 - 4)$, which is negative.

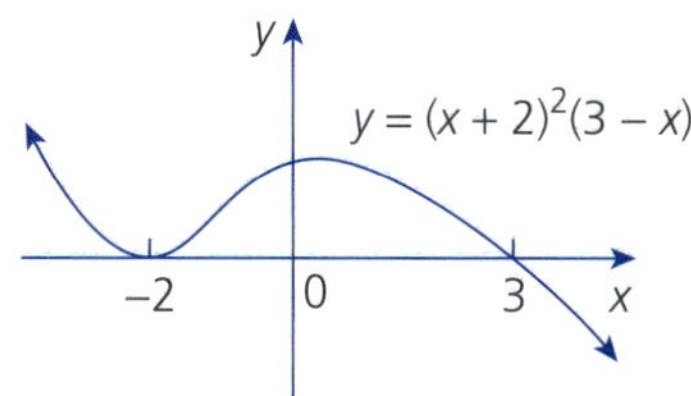

3 Sketch $y = (x - 1)(x + 2)^3$.

Solving $(x - 1)(x + 2)^3 = 0$, $x = 1$, and $x = -2$ is a triple root.

- At a triple root the curve has a kink.
- We substitute, say, $x = 2$,
  $\therefore y(2) = (2 - 1)(2 + 2)^3 > 0$.

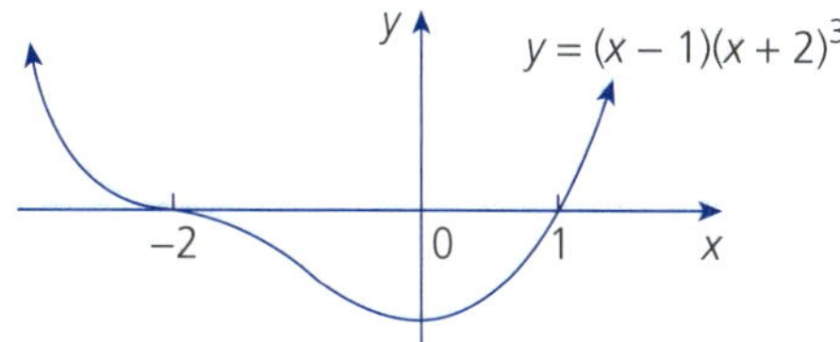

4 Solve the equation $x^3 + 8x^2 + 17x + 10 = 0$ and sketch the graph of $y = x^3 + 8x^2 + 17x + 10$.

Let $P(x) = x^3 + 8x^2 + 17x + 10$.

Try a few values of x:

$$\begin{aligned} P(-1) &= (-1)^3 + 8(-1)^2 + 17(-1) + 10 \\ &= -1 + 8 - 17 + 10 \\ &= 0 \end{aligned}$$

The equation had all positive terms so the roots must be negative.

$\therefore x = -1$ is a root, and $(x + 1)$ is a factor.

Found it on our first try!

Now divide $P(x)$ by $(x + 1)$.

$$\begin{array}{r} x^2 + 7x + 10 \\ x + 1\overline{)\,x^3 + 8x^2 + 17x + 10} \\ \underline{x^3 + x^2 \qquad\qquad} \\ 7x^2 + 17x \qquad \\ \underline{7x^2 + 7x \qquad} \\ 10x + 10 \\ \underline{10x + 10} \\ 0 \end{array}$$

$$\begin{aligned} \therefore P(x) &= x^3 + 8x^2 + 17x + 10 \\ &= (x + 1)(x^2 + 7x + 10) \\ &= (x + 1)(x + 5)(x + 2) \\ &= 0 \end{aligned}$$

$\therefore x = -1, -5, -2$

Therefore the zeros are $-1, -5, -2$.

Now, for $y = x^3 + 8x^2 + 17x + 10$
choose $x = 0$,
$\therefore y(0) = (0 + 1)(0 + 5)(0 + 2)$
$= 10 > 0$

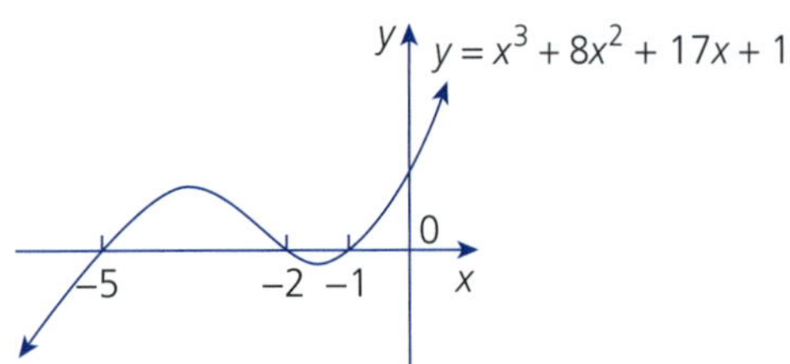

5 Give values of a and b such that the graph of $y = (ax - 5)(x - b)^2$ cuts the x-axis at $x = 2.5$ and touches the x-axis at $x = 3$.

S

- Cuts at $x = 2.5$

$$x = \frac{5}{2}$$
$$2x = 5$$
$$2x - 5 = 0$$

Refer back to $(ax - 5)(x - b)^2$

$\therefore (2x - 5)$ is a factor,
$\therefore 2x - 5 = ax - 5$, therefore $a = 2$.

- Touches at $x = 3$

$$x - 3 = 0$$

Therefore a double root exists,
$\therefore (x - 3)^2$ is a factor, that is, $x - 3 = x - b$, therefore $b = 3$.

Thus $a = 2$, $b = 3$.

6 Write down the equations of the polynomials with the following graphs:

a

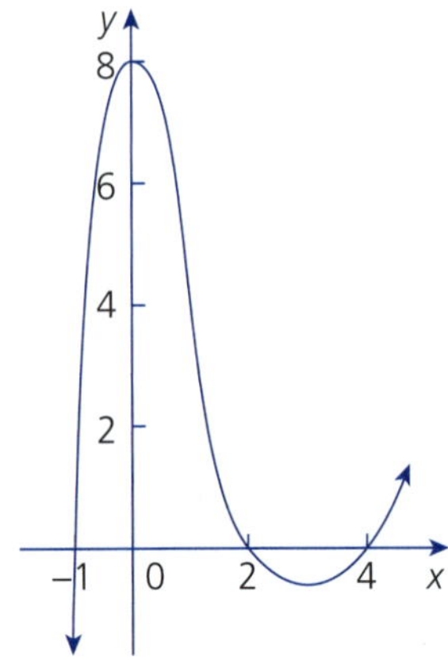

S $y = a(x + 1)(x - 2)(x - 4)$

Substitute in (0, 8)

$\therefore 8 = a(1)(-2)(-4)$
$\therefore 8 = 8a$
$a = 1$
$\therefore y = (x + 1)(x - 2)(x - 4)$

b

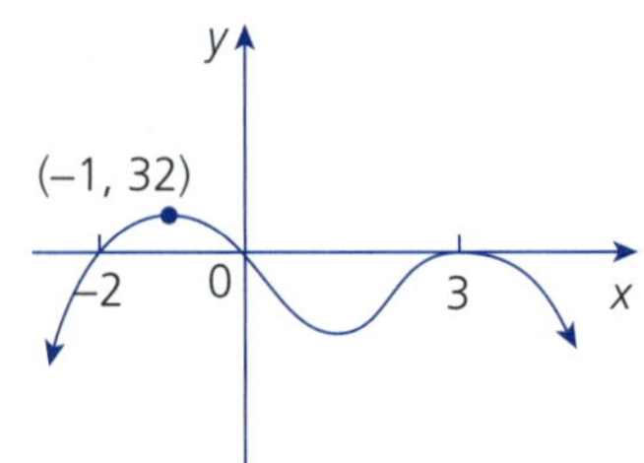

S $y = a(x + 2) \cdot x \cdot (x - 3)^2$
$= ax(x + 2)(x - 3)^2$

Substitute in (−1, 32)

$\therefore 32 = -a(1)(16)$
$-16a = 32$
$a = -2$
$\therefore y = -2x(x + 2)(x - 3)^2$

c

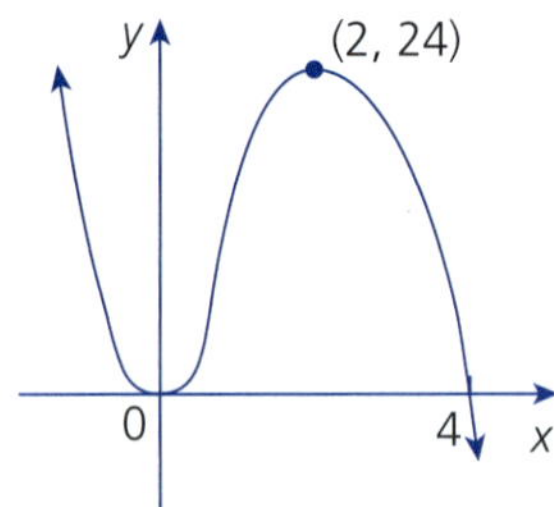

S $y = ax^2(x - 4)$

Substitute in (2, 24)

$\therefore 24 = a(2)^2(2 - 4)$
$24 = -8a$
$a = -3$
$\therefore y = -3x^2(x - 4)$

Practise Practise

1 For the following polynomials, find the degree, the leading term, leading coefficient and constant term:

a $P(x) = 3x^4 - 7x + 2$

b $P(x) = 4 - 2x^5$

c $P(x) = 3x^5 - 2x^3 + 4x$

d $P(x) = 7$ pp. 229–230

2 Determine whether the following are polynomials in x:

a $\frac{4}{x} - 3x + 2$

b $\frac{5x^4 - x}{x}$

c $\frac{3x}{4} + 2$

d $4x^4 + \sqrt{x} - 2$

e $(x + 4)^2 - 2x + 3$ pp. 229–230

3 Complete the following:

a $\begin{array}{l} 4x^4 - 2x^2 + 3 \quad + \\ \underline{3x^4 - 5x^2 - 7} \end{array}$

b $\begin{array}{l} 3x^2 - 4x + 1 \quad + \\ \underline{5x^2 + 2x - 5} \end{array}$

c $\begin{array}{l} 4x^3 + 12x^2 + 3x + 7 \quad + \\ \underline{5x^3 \qquad\qquad + \ x + 3} \end{array}$

d $\begin{array}{l} 4x^5 - 21x + 2 \quad + \\ \underline{3x^5 + \ 2x - 1} \end{array}$

e $\begin{array}{l} 5x^3 + 2x^2 - x - 1 \quad - \\ \underline{3x^3 - \ x^2 - x - 5} \end{array}$

f $\begin{array}{l} 4x^4 - 2x^2 + 1 \quad - \\ \underline{x^4 - 5x^2 - 7} \end{array}$

g $\begin{array}{l} 4x^3 \qquad\quad - x + 2 \quad - \\ \underline{3x^3 + 2x^2 \qquad - 1} \end{array}$

h $\begin{array}{l} 5x^4 \qquad - 2x^2 \qquad + 1 \quad - \\ \underline{\qquad 3x^3 + \ x^2 - 5x + 2} \end{array}$ pp. 230–231

4 Find the degree of $P(x) \cdot Q(x)$ if:

a $P(x) = 3x^2 - 4x + 1$ and $Q(x) = 4 - x + x^2$

b $P(x) = 5x^4 - 2x - 1$ and $Q(x) = 3x - 5x^2 + x^3$ p. 231

5 Find the product of the following polynomials:

a $2x^3 - 4$ and $5x^2 - x + 2$

b $6x^2 - 4x$ and $x^3 + x - 1$ p. 231

6 Complete the following divisions, leaving your answer in the form:

Dividend = Divisor × Quotient + Remainder

a $(x^3 - 3x^2 + 7x - 5) \div (x - 3)$

b $(-x^3 + 2x^2 - 5x - 1) \div (x + 2)$

c $(x^5 + x^3 - x) \div (x^2 + x)$

d $(x^4 - 1) \div (x^2 - 3)$ pp. 231–232

7 If $P(x) = x^3 + 2x^2 - x + 1$ and $Q(x) = x - 1$, find:

a $P(x) + Q(x)$

b $P(x) - Q(x)$

c $P(x) \cdot Q(x)$

d $\frac{P(x)}{Q(x)}$ pp. 230–232

8 Find the remainder after the following divisions:

a $(x^3 + 4x^2 + x - 1) \div (x + 2)$

b $(x^3 - 4x + 2) \div (x - 1)$ pp. 232–233

9 If $x^3 + 3x^2 - 7x + k$ is divided by $(x + 2)$, the remainder is 5. Find the value of k. pp. 232–233

10 When $3x^4 - 2x^3 + 2k$ is divided by $(x + 1)$, the remainder is 3. Find the value of k. pp. 232–233

11 If $x^2 - 7x + 9$ is divided by $(x - k)$, the remainder is 3. Find the values of k. pp. 232–233

12 If $(x - 5)$ is a factor of $x^3 + kx - 10$, find the value of k. pp. 233–234

13 Show that $(x - 1)$ is a factor of $x^3 - 2x^2 - 5x + 6$, and hence solve the equation $x^3 - 2x^2 - 5x + 6 = 0$. pp. 233–234

14 Factorise $2x^3 - 5x^2 - 4x + 3$. pp. 233–234

15 Solve for x, the equation
$x^3 + 2x^2 - 9x - 18 = 0$. pp. 233–234

16 If $x = -3$ and $x = 1$ are roots of the equation $x^3 - 2x^2 - ax + b$, find the values of a and b, and hence find the third root of the equation. pp. 233–234

17 Sketch the graphs of the following polynomials:

a $y = (x - 4)(x - 1)(x + 2)$

b $y = x(x - 2)^2(4 - x)$

c $y = (3 - x)(x + 1)^3(4 - x)$

d $y = x^2(1 - x)^2$

e $y = (1 + x)^3$ pp.234–236

18 Factorise the following and then sketch their graphs, showing clearly the zeros of each function:

a $y = x^3 - 4x^2 + x + 6$

b $y = x^3 - x^2 - x + 1$

c $y = x^3 - 3x^2 + 3x - 1$

d $y = -x^3 + 12x + 16$ pp. 234–236

19 Match the polynomial function to the correct sketch:

a $y = x(x - 1)^2(4 - x)$

b $y = x^2(x - 1)(x + 4)$

c $y = x(x - 1)^3(4 - x)$

d $y = x(x + 1)(4 + x)$

i

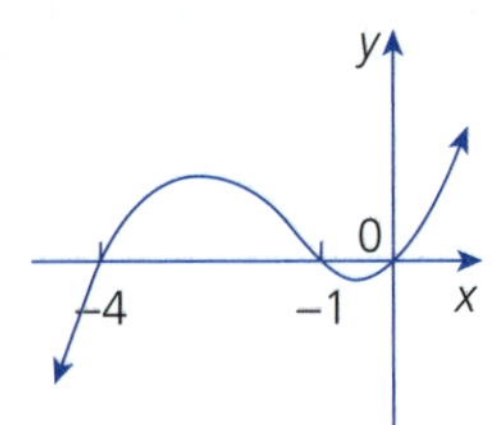

ii

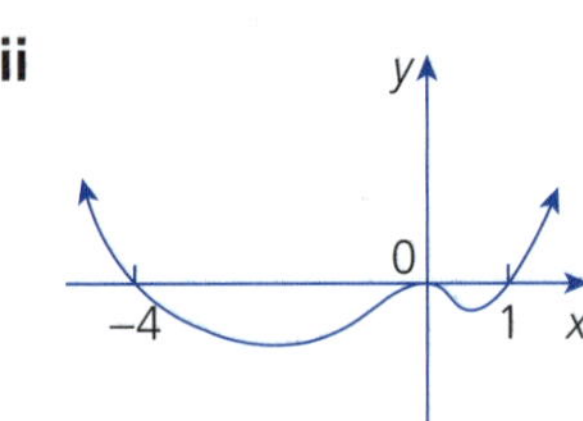

iii

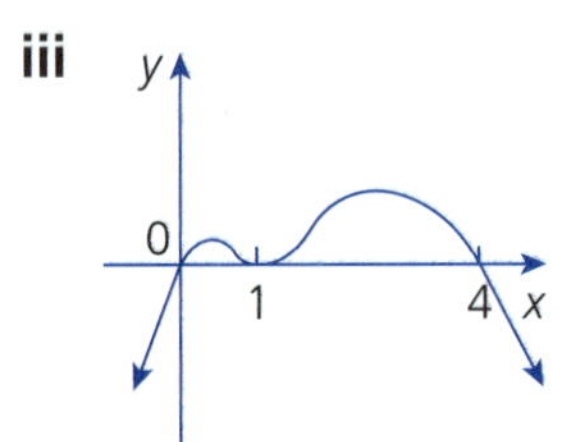

iv

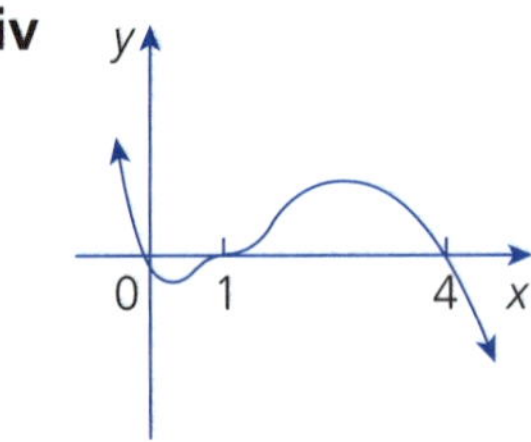

pp. 234–236

Go to p. 302 for **Quick Answers** or to pp. 348–350 for **Worked Solutions**

For a complete understanding of this topic, you must be able to:

✓	Find degree, leading coefficient, and constant term of a polynomial		pp. 229–230
✓	Add, subtract, multiply and divide polynomials		pp. 230–232
✓	Find the remainder using the remainder theorem		pp. 232–233
✓	Use the factor theorem to solve polynomial equations		pp. 233–234
✓	Graph polynomial functions.		pp. 234–236

Now you are ready to do the tests!

Intermediate Test

Polynomials

(20 marks)

1 For the polynomial $P(x) = 3x^4 - 7x^2 + 2$, find the:

a degree of the polynomial (1 mark)

b constant term. (1 mark)

2 If $P(x) = Q(x) \times S(x)$, find the degree and constant term of $P(x)$:

a $Q(x) = 3x - 1$ and $S(x) = 2x^2 + x - 1$ (2 marks)

b $Q(x) = 4x^3 - 1$ and $S(x) = 2 - 7x^2$ (2 marks)

3 Complete the division $(x^3 - 2x^2 + 6x - 1) \div (x - 1)$, leaving your answer in the form:

Dividend = Divisor × Quotient + Remainder

(2 marks)

4 If $P(x) = 2x^3 - 4x^2 + 5x - 1$, find:

a $P(0)$ (1 mark)

b $P(3)$ (1 mark)

c $P(-2)$ (1 mark)

5 Show that $(x + 1)$ is a factor of $x^3 - 4x^2 + x + 6$. (2 marks)

6 Factorise $x^3 + 3x^2 - 6x - 8$. (3 marks)

7 Sketch the following graphs:

a $y = (x - 2)(x + 1)(x - 3)$ (1 mark)

b $y = x^2(2 - x)(x + 1)^3$ (1 mark)

8 Write the equation of the graph:

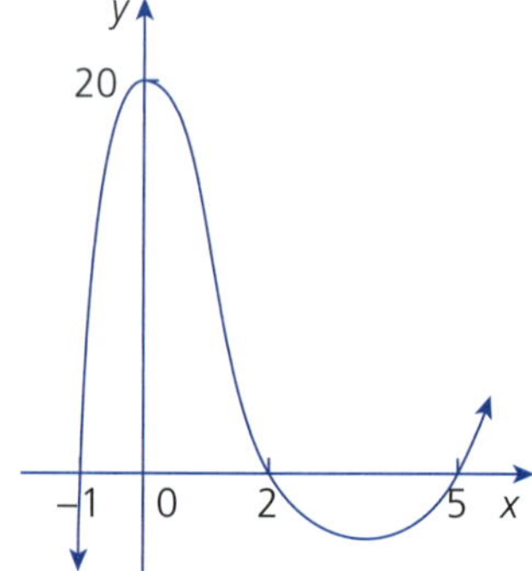

(2 marks)

Your Feedback

☐ × 100% = ☐ %

QA PAGE 307

WS PAGE 378

Advanced Test

Polynomials

(20 marks)

1 Complete the division $(x^3 - 4x^2 + 5x + 2) \div (x - 3)$, leaving your answer in the form:

Dividend = Divisor × Quotient + Remainder

(2 marks)

2 Divide $2x^3 - x - 1$ by $x - 1$. (2 marks)

3 Show that $(x - 5)$ is a factor of $P(x)$ if $P(x) = x^3 - 6x^2 - x + 30$. (2 marks)

4 For what value of k will $(x^3 - 3x^2 + kx - 2) \div (x + 1)$ have a remainder of 5? (2 marks)

5 If $P(x) = x^3 + ax^2 + bx - 3$ and $P(1) = -1$, $P(-1) = -9$, find the values of a and b. (2 marks)

6 If $x^3 - x^2 - 14x + 24 = (x - p)(x - q)(x - r)$, find the values of p, q and r. (3 marks)

7 Solve the equation $x^3 - 7x + 6 = 0$. (3 marks)

8 Sketch the following graphs:

a $y = x^3 + x^2 - 12x$ (2 marks)

b $y = x^3 + x^2 - 16x + 20$ (2 marks)

Your Feedback

☐ × 100% = ☐ %

QA PAGE 307

WS PAGE 379

Chapter 14
Functions and Logarithms

Function definition

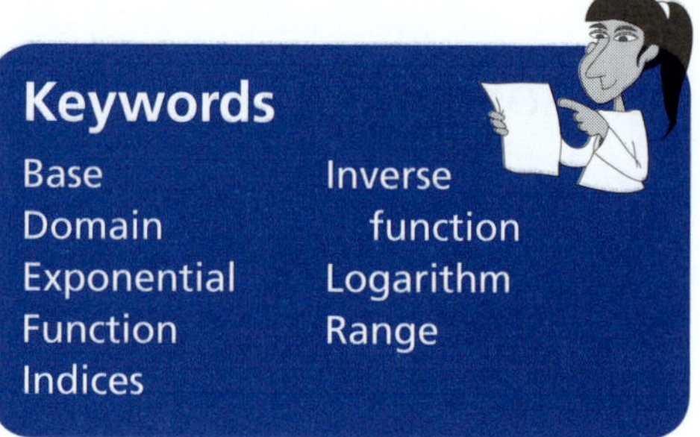

A **function** is a rule or relationship where for each input there is only one output, that is, for each x value there is only one y value.

If a vertical line can cut the graph of a relationship at more than one point then it is **not** a function.

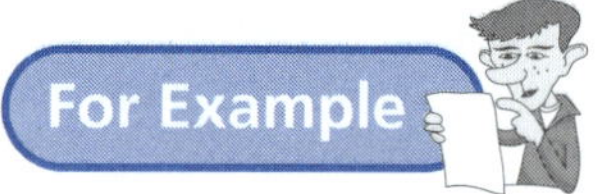

1

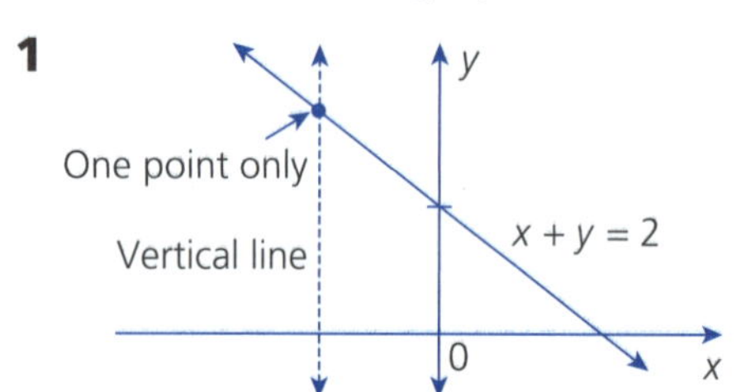

Function

2

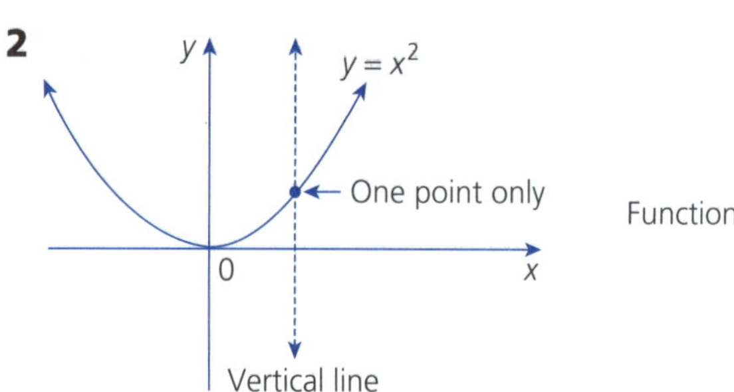

Function

3

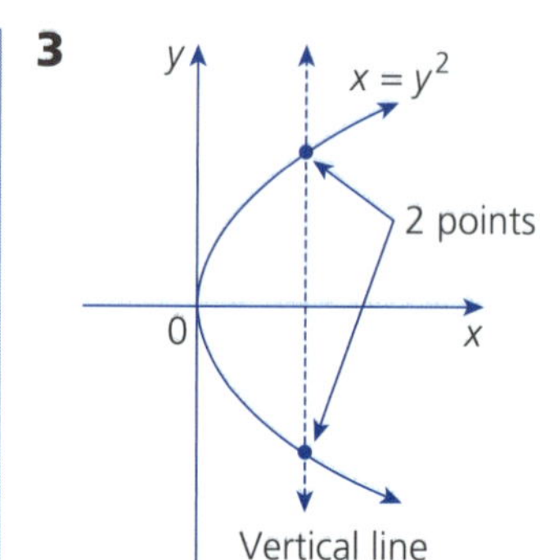

Not a function

4

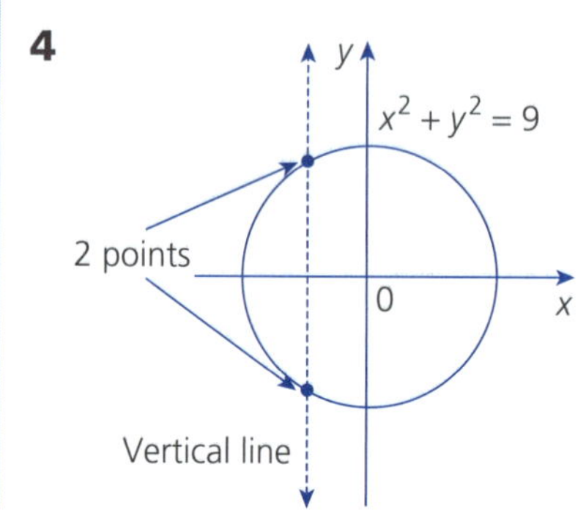

Not a function

Function notation

The notation $f(x)$ is often used to indicate a function, for example:

$f(x) = x^2 + x$

$f(x) = 7 - 2x$

The notation used does not always have to be $f(x)$ (although it usually is)—$F(x)$, $\phi(x)$, $g(x)$, $h(x)$, $C(t)$ and many others can be used. For example:

$C(t) = t^2 - 4$

defines a function of t.

The notation implies substitution into a function. For example, if $f(x) = x^2 - 4$, evaluate $f(2)$. This is the same question as 'given $y = x^2 - 4$, calculate the value of y when $x = 2$'. The notation simplifies the expression of the question:

$f(x) = x^2 - 4$

$f(2) = (2)^2 - 4 = 4 - 4 = 0$

For Example

1 For $f(x) = 3x - 5$, find the following:

a $f(2)$

S $f(2) = 3(2) - 5 = 6 - 5 = 1$

b $f(-2)$

S $f(-2) = 3(-2) - 5 = -6 - 5 = -11$

c $f(a)$

S $f(a) = 3(a) - 5 = 3a - 5$

d $f(a + 2)$

S $f(a + 2) = 3(a + 2) - 5 = 3a + 6 - 5 = 3a + 1$

e $f(4a)$

S $f(4a) = 3(4a) - 5 = 12a - 5$

f $f(2 - a^2)$

S $f(2 - a^2) = 3(2 - a^2) - 5 = 6 - 3a^2 - 5 = 1 - 3a^2$

g $f(x - 1)$

S $f(x - 1) = 3(x - 1) - 5 = 3x - 3 - 5 = 3x - 8$

2 A stone is thrown from the top of a tower such that the distance of the stone above ground level is a function of the elapsed time. The function is given by $f(t) = 60 + 5t - 5t^2$, where t is the time in seconds and $f(t)$ the distance in metres.

a What does the value $f(0)$ represent?

S $f(0) = 60$. This represents the height of the stone above ground level when $t = 0$, that is, the initial height, the starting height, the height of the tower.

b What does $f(0.5)$ represent? What does this value mean?

S $f(0.5) = 60 + 5(0.5) - 5(0.5)^2$
$= 60 + 2.5 - 1.25 = 61.25$

This represents the height of the stone $\frac{1}{2}$ second after the stone was thrown.

The stone has been thrown upwards.

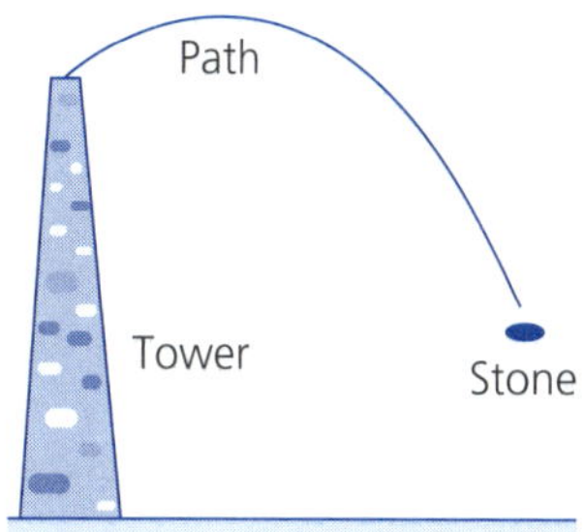

c Find the height of the stone after 1 second, 2 seconds and 3 seconds. What does this mean?

S $f(1) = 60 + 5(1) - 5(1)^2 = 60 + 5 - 5 = 60$
$f(2) = 60 + 5(2) - 5(2)^2 = 60 + 10 - 20 = 50$
$f(3) = 60 + 5(3) - 5(3)^2 = 60 + 15 - 45 = 30$

The stone rose and has now begun to fall. When it has reached the height, $f(t) = 0$, then the stone has reached ground level.

d Calculate the time taken for the stone to reach ground level.

S
$$f(t) = 0$$
$$\therefore\ 60 + 5t - 5t^2 = 0$$
i.e. $t^2 - t - 12 = 0$ (divide by -5)
$$(t - 4)(t + 3) = 0$$
$$\therefore\ t = 4 \quad or \quad -3$$

Negative time has no meaning, thus $t = 4$ seconds.

The particle reaches ground level after 4 seconds.

Domain and range

Let a function be defined by $y = f(x)$. The **domain** of a function is the set of all possible x values (inputs). The **range** of a function is the set of all possible y values (outputs).

1 For the given sketches, state whether or not they are functions and find the domain and range.

a

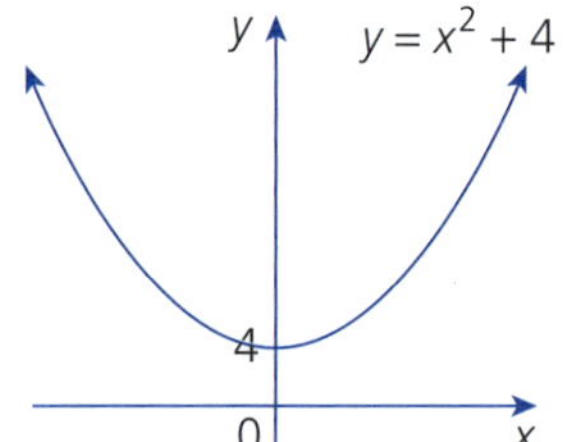

S

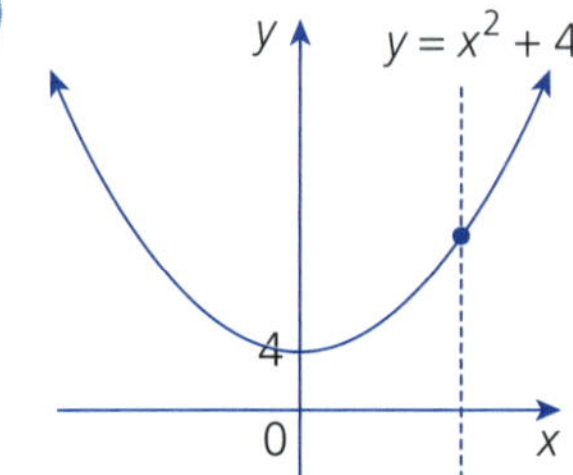

By vertical line test this represents a function.

Domain: All values of x. (No restriction.)

Range: Only y values greater or equal to 4 are outputs, that is, $y \geq 4$.

b

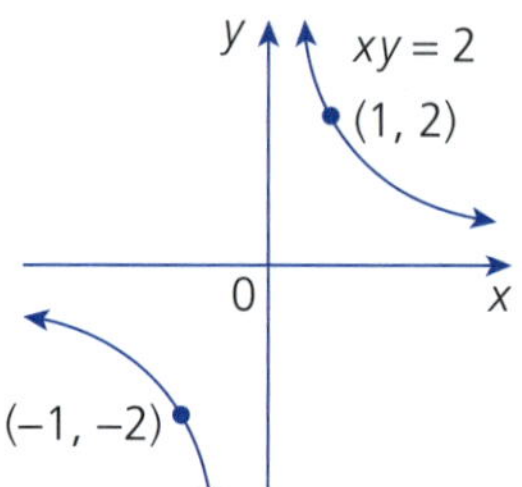

S

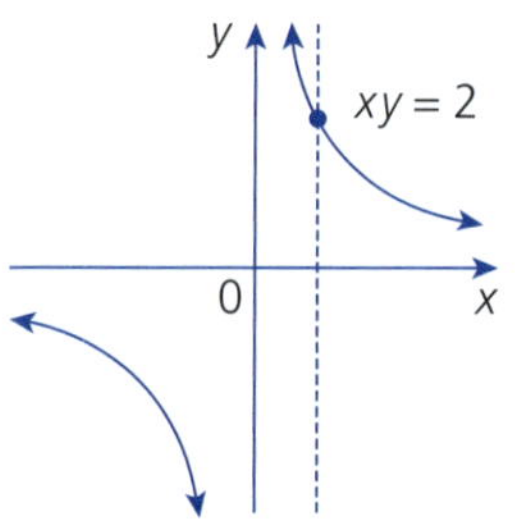

By vertical line test this represents a function.

Domain: All x values except 0 can be used, that is, $x \neq 0$.

Range: All y values except 0 can be outputs, that is, $y \neq 0$.

Inverse functions

If in the equation of a function the x and y are interchanged the inverse function is generated, that is, the x replaces the y and the y replaces the x. For example:

Original function $y = 2x - 3$. Interchange x and y gives $x = 2y - 3$.

Make y the subject $\quad 2y = x + 3$

$$y = \frac{1}{2}x + \frac{3}{2}$$

$\therefore$ Inverse function

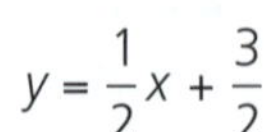

$$y = \frac{1}{2}x + \frac{3}{2}$$

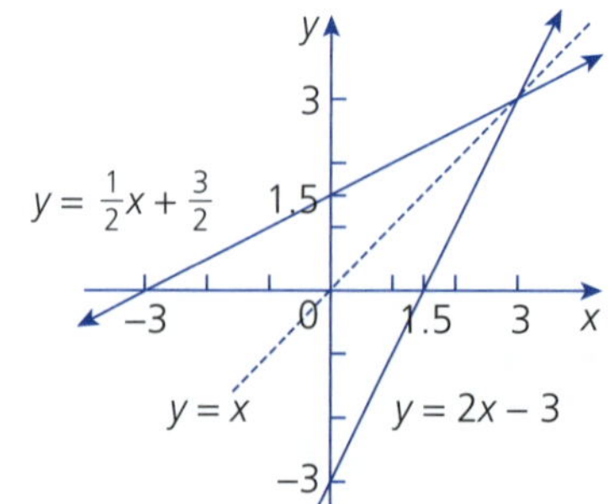

Graphically:

By drawing graphs of $y = 2x - 3$ and $y = \frac{1}{2}x + \frac{3}{2}$ together with $y = x$, it can be seen that the two functions are mirror images through $y = x$.

The graphs of inverse functions are mirror images through $y = x$.

Some functions can only have an inverse function if the range is restricted. For example:

Function: $y = x^2$. Interchange x and y gives $x = y^2$, $\therefore\ y = \pm\sqrt{x}$.

Now when $y = \pm\sqrt{x}$ is graphed:

Clearly $y = \pm\sqrt{x}$ (or $x = y^2$) does not represent a function (vertical line test).

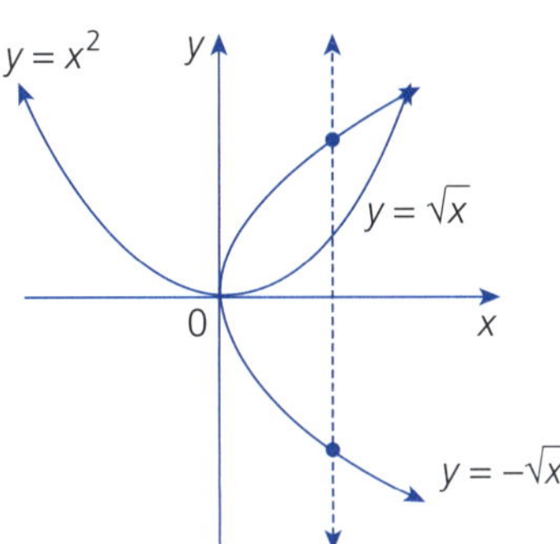

If the domain of $y = x^2$ is restricted such that $x \geq 0$, then the inverse function will have a range such that $y \geq 0$, that is, $y = \sqrt{x}$ and the function has an inverse.

$\therefore$ Function: $y = x^2$ Domain: $x \geq 0$

Inverse: $y = \sqrt{x}$ Range: $y \geq 0$

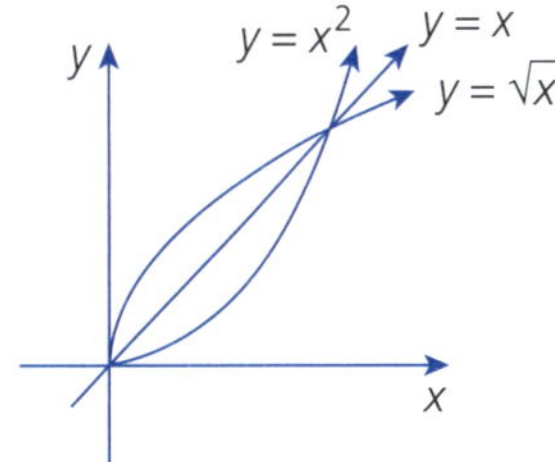

The range of the function becomes the domain of the inverse function and vice versa.

Using graphs to draw other graphs

Given the graph of $y = f(x)$,

then $y = f(x) + a$ represents a shift of the graph up by 'a' units

$y = f(x) - a$ represents a shift of the graph down by 'a' units

$y = f(x + a)$ represents a shift of the graph to the left by 'a' units

$y = f(x - a)$ represents a shift of the graph to the right by 'a' units.

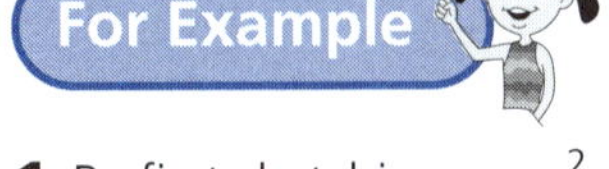

1 By first sketching $y = x^2$, draw sketches of the following:

a $y = x^2 + 2$

S

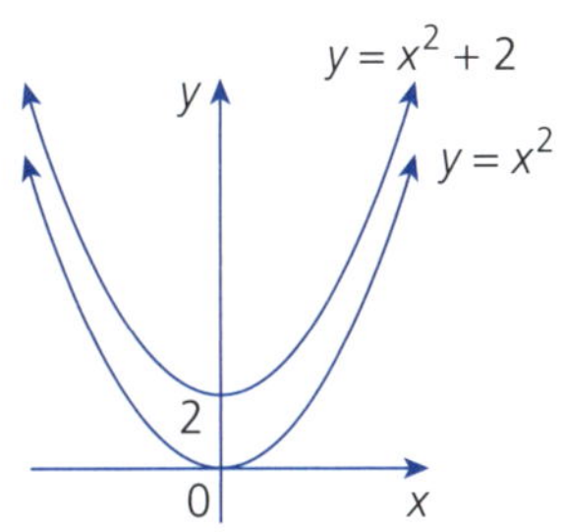

Graph of $y = x^2$ moved up by 2.

b $y = x^2 - 2$

S

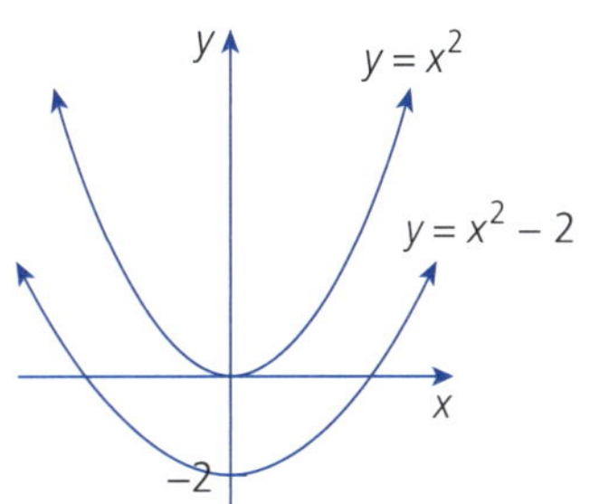

Graph of $y = x^2$ moved down by 2.

c $y = (x + 2)^2$

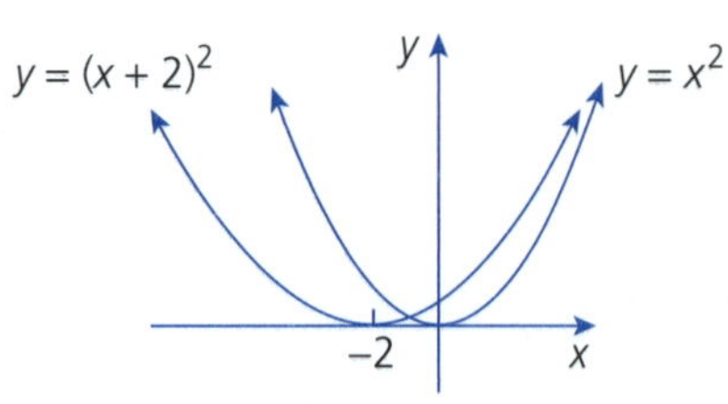

Graph of $y = x^2$ moved left by 2 units.

d $y = (x - 2)^2$

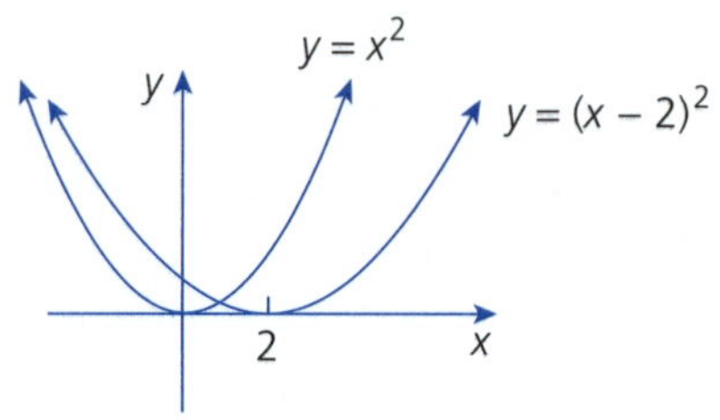

Graph of $y = x^2$ moved right by 2 units.

Review of indices

Index laws

- $x^a \times x^b = x^{a+b}$
- $x^a \div x^b = x^{a-b}$
- $(x^a)^b = x^{ab}$

Consider the following:

- $2^4 \times 2^3 = 2^7$
- $4^7 \div 4^3 = 4^4$
- $(2^4)^3 = 2^{12}$
- $4^2 \times 2^3 = (2^2)^2 \times 2^3 = 2^4 \times 2^3 = 2^7$
- $9^4 \div 3^5 = (3^2)^4 \div 3^5 = 3^8 \div 3^5 = 3^3$
- $(5^2)^3 \div 25^2 = 5^6 \div (5^2)^2 = 5^6 \div 5^4 = 5^2$

Special indices

Zero index

Any non-zero number when raised to zero power will give an answer of 1.

$x^0 = 1, (x \neq 0)$

Consider: $2^0 = 1$, $5^0 = 1$, $\frac{3}{4}^{\,0} = 1$

Fractional index

Index $\frac{1}{n}$ means the nth root. Consider the following:

- $x^{\frac{1}{2}} = \sqrt{x}$, $4^{\frac{1}{2}} = \sqrt{4} = 2$
- $x^{\frac{1}{3}} = \sqrt[3]{x}$, $27^{\frac{1}{3}} = \sqrt[3]{27} = 3$
- $x^{\frac{2}{3}} = \sqrt[3]{x^2}$, or $x^{\frac{2}{3}} = (\sqrt[3]{x})^2$

  $x^{\frac{2}{3}} = (x^2)^{\frac{1}{3}}$, or $= (x^{\frac{1}{3}})^2$
- $27^{\frac{2}{3}} = (\sqrt[3]{27})^2 = (3)^2 = 9$

It is usually easier to take the root, then calculate the power.

Negative index

The minus sign in the index indicates that the reciprocal should be taken. Consider the following:

- $x^{-n} = \frac{1}{x^n}$
- $2^{-3} = \frac{1}{2^3} = \frac{1}{8}$
- $5^{-2} = \frac{1}{5^2} = \frac{1}{25}$
- $a^{-\frac{1}{n}} = \frac{1}{a^{\frac{1}{n}}} = \frac{1}{\sqrt[n]{a}}$
- $25^{-\frac{1}{2}} = \frac{1}{25^{\frac{1}{2}}} = \frac{1}{\sqrt{25}} = \frac{1}{5}$
- $27^{-\frac{2}{3}} = \frac{1}{27^{\frac{2}{3}}} = \frac{1}{(\sqrt[3]{27})^2} = \frac{1}{3^2} = \frac{1}{9}$

Note that the number part of the index is unchanged.

Tables of powers

In this chapter it will be useful to have drawn up the powers of the following numbers. Use your calculator to check these tables.

Powers of 2	
2^0	1
2^1	2
2^2	4
2^3	8
2^4	16
2^5	32
2^6	64
2^7	128
2^8	256
2^9	512
2^{10}	1 024
2^{11}	2 048
2^{12}	4 096

Powers of 3	
3^0	1
3^1	3
3^2	9
3^3	27
3^4	81
3^5	243
3^6	729
3^7	2 187
3^8	6 561

Powers of 5	
5^0	1
5^1	5
5^2	25
5^3	125
5^4	625
5^5	3 125
5^6	15 625

Powers of 10	
10^0	1
10^1	10
10^2	100
10^3	1 000
10^4	10 000
10^5	100 000
10^6	1 000 000

Using the index laws

We can use the index laws to simplify numerical expressions without the use of a calculator.

1 Simplify:

a $\frac{10^{-2} \times 10^5}{10^2}$

S $\frac{10^{-2} \times 10^5}{10^2} = 10^{-2+5-2} = 10^1 = 10$

b $16^{\frac{3}{4}}$

S $16^{\frac{3}{4}} = (16^{\frac{1}{4}})^3 = 2^3 = 8$

c $125^{-\frac{1}{3}}$

S $125^{-\frac{1}{3}} = \frac{1}{125^{\frac{1}{3}}} = \frac{1}{\sqrt[3]{125}} = \frac{1}{5}$

2 Solve for x:

① Express both sides as a power of the same base number.

② Then the indices must be equal.

a $2^x = 64$

S
$$2^x = 64$$
$$2^x = 2^6$$
$$\therefore x = 6$$

Both as powers of 2. Indices are equal.

b $3^{x+2} = 81$

S
$$3^{x+2} = 81$$
$$3^{x+2} = 3^4$$
$$\therefore x + 2 = 4$$
$$x = 2$$

Both as powers of 3. Indices are equal.

c $25^{2x-1} = 3125$

S
$$25^{2x-1} = 3125$$
$$(5^2)^{2x-1} = 5^5$$
$$5^{4x-2} = 5^5$$
$$\therefore 4x - 2 = 5$$
$$4x = 7$$
$$x = \frac{7}{4}$$

Powers of 5.

3 By using the tables of powers, simplify the following calculations:

a 27×243

S
$$27 \times 243$$
$$3^3 \times 3^5 = 3^8 = 6561$$

b $\dfrac{125 \times 3125}{625}$

S
$$\frac{125 \times 3125}{625}$$
$$\frac{5^3 \times 5^5}{5^4} = \frac{5^8}{5^4} = 5^4 = 625$$

c $\sqrt{16 \times 32 \times 128}$

S
$$\sqrt{16 \times 32 \times 128}$$
$$\sqrt{2^4 \times 2^5 \times 2^7} = \sqrt{2^{16}}$$
$$= (2^{16})^{\frac{1}{2}}$$
$$= 2^8$$
$$= 256$$

d $(32)^2$

S
$$(32)^2$$
$$(2^5)^2 = 2^{10} = 1024$$

Logarithms

Definition of a logarithm

The logarithm (log) of a number is the index or the power to which the base must be raised to give that number.

If $N = a^x$, ⇐ where a is the base, x is the index

then $\log_a N = x$ ⇐ where a is the base of the logarithm.

Note:
1 Base of index becomes base of log.
2 Index becomes subject of log equation.
3 Number N goes into the logarithm.

or

If $\log_a N = x$
then $N = a^x$

You must be able to change readily from index form to logarithm form and vice versa. The following statements are identical:

- If $2^5 = 32$, then $\log_2 32 = 5$.
- If $5^4 = 625$, then $\log_5 625 = 4$.
- If $10^3 = 1000$, then $\log_{10} 1000 = 3$.

For Example

1 Write in logarithm form:

a $3^6 = 729$

S If $3^6 = 729$, then $\log_3 729 = 6$.

b $2^{-2} = \frac{1}{4}$

S If $2^{-2} = \frac{1}{4}$, then $\log_2 \frac{1}{4} = -2$.

2 Write in index form:

a $\log_{10} 10\,000 = 4$

S If $\log_{10} 10\,000 = 4$, then $10^4 = 10\,000$.

b $\log_3 \frac{1}{27} = -3$

S If $\log_3 \frac{1}{27} = -3$, then $3^{-3} = \frac{1}{27}$.

Logarithmic equations

Logarithmic equations are solved by changing to equivalent index expressions and solving the index equation.

For Example

1 Solve the logarithmic equations:

a $\log_2 256 = x$

S If $\log_2 256 = x$
then $2^x = 256$
that is $2^x = 2^8$
$x = 8$

> **Remember:**
> **1** Express in index form with the same base.
> **2** Equate indices.

b $\log_3 x = 2$

S If $\log_3 x = 2$
then $3^2 = x$
$x = 9$

c $\log_x \frac{1}{8} = -3$

S If $\log_x \frac{1}{8} = -3$
then $x^{-3} = \frac{1}{8}$
$x^{-3} = 2^{-3}$
$x = 2$

Exponential and logarithm graphs

Using your calculator

Your calculator can be used to evaluate powers, both integral and fractional. The buttons to be used are:

[x^y] or $\overset{x^y}{[\ \ ]}$

and [$x^{\frac{1}{y}}$] or $\overset{x^{\frac{1}{y}}}{[\ \ ]}$

Some calculators use [y^x]

To calculate 4^3:

Press [4] [x^y] 3 [=]

or 4 [INV] $\overset{x^y}{[\ \ ]}$ 3 [=]

To calculate $2^{\frac{1}{3}}$:

Press [2] [$x^{\frac{1}{y}}$] 3 [=]

or 2 [INV] $\overset{x^{\frac{1}{y}}}{[\ \ ]}$ [=]

Exponential graphs

All exponential graphs of the form $y = a^x$ pass through (0, 1).

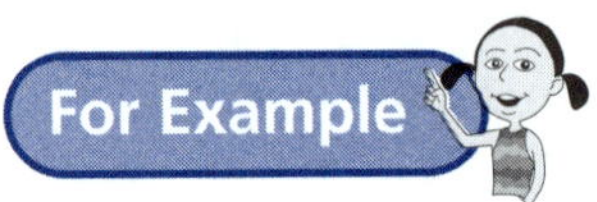

1 Graph $y = 2^x$ by first completing a table of values:

x	−2	−1	0	1	2	3
y						

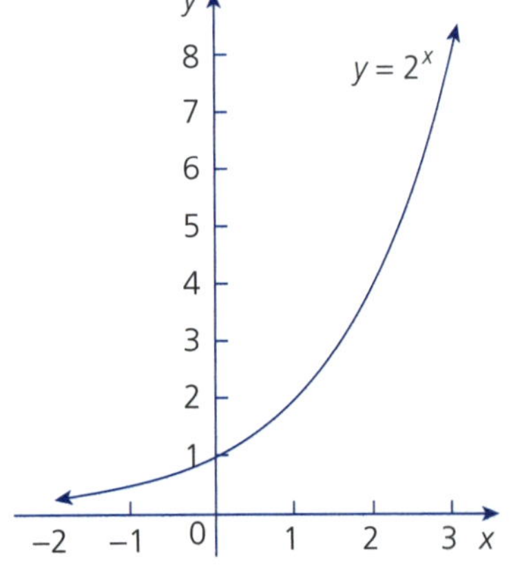

x	−2	−1	0	1	2	3
y	$\frac{1}{4}$	$\frac{1}{2}$	1	2	4	8

Using $y = 2^x$

$x = -2, y = 2^{-2} = \frac{1}{2^2} = \frac{1}{4}$

$x = 0, y = 2^0 = 1$

$2^0 = 1$, but so is 3^0, 4^0, etc.

Note:
1 The graph cuts the y-axis at 1.
2 All y-values are greater than 0.
3 The graph runs along the x-axis but does not cut the x-axis.

These are properties of all exponential graphs of the form $y = a^x$ $(a > 1)$.

Logarithmic graphs

All logarithmic graphs of the form $y = \log_a x$ pass through (1, 0).

Graph $y = \log_2 x$ by first expressing this as an index equation.

If $y = \log_2 x$, then $x = 2^y$.

x	$\frac{1}{4}$	$\frac{1}{2}$	1	2	4	8
y	−2	−1	0	1	2	3

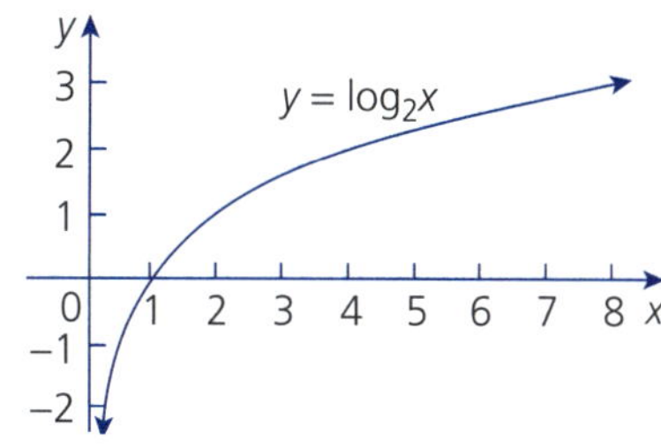

This time we give y-values and calculate the x-values.

Using $x = 2^y$

when $y = -2$, $x = 2^{-2} = \frac{1}{2^2} = \frac{1}{4}$

$y = 0$, $x = 2^0 = 1$

$\log_2 1 = 0$ because $2^0 = 1$.
Similarly: $\log_3 1 = 0$, $\log_5 1 = 0$, etc.

Note:

1 The graph cuts x-axis at 1.
2 All x-values are greater than zero.
3 The graph runs along the y-axis but does not cut the y-axis.

These are properties of all logarithmic graphs of the form: $y = \log_a x$ $(a > 1)$

$y = 2^x$ and $y = \log_2 x$ are **inverse functions**.

The graph of $y = \log_{10} x$ is readily obtained by using the LOG button on your calculator.

Laws for logarithms

The laws for logarithms are as follows:

These relate to the index laws.

1 $\log_a MN = \log_a M + \log_a N$

For example:

$\log_2(3 \times 5) = \log_2 3 + \log_2 5$, or

$\log_{10} 7 + \log_{10} 3 = \log_{10} 21$

2 $\log_a \frac{M}{N} = \log_a M - \log_a N$

For example:

$\log_3 \frac{20}{4} = \log_3 20 - \log_3 4$, or

$\log_{10} 63 - \log_{10} 9 = \log_{10} \frac{63}{9} = \log_{10} 7$

3 $\log_a M^r = r\log_a M$

For example:

$\log_3 4^2 = 2\log_3 4$, or

$\frac{1}{2}\log_5 16 = \log_5 16^{\frac{1}{2}} = \log_5 4$

$16^{\frac{1}{2}} = \sqrt{16}$

There are also the following important cases:

4 $\log_a 1 = 0$

For example:

$\log_5 1 = 0$
$\log_{10} 1 = 0$
$\log_x 1 = 0$

This corresponds to:
$5^0 = 1$
$10^0 = 1$
$x^0 = 1$

5 $\log_a a = 1$

For example:

$\log_2 2 = 1$
$\log_{10} 10 = 1$
$\log_x x = 1$

This corresponds to:
$2^1 = 2$
$10^1 = 10$
$x^1 = x$

For Example

1 Simplify the following:

a $\log_2 x + \log_2 y$

S $\log_2 x + \log_2 y = \log_2 xy$

b $\log_a S - \log_a T$

S $\log_a S - \log_a T = \log_a \dfrac{S}{T}$

c $\dfrac{1}{2}\log_2 x$

S $\dfrac{1}{2}\log_2 x = \log_2 x^{\frac{1}{2}} = \log_2 \sqrt{x}$

d $3\log_{10} x + 4\log_{10} y$

S $3\log_{10} x + 4\log_{10} y = \log_{10} x^3 + \log_{10} y^4$
$= \log_{10}(x^3y^4)$

e $\dfrac{1}{2}\log_a x - 2\log_a y$

S $\dfrac{1}{2}\log_a x - 2\log_a y = \log_a x^{\frac{1}{2}} - \log_a y^2$
$= \log_a \sqrt{x} - \log_a y^2$
$= \log_a \dfrac{\sqrt{x}}{y^2}$

2 Expand the following expressions:

a $\log_2 (ab)$

S $\log_2 ab = \log_2 a + \log_2 b$

b $\log_x \dfrac{p}{q}$

S $\log_x \dfrac{p}{q} = \log_x p - \log_x q$

c $\log_3 a^4$

S $\log_3 a^4 = 4\log_3 a$

d $\log_{10} \sqrt[3]{t}$

S $\log_{10} \sqrt[3]{t} = \log_{10} t^{\frac{1}{3}} = \dfrac{1}{3}\log_{10} t$

e $\log_n \dfrac{a^4}{b^2}$

S $\log_n \dfrac{a^4}{b^2} = \log_n a^4 - \log_n b^2$
$= 4\log_n a - 2\log_n b$

f $\log_e \dfrac{\sqrt{xy^3}}{z^2}$

S $\log_e \dfrac{\sqrt{xy^3}}{z^2}$

$= \log_e \dfrac{(xy^3)^{\frac{1}{2}}}{z^2}$

$= \log_e \dfrac{\left(x^{\frac{1}{2}}y^{\frac{3}{2}}\right)}{z^2}$

$= \log_e x^{\frac{1}{2}} + \log_e y^{\frac{3}{2}} - \log_e z^2$

$= \frac{1}{2}\log_e x + \frac{3}{2}\log_e y - 2\log_e z$

3 Evaluate the following expressions:

a $\log_3 27$

Remember: $\log_a a = 1$, $\log_2 2 = 1$, etc.

S $\log_3 27 = \log_3 3^3 = 3\log_3 3 = 3$

b $\log_a a^2$

S $\log_a a^2 = 2\log_a a = 2$

c $\log_2 \sqrt{2}$

S $\log_2 \sqrt{2} = \log_2 2^{\frac{1}{2}} = \frac{1}{2}\log_2 2 = \frac{1}{2}$

d $\log_n \sqrt[3]{n}$

S $\log_n \sqrt[3]{n} = \log_n n^{\frac{1}{3}} = \frac{1}{3}\log_n n = \frac{1}{3}$

e $\log_3 \frac{1}{3}$

S $\log_3 \frac{1}{3} = \log_3 3^{-1} = -1\log_3 3 = -1$

f $\log_3 \frac{1}{81}$

S $\log_3 \frac{1}{81} = \log_3 \frac{1}{3^4} = \log_3 3^{-4}$
$= -4\log_3 3 = -4$

g $\log_2 4\sqrt{2}$

S $\log_2 4\sqrt{2} = \log_2 (2^2 \cdot 2^{\frac{1}{2}}) = \log_2 2^{2.5}$
$= 2.5\log_2 2 = 2.5$

h $\log_a (a^2 + a) - \log_a (a + 1)$

S $\log_a (a^2 + a) - \log_a (a + 1)$

$= \log_a \dfrac{a^2 + a}{a + 1} = \log_a \dfrac{a\cancel{(a+1)}}{\cancel{(a+1)}}$

$= \log_a a = 1$

i $\log_{10} b + \log_{10} \frac{1}{b}$

S $\log_{10} b + \log_{10} \frac{1}{b} = \log_{10} b + \log_{10} b^{-1}$
$= \log_{10} b - \log_{10} b$
$= 0$

j $\log_a [\log_e e]$

S $\log_a [\log_e e] = \log_a 1 = 0$

k $\log_2 18 - \log_2 9$

S $\log_2 18 - \log_2 9 = \log_2 \frac{18}{9} = \log_2 2 = 1$

l $\log_{10} 250 - \log_{10} 4 + \log_{10} 16$

S $\log_{10} 250 - \log_{10} 4 + \log_{10} 16$

$= \log_{10} \dfrac{250 \times \cancel{16}}{\cancel{4}}$

$= \log_{10} 1000$
$= \log_{10} 10^3$
$= 3\log_{10} 10$
$= 3$

4 If $p = \log_a 2$ and $q = \log_a 5$, then write in terms of p and q:

a $\log_a 8$

S $\log_a 8 = \log_a 2^3 = 3\log_a 2 = 3p$

b $\log_a \sqrt{5}$

S $\log_a \sqrt{5} = \log_a 5^{\frac{1}{2}} = \frac{1}{2}\log_a 5 = \frac{1}{2}q$

c $\log_a 10$

S $\log_a 10 = \log_a (2 \times 5)$
$= \log_a 2 + \log_a 5 = p + q$

d $\log_a 0.4$

S $\log_a 0.4 = \log_a \frac{2}{5}$
$= \log_a 2 - \log_a 5 = p - q$

e $\log_a 0.2$

S $\log_a 0.2 = \log_a \frac{1}{5} = \log_a 5^{-1}$
$= -1 \log_a 5 = -q$

f $\log_a 50$

S $\log_a 50 = \log_a (2 \times 25) = \log_a (2 \times 5^2)$
$= \log_a 2 + \log_a 5^2$
$= \log_a 2 + 2\log_a 5$
$= p + 2q$

g $\log_a \sqrt{50}$

S $\log_a \sqrt{50} = \log_a 50^{\frac{1}{2}} = \frac{1}{2}\log_a 50$
$= \frac{1}{2}(p + 2q) = \frac{1}{2}p + q$

h $\log_a 40$

S $\log_a 40 = \log_a (8 \times 5)$
$= \log_a 8 + \log_a 5$
$= \log_a 2^3 + \log_a 5$
$= 3\log_a 2 + \log_a 5$
$= 3p + q$

i $\log_a \frac{8}{5}$

S $\log_a \frac{8}{5} = \log_a 8 - \log_a 5$
$= \log_a 2^3 - \log_a 5$
$= 3 \log_a 2 - \log_a 5$
$= 3p - q$

j $\log_a 0.625$

S $\log_a 0.625 = \log_a \frac{5}{8} = \log_a 5 - \log_a 8$
$= \log_a 5 - \log_a 2^3$
$= \log_a 5 - 3\log_a 2$
$= q - 3p$

5 If $a = \log_{10} 3$ and $b = \log_{10} 4$, write simple log expressions for:

a $a + b$

S $a + b = \log_{10} 3 + \log_{10} 4 = \log_{10} (3 \times 4)$
$= \log_{10} 12$

b $a - b$

S $a - b = \log_{10} 3 - \log_{10} 4 = \log_{10} \frac{3}{4}$
$= \log_{10} 0.75$

c $\frac{1}{2}b$

S $\frac{1}{2}b = \frac{1}{2}\log_{10} 4 = \log_{10} \sqrt{4} = \log_{10} 2$

d $3a$

S $3a = 3\log_{10} 3 = \log_{10} 3^3 = \log_{10} 27$

e $3a + 2b$

S $3a + 2b = 3\log_{10} 3 + 2\log_{10} 4$
$= \log_{10} 3^3 + \log_{10} 4^2$
$= \log_{10} 27 + \log_{10} 16$
$= \log_{10} (27 \times 16)$
$= \log_{10} 432$

f $2a - b$

S $2a - b = 2\log_{10} 3 - \log_{10} 4$
$= \log_{10} 3^2 - \log_{10} 4$
$= \log_{10} 9 - \log_{10} 4$
$= \log_{10} \frac{9}{4}$
$= \log_{10} 2.25$

g Show that $2a + b - 1 = \log_{10} 3.6$

S $2a + b - 1 = 2\log_{10} 3 + \log_{10} 4 - \log_{10} 10$
$= \log_{10} 3^2 + \log_{10} 4 - \log_{10} 10$
$= \log_{10} 9 + \log_{10} 4 - \log_{10} 10$
$= \log_{10} \frac{9 \times 4}{10} = \log_{10} \frac{36}{10}$
$= \log_{10} 3.6$

6 Write an expression without logarithms for y, given:

a $\log_a y = \frac{1}{2}\log_a x$

S
$$\log_a y = \frac{1}{2}\log_a x$$
$$= \log_a \sqrt{x}$$
$$\therefore y = \sqrt{x}$$

b $\log_a y = \log_a x - \log_a z$

S
$$\log_a y = \log_a x - \log_a z$$
$$= \log_a \frac{x}{z}$$
$$\therefore y = \frac{x}{z}$$

c $\log_a y = \log_a a - 2\log_a x$

S
$$\log_a y = \log_a a - 2\log_a x$$
$$= \log_a a - \log_a x^2$$
$$= \log_a \frac{a}{x^2}$$
$$\therefore y = \frac{a}{x^2}$$

d $\log_a y = \frac{1}{2}\log_a x - 2\log_a z$

S
$$\log_a y = \frac{1}{2}\log_a x - \log_a z$$
$$= \log_a x^{\frac{1}{2}} - \log_a z^2$$
$$= \log_a \sqrt{x} - \log_a z^2$$
$$= \log_a \frac{\sqrt{x}}{z^2}$$
$$\therefore y = \frac{\sqrt{x}}{z^2}$$

e $\log_a y = \log_a x - 1$

S
$$\log_a y = \log_a x - 1$$
$$= \log_a x - \log_a a$$ ← Note again
$$= \log_a \frac{x}{a}$$
$$\therefore y = \frac{x}{a}$$

7 Solve for x:

a $\log_2 x = \log_2 5 + \log_2 3$

S
$$\log_2 x = \log_2 5 + \log_2 3$$
$$= \log_2 (5 \times 3)$$
$$= \log_2 15$$
$$\therefore x = 15$$

b $\log_2 24 - \log_2 8 = \log_2 x$

S
$$\log_2 24 - \log_2 8 = \log_2 x$$
$$\therefore \log_2 \frac{24}{8} = \log_2 x$$
$$\therefore \log_2 3 = \log_2 x$$
$$\therefore x = 3$$

c $\log_2 3x + \log_2 7 = \log_2 42$

S
$$\log_2 3x + \log_2 7 = \log_2 42$$
$$\log_2 (3x \times 7) = \log_2 42$$
$$\log_2 21x = \log_2 42$$
$$\therefore 21x = 42$$
$$\therefore x = 2$$

d $2\log_2 3 + \log_2 x = 3\log_2 3$

S
$$2\log_2 3 + \log_2 x = 3\log_2 3$$
$$\log_2 x = \log_2 3$$
$$\therefore x = 3$$

e $\log_2 (x + 1) + \log_2 (x + 3) = 3$

S
$$\log_2 (x + 1) + \log_2 (x + 3) = 3$$
$$\therefore \log_2 (x + 1)(x + 3) = 3\log_2 2$$
$$\log_2 (x^2 + 4x + 3) = \log_2 2^3$$
$$= \log_2 8$$
$$\therefore x^2 + 4x + 3 = 8$$
$$x^2 + 4x - 5 = 0$$
$$(x + 5)(x - 1) = 0$$
$$\therefore x = 1 \text{ or } -5.$$

Now $x = -5$ is inadmissible as $\log_2 -4$ does not exist, $\therefore x = 1$.

f $\log_2 (x + 4) - \log_2 (x - 2) = 1$

S
$$\log_2 (x + 4) - \log_2 (x - 2) = 1$$
$$\therefore \log_2 \frac{x + 4}{x - 2} = \log_2 2$$
$$\therefore \frac{x + 4}{x - 2} = 2$$
$$x + 4 = 2x - 4$$
$$8 = x$$
$$\therefore x = 8$$

Solving exponential equations

Exponential equations can be solved by taking logarithms of both sides.

1 Solve for x, giving the answer correct to two decimal places:

a $4^x = 12$

S Take logs of both sides—here logs to base 10 are taken:

$$\begin{aligned}\log_{10} 4^x &= \log_{10} 12\\ x\log_{10} 4 &= \log_{10} 12\\ x &= \frac{\log_{10} 12}{\log_{10} 4}\\ &= 1.792\,481\,25\ \ldots\\ &= 1.79 \text{ (2 dec. pl.)}\end{aligned}$$

b $5^x = 3^{2x-1}$

S

$$\begin{aligned}\log_{10} 5^x &= \log_{10} 3^{2x-1}\\ x\log_{10} 5 &= (2x-1)\log_{10} 3\\ x\log_{10} 5 &= 2x\log_{10} 3 - \log_{10} 3\\ 2x\log_{10} 3 - x\log_{10} 5 &= \log_{10} 3\\ x(2\log_{10} 3 - \log_{10} 5) &= \log_{10} 3\\ x &= \frac{\log_{10} 3}{2\log_{10} 3 - \log_{10} 5}\\ &= 1.869\,066\,371\ \ldots\\ &= 1.87 \text{ (2 dec. pl.)}\end{aligned}$$

Practise Practise

1 By using a vertical line test decide whether these graphs represent functions:

a

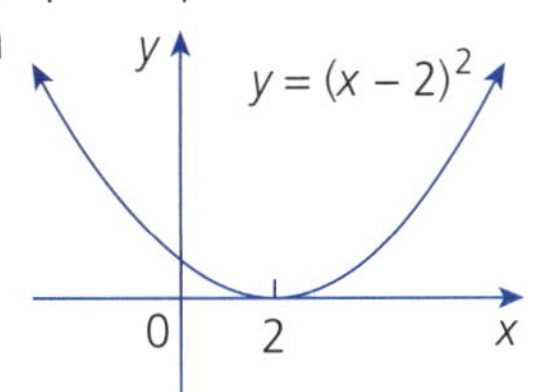

b

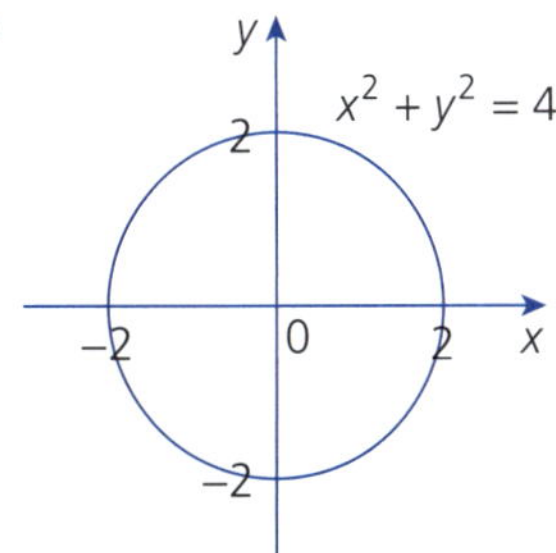

c

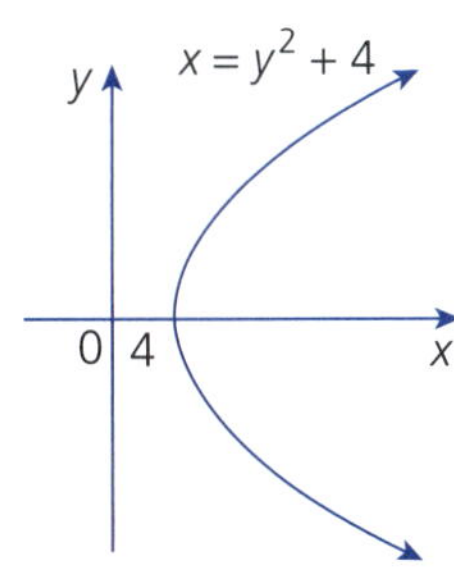

d

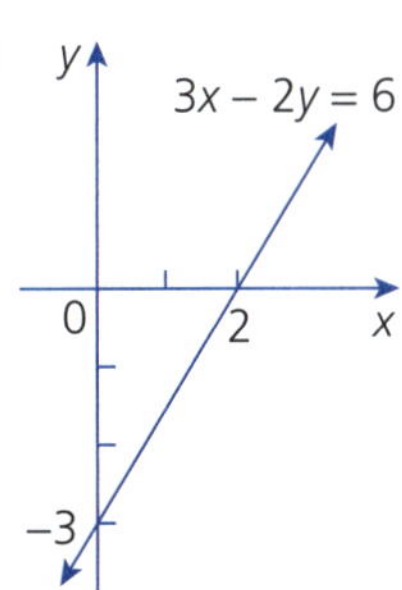

e

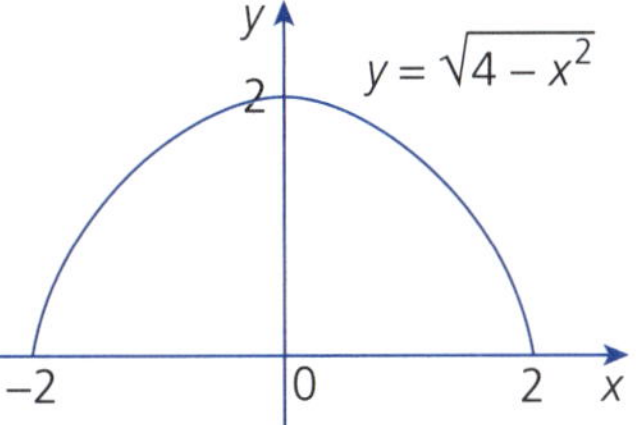

f

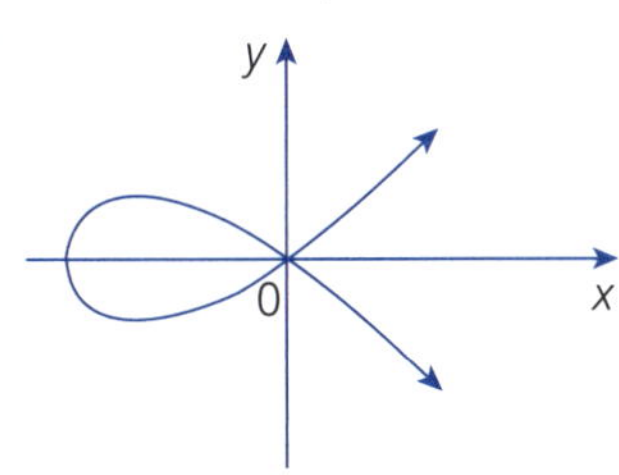

p. 242

2 Given that $f(x) = 6x - 11$, evaluate:

a $f(0)$ b $f(2)$

c $f(-2)$ d $f(5)$

pp. 242–243

3 If $f(x) = x^2 - 4x$, evaluate:

a $f(0)$ b $f(1)$

c $f(-1)$ d $f(3)$

pp. 242–243

4 For the following functions of t, evaluate $f(0)$, $f(2)$ and $f(-2)$:

a $f(t) = 1 - t^2$ b $f(t) = \dfrac{1+t}{1-t}$

c $f(t) = t^2 - t^3$ d $f(t) = (t-3)^2$

e $f(t) = (3-t)^2$ f $f(t) = 3^t$

pp. 242–243

5 When $\phi(x) = x^2 + x$, find expressions (in the simplest form) for:

a $\phi(h)$ b $\phi(x+2)$

c $\phi(x+2) - \phi(2)$ d $\phi(-x)$

pp. 242–243

6 a For $g(x) = x^4$, show that $g(x) = g(-x)$.

b For $H(x) = x^3$, show that $H(-x) = -H(x)$.

pp. 242–243

7 From the given sketches state the domain and range for the functions shown:

a

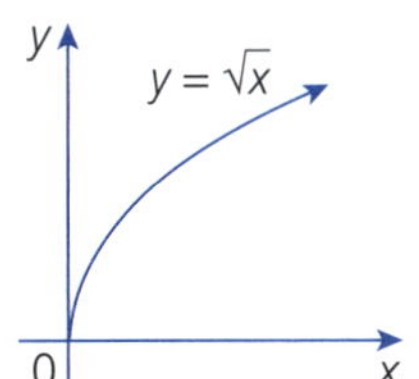

b

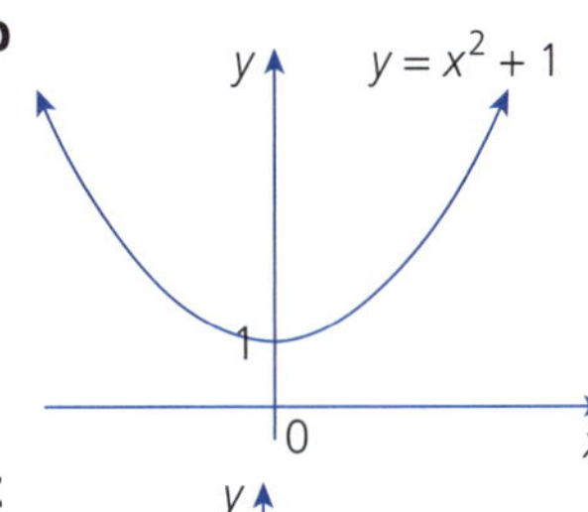

c

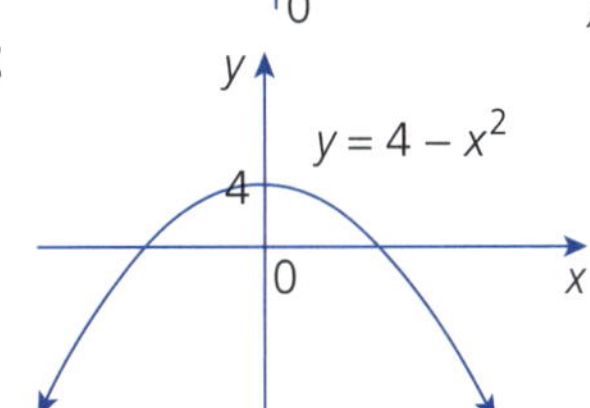

d

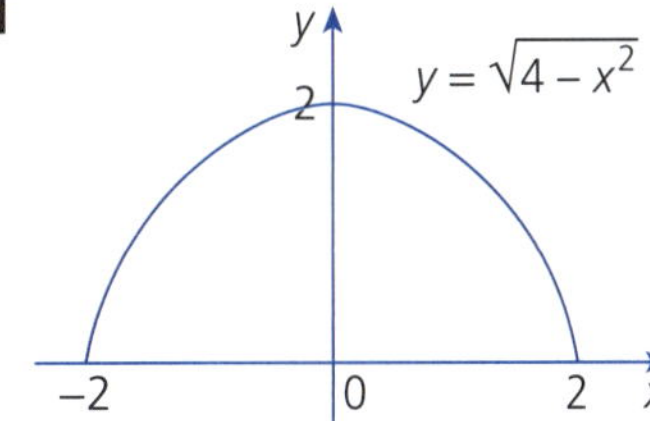

p. 244

8 For the equation $y = 2x - 3$:
a sketch the line representing $y = 2x - 3$
b derive the equation for the inverse function of $y = 2x - 3$, and sketch this inverse function on the same number plane
c By drawing the line $y = x$, verify that $y = 2x - 3$ and its inverse are mirror images through the line $y = x$. pp. 244–245

9 For the equation $y = 4 - 2x$:
a sketch the line representing $y = 4 - 2x$
b derive the equation for the inverse function of $y = 4 - 2x$ and sketch this inverse function on the same number plane. pp. 244–245

10 Find the inverse function of $y = \sqrt{x}$. pp. 244–245

11 Repeat Question 8 for:
a $x - 2y = 4$ **b** $y = -x + 2$
c $y = x + 2$ pp. 244–245

12 By first sketching $y = x^2$, make sketches of:
a $y = x^2 - 4$ **b** $y = (x - 4)^2$ pp. 245–246

13 Sketch the following functions:
a $y = -x^2$ **b** $y = 4 - x^2$
c $y = -(x - 4)^2$ pp. 245–246

14 By first sketching $y = x^3$, sketch the following:
a $y = x^3 + 2$ **b** $y = x^3 - 2$
c $y = (x - 2)^3$ **d** $y = (x + 2)^3$ pp. 245–246

15 Simplify:
a $\dfrac{a^4 \times a^7}{a^9}$ **b** $\dfrac{5a^{12} \times 3a^3}{15a^5}$
c $(2a^5)^4$ **d** $\dfrac{(3a^2)^4}{(2a^3)^2}$
e $\dfrac{a^{\frac{3}{4}} \sqrt{a^{\frac{1}{2}}}}{a^{\frac{1}{4}}}$ **f** $\dfrac{(a^{\frac{2}{3}})^4 \times (a^{\frac{3}{4}})}{(a^{\frac{3}{2}})^3 \times a^{\frac{1}{4}}}$ p. 246

16 Evaluate:
a $81^{\frac{1}{2}}$ **b** $81^{\frac{1}{4}}$
c $2^3 \times 64^{\frac{1}{6}}$ **d** $256^{\frac{1}{2}} \times 16^{-\frac{1}{2}}$
e $2187^{\frac{4}{7}}$ **f** $125^{-\frac{2}{3}}$
g $1000^{\frac{1}{3}} \times 16^{-\frac{3}{4}}$ pp. 246–278

17 Solve for a, where $a > 0$:
a $3^a = 81$ **b** $a^2 = 1024$
c $5^{2a} = 625$ **d** $10^{3a - 2} = 10\,000$
e $2^a = \dfrac{1}{4}$ **f** $5^{a + 1} = \dfrac{1}{125}$
g $(\sqrt{3})^a = 27$ **h** $\left(\dfrac{1}{\sqrt{5}}\right)^{4a} = 625^{2a - 15}$ pp. 247–248

18 Simplify the following expressions using the tables of powers. Express your answer without an index:
a $9 \times 27 \times 27$ **b** $\dfrac{2 \times 16 \times 64 \times 128}{256}$
c $\dfrac{\sqrt{729}}{9}$ **d** $\dfrac{\sqrt{3125 \times 125}}{625}$
e $\dfrac{(81)^2 \times (9)^3}{(27)^4}$ pp. 247–248

19 Write in logarithm form:
a $5^3 = 125$ **b** $3^{-3} = \dfrac{1}{27}$
c $\sqrt{8} = 2^{\frac{3}{2}}$ pp. 248–249

20 Write in index form:
a $\log_{10} 1000 = 3$ **b** $\log_2 16 = 4$
c $\log_2 4\sqrt{2} = 2.5$ **d** $\log_5 \dfrac{1}{25} = -2$ p. 249

21 Solve the following logarithmic equations:
a $\log_2 64 = x$ **b** $\log_5 125 = x$
c $x = \log_3 \sqrt{27}$ **d** $\log_{10} 1\,000\,000 = x$
e $\log_2 \dfrac{1}{4} = x$ **f** $\log_2 x = 3$
g $\log_5 x = 2$ **h** $\log_3 x = \dfrac{1}{2}$
i $\log_{10} x = -1$ **j** $\log_2 x = -3$
k $\log_8 x = -\dfrac{1}{3}$ **l** $\log_x 256 = 8$
m $\log_x \dfrac{1}{9} = -2$ **n** $\log_x 4 = \dfrac{2}{3}$ p. 249

22 By first completing the following table of values:

x	–2	–1	0	0.5	1	1.5	2
y							

sketch the graph of $y = 3^x$. p. 250

23 Complete the following table for $y = 10^x$ and then sketch the curve:

x	–1	–0.5	0	0.2	0.5	0.8	1
y							

p. 250

24 Rewrite $y = \log_3 x$ in index form and complete the following table:

x							
y	–2	–1	0	0.5	1	1.5	2

Sketch the curve $y = \log_3 x$. p. 251

25 Rewrite $y = \log_{10} x$ in index form and complete the following table:

x							
y	-1	-0.5	0	0.2	0.5	0.8	1

Sketch the curve $y = \log_{10} x$. p. 251

26 Simplify the following logarithmic expressions:

a $\log_a N + \log_a M$
b $\log_a N - \log_a M$
c $i \log_a N$
d $i \log_a N + j \log_a M$
e $\frac{1}{2}\log_a N$
f $\frac{1}{2}\log_a N - 2\log_a M$

pp. 251–255

27 Expand the following logarithmic expressions:

a $\log_a xy$
b $\log_a \frac{x}{y}$
c $\log_a \frac{1}{y}$
d $\log_a \sqrt{x}$
e $\log_a \frac{x^2}{y}$
f $\log_a \frac{x^2\sqrt{y}}{z^4}$

pp. 251–255

28 Use the laws of logarithms to evaluate the following expressions:

a $\log_2 64$
b $\log_4 64$
c $\log_3 \sqrt{3}$
d $\log_5 5\sqrt{5}$
e $\log_5 \frac{1}{\sqrt{5}}$
f $\log_a a^3$
g $\log_a \sqrt{a}$
h $\log_a a + \log_a \sqrt{a}$
i $\log_a (\log_2 2)$
j $\log_x (x^2 - 2x) - \log_x (x - 2)$
k $\log_5 40 - \log_5 8$
l $\log_3 54 - \log_3 2$

pp. 251–255

29 Given $\log_{10} 3 = 0.477$ and $\log_{10} 5 = 0.699$, evaluate the following correct to three decimal places:

a $\log_{10} 15$
b $\log_{10} \frac{5}{3}$
c $\log_{10} \sqrt{3}$
d $\log_{10} 3\sqrt{3}$
e $\log_{10} 5\sqrt{3}$
f $\log_{10} \frac{1}{3}$
g $\log_{10} 0.2$
h $\log_{10} 45$
i $\log_{10} 75$
j $\log_{10} 5.4$

pp. 251–255

30 Given $\log_a 7 = 2.8$ and $\log_a 5 = 2.3$, evaluate the following correct to one decimal place:

a $\log_a 35$
b $\log_a \frac{7}{5}$
c $\log_a 1.4$
d $\log_a \sqrt{7}$
e $\log_a \frac{1}{7}$
f $\log_a \frac{1}{25}$
g $\log_a \frac{1}{\sqrt{7}}$
h $\log_a 175$
i $\log_a \sqrt{35}$
j $\log_a 9.8$

pp. 251–255

31 If $x = \log_n 2$ and $y = \log_n 3$, write the following in terms of x and y:

a $\log_n 6$
b $\log_n \frac{2}{3}$
c $\log_n 1.5$
d $\log_n \sqrt{2}$
e $\log_n \sqrt{6}$
f $\log_n 24$
g $\log_n (6\sqrt{2})$
h $\log_n 4.5$
i $\log_n \frac{16}{3}$
j $\log_n 0.5$

pp. 251–255

32 If $N = \log_a 10$ and $M = \log_a 3$, write simple logarithmic expressions for:

a $N + M$
b $N - M$
c $\frac{1}{2}N$
d $\frac{1}{2}M - \frac{1}{2}N$
e $2N + 3M$
f $4M - 2N$

pp. 251–255

33 Express y in terms of the other variables:

a $\log_2 y = \log_2 a + \log_2 b$
b $\log_2 y = \frac{1}{3}\log_2 a$
c $3\log_2 y = -\log_2 x$
d $\log_2 y = 1 - \frac{1}{2}\log_2 x$

pp. 251–255

34 Solve the following logarithmic equations for a:

a $\log_{10} a = \log_{10} 3 + \log_{10} 8$
b $\log_{10} a = \frac{1}{2}\log_{10} 81$
c $\log_{10} a = \log_{10} 72 - \log_{10} 9$
d $\log_{10} a = 3\log_{10} 2 - 2\log_{10} 5$
e $\log_{10} a = 1 - 2\log_{10} 5$
f $\log_{10} 16a - \log_{10} (a + 6) = 1$
g $\log_2 a - \log_2 (a - 2) = 3$

pp. 251–255

35 Solve the following for y, giving the answer correct to two decimal places:

a $3^x = 5$
b $3^x = 10$
c $3^{2x} = 5$
d $3^x = \sqrt{5}$
e $3^x = \frac{1}{5}$
f $3^x = 2^{3x-5}$
g $2^{3-x} = 5^{2x+1}$

p. 256

Go to p. 302 for **Quick Answers** or to pp. 350–356 for **Worked Solutions**

For a complete understanding of this topic, you must be able to:

✓	Identify functions		p. 242
✓	Use the notation $f(c)$ when using $f(x)$ and $x = c$		pp. 242–243
✓	Find the permissible x values (domain) and y values (range) for a variety of functions		p. 244
✓	Determine the inverse functions for a variety of functions		pp. 244–245
✓	Sketch the graphs of $y = f(x) + a$, $y = f(x + a)$		pp. 245–246
✓	Recall the index laws		pp. 246–248
✓	Evaluate simple expressions involving logarithms		pp. 248–249
✓	Solve equations involving logarithms		p. 249
✓	Graph exponential and logarithmic functions		p. 250
✓	Use the laws of logarithms to evaluate simple expressions		pp. 251–255
✓	Solve equations involving exponentials.		p. 256

Now you are ready to do the tests!

Intermediate Test

Functions and Logarithms

(30 marks)

1 Which of the following is NOT a function?

A

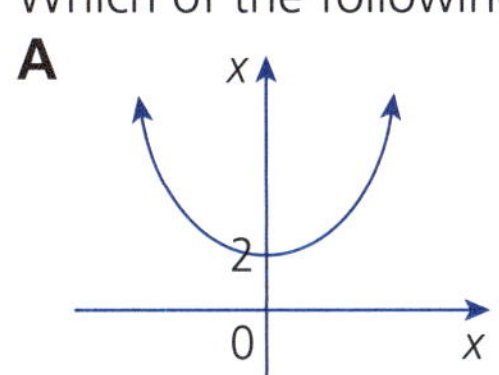

B

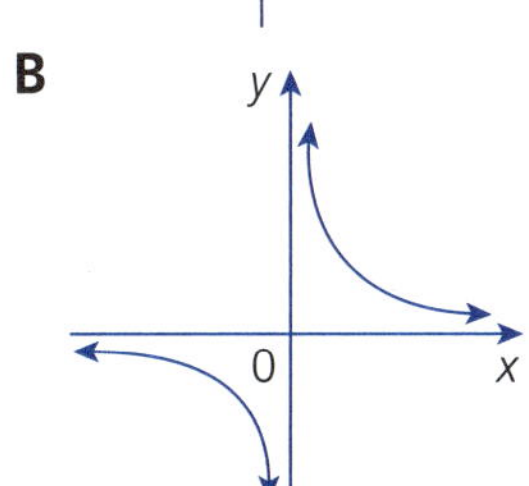

C

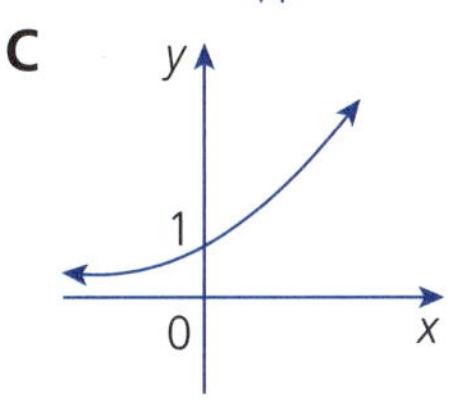

D

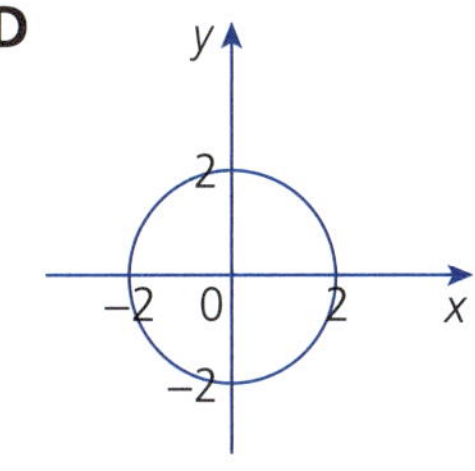

(1 mark)

2 If $f(x) = 2x^2 - 4x$, find $f(-3)$. (2 marks)

3 If $g(x) = 4 - x$, find $g(x + 2)$. (2 marks)

4 For the following functions, find the domain (possible x values):

a $y = \dfrac{1}{x}$ (1 mark)

b $y = \sqrt{x}$ (1 mark)

5 Find the domain (possible x values) and range (possible y values) of:

a $y = 2x + 1$ (2 marks)

b $y = x^2 + 3$ (2 marks)

6 **a** Find the inverse function of $y = 3x - 2$. (1 mark)

b Sketch $y = 3x - 2$ and its inverse on the same number plane. (2 marks)

7 The graph of $y = f(x)$ is drawn on the number plane below:

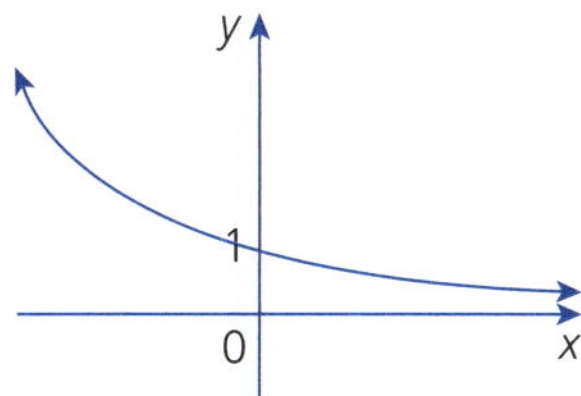

Copy the graph of $y = f(x)$ on to your page and then graph $y = f(x) + 2$ and $y = -f(x)$ on the same number plane. (2 marks)

8 If $\log_2 x = 3$, then $x = ?$ (1 mark)

9 On the same number plane graph $y = \log_{10} x$ and $y = 10^x$. (2 marks)

10 If $\log_a 2 = 0.4176$ and $\log_a 3 = 0.6413$, find:

a $\log_a 12$ (2 marks)

b $\log_a 1.5$ (2 marks)

c $\log_a \sqrt{6}$ (2 marks)

11 Simplify:

a $\log_5 15 - \log_5 3$ (1 mark)

b $2\log_2 8 + \log_2 4$ (1 mark)

c $\dfrac{\log_3 27}{\log_3 3}$ (1 mark)

12 Solve:

a $\log_{10} 10\,000 = x$ (1 mark)

b $3^{x+2} = 81$ (1 mark)

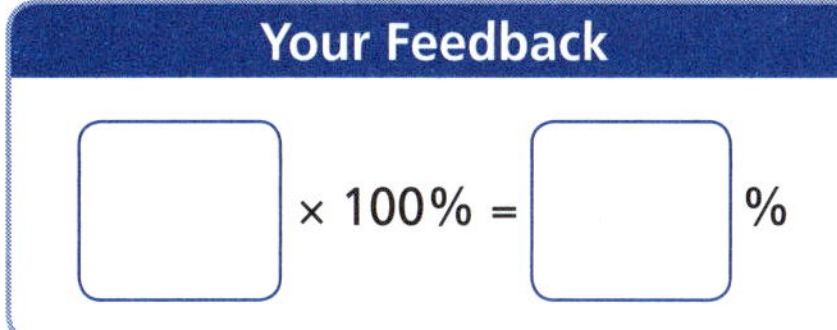

QA PAGE 308
WS PAGE 379

Advanced Test

Functions and Logarithms

(30 marks)

1 If $f(x) = x^2 - 4x - 6$:

a find $f(-2)$ (1 mark)

b find $f(x + 2)$ (1 mark)

c solve $f(x) = 6$. (2 marks)

2 For the following functions, find the domain (possible x values):

a $y = \frac{1}{x - 2}$ (1 mark)

b $y = \frac{1}{2x^2 + 5x - 12}$ (1 mark)

c $y = \sqrt{3 - x}$ (1 mark)

d $y = \sqrt{9 - x^2}$ (2 marks)

3 Find the domain (possible x values) and range (possible y values):

a $y = 4 - x^2$ (2 marks)

b $y = \frac{1}{2x + 1}$ (2 marks)

c $y = 3^x + 1$ (2 marks)

4 The graph of $y = g(x)$ is drawn on the number plane below:

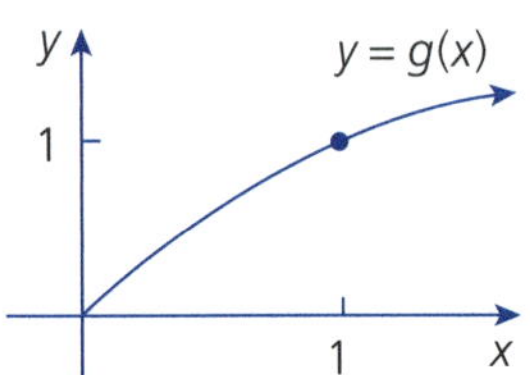

Copy the graph of $y = g(x)$ and graph its inverse function $y = h(x)$, on the same number plane.
(1 mark)

5 Evaluate:

a $\log_2 4\sqrt{2}$ (1 mark)

b $\log_5 b + \log_5 \frac{1}{b}$ (1 mark)

c $\log_2 36 - \log_2 9$ (1 mark)

6 If $x = \log_b 3$ and $y = \log_b 4$, write a simple expression for:

a $x + y$ (1 mark)

b $2x - y$ (2 marks)

7 Solve:

a $\log_2 3x - \log_2 4 = \log_2 6$ (2 marks)

b $\log_2 (x + 2) + \log_2 (x - 5) = 3$ (3 marks)

8 Evaluate:

$\log_3 (\log_4 64)$ (1 mark)

9 Solve:

$4^{x+2} = \frac{1}{2\sqrt{2}}$ (2 marks)

$\square \times 100\% = \square\ \%$

QA PAGE 308

WS PAGE 380

Chapter 15
Circle Geometry

Some important points

- Always draw a clear diagram and mark on the diagram all given information.
- Work from the known towards the desired result.
- If an angle is required, mark sizes of angles on the diagram as you find them. If it is a deductive question it is generally useful to label one of the angles with a pronumeral (one of the angles in the result).
- If sides are involved, consider isosceles triangles or congruent triangles.
- If you become lost, check that all the given information has been used.
- Use any *hints* given in the question.

Keywords

Alternate, Arc, Bisector, Centre, Chord, Circumference, Collinear, Common chord, Concyclic, Converse, Cyclic, Deductive reasoning, Diameter, Intersect, Intersection, Perpendicular, Point of contact, Radius, Reflex, Secant, Sector, Segment, Semicircle, Subtend, Tangent

Some parts of the circle

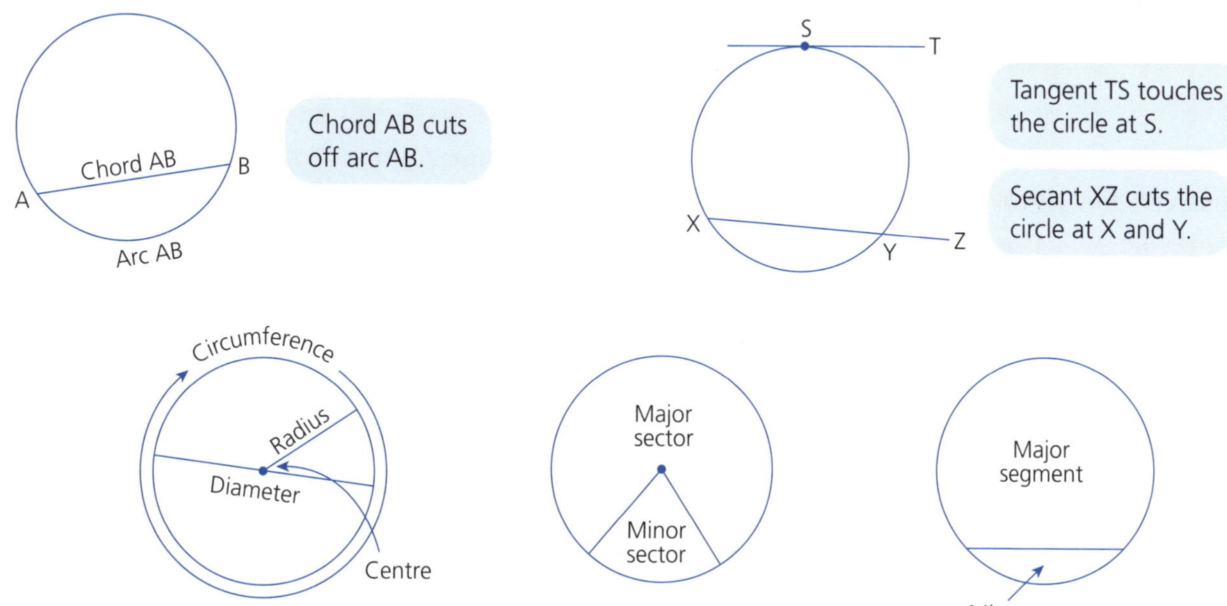

Properties of chords

All properties in this section rely on the use of congruent triangles.

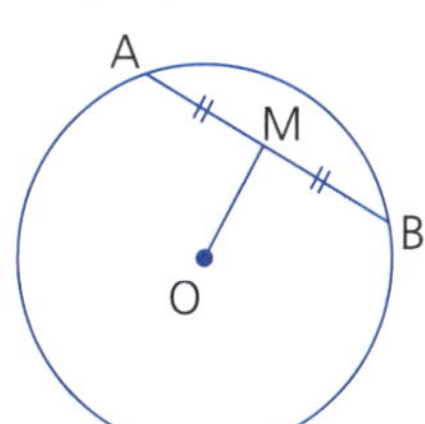

1 A line from the centre of a circle to the midpoint of a chord meets the chord at right-angles.

Given AM = MB, then OM ⊥ AB.

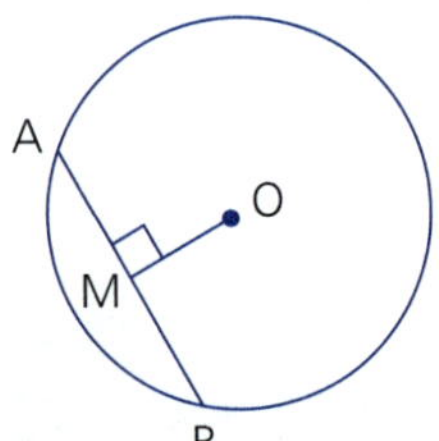

2 A perpendicular drawn to a chord from the centre of a circle bisects the chord.

Given OM ⊥ AB, then AM = MB.

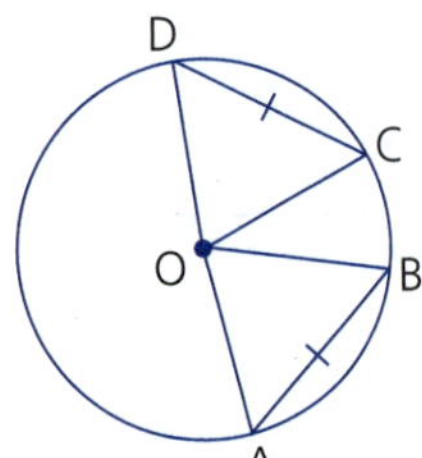

3 Equal chords of a circle subtend equal angles at the centre.
Given AB = DC, then ∠AOB = ∠DOC.

Other interesting properties arising from these first three properties are:

a The perpendicular bisector of a chord of a circle must pass through the centre of the circle. If AM = MB and NM ⊥ AB, then NB passes through O.

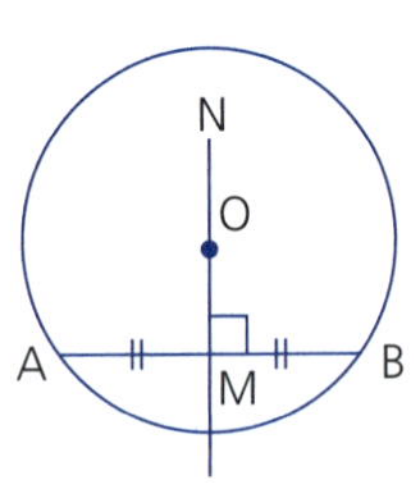

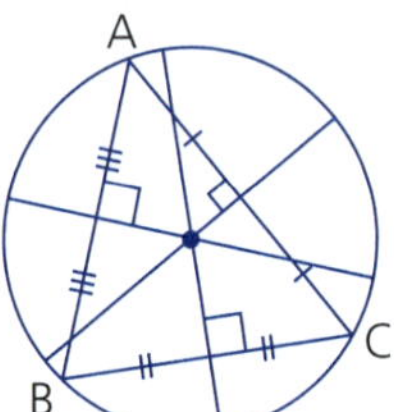

Note: this property is used to locate the centre of a circle passing through any three non-collinear points. The centre is the point of intersection of the perpendicular bisectors of lines joining AB, BC and AC.

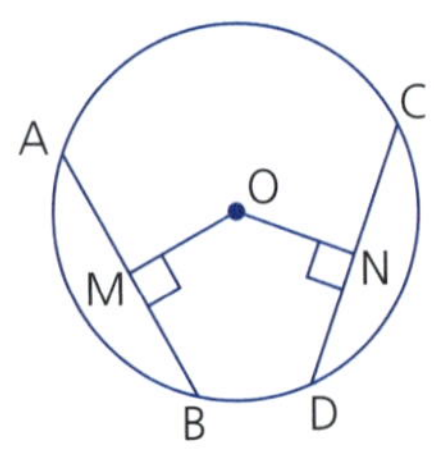

b Equal chords of a circle are equidistant from the centre of the circle. Given AB = CD, then OM = ON.

For Example

1 Find the value of the pronumeral in the following questions:

a Given OM ⊥ AB, O is the centre, AB = 12 cm, find x.

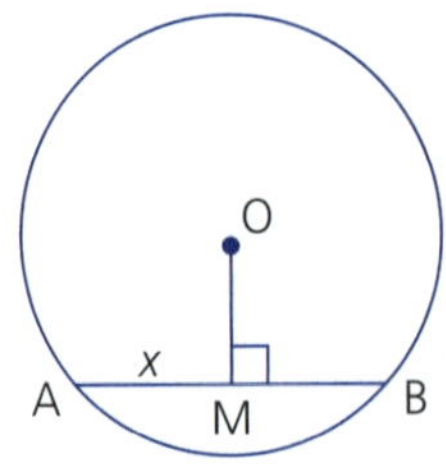

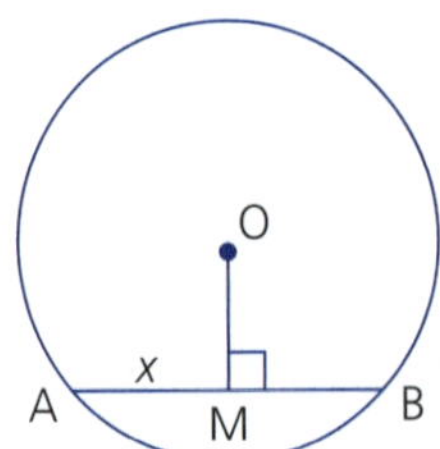
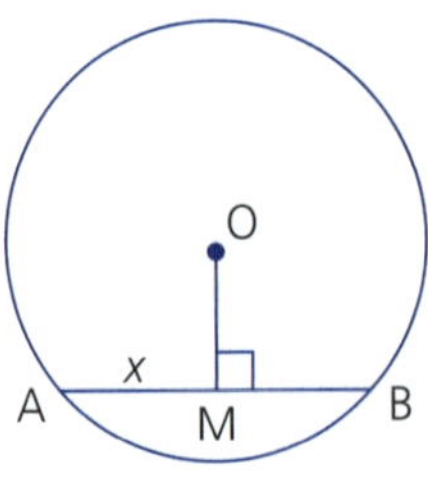

S AM = MB (line from centre ⊥ chord)
∴ AM = 6 cm
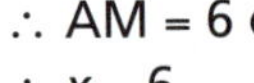
∴ $x = 6$

b Given O is the centre, AM = MB = 4 cm, find x.

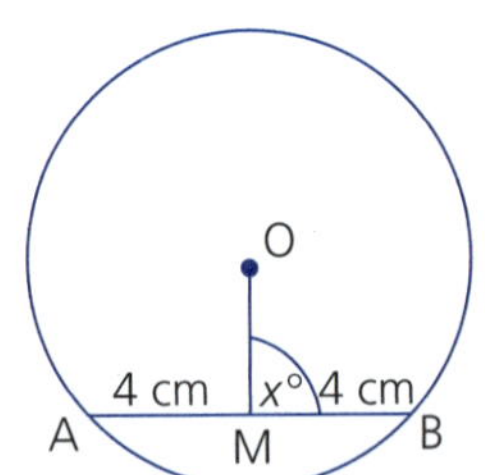

S OM ⊥ AB
∴ $x = 90$

c O is the centre.
Find y.

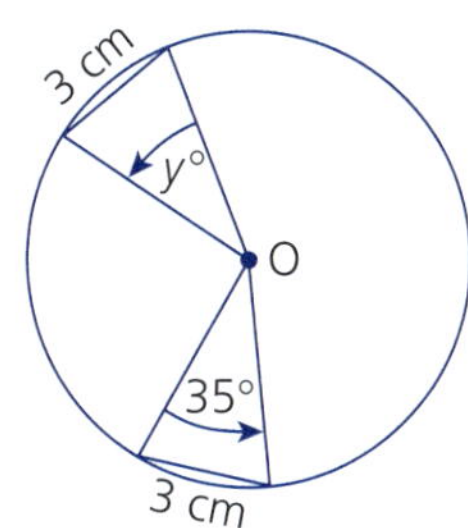

S $y = 35$ (Equal chords subtend equal angles at the centre.)

d O is the centre.
Find m.

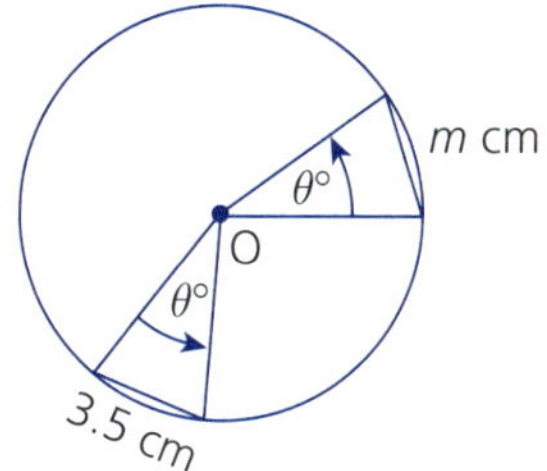

S $m = 3.5$ (Equal chords subtend equal angles at the centre.)

e O is the centre,
AB = 14 cm,
CD = 14 cm.
Find x.

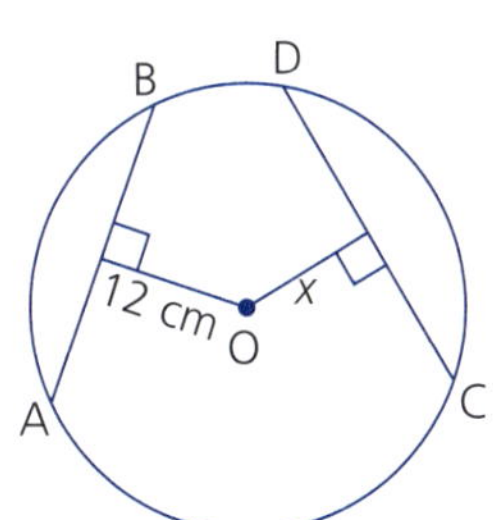

S $x = 12$ (Equal chords are equidistant from the centre.)

f O is the centre,
EO = 12 cm,
OF = 12 cm,
FC = 5 cm,
AE = x cm.
Find x.

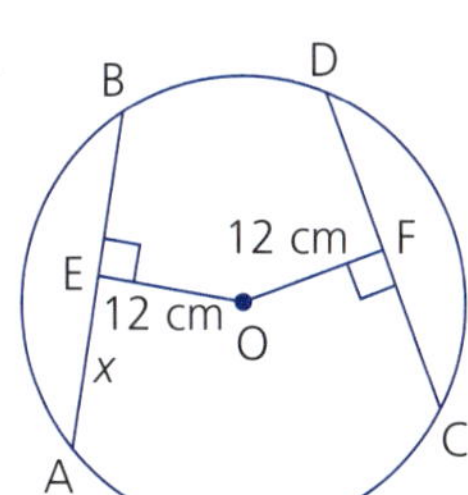

S DC = 10 cm
(perpendicular from centre to a chord bisects the chord)
$\therefore$ AB = 10 cm
(equal chords are equidistant from the centre)
$\therefore x = 5$
(perpendicular from centre to chord bisects the chord)

Examples involving Pythagoras' Theorem

g Given that a chord AB, 10 cm long, is 12 cm from the centre O, find the radius of the circle.

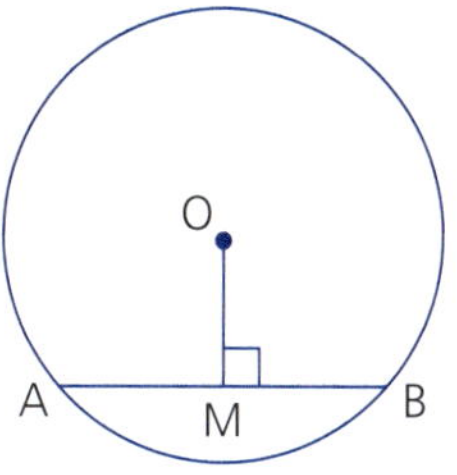

S Join OB and mark the information on the diagram.

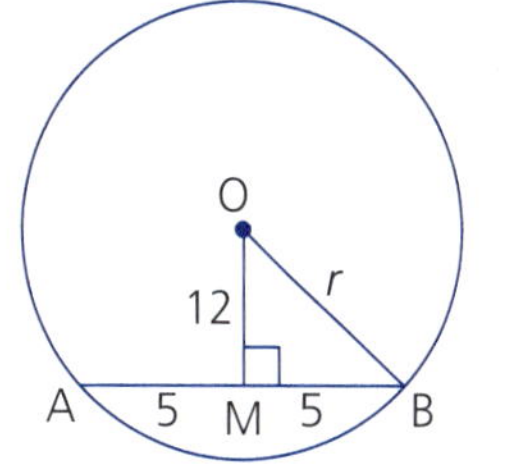

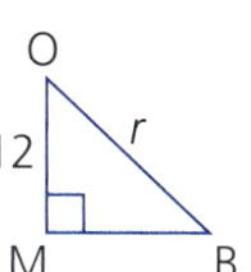

Using Pythagoras' Theorem:

$$\begin{aligned} r^2 &= 12^2 + 5^2 \\ &= 169 \\ \therefore r &= \sqrt{169} \\ &= 13 \end{aligned}$$

The radius of the circle is 13 cm.

h A chord is drawn 12 cm from the centre of concentric circles with radii 13 cm and 15 cm. Find the length of DB in the diagram.

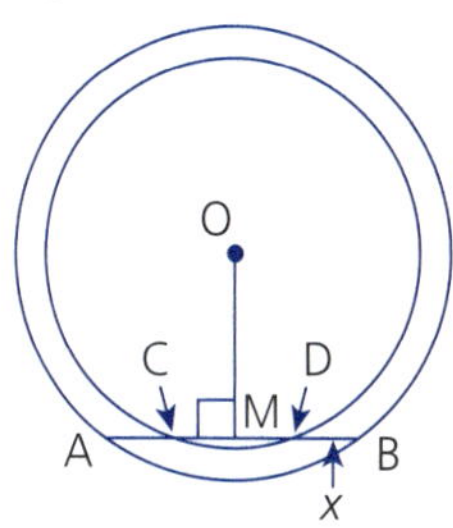

S Join OD, OB and mark the information on the diagram.

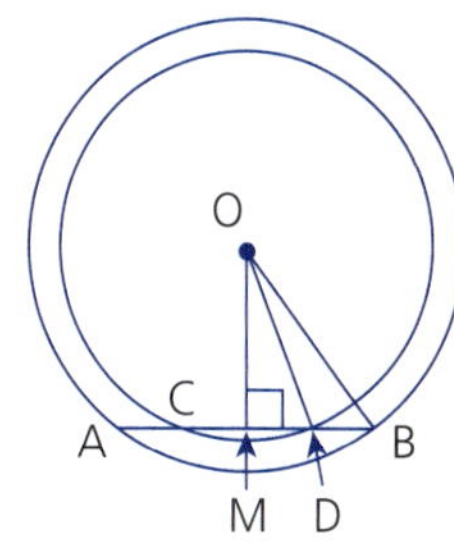

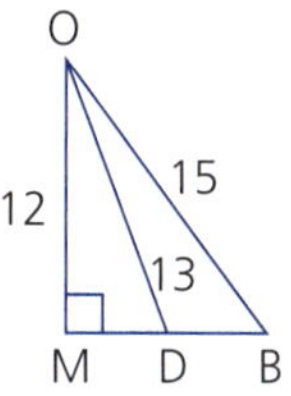

Using Pythagoras' Theorem:

From $\triangle$OMD
$MD^2 + 12^2 = 13^2$
$\therefore MD^2 = 13^2 - 12^2$
$= 25$
$\therefore MD = 5$

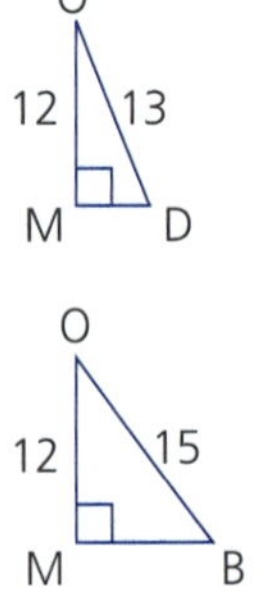

From $\triangle$OMD
$MB^2 + 12^2 = 15^2$
$\therefore MB^2 = 15^2 - 12^2$
$= 81$
$\therefore MB = 9$

Now DB = BM – MD
= 9 – 5
= 4

DB is 4 cm.

Properties of angles

1 The angle at the centre of a circle is twice the angle at the circumference standing on the same arc.

Given O is the centre, then:
$\angle AOB = 2 \times \angle ACB$, or

$\angle ACB = \frac{1}{2} \times \angle AOB$

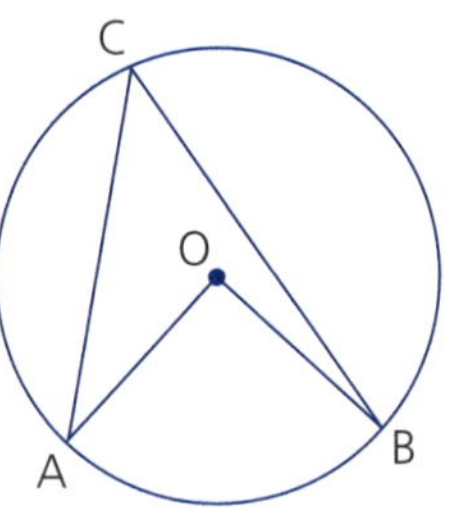

Some important notes

Arcs and chords

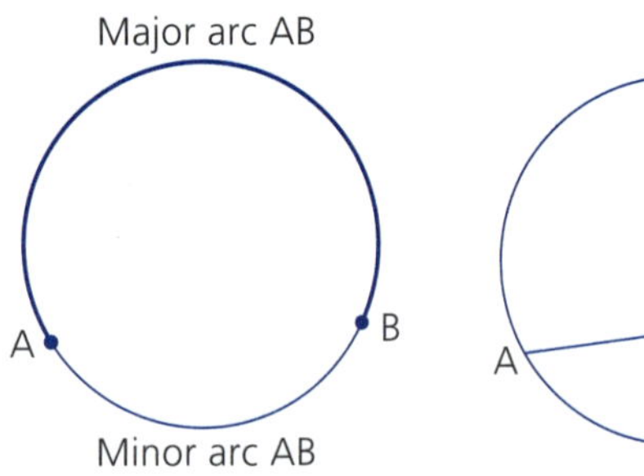

Any chord (AB in this diagram) divides the circumference into two sections. In this diagram the minor arc is ASB and the major arc is ATB.

Reflex angle at the centre

There are always two angles at the centre; one stands on the major arc—the reflex angle, while the other stands on the minor arc—the acute or obtuse angle. This second type is the most common and the easiest to see.

For example:

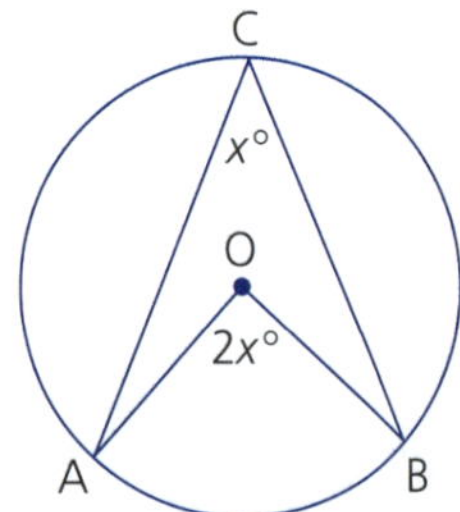

Normal case

$\angle$AOB is twice $\angle$ACB
or
$\angle$ACB is half $\angle$AOB

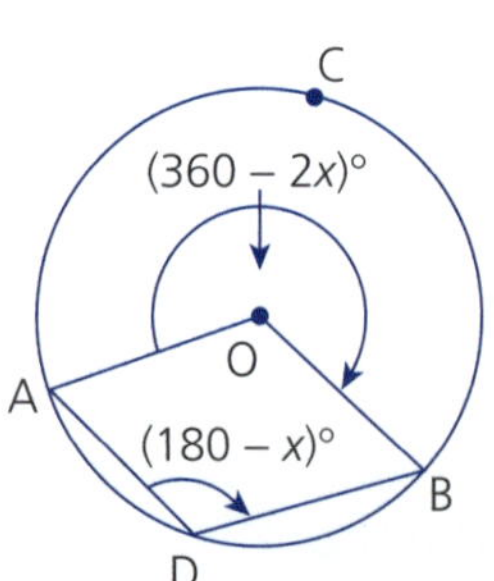

Reflex angle at the centre case

Reflex $\angle$AOB is twice $\angle$ADB.
(Both stand on the major arc ACB.)
If obtuse $\angle AOB = 2x°$, then
reflex $\angle AOB = (360° - 2x)°$

and $\angle ADB = \frac{1}{2}(360 - 2x)°$
$= (180 - x)°$

2 The angle in a semicircle is a right-angle.

Given AB is a diameter, ∠ACB = 90°.

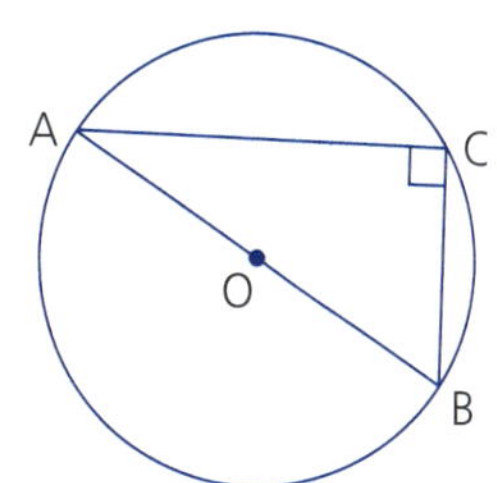

3 Angles in the same segment are equal (that is, angles at the circumference standing on the same arc).

Given ∠ADB, ∠ACB standing on arc AB, then ∠ADB = ∠ACB.

Or, angles at the circumference standing on the same arc are equal.

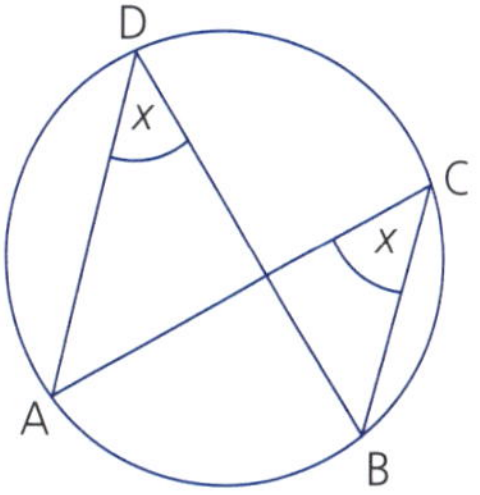

Cyclic quadrilaterals

1 A **cyclic quadrilateral** is a quadrilateral with all four vertices on the circumference of a circle.

Concyclic points are points through which a circle can be drawn, that is, all points lie on the circumference of a circle.

The four vertices of a cyclic quadrilateral are concyclic points.

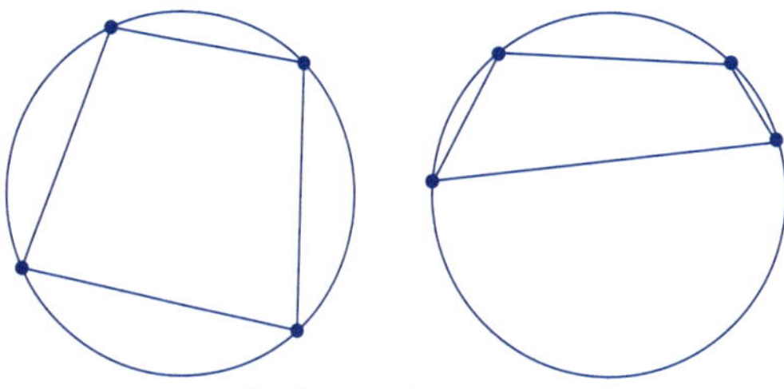
Cyclic quadrilaterals

2 Opposite angles in a cyclic quadrilateral are supplementary.

∠DAB + ∠BCD = 180°, ∠ABC + ∠ADC = 180°

Conversely, **if the opposite angles of a quadrilateral are supplementary then the quadrilateral is cyclic**.

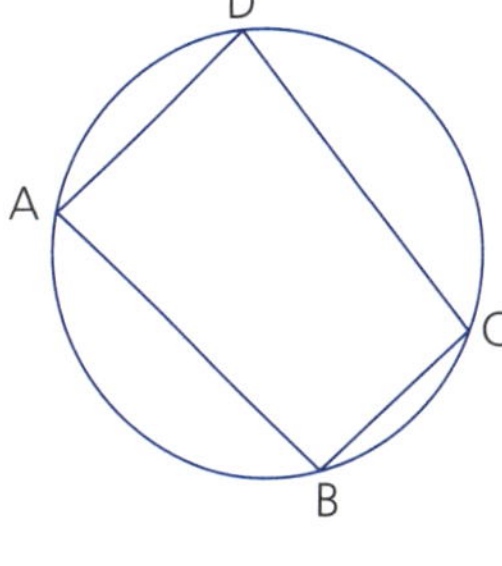

3 The exterior angle of a cyclic quadrilateral is equal to the interior opposite angle.

Given that ABCD is a cyclic quadrilateral, ∠CBE = ∠ADC.

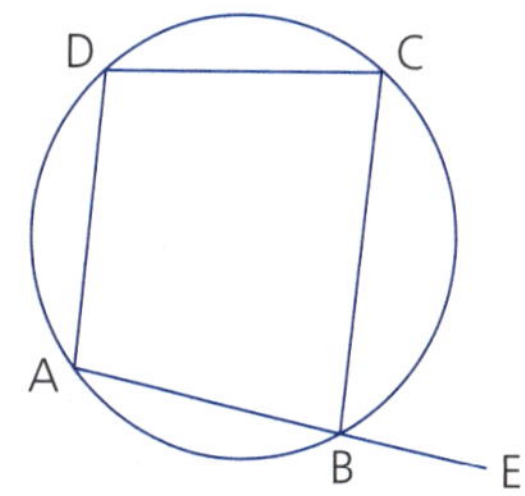

1 Find the value of the pronumeral in each diagram, giving adequate reasons. (O is the centre in each diagram.)

a ∠AOB = 112°
∠ACB = a°

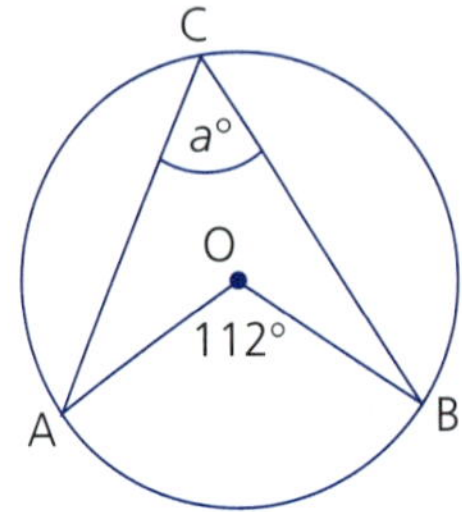

S $a = \frac{1}{2} \times 112 = 56$ (∠ at the circum. is half ∠ at the centre on the same arc)

b ∠ACB = 66°
∠AOB = b°

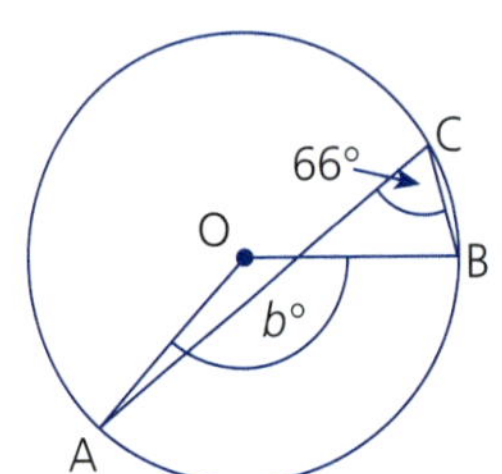

S $b = 2 \times 66 = 132$
(∠ at centre is twice ∠ at the circumference on the same arc)

c AB is diameter

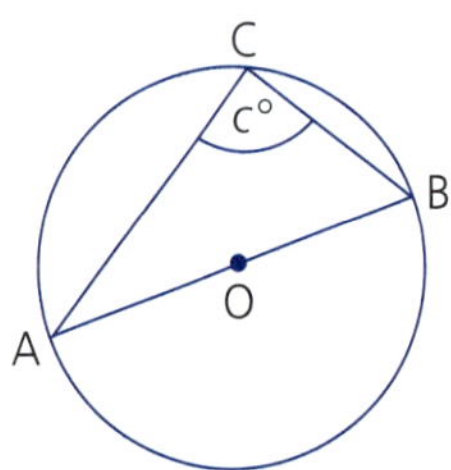

S $c = 90$
(angle in a semicircle)

d ∠ADB = 47°
∠ACB = x°

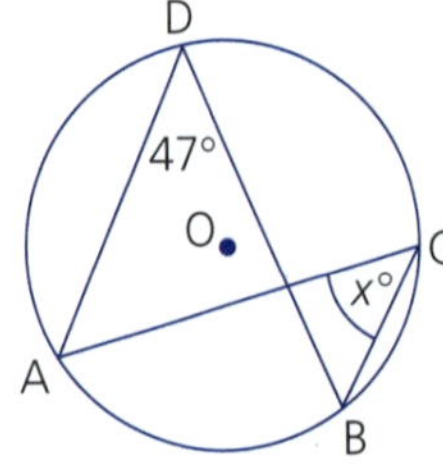

S $x = 47$
(angles in the same segment)

e ∠ADB = 46°
∠DBC = 49°
∠DAC = y°
∠ACB = x°
∠AED = z°

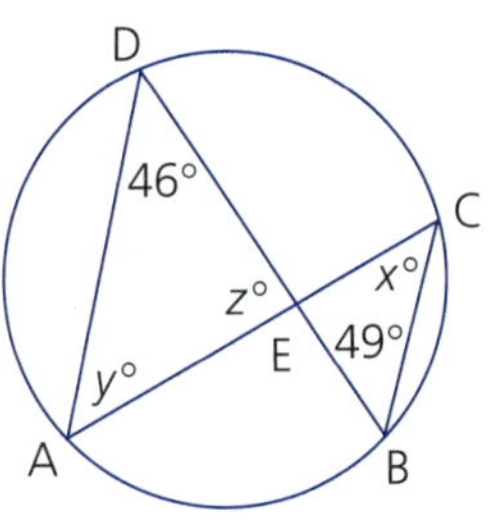

S $x = 46$
(angles in the same segment standing on arc AB)

$y = 49$
(angles in the same segment, on arc DC)

$z = 180 - (46 + 49)$ (angle sum of △ADE)
$= 85$

f ∠DAB = 78°
∠DCB = a°

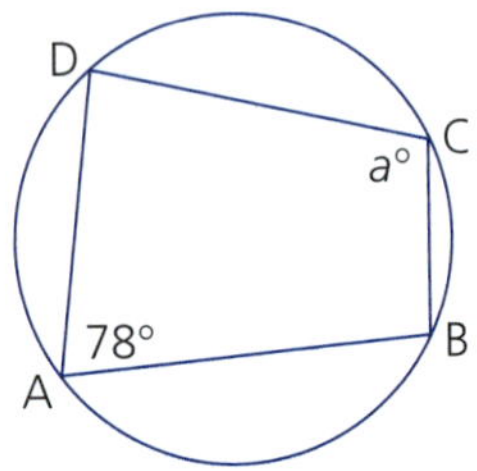

S $a = 180 - 78$ (opposite angles of a
$= 102$ cyclic quadrilateral)

g AE is a straight line.
∠CBE = 108°
∠ADC = x°

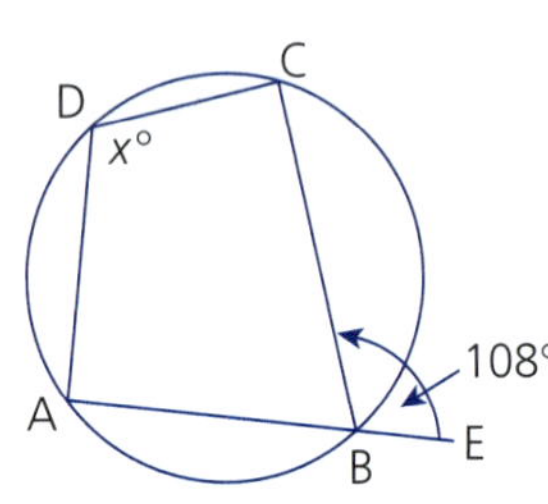

S $x = 108$
(exterior angle of a cyclic quadrilateral)

h AC is a diameter.
∠ACB = 46°
∠CAB = x°
∠CBO = y°

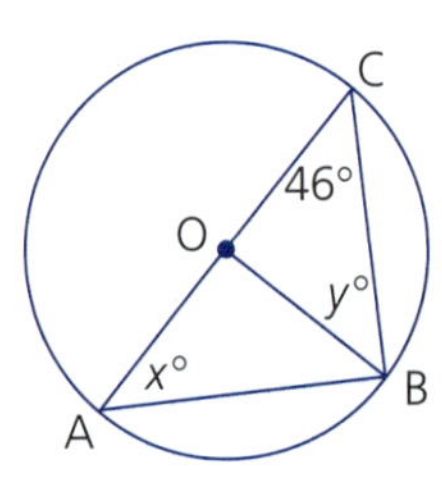

S OC = OB (equal radii)
∴ $y = 46$ (base angles, isosceles △)
But ∠ABC = 90° (∠ in a semicircle)
∴ ∠ABO = 90° − 46° = 44°
Also AO = OB (equal radii)
∴ $x = 44$ (base angles, isosceles △)

i AB ∥ ED
AC is a straight line.
∠DBC = 100°

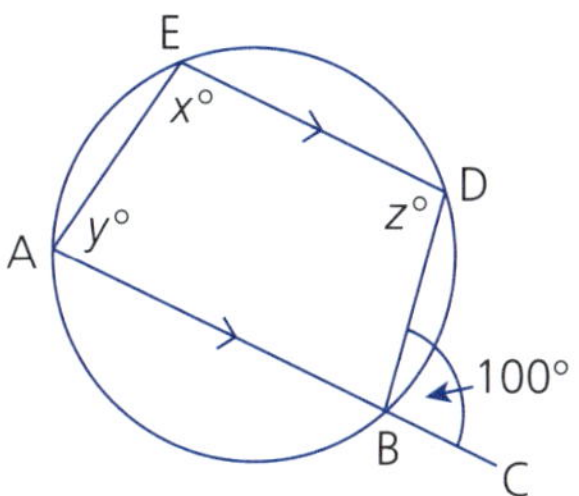

S $x = 100$
(exterior angle of a cyclic quadrilateral)

$z = 100$ (alternate angles, AB ∥ ED)

$\therefore y = 180 - 100$ (opposite angles of
$= 80$ a cyclic quadrilateral)

j ∠POR = 146°
∠PQR = x°

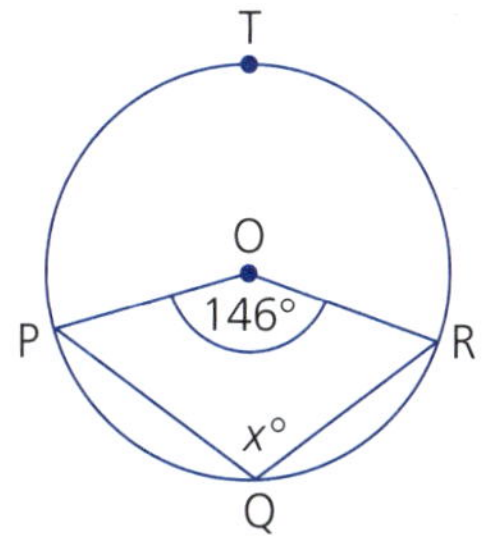

Note: PQRO is not a cyclic quadrilateral. Only three vertices lie on the circle. The other vertex is at the centre.

S Reflex ∠POR stands on major arc PTR.
∠PQR is the angle at the circumference on major arc PTR.

∠POR = 360° − 146° (angles at a point)
= 214°

$\therefore x = 107$ (∠ at circumference is half ∠ at centre on major arc PTR)

Compare this with the following case:

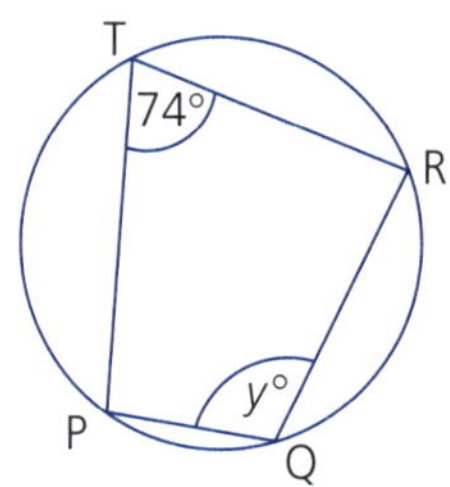

This is a cyclic quadrilateral as all four vertices lie on the circle.

∠PTR = 74°, ∠PQR = y°,
$y = 180 - 74$ (opposite angles of a
$\therefore y = 106$ cyclic quadrilateral)

k ∠XYZ = 112°
∠XOZ = θ°

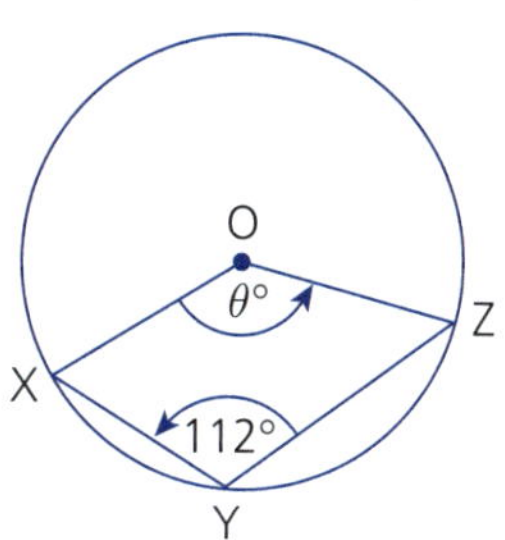

S Reflex ∠XOZ = 224° 2 × 112°

(∠ at centre is twice ∠ at circumference on major arc XZ)

$\therefore \theta = 360 - 224$ (angles at a point)
$= 136$

l ∠AOC = 140°,
∠ADC = a°,
∠ABC = b°

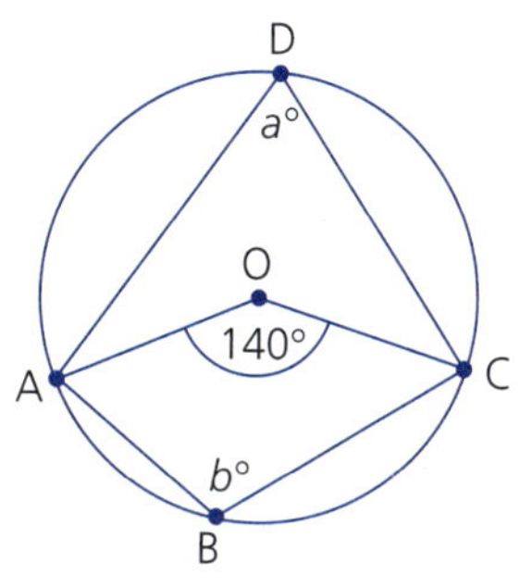

S $a = 70$ (∠ at circumference is half ∠ at centre on arc AC)

$\therefore b = 180 - 70$ (opposite angles of cyclic
$= 110$ quadrilateral ABCD)

m ∠OPR = 36°
∠PSR = x°

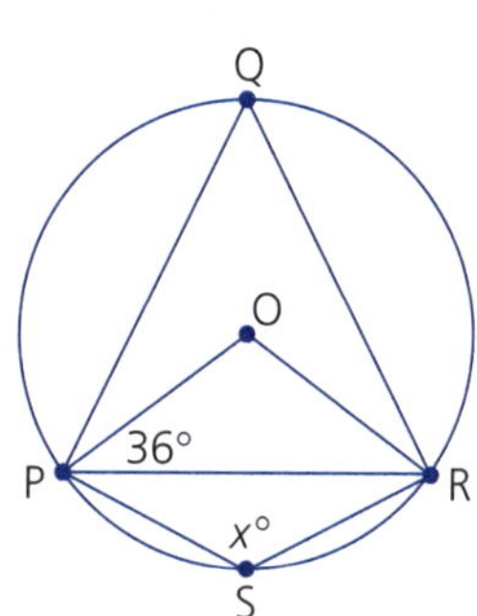

S ∠ORP = 36° (base ∠s of isos. △OPR)
∠POR = 108° (∠ sum of △)
∠PQR = 54° (∠ at circumference is half ∠ at centre on same arc)
x = 126° (opposite angles of cyclic quadrilateral PSRQ)

Properties of tangents

1 The angle between a tangent and a radius drawn to the point of contact is a right-angle.

PT is a tangent at T, O is the centre, then ∠OTP = 90°.

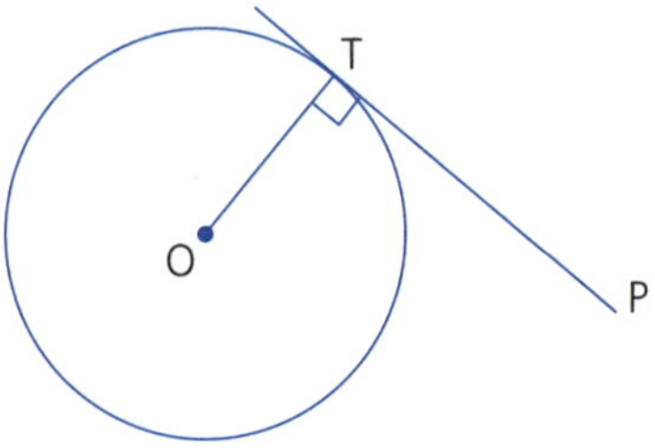

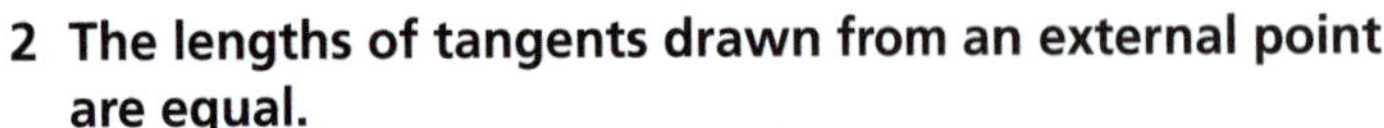

2 The lengths of tangents drawn from an external point are equal.

PT and PS are tangents, then PT = PS.

Congruent triangles are used after joining OT and OS.

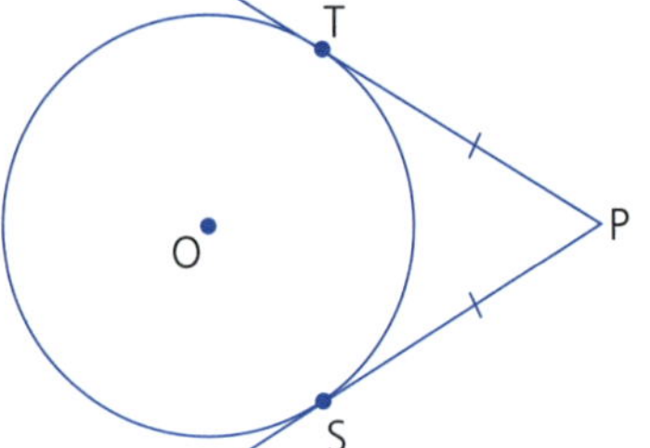

For Example

1 In Examples **a** to **c**, find the value of x.

a

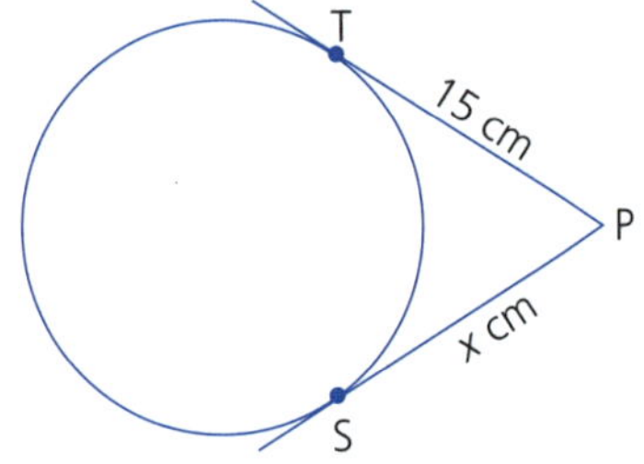

PT, PS are tangents, PT = 15 cm.

S x = 15 cm (tangents drawn from external point)

b

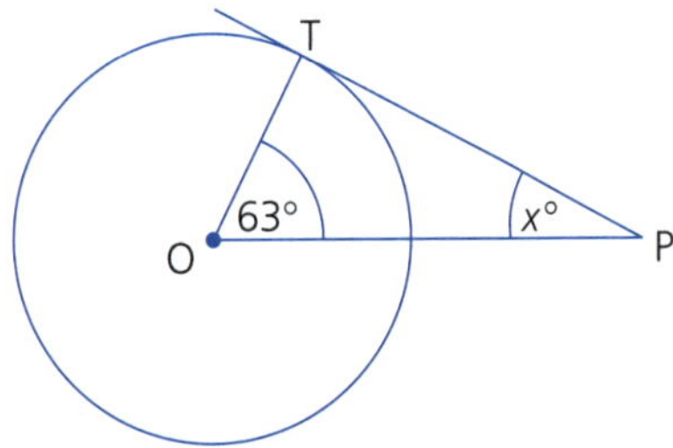

O is centre. PT is a tangent touching at T.
∠TOP = 63°, ∠OPT = x°

S ∠OTP = 90°
(radius drawn to a point of contact)

$\therefore x = 90 - 63$ (∠ sum of △)
$x = 27$

c

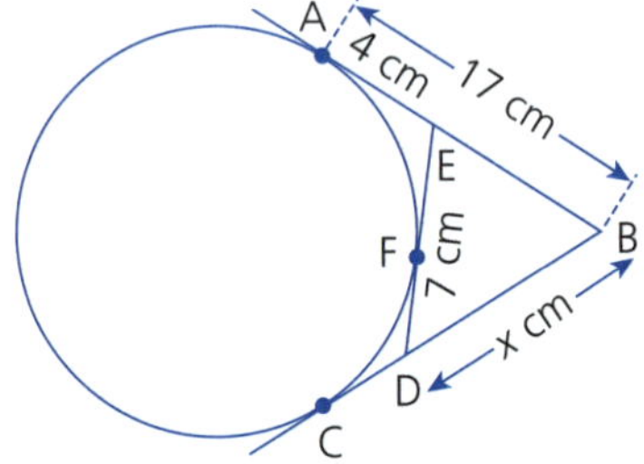

AB, BC and ED are tangents touching at A, C and F respectively.
AB = 17 cm, AE = 4 cm, ED = 7 cm, BD = x cm. Find x.

Draw clear diagrams. Mark all known information on the diagram. Use all the given data.

S Look for:

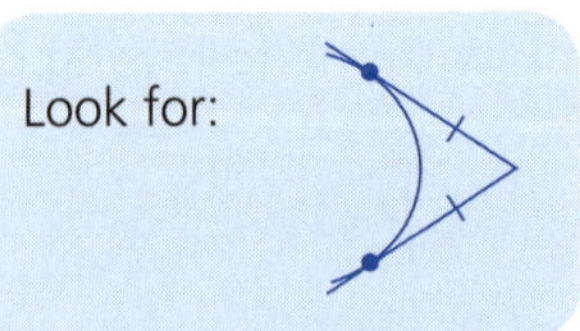

AB = BC (tangents from point B)
∴ BC = 17 cm
Also AE = EF (tangents from point E)
∴ EF = 4 cm
Then FD = 3 cm (ED = 7 cm)
But FD = DC (tangents from point D)
∴ DC = 3 cm
Then BD = 14 cm (BC = 17 cm)
∴ $x = 14$

More applications of Pythagoras' Theorem

Remember that whenever a right-angle occurs in a triangle, there are opportunities to use either Pythagoras' Theorem or trigonometry. Here Pythagoras' Theorem is considered.

1

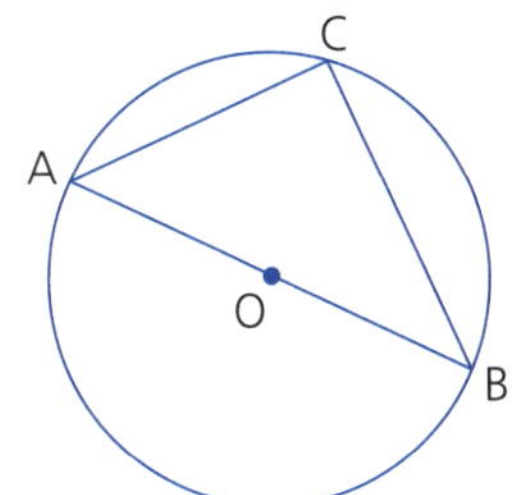

AB is a diameter of a circle, centre O, and radius 20 cm. Given BC = 32 cm, find the length of chord AC.

Let chord AC = x cm
Now AB = 40 cm (radius = 20 cm)
Also $\angle$ACB = 90° (angle in a semicircle)

Now, using Pythagoras' Theorem:

$$\begin{aligned} AB^2 &= BC^2 + AC^2 \\ 40^2 &= 32^2 + x^2 \\ \therefore x^2 &= 40^2 - 32^2 \\ &= 576 \\ \therefore x &= \sqrt{576} \\ &= 24 \end{aligned}$$

$\therefore$ AC is 24 cm long.

2

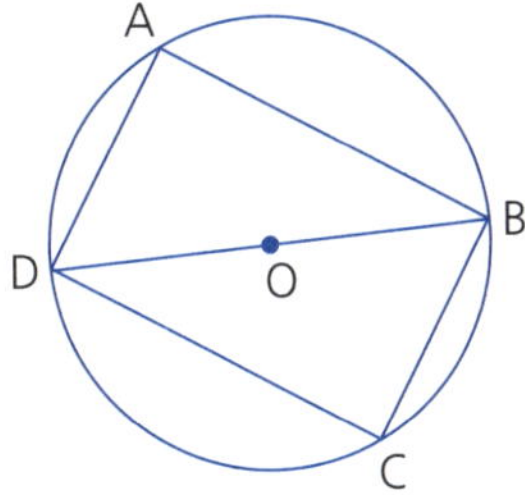

BD is diameter of circle, centre O, and radius 10 cm. Given AD = 12 cm, find the perimeter of ABCD.

$\angle$DAB is a right-angle.

By Pythagoras:

$$\begin{aligned} AB^2 &= 20^2 - 12^2 \\ &= 256 \\ AB &= \sqrt{256} \\ &= 16 \end{aligned}$$

$$\begin{aligned} \text{Perimeter of rectangle ABCD} &= 2(16 + 10) \\ &= 52 \end{aligned}$$

$\therefore$ Perimeter is 52 cm.

3

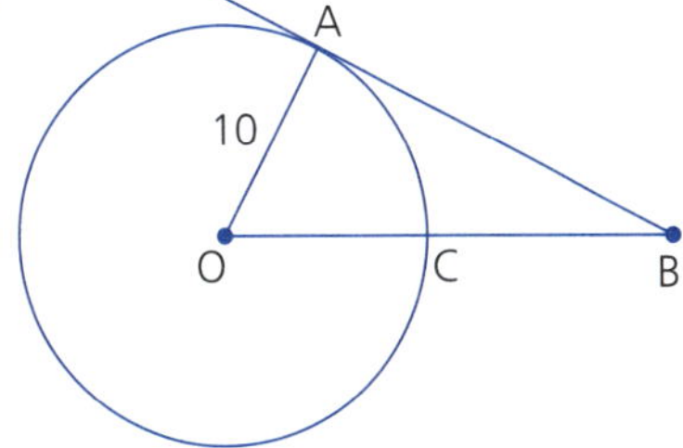

AB is a tangent to a circle, centre O, radius 10 cm. If AB is 24 cm, find the length of BC.

To find BC, we must first find BO.
$\angle$OAB = 90°
(angle between the radius and the tangent to the point of contact)

Let OB = y cm.

Using Pythagoras' Theorem in $\triangle$OAB:

$$\begin{aligned} OB^2 &= OA^2 + AB^2 \\ y^2 &= 10^2 + 24^2 \\ &= 676 \\ \therefore y &= \sqrt{676} \\ &= 26 \end{aligned}$$

OB = 26 cm

$$\begin{aligned} \text{Now } BC &= OB - OC \\ &= 26 - 10 \\ &= 16 \end{aligned}$$

(OC is a radius and is thus 10 cm.)
$\therefore$ BC is 16 cm long.

Ratio theorems (proved using similar triangles)

1 **The products of intercepts of intersecting chords are equal.** That is:

AE · EB = DE · EC

AE × EB = DE × EC

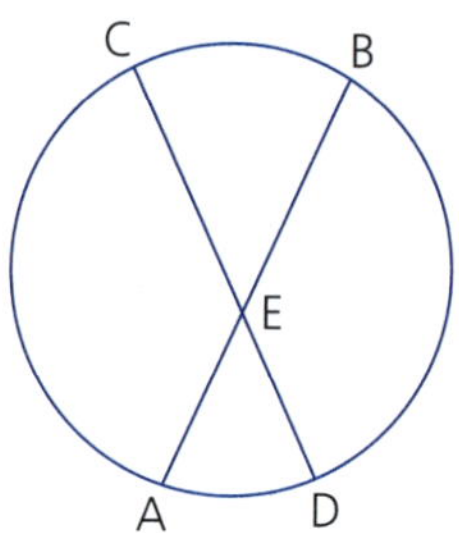

2 Here an extension of Example **1** concerning secants from an external point B is illustrated in the diagram.

This is a case of chords produced intersecting at an external point. Chords AD and CE are extended to meet at B. Then AB × BD = CB × BE.

(End point A to intersection B × intersection B to other end point D.)

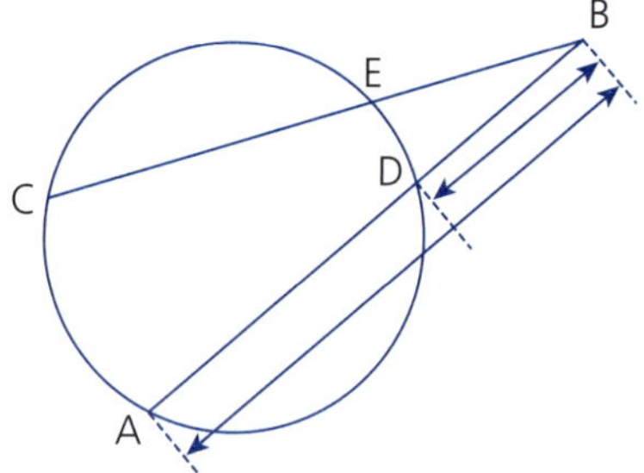

3 **The square of the length of the tangent is equal to the product of the intercepts of a secant drawn from an external point.**

PT^2 = QP × PR

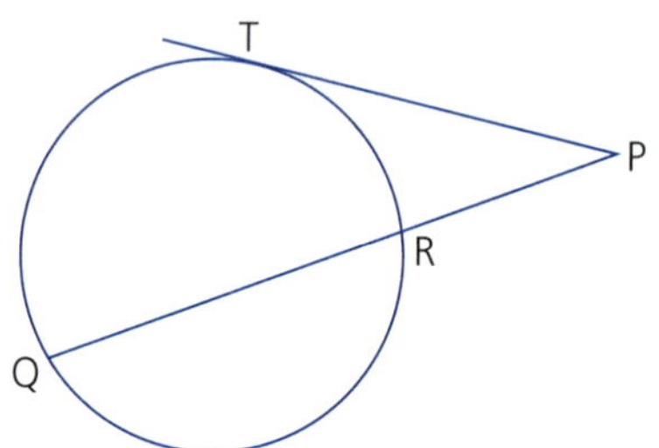

For Example

Find the value of the pronumeral in each diagram. All lengths are in cm.

1 AE = 8
EB = 10
DE = 12
EC = x

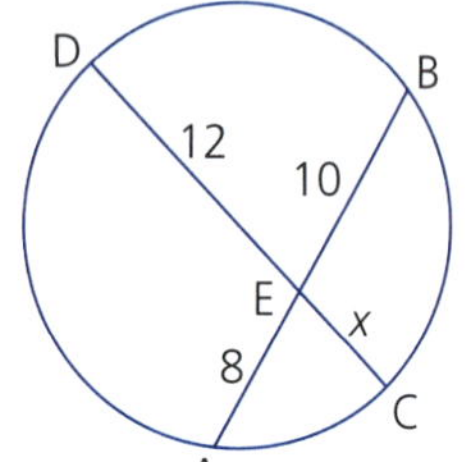

S $12 \times x = 8 \times 10$
(intercepts of intersecting chords)
$12x = 80$
$x = \frac{80}{12} = 6\frac{2}{3}$

2 XY = 25,
XP = 15,
PT = y.
Find the exact value.

Note: YP = 40

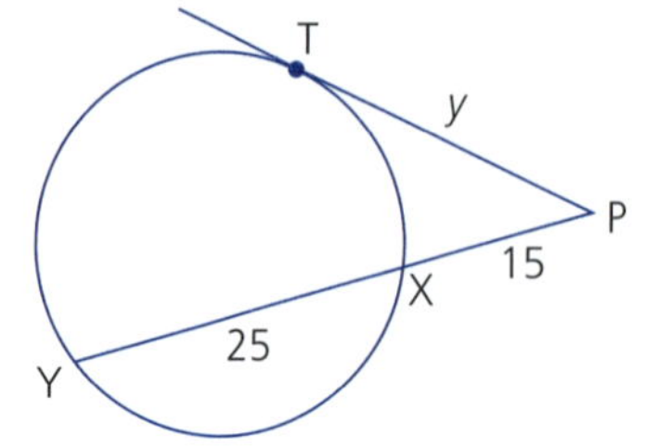

S $y^2 = 40 \times 15$
(product of intercepts of secant = square of tangent)
$= 600$
$\therefore y = \sqrt{600}$
$= 10\sqrt{6}$

3 AB = x
BC = 8
DE = 10
EC = 6

Note:
AC = $x + 8$
DC = 16

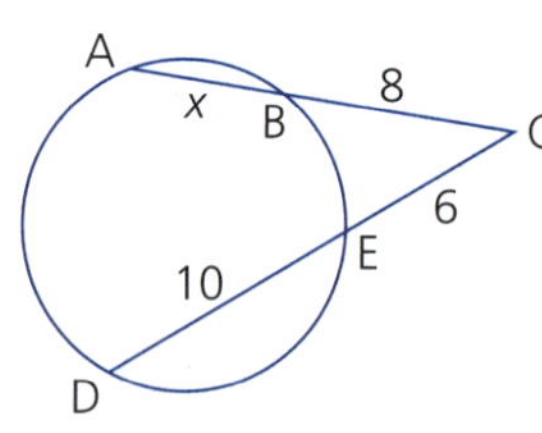

S $(x + 8) \times 8 = 16 \times 6$
(intercepts of intersecting chords)
$\therefore x + 8 = 12$
$x = 4$

4 AB = x, BP = 12, CD = 24, DP = y, and TP = 16.

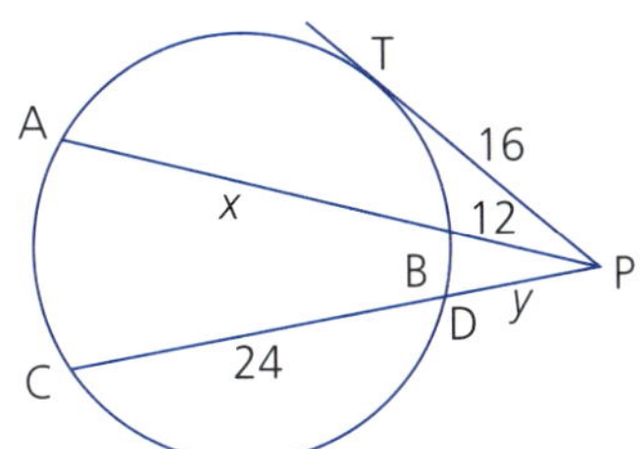

S $16^2 = (x + 12)12$
(product of intercepts of secant = square of the tangent)

$$\frac{256}{12} = x + 12$$

$$21\frac{1}{3} = x + 12$$

$$\therefore x = 9\frac{1}{3}$$

$$16^2 = (y + 24)y$$

(product of intercepts of secant = square of the tangent)

$$\therefore 256 = y^2 + 24y,$$

That is $y^2 + 24y - 256 = 0$

$$(y + 32)(y - 8) = 0$$

$$\therefore y = -32 \text{ or } 8$$

That is $y = 8$

Length must be positive.

Angle in the alternate segment

1 PT is a tangent at T. AT is a chord drawn to the point of contact. ∠TSA is the angle at the circumference subtended by chord AT.

∠TSA is the angle in the alternate segment relative to ∠PTA.

2 **An angle formed by a tangent to a circle with a chord drawn to the point of contact is equal to any angle in the alternate segment.**

From the diagram, ∠PTA = ∠TSA.

Also note that ∠RTS = ∠TAS for the same reason (tangent RT, chord TS).

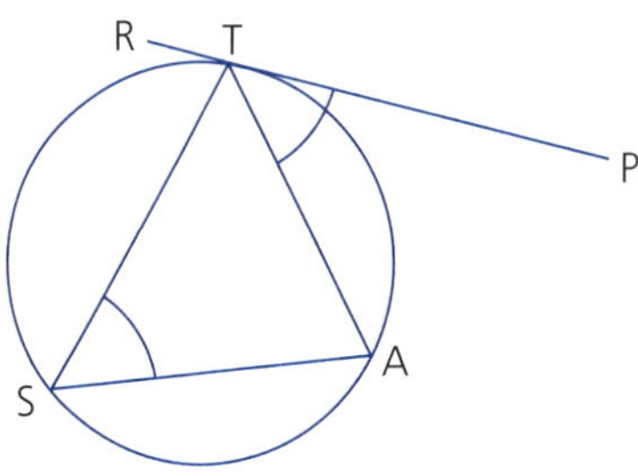

Circles that touch

Intersection of two circles

When two circles intersect, the line joining their centres bisects their common chord at right-angles.

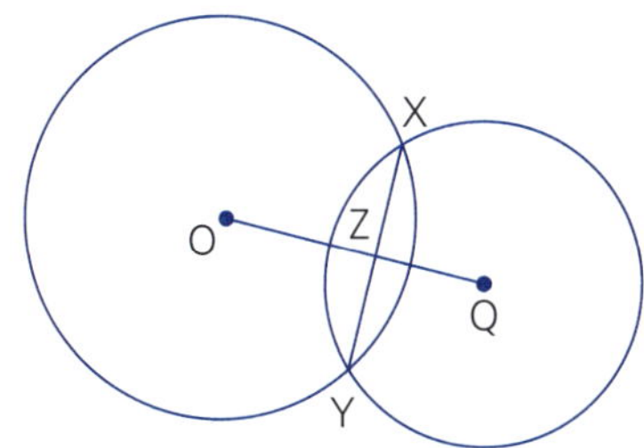

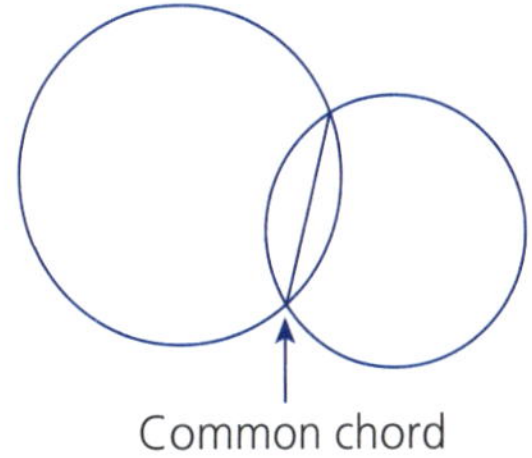

Two circles, centres O and Q, intersect at X and Y.
The line joining the centres OQ intersect the common chord XY at Z.
Then XZ = ZY and XY ⊥ OQ.

This is proven using congruent triangles.

Point of contact of two circles

When two circles touch, their centres and the point of contact are collinear.

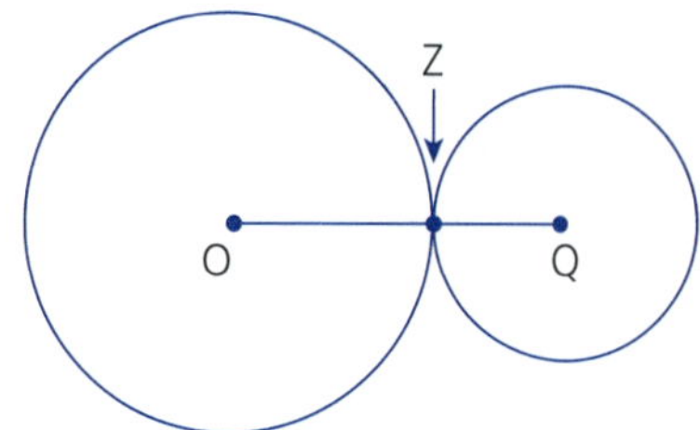

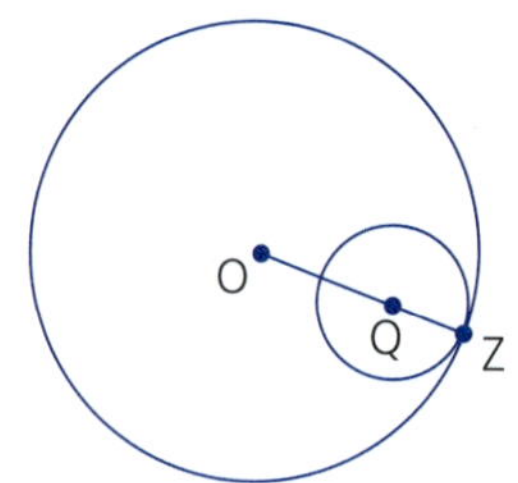

Two circles with centres O and Q touch at Z.
OQ is a straight line and Z lies on OQ (or OQ produced), that is, O, Q and Z are collinear.

This is proved by drawing the common tangent at Z.

Worked examples using deductive reasoning

For worked examples using deductive reasoning, use the following guidelines:

- Always draw a clear diagram.
- Mark on the diagram all the given information.
- Work from the known towards the unknown.
- If you become lost, check that you have used all the data.
- Any known facts can be used—you cannot use the fact that you are asked to prove.

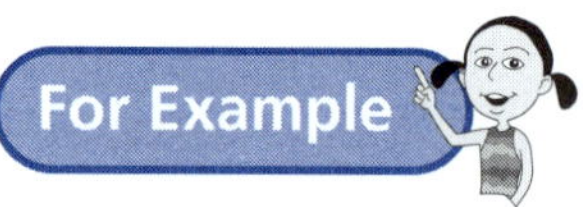

In all diagrams, O is the centre of the circle.

1 $\angle OAB = 46°$.
Find $\angle ACB$.

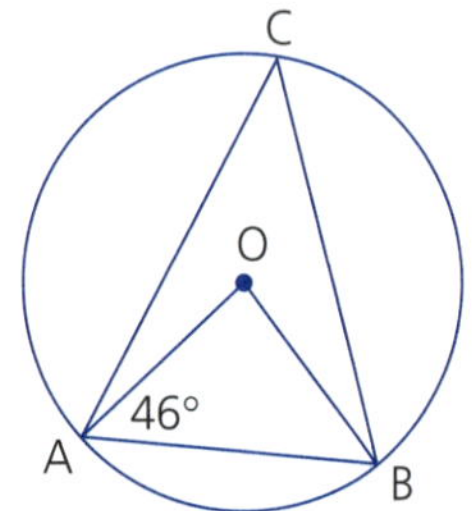

AO = OB (equal radii)
$\therefore \angle ABO = 46°$ (base angles of isosceles $\triangle$)
$\therefore \angle AOB = 180° - (46° + 46°)$
$= 88°$ (angle sum of $\triangle$)
$\therefore \angle ACB = 44°$
(angle at circumference is half $\angle$ at centre on arc AB)

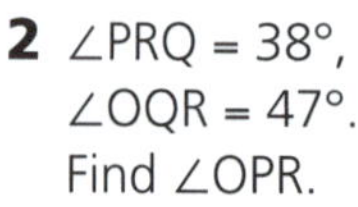

2 $\angle PRQ = 38°$,
$\angle OQR = 47°$.
Find $\angle OPR$.

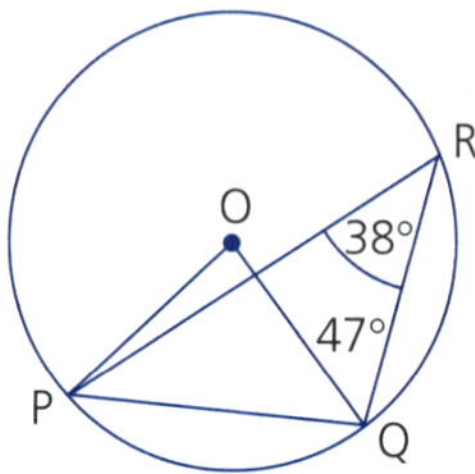

$\angle POQ = 76°$
($\angle$ at centre = is twice $\angle$ at circumference on arc PQ)
But OP = OQ (equal radii)
$\therefore \angle OPQ = \angle OQP$
(base angles, isosceles $\triangle$)
$\angle OQP = \angle OPQ$
$= \frac{1}{2}(180° - 76°)$
$= 52°$
Then $\angle PQR = \angle OQP + \angle OQR$
$= 52° + 47°$
$= 99°$
Then $\angle RPQ = 43°$ ($\angle$ sum of $\triangle PRQ$)
But $\angle OPR = \angle OPQ - \angle RPQ$
$= 52° - 43°$
$= 9°$

3 AC ∥ DB.
Prove that ∠ACE = ∠CAE.

S Let ∠ACE = $\theta°$
∴ ∠EDB = $\theta°$
(alternate angles, AC ∥ DB)
and ∠CAB = $\theta°$
(angles in the same segment on arc CB)
Then ∠ACE = ∠CAE (both $\theta°$).

4 Given that AB, CD are diameters, ∠CAB = 49°. Find ∠ADC.

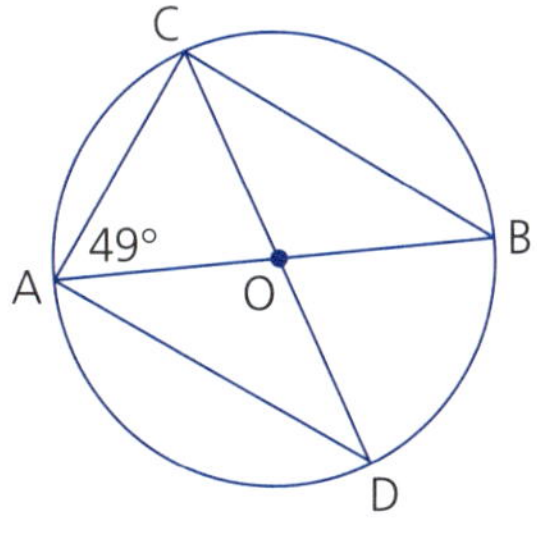

S AO = OC (equal radii)
∠ACO = 49°
(base angles of isosceles △)
But ∠ACB = 90° (angle in a semicircle)
∴ ∠OCB = 41° (90° – 49°)
Also OC = OB (equal radii)
∴ ∠OBC = 41°
(base angles of isosceles △)
and ∠ADC = 41°
(angles in the same segment)

5 Given that AB ∥ DC and ABCD is a cyclic quadrilateral, prove that ∠ADC = ∠BCD.

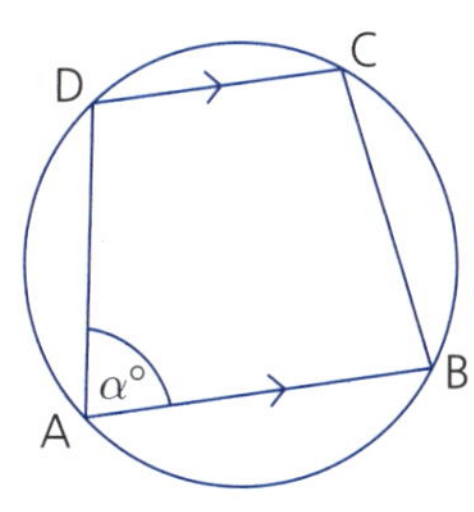

S Let ∠DAB = $\alpha°$
∴ ∠ADC = $(180 - \alpha)°$
(co-interior angles, AB ∥ DC)
and ∠DCB = $(180 - \alpha)°$
(opposite angles of a cyclic quadrilateral)
∴ ∠ADC = ∠BCD [both $(180 - \alpha)°$]

6 Given that AF ∥ BE, and ABEF is a cyclic quadrilateral, prove that ∠CBE = ∠DEB.

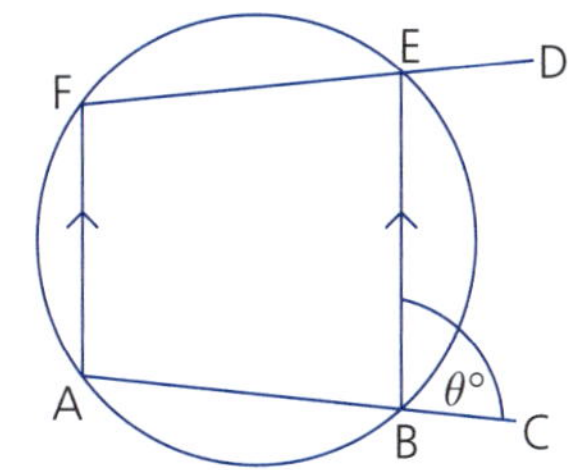

S Let ∠EBC = $\theta°$
∴ ∠FAB = $\theta°$
(corresponding angles, AF ∥ BE)
and ∠DEB = $\theta°$
(external ∠ of a cyclic quadrilateral)
∴ ∠CBE = ∠DEB (both $\theta°$)

7 Given that AB is tangent at C and ∠ACD = ∠BCE, prove that DE ∥ AB.

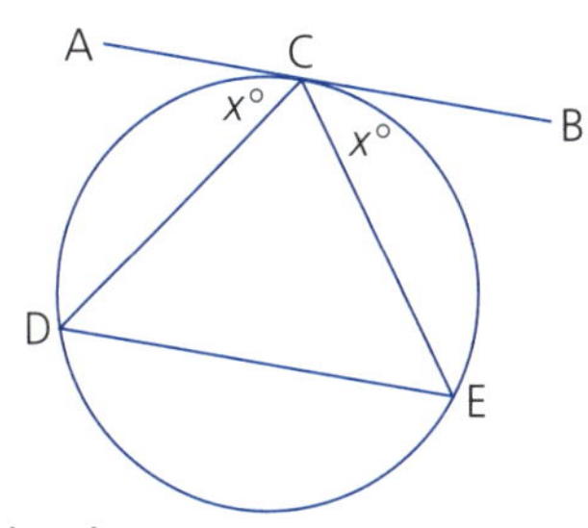

S Let ∠ACD = $x°$
Then ∠BCE = $x°$ (data)
But ∠BCE = ∠CDE
(angle in the alternate segment)
∴ ∠CDE = $x°$
Then ∠ACD = ∠CDE (both $x°$)
∴ AB ∥ DE
(a pair of alternate angles are equal)

8 Given that TR is a tangent at S, prove that ∠POS = 2 × ∠PST.

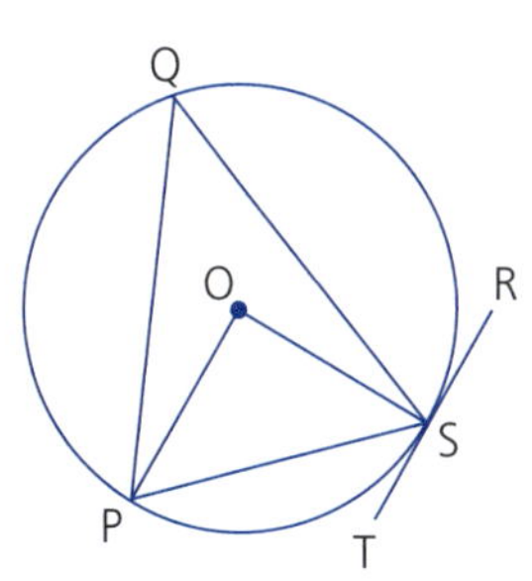

S Let ∠PST = $\theta°$
Now ∠PQS = ∠PST
(angle in alternate segment)
∴ ∠PQS = $\theta°$
But ∠POS = 2 × ∠PQS = $2\theta°$
(angle at centre is twice angle at circum. on arc PS)
∴ ∠POS = 2 × ∠PST

9 Given that ABCD is a cyclic quadrilateral and ∠CAB = $a°$, ∠ADB = $b°$, ∠DBC = $c°$, ∠ACD = $e°$, prove that $a + b + c + e = 180°$.

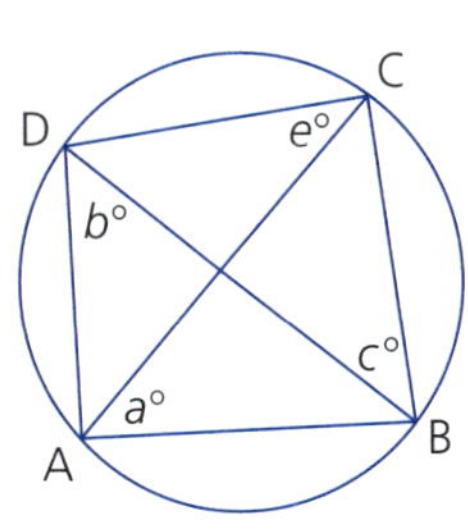

S ∠BDC = $a°$
(angles in same segment)
and ∠ABD = $e°$
(angles in same segment)
But ∠ADC + ∠ABC = 180°
(opposite angles of a cyclic quadrilateral)
∴ $a + b + c + e = 180$

Summary of circle properties

Chord properties

1 Chords of equal length in a circle subtend equal angles at the centre and are equidistant from the centre.

2 The perpendicular from the centre of a circle to a chord bisects the chord.

3 Conversely, the line from the centre of a circle to the midpoint of a chord is perpendicular to the chord.

4 The perpendicular bisector of a chord of a circle passes through the centre.

5 Given any three non-collinear points, the point of intersection of the perpendicular bisectors of any two sides of the triangle formed by the three points is the centre of the circle through all three points.

6 When two circles intersect, the line joining their centres bisects their common chord at right-angles.

Angle properties

1 The angle at the centre of a circle is twice the angle at the circumference standing on the same arc.

2 The angle in a semicircle is a right-angle.

3 Angles at the circumference, standing on the same arc, are equal.

4 The opposite angles of cyclic quadrilaterals are supplementary.

5 An exterior angle at a vertex of a cyclic quadrilateral is equal to the interior opposite angle.

Tangent and secant properties

1 The angle between a tangent and a radius drawn to the point of contact is a right-angle.

2 The two tangents drawn to a circle from an external point are equal in length.

3 The angle between a tangent and a chord drawn to the point of contact is equal to the angle in the alternate segment.

4 When two circles touch, their centres and the point of contact are collinear.

5 The products of the intercepts of two intersecting chords of a circle are equal.

6 The products of the intercepts of two intersecting secants to a circle from an external point are equal.

7 The square of a tangent to a circle from an external point equals the product of the intercepts of any secants from the point.

Practise Practise

Note: when O is used it always indicates the centre of the circle.

1

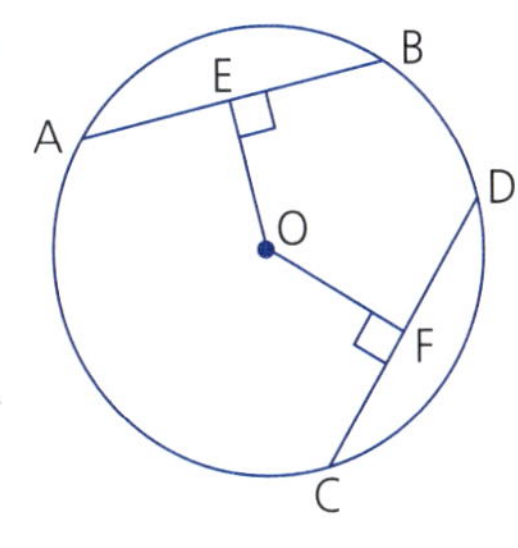

Prove that equal chords are equidistant from the centre, that is, given AB = CD, prove that OE = OF, where OE ⊥ AB and OF ⊥ CD. pp. 263–266

2 Find the value of the pronumeral in the following questions. (All lengths are in cm.) Give adequate reasons.

a Given: RQ = 7 cm

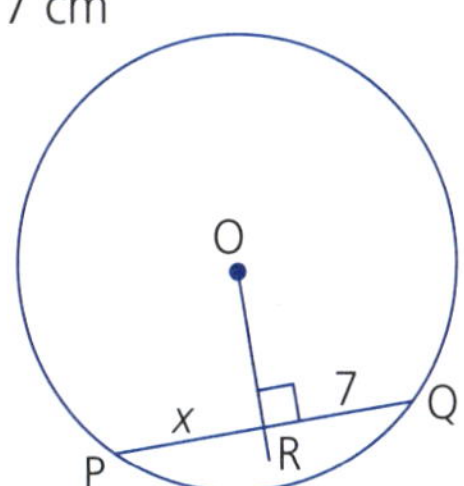

b Given: RT = TS

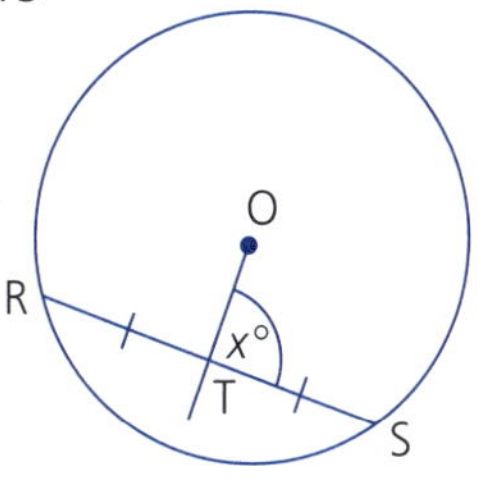

c Given: ∠AOB = ∠COD, AB = 8.4 cm

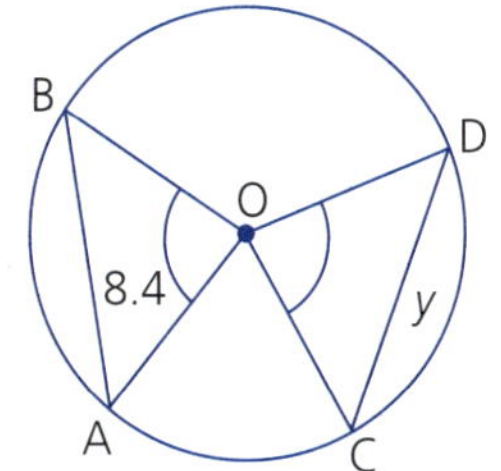

d Given: PQ = RS, ∠POQ = 110°, ∠ROS = θ°

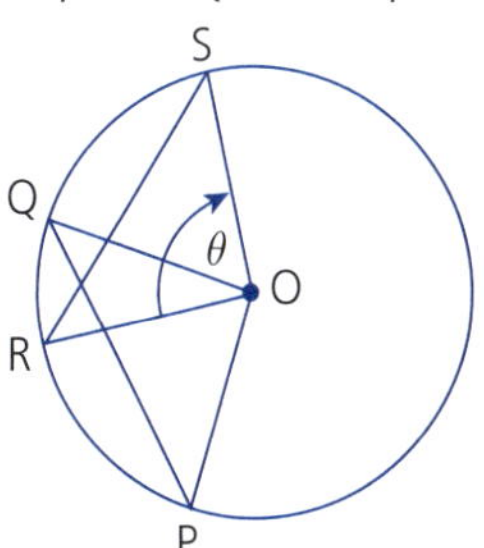

pp. 263–266

3 Use properties of the circle and Pythagoras' Theorem in the following questions:

a AB is a 16 cm chord of a circle with radius 10 cm. Find the length OC if OC ⊥ AB.

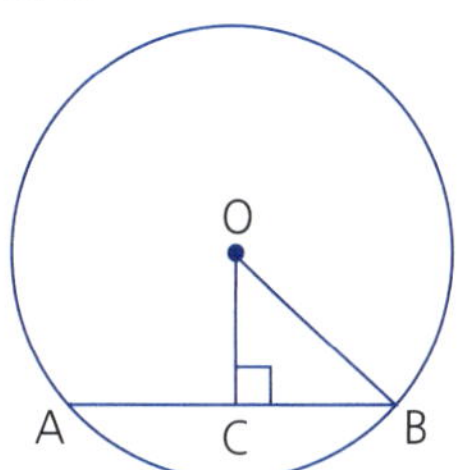

b Given that AM = MB, OM = 10 cm, and the circle has radius 26 cm, find the length of chord AB.

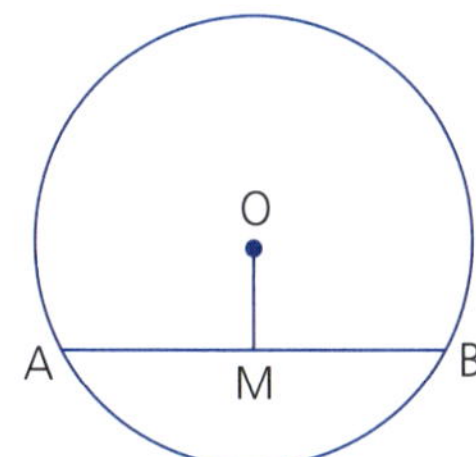

c A chord 15 cm long is 4 cm from the centre of a circle. Calculate the radius of the circle correct to one decimal place.

d Two parallel chords 32 cm long and 24 cm long are drawn in a circle of radius 20 cm as shown in the figure. Find the distance between the two chords if they are both on the same side of the centre.

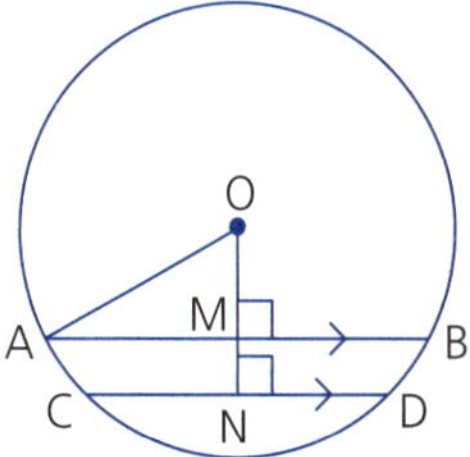

Given AB || CD, AB = 32 cm, CD = 24 cm, OA = 20 cm, find the length MN. pp. 263–266

4 Given that O is the centre of the circle, prove that $\angle AOB = 2 \times \angle ACB$.

Construction: join CO and produce it to X. Let $\angle OAC = x°$, $\angle OBC = y°$ and remember that radii are equal.

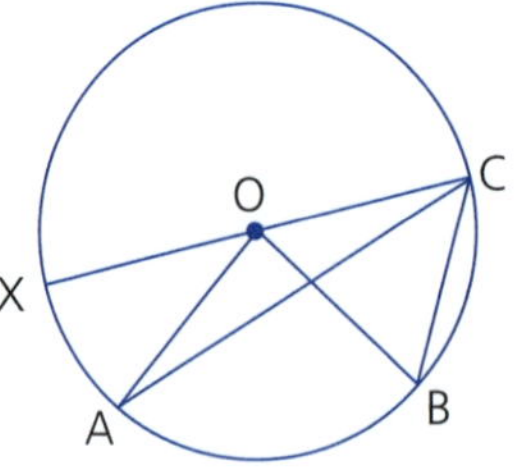

pp. 266–267

5 Given that PQRS is a cyclic quadrilateral, prove that $\angle SPQ = \angle QRT$.

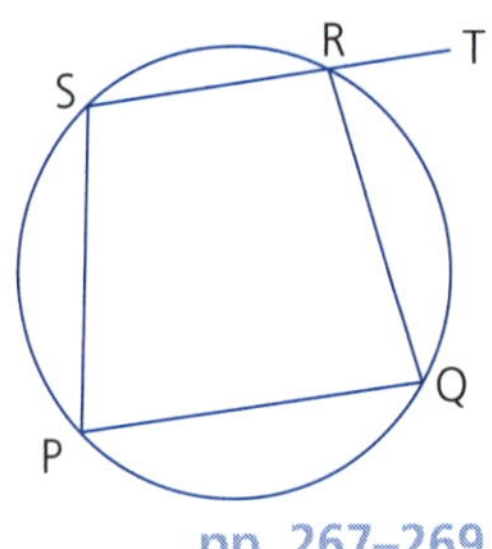

pp. 267–269

6 Find the value of the pronumeral(s) in each of the following diagrams, giving adequate reasons in each case.

a Given $\angle AOB = 86°$, $\angle ACB = a°$.

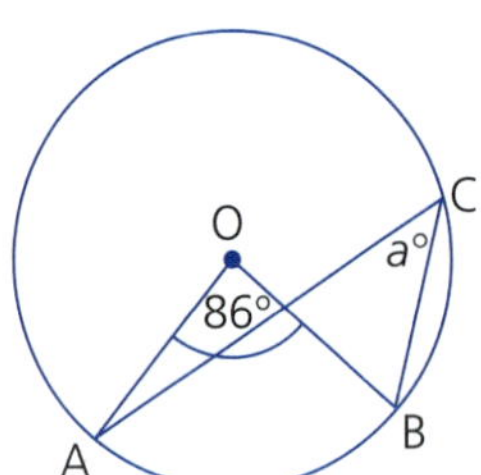

b Given PQ is a diameter, $\angle RPQ = 52°$, $\angle RQP = b°$.

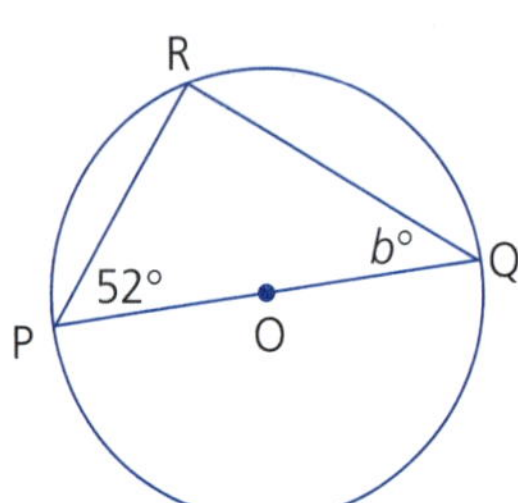

c Given $\angle XAY = 38°$, $\angle XZY = y°$.

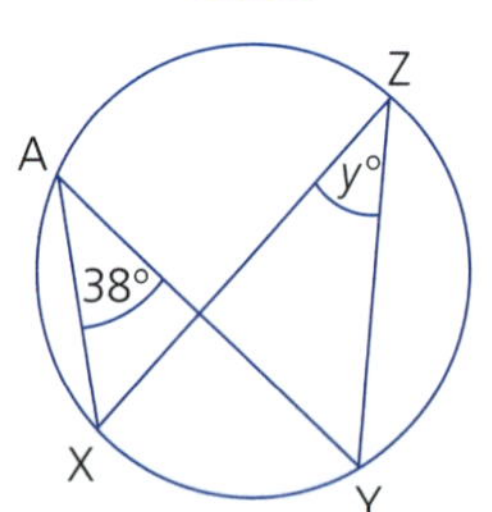

d Given $\angle ADB = 42°$, $\angle DBC = 37°$, $\angle DAC = x°$, $\angle DTA = y°$.

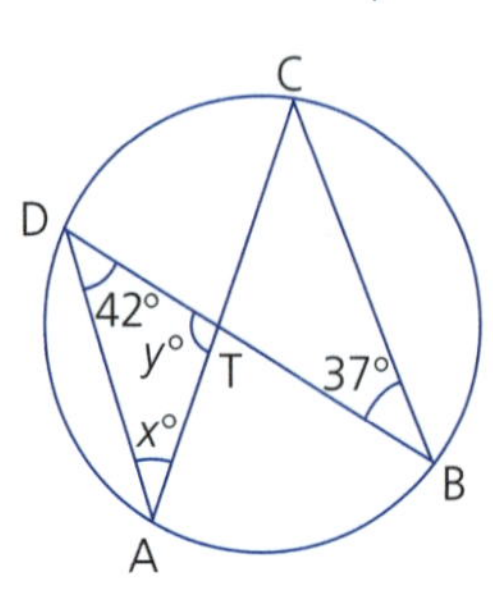

e Given $\angle DAB = 84°$, $\angle DCB = g°$.

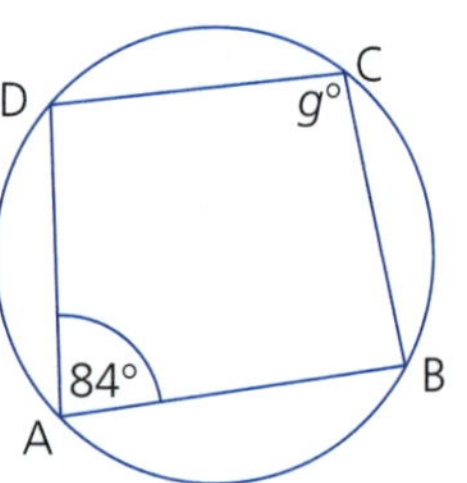

f $\angle CBE = 114°$, $\angle ADC = h°$.

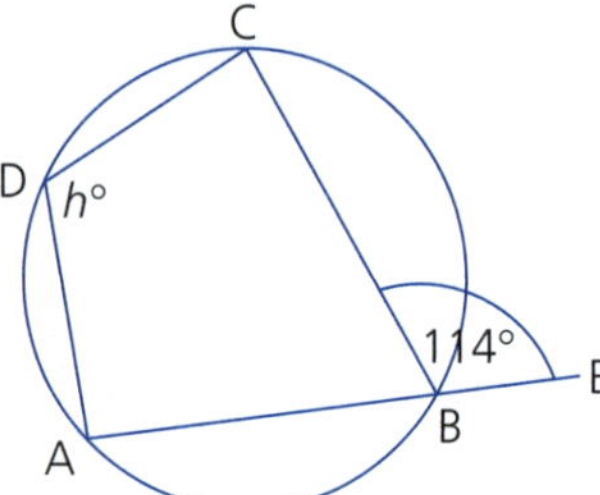

g AD || BC, $\angle CBE = 114°$, $\angle BAD = a°$, $\angle ADC = b°$, $\angle DCB = c°$.

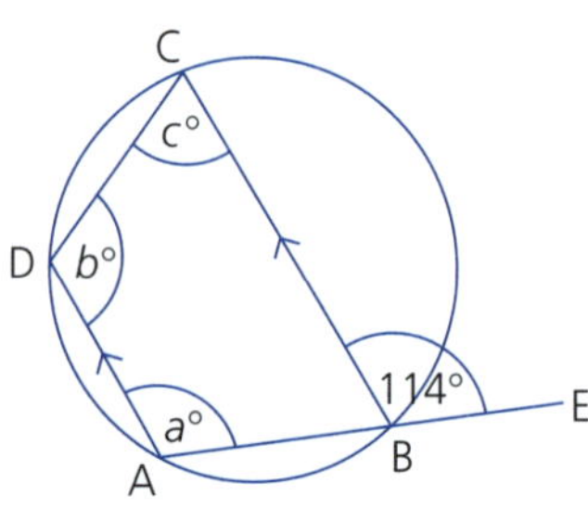

h AC is a diameter, $\angle BCO = 48°$, $\angle OBC = y°$, $\angle ABO = x°$, $\angle BAC = z°$.

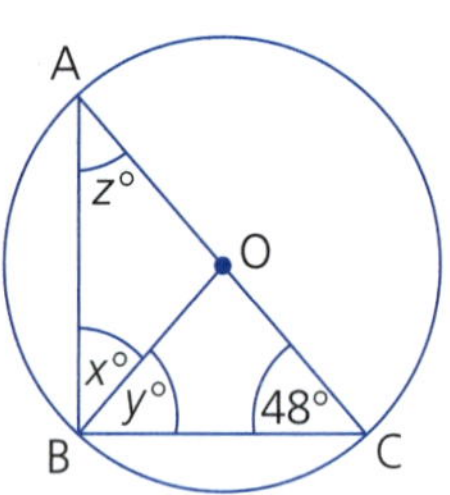

i $\angle AOC = 138°$, $\angle ABC = \theta°$.

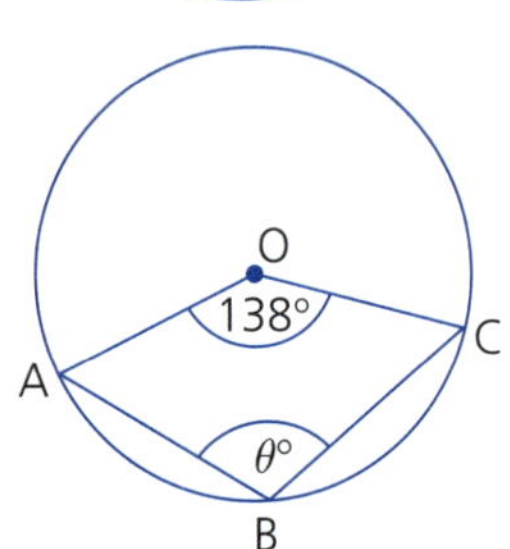

j $\angle PQR = 88°$, $\angle POR = \phi°$.

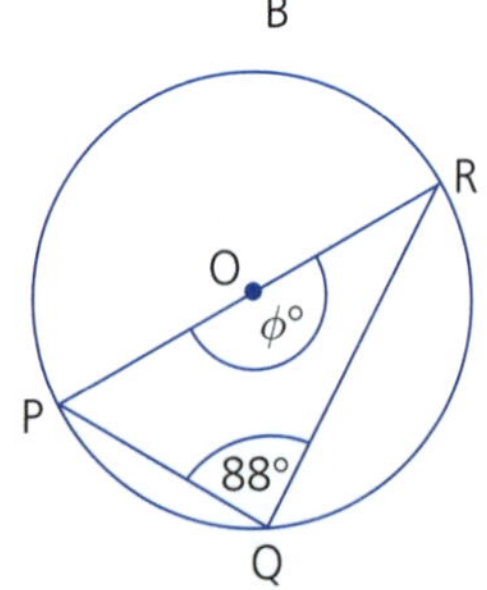

k AB || OD, $\angle ABC = 124°$, $\angle AOC = x°$, $\angle BCD = y°$, $\angle BAO = z°$.

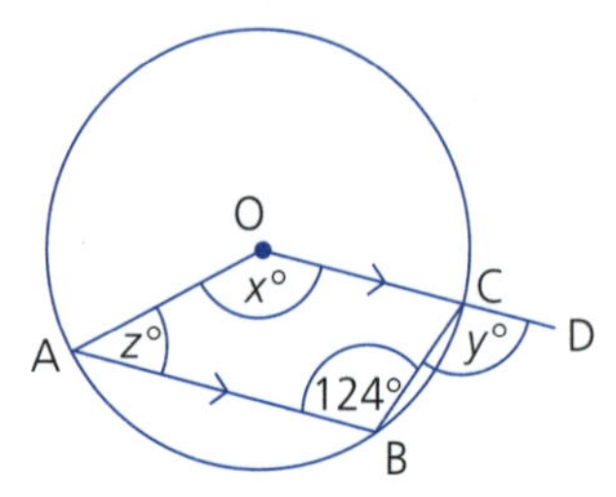

l ∠PSR = 48°,
∠POR = a°,
∠PQR = b°.

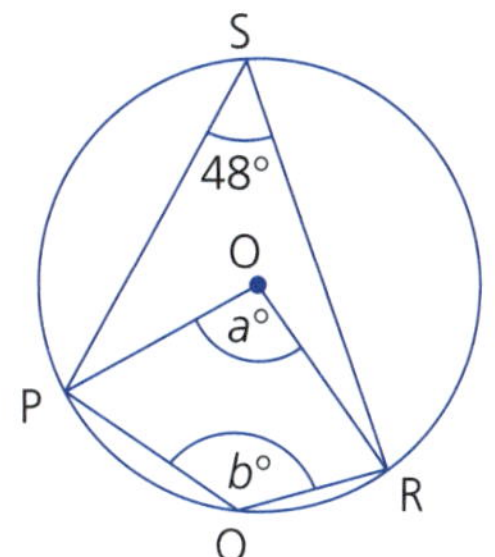

m AD ∥ BC,
∠DAC = 42°,
∠DBC = x°,
∠ADB = y°,
∠CTB = z°.

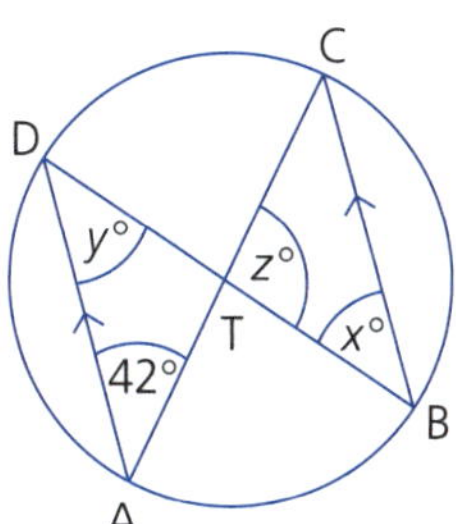

n ST is a tangent at T,
∠ABT = 44°,
∠ATS = y°.

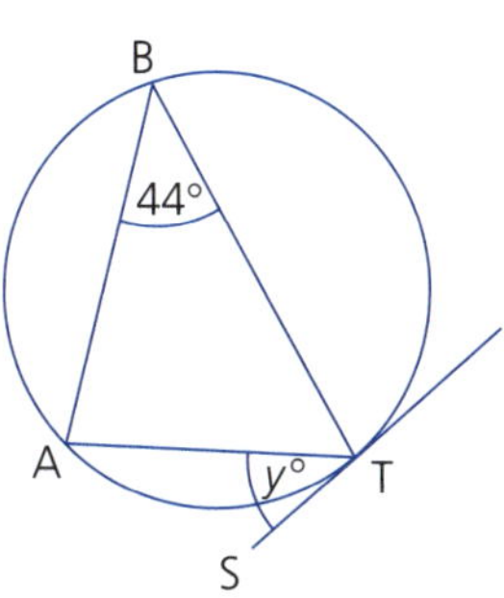

o ST is a tangent at T,
∠ATS = 52°,
∠TBA = x°,
∠TOA = y°.

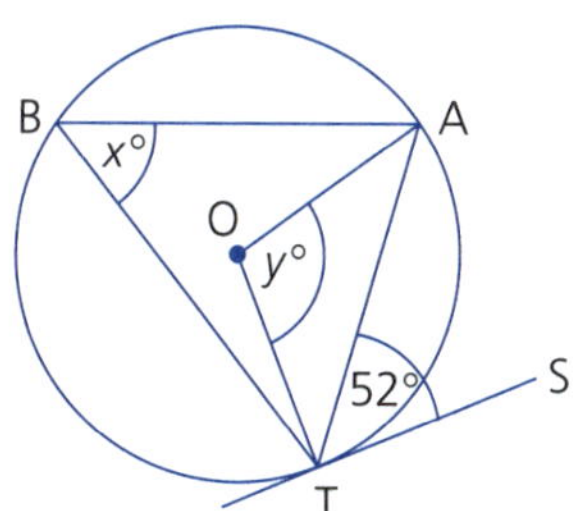

p AS and AT are tangents at S and T,
∠SAT = 82°,
∠STA = a°.

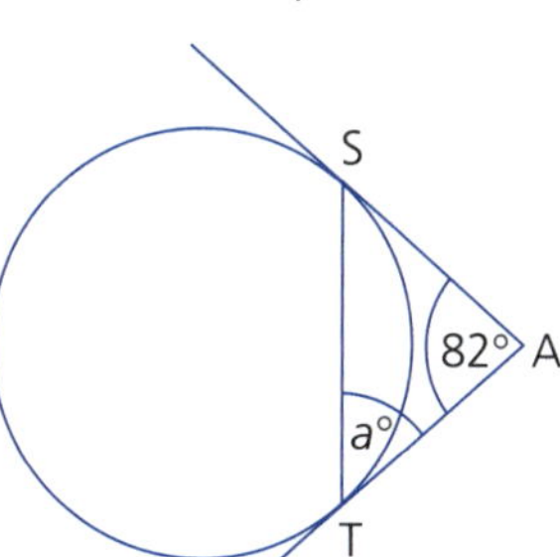

q ST is a tangent at T,
∠ATS = 46°,
∠OTA = x°,
∠TOA = y°.

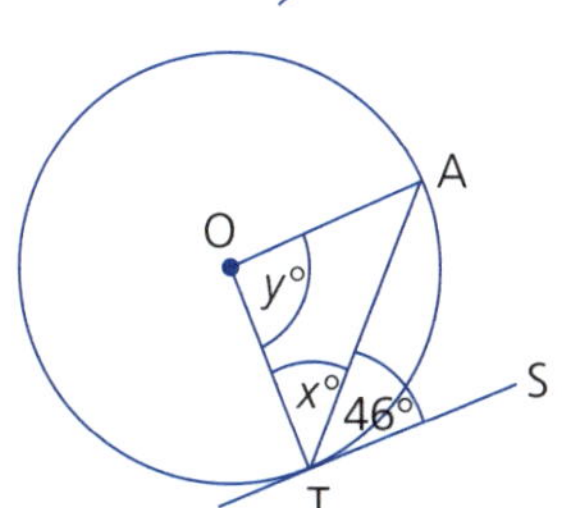

r ST is a tangent at T,
∠TBA = 38°,
∠ATS = x°,
∠OAT = y°.

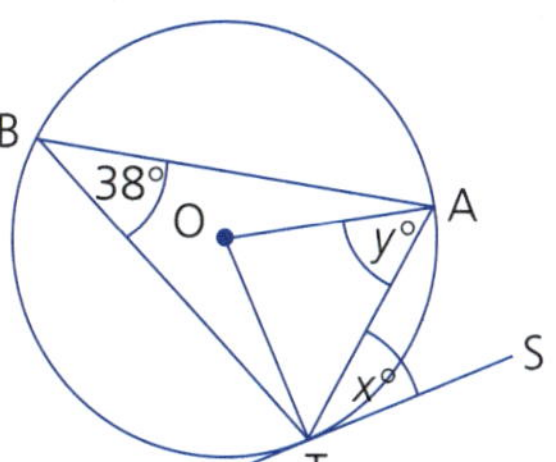

s PT is a tangent at T,
∠OPT = 26°,
∠TOP = ϕ°.

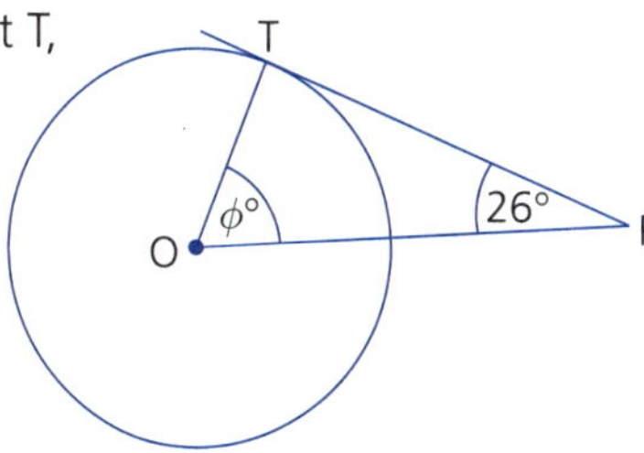

t

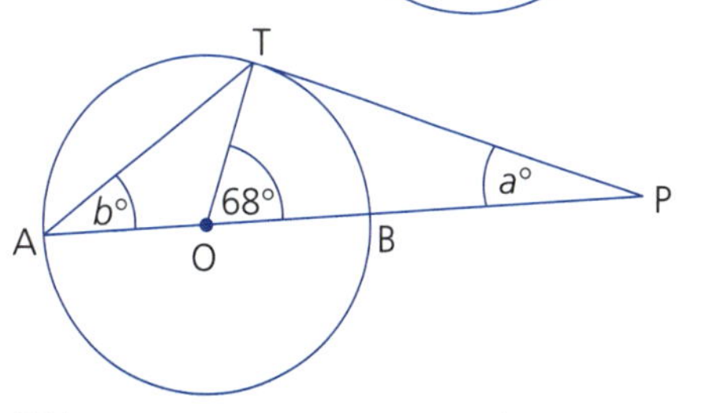

PT is a tangent at T, AB is a diameter,
∠TOB = 68°, ∠TPB = a°, ∠TAO = b°.

pp. 266–273

7 Find the value of h in each diagram.
All measurements are in cm.

a AB is the diameter of a circle of radius 10 cm, while AC = 16 cm, and BC = h cm.

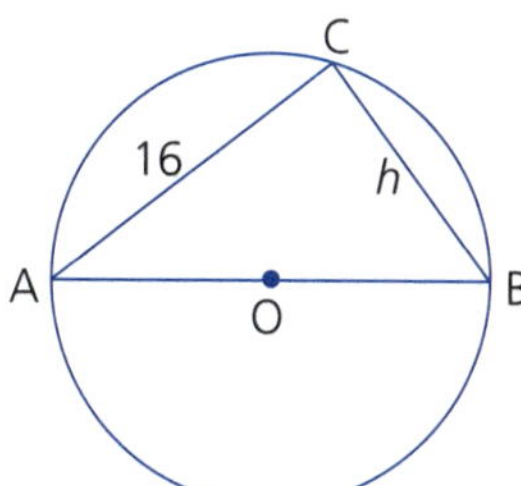

b

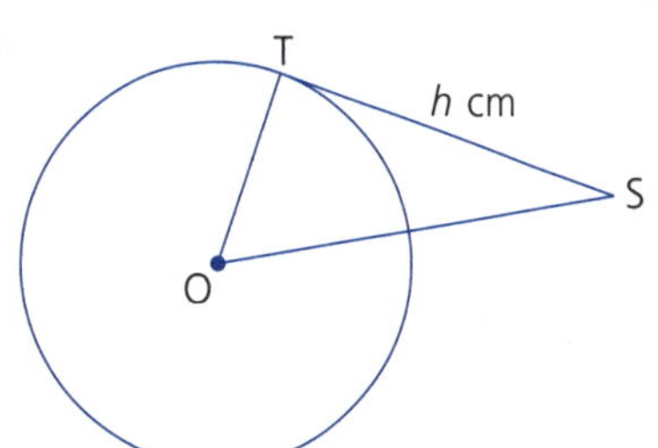

ST is a tangent at T to the circle of radius 15 cm.
If OS is 25 cm, calculate h.

c

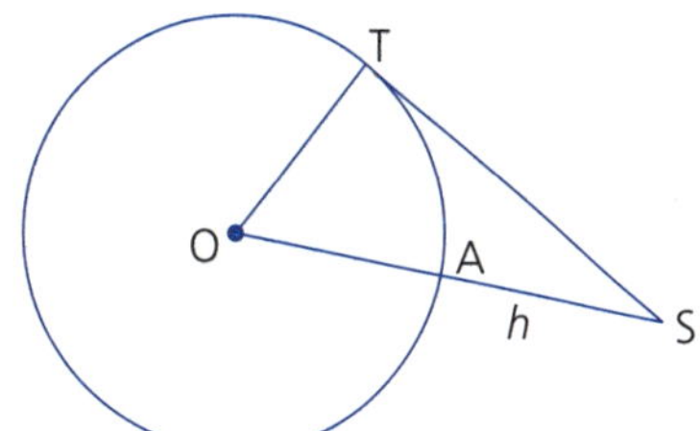

ST is a tangent at T to the circle of radius 2.7 cm.
Given that ST = 3.6 cm and AS = h cm, calculate h.

d

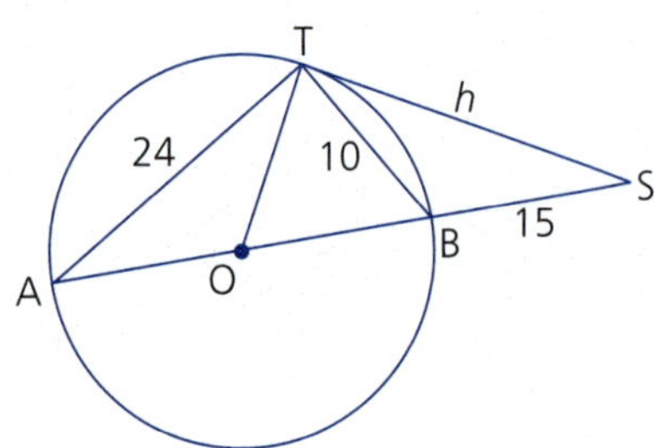

ST is a tangent at T. AT = 24 cm, BT = 10 cm, BS = 15 cm and ST = h cm. Calculate h correct to one decimal place. p. 271

8 Calculate the value of the pronumeral in each diagram. All lengths are in cm. Answer correct to one decimal place where necessary.

a AE = 5 cm,
EB = 8 cm,
ED = 4 cm,
EC = x cm.

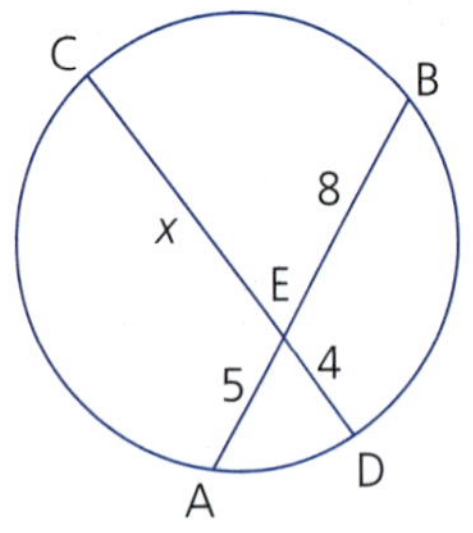

b AB = 7 cm,
BS = 8 cm,
TS = t cm.

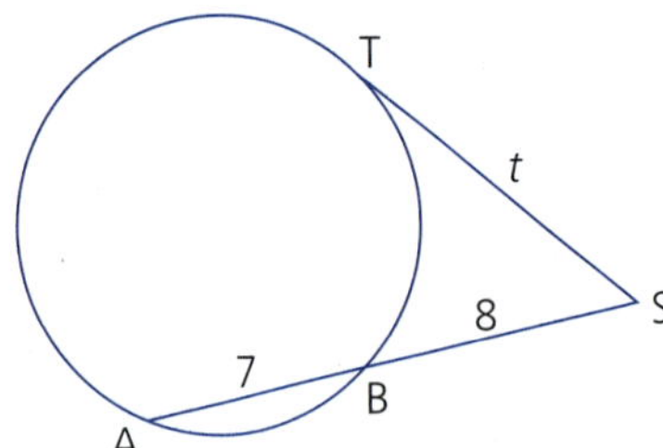

c BC = 6 cm,
DC = 8 cm,
ED = 10 cm,
AB = y cm.

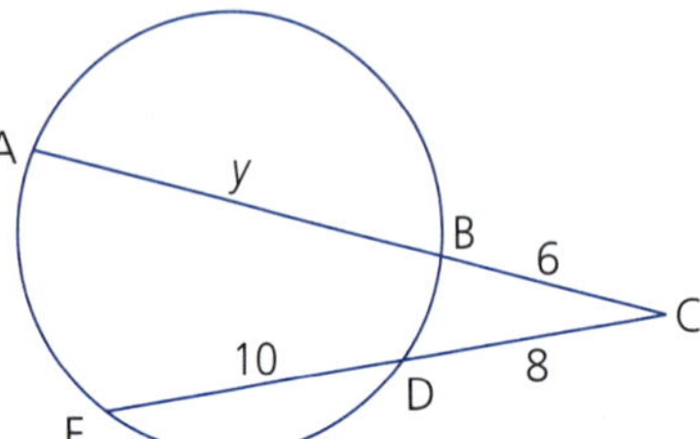

d ST = 12 cm,
SB = 8 cm,
AB = a cm,
CD = 7 cm,
DS = y cm.

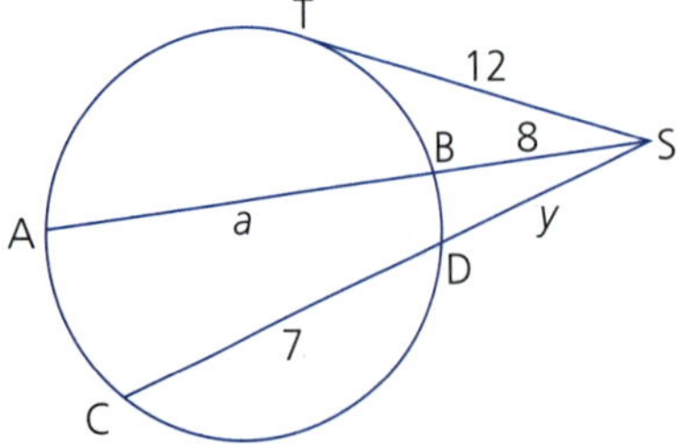

e ST is a tangent at T.
ED = 5 cm,
EA = 3 cm,
EC = 9 cm,
EB = x cm,
BS = 18 cm,
TS = t cm.
(Leave t in exact form.)

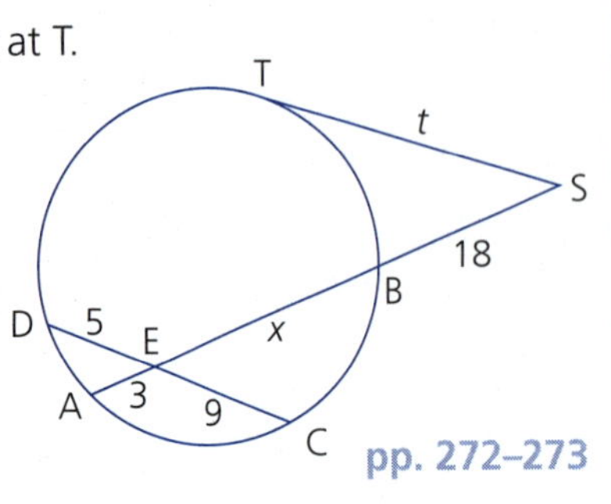

pp. 272–273

9 Answer each part of this question with a logical sequence of steps, giving adequate reasons. The point O is always the centre of the circle.

a Given ∠QOP = 74°,
∠OPR = 28°,
find ∠RQO.

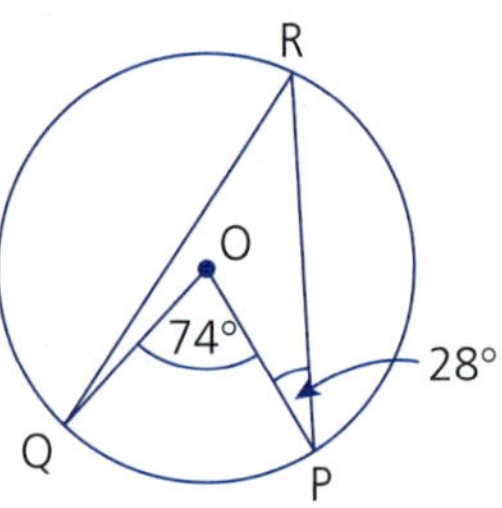

b Given AB = AC,
∠BOC = 116°,
find ∠ABO.

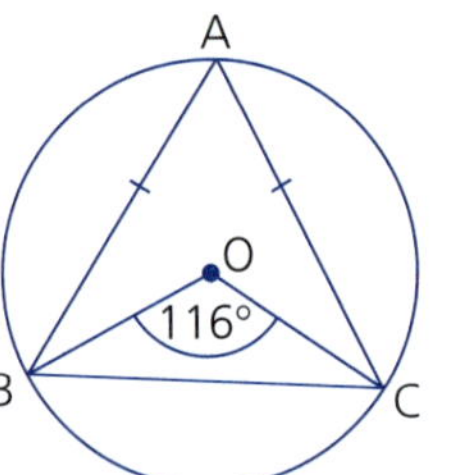

c Given ∠POR = 142°,
∠POQ = 94°,
find ∠ORQ.

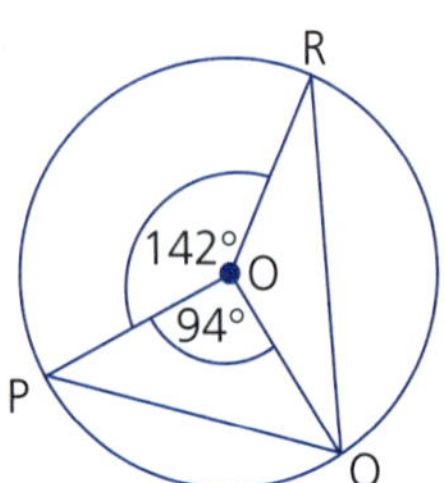

d Given ∠OPR = ∠QPR,
∠SRO = θ°,
prove ∠ROS = 2θ°.

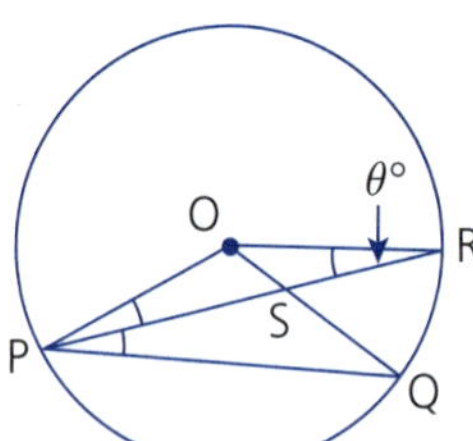

e Given that ∠OXZ = θ°,
and ∠XYZ = θ°,
prove that ∠XOZ = 90°.

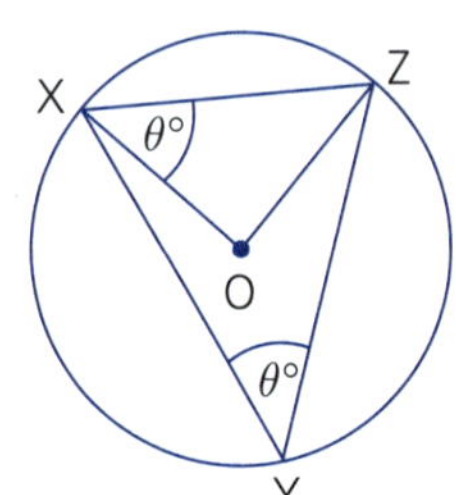

f Given DO ∥ BC,
∠DAB = 118°,
find x.

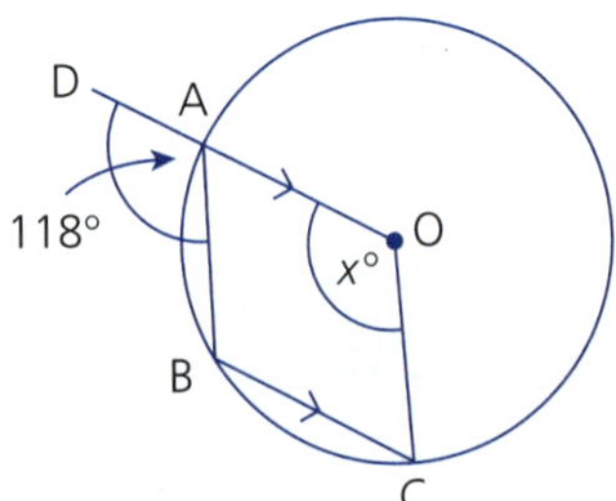

g Given ∠XAY = 47°, ∠YPZ = 90°, find θ.

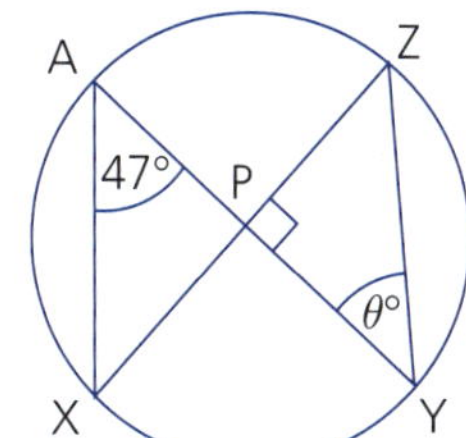

h PQRS is a parallelogram, and PQRT is a cyclic quadrilateral. Prove that PT = PS. (Call ∠PQR = α°.)

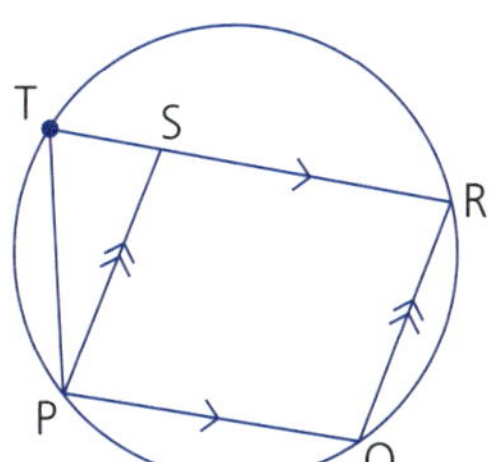

i Given ∠BAE = 27°, ∠DCA = 36°, find ∠DBC.

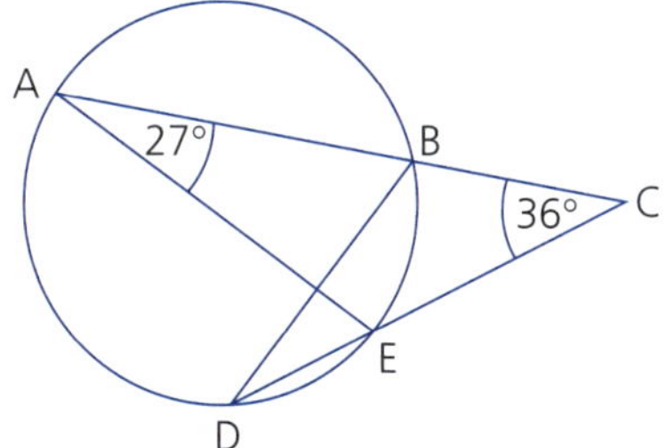

j Given PR = PQ, ∠PQS = x°, ∠SQR = y°, prove that ∠STR = x°.

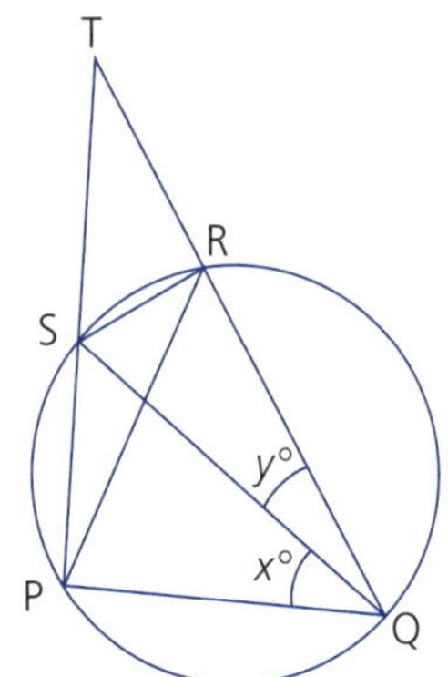

k Given ∠TPS = 94°, ∠RQU = 92°, ∠QSR = 48°, find ∠SVR.

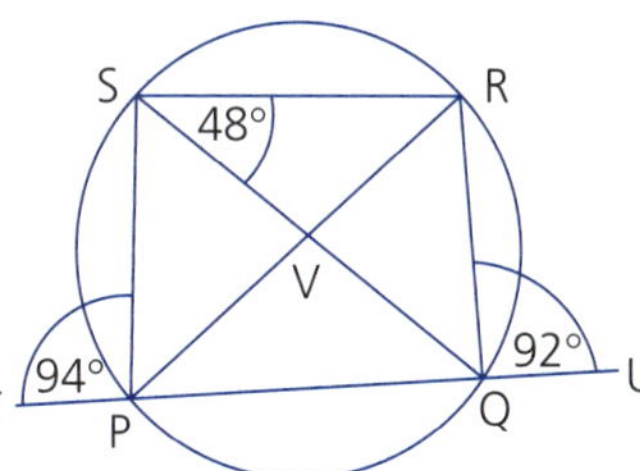

l

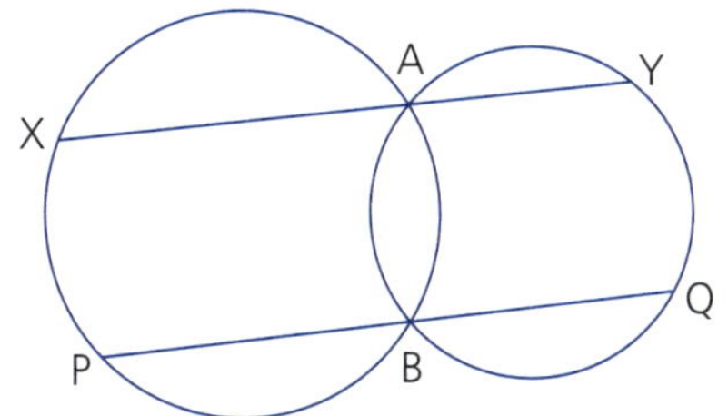

Circles intersect at A and B. XY and PQ are straight lines. Prove that PX || QY.

m Given that PT is a tangent at T and XP ⊥ PT, prove that ∠OTX = ∠TXP.

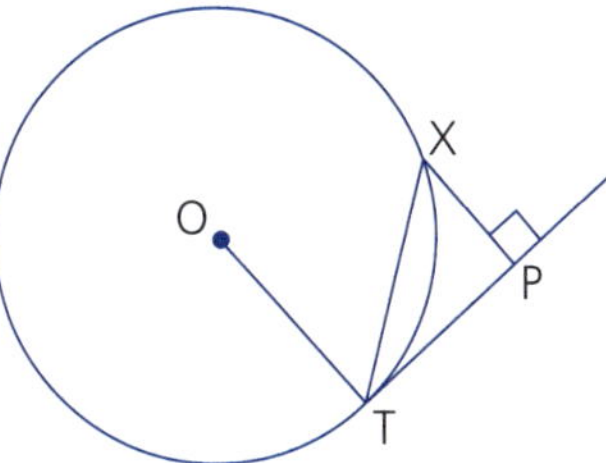

n Given that TB is a tangent at B and ∠ABC = θ°. AT is a straight line. Prove that ∠ATB = θ°.

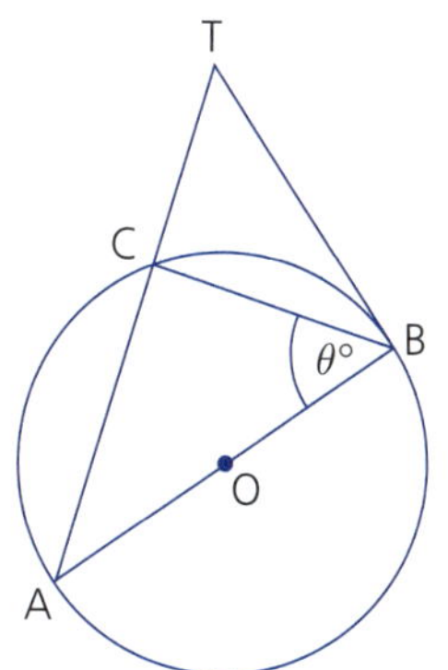

o Given that SP || AB, PT is a tangent at T and ∠PTB = α°, prove that △ABT is isosceles.

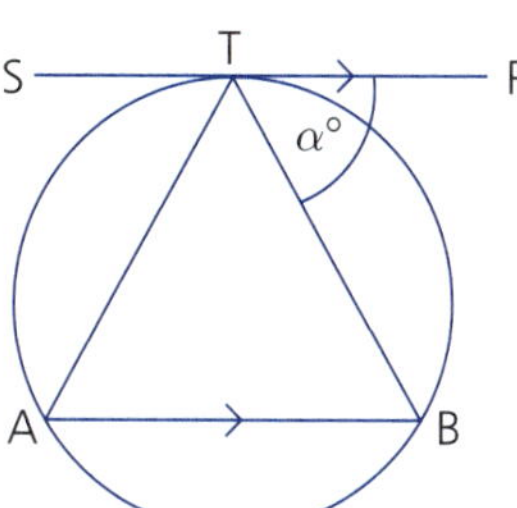

p Given that AC is a straight line, XY is a diameter and ∠XYA = θ°, prove that ∠XYA + ∠YBC = 90°.

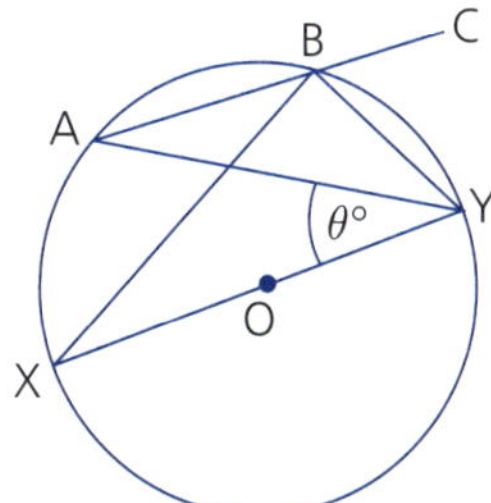

q

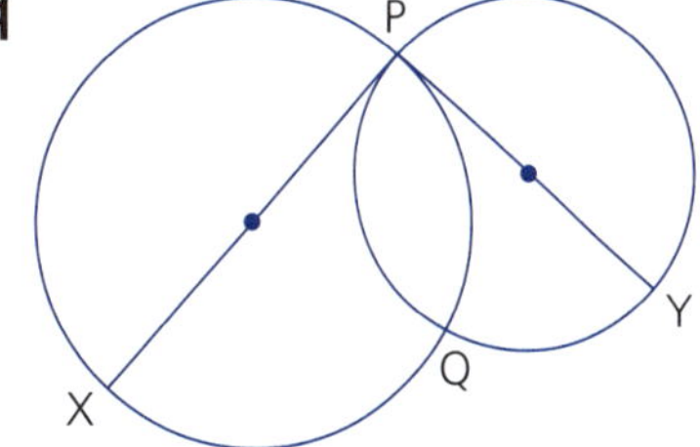

Two circles intersect at P and Q. PX and PY are the respective diameters. Prove that X, Q and Y are collinear.

pp. 264–276

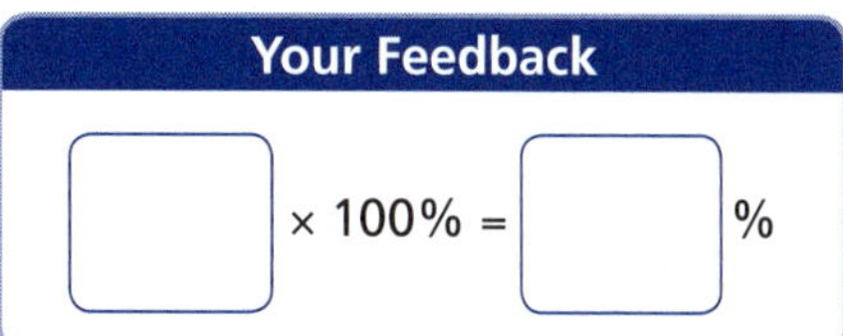

QA PAGE 303

WS PAGE 356

Your Checklist

For a complete understanding of this topic, you must be able to:

✓	Recognise various parts of the circle		p. 263
✓	Use chord properties of the circle		pp. 263–266
✓	Use angle properties of the circle		pp. 266–267
✓	Use cyclic quadrilateral properties of the circle		pp. 267–269
✓	Use tangent properties of the circle		pp. 269–270
✓	Apply ratio properties of tangents and secants		pp. 272–273
✓	Apply the alternate segment theorem		p. 273
✓	Use two circle properties		pp. 273–274
✓	Apply circle geometry knowledge to deductive geometry.		pp. 274–276

Now you are ready to do the tests!

Intermediate Test

Circle Geometry

(35 marks)

1 Complete the statements:

a The perpendicular bisector of a chord of a circle ____________________

b Equal chords of a circle subtend ____________________

c The opposite angles in a cyclic quadrilateral ____________________

d The angle in a semicircle is ____________________

e The angle between a tangent and a radius drawn to the point of contact ____________________

(5 marks)

2 Name the parts of the circle:

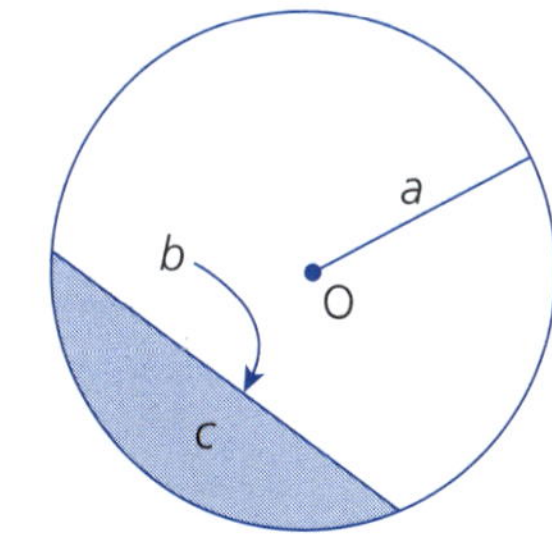

a ____________________

b ____________________

c ____________________ (3 marks)

3 Find the value of the letter in each diagram (O is the centre). Give the main reason for each answer:

a

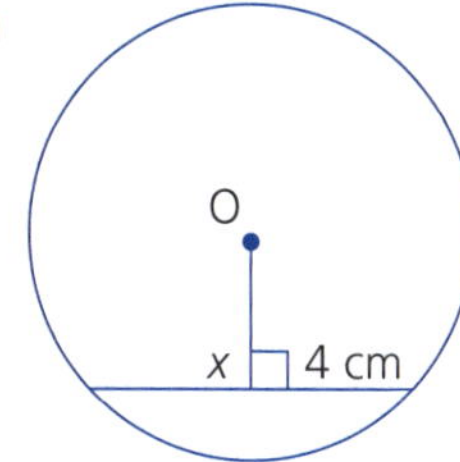

$x =$ __________

b

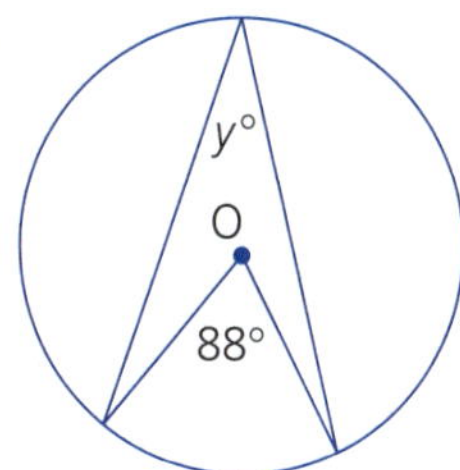

$y =$ __________

c

12 cm

a cm

$a =$ __________

d

23°

O

b°

$b =$ __________

e

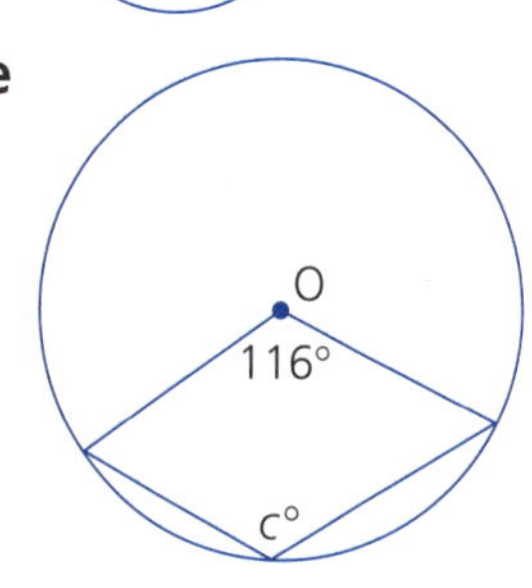

$c =$ __________

f

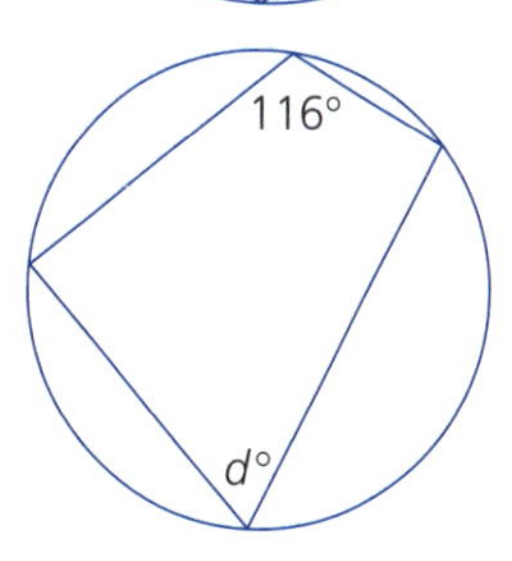

$d =$ __________

(12 marks)

4 A, B, C and D lie on the circle centre O. Let $\angle AOC = x°$. Giving reasons for each step prove that $\angle ABC = 180° - \frac{1}{2}x$Υ.

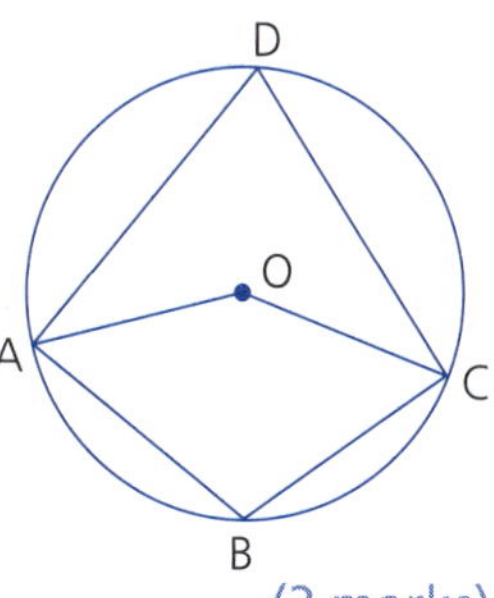

(3 marks)

5

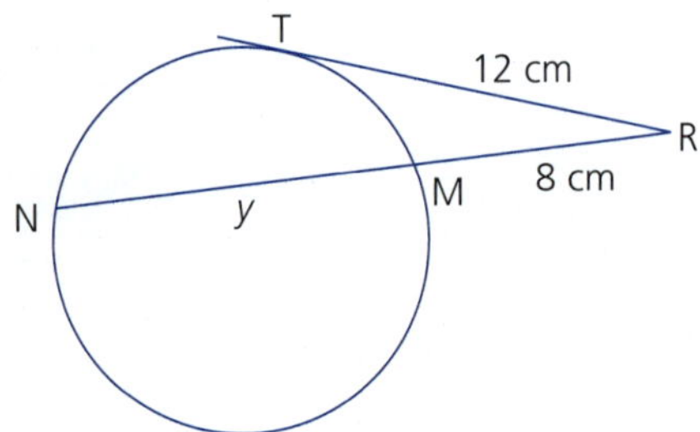

RT is a tangent to the circle, RN cuts the circle at M. If RT = 12 cm, RM = 8 cm, NM = y cm, calculate the value of y. (2 marks)

6 AB is a tangent to the circle at C, AB || DE. Show that:

a $\angle ACD = \angle BCE$

b $\triangle DEC$ is isosceles, giving reasons for all working.

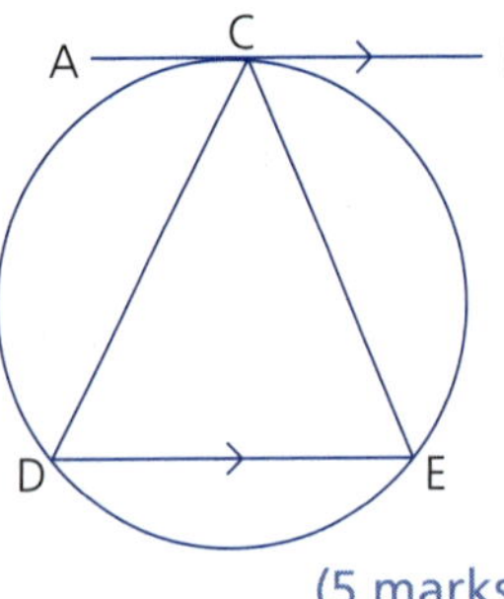

(5 marks)

7

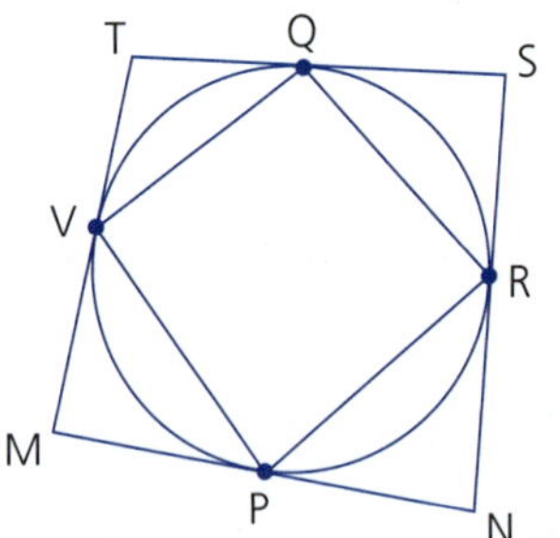

MN, NS, ST and TM are tangents to the circle with points of contact P, R, Q and V respectively.
VP, PR, RQ and QV are joined.
If $\angle MPV = x°$, $\angle NRP = y°$
$\angle SQR = z°$, $\angle TVQ = w°$,
show that $x + y + z + w = 180$. (5 marks)

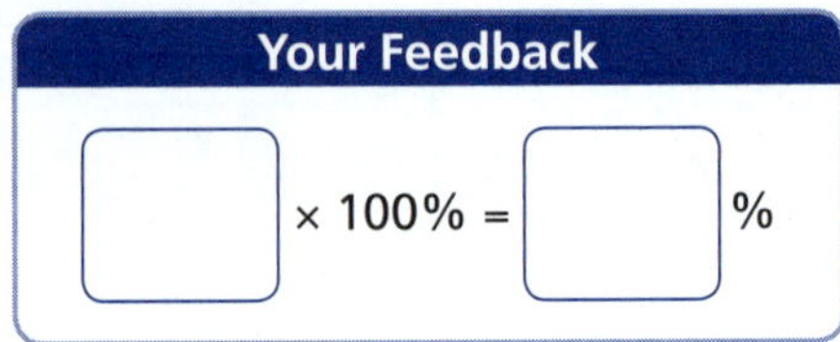

QA PAGE 308

WS PAGE 381

Advanced Test

Circle Geometry

(35 marks)

1 Complete the following statements:

a The square of a tangent to a circle from an external point equals ______________________________

b When two circles touch, their centres ______________________________

c The angle between a tangent and a chord drawn to the point of contact is ______________________________

d When two circles intersect, the line joining their centres ______________________________

e The angle at the centre of a circle is ______________________________

(5 marks)

2 Name the parts of the circle:

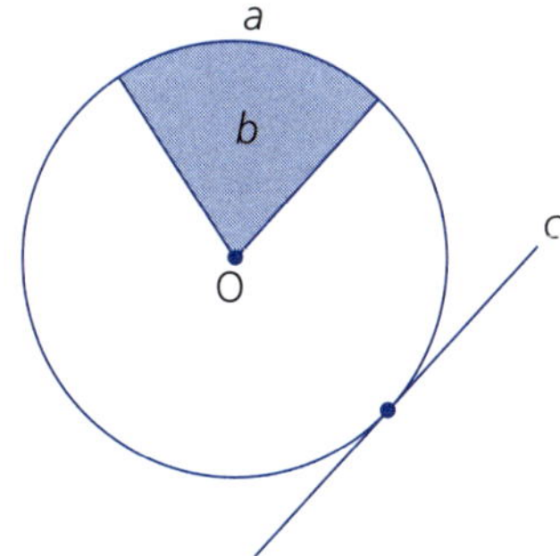

a ______________________

b ______________________

c ______________________ (3 marks)

3 Find the value of the letter in each diagram giving the main reason for your answer (O is the centre of the circle).

a

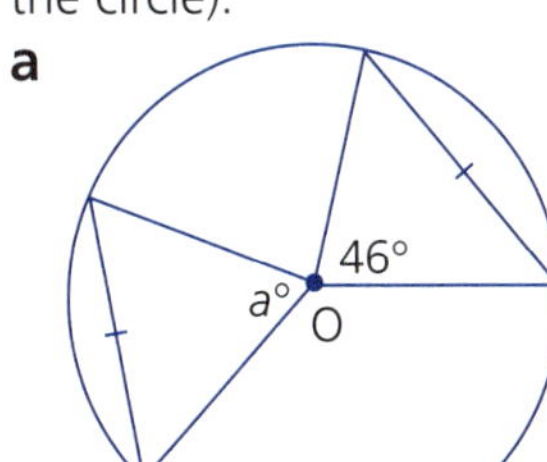

$a =$ __________

b

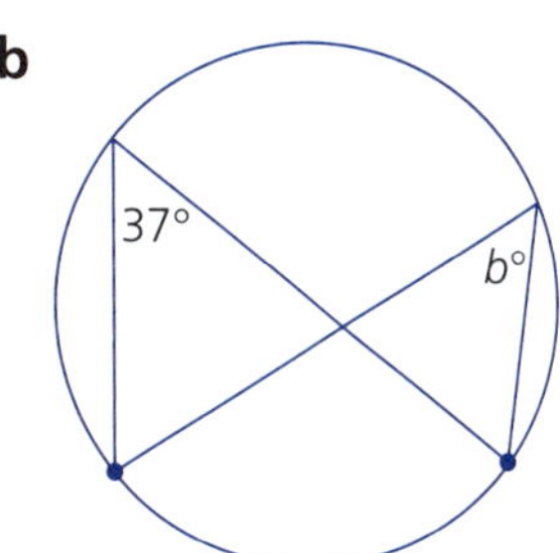

$b =$ __________

c

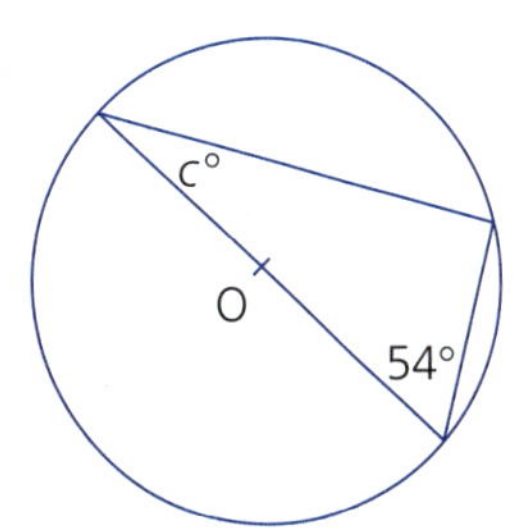

$c =$ __________

d

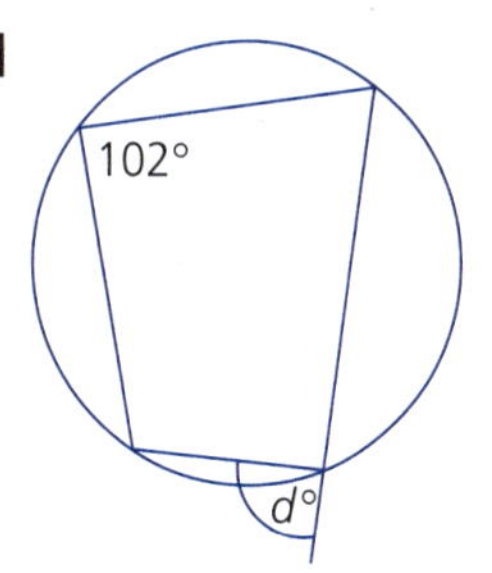

$d =$ __________

e

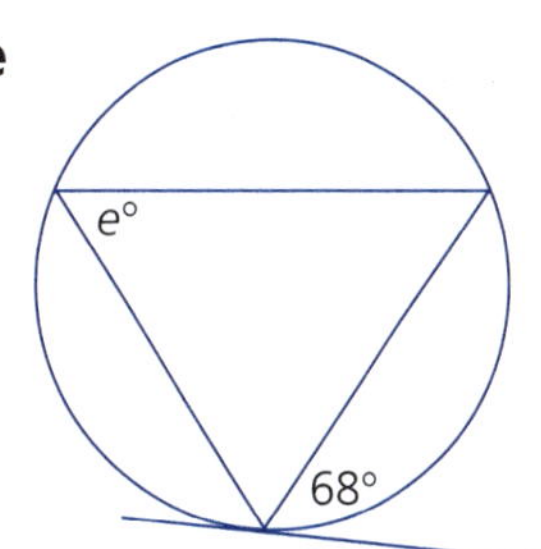

$e =$ __________

f

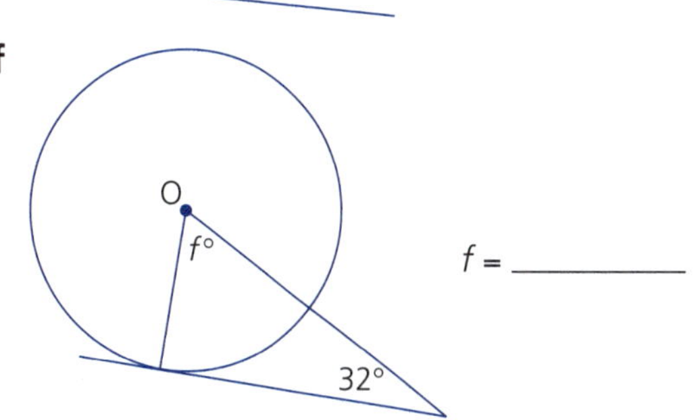

$f =$ __________

(12 marks)

4

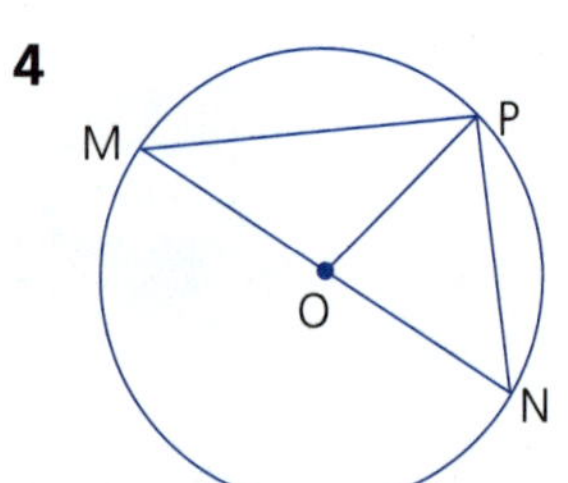

O is the centre of a circle with M, N and P on the circumference. If ∠ONP = 43°, find the size of ∠MPO and ∠MOP. (4 marks)

5

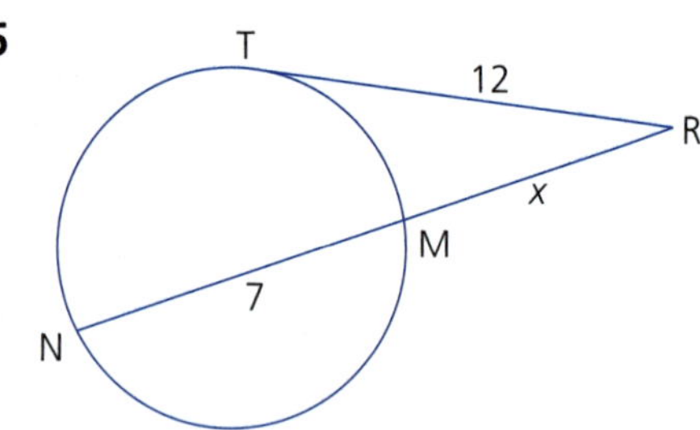

RT is a tangent to the circle. RN cuts the circle at M. If TR = 12 cm, NM = 7 cm and MR = x cm, calculate the value of x. (3 marks)

6

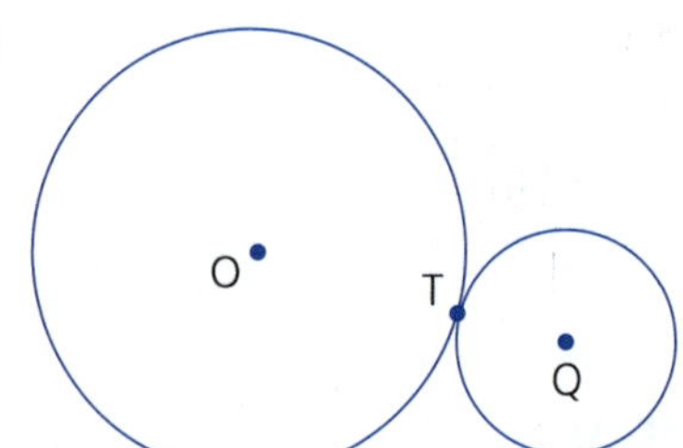

Two circles with centres O and Q touch at T. By drawing the common tangent at T, prove that O, T and Q are collinear. (3 marks)

7

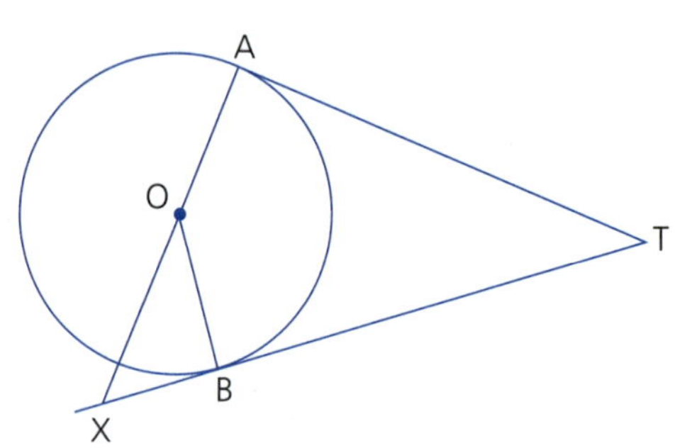

TA and TB are tangents at A and B respectively. AX is a straight line. Prove that ∠BOX = ∠ATB and that ∠ATB = 2 × ∠OAB. (5 marks)

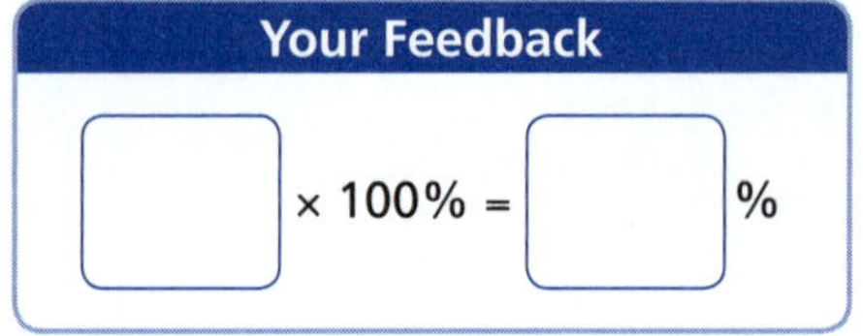

QA PAGE 308

WS PAGE 382

Sample
Examination 1

(120 marks)

Financial Mathematics

1 a

Taxable Income	Tax on this income
0–\$18 200	Nil
\$18 201–\$37 000	19c for each \$1 over \$18 200
\$37 001–\$80 000	\$3572 plus 32.5c for each \$1 over \$37 000
\$80 001–\$180 000	\$17 547 plus 37c for each \$1 over \$80 000
\$180 001 and over	\$54 547 plus 45c for each \$1 over \$180 000

Use the table to find the amount of tax Jannah pays on a taxable income of \$72 340. (1 mark)

b Oliver invested \$4200 in a simple interest account and earned interest of \$1134. How long was the money invested, if the interest rate is 4.5% p.a.? (2 marks)

c Sam decided to buy a car originally priced at \$24 000. As Sam did not have a trade-in he received a discount of 15%. The dealership also had an 'end-of-month sale' of 5% off all cars. How much did Sam pay for the new car? (1 mark)

d Zach invests \$6500 in an account earning 3.75% p.a. compound interest. What amount of compound interest will he receive if the investment is for 4 years? (2 marks)

e A caravan depreciates in value at the rate of 12% per annum. If the value of the caravan in 2010 was \$52 000, what is its value in 2015? (2 marks)

Algebraic Techniques

2 a Expand and simplify $(x + 5)^2 - (x + 5)(x - 5)$. (2 marks)

b Factorise:

i $x^2 - x - 6$ (1 mark)

ii $2x^2 - x - 3$ (1 mark)

c Simplify:

i $\dfrac{2x - 2}{x^2 - 1}$ (2 marks)

ii $\dfrac{2a - 5}{3} - \dfrac{4 - a}{2}$ (2 marks)

Surds and Indices

3 a Simplify $\sqrt{12} + \sqrt{27}$. (1 mark)

b If $\dfrac{2 - \sqrt{3}}{2 + \sqrt{3}} = a + b\sqrt{3}$, find the values of a and b. (2 marks)

c Expand and simplify $(\sqrt{2} - 1)^2 - (\sqrt{2} + 1)^2$. (2 marks)

d Simplify $8^{-\frac{2}{3}} \times \left(\dfrac{1}{2}\right)^{-2}$. (1 mark)

e Solve, by equating powers: $4^{x - 2} = 8^x$ (2 marks)

Equations

4 a Make y the subject of $3x - 4y = 5$. (1 mark)

b Solve $x^2 - 6x - 16 = 0$ (1 mark)

c Solve, leaving your answer as a surd, $3x^2 - 5x - 1 = 0$. (2 marks)

d Find the points of intersection of the line $2x - y = 5$ and the parabola $y = x^2 - 4x$. (3 marks)
e Solve $x^4 - 2x^2 + 1 = 0$. (1 mark)

Linear Relationships

5 a For the points A(−1, 2) and B(3, −4), find:
  i distance AB (1 mark)
  ii midpoint AB (1 mark)
  iii gradient AB. (1 mark)
b For the line $3x - 2y = 4$, find the:
  i gradient (1 mark)
  ii y-intercept. (1 mark)
c Find the equation of the line passing through (−2, 4) and perpendicular to $3x - 4y = 2$. (2 marks)
d The point $(m + 2, 4)$ lies on the line $2x - 3y = 4$. Find the value of m. (1 mark)

Non-Linear Relationships

6 a For the parabola $y = (x - 4)(2 - x)$:
  i find the axis of symmetry (1 mark)
  ii find the maximum value (1 mark)
  iii hence, sketch the curve. (1 mark)
b Write down the equation of the parabola:

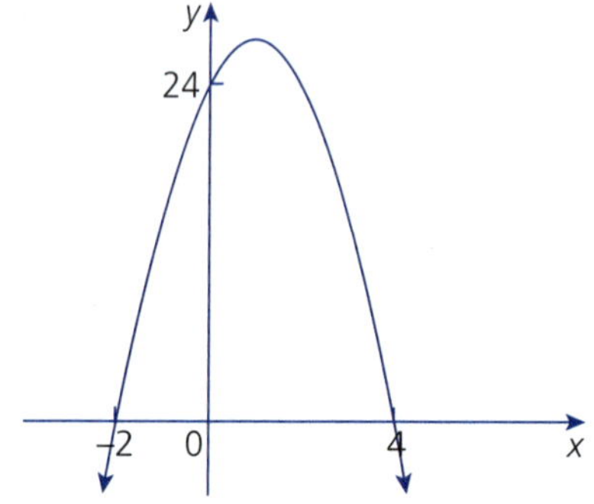

(2 marks)

c Find the coordinates of the centre and the radius of the circle with equation $x^2 + 4x + y^2 - 6y = 3$. (2 marks)
d Sketch the curve $y = 3^x$. (1 mark)

Rates and Proportion

7 a Convert 30 grams/m^2 to kg/hectare. (1 mark)
b It is known that y is directly proportional to x and if $x = 6$ then $y = 18$.
  i Write an equation that links y and x using the constant of proportionality k. (1 mark)
  ii Find the value of k. (1 mark)
  iii What is the value of y when $x = 20$? (1 mark)
  iv Find the value of x when $y = 96$. (1 mark)
c Anna's wage is directly proportional to the number of hours she works. When she works 24 hours she is paid $528. Let n = number of hours worked and W = wages in dollars. By finding the constant of proportionality, write an equation that links W and n. (2 marks)
d

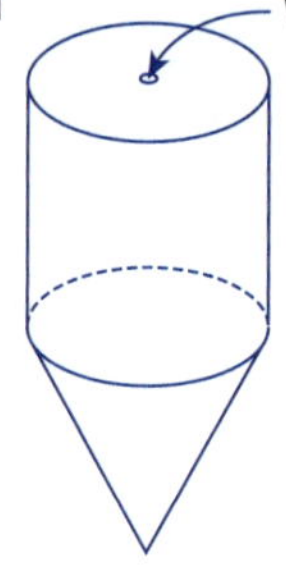

Water is poured into an empty container at a constant rate.

Which of the following represents the graph showing the relationship between depth (d) and time (t)?

A 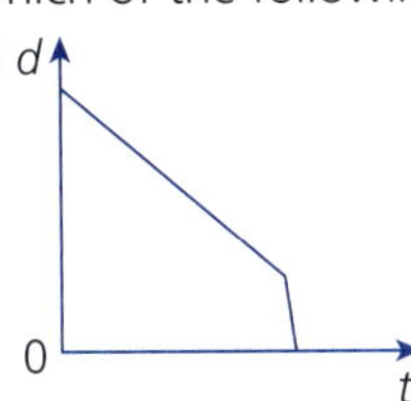

B 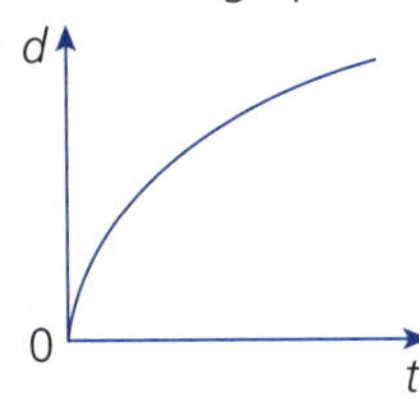

C 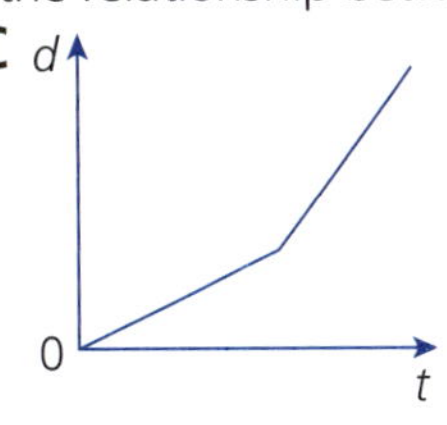

D 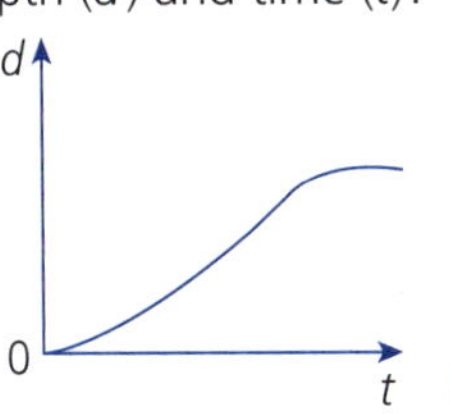

(1 mark)

Trigonometry and Pythagoras' Theorem

8 a A ladder reaches 2.3 metres up a vertical wall. If the ladder is 3 metres long, how far is the base of the ladder from the base of the wall? Give your answer correct to three significant figures. (2 marks)

b Write the exact value of $\tan 150°$. (1 mark)

c If $\cos\theta = \frac{3}{7}$, and θ is acute, write the exact value of $\tan\theta$. (1 mark)

d Find the exact length of x:

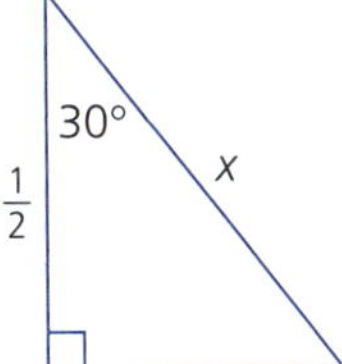

(2 marks)

e Find the smallest angle in the triangle with sides 3 cm, 5 cm and 6 cm. Give your answer to the nearest minute. (2 marks)

Measurement

9 a Find the volume, in terms of π:

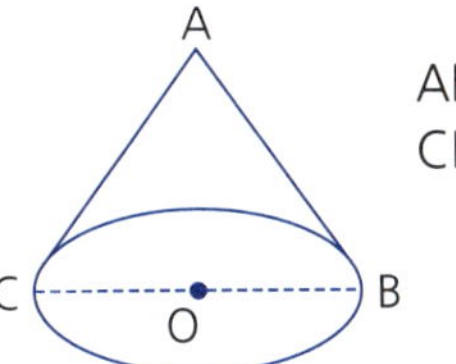

AB = 10 cm
CB = 16 cm

(2 marks)

b 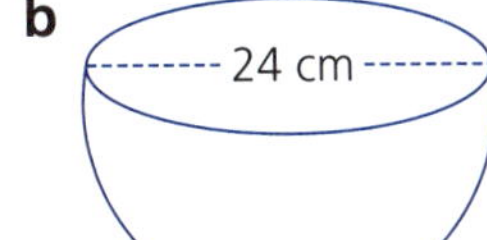

For the hemisphere, left, find the volume and surface area, correct to two decimal places. (4 marks)

c Find the volume and surface area of the square-based pyramid:
PQ = 12 cm
QM = MR
AM = 10 cm

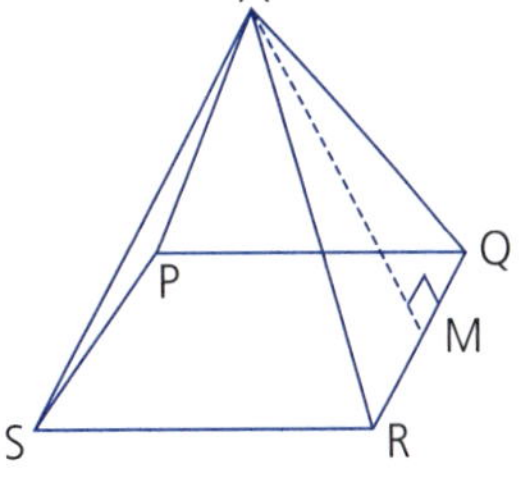

(2 marks)

Geometry

10 a M is midpoint of BC and AC = AB.

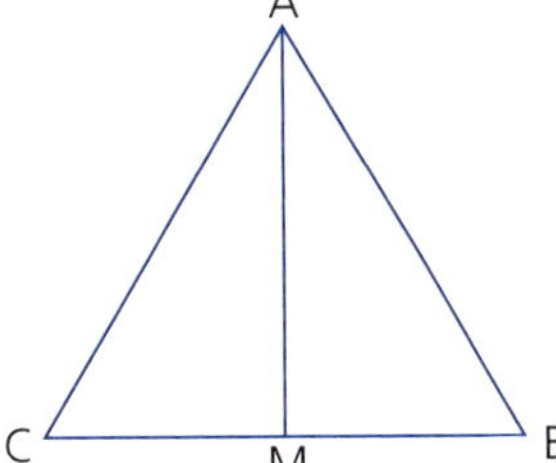

Prove that:

i $\triangle ACM \equiv \triangle ABM$. (2 marks)

ii Hence, prove $\angle CAM = \angle BAM$. (1 mark)

b $\angle PQR = 90°$,
T is the midpoint of PQ and
TS ∥ QR.

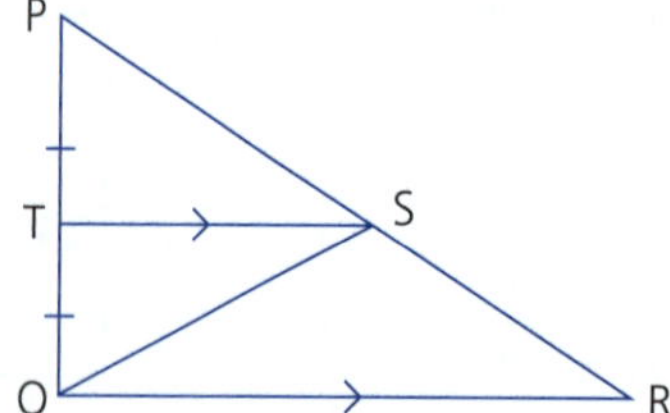

i Prove $\triangle PTS \equiv \triangle QTS$. (2 marks)
ii Prove QS = SR. (2 marks)

c Two similar cones have curved surface areas of 25.2 cm^2 and 56.7 cm^2. Find the ratio of their diameters. (1 mark)

Statistics

11 a A box plot is drawn to show the distribution of marks for a class of 24 students. The test was marked out of 20:

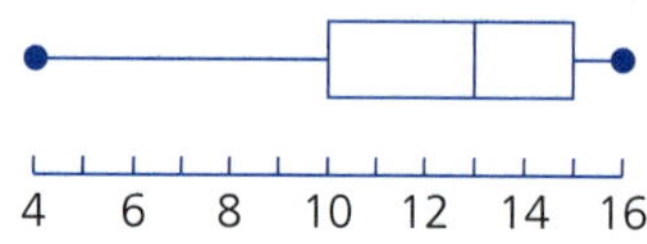

i What percentage of students scored 15 or more? (1 mark)
ii How many students scored 50% or less? (1 mark)
iii Comment on the skew of the distribution. (1 mark)

b Find the standard deviation of the following sets of scores:
i 12, 16, 10, 11, 19, 15 (1 mark)
ii

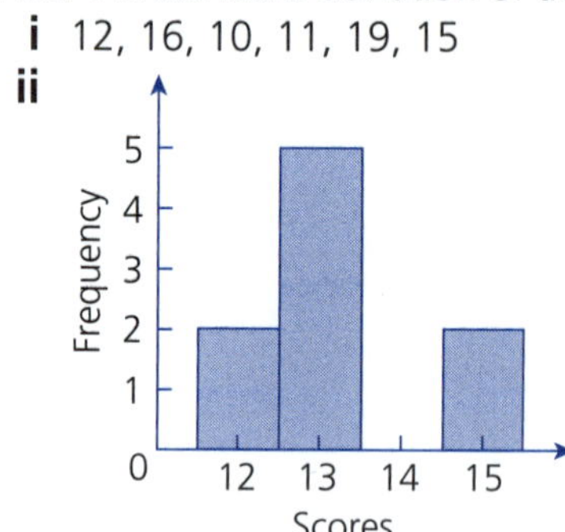

(1 mark)

c The scatter plot shows the relationship between the student results in a geography test (x) and a history test (y):

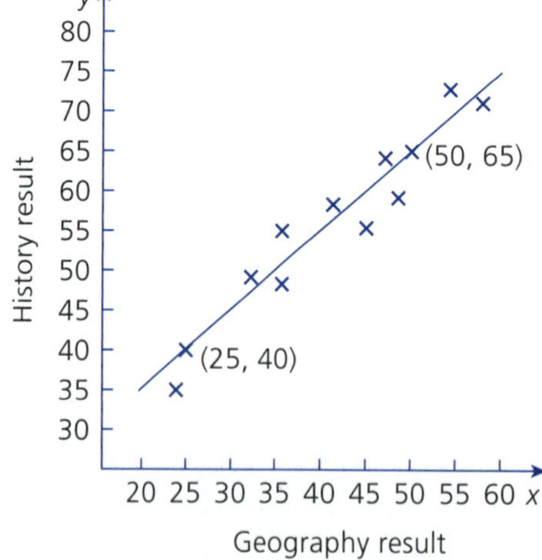

i Use the scatter plot to find an equation linking the history results to the geography results. (2 marks)
ii Use the equation to estimate the history results for a student who scored 52 in the geography test. (1 mark)

Probability

12 a A coin is tossed three times.

  i Draw a tree diagram to list the possible outcomes. (2 marks)

  ii Find the probability of tossing:

   1 three heads (1 mark)

   2 a head and two tails (1 mark)

   3 at least one tail. (1 mark)

b Cards with the numbers from 1 to 15 are placed in a bag. A card is chosen at random from the bag.

  i Find the probability that the number is less than 10. (1 mark)

  ii If the number is more than 5, what is the probability that it is even? (1 mark)

  iii The number is a factor of 15. What is the probability that it is prime? (1 mark)

Polynomials

13 a Divide $(2x^3 - 5x^2 + 3x - 2)$ by $(x - 1)$, leaving your answer in the form
Dividend = Divisor × Quotient + Remainder. (2 marks)

b Show that $(x - 3)$ is a factor of $P(x)$ if $P(x) = x^3 - 2x^2 - 5x + 6$. (2 marks)

c For what value of k will $(x^4 + kx^2 - 2x + 1) \div (x + 2)$ have a remainder of 9? (2 marks)

d Sketch $y = x^3 - 6x^2 + 9x$ on a number plane. (2 marks)

Functions and Logarithms

14 a If $f(x) = x^2 - 4x + 5$, find $f(x - 2)$. (2 marks)

b The number plane shows the graph of $y = f(x)$.
On the same number plane graph $y = f(x) + 2$
and $y = -f(x)$. (2 marks)

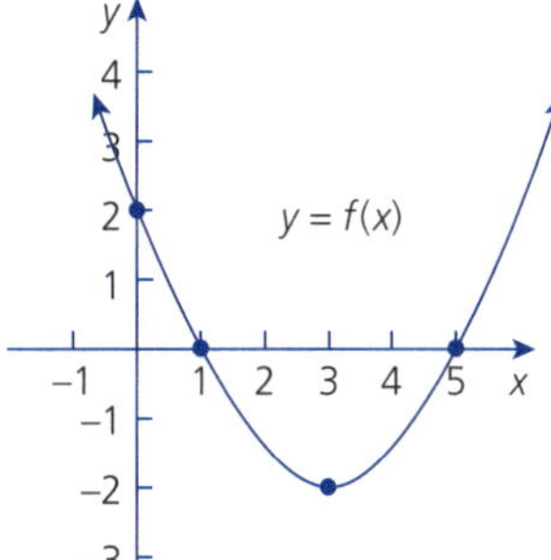

c If $\log_5 x = 3$, find the value of x. (1 mark)

d If $\log_a 2 = 0.1476$ and $\log_a 3 = 0.3416$, find the value of $\log_a 18$. (2 marks)

e Solve $2^x = 11$, leaving your answer correct to two decimal places. (1 mark)

Circle Geometry

15 a

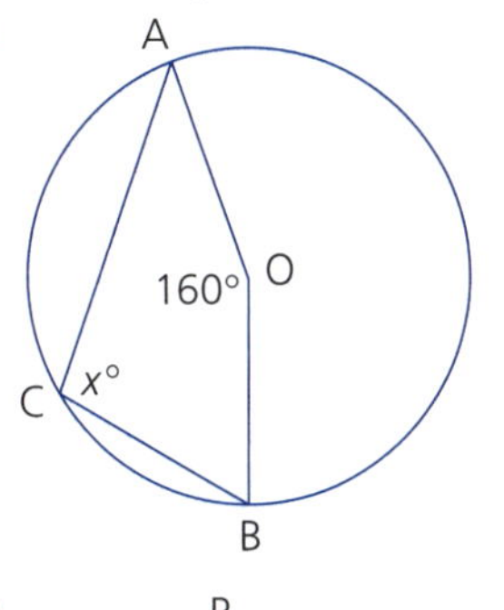

The points A, B and C lie on a circle with centre O.
Find the value of x, giving reasons. (2 marks)

b

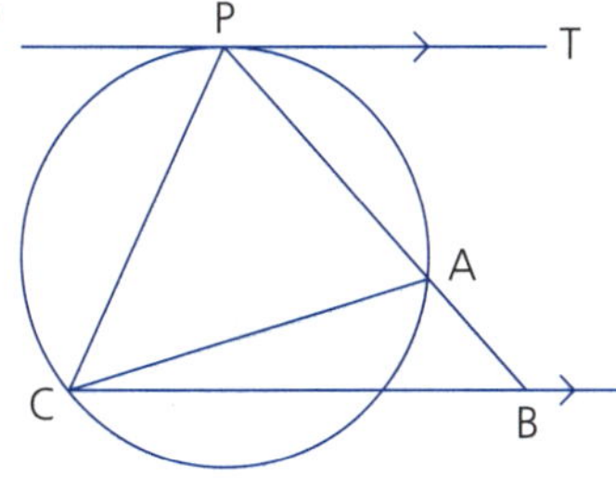

The tangent from T touches a circle at P. BC is parallel to TP.
Show that ∠PCA = ∠PBC. (2 marks)

c

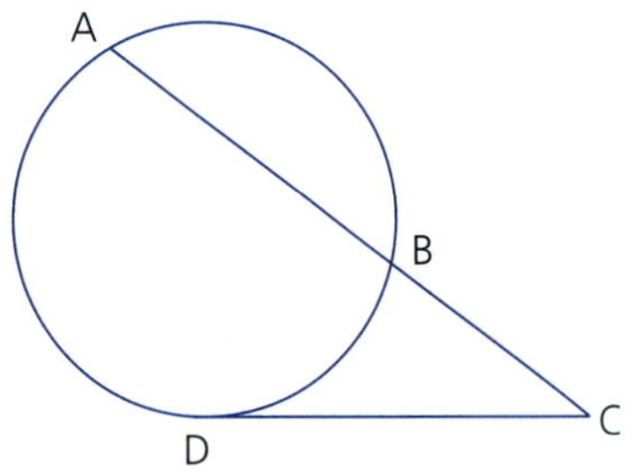

If $AB = 6$ cm, $DC = 5$ cm, find the length of BC, correct to two decimal places. (2 marks)

d

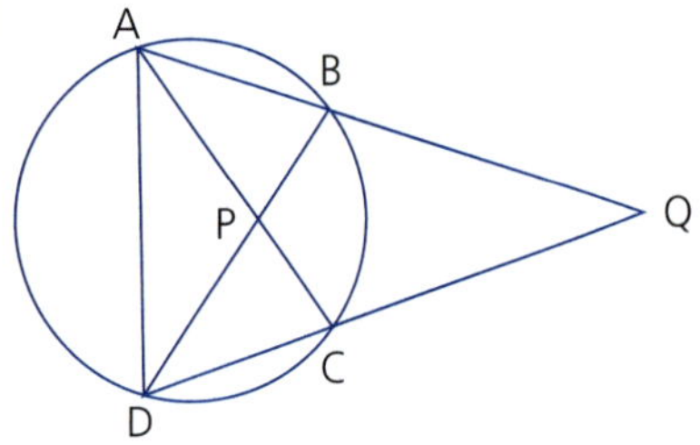

QB is produced to A and QC is produced to D.
A, B, C and D lie on a circle. PBQC is a cyclic quadrilateral.
Find the size of ∠ABD. (2 marks)

Your Feedback

☐ × 100% = ☐ %

QA PAGE 309

WS PAGE 383

Sample Examination 2

Part A (Multiple-Choice Questions) (30 marks)

- Part A: 30 multiple-choice questions—allow 30 minutes
- Part B: Single- and free-response questions—allow 50 minutes

Select the answer that best fits the question.

1 If $(x - 2)(3x + 1) = 0$, then x equals:

A 2 or −1 **B** $2 \text{ or } -\frac{1}{3}$ **C** −2 or −1 **D** $-2 \text{ or } \frac{1}{3}$ (1 mark)

2 The gradient of the line $3x + y = 2$ is:

A 2 **B** 3 **C** −3 **D** $\frac{2}{3}$ (1 mark)

3 Express $\frac{1}{\sqrt{11} - 3}$ with a rational denominator:

A $\frac{\sqrt{11} + 3}{2}$ **B** $\frac{\sqrt{11} + 3}{20}$ **C** $\frac{\sqrt{11} - 3}{2}$ **D** $\frac{\sqrt{11} - 3}{20}$ (1 mark)

4 It is known that an ant travels 10 cm in 20 seconds. Its speed in kilometres per hour is:

A 0.018 km/h **B** 0.18 km/h **C** 1.8 km/h **D** 18 km/h (1 mark)

5 A packet contains four balloons: two blue and two red. What is the probability that Fiona selects two balloons at random that are the same colour?

A $\frac{1}{4}$ **B** $\frac{1}{3}$ **C** $\frac{1}{2}$ **D** $\frac{2}{3}$ (1 mark)

6 Expand and simplify $4 - 2(5 - 3x)$:

A $10 - 6x$ **B** $-6 - 3x$ **C** $-6 + 3x$ **D** $-6 + 6x$ (1 mark)

7 If $y = \frac{4x + 3}{2x}$, then:

A $x = \frac{4y + 3}{2y}$ **B** $x = \frac{y - 3}{2}$ **C** $x = \frac{3}{2y - 4}$ **D** $x = \frac{2y - 4}{3}$ (1 mark)

8

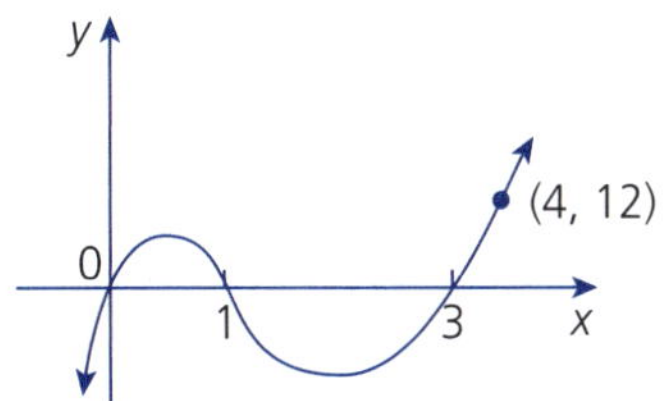

The equation of the graph above is:

A $y = x(x - 1)(x - 3)$ **B** $y = x(x + 1)(x + 3)$ **C** $y = -x(x + 1)(x - 3)$ **D** $y = -x(x - 1)(x - 3)$ (1 mark)

9 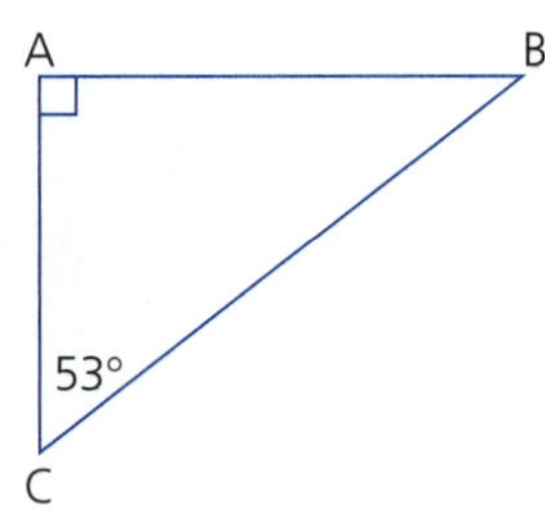

The ratio of AB to BC is closest to:

A 0.7986 **B** 0.6018 **C** 1.327 **D** 1.2522 (1 mark)

10 The term $a^{-\frac{2}{3}}$ can be rewritten as:

A $\frac{-2a}{3}$ **B** $\frac{\sqrt[3]{a}}{2}$ **C** $\frac{1}{\sqrt{a^3}}$ **D** $\frac{1}{\sqrt[3]{a^2}}$ (1 mark)

11 For $2x^2 - 3x - 4 = 0$, the values of x are:

A $\frac{-3 \pm \sqrt{41}}{4}$ **B** $\frac{3 \pm \sqrt{41}}{4}$ **C** $\frac{-3 \pm \sqrt{24}}{4}$ **D** $\frac{3 \pm \sqrt{24}}{4}$ (1 mark)

12 AB is parallel to DE. The length of CD is:

A 10 cm
B 11 cm
C 15 cm
D Not enough information.

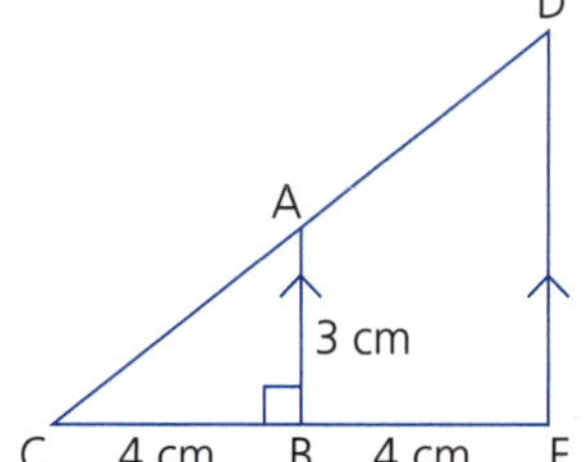

(1 mark)

13

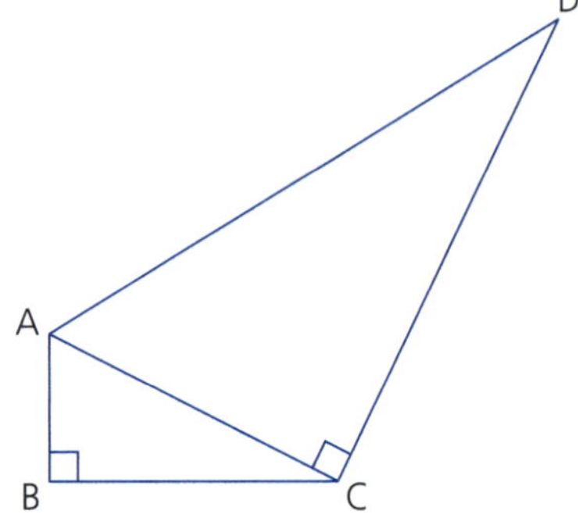

AB = 3, BC = 4, AD = 13. What is the length of CD?

A $\sqrt{34}$ **B** 8 **C** $\sqrt{120}$ **D** 12 (1 mark)

14 If $(x - 2)$ divides $(x^3 - 2x^2 - x + 2)$, what is the remainder?

A 0 **B** 4 **C** 8 **D** 16 (1 mark)

15 The smallest angle in a triangle with sides 6 cm, 7 cm and 8 cm is about:

A 43° **B** 47° **C** 36° **D** 54° (1 mark)

16 Simplify $\frac{3x}{5} - \frac{x}{2}$:

A $\frac{x}{10}$ **B** $\frac{2x}{3}$ **C** 1 **D** $\frac{x}{3}$ (1 mark)

17 Given that $f(x) = 3x^2 - x$, then $f(-2)$ equals:

A –10 **B** 38 **C** 14 **D** 10 (1 mark)

18 The area of the triangle is closest to:

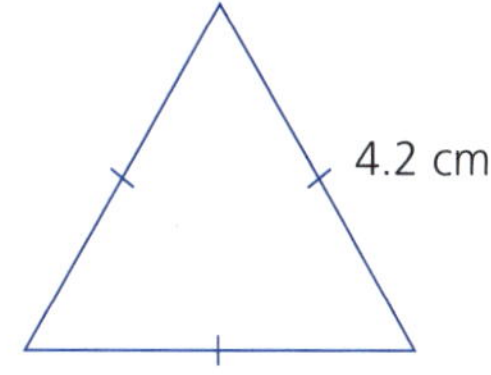

A 4.41 cm^2 **B** 7.64 cm^2 **C** 8.82 cm^2 **D** 15.28 cm^2 (1 mark)

19 The solution of $4 - 2x \geq 6$ is represented on the number line as:

A 

B –2 –1 0 1 2 3 4 5 6

C –2 –1 0 1 2 3 4 5 6

D –2 –1 0 1 2 3 4 5 6

(1 mark)

20 Solve these simultaneous equations: $3x - 2y = 0$, $x + y = 5$

A $x = 10$ **B** $x = 3$ **C** $x = -2$ **D** $x = 2$ (1 mark)

21 ABCD is a parallelogram. The value of x is:

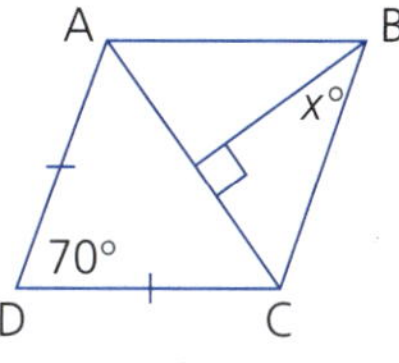

A 20 **B** 35 **C** 40 **D** 45 (1 mark)

22 The centre of the circle which equation $x^2 - 4x + y^2 + 6y - 2 = 0$ is:

A (2, –3) **B** (–2, 3) **C** (–4, 6) **D** (4, –6) (1 mark)

23 The value of $\sin\theta$ is:

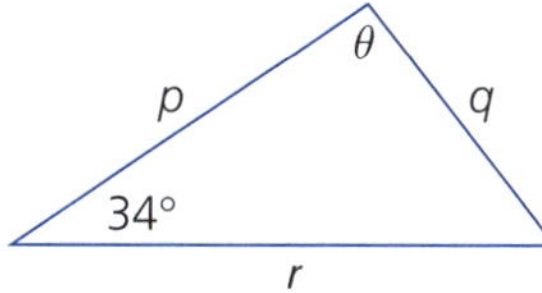

A $\frac{p\sin 34\Upsilon}{r}$ **B** $\frac{q\sin 34\Upsilon}{r}$ **C** $\frac{r\sin 34\Upsilon}{q}$ **D** $\frac{r\sin 34\Upsilon}{p}$ (1 mark)

24 Find the value of k: $\sqrt{8} + \sqrt{18} = \sqrt{k}$

A $k = 26$ **B** $k = 10$ **C** $k = 25$ **D** $k = 50$ (1 mark)

25 Simplify $\frac{a^3b^2}{(a^2b)^2}$:

A ab **B** a **C** $\frac{1}{a}$ **D** $\frac{1}{ab}$ (1 mark)

26 There are two ways to hang nappies on a clothes line using pegs. Method B involves overlapping one nappy onto another and then pegging them on the line. If n nappies are to be hung on a clothes line, how many more pegs would be required using Method A than using Method B?

Peg
Nappy
Method A **Method B**

A $n - 1$ **B** $n + 1$ **C** $2n - 1$ **D** $2n + 1$ (1 mark)

27 Which of the following is most likely to represent the graph?

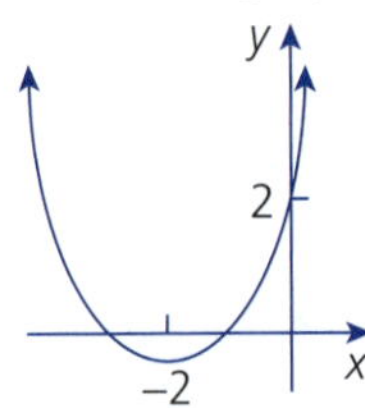

A $y = x^2 - 4x - 2$ **B** $y = x^2 + 4x - 2$ **C** $y = x^2 - 4x + 2$ **D** $y = x^2 + 4x + 2$ (1 mark)

28 The following solids are known to be similar.
If the surface area of the smaller solid is 16 cm^2,
the surface area of the larger solid is:
A 17.9685 cm^2 **B** 43.56 cm^2
C 71.874 cm^2 **D** 95.832 cm^2

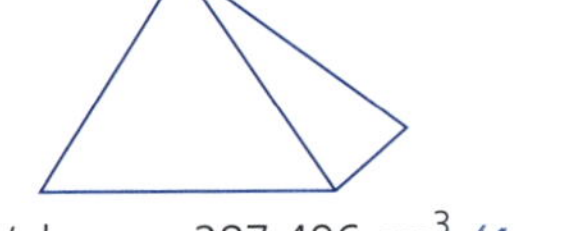

Volume = 64 cm^3 Volume = 287.496 cm^3 (1 mark)

29 The volume of a hemisphere with radius of 3 cm is:

A 36π cm^3 **B** 18π cm^3 **C** $\frac{8}{3}\neq$ cm^3 **D** 24π cm^3 (1 mark)

30 If $x^2 + 8x + p = (x + q)^2$, what is the value of p and q?
A $p = 4$ and $q = 16$ **B** $p = 16$ and $q = 16$ **C** $p = 16$ and $q = 4$ **D** $p = 8$ and $q = 4$ (1 mark)

Part B (Short-Answer Questions) (45 marks)

31 **a** Calculate the distance between (2, −1) and (−3, 4). (1 mark)
b Solve $4x^3 = 32$. (1 mark)
c Solve $3x = \frac{x}{2} + 5$. (1 mark)
d Simplify $\frac{x^2 - 5x + 6}{x^2 - 9}$. (1 mark)
e Find the degree of $P(x) \cdot Q(x)$ if $P(x) = 3x^2 - 4x^3 + 7$ and $Q(x) = 2x^4 - 7x^2 + 1$. (1 mark)
f Given $v = u + at$, find t when $v = 60$, $u = 24$, $a = 6$. (1 mark)
g The surface area of this rectangular prism is 62 cm^2. Find the value of x.

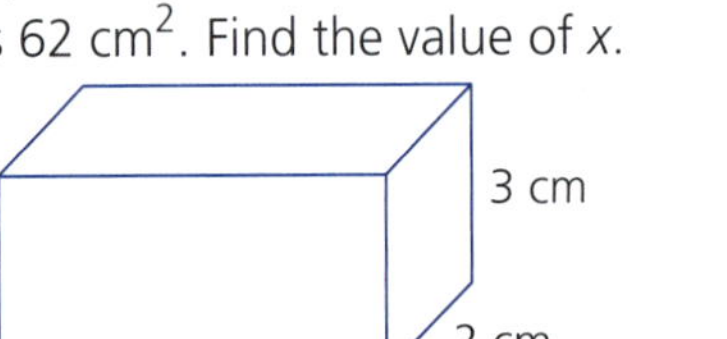

(1 mark)

h

The box plot shows the marks scored by 24 students in a maths test marked out of 60.
How many students scored 80% or less? (1 mark)

i Find the equation of line l.

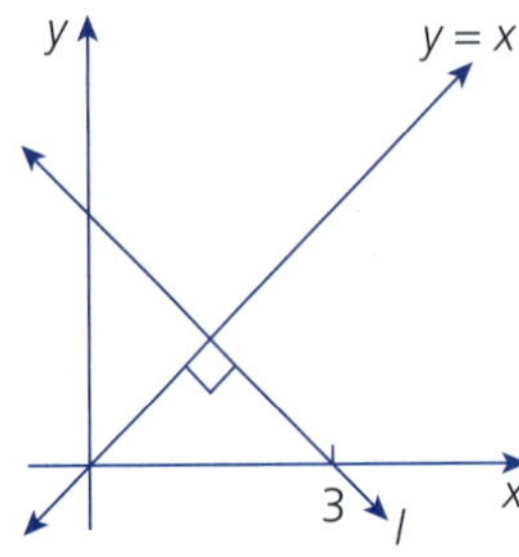

(1 mark)

j Solve $2x^2 = 4x$. (1 mark)

k Factorise $6x^2 - 5x - 4$. (1 mark)

l

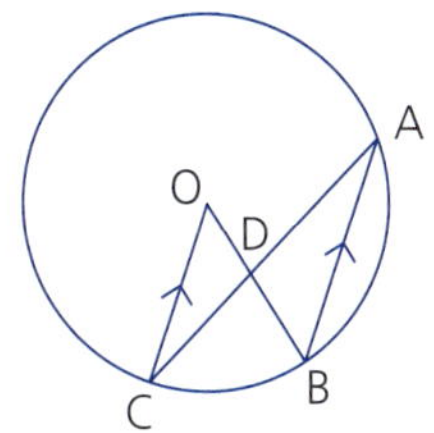

A, B and C lie on the circle centre O. OC is parallel to AB.
If $\angle CAB = 32°$, find the size of $\angle BDA$. (1 mark)

m It is known that $M \propto n$ and when $n = 8$ then $M = 128$.
By finding the constant of proportionality k, write an equation that links M and n. (1 mark)

n Given $\log_a 3 = 1.4$ and $\log_a 2 = 0.9$, evaluate $\log_a 12$. (1 mark)

o The sum of two whole numbers is 6 but their product is less than 5. Give a possible pair of numbers. (1 mark)

32 A farmer wishes to fence a rectangular enclosure using an existing river as the fourth side. He has 80 metres of fencing material to use. Let the width of the enclosure be x metres.

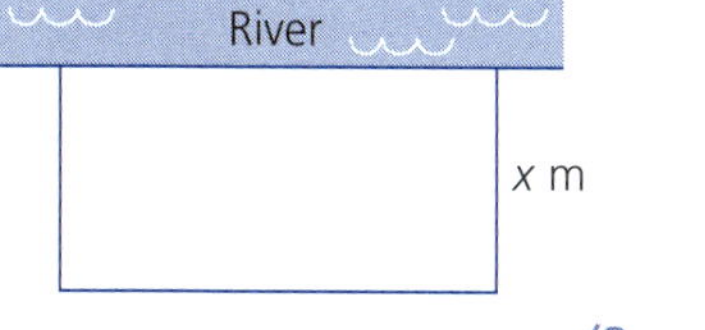

a Find an expression for the length of the rectangle in terms of x.

b If the length of the enclosure is chosen to be 50 metres, find the width, and hence the area, of the enclosure. (3 marks)

33 An icecream shop records the maximum temperature and the number of icecreams sold each day and the data is recorded in the scatter plot below:

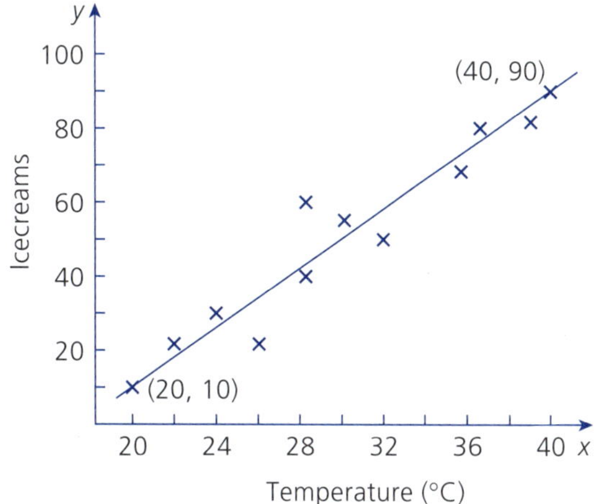

a Find the equation of the line of best fit. (2 marks)

b Predict the sales of icecreams on a day when the temperature reaches a maximum of 38°C. (1 mark)

34 The diagram shows the parabola $y = x^2 - 3x$ and the line $y = 2x$.

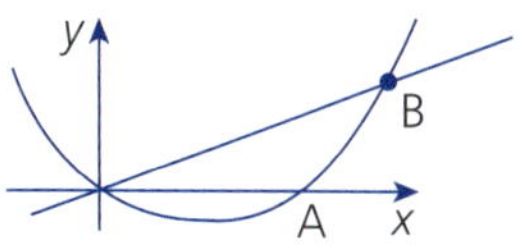

a Find the coordinates of A. (1 mark)

b Find the coordinates of B. (2 marks)

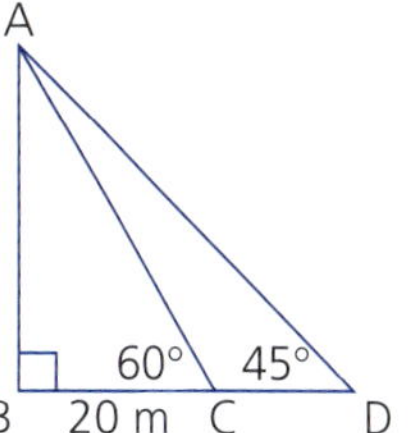

35 From a point D, the angle of elevation of a tower AB is 45°. Moving to a point C, 20 metres from the base of the tower, the angle of elevation is 60°.

a Find the height AB of the tower.

b Find the distance between C and D.

Give your answers correct to two decimal places. (3 marks)

36 A stone is thrown from a cliff and lands in water at B. For the axes on the diagram, the flight path of the ball is the parabola $h = 32 + 4x - x^2$, where x and h are in metres.

a How far from the base of the cliff is the stone when it hits the water?

b What was the maximum height (above the water) that the stone attained?

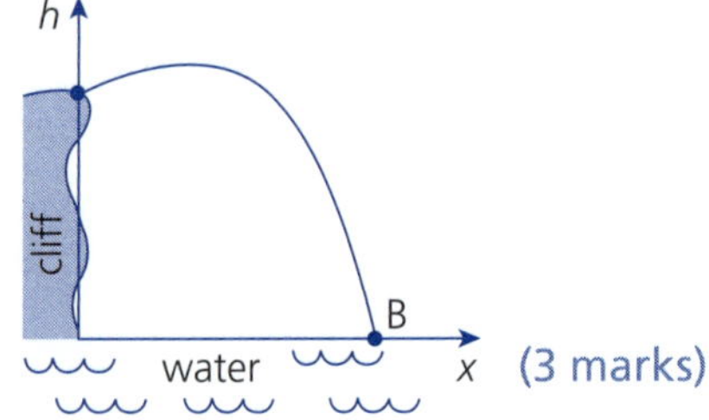

(3 marks)

37 Given AB = BC, BE bisects ∠CBD.
Prove AC || BE.

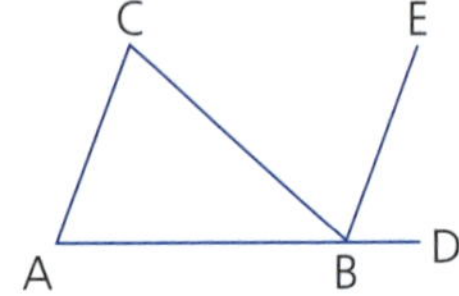

(3 marks)

38 ABCD is cyclic quadrilateral. TD is a tangent to the circle. ∠BDT = 40° and ∠DBC = 15°. Find ∠CAB, giving reasons.

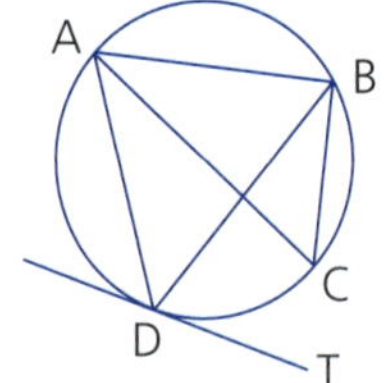

(3 marks)

39 It is known that a certain parabola cuts the x-axis at 0 and 6.

a Write down equations of two parabolas that fit this description, and sketch them on the number plane.

b On further investigation, it is discovered that the parabola has a minimum value of –18. Find the equation of the parabola.

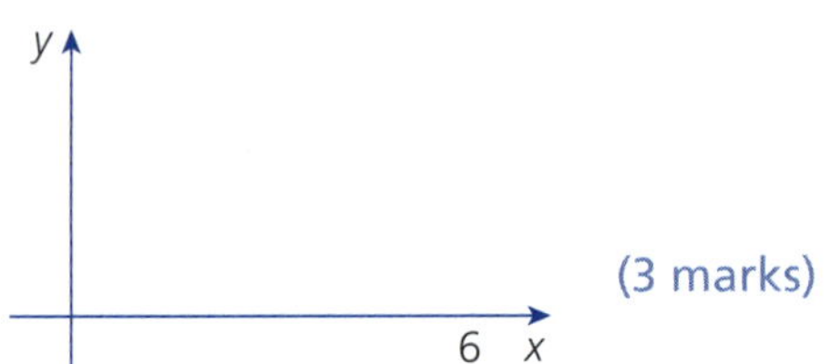

(3 marks)

40 Shape A represents a cone surmounted on a cylinder, shape B represents a sphere. Water is poured into the shapes at a constant identical rate.

The graph shows the relationship between the volume and the height of the water in each shape.

a On separate graphs, sketch the relationship between the volume and the height of the water in each shape.

b Explain why the capacity of each shape is the same.

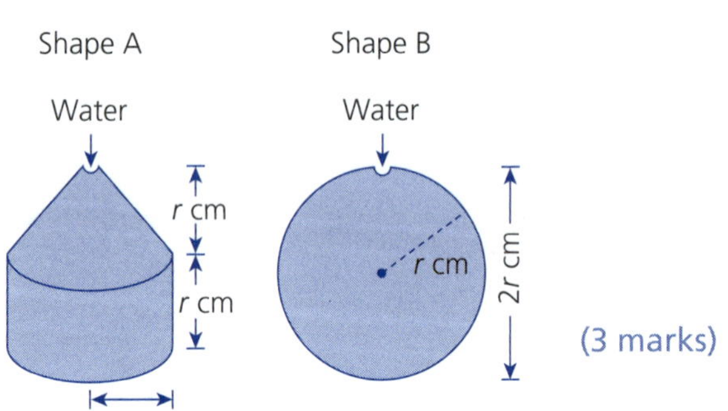

(3 marks)

41 The line l touches the circle $x^2 + y^2 = 25$ at P.
It is known that the point P has coordinates $(3, a)$.

a Find the value of a.

b The line l is written in the form $3x + 4y + k = 0$.
Find the value of k.

c Find the area of the triangle OXY.

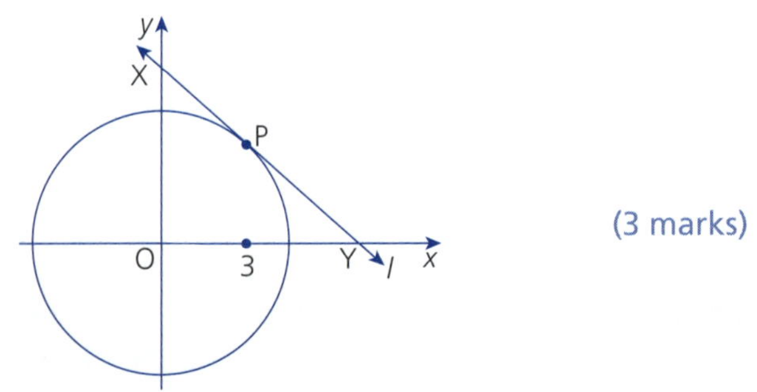

(3 marks)

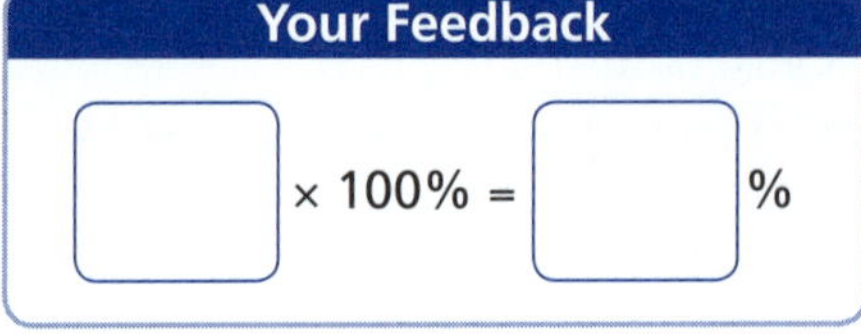

QA PAGE 309

WS PAGE 386

Quick Answers to Practise Practise

Chapter 1 Financial Mathematics . . . . pp. 8–10

1 a Nil tax b \$2133.70 c \$13 807.88 d \$36 944.99 e \$60 614.35 **2** a \$23 000 b \$100 000 c \$500 000 **3** \$92 201.35 **4** a \$14 947 b \$1625 **5** a \$630 b \$114.75 c \$162 d \$10.20 e \$480 **6** a 6% p.a. b 4.5% p.a. c 3.75% p.a. d 3.8% p.a. **7** a \$2500 b \$2300 c \$8400 d \$10 000 **8** a 3 years b 6 years c 9 years d 15 months **9** 20 years **10** \$264 **11** \$1176 **12** a \$27 200 b \$566.67 **13** a \$352 800 b \$72 800 **14** a \$2542 b \$1047 **15** a \$30 b \$27 c \$28 d \$30 ∴ B is the cheapest buy **16** a \$21.60 b \$261.60 c \$16.80 **17** \$35.14 **18** \$1044 **19** A **20** \$68.28 **21** 19% **22** \$200 **23** \$1000 **24** \$1042 **25** \$36 000 **26** a \$1222.50 b \$2467.92 c \$330.62 d \$4435.33 **27** \$20 755.57 after 10 years **28** 23 534 **29** a \$2400 b \$4500 c \$25 000 **30** \$20 000 **31** a 3.0% b 4.5% c 5.3% **32** \$36 971.90 **33** 3.9% **34** a 7 years b 3 years c 15 years **35** 25 years **36** \$982.60 **37** \$6043.06 **38** 25% p.a. **39** 28% p.a. **40** 2021

Chapter 2 Algebraic Techniques . . . . pp. 29–30

1 $(x-y)$ km/hour **2** a $(14+x)$ years old b $(14-y)$ years old c $(p+14)$ years old d $(p+14+q)$ years old **3** a xy m^2 b mq m^2 c rmq m^2 d $(xy-rmq)$ m^2 **4** a $10k$ b $1440p$ c $1000k$ d $\frac{r}{100}$ e $\frac{v}{1000}$ **5** $(16\,400-1000x+y)$ grams **6** $(100x+y)$ cents **7** $\frac{\$3c}{20}$ **8** $180° - (p° + q°)$ **9** a st km b $\frac{k}{p}$ km/h c $\frac{b}{c}$ hours **10** a 30 b −13 c $-\frac{5}{6}$ d 1 e −5 f 17 **11** a $6y+4x$ b $8y^2-6y$ c $9xy$ d $-6ab$ e $25y^2$ f $7a^3$ g $-24x^2y$ h $2b$ i $-3y$ j 2 **12** a $\frac{7x}{6}$ b $\frac{3a}{4}$ c $\frac{11x}{26}$ d $\frac{2}{x}$ e $-\frac{1}{2a}$ f $\frac{ad-bc}{bd}$ g $-\frac{2a}{15b}$ h 4 i $\frac{2}{3m}$ j 1 k $7\frac{1}{2}$ l $\frac{1}{6}$ m $1\frac{7}{9}$ **13** a $8x-28$ b $-15a+36$ c $12y-1$ d $12-5x$ e $4-3x$ f $6x-14$ g $16a-49$ h $x^2-9x+14$ **14** a $\frac{7x-9}{4}$ b $\frac{17x-18}{6}$ c $\frac{3a+1}{3x}$ **15** a $3x-7$ b $14y-9$ c $8x-8x^2-2$ **16** a a^2+2a-8 b $2x^2+5x-3$ c $4x^2+25x-56$ d $11x+4-3x^2$ e $x^2-8x+16$ f $16x^2+8x+1$ g $4y^2-1$ h $25y^2-49$ **17** a $(2x^2+7x-4)$ cm^2 b $(9y^2+24y+16)$ m^2 **18** a $8x-16$ b $-4x$ c $x^2+2+\frac{1}{x^2}$ d $2ab-2a^2-2b^2$ e $3x^2-24x+48$ f $-4a+4$ **19** a 1681 b 841 **20** a $5(x-2y)$ b $2(4a-3)$ c $4y(3x-2)$ d $5y(3y-2)$ e $-2(5a+9)$ f $2yz(2y-3)$ g $3(x^2-3x-16)$ h $\pi r(r-2)$ i $2(l+b)$ j $(x+y)(3+x)$ k $(a-7)(a-b)$ **21** a $(x+y)(2+a)$ b $-(a-b)(2+c)$ c $(c+6)(a-t)$ d $(g-3)(g^2-2)$ e $(x-y)(y-7)$ **22** a $(u-6)(u+6)$ b $(1-5p)(1+5p)$ c $(2a-3b)(2a+3b)$ d $(5k-12m)(5k+12m)$ e $(4x^2+1)(2x+1)(2x-1)$ f $(1+x^2y^2)(1-xy)(1+xy)$ g $2(a-1)(a+1)$ h $5(a-5)(a+5)$ **23** a $(x+4)(x-2)$ b $(k-11)(k-1)$ c $(h+16)(h-3)$ d $(b-11)(b+8)$ e $(x-2y)(x-4y)$ f $(x-y)(x+4y)$ **24** a $(2x+1)(x-1)$ b $(3p+8)(p+1)$ c $(3c-4)(c+3)$ d $(9x+8)(x+1)$ e $(8v+7)(v-1)$ f $(3b+c)(b+5c)$ **25** a $\frac{4}{a-3}$ b $\frac{1}{3}$ c $\frac{a-4}{6}$ d $\frac{a-3}{a+3}$ e $\frac{x+3}{x-4}$ f $\frac{p+4}{p-q}$ **26** a $\frac{1}{2}$ b 5 c $\frac{(a-7)(a+1)}{a(a-1)}$ **27** a $\frac{9x+43}{(x+3)(x+7)}$ b $\frac{6x-22}{(x-3)(x-5)}$ c $\frac{-2x+10}{(x-2)(x+4)}$ d $\frac{6c+34}{(c-7)(c+7)(c+3)}$ e $\frac{7y+8}{(y+3)(y-1)(y+4)}$

Chapter 3 Surds and Indices . . . . pp. 44–45

1 a $\sqrt{7}$, 3, $\sqrt{11}$, $\sqrt{12}$, 5 b $\sqrt{24}$, 5, $\sqrt{33}$, 6, $\sqrt{40}$ **2** a $\sqrt{17}$, 4, $\sqrt{15}$, 3, $\sqrt{8}$ b $\sqrt{70}$, 8, $\sqrt{56}$, $\sqrt{50}$, 7 **3** a $2\sqrt{2}$ b $3\sqrt{3}$ c $8\sqrt{2}$ d $4\sqrt{5}$ e $10\sqrt{2}$ f $40\sqrt{3}$ g $\frac{2}{3}$ h $\frac{9}{10}$ i $\frac{\sqrt{3}}{2}$ j 3 k 20 l 48 **4** a $\sqrt{3}$ b $6\sqrt{5}$ c $4\sqrt{2}-\sqrt{3}$ d $\sqrt{7}$ e $10\sqrt{2}$ f $2\sqrt{x}(x-2)$ **5** a $\sqrt{15}$ b 2 c $8\sqrt{14}$ d 10 e −12 f 12 **6** a $3\sqrt{y-2}$ b $2\sqrt{2a+1}$ c $x\sqrt{x-1}$ **7** a 4 b 2 c $\frac{2}{3}$ d $\frac{\sqrt{2}}{2}$ e 4 f $\sqrt{3}$ **8** a $2-\sqrt{2}$ b $15+2\sqrt{5}$ c $8\sqrt{3}-12$ d $14-\sqrt{14}$ e $6+10\sqrt{6}$ f $x\sqrt{x}-x^2$ **9** a $1+\sqrt{3}$ b $\frac{2-3\sqrt{2}}{2}$ c $\frac{4+\sqrt{5}}{3}$ d $\frac{6+\sqrt{2}}{4}$ e $\frac{\sqrt{3}-\sqrt{6}}{3}$ f $\sqrt{3}+2\sqrt{2}$ **10** a $7+4\sqrt{3}$ b $9-4\sqrt{2}$ c $21+6\sqrt{6}$ d 1 e 1 f $\sqrt{6}+2\sqrt{2}-\sqrt{3}-2$ g $x-2\sqrt{x}+1$ h $x^3+2x\sqrt{x}+1$ i $x-2+\frac{1}{x}$ **11** a $\frac{\sqrt{6}}{6}$ b $\frac{3\sqrt{5}}{10}$ c $\frac{\sqrt{6}}{3}$ d $\frac{\sqrt{14}-\sqrt{7}}{7}$ e $2-\sqrt{3}$ f $3+2\sqrt{2}$

g $3(\sqrt{7} - \sqrt{3})$ h $\dfrac{\sqrt{15} + \sqrt{6}}{6}$ i $5 - 2\sqrt{6}$ **12** a y^7 b x^7 c $-12q^5$ d a^4 e m^2 f $8b^3$ g 27 h 5 i 216 j x^{14} k $8a^9$ l a^2 m 3 n 1 o 3 **13** a $\frac{1}{9}$ b $\frac{1}{8}$ c 4 d $2\frac{7}{9}$ e $\dfrac{4}{x^2}$ f $\dfrac{1}{16x^2}$ g $\dfrac{1}{4a^8}$ h x^{-5} i x j $\dfrac{1}{y^4}$ k $\dfrac{a}{b^2}$ l $\dfrac{2}{x^2}$ **14** a 5 b 3 c 4 d 100 e 8 f $\frac{1}{2}$ g 2 h $5\sqrt{a}$ i $4y^2$ j $3x^2$ **15** a $(2y - 1)^{\frac{1}{2}}$ b $(x + 5)^{\frac{1}{3}}$ c $(5y + 2)^{-1}$ d $2a(3a - 1)^{-2}$ e $(7b - 3)^{-\frac{1}{2}}$ f $(1 - 8p)^{-\frac{1}{3}}$ g $(2y + 3)(6y - 5)^{-\frac{1}{2}}$ h $x^{-\frac{3}{2}}$ i $(8y - 7)^{-\frac{2}{3}}$ **16** a 19 b 729 c $1\frac{1}{2}$ **17** a $x = 4$ b $x = 7$ c $x = 3$ d $x = 2$ e $x = 1$ f $x = 2$ **18** a $0.\dot{4}2857\dot{1}$ b $0.\dot{5}\dot{4}$ c $0.\dot{3}\dot{1}$ **19** a $\frac{97}{99}$ b $\frac{127}{333}$ c $1\frac{23}{90}$ **20** a 3.823×10^4 b 5×10^9 c 2.37×10^{-1} d 3×10^{-7} e 9×10^{-2} f 1.007×10^{-3} **21** a 42 000 b 0.0107 c 0.0008 **22** a 4.28×10^4 b 1.81×10^5 c 1.71×10^{-2} **23** a 7.45×10^8 b 2.33×10^2 c 5.46×10^{-2}

Chapter 4 Equations pp. 63–65

1 a $x = 18$ b $y = 8$ c $y = -\frac{5}{8}$ d $x = 4\frac{1}{4}$ e $y = 0$ f $a = 3\frac{11}{19}$ g $x = 3\frac{5}{12}$ h $x = 3\frac{1}{3}$ i $y = 1\frac{1}{3}$ j $y = \frac{1}{8}$ k $a = 1\frac{2}{15}$
2 a $a > 5$ b $y \geq -8$ c $y \leq -3\frac{1}{2}$ d $a \geq -\frac{3}{4}$ e $x \leq 0$ f $x \geq -\frac{3}{4}$ **3** a 

b 

4 a $-2 < x \leq 4$ b $x \leq -4$, $x > -2$ **5** a $x \geq -5$ 

b $x > \frac{2}{5}$ 

6 $x \leq 10$ 

7 144 **8** 1500 **9** 158.4 **10** a $y = 2x + 2$ b $y = x^2 + 4$ **11** a $C = \dfrac{3t}{2} + 2$ b \$11
12 a $A = 4x^2 + \frac{1}{2}{\neq}x^2$ b $A = 11x^2$ **13** $A = (20 + 2t)(8 + 2t)$ **14** a $y = 8 - 2x$ b $y = \dfrac{5x - 11}{3}$ c $y = \dfrac{11 - x}{2x}$ d $y = \dfrac{7}{5 - 2x}$ e $y = \dfrac{4x}{A}$ f $y = \dfrac{4P^2}{x}$ g $y = \dfrac{1}{1 - k}$ h $y = \dfrac{2x^2}{x^2 - 1}$ **15** a $h = 5$ b $u = 44.1$ **16** a $d = \dfrac{v^2}{70.56}$ b 25 m **17** a $x = -2, y = 3$ b $x = 6, y = 2$ c $x = 1, y = 1$ d $x = 2, y = -6$ **18** a 9 b 7 and 5 c i $a = 3, b = 1$ ii $a = 2, b = -1$ d \$1.05
19 a $\frac{1}{3}$ or -2 b 0 or 5 c 0 or 3 d 2 e 2 or 3 f ± 7 g 3 or -3 h $-\frac{3}{5}$ or 1 i $-\frac{2}{3}$ or $\frac{8}{3}$ j $\frac{3}{4}$ or -4 k $\frac{1}{2}$ or 2 l 7 or -1 m -1 or 4 **20** a $-2 \pm \sqrt{3}$ b $3 \pm \sqrt{7}$ **21** a $2 \pm \sqrt{5}$ b $4 \pm \sqrt{21}$ **22** a -0.736 or -2.264 b 1.117 or -0.717
23 a -6 and -5 or 5 and 6 b 6 and 4 c i Length $= (x + 8)$ cm ii 20 cm and 12 cm d 9 cm e 84°, 84°, 12° f $(-5, -7)$ or $(1, 5)$ g 5 **24** a $x = 1, y = 2$ or $x = 2, y = 1$ b $x = -5, y = 42$ or $x = 2, y = 7$ c $x = -2, y = -1$ or $x = 1, y = 2$
25 Points are $(-5, -7)$ and $(1, 5)$ **26** Dimensions are 8 cm by 4 cm **27** 7 and 3 or -7 and -3 **28** a $x = 1$ b $x = 2$ c $x = -5$
29 a 3, 1, -1, -3 b $-\frac{1}{5}$ or $\frac{1}{2}$

Chapter 5 Linear Relationships pp. 81–83

1 a $2\sqrt{17}$ b 2 c 5 d $3\sqrt{2}$ **2** a $(3, -2)$ b $(-1\frac{1}{2}, -1)$ c $(1, 4)$ d $(2, -2)$ **3** a $m = \frac{1}{4}$ b $m = \frac{3}{2}$ c $m = -1$ d $m = 0$ [Line || x-axis] **4** a $m = -\frac{3}{2}$ b $m = -2$ c $m = -\frac{5}{3}$ d $m = \frac{3}{4}$ e $m = 1$ **5** a See solution b $-\frac{1}{3}$ c $(-1, 1)$ d $(2, 0)$ See solution **6** a See solution b See solution c 8.5 units^2 d AB = BC, △ABC is isosceles **7** a -3 b $\frac{1}{4}$ c 3 d $-\frac{1}{5}$ **8** $x - 2y - 4 = 0$ **9** $3x + y - 4 = 0$ **10** a $2x + y + 1 = 0$ b $x - 2y = 0$ c $3x - 4y + 13 = 0$ **11** $2x - 3y - 19 = 0$ **12** a $5x - y - 6 = 0$ b $x - y + 3 = 0$ c $x - 6y + 8 = 0$ d $y = -5$ **13** A and C **14** See solutions **15** a Neither b Perpendicular c Perpendicular **16** a $4x - y - 4 = 0$ b $2x - 3y - 1 = 0$ c $3x - y - 11 = 0$ d $2x - y + 7 = 0$ **17** $k = -2\frac{2}{3}$
18 $2x - 3y + 5 = 0$ **19** a $x + 2y + 2 = 0$ b $m_{BC} = 2$ c $2x - y + 9 = 0$ d $2\sqrt{5}$ e 15 units^2 f $11x + 2y + 12 = 0$
20 a $2x + 3y + 5 = 0$ b $2x + 3y - 9 = 0$ c $(6, -1)$ d See solution **21** $2x - y - 3 = 0$ **22** $x - 4y + 13 = 0$ **23** See solution
24 a $(1, 4)$, $(2, 1)$, $(-1, 3)$ b $y = 4x$, $5x + 4y = 14$, $x + 5y = 14$ c See solution d See solution e $3x - 2y = 0$
25 $d = 3\frac{1}{4}$ **26** Rectangle **27** $y = 5x$ **28** a $3\sqrt{5}$ b $x - 2y + 12 = 0$ c $x - 2y + 2 = 0$ d $2x + y - 6 = 0$ e $(3, 5)$ f $y = 5$
29 See solution **30** See solution

Chapter 6 Non-linear Relationships pp. 98–99

1 See solutions **2** a Concave up b Concave down c Concave up d Concave up **3** a x-int. of -2 and 1, y-int. of -2 b x-int. of 0, 1 y-int. of 0 c x-int. of 2 and -2, y-int. of -4 d x-int. of 3 and -3, y-int. of 9 **4** a $x = 1$ b $x = -2\frac{1}{2}$ c $x = 1$ d $x = 0$ **5** a $4\frac{1}{4}$ b 6 **6** a 3 b 2 **7** See solutions **8** a $(2\frac{1}{2}, -12\frac{1}{4})$ b $(-1, -9)$ c $(\frac{3}{4}, -\frac{1}{8})$ d $(-1, 6)$ **9** a $y = 2(x + 1)(x - 3)$ b $y = x(x - 4)$ c $y = -(x + 2)(x - 2)$ d $y = 3(x + 2)(x - 4)$ **10** See solutions **11** a $y = (x - 2)^3$ b $y = 2 - x^3$ **12** See solutions
13 a G b F c J d B e E f C g I h D i A j H **14** See solution **15** See solutions **16** a $C(0, 0)$, 7 units b $C(0, 0)$, $2\sqrt{3}$ units c $C(0, 0)$, $\sqrt{5}$ units **17** a $C(-3, 1)$, 4 units b $C(1, -4)$, $4\sqrt{3}$ units c $C(0, -1)$, $2\sqrt{6}$ units d $C(-4, 5)$, $\sqrt{61}$ units

e $C(3, -8)$, $\sqrt{89}$ units f $C(1\frac{1}{2}, -5)$, $\frac{\sqrt{141}}{2}$ units g $C(0, -1)$, $\sqrt{19}$ units **18** See solutions **19** a $(-3, 9)$ b $(2, 6)$, $(1, 3)$ c $(-3, 13)$, $(2, 18)$ d $(-4, -1)$, $(1, 4)$ e $(2, 2)$, $(-2, 2)$ f $(-1, -1)$

Chapter 7 Rates and Proportion

1 a $27\frac{7}{9}$ m/s b 144 km/h c 5L/100 km d 10 km/L e 1.008 km/h f 0.125 kg/\$ g 62.5 L/\$100 h $3\frac{1}{3}$ h/L i 1.8 s/mL **2** a $C = 12d$ b Increases c \$420 d 42 metres e See solutions f 12 **3** a $W = 16.5n$ b Increases c \$495 d 36 hours e See solutions f 16.5 **4** a $D = 9n$ b 315 km c $\frac{1}{4}$ of petrol remains **5** a See solutions b i 100°F ii 15°C **6** a See solutions b i 32 km ii 25 miles **7** a

Kilograms	0	5	10	15	20
Pounds	0	11	22	33	44

b See solutions c i 17.6 pounds ii 330 pounds

8 a i €60 ii £80 b i €36 ii \$70 iii £48 iv €25 c i €3000 ii \$A4000 iii £600 iv €125 d i £48 ii \$A125 **9** a $y = kx$ b $y = 3x$ c $y = 60$ d $x = 6$ **10** a $P = kt$ b $P = 5t$ c $P = 80$ d $t = 6$ **11** a $W = 18n$ b \$486 c 36 hours **12** a $P = 5.2m$ b \$18.72 c 2.8 kg **13** a $y = 3.6x$ b

x	1.3	2.9	4.5
y	4.68	10.44	16.2

14 a $F = 12m$ b 74.4 Newtons c 3.6 kg **15** a $y = k\sqrt{x}$ b $y = 8\sqrt{x}$ c $y = 32$ d $x = 900$ **16** a $y = kx^2$ b $y = 3x^2$ c $y = 108$ d $x = 14$ **17** See solutions **18** a $y = \frac{k}{x}$ b $k = 24$ c $y = 1.5$ d $x = 0.3$ **19** a $M = \frac{k}{a}$ b $k = 32$ c $M = 4$ d $a = 0.5$ **20** a $T = \frac{100}{n}$ b 50 minutes c 4 tankers **21** a $b = \frac{4}{a}$ b

a	2	4	8
b	2	1	0.5

22 See solutions **23** See solutions

Chapter 8 Pythagoras and Trigonometry

1 a $n = \sqrt{34}$ b $m = \sqrt{301}$ c $r = 2\sqrt{33}$ d $s = 3\sqrt{33}$ **2** $4\sqrt{41}$ cm **3** 4.2 cm **4** 28 cm **5** 2.8 m **6** 10.96 m **7** 12.7 cm **8** a Right-angled b Not right-angled c Right-angled **9** a 25 b 15 c 24 d 40 e 84 f 75 **10** a $x = 20$ b $y = 85$ **11** 1 **12** a $y = 21.0$ b $y = 23.3$ c $y = 2.3$ d $y = 48.1$ e $y = 44.6$ f $y = 38.9$ g $y = 12.2$ h $y = 56.3$ **13** a $\theta = 28°4'$ b $x = 69°47'$ c $\phi = 41°25'$ d $\theta = 28°4'$ **14** a $\theta = 30°$ b $\theta = 55°$ c $f = 34.8°$ **15** $\frac{3}{5}, \frac{3}{4}$ **16** $\frac{3}{5}, \frac{4}{5}$ **17** 9.8 m **18** 19°17' **19** 8.0 m **20** 3.1 m, 30°51' **21** 35° **22** a 21°48' b 14°56' **23** 300° **24** a 9.4 m b 56°19' **25** 60° **26** $\frac{1}{\sqrt{2}}, \frac{1}{\sqrt{2}}$, 45° **27** See solution **28** See solution **29** 26°34' **30** a 108°27' b 127°53' d 157°4' **31** a 19.8 b 24.8 c 21.7 **32** 10.3 m **33** 40.5 km **34** a i 10.9 ii 16.4 iii 4.1 b i 40.8 units^2 ii 35.6 units^2 iii 7.3 units^2 **35** a 72°33' b 42°11', 19°37' **36** 12.0 km **37** 17°39', 119°19', 62.8 cm^2 **38** 57° 55', 122° 5', 61.4 cm **39** 530 m **40** 2.9 km **41** 98 m **42** 1455 m, 363 m **43** 3 h 47 min **44** 1300 m **45** 197 m **46** a 153°26' b 18°26' c 56°19'

Chapter 9 Measurement

1 a 96 mm^2 b $12x^2$ units^2 c 102 cm^2 d $3a^2$ units^2 **2** a 58 cm^2, 38 cm b 24 cm^2, 20 cm c 127.27 cm^2, 45.71 cm **3** a 320 cm^3 b 1884.96 cm^3 c 14137.17 cm^3 d 3082.95 cm^3 **4** 8181.23 mL **5** a 679 cm^3 b 3670 cm^3 c 378 cm^3 d 396 cm^3 **6** 1152π cm^3 **7** 1361.36 cm^3 **8** a 904.779 cm^3 b 1357.168 cm^3 **9** $\sqrt[3]{\frac{6}{\neq}}$ cm **10** 1.8 cm **11** 2.5 cm **12** a 5832 cm^3 b 1944 cm^3 c 1526.81 cm^3 d $33\frac{1}{3}$%, 26.18% **13** 48π cm^2 **14** a 1105.84 cm^2 b 340 cm^2 c 1809.56 cm^2 d 1256.64 cm^2 **15** a 360 cm^2 b 628.32 cm^2 **16** a 600π cm^2 b 160π cm^2 **17** 192π cm^2 **18** a 502.655 cm^3 b 503 mL c 351.86 cm^2 d 251.33 cm^2 **19** 3468.32 cm^2

Chapter 10 Geometry

1 $x = 50$ **2** $a = 81$ **3** $b = 72$ **4** $c = 128, d = 52$ **5** $e = 47$ **6** $g = 95, f = 22$ **7** $x = 6$, 46° **8** $h = 124$ **9** $k = 126$ **10** $m = 48$ **11** $p = 102$ **12** $m = 104, n = 76$ **13** $x = 102$ **14** $y = 15$ **15** $a = 140$ **16** $b = 55$ **17** $c = 62$ **18** $a = 45$ **19** $d = 62$ **20** $x = 15$ **21** $y = 128$ **22** $b = 39$ **23** $y = 44$ **24** $d = 45$ **25** $e = 121$ **26** $f = 142$ **27** $x = 39, y = 87$ **28** $z = 20$ **29** $g = 147, h = 99$ **30** $x = 62$ **31** $y = 95$ **32** a 540° b 1800° c 3240° **33** a 135° b 150° c 168.75° **34** a 60° b 36° c 18° **35** $\frac{360}{n}$ **36** See solution **37** See solution **38** See solution **39** See solution **40** See solution

41 See solution **42** See solution **43** See solution **44** See solution **45** See solution **46** See solution **47** 64° **48** 102° **49** 86° **50** 85° **51** 68° **52** 104° **53** 70° **54** 286° **55** 91° **56** See solution **57** See solution **58** See solution **59** See solution **60** See solution **61** See solution **62** See solution **63** See solution **64** a See solution b 5.6 m **65** a 1 : 2 b 1 : 4 **66** 16.2 cm^2 **67** a 4 : 5 b 25.6 m c 480 m^2, 307.2 m^2 **68** a i 4 : 25 ii 8 : 125 b 3.84 cm^3 c 87.5 cm^2 **69** a 2 : 3 b 4 : 9 **70** a 1 : 32 b 1 : 1024 c 46.875 mL d 250 L

Chapter 11 Statistics

pp. 203–206

1 a i 6 ii 5, 8 iii 6.5 b 1, 0.5 c 3 **2** a 5.3, 5 b 5.8, 5.5 **3** 95 **4** See solution **5** See solution **6** See solution **7** a See solution b Range = 38, Mode = 39, 47, Median = 48.5 **8** a 6.5 b 7 c 16 d 10 **9** a 38 b 124 **10** a Range 4, 5.5, 6.5, 8, 9, see solutions b 5, 6, 7, 9, 9, see solutions c 12, 14, 16, 18, 19, see solutions d 7, 8, 10, 12, 14, see solutions **11** a i 16 ii 63 b i 3 ii 16 **12** a 38, 24, 47, 10, 54, see solutions b 124, 114.5, 135.5, 105, 147, see solutions **13** a See solutions b 167.5, 161, 172, 11 c See solutions **14** a 12 b 3.74 **15** a $\bar{x} = 19.5$, $\sigma = 3.82$ b $\bar{x} = 104$, $\sigma = 4.46$ c $\bar{x} = 1.64$, $\sigma = 0.0459$ **16** a See solution b See solution **17** a See solution b See solution c See solution **18** Range = 9, IQR = 6, Standard deviation = 3.04 **19** a Positively skewed b Normally distributed (symmetrical) c Positively skewed d Negatively skewed e Negatively skewed f Symmetrical—without knowing mean and mode we cannot say that it is normally distributed **20** a See solution b Mean = $4\frac{4}{15}$, Mode = 3, Median = 4 (8th score) c Positively skewed **21** a Negatively skewed b Positively skewed **22** a Normally distributed b Median = 8, Mode = 8, Mean = 8 (as the data is normally distributed, it will be the same as the median and mode) **23** The trigonometry test is positively skewed and the algebra test is negatively skewed **24** a See solutions b See solutions c See solutions **25** a *r* b *n* c *s* **26** a Weak positive b Strong negative **27** a See solutions b Strong positive relationship c About 166–167 cm d The length of the arm span increases as the height increases **28** a See solutions b About 41 cm **29** a See solution b Negative c Strong **30** a $y = 5x$ b 13 **31** a $y = \dfrac{6}{5}x - 129$ b i 63 kg ii 175 cm

Chapter 12 Probability

pp. 220–223

1 a $\frac{1}{13}$ b $\frac{1}{13}$ c $\frac{1}{4}$ d $\frac{1}{4}$ e $\frac{1}{2}$ f $\frac{3}{13}$ g $\frac{10}{13}$ h $\frac{5}{13}$ or $\frac{6}{13}$ i $\frac{1}{52}$ j $\frac{7}{13}$ k $\frac{1}{26}$ **2** $\frac{3}{10}$ **3** a $\frac{7}{15}$ b $\frac{8}{15}$ c $\frac{1}{3}$ d $\frac{2}{3}$ e 0 **4** a $\frac{1}{10}$ b $\frac{1}{2}$ c $\frac{3}{5}$ d $\frac{3}{10}$ **5** $\frac{2}{5}$ **6** a $\frac{9}{50}$ b $\frac{17}{100}$ c $\frac{11}{25}$ **7** a 33 b i $\frac{8}{33}$ ii $\frac{4}{33}$ iii $\frac{16}{33}$ **8** a $\frac{1}{3}$ b $\frac{1}{12}$ c $\frac{13}{36}$ **9** See solution **10** a 20 b i $\frac{4}{5}$ ii $\frac{1}{5}$ iii $\frac{3}{10}$ **11** a 30 b i $\frac{3}{5}$ ii $\frac{3}{10}$ iii $\frac{1}{6}$ iv $\frac{2}{5}$ **12** a See solution b i $\frac{1}{12}$ ii $\frac{1}{4}$ **13** a $\frac{1}{9}$ b $\frac{1}{9}$ c $\frac{1}{2}$ d $\frac{1}{6}$ e $\frac{5}{6}$ f $\frac{8}{9}$ g $\frac{11}{18}$ h $\frac{1}{18}$ **14** a $\frac{1}{18}$ b $\frac{1}{4}$ c $\frac{19}{36}$ d $\frac{23}{36}$ e $\frac{5}{9}$ f $\frac{5}{36}$ g $\frac{2}{3}$ **15** a $\frac{11}{36}$ b $\frac{1}{36}$ c $\frac{25}{36}$ d $\frac{2}{3}$ e $\frac{7}{12}$ f $\frac{17}{36}$ g $\frac{1}{12}$ h $\frac{35}{36}$ i $\frac{4}{9}$ j 0 **16** a See worked solutions b i $\frac{7}{25}$ ii $\frac{23}{100}$ c i $\frac{3}{10}$ ii $\frac{7}{10}$ **17** See solutions **18** See solutions **19** See solutions **20** a $\frac{1}{8}$ b $\frac{3}{8}$ c $\frac{3}{8}$ d $\frac{1}{8}$ e $\frac{7}{8}$ **21** a $\frac{3}{10}$ b $\frac{1}{10}$ c $\frac{3}{5}$ d $\frac{2}{5}$ **22** a $\frac{5}{18}$ b $\frac{5}{9}$ c $\frac{5}{18}$ d $\frac{1}{6}$ e $\frac{5}{6}$ **23** a $\frac{1}{50}$ b $\frac{1}{4950}$ c $\frac{49}{2475}$ d $\frac{197}{4950}$ e $\frac{4753}{4950}$ f $\frac{197}{4950}$ **24** a $\frac{1}{3}$ b $\frac{2}{3}$ **25** a $\frac{5}{26}$ b $\frac{1}{5}$ c $\frac{1}{21}$ **26** a $\frac{6}{25}$ b $\frac{10}{27}$ c $\frac{8}{13}$ **27** a See solutions b i $\frac{14}{65}$ ii $\frac{12}{65}$ c i $\frac{11}{30}$ ii $\frac{3}{5}$ d $\frac{7}{20}$ **28** a $\frac{1}{2}$ b $\frac{1}{2}$ c $\frac{1}{2}$ **29** a $\frac{11}{23}$ b $\frac{4}{7}$ c $\frac{1}{2}$ **30** a See solution b $\frac{9}{17}$ c $\frac{4}{9}$

Chapter 13 Polynomials

pp. 237–238

1 a 4, $3x^4$, 3, 2 b 5, $-2x^5$, −2, 4 c 5, $3x^5$, 3, 0 d 0, 7, 7, 7 **2** a Not polynomial b Polynomial c Polynomial d Not polynomial e Polynomial **3** a $7x^4 - 7x^2 - 4$ b $8x^2 - 2x - 4$ c $9x^3 + 12x^2 + 4x + 10$ d $7x^5 - 19x + 1$ e $2x^3 + 3x^2 + 4$ f $3x^4 + 3x^2 + 8$ g $x^3 - 2x^2 - x + 3$ h $5x^4 - 3x^3 - 3x^2 + 5x - 1$ **4** a 4 b 7 **5** a $10x^5 - 2x^4 + 4x^3 - 20x^2 + 4x - 8$ b $6x^5 - 4x^4 + 6x^3 - 10x^2 + 4x$ **6** a $x^3 - 3x^2 + 7x - 5 = (x - 3)(x^2 + 7) + 16$ b $-x^3 + 2x^2 - 5x - 1 = (x + 2)(-x^2 + 4x - 13) + 25$ c $x^5 + x^3 - x = (x^2 + x)(x^3 - x^2 + 2x - 2) + x$ d $x^4 - 1 = (x^2 - 3)(x^2 + 3) + 8$ **7** a $x^3 + 2x^2$ b $x^3 + 2x^2 - 2x + 2$ c $x^4 + x^3 - 3x^2 + 2x - 1$ d $(x^2 + 3x + 2) + 3$ **8** a 5 b −1 **9** $k = -13$ **10** $k = -1$ **11** $k = 6$ or 1 **12** $k = -23$ **13** See solutions, $x = 1, 3, -2$ **14** $(x + 1)(2x - 1)(x - 3)$ **15** $x = -2, 3, -3$ **16** $a = 11$, $b = 12$, $x = 4$ **17** See solutions **18** See solutions a $y = (x + 1)(x - 3)(x - 2)$ b $y = (x - 1)^2(x + 1)$ c $y = (x - 1)^3$ d $y = -(x + 2)^2(x - 4)$ **19** a iii b ii c iv d i

Chapter 14 Functions and Logarithms

pp. 257–259

1 a Function b Not function c Not function d Function e Function f Not function **2** a −11 b 1 c −23 d 19 **3** a 0 b −3 c 5 d −3 **4** a 1, −3, −3 b 1, −3, $-\frac{1}{3}$ c 0, −4, 12 d 9, 1, 25 e 9, 1, 25 f 1, 9, $\frac{1}{9}$ **5** a $h^2 + h$ b $x^2 + 5x + 6$ c $x^2 + 4x - 2$ d $x^2 - x$ **6** See solutions **7** a D: $x \geq 0$, R: $y \geq 0$ b D: all real x, R: $y \geq 4$ c D: all real x, R: $y \leq 4$ d D: $-2 \leq x \leq 2$, R: $0 \leq y \leq 2$ **8** a See solutions b $y = \frac{1}{2}x + \frac{3}{2}$, see solutions c See solutions **9** a See solutions b $y = 2 - \dfrac{x}{2}$, see solutions

10 $y = x^2$, where $x \geq 0$ **11** a $y = 2x + 4$, see solutions b $y = -x + 2$, see solutions c $y = x - 2$, see solutions **12** See solutions **13** See solutions **14** See solutions **15** a a^2 b a^{10} c $16a^{20}$ d $\frac{81}{4}a^2$ e 1 f $a^{-\frac{4}{3}}$ **16** a 9 b 3 c 16 d 4 e 81 f $\frac{1}{25}$ g $1\frac{1}{4}$ **17** a $a = 4$ b $a = 32$ c $a = 2$ d $x = 2$ e $x = -2$ f $x = -4$ g $x = 6$ h $x = 6$ **18** a 6561 b 1024 c 3 d 9 e 1 f 9 **19** a $\log_5 125 = 3$ b $\log_3(\frac{1}{27}) = -3$ c $\log_2 \sqrt{8} = \frac{3}{2}$ **20** a $10^3 = 1000$ b $2^4 = 16$ c $2^{2.5} = 4\sqrt{2}$ d $5^{-2} = \frac{1}{25}$ **21** a $x = 6$ b $x = 3$ c $x = \frac{3}{2}$ d $x = 6$ e $x = -2$ f $x = 8$ g $x = 25$ h $x = \sqrt{3}$ i $x = \frac{1}{10}$ j $x = \frac{1}{8}$ k $x = \frac{1}{2}$ l $x = 2$ m $x = 3$ n $x = 8$ **22** See solution **23** See solution **24** See solution **25** See solution **26** a $\log_a NM$ b $\log_a \frac{N}{M}$ c $\log_a N^i$ d $\log_a(N^iM^j)$ e $\log_a \sqrt{N}$ f $\log_a \frac{\sqrt{N}}{M^2}$ **27** a $\log_a x + \log_a y$ b $\log_a x - \log_a y$ c $-\log_a y$ d $\frac{1}{2}\log_a x$ e $2\log_a x - \log_a y$ f $2\log_a x + \frac{1}{2}\log_a y - 4\log_a z$ **28** a 6 b 3 c $\frac{1}{2}$ d 1.5 e $-\frac{1}{2}$ f 3 g $\frac{1}{2}$ h 1.5 i 0 j 1 k 1 l 3 **29** a 1.176 b 0.222 c 0.239 d 0.716 e 0.938 f −0.477 g −0.699 h 1.653 i 1.875 j 0.732 **30** a 5.1 b 0.5 c 0.5 d 1.4 e −2.8 f −4.6 g −1.4 h 7.4 i 2.6 j 3.3 **31** a $x + y$ b $x - y$ c $y - x$ d $\frac{1}{2}x$ e $\frac{1}{2}(x + y)$ f $3x + y$ g $\frac{3}{2}x + y$ h $2y - x$ i $4x - y$ j $-x$ **32** a $\log_a 30$ b $\log_a(\frac{10}{3})$ c $\log_a \sqrt{10}$ d $\log_a \sqrt{0.3}$ e $\log_a 2700$ f $\log_a 0.81$ **33** a $y = ab$ b $y = \sqrt[3]{a}$ c $y = \frac{1}{\sqrt[3]{x}}$ d $y = \frac{2}{\sqrt{x}}$ **34** a $a = 24$ b $a = 9$ c $a = 8$ d $a = \frac{8}{25}$ e $a = \frac{2}{5}$ f $a = 10$ g $a = -2\frac{2}{7}$ **35** a $x = 1.46$ b $x = 2.10$ c $x = 0.73$ d $x = 0.73$ e $x = -1.46$ f $x = 3.53$ g $x = 0.12$

Chapter 15 Circle Geometry pp. 277–281

1 See solutions **2** a $x = 7$ b $x = 90$ c $y = 8.4$ d $\theta = 110$ **3** a 6 cm b 48 cm c 8.5 cm d 4 cm **4** See solutions **5** See solutions **6** a $a = 43$ b $= 38$ c $y = 38$ d $x = 37, y = 101$ e $g = 96$ f $h = 114$ g $a = 114, b = 114, c = 66$ h $y = 48, x = 42, z = 42$ i $\theta = 111$ j $\phi = 184$ k $y = 124, x = 112, z = 68$ l $a = 96, b = 132$ m $x = 42, y = 42, z = 96$ n $y = 44$ o $x = 52, y = 104$ p $a = 49$ q $x = 44, y = 92$ r $x = 38, y = 52$ s $\phi = 64$ t $a = 22, b = 34$ **7** a $h = 12$ b $h = 20$ c $h = 1.8$ d $h = 24.8$ **8** a $x = 10$ b $t = 11.0$ c $y = 18$ d $a = 10, y = 9$ e $x = 15, t = 18\sqrt{2}$ **9** a 9° b 29° c 28° d See solutions e See solutions f $x = 124$ g $\theta = 43$ h See solutions i 117° j See solutions k 82° l See solutions m See solutions n See solutions o See solutions p See solutions q See solutions

Quick Answers to Tests

Chapter 1 Financial Mathematics

Intermediate Test .... p. 12

1 a \$4027 b i \$90 000 ii \$21 247 **2** \$189 **3** \$2400 **4** \$4830 **5** \$58.50 **6** a \$3840 b \$330 **7** \$179.90 **8** a \$4600 b \$1069.50 **9** \$945.75 **10** \$8852.18 **11** \$7025.96 **12** Compound interest account by \$101.89 **13** \$16 000

Advanced Test .... p. 13

1 a \$40 600 b \$127 200 **2** 6% p.a. **3** 7 years **4** \$80 **5** 44 **6** \$766.67 **7** \$1761.52 **8** 5.5% p.a. **9** 3.1808 (or 4 years as interest would be added at the end of the year) **10** 2 years **11** 449 458

Chapter 2 Algebraic Techniques

Intermediate Test .... p. 32

1 a $8x$ b $6xy$ c $400x$ cents **2** a -48 b 3 c 20 d $\frac{1}{5}$ **3** a $10m$ b $36y^2$ c $-7ab$ d $16y^2$ e $3y$ f $\frac{b}{2a}$ **4** a $\frac{6x}{y}$ b $\frac{-5m}{6}$ c $\frac{19}{6x}$ d x^2 e $\frac{4}{25}$ **5** a $15x - 10y$ b $15 - 6x$ c 7 **6** a $\frac{10x-3}{12}$ b $\frac{16x-15}{6}$ **7** a $x^2 + 2x - 15$ b $4x^2 - 25$ c $4x^2 + 20x + 25$ **8** a $7(y - 3)$ b $5m(m + 3n)$ c $(m + 3)(m - 3)$ d $(a + 8)(a - 5)$ e $(y - 9)(y - 2)$ f $(y + 9)(y + 2x)$ g $(x + 5)(2x - 1)$ h $7(x - 2)(x + 2)$ **9** a $\frac{3}{x+y}$ b $\frac{m+2n}{4}$ c $\frac{x^2}{x+5}$ d $\frac{8x+1}{x(x+1)}$

Advanced Test .... p. 33

1 a $3y$ b $24a^2$ c $\$\frac{3x}{100}$ **2** a $-2t$ b $-42mn$ c $-17m^2$ d $3b$ e $27m^2n$ f $10a - 10c$ **3** a $5b^2$ b $\frac{19a}{10}$ c $\frac{16}{3m}$ d $\frac{9}{y}$ e $\frac{5y}{2}$ **4** a $35m - 21m^2$ b $-t + 40$ c $-3a + 40$ **5** a $\frac{23x+4}{12}$ b $\frac{21m}{10}$ **6** a $3x^2 + 13x - 10$ b $25x^2 - 9$ c $25x^2 - 30x + 9$ d $-x$ e $-8x$ **7** a $9(6 - a)$ b $(m - 8)(m + 8)$ c $9y(3y - 1)$ d $(p + 7)(p - 6)$ e $(p - 9)(p + 8)$ f $(3x - 2)(2x + 1)$ g $(m - 8)(m - 7t)$ h $(a^2 + b^2)(a + b)(a - b)$ **8** a $\frac{7}{x}$ b $\frac{-10}{(2x+1)(2x-1)}$ c $\frac{9a}{b(3a-1)}$ d $\frac{1}{4y}$ e $\frac{1}{x(x+4)}$

Chapter 3 Surds and Indices

Intermediate Test .... p. 47

1 a $\frac{3}{5}$ b 2 **2** a $6 - 9\sqrt{2}$ b $3 - 2\sqrt{2}$ **3** $a = 7$ **4** $2 - \sqrt{3}$ **5** $\frac{12-\sqrt{6}}{9}$ **6** $\frac{2}{x^3}$ **7** $\frac{1}{9}$ **8** $x = 4$ **9** a x^2y^3 b $2\sqrt[3]{x}$ **10** $\frac{47}{99}$ **11** 3.17×10^2

Advanced Test .... p. 48

1 $2a\sqrt{a}$ **2** 121 **3** $a = 17, b = -12$ **4** $x = 20, y = 8$ **5** -1 **6** $\frac{4}{27}$ **7** $\frac{1}{a^7}$ **8** a $x = \frac{1}{2}$ b $x = 0$ **9** $m = 1, n = 4$ **10** $5(1 - 3x + x^2)^{-\frac{1}{3}}$

Chapter 4 Equations

Intermediate Test p. 67

1 a $9\frac{1}{3}$ b 42 **2** $-1 < x \leq 2$ **3** $x < -2$ 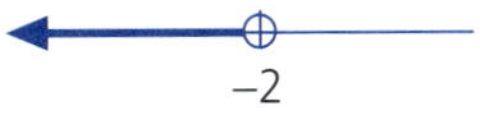 **4** 68 **5** $y = \frac{7x - 9}{2}$ **6** $x = 2$ **7** $x = 3, y = -5$ **8** a $x = 8$ or -1
b $x = 0$ or $\frac{4}{5}$ **9** $y = \frac{5 \pm \sqrt{33}}{2}$ **10** $x = 6$ or 2 **11** $(-7, 28), (2, 10)$ **12** $\frac{1}{2}$ cm and 4 cm

Advanced Test p. 68

1 a 2 b $\frac{4}{9}$ **2**  **3** $x > 2$ **4** $-\frac{1}{2}, -2$ **5** a Roots are real b Roots are not real **6** a $y = \frac{M + bx}{x}$
b $y = \frac{nx + mx - Pmn}{m - n}$ **7** $x = 3, y = 4$ **8** a $\frac{5}{2}, -\frac{2}{3}$ b $7, -5$ **9** 4.79, 0.21 **10** $(1\frac{1}{2}, 2), (-1, -3)$ **11** 10, 11

Chapter 5 Linear Relationships

Intermediate Test p. 85

1 a $\sqrt{89}$ b $(1, -\frac{3}{2})$ c $-\frac{5}{8}$ **2** $-\frac{4}{3}, y = -\frac{4}{3}x + 4$ **3** $\frac{1}{2}, -\frac{5}{4}$ **4** $y = 2x - 4$ **5** $x + y - 3 = 0$ **6** See solutions **7** $x - 2y + 6 = 0$
8 a $3\sqrt{5}$ b 3 c $(-2, 1)$ d $3x - y + 7 = 0$ e $x + 3y - 1 = 0$ f $8x + 2y + 1 = 0$ **9** $k = 2$

Advanced Test p. 86

1 a $\sqrt{65}$ b -6 c $(1, -4)$ d $(-\frac{1}{2}, -4\frac{1}{2})$ **2** $-2, y = -2x - 4$ **3** $3, 3\frac{1}{2}$ **4** $3x - 4y - 18 = 0$ **5** $x + y + 2 = 0$ **6** $2x - y + 6 = 0$
7 $m = -2$ **8** a $p = 2, q = -5$ b 5 **9** See solutions, (2, 2) point of intersection **10** a $5\sqrt{2}$ b $\frac{1}{7}$ c $x - 7y - 10 = 0$
d $x - 7y + 16 = 0$ e $(-3, 0)$ f $(5, 3)$ g $5\sqrt{2}$ **11** See solutions

Chapter 6 Non-linear Relationships

Intermediate Test p. 101

1 D **2** B **3** See solutions **4** a $x = -1$ b -4 c See solutions **5** $x = -1$, max. of 5 **6** $y = (x + 2)(x - 4)$ **7** See solutions
8 $x^2 + y^2 = 16$ **9** $x = 5, -4$ and $y = -20$ **10** $x = 0$ **11** a See solutions b See solutions **12** a See solutions
b See solutions c See solutions **13** a $y = 3 - x^2$ b $y = (x + 2)(x - 4)$ c $y = x(x - 3)$ **14** a See solutions b See solutions
15 a $(x - 5)^2 + (y + 2)^2 = 16$ b $x^2 + y^2 = 17$ c $(x - 2)^2 + (y - 2)^2 = 4$ **16** a See solutions b See solutions **17** (3, 9) and (−1, 1)

Advanced Test p. 102

1 C **2** B **3** See solutions **4** a $x = 3\frac{1}{2}$ b $2\frac{1}{4}$ **5** $x = 1$ and min. of 3 **6** a $y = -3x(x - 4)$ b $y = 2(x + 4)(x - 6)$
c $y = 4(x + 2)(x - 8)$ **7** A **8** See solutions **9** $4\sqrt{3}$ units **10** a $(-2, -4)$ b $(1, -16)$ c $(1, -1)$ **11** a $y = 3(x + 1)(x - 3)$
b $y = -2x(x - 6)$ c $y = 4(x - 3)(x + 3)$ **12** a See solutions b See solutions **13** $C(2, 6)$, radius 7 units **14** See solutions
15 See solutions **16** (2, 6) and (−2, −6) **17** See solutions

Chapter 7 Rates and Proportion

Intermediate Test p. 114

1 14.6 cm/year **2** a 200 kg/ha b $800/m^3$ **3** Cheetah is faster by 23 km/h **4** a

h	0	1	2	3	5
D	0	90	180	270	360

b $D = 90h$ c 90 km/h d 112.5 km e See solutions f First car travelled 105 km more g Second car takes 1 hour longer
5 a $C = 1175A$ b $364 250 **6** a $d = 3.2\sqrt{h}$ b 6.25 metres c As 13.58 > 12, then the person can see the island
7 See solutions **8** See solutions

Advanced Test p. 115

1 160 g/L **2** If x is doubled, then y is doubled **3** If x is doubled, then y is multiplied by 4 **4** a $k = 2$
b

x	4	16	49	144
y	4	8	14	24

5 a Truck A: $D = 100t$; Truck B: $D = 100t$ b 80 km more c After 3 hours **6** a $C = 7.2n^2$
b 2.5 cm c $133.13 **7** 4.32 minutes **8** See solutions **9** See solutions **10** See solutions

Chapter 8 Pythagoras and Trigonometry

Intermediate Test pp. 140–141

1 a 11.40 b 5.85 **2** Right-angled **3** $\frac{5}{12}, \frac{5}{13}$ **4** 20 **5** 12 **6** a 16.3 b 17.1 c 29.9 **7** a 73°37′ b 37°40′ **8** 4.9 m
9 a 13.6 b 12.6 **10** 36.6 cm^2 **11** 36°52′ **12** 134°50′, 45°10′

Advanced Test pp. 142–143

1 24 m **2** 50π cm^2 **3** See solutions **4** $\frac{15}{17}, \frac{15}{8}$ **5** a $\frac{1}{\sqrt{2}}$ b $\frac{-1}{\sqrt{3}}$ c 1 **6** a 3.7 cm b 3.0 cm c 0.7 cm **7** a 65° b 51°20′
8 26.4 **9** 383.4 m **10** a 11.9 cm b 11.3 cm **11** 114°57′ **12** 99.7 cm^2 **13** 63°26′, 53°8′, 63°26′ **14** a See solutions
b 1467 km c 155°

Chapter 9 Measurement

Intermediate Test pp. 155–156

1 a 320 cm^3 b 1017.88 cm^3 c 288π cm^3 d 4817.11 cm^3 **2** a 1130.97 cm^2 b 260 cm^2 c 1130.97 cm^2 d 256π cm^2
3 a 838 cm^3 b $\frac{1}{5}$ **4** 116.58 cm^2 **5** 40π cm^2 **6** a 4 cm b 36π cm^2 **7** 60 cm

Advanced Test pp. 157–158

1 a 400 cm^3 b 96π cm^3 c 8578.64 cm^3 **2** a 2144.66 cm^3 b 2144.66 cm^3 **3** 160π cm^2 **4** a 384.96 cm^2 b 804.89 cm^2
c 1592.79 cm^2 **5** a $\frac{3}{4}$ b $\frac{2}{7}$ **6** Surface area is four times the original **7** 52% of cube taken by basketball **8** 80 cm

Chapter 10 Geometry

Intermediate Test p. 184

1 a 74 b 126 c 86 d 106 e 78 f 39 **2** See solutions **3** See solutions **4** See solutions **5** a See solutions
b See solutions c $5\frac{1}{3}$ cm **6** 144° **7** a i 3 : 4 ii 9 : 16 iii 27 : 64 b 270 cm^3

Advanced Test .... pp. 185–186

1 a 30 b 122 c 66 d 68 e 66 **2** a, b, c, see solutions **3** a, b, c, see solutions **4** a, b see solutions **5** a See solutions b $2y$ **6** a, b, see solutions **7** 3240° **8** a 3 cm b 160 cm^2

Chapter 11 Statistics

Intermediate Test .... p. 208

1 C **2** a Mean number of points is 31 b Team needs to score 31 points c Mean after six games is 32 d Team lost by 10 points **3** a Mean = 4, Mode = 5, Range = 5 b Median = 4, Lower quartile = 3, Upper quartile = 5, IQR = 2 c See solutions **4** a Mean = 31.5, Standard deviation = 4.61 b Mean = 2.8, Standard deviation = 2.48 **5** a Highest was 24 out of 25; lowest was 3 out of 25 b i 50% ii 75% iii 25% c 7 students d Bottom half e Negatively skewed **6** a i Wildfires has a bigger range ii Diamonds has a bigger mode iii Diamonds has bigger median b Wildfires is positively skewed; Diamonds is negatively skewed **7** a See solution b y c Positive d Strong **8** 120

Advanced Test .... pp. 209–210

1 a Normally distributed b Bi-modal c Negatively skewed d Positively skewed **2** a Mode = 10; Range = 35 b 18 c IQR = 21 d See solution **3** a Negatively skewed b Standard deviation = 1.04 c The class mean is 8.7. This means that this class' mean is higher than the year mean **4** a \$10 b 75% c 100 books d Shop A had IQR of 25, which is higher than Shop B's IQR of 15 **5** Mean = 30, Standard deviation is unchanged **6** a See solution b Positive c Strong relationship. The hotter the temperature, the more people want to go to the beach **7** IQR = 2.7 **8** a Weak b 50 c $a = 45$; $b = 40$ **9** a $y = 35x - 650$ b 645

Chapter 12 Probability

Intermediate Test .... pp. 225–226

1 a $\frac{1}{4}$ b $\frac{3}{4}$ **2** a See solutions b i $\frac{25}{64}$ ii $\frac{15}{32}$ **3** a $\frac{1}{12}$ b $\frac{1}{3}$ **4** a $\frac{27}{100}$ b $\frac{16}{25}$ c $\frac{137}{200}$ **5** a $\frac{37}{80}$ b $\frac{43}{80}$ **6** a $\frac{1}{12}$ b $\frac{1}{4}$ c $\frac{1}{6}$ **7** a See solutions b i $\frac{1}{9}$ ii $\frac{1}{9}$ iii $\frac{5}{9}$ iv $\frac{1}{3}$ v $\frac{2}{3}$ **8** a See solutions b i $\frac{2}{7}$ ii $\frac{2}{7}$ iii $\frac{1}{7}$ **9** a $\frac{1}{4}$ b $\frac{1}{2}$ c $\frac{3}{5}$

Advanced Test .... pp. 227–228

1 a See solutions b i $\frac{8}{15}$ ii $\frac{13}{15}$ **2** a 0.45 b 0.425 **3** a i 0.61 ii 0.31 b i $\frac{5}{9}$ ii $\frac{8}{27}$ **4** a See solutions b $\frac{1}{5}$ c $\frac{3}{11}$ **5** a $\frac{1}{5}$ b $\frac{2}{5}$ **6** a See solutions b i $\frac{1}{8}$ ii $\frac{3}{8}$ iii $\frac{7}{8}$ **7** a See solutions b i $\frac{15}{77}$ ii $\frac{39}{77}$ iii $\frac{38}{77}$ iv $\frac{62}{77}$ **8** a See solutions b i 0.504 ii 0.054 iii 0.398

Chapter 13 Polynomials

Intermediate Test .... p. 240

1 a 4 b 2 **2** a Degree = 3, constant term = 1 b Degree = 5, constant term = −2 **3** $x^3 - 2x^2 + 6x - 1 = (x - 1)(x^2 - x + 5) + 4$ **4** a −1 b 32 c −43 **5** See solutions **6** $(x + 1)(x + 4)(x - 2)$ **7** a See solutions b See solutions **8** $y = 2(x + 1)(x - 2)(x - 5)$

Advanced Test .... p. 241

1 $x^3 - 4x^2 + 5x + 2 = (x - 3)(x^2 - x + 2) + 8$ **2** $2x^3 - x - 1 = (x - 1)(2x^2 + 2x + 1)$ **3** See solutions **4** −11 **5** $a = -2$, $b = 3$ **6** 2, −4, 3 **7** 1, −3, 2 **8** a See solutions b See solutions

Chapter 14 Functions and Logarithms

Intermediate Test p. 261

1 D **2** 30 **3** $2 - x$ **4** a $x \neq 0$ b $x \geq 0$ **5** a Dom: all real x; Range: all real y b Dom: all real x; Range: $y \geq 3$ **6** a $y = \frac{x+2}{3}$ b See solutions **7** See solutions **8** 8 **9** See solutions **10** a 1.4765 b 0.2237 c 0.52945 **11** a 1 b 8 c 3 **12** a 4 b 2

Advanced Test p. 262

1 a 6 b $x^2 - 10$ c $x = 6, -2$ **2** a $x \neq 2$ b $x \neq \frac{3}{2}, -4$ c $x \leq 3$ d $-3 \leq x \leq 3$ **3** a Dom: all real x; Range: $y \leq 4$ b Dom: all real x, except $x = -\frac{1}{2}$; Range: all real y, except $y = 0$ c Dom: all real x; Range: $y > 1$ **4** See solutions **5** a $2\frac{1}{2}$ b 0 c 2 **6** a $\log_b 12$ b $\log_b \frac{9}{4}$ **7** a $x = 8$ b $x = 6$ **8** 1 **9** $x = -2\frac{3}{4}$

Chapter 15 Circle Geometry

Intermediate Test pp. 283–284

1 a Passes through the centre of the circle b Equal angles at the centre c Are supplementary d A right-angle e Is a right-angle **2** a Radius b Chord c Minor segment **3** a 4 b 44 c 12 d 67 e 122 f 64 **4** See solutions **5** 10 **6** a See solutions b See solutions **7** See solutions

Advanced Test pp. 285–286

1 a … the product of the intercepts of any secants from the point b … and the point of contact are collinear c … equal to the angle in the alternate segment d … bisects their common chord at right angles e … twice the angle at the circumference standing on the same arc **2** a Arc b Sector c Tangent **3** a 46 b 37 c 36 d 102 e 68 f 58 **4** 47°, 86° **5** 9 **6** See solutions **7** See solutions

Quick Answers to Sample Examinations

Examination Paper 1 pp. 287–292

1 a \$15 057.50 b 6 years c \$19 380 d \$1031.23 e \$27 442.06 **2** a $10x + 50$ b i $(x - 3)(x + 2)$ ii $(2x - 3)(x + 1)$ c i $\frac{2}{x+1}$ ii $\frac{7a - 22}{6}$ **3** a $5\sqrt{3}$ b $a = 7, b = -4$ c $-4\sqrt{2}$ d 1 e -4 **4** a $y = \frac{3x - 5}{4}$ b $x = 8, -2$ c $\frac{5 \pm \sqrt{37}}{6}$ d (5, 5), (1, −3) e $x = \pm 1$ **5** a i $2\sqrt{13}$ units ii (1, −1) iii $-\frac{3}{2}$ b i $\frac{3}{2}$ ii −2 c $4x + 3y - 4 = 0$ d 6 **6** a i $x = 3$ ii 1 iii See solution b $y = -3(x + 2)(x - 4)$ c C(−2, 3) and radius is 4 units d See solution **7** a 300 kg/ha b i $y = kx$ ii $y = 3x$ iii 60 iv 32 c $W = 22n$ d B **8** a 1.93 m b $-\frac{1}{\sqrt{3}}$ c $\frac{2\sqrt{10}}{3}$ d $\frac{1}{\sqrt{3}}$ e 29°56′ **9** a 128π cm^3 b Volume is 3619.11 cm^3, surface area is 1357.17 cm^2 c 384 cm^3 **10** a i See solution ii See solution b i See solution ii See solution c 2 : 3 **11** a i 25% ii 6 iii Negatively skewed b i 3.13 ii 1.03 c i $y = x + 15$ ii 67 **12** a i See solution ii $1\frac{1}{8}$ **2** $\frac{3}{8}$ **3** $\frac{7}{8}$ b i $\frac{3}{5}$ ii $\frac{1}{2}$ iii $\frac{1}{2}$ **13** a $(2x^3 - 5x^2 + 3x - 2) = (x - 1)(2x^2 - 3x) - 2$ b See solution c −3 d See solution **14** a $x^2 - 8x + 17$ b See solution c 125 d 0.8308 e 3.46 **15** a $x = 100$ b See solution c 2.83 cm d 90°

Examination Paper 2 pp. 293–298

Part A

1 B **2** C **3** A **4** A **5** B **6** D **7** C **8** A **9** A **10** D **11** B **12** A **13** D **14** A **15** B **16** A **17** C **18** B **19** B **20** D **21** B **22** A **23** C **24** D **25** C **26** A **27** D **28** B **29** B **30** C

Part B

31 a $5\sqrt{2}$ units b $x = 2$ c $x = 2$ d $\frac{x - 2}{x + 3}$ e 7 f $t = 6$ g $x = 5$ h 12 i $x + y - 3 = 0$ j $x = 0, 2$ k $(3x - 4)(2x + 1)$ l 84° m $M = 16n$ n 3.2 o 6 and 0 **32** a $(80 - 2x)$ m b Width = 15 m; area = 750 m^2 **33** a $y = 4x - 70$ b 82 **34** a A(3, 0) b B(5, 10) **35** a 34.64 m b 14.64 m **36** a 8 m b 36 m **37** See solutions **38** 25° (see solution) **39** a See solutions b $y = 2x(x - 6)$ **40** a See solutions b See solutions **41** a $a = 4$ b $k = -25$ c $26\frac{1}{24}$ units^2

Worked Solutions to Practise Practise

Chapter 1 . . . . pp. 8–10

Financial Mathematics

1 a Nil tax

b $0.19 \times (29430 - 18200)$
$= 2133.7$
∴ Tax of \$2133.70.

c $3572 + 0.325 \times (68495 - 37000)$
$= 13807.875$
$= 13807.88$ (2 dec. pl.)
∴ Tax of \$13 807.88.

d $17547 + 0.37 \times (132427 - 80000)$
$= 36944.99$
∴ Tax of \$36 944.99.

e $54547 + 0.45 \times (193483 - 180000)$
$= 60614.35$
∴ Tax of \$60 614.35.

2 a \$912 tax means the taxable income is between \$18 200 and \$37 000.
As $912 \div 0.19 = 4800$, the income is \$4800 more than \$18 200.
∴ Taxable income is \$23 000.

b \$24 947 tax means the taxable income is between \$80 000 and \$180 000.
Amount over 17 547
$= 24947 - 17547$
$= 7400$
As $7400 \div 0.37 = 20000$, the income is \$20 000 more than \$80 000.
∴ Taxable income is \$100 000.

c \$198 547 tax means the taxable income is over \$180 000.
Amount over 54 547
$= 198547 - 54547$
$= 144000$
As $144000 \div 0.45 = 320000$, the income is \$320 000 more than \$180 000.
∴ Taxable income is \$500 000.

3 Aiden:
Tax $= 3572 + 0.325 \times (59950 - 37000)$
$= 11030.75$
∴ Aiden pays \$11 030.75 tax.
∴ Emma pays \$22 061.50 tax.
Emma: Tax of \$22 061.50 means a taxable income between \$80 001 and \$180 000.
Firstly, $22061.5 - 17547 = 4514.5$.
Also, $4514.5 \div 0.37 = 12201.35$ (2 dec. pl.)
∴ Emma earns \$12 201.35 more than \$80 000.
∴ She earns \$92 201.35.

4 a $3572 + 0.325 \times (72000 - 37000)$
$= 14947$
∴ Tax of \$14 947.

b After pay rise of \$5000, Jacob is still in the same tax bracket.
Tax on extra \$5000
$= 0.325 \times 5000$
$= 1625$
∴ He pays \$1625 of the bonus in tax.

5 a $I = Prn$
$= 1800 \times 0.07 \times 5$
$= 630$
∴ Simple interest of \$630.

b $I = 450 \times 0.085 \times 3$
$= 114.75$
∴ Simple interest of \$114.75.

c $I = 1200 \times 0.09 \times 1.5$
$= 162$
∴ Simple interest of \$162.

d $I = 680 \times 0.005 \times 3$
$= 10.2$
∴ Simple interest of \$10.20.

e $I = 10000 \times 0.012 \times 4$
$= 480$
∴ Simple interest of \$480.

6 a $I = Prn$
$840 = 3500 \times r \times 4$
$14000r = 840$
$r = \dfrac{840}{14000}$
$= 0.06$
∴ Rate of 6% p.a.

b $I = Prn$
$1921.5 = 6100 \times r \times 7$
$42700r = 1921.5$
$r = \dfrac{1921.5}{42700}$
$= 0.045$
∴ Rate of 4.5% p.a.

c $I = Prn$
$2250 = 10000 \times r \times 6$
$60000r = 2250$
$r = \dfrac{2250}{60000}$
$= 0.0375$
∴ Rate of 3.75% p.a.

d $I = Prn$
$410.4 = 2700 \times r \times 4$
$10800r = 410.4$
$r = \dfrac{410.4}{10800}$
$= 0.038$
∴ Rate of 3.8% p.a.

7 a $I = Prn$
$500 = P \times 0.04 \times 5$
$0.2P = 500$
$P = \dfrac{500}{0.2}$
$= 2500$
∴ \$2500 invested.

b $I = Prn$
$680.8 = P \times 0.037 \times 8$
$0.296P = 680.8$
$P = \dfrac{680.8}{0.296}$
$= 2300$
∴ \$2300 invested.

c $I = Prn$
$3570 = P \times 0.0425 \times 10$
$0.425P = 3570$
$P = \dfrac{3570}{0.425}$
$= 8400$
∴ \$8400 invested.

d $I = Prn$
$280 = P \times 0.004 \times 7$
$0.028P = 280$
$P = \dfrac{280}{0.028}$
$= 10000$
∴ \$10 000 invested.

8 a $I = Prn$
$336 = 2800 \times 0.04 \times n$
$112n = 336$
$n = \dfrac{336}{112}$
$= 3$
∴ Invested for 3 years.

b $I = Prn$
$940.5 = 5700 \times 0.0275 \times n$
$156.75n = 940.5$
$n = \dfrac{940.5}{156.75}$
$= 6$
∴ Invested for 6 years.

c $I = Prn$
$1782 = 3600 \times 0.055 \times n$
$198n = 1782$
$n = \dfrac{1782}{198}$
$= 9$
∴ Invested for 9 years.

d $I = Prn$
$225 = 5000 \times 0.003 \times n$
$15n = 225$
$n = \frac{225}{15}$
$= 15$
$\therefore$ Invested for 15 months.

9 Let Interest = Principal = x
$I = Prn$
$x = x \times 0.05 \times n$
$0.05n = 1$
$n = \frac{1}{0.05}$
$= 20$
$\therefore$ Will take 20 years.

10 Interest = $1200 \times 0.11 \times 2$
= 264 $\therefore$ \$264

11 Amount paid for the sofa
= 49×24
= 1176 $\therefore$ \$1176

12 a Interest charged = \$7200
Amount to be repaid
= Amount borrowed + Interest
= 20 000 + 7200
= 27 200 $\therefore$ \$27 200

b Monthly repayment
= $27\,200 \div 48$
= 566.67 (to 2 dec. pl.) $\therefore$ \$566.67

13 a Total amount he will pay
= $1400 \times 12 \times 21$
= 352 800 $\therefore$ \$352 800

b Interest he will pay over 21 years
= 352 800 − 280 000
= 72 800 $\therefore$ \$72 800

14 a Cost on terms
= $150 + 11.50 \times 208$
= 2542 $\therefore$ \$2542

b Amount of interest charged
= 2542 − 1495
= 1047 $\therefore$ \$1047

15 A Total cost
= 250×12
= 3000 $\therefore$ \$3000

B Total cost
= $1500 + 100 \times 12$
= 2700 $\therefore$ \$2700

C Total cost
= $1000 + 150 \times 12$
= 2800 $\therefore$ \$2800

D Total cost
= $0.2 \times 2400 + 210 \times 12$
= 3000 $\therefore$ \$3000

$\therefore$ **B** is the cheapest buy.

16 Cash price = \$240
Deposit = \$60
Balance = Cash price − Deposit
= \$240 − \$60
= \$180

a $I = 0.12 \times 180$
= 21.60 $\therefore$ \$21.60

b Total amount to be repaid
= Cash price + Interest
= 240 + 21.60
= 261.60 $\therefore$ \$261.60

c Deposit = \$60
Amount owed = \$261.60 − \$60
= \$201.60
Instalment = $201.60 \div 12$
= 16.80 $\therefore$ \$16.80

17 Cash price = \$670
Deposit = 20% of \$670
= \$134
Balance = \$670 − \$134
= \$536

Interest charged = $\frac{\$536 \times 12 \times 1.5}{100}$
= \$96.48

Total amount to be repaid
= Cash price + Interest
= \$670 + \$96.48
= \$766.48
Amount still owed = \$766.48 − \$134
= \$632.48
Monthly repayment = $632.48 \div 18$
= 35.14 $\therefore$ \$35.14

18 Price = $0.8 \times 0.9 \times 1450$
= 1044
$\therefore$ The price paid is \$1044.

19 Let product cost \$100
$\therefore$ A: $0.85 \times 0.75 \times 100 = 63.75$
B: $0.65 \times 100 = 65$
$\therefore$ A is a better deal as the product will cost \$63.75 compared to \$65 for deal B.

20 Cost = $0.85 \times 0.9 \times 11.90 \times 7.5$
= 68.27625
= 68.28 (to 2 dec. pl.)
Steak costs \$68.28.

21 Let original price be \$100.
New price = $100 \times 0.90 \times 0.90$
= 81
As 100 − 81 = 19, the total discount is 19%.

22 Original price = $150 \div 0.75$
= 200
$\therefore$ Price was \$200.

23 Original price = $902.5 \div 0.95 \div 0.95$
= 1000
$\therefore$ Price was \$1000.

24 Price = 1200×0.98^7
= 1041.750 64…
= 1042 (nearest whole)
$\therefore$ Price is \$1042.

25 January value = $25\,977.6 \div 0.88 \div 0.82$
= 36 000
$\therefore$ Value was \$36 000.

26 a Interest = $6400 \times 1.063^3 - 6400$
= 1222.5024
= 1222.50 (to 2 dec. pl.)
$\therefore$ \$1222.50

b Interest = $4900 \times 1.085^5 - 4900$
= 2467.917 782
= 2467.92 (to 2 dec. pl.)
$\therefore$ \$2467.92

c Monthly: 6% pa = 0.5% per month
2 years = 24 months
$\therefore$ Interest = $2600 \times 1.005^{24} - 2600$
= 330.615 4181
= 330.62 (to 2 dec. pl.)
$\therefore$ \$330.62

d Six monthly:
7% p.a. = 3.5% per 6 months
4 years = 48 months
$\therefore$ Interest
= $14\,000 \times 1.035^8 - 14\,000$
= 4435.326 517
= 4435.33 (to 2 dec. pl.)
$\therefore$ \$4435.33

27 Quarterly: 8% p.a. = 2% per quarter
10 years = 40 quarters
$\therefore$ Amount = 9400×1.02^{40}
= 20 755.572 84
= 20 755.57 (to 2 dec. pl.)
$\therefore$ \$20 755.57 after 10 years.

28 Pop = $11\,240 \times 1.03^{25}$
= 23 534.063 93
= 23 534 (to nearest whole number)
$\therefore$ The population is 23 534.

29 a $A = P(1 + r)^n$
$2920 = P(1.04)^5$
$P = \frac{2920}{1.04^5}$
= 2400.027 152…
= 2400 (nearest whole)
$\therefore$ Investment of \$2400.

b $A = P(1 + r)^n$
$5926 = P(1.035)^8$
$P = \frac{5926}{1.035^8}$
= 4500.272 8882…
= 4500 (nearest whole)
$\therefore$ Investment of \$4500.

c $A = P(1 + r)^n$
$39\,763 = P(1.0475)^{10}$
$P = \frac{39763}{1.0475^{10}}$
= 24 999.931 397…
= 25 000 (nearest whole)
$\therefore$ Investment of \$25 000.

30 As $4.8 \div 12 = 0.4$, then 0.4% per month for 72 months:
$A = P(1 + r)^n$
$26\,660 = P(1.004)^{72}$
$P = \frac{26660}{1.004^6}$
= 20 000.1302…
= 20 000 (nearest whole)
$\therefore$ Investment of \$20 000.

31 a $2400 + 466 = 2866$

$A = P(1 + r)^n$

$2866 = 2400(1 + r)^6$

$(1 + r)^6 = \frac{2866}{2400}$

$1 + r = \left(\frac{2866}{2400}\right)^{\frac{1}{6}}$

$= 1.030016442\ldots$

$r = 0.030016422\ldots$

$= 0.030$ (3 dec. pl.)

$\therefore$ Interest rate of 3.0% (1 dec. pl.)

b $8000 + 4424 = 12424$

$A = P(1 + r)^n$

$12424 = 8000(1 + r)^{10}$

$(1 + r)^{10} = \frac{12424}{8000}$

$1 + r = \left(\frac{12424}{8000}\right)^{\frac{1}{10}}$

$= 1.045002058\ldots$

$r = 0.045002058\ldots$

$= 0.045$ (3 dec. pl.)

$\therefore$ Interest rate of 4.5%.

c $25000 + 12646 = 37646$

$A = P(1 + r)^n$

$37646 = 25000(1 + r)^8$

$(1 + r)^8 = \frac{37646}{25000}$

$1 + r = \left(\frac{37646}{25000}\right)^{\frac{1}{8}}$

$= 1.052500604\ldots$

$r = 0.052500604\ldots$

$= 0.0525$ (4 dec. pl.)

$\therefore$ Interest rate of 5.3% (1 dec. pl.).

32 6% p.a. = 0.5% per month for 120 months

$A = P(1 + r)^n$

$= 20000(1.005)^{120}$

$= 36387.93$ (2 dec. pl.)

9% p.a. = 0.75% per month for 60 months

$A = P(1 + r)^n$

$= 36387.93(1.0075)^{60}$

$= 56971.90$ (2 dec. pl.)

$56971.90 - 20000 = 36971.90$

$\therefore$ Interest of \$36971.90.

33 $A = P(1 + r)^n$

$12223 = 9000(1 + r)^8$

$(1 + r)^8 = \frac{12223}{9000}$

$1 + r = \left(\frac{12223}{9000}\right)^{\frac{1}{8}}$

$= 1.039003266\ldots$

$r = 0.039003266\ldots$

$= 0.039$ (2 sig. fig.)

$\therefore$ Interest rate of 3.9%.

34 a $5000 + 2036 = 7036$

$A = P(1 + r)^n$

$7036 = 5000(1.05)^n$

$1.05^n = \frac{7036}{5000}$

$1.05^n = 1.4072$

Use 'guess and check' strategy:

$n = 7 \therefore 1.05^7 = 1.4072$

$\therefore$ Invested for 7 years.

b $5000 + 955 = 5955$

$A = P(1 + r)^n$

$5955 = 5000(1.06)^n$

$1.06^n = \frac{5955}{5000}$

$1.06^n = 1.191$

Use 'guess and check' strategy:

$n = 3 \therefore 1.06^3 = 1.191\ldots$

$\therefore$ Invested for 3 years.

c $5000 + 10861 = 15861$

$A = P(1 + r)^n$

$15861 = 5000(1.08)^n$

$1.08^n = \frac{15861}{5000}$

$1.08^n = 3.1722$

Use 'guess and check' strategy:

$n = 15 \therefore 1.08^{15} = 3.17216\ldots$

$\therefore$ Invested for 15 years.

35 $A = P(1 + r)^n$

$33487 = 7500(1.005)^n$

$1.005^n = \frac{33487}{7500}$

$1.005^n = 4.464933333\ldots$

Use 'guess and check' strategy:

$n = 300 \therefore 1.005^{300} = 4.46496\ldots$

$\therefore$ Invested for 300 months or 25 years.

36 Value $= 1600 \times 0.85^3$

$= 982.6$

$\therefore$ The computer is valued at \$982.60.

37 Time = 5 years

$\therefore$ Value $= 16300 \times 0.82^5$

$= 6043.059444$

$= 6043.06$ (to 2 dec. pl.)

$\therefore$ The motorbike is valued at \$6043.06.

38 $A = P(1 - r)^n$

$17930 = 42500(1 - r)^3$

$(1 - r)^3 = \frac{17930}{42500}$

$1 - r = \left(\frac{17930}{42500}\right)^{\frac{1}{3}}$

$r = 1 - \left(\frac{17930}{42500}\right)^{\frac{1}{3}}$

$= 0.249995642\ldots$

$= 0.25$ (to 2 dec. pl.)

$\therefore$ Rate is 25% per annum.

39 $A = P(1 - r)^n$

$5644 = 21000(1 - r)^4$

$(1 - r)^4 = \frac{5644}{21000}$

$1 - r = \left(\frac{5644}{21000}\right)^{\frac{1}{4}}$

$r = 1 - \left(\frac{5644}{21000}\right)^{\frac{1}{4}}$

$= 0.279984364\ldots$

$= 0.28$ (to 2 dec. pl.)

$\therefore$ Rate is 28% per annum.

40 Let $P = 4$ and $A = 1$:

$A = P(1 - r)^n$

$1 = 4(1 - 0.2)^n$

$0.25 = 0.8^n$

$0.8^n = 0.25$

Use 'guess and check' strategy:

$n = 6.2 \therefore 0.8^{6.2} = 0.25070\ldots$

$\therefore$ after 6.2 years

$\therefore$ One-quarter cost is in 2021.

Chapter 2 . . . . pp. 29–30

Algebraic Techniques

1 $(x - y)$ km/hr

2 a $(14 + x)$ years old

b $(14 - y)$ years old

c $(p + 14)$ years old

d $(p + 14 + q)$ years old

3 a xy m^2

b mq m^2

c rmq m^2

d $(xy - rmq)$ m^2

4 a $10k$

b $60 \times 24 \times p = 1440p$

c $1000k$

d $\frac{r}{100}$

e $\frac{y}{1000}$

5 $(16400 - 1000x + y)$ g

6 $(100x + y)$ cents

7 $\$\frac{3c}{20}$

8 $[180 - (p + q)]°$

9 a st km

b $\frac{k}{p}$ km/h

c $\frac{b}{c}$ h

10 a $-2 \times 3 \times -5 = 30$

b $3 \times (-5) - (-2) = -13$

c $\frac{-1}{2} - \frac{1}{3} = \frac{-5}{6}$

d $\frac{-2 \times 3}{-5 - 1} = \frac{-6}{-6} = 1$

e $(-2)^2 - 3^2 = 4 - 9 = -5$

f $(-5)^2 - 2(-2)^2 = 25 - 2(4) = 17$

11 **a** $6y + 4x$

b $8y^2 - 6y$

c $9xy$

d $-6ab$

e $25y^2$

f $7a^3$

g $-24x^2y$

h $2b$

i $-3y$

j 2

12 **a** $\frac{4x + 3x}{6} = \frac{7x}{6}$

b $\frac{5a - 2a}{4} = \frac{3a}{4}$

c $\frac{24x - 13x}{26} = \frac{11x}{26}$

d $\frac{2}{x}$

e $\frac{5 - 6}{2a} = \frac{-1}{2a}$

f $\frac{ad - bc}{bd}$

g $\frac{10a - 12a}{15b} = \frac{-2a}{15b}$

h $\frac{6}{y} \times \frac{2y}{3} = \frac{12y}{3y} = 4$

i $\frac{6m}{5} \times \frac{10}{18m^2} = \frac{60m}{90m^2} = \frac{2}{3m}$

j $\frac{3}{2m} \times \frac{4m}{6} = \frac{12m}{12m} = 1$

k $\frac{6a}{5} \times \frac{25}{4a} = \frac{150a}{20a} = \frac{15}{2} = 7\frac{1}{2}$

l $\frac{3a}{8bc} \times \frac{b}{9a} \times \frac{4c}{1} = \frac{\cancel{12abc}^{1}}{\cancel{72abc}_{6}} = \frac{1}{6}$

m $\frac{4}{\cancel{x}} \times \frac{\cancel{x}}{3} \times \frac{4}{3} = \frac{16}{9} = 1\frac{7}{9}$

13 **a** $8x - 28$

b $-15a + 36$

c $2 + 12y - 3 = 12y - 1$

d $7 - 5x + 5 = 12 - 5x$

e $-3x + 4$

f $4x - 4 + 2x - 10 = 6x - 14$

g $10a - 25 - 24 + 6a = 16a - 49$

h $x^2 - 7x - 2x + 14 = x^2 - 9x + 14$

14 **a** $\frac{2(3x - 4) + x - 1}{4}$
$= \frac{6x - 8 + x - 1}{4}$
$= \frac{7x - 9}{4}$

b $\frac{3(5x - 4) - 2(3 - x)}{6}$
$= \frac{15x - 12 - 6 + 2x}{6}$
$= \frac{17x - 18}{6}$

c $\frac{3(4a - 2) - 2(3a - 4)}{6x}$
$= \frac{12a - 6 - 6a + 8}{6x}$
$= \frac{6a + 2}{6x}$
$= \frac{3a + 1}{3x}$

15 **a** $4x - 2 - (x + 5)$
$= 4x - 2 - x - 5$
$= 3x - 7$

b $12y - 4 - (5 - 2y)$
$= 12y - 4 - 5 + 2y$
$= 14y - 9$

c $3x - x^2 - (7x^2 - 5x + 2)$
$= 3x - x^2 - 7x^2 + 5x - 2$
$= 8x - 8x^2 - 2$

16 **a** $a^2 + 4a - 2a - 8 = a^2 + 2a - 8$

b $2x^2 + 6x - x - 3 = 2x^2 + 5x - 3$

c $4x^2 + 32x - 7x - 56$
$= 4x^2 + 25x - 56$

d $12x + 4 - 3x^2 - x$
$= 11x + 4 - 3x^2$

e $x^2 - 8x + 16$

f $16x^2 + 8x + 1$

g $4y^2 - 1$

h $25y^2 - 49$

17 **a** $A = (x + 4)(2x - 1)\text{ cm}^2$
$= (2x^2 - x + 8x - 4)\text{ cm}^2$
$= (2x^2 + 7x - 4)\text{ cm}^2$

b $A = (3y + 4)^2\text{ m}^2$
$= (9y^2 + 24y + 16)\text{ m}^2$

18 **a** $x^2 - (x^2 - 8x + 16)$
$= x^2 - x^2 + 8x - 16)$
$= 8x - 16$

b $x^2 - 2x + 1 - (x^2 + 2x + 1)$
$= x^2 - 2x + 1 - x^2 - 2x - 1$
$= -4x$

c $x^2 + 2x\ \frac{1}{x}\ + \ \frac{1}{x}^{\ 2}$
$= x^2 + 2 + \frac{1}{x^2}$

d $ab - a^2 - b^2 + ab - a^2 - b^2$
$= 2ab - 2a^2 - 2b^2$

e $3(x^2 - 8x + 16)$
$= 3x^2 - 24x + 48$

f $2(a^2 - 2a + 1) -2(a^2 - 1)$
$= 2a^2 - 4a + 2 - 2a^2 + 2$
$= -4a + 40$

19 **a** $41^2 = (40 + 1)^2$
$= 40^2 + 2(40)(1) + 1^2$
$= 1600 + 80 + 1$
$= 1681$

b $29^2 = (30 - 1)^2$
$= 30^2 - 2(30)(1) + 1$
$= 900 - 60 + 1$
$= 841$

20 **a** $5(x - 2y)$

b $2(4a - 3)$

c $4y(3x - 2)$

d $5y(3y - 2)$

e $-2(5a + 9)$

f $2yz(2y - 3)$

g $3(x^2 - 3x - 16)$

h $\pi r(r - 2)$

i $2(l + b)$

j $(3 + x)(x + y)$

k $(a - b)(a - 7)$

21 **a** $2(x + y) + a(x + y)$
$= (x + y)(2 + a)$

b $-2(a - b) - c(a - b)$
$= (a - b)(-2 - c)$
$= -(a - b)(2 + c)$

c $6(a - t) + c(a - t)$
$= (a - t)(6 + c)$

d $g^2(g - 3) -2(g - 3)$
$= (g - 3)(g^2 - 2)$

e $y(x - y) - 7(x - y)$
$= (x - y)(y - 7)$

22 **a** $(u - 6)(u + 6)$

b $(1 - 5p)(1 + 5p)$

c $(2a - 3b)(2a + 3b)$

d $(5k - 12m)(5k + 12m)$

e $(4x^2 - 1)(4x^2 + 1)$
$= (2x - 1)(2x + 1)(4x^2 + 1)$

f $(1 - x^2y^2)(1 + x^2y^2)$
$= (1 - xy)(1 + xy)(1 + x^2y^2)$

g $2(a^2 - 1)$
$= 2(a - 1)(a + 1)$

h $5(a^2 - 25)$
$= 5(a + 5)(a - 5)$

23 a $(x+4)(x-2)$

b $(k-11)(k-1)$

c $(h+16)(h-3)$

d $(b-11)(b+8)$

e $(x-4y)(x-2y)$

f $(x+4y)(x-y)$

24 a $(2x+1)(x-1)$

b $\dfrac{^{p+1}\cancel{(3p+3)}(3p+8)}{\cancel{3}}$ (24)

$= (p+1)(3p+8)$

c $\dfrac{^{c+3}\cancel{(3c+9)}(3c-4)}{\cancel{3}}$ (−36)

$= (c+3)(3c-4)$

d $\dfrac{^{x+1}\cancel{(9x+9)}(9x+8)}{\cancel{9}_1}$ (72)

$= (x+1)(9x+8)$

e $\dfrac{^{v-1}\cancel{(8v-8)}(8v+7)}{\cancel{8}}$ (−56)

$= (v-1)(8v+7)$

f $\dfrac{^{b+5c}\cancel{(3b+15c)}(3b+1c)}{\cancel{3}}$ (15)

$= (b+5c)(3b+c)$

25 a $\dfrac{^4\cancel{8}}{_1\cancel{2}(a-3)} = \dfrac{4}{a-3}$

b $\dfrac{\cancel{3y-4}^1}{3\cancel{(3y-4)}_1} = \dfrac{1}{3}$

c $\dfrac{^1\cancel{a}(a-4)}{6\cancel{a}_1} = \dfrac{a-4}{6}$

d $\dfrac{^1\cancel{(a-3)}(a-3)}{_1\cancel{(a-3)}(a+3)} = \dfrac{a-3}{a+3}$

e $\dfrac{^1\cancel{(x+4)}(x+3)}{_1\cancel{(x+4)}(x-4)} = \dfrac{x+3}{x-4}$

f $p^2 + 4p + pq + 4q$

$= p(p+4) + q(p+4)$

$= (p+q)(p+4)$

Then $\dfrac{^1\cancel{(p+q)}(p+4)}{_1\cancel{(p+q)}(p-q)} = \dfrac{p+4}{p-q}$

26 a $\dfrac{^1\cancel{(y-6)}}{_1\cancel{(y+6)}\,\cancel{(y-6)}_1} \times \dfrac{\cancel{(y+6)}^1}{2} = \dfrac{1}{2}$

b $\dfrac{^1\cancel{(x+1)}\,\cancel{(x-1)}^1}{\cancel{(x-1)}_1} \times \dfrac{5}{\cancel{(x+1)}} = 5$

c $\dfrac{(a-7)(a+1)}{(a-1)\cancel{(a+1)}_1} \times \dfrac{^1\cancel{a}\,\cancel{(a+1)}^1}{a^{\cancel{2}^1}}$

$= \dfrac{(a-7)(a+1)}{a(a-1)}$

27 a $\dfrac{4(x+7)+5(x+3)}{(x+3)(x+7)}$

$= \dfrac{4x+28+5x+15}{(x+3)(x+7)}$

$= \dfrac{9x+43}{(x+3)(x+7)}$

b $\dfrac{2(x-5)+4(x-3)}{(x-5)(x-3)}$

$= \dfrac{2x-10+4x-12}{(x-5)(x-3)}$

$= \dfrac{6x-22}{(x-5)(x-3)}$

c $\dfrac{^1\cancel{x-3}}{_1\cancel{(x-3)}(x-2)} - \dfrac{3\cancel{x}^1}{_1\cancel{x}(x+4)}$

$= \dfrac{1}{x-2} - \dfrac{3}{x+4}$

$= \dfrac{x+4-3x+6}{(x-2)(x+4)}$

$= \dfrac{-2x+10}{(x-2)(x+4)}$

d $\dfrac{2}{(c-7)(c+7)} + \dfrac{4}{(c-7)(c+3)}$

$= \dfrac{2(c+3)+4(c+7)}{(c-7)(c+7(c+3)}$

$= \dfrac{2c+6+4c+28}{(c-7)(c+7)(c+3)}$

$= \dfrac{6c+34}{(c-7)(c+7)(c+3)}$

e $\dfrac{y+2}{(y+3)(y-1)} - \dfrac{y}{(y+3)(y+4)}$

$= \dfrac{(y+2)(y+4)-y(y-1)}{(y+3)(y-1)(y+4)}$

$= \dfrac{y^2+6y+8-y^2+y}{(y+3)(y-1)(y+4)}$

$= \dfrac{7y+8}{(y+3)(y-1)(y+4)}$

Chapter 3 . . . . pp. 44–45

Surds and Indices

1 a $\sqrt{11}, \sqrt{7}, \sqrt{12}, 3 = \sqrt{9}, 5 = \sqrt{25}$

$\therefore \sqrt{7}, 3, \sqrt{11}, \sqrt{12}, 5$

b $6 = \sqrt{36}, \sqrt{33}, \sqrt{24}, 5 = \sqrt{25}, \sqrt{40}$

$\therefore \sqrt{24}, 5, \sqrt{33}, 6, \sqrt{40}$

2 a $\sqrt{15}, \sqrt{8}, \sqrt{17}, 3 = \sqrt{9}, 4 = \sqrt{16}$

$\therefore \sqrt{17}, 4, \sqrt{15}, 3, \sqrt{8}$

b $8 = \sqrt{64}, \sqrt{70}, \sqrt{56}, 7 = \sqrt{49}, \sqrt{50}$

$\therefore \sqrt{70}, 8, \sqrt{56}, \sqrt{50}, 7$

3 a $\sqrt{8} = \sqrt{4} \times \sqrt{2}$

$= 2\sqrt{2}$

b $\sqrt{27} = \sqrt{9} \times \sqrt{3}$

$= 3\sqrt{3}$

c $\sqrt{128} = \sqrt{64} \times \sqrt{2}$

$= 8\sqrt{2}$

d $\sqrt{80} = \sqrt{16} \times \sqrt{5}$

$= 4\sqrt{5}$

e $2\sqrt{50} = 2 \times \sqrt{25} \times \sqrt{2}$

$= 10\sqrt{2}$

f $4\sqrt{300} = 4 \times \sqrt{100} \times \sqrt{3}$

$= 40\sqrt{3}$

g $\sqrt{\dfrac{4}{9}} = \dfrac{\sqrt{4}}{\sqrt{9}}$

$= \dfrac{2}{3}$

h $\sqrt{\dfrac{81}{100}} = \dfrac{\sqrt{81}}{\sqrt{100}}$

$= \dfrac{9}{10}$

i $\sqrt{\dfrac{3}{4}} = \dfrac{\sqrt{3}}{2}$

j $(\sqrt{3})^2 = \sqrt{3} \times \sqrt{3}$

$= \sqrt{9}$

$= 3$

k $(2\sqrt{5})^2 = 2\sqrt{5} \times 2\sqrt{5}$

$= 4\sqrt{25}$

$= 20$

l $(4\sqrt{3})^2 = 4\sqrt{3} \times 4\sqrt{3}$

$= 16\sqrt{9}$

$= 48$

4 a $2\sqrt{3} - \sqrt{3} = \sqrt{3}$

b $12\sqrt{5} + \sqrt{5} - 7\sqrt{5} = 6\sqrt{5}$

c $\sqrt{2} - \sqrt{3} + 3\sqrt{2} = 4\sqrt{2} - \sqrt{3}$

d $\sqrt{63} - \sqrt{28} = 3\sqrt{7} - 2\sqrt{7} = \sqrt{7}$

e $2\sqrt{2} + \sqrt{128} = 2\sqrt{2} + 8\sqrt{2} = 10\sqrt{2}$

f $x\sqrt{x} + \sqrt{x^3} - 4\sqrt{x}$

$= x\sqrt{x} + x\sqrt{x} - 4\sqrt{x}$

$= 2x\sqrt{x} - 4\sqrt{x}$

$= 2\sqrt{x}(x-2)$

5 a $\sqrt{3} \times \sqrt{5} = \sqrt{15}$

b $\sqrt{2} \times \sqrt{2} = 2$

c $2\sqrt{2} \times 4\sqrt{7} = 8\sqrt{14}$

d $\sqrt{5} \times 2\sqrt{5} = 2\sqrt{25} = 10$

e $3\sqrt{2} \times -2\sqrt{2} = -6\sqrt{4} = -12$

f $(2\sqrt{3})^2 = 4 \times 3 = 12$

6 a $\sqrt{9y-18} = \sqrt{9(y-2)}$

$= 3\sqrt{y-2}$

b $\sqrt{8a+4} = \sqrt{4(2a+1)}$

$= 2\sqrt{2a+1}$

c $\sqrt{x^3 - x^2} = \sqrt{x^2(x-1)}$
$= x\sqrt{x-1}$

7 a $\sqrt{\frac{48}{3}} = \sqrt{16}$
$= 4$

b $\frac{\sqrt{24}}{\sqrt{6}} = \sqrt{\frac{24}{6}}$
$= \sqrt{4}$
$= 2$

c $\frac{2\sqrt{5}}{\sqrt{45}} = \frac{2}{\sqrt{9}}$
$= \frac{2}{3}$

d $\frac{3\sqrt{10}}{6\sqrt{5}} = \frac{\sqrt{2}}{2}$

e $\frac{4\sqrt{3} \times \sqrt{2}}{\sqrt{6}} = \frac{4\sqrt{6}}{\sqrt{6}}$
$= 4$

f $\frac{2\sqrt{3} \times \sqrt{2}}{\sqrt{8}} = \frac{2\sqrt{3} \times \sqrt{2}}{2\sqrt{2}}$
$= \sqrt{3}$

8 a $\sqrt{2}(\sqrt{2} - 1) = 2 - \sqrt{2}$

b $\sqrt{5}(3\sqrt{5} + 2) = 15 + 2\sqrt{5}$

c $4\sqrt{3}(2 - \sqrt{3}) = 8\sqrt{3} - 12$

d $\sqrt{7}(2\sqrt{7} - \sqrt{2}) = 14 - \sqrt{14}$

e $2\sqrt{3}(\sqrt{3} + 5\sqrt{2}) = 6 + 10\sqrt{6}$

f $\sqrt{x}(x - x\sqrt{x}) = x\sqrt{x} - x^2$

9 a $\frac{3 + 3\sqrt{3}}{3} = \frac{3(1 + \sqrt{3})}{3}$
$= 1 + \sqrt{3}$

b $\frac{4 - 6\sqrt{2}}{4} = \frac{2(2 - 3\sqrt{2})}{4}$
$= \frac{2 - 3\sqrt{2}}{2}$

c $\frac{12 + 3\sqrt{5}}{9} = \frac{3(4 + \sqrt{5})}{9}$
$= \frac{4 + \sqrt{5}}{3}$

d $\frac{12 + \sqrt{8}}{8} = \frac{2(6 + \sqrt{2})}{8}$
$= \frac{6 + \sqrt{2}}{4}$

e $\frac{\sqrt{12} - \sqrt{24}}{6} = \frac{2\sqrt{3} - 2\sqrt{6}}{6}$
$= \frac{2(\sqrt{3} - \sqrt{6})}{6}$
$= \frac{\sqrt{3} - \sqrt{6}}{3}$

f $\frac{\sqrt{48} + \sqrt{128}}{4} = \frac{4\sqrt{3} + 8\sqrt{2}}{4}$
$= \frac{4(\sqrt{3} + 2\sqrt{2})}{4}$
$= \sqrt{3} + 2\sqrt{2}$

10 a $(\sqrt{3} + 2)^2 = 3 + 4\sqrt{3} + 4$
$= 7 + 4\sqrt{3}$

b $(2\sqrt{2} - 1)^2 = 8 - 4\sqrt{2} + 1$
$= 9 - 4\sqrt{2}$

c $(3\sqrt{2} + \sqrt{3})^2 = 18 + 6\sqrt{6} + 3$
$= 21 + 6\sqrt{6}$

d $(\sqrt{5} - 2)(\sqrt{5} + 2) = 5 - 4$
$= 1$

e $(3 - 2\sqrt{2})(3 + 2\sqrt{2}) = 9 - 8$
$= 1$

f $(\sqrt{2} - 1)(\sqrt{3} + 2)$
$= \sqrt{6} + 2\sqrt{2} - \sqrt{3} - 2$

g $(\sqrt{x} - 1)^2 = x - 2\sqrt{x} + 1$

h $(x\sqrt{x} + 1)^2 = x^3 + 2x\sqrt{x} + 1$

i $\left(\sqrt{x} - \frac{1}{\sqrt{x}}\right)^2 = x - 2 + \frac{1}{x}$

11 a $\frac{1}{\sqrt{6}} = \frac{1}{\sqrt{6}} \times \frac{\sqrt{6}}{\sqrt{6}}$
$= \frac{\sqrt{6}}{6}$

b $\frac{3}{2\sqrt{5}} = \frac{3}{2\sqrt{5}} \times \frac{\sqrt{5}}{\sqrt{5}}$
$= \frac{3\sqrt{5}}{10}$

c $\frac{\sqrt{2}}{\sqrt{3}} = \frac{\sqrt{2}}{\sqrt{3}} \times \frac{\sqrt{3}}{\sqrt{3}}$
$= \frac{\sqrt{6}}{3}$

d $\frac{\sqrt{2} - 1}{\sqrt{7}} = \frac{\sqrt{2} - 1}{\sqrt{7}} \times \frac{\sqrt{7}}{\sqrt{7}}$
$= \frac{\sqrt{14} - \sqrt{7}}{7}$

e $\frac{1}{2 + \sqrt{3}} = \frac{1}{2 + \sqrt{3}} \times \frac{2 - \sqrt{3}}{2 - \sqrt{3}}$
$= \frac{2 - \sqrt{3}}{4 - 3}$
$= 2 - \sqrt{3}$

f $\frac{1}{3 - \sqrt{8}} = \frac{1}{3 - \sqrt{8}} \times \frac{3 + \sqrt{8}}{3 + \sqrt{8}}$
$= \frac{3 + \sqrt{8}}{9 - 8}$
$= 3 + 2\sqrt{2}$

g $\frac{12}{\sqrt{7} + \sqrt{3}} = \frac{12}{\sqrt{7} + \sqrt{3}} \times \frac{\sqrt{7} - \sqrt{3}}{\sqrt{7} - \sqrt{3}}$
$= \frac{12(\sqrt{7} - \sqrt{3})}{7 - 3}$
$= \frac{12(\sqrt{7} - \sqrt{3})}{4}$
$= 3(\sqrt{7} - \sqrt{3})$

h $\frac{\sqrt{5} + \sqrt{2}}{2\sqrt{3}} = \frac{\sqrt{5} + \sqrt{2}}{2\sqrt{3}} \times \frac{\sqrt{3}}{\sqrt{3}}$
$= \frac{\sqrt{15} + \sqrt{6}}{6}$

i $\frac{\sqrt{3} - \sqrt{2}}{\sqrt{3} + \sqrt{2}} = \frac{\sqrt{3} - \sqrt{2}}{\sqrt{3} + \sqrt{2}} \times \frac{\sqrt{3} - \sqrt{2}}{\sqrt{3} - \sqrt{2}}$
$= \frac{3 - 2\sqrt{6} + 2}{3 - 2}$
$= 5 - 2\sqrt{6}$

12 a $y^4 \times y^3 = y^7$

b $x^2 \times x^4 \times x = x^7$

c $4q^2 \times (-3q^3) = -12q^5$

d $a^7 \div a^2 \div a = a^4$

e $\frac{m^5}{m^3} = m^2$

f $16b^6 \div 2b^3 = 8b^3$

g $3^2 \times 3 = 3^3$
$= 27$

h $5^8 \div 5^7 = 5$

i $6^4 \div 6^2 \times 6 = 6^3$
$= 216$

j $(x^7)^2 = x^{14}$

k $(2a^3)^3 = 8a^9$

l $(a^4)^{\frac{1}{2}} = a^2$

m $3a^0 = 3$

n $(4a^3 + 2a^3)^0 = 1$

o $(2y)^0 + 2y^0 = 1 + 2$
$= 3$

13 a $3^{-2} = \frac{1}{9}$

b $2^{-3} = \frac{1}{8}$

c $\left(\frac{1}{2}\right)^{-2} = 2^2$
$= 4$

d $\left(\frac{3}{5}\right)^{-2} = \frac{25}{9}$
$= 2\frac{7}{9}$

e $4x^{-2} = \frac{4}{x^2}$

f $(4x)^{-2} = \frac{1}{16x^2}$

g $(2a^4)^{-2} = \frac{1}{(2a^4)^2}$
$= \frac{1}{4a^8}$

h $x^{-2} \times x^{-3} = x^{-5}$

i $x^{-2} \div x^{-3} = x$

j $\frac{y^3}{y^7} = \frac{1}{y^4}$

k $\frac{a^2b}{ab^3} = \frac{a}{b^2}$

l $\frac{10x^5}{5x^7} = \frac{2}{x^2}$

14 a $25^{\frac{1}{2}} = \sqrt{25}$
$= 5$

b $27^{\frac{1}{3}} = \sqrt[3]{27}$
$= 3$

c $\left(8^{\frac{1}{3}}\right)^2 = 2^2$
$= 4$

d $1000^{\frac{2}{3}} = 10^2$
$= 100$

e $4^{\frac{3}{2}} = 2^3$
$= 8$

f $32^{-\frac{1}{5}} = 2^{-1}$
$= \frac{1}{2}$

g $\frac{1}{4}^{-\frac{1}{2}} = 4^{\frac{1}{2}}$
$= 2$

h $(25a)^{\frac{1}{2}} = 5\sqrt{a}$

i $(64y^6)^{\frac{1}{3}} = 4y^2$

j $\frac{1}{9x^4}^{-\frac{1}{2}} = (9x^4)^{\frac{1}{2}}$
$= 3x^2$

15 a $\sqrt{2y-1} = (2y-1)^{\frac{1}{2}}$

b $\sqrt[3]{x+5} = (x+5)^{\frac{1}{3}}$

c $\frac{1}{5y+2} = (5y+2)^{-1}$

d $\frac{2a}{(3a-1)^2} = 2a(3a-1)^{-2}$

e $\frac{1}{\sqrt{7b-3}} = (7b-3)^{-\frac{1}{2}}$

f $\frac{1}{\sqrt[3]{1-8p}} = (1-8p)^{-\frac{1}{3}}$

g $\frac{2y+3}{\sqrt{6y-5}} = (2y+3)(6y-5)^{-\frac{1}{2}}$

h $\frac{1}{x\sqrt{x}} = \frac{1}{x^{\frac{3}{2}}}$
$= x^{-\frac{3}{2}}$

i $\frac{1}{\sqrt[3]{(8y-7)^2}} = (8y-7)^{-\frac{2}{3}}$

16 a $361^{\frac{1}{2}} = 19$

b $81^{\frac{3}{2}} = 729$

c $\frac{4}{9}^{-\frac{1}{2}} = 1\frac{1}{2}$

17 a $3^x = 81$
$3^x = 3^4$
$x = 4$

b $2^{x+3} = 1024$
$2^{x+3} = 2^{10}$
$x + 3 = 10$
$x = 7$

c $5^{2x-3} = 125$
$5^{2x-3} = 5^3$
$2x - 3 = 3$
$2x = 6$
$x = 3$

d $8^{2x-1} = 512$
$8^{2x-1} = 8^3$
$2x - 1 = 3$
$2x = 4$
$x = 2$

e $9^{x+1} = 3^{4x}$
$(3^2)^{x+1} = 3^{4x}$
$3^{2x+2} = 3^{4x}$
$4x = 2x + 2$
$2x = 2$
$x = 1$

f $64^{\frac{x}{2}} = 4^{x+1}$
$8^x = 4^{x+1}$
$(2^3)^x = (2^2)^{x+1}$
$2^{3x} = 2^{2x+2}$
$3x = 2x + 2$
$x = 2$

18 a $\frac{3}{7} = 0.428\,571\,4285\ldots$
$= 0.\dot{4}2857\dot{1}$

b $\frac{6}{11} = 0.545\,454\,54\ldots$
$= 0.\dot{5}\dot{4}$

c $\frac{31}{99} = 0.313\,131\,3131\ldots$
$= 0.\dot{3}\dot{1}$

19 a Let $x = 0.979\,797\,979\ldots$
$100x = 97.979\,797\,979\ldots$
$\therefore 99x = 97$
$x = \frac{97}{99}$
$\therefore 0.\dot{9}\dot{7} = \frac{97}{99}$

b Let $x = 0.381\,381\,381\ldots$
$1000x = 381.381\,381\,381\ldots$
$\therefore 999x = 381$
$x = \frac{381}{999}$
$= \frac{127}{333}$
$\therefore 0.\dot{3}8\dot{1} = \frac{127}{333}$

c Let $x = 1.255\,555\,55\ldots$
$10x = 12.555\,555\,55\ldots$
$\therefore 9x = 11.3$
$x = \frac{11.3}{9}$
$= \frac{113}{90}$
$= 1\frac{23}{90}$
$\therefore 1.2\dot{5} = 1\frac{23}{90}$

20 a $38\,230 = 3.823 \times 10^4$

b $5\,000\,000\,000 = 5 \times 10^9$

c $0.237 = 2.37 \times 10^{-1}$

d $0.000\,0003 = 3 \times 10^{-7}$

e $0.09 = 9 \times 10^{-2}$

f $0.001\,007 = 1.007 \times 10^{-3}$

21 a $4.2 \times 10^4 = 42\,000$

b $1.07 \times 10^{-2} = 0.0107$

c $8 \times 10^{-4} = 0.0008$

22 a $\frac{125.6 \times 1.09}{0.0032} = 42\,782.5$
$= 4.28 \times 10^4$

b $\frac{8.038}{0.003 + 0.1392}^3$
$= 180\,611.4268$
$= 1.81 \times 10^5$

c $\frac{\sqrt{10.857 - 9.8}}{3.92^3} = 0.017\,067\,869\ldots$
$= 1.71 \times 10^{-2}$

23 a $(3.76 \times 10^3) \times (1.982 \times 10^5)$
$= 7.45 \times 10^8$

b $\frac{2.35 \times 10^5}{1.0089 \times 10^3} = 2.33 \times 10^2$

c $\sqrt{2.9812 \times 10^{-3}} = 5.46 \times 10^{-2}$

Chapter 4 . . . . pp. 63–65

Equations

1 a $4x - 3x = 6 + 12$
$x = 18$

b $5y - 3y = 18 - 2$
$2y = 16$
$y = \frac{16}{2}$
$= 8$

c $14y - 28 = 6y - 33$
$14y - 6y = -33 + 28$
$8y = -5$
$y = -\frac{5}{8}$

d $4x - 2 = 15$
$4x = 17$
$x = \frac{17}{4}$
$= 4\frac{1}{4}$

e $4 - 2y = 4(3y + 1)$
$4 - 2y = 12y + 4$
$4 - 4 = 12y + 2y$
$0 = 14y$
$y = 0$

f $3(7a - 12) - 2(a + 1) = 5 \times 6$
$21a - 36 - 2a - 2 = 30$
$19a - 38 = 30$
$19a = 68$
$a = \frac{68}{19}$
$= 3\frac{11}{19}$

g $2(12x - 4) = 3(4x + 11)$
$24x - 8 = 12x + 33$
$12x = 41$
$x = \frac{41}{12}$
$= 3\frac{5}{12}$

h $20 = 6x$
$x = \frac{20}{6}$
$= 3\frac{1}{3}$

i $6(2y - 1) = 2(3y + 1)$
$12y - 6 = 6y + 2$
$6y = 8$
$y = \frac{8}{6}$
$= 1\frac{1}{3}$

j $3 + 4(2y - 1) = 0$
$3 + 8y - 4 = 0$
$8y = 1$
$y = \frac{1}{8}$

k $5(5a - 3) - 10(a - 1) = \frac{3}{5} \times 20$
$25a - 15 - 10a + 10 = 12$
$15a - 5 = 12$
$15a = 17$
$a = \frac{17}{15}$
$= 1\frac{2}{15}$

2 a $5a - 2a > 12 + 3$
$3a > 15$
$a > 5$

b $7y - 9y \le 12 + 4$
$-2y \le 16$
$y \ge \frac{16}{-2}$
$y \ge -8$

c $4 - 2y \ge 11$
$-2y \ge 7$
$y \le \frac{7}{-2}$
$y \le -3\frac{1}{2}$

d $3 \le 4a + 6$
$-3 \le 4a$
$-\frac{3}{4} \le a$
$a \ge -\frac{3}{4}$

e $4 - 2x + 8 \ge 12$
$12 - 2x \ge 12$
$0 \ge 2x$
$x \le 0$

f $3 - 4x \le 6$
$-4x \le 3$
$x \ge -\frac{3}{4}$

3 a

b

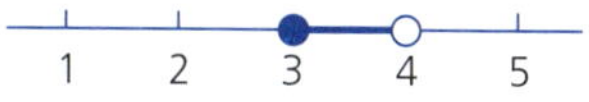

4 a $-2 < x \le 4$

b $x \le -4, x > -2$

5 a $3x - 2x \ge -10 + 5$
$x \ge -5$

b $4 - 5x < 2$
$-5x < -2$
$x > \frac{-2}{-5}$
$x > \frac{2}{5}$

6 $\frac{4x - 1}{3} \le \frac{2 + 5x}{4}$
$4(4x - 1) \le 3(2 + 5x)$
$16x - 4 \le 6 + 15x$
$16x - 15x \le 6 + 4$
$x \le 10$

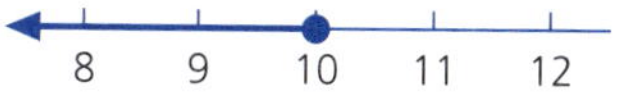

7 $P = 16(11 - 2)$
$= 16 \times 9$
$= 144$

8 $S = \frac{100(2^4 - 1)}{2 - 1}$
$= \frac{100(15)}{1}$
$= 1500$

9 $I = \frac{160 \times 11 \times 9}{100}$
$= 158.4$

10 a $y = 2x + 2$

b $y = x^2 + 4$

11 a $C = \frac{3t}{2} + 2$

b $C = \frac{3 \times 6}{2} + 2 = 11$
Cost = \$11

12 a $A = 2x \times 2x + \frac{1}{2} \times \pi \times x^2$
$= 4x^2 + \frac{1}{2}\pi x^2$

b $A = 4x \times 3x - x \times x$
$= 11x^2$

13

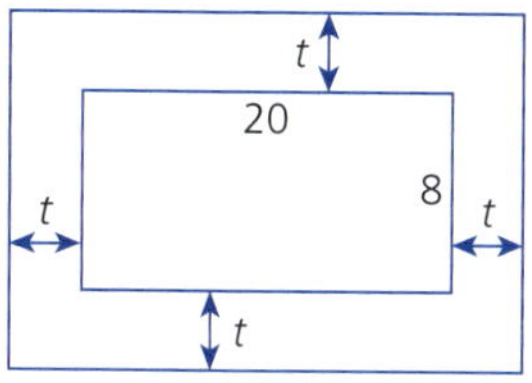

Dimensions are $20 + 2t$ and $8 + 2t$
$A = (20 + 2t)(8 + 2t)$

14 a $2y = -4x + 16$
$y = -2x + 8$ or $y = 8 - 2x$

b $5x = 3y + 11$
$5x - 11 = 3y$
$\frac{5}{3}x - \frac{11}{3} = y$ or $y = \frac{5x - 11}{3}$

c $2xy = 11 - x$
$y = \frac{11 - x}{2x}$

d $y(5 - 2x) = 7$
$y = \frac{7}{5 - 2x}$

e $Ay = 4x$
$y = \frac{4x}{A}$

f $2P = \sqrt{xy}$
$4P^2 = xy$
$y = \dfrac{4P^2}{x}$

g $yk = y - 1$
$yk - y = -1$
$y(k - 1) = -1$
$y = \dfrac{-1}{k - 1}$
OR $y = \dfrac{1}{1 - k}$

h $(y - 2)x^2 = y$
$yx^2 - 2x^2 = y$
$yx^2 - y = 2x^2$
$y(x^2 - 1) = 2x^2$
$y = \dfrac{2x^2}{x^2 - 1}$

15 a $770 = \dfrac{22}{\not{7}_1} \times \not{7}^1 \times 7 \times h$
$h = \dfrac{770}{22 \times 7}$
$= 5$

b $345.5 = 5u + \dfrac{1}{2} \times 10 \times 25$
$345.5 = 5u + 125$
$220.5 = 5u$
$u = \dfrac{220.5}{5}$
$= 44.1$

16 a $v = 8.4\sqrt{d}$
$\dfrac{v}{8.4} = \sqrt{d}$
$d = \dfrac{v^2}{8.4^2}$
$= \dfrac{v^2}{70.56}$

b $d = \dfrac{42^2}{70.56} = 25$
Object falls 25 m.

17 a
$x + 2y = 4$ (1)
$3x + 14y = 36$ (2)
(1) × 3 $3x + 6y = 12$ (3)
(2) − (3) $8y = 24$
$y = 3$
Sub. into (1) $x + 6 = 4$
$x = -2$
Solution is $x = -2$, $y = 3$.

b
$3x - 8y = 2$ (1)
$4x + 10y = 44$ (2)
(1) × 5 $15x - 40y = 10$ (3)
(2) × 4 $16x + 40y = 176$ (4)
(3) + (4) $31x = 186$
$x = 6$
Sub. into (2) $24 + 10y = 44$
$10y = 20$
$y = 2$
Solution is $x = 6$, $y = 2$.

c
$3x = 5 - 2y$ (1)
$4x = 1 + 3y$ (2)
(1) × 4 $12x = 20 - 8y$ (3)
(2) × 3 $12x = 3 + 9y$ (4)
Substitute (3) into (4)
$20 - 8y = 3 + 9y$
$17 = 17y$
$y = 1$
Sub. into (1) $3x = 5 - 2$
$x = 1$
Solution is $x = 1$, $y = 1$.

d
$x - y = 8$ (1)
$3x + 2y = -6$ (2)
(1) × 2 $2x - 2y = 16$ (3)
(2) + (3) $5x = 10$
$x = 2$
Sub. into (1) $2 - y = 8$
$y = 2 - 8$
$y = -6$
Solution is $x = 2$, $y = -6$.

18 a Let number be x.
$\therefore\ 2x - 5 = x + 4$
$2x - x = 4 + 5$
$x = 9$
∴ The number is 9.

b Let the numbers be x and y.
$\therefore\ x + y = 12$ (1)
$x - y = 2$ (2)
(1) − (2)
$\therefore\ 2y = 10$
$y = 5$
Substitute in (1)
$\therefore\ x + 5 = 12$
$x = 7$
∴ Numbers are 7 and 5.

c i Equilateral triangle
$\therefore\ 3a - 2b = 7$ (1)
$2a + b = 7$ (2)
2 × (2) $4a + 2b = 14$ (3)
(1) + (3) $7a = 21$
$a = 3$
Substitute in (2)
$\therefore\ 6 + b = 7$
$b = 7 - 6$
$b = 1$
$\therefore\ a = 3, b = 1$

ii Opposite sides of rectangle equal.
$\therefore\ 3a - 1 = 3 - 2b$
$5a - 7 = b + 4$
Rewriting these equations:
$\therefore\ 3a + 2b = 4$ (1)
$5a - b = 11$ (2)
2 × (2) $10a - 2b = 22$ (3)
(1) + (3)
$\therefore\ 13a = 26$
$a = 2$
Substituting in (2):
$10 - b = 11$
$-b = 11 - 10$
$-b = 1$
$\therefore\ b = -1$
$\therefore\ a = 2$ and $b = -1$

d Let cost of a pen = x cents.
Cost of a pencil = y cents.
$\therefore\ 6x + 5y = 250$ (1)
$3x + 2y = 115$ (2)
2 × (2) $6x + 4y = 230$ (3)
(1) − (2)
$\therefore\ y = 20$
Substitute in (2)
$\therefore\ 3x + 40 = 115$
$3x = 115 - 40$
$3x = 75$
$x = 25$
∴ Pens cost 25c while pencils cost 20c.
∴ Pen and four pencils cost $1.05.

19 a $x = \dfrac{1}{3}$ or -2

b $x(x - 5) = 0$
$x = 0$ or 5

c $x^2 - 3x = 0$
$x(x - 3) = 0$
$\therefore\ x = 0$ or 3

d $(x - 2)(x - 2) = 0$
$\therefore\ x = 2$

e $x^2 - 5x + 6 = 0$
$(x - 3)(x - 2) = 0$
$\therefore\ x = 2$ or 3

f $x = 7$ or -7

g $x^2 = 9$
$x = 3$ or -3

h $(5x + 3)(x - 1) = 0$
$x = -\dfrac{3}{5}$ or 1

i $9x^2 = 18x + 16$
$9x^2 - 18x - 16 = 0$
$(3x - 8)(3x + 2) = 0$
$\therefore\ x = \dfrac{8}{3}$ or $-\dfrac{2}{3}$

j $4x^2 - 12 + 13x = 0$
$4x^2 + 13x - 12 = 0$
$(x + 4)(4x - 3) = 0$
$\therefore\ x = -4$ or $\dfrac{3}{4}$

k $2x^2 - 5x + 2 = 0$
$(x - 2)(2x - 1) = 0$
$\therefore\ x = 2$ or $\dfrac{1}{2}$

l $x - 3 = 4$ or -4
$\therefore\ x = 7$ or -1

m $3x + 4 = x^2$
$\therefore\ 0 = x^2 - 3x - 4$
$0 = (x - 4)(x + 1)$
$\therefore\ x = 4$ or -1

20 a $x^2 + 4x = -1$
$x^2 + 4x + 4 = -1 + 4$
$(x + 2)^2 = 3$
$x + 2 = \pm\sqrt{3}$
$x = -2 \pm \sqrt{3}$

b
$$\begin{aligned} x^2 - 6x &= -2 \\ x^2 - 6x + 9 &= -2 + 9 \\ (x-3)^2 &= 7 \\ x - 3 &= \pm\sqrt{7} \\ x &= 3 \pm \sqrt{7} \end{aligned}$$

21 Using $x = \dfrac{-b \pm \sqrt{b^2 - 4ac}}{2a}$

a
$$\begin{aligned} x &= \frac{4 \pm \sqrt{(-4)^2 - 4(1)(-1)}}{2} \\ &= \frac{4 \pm \sqrt{20}}{2} \\ &= \frac{4 \pm 2\sqrt{5}}{2} \\ &= 2 \pm \sqrt{5} \end{aligned}$$

b
$$\begin{aligned} x &= \frac{8 \pm \sqrt{(-8)^2 - 4(1)(-5)}}{2} \\ &= \frac{8 \pm \sqrt{84}}{2} \\ &= \frac{8 \pm 2\sqrt{21}}{2} \\ &= 4 \pm \sqrt{21} \end{aligned}$$

22 a
$$\begin{aligned} x &= \frac{-9 \pm \sqrt{9^2 - 4(3)(5)}}{6} \\ &= \frac{-9 \pm \sqrt{81 - 60}}{6} \\ &= \frac{-9 \pm \sqrt{21}}{6} \\ &= \frac{-9 + \sqrt{21}}{6} \text{ or } \frac{-9 - \sqrt{21}}{6} \\ &= -0.736 \text{ or } -2.264 \end{aligned}$$

b $5x^2 - 2x - 4 = 0$
$$\begin{aligned} x &= \frac{2 \pm \sqrt{(-2)^2 - 4(5)(-4)}}{10} \\ &= \frac{2 \pm \sqrt{4 + 80}}{10} \\ &= \frac{2 + \sqrt{84}}{10} \text{ or } \frac{2 - \sqrt{84}}{10} \\ &= 1.117 \text{ or } -0.717 \end{aligned}$$

23 a Let numbers be x and $x + 1$.
$$\begin{aligned} \therefore \quad x(x+1) &= 30 \\ x^2 + x - 30 &= 0 \\ (x+6)(x-5) &= 0 \\ \therefore \quad x &= 5 \text{ or } -6 \end{aligned}$$
Numbers are 5, 6 or −6 and −5.

b Let numbers be x and $10 - x$ (sum = 10)
Then:
$$\begin{aligned} x^2 + (10-x)^2 &= 52 \\ x^2 + 100 - 20x + x^2 &= 52 \\ 2x^2 - 20x + 48 &= 0 \\ x^2 - 10x + 24 &= 0 \\ (x-6)(x-4) &= 0 \\ \therefore \quad x &= 6 \text{ or } 4 \end{aligned}$$
Numbers are 6 and 4.

c **i** $x + 8$ cm

ii
$$\begin{aligned} A &= x(x+8) \\ &= x^2 + 8x \\ \therefore \quad x^2 + 8x &= 240 \\ x^2 + 8x - 240 &= 0 \\ (x+20)(x-12) &= 0 \\ \therefore \quad x &= 12 \text{ or } -20 \end{aligned}$$
(cannot have negative length)
$\therefore$ width = 12 cm, length = 20 cm

d Let height = h cm
$\therefore$ Base = $h + 5$
$$\begin{aligned} \text{Area} &= \frac{1}{2} \times (h+5)h \\ &= \frac{h^2 + 5h}{2} \\ \therefore \quad \frac{h^2 + 5h}{2} &= 18 \\ h^2 + 5h &= 36 \\ h^2 + 5h - 36 &= 0 \\ (h+9)(h-4) &= 0 \\ \therefore \quad h &= 4 \text{ or } -9 \end{aligned}$$
(cannot have negative length)
$\therefore$ height = 4 cm, base = 9 cm

e
$$\begin{aligned} 4x + 12 &= 2x + 48 \\ 2x &= 36 \\ x &= 18 \end{aligned}$$
Then $4x + 12 = 4(18) + 12 = 84$
obviously $2x + 48 = 84$ (equal lengths)
Then 3rd angle = 180 − 2(84)
= 12
Angles are 84°, 84°, 12°.

f $y = x^2 + 6x - 2$ (1)
$y = 2x + 3$ (2)
Let (1) = (2)
$$\begin{aligned} \therefore x^2 + 6x - 2 &= 2x + 3 \\ x^2 + 4x - 5 &= 0 \\ (x+5)(x-1) &= 0 \\ x &= -5, 1 \end{aligned}$$
Sub. $x = -5$ in (2)
$\therefore y = 2(-5) + 3$
$= -7$
Sub. $x = 1$ in (2)
$\therefore y = 2(1) + 3$
$= 5$
$\therefore$ (−5, −7) or (1, 5)

g Let integer be x.
$$\begin{aligned} x + 20 &= x^2 \\ x^2 - x - 20 &= 0 \\ (x-5)(x+4) &= 0 \\ \therefore x &= 5, -4 \end{aligned}$$
But $x > 0$
$\therefore$ The number is 5.

24 a
$$\begin{aligned} x + y &= 3 \\ \therefore y &= 3 - x \end{aligned}$$
Subs. in $xy = 2$
$$\begin{aligned} x(3-x) &= 2 \\ 3x - x^2 &= 2 \\ x^2 - 3x + 2 &= 0 \\ (x-1)(x-2) &= 0 \\ x &= 1, 2 \end{aligned}$$
$\therefore$ Subs. $x = 1$ in $xy = 2$: $\therefore y = 2$
Subs. $x = 2$ in $xy = 2$: $\therefore y = 1$
$\therefore x = 1, y = 2$ or $x = 2, y = 1$

b
$$\begin{aligned} 2x^2 + x - 3 &= x^2 - 2x + 7 \\ \therefore x^2 + 3x - 10 &= 0 \\ (x+5)(x-2) &= 0 \\ x &= -5, 2 \end{aligned}$$
Subs. in $y = x^2 - 2x + 7$
For $x = -5$: $y = (-5)^2 - 2(-5) + 7$
$= 42$ $\therefore y = 42$
For $x = 2$: $y = 2^2 - 2(2) + 7$
$= 7$ $\therefore y = 7$
$\therefore x = -5, y = 42$ or $x = 2, y = 7$

c
$$\begin{aligned} \frac{2}{x} &= x + 1 \\ \therefore 2 &= x^2 + x \\ x^2 + x - 2 &= 0 \\ (x+2)(x-1) &= 0 \\ x &= -2, 1 \end{aligned}$$
Subs. in $y = x + 1$:
For $x = -2$: $y = -2 + 1$
$= -1$
$\therefore y = -1$
For $x = 1$: $y = 1 + 1$
$= 2$
$\therefore y = 2$
$\therefore x = -2, y = -1$ or $x = 1, y = 2$

25
$$\begin{aligned} x^2 + 6x - 2 &= 2x + 3 \\ x^2 + 4x - 5 &= 0 \\ (x+5)(x-1) &= 0 \\ x &= -5, 1 \end{aligned}$$
Subs. in $y = 2x + 3$:
For $x = -5$: $y = 2(-5) + 3$
$= -7$
$\therefore y = -7$
For $x = 1$: $y = 2(1) + 3$
$= 5$
$\therefore y = 5$
$\therefore$ Points are (−5, −7) and (1, 5).

26 Let length = x and width = y
$$\begin{aligned} 2(x+y) &= 24 \\ \therefore x + y &= 12 \\ \therefore y &= 12 - x \\ xy &= 32 \\ \therefore x(12-x) &= 32 \\ 12x - x^2 &= 32 \\ x^2 - 12x + 32 &= 0 \\ (x-8)(x-4) &= 0 \\ x &= 8, 4 \end{aligned}$$
Subs. in $y = 12 - x$:
For $x = 8$: $y = 12 - 8$
$= 4$
$\therefore y = 4$
For $x = 4$: $y = 12 - 4$
$= 8$
$\therefore y = 8$
$\therefore$ Dimensions are 8 cm by 4 cm.

27 Let the numbers be x and y.

$$x - y = 4$$
$$y = x + 4$$
$$xy = 21$$
$$x(x + 4) = 21$$
$$x^2 + 4x - 21 = 0$$
$$(x + 7)(x - 3) = 0$$
$$x = -7, 3$$

Subs. in $y = x + 4$:

For $x = -7$: $y = -7 + 4$
$= -3$

$\therefore y = -3$

For $x = 3$: $y = 3 + 4$
$= 7$

$\therefore y = 7$

$\therefore$ The numbers are either 7 and 3, or -7 and -3.

28 **a** $4x^3 = 4$
$x^3 = 1$
$x = 1$

b $8x^3 = 64$
$x^3 = 8$
$x = 2$

c $2x^3 + 250 = 0$
$2x^3 = -250$
$x^3 = -125$
$x = -5$

29 **a** $(x^2)^2 - 10x^2 + 9 = 0$

Put $y = x^2$

$y^2 - 10y + 9 = 0$
$(y - 9)(y - 1) = 0$
$\therefore y = 9$ or 1

Replace x^2

$x^2 = 9$ or $x^2 = 1$

$\therefore x = 3, -3$ or $x = 1, -1$

$x = 3, 1, -1, -3$

b $\frac{1}{x}^{2} + 3\,\frac{1}{x} - 10 = 0$

Put $w = \frac{1}{x}$

$w^2 + 3w - 10 = 0$

$\therefore (w + 5)(w - 2) = 0$

$\therefore w = -5$ or 2

Replace $\frac{1}{x}$

$\frac{1}{x} = -5$ or $\frac{1}{x} = 2$

$\therefore 1 = -5x$ or $1 = 2x$

$\therefore x = \frac{-1}{5}$ or $x = \frac{1}{2}$

Chapter 5 . . . . pp. 81–83

Linear Relationships

1 $d = \sqrt{(x_2 - x_1)^2 + (y_2 - y_1)^2}$ is used in each part.

a $d = \sqrt{(1 + 1)^2 + (-4 - 4)^2}$
$= \sqrt{2^2 + (-8)^2}$
$= \sqrt{68}$
$= 2\sqrt{17}$

b $d = \sqrt{(4 - 2)^2 + (-1 + 1)^2}$
$= \sqrt{2^2 + 0}$
$= \sqrt{4}$
$= 2$

c $d = \sqrt{(1 + 3)^2 + (1 + 2)^2}$
$= \sqrt{4^2 + 3^2}$
$= \sqrt{25}$
$= 5$

d $d = \sqrt{(-3 - 0)^2 + (7 - 4)^2}$
$= \sqrt{9 + 9}$
$= \sqrt{18}$
$= 3\sqrt{2}$

2 MP $= \frac{x_1 + x_2}{2}, \frac{y_1 + y_2}{2}$ is used in each part.

a MP $= \frac{2 + 4}{2}, \frac{-1 - 3}{2} = (3, -2)$

b MP $= \frac{-4 + 1}{2}, \frac{2 - 4}{2}$
$= \frac{-3}{2}, \frac{-2}{2} = 1\frac{1}{2}, -1$

c MP $= \frac{5 + -3}{2}, \frac{3 + 5}{2}$
$= \frac{2}{2}, \frac{8}{2} = (1, 4)$

d MP $= \frac{7 + -3}{2}, \frac{-2 + 2}{2}$
$= \frac{4}{2}, \frac{-4}{2} = (2, -2)$

3 $m = \frac{y_2 - y_1}{x_2 - x_1}$ is used in each part.

a $m = \frac{-1 - 1}{-4 - 4} = \frac{-2}{-8} = \frac{1}{4}$

b $m = \frac{1 - 4}{-1 - 1} = \frac{-3}{-2} = \frac{3}{2}$

c $m = \frac{-3 + 2}{-2 + 3} = \frac{-1}{1} = -1$

d $m = \frac{-1 + 1}{3 - 4} = 0$ (line || x-axis)

4 Gradient $= \frac{\text{Rise}}{\text{Run}}$

a $m = \frac{-3}{2}$ (↘)

b $m = -\frac{4}{2} = -2$ (↘)

c $m = -\frac{5}{3}$ (↘)

d $m = \frac{3}{4}$ (↗)

e Gradient $= \frac{\text{Rise}}{\text{Run}} = \frac{3}{3} = 1$, i.e. $m = 1$.

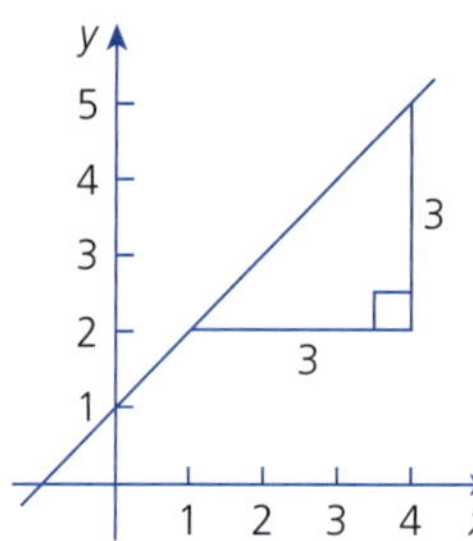

5 **a** AB $= \sqrt{40} = 2\sqrt{10}$
AC $= \sqrt{40} = 2\sqrt{10}$
$\therefore$ AB = AC
△ABC is isosceles.

b $m_{AB} = \frac{2 - 4}{4 + 2}$
$= -\frac{2}{6}$
$= -\frac{1}{3}$

c Midpoint AC = (−1, 1) (Call it P.)

d Midpoint BC = (2, 0) (Call it Q.)

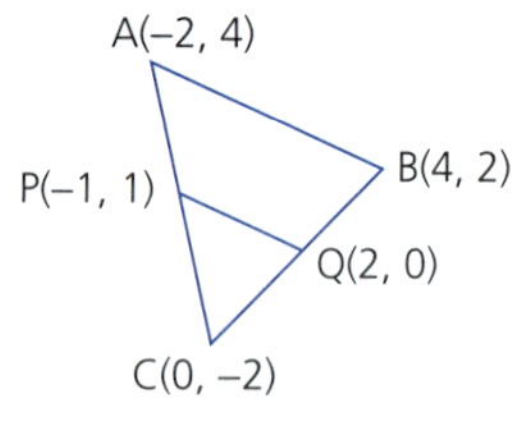

$m_{PQ} = -\frac{1}{3}$

$\therefore m_{PQ} = m_{AB}$

i.e. PQ || AB

Distance PQ $= \sqrt{10} = \frac{1}{2}$AB

PQ || AB and equal to half AB.

6 **a** $m_{AB} = 4$, $m_{BC} = -\frac{1}{4}$

As $m_{AB} \times m_{BC} = 4 \times -\frac{1}{4} = -1$

then AB ⊥ BC,

i.e. △ABC is right-angled at B.

b AC $= \sqrt{34}$, AB $= \sqrt{17}$, BC $= \sqrt{17}$

$\therefore$ AC^2 = 34, AB^2 = 17, BC^2 = 17

i.e. AC^2 = AB^2 + BC^2.

c

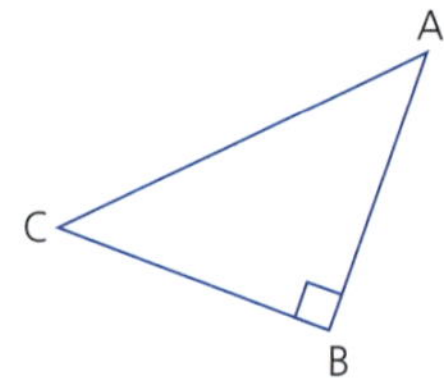

$$\text{Area} = \frac{1}{2}bh$$
$$= \frac{1}{2} \times \sqrt{17} \times \sqrt{17}$$
$$= \frac{17}{2}$$
$$= 8.5 \text{ units}^2$$

d As AB = BC, $\triangle$ABC is isosceles.

7 Put equations into form $y = mx + b$ where m = gradient.

a $m = -3$

b $y = \frac{1}{4}x + \frac{2}{4}$

$m = \frac{1}{4}$

c $3x - y = 7$

$3x - 7 = y$

$m = 3$

d $x + 5y - 8 = 0$

$5y = -x + 8$

$y = -\frac{1}{5}x + \frac{8}{5}$

$m = -\frac{1}{5}$

8 Using $y = mx + b$

$y = \frac{1}{2}x - 2$

$2y = x - 4$

$x - 2y - 4 = 0$

9 $y = -3x + 4$

Or $3x + y - 4 = 0$

10 Using $y - y_1 = m(x - x_1)$

a $y - 1 = -2(x + 1)$

$y - 1 = -2x - 2$

$\therefore 2x + y + 1 = 0$

b $y - 2 = \frac{1}{2}(x - 4)$

$2y - 4 = x - 4$

$\therefore x - 2y = 0$

c $y - 1 = \frac{3}{4}(x + 3)$

$4y - 4 = 3x + 9$

$\therefore 3x - 4y + 13 = 0$

11 $y + 5 = \frac{2}{3}(x - 2)$

$3y + 15 = 2x - 4$

$\therefore 2x - 3y - 19 = 0$

12 Using $\frac{y - y_1}{x - x_1} = \frac{y_2 - y_1}{x_2 - x_1}$ in each question.

a $\frac{y + 1}{x - 1} = \frac{4 + 1}{2 - 1} = \frac{5}{1}$

$\therefore \quad y + 1 = 5x - 5$

$\therefore \quad 5x - y - 6 = 0$

b $\frac{y - 3}{x - 0} = \frac{0 - 3}{-3 - 0} = \frac{1}{1}$

$y - 3 = x$

$\therefore \quad x - y + 3 = 0$

c $\frac{y - 1}{x + 2} = \frac{2 - 1}{4 + 2} = \frac{1}{6}$

$\therefore \quad 6y - 6 = x + 2$

$\therefore \quad x - 6y + 8 = 0$

d $\frac{y + 5}{x - 1} = \frac{-5 + 5}{-1 - 1} = \frac{0}{-2} = 0$

Consider the points now that you have seen $m = 0$.

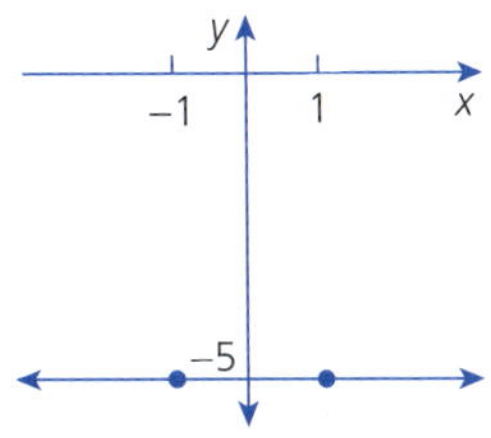

The required line is parallel to the x-axis and passes through points with y-coordinate -5. The equation is $y = -5$.

13 $2x + y = 1$

Substitute each point.

A$(-3, 7)$, LHS $= -6 + 7 = 1$ = RHS

$(-3, 7)$ lies on $2x + y = 1$.

B$(3, -7)$, LHS $= 6 - 7 = -1$

$(3, -7)$ does not lie on the line.

C$(-11, 23)$, LHS $= -22 + 23 = 1$ = RHS

$(-11, 23)$ lies on $2x + y = 1$.

A and C lie on $2x + y = 1$.

14 a $y = x + 2$

x	0	−2
y	2	0

$\therefore$ x-int.: -2

y-int.: 2

b $2x + y = 6$

x	0	3
y	6	0

$\therefore$ x-int.: 3

y-int.: 6

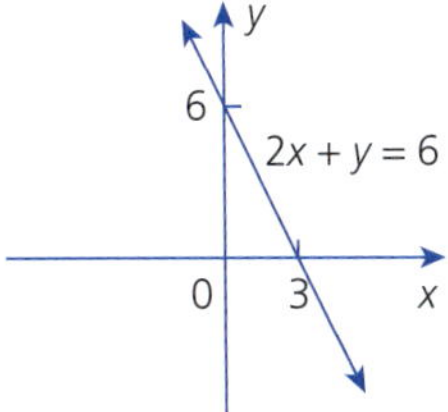

c $x - 3y - 9 = 0$

x	0	9
y	−3	0

$\therefore$ x-int.: 9

y-int.: -3

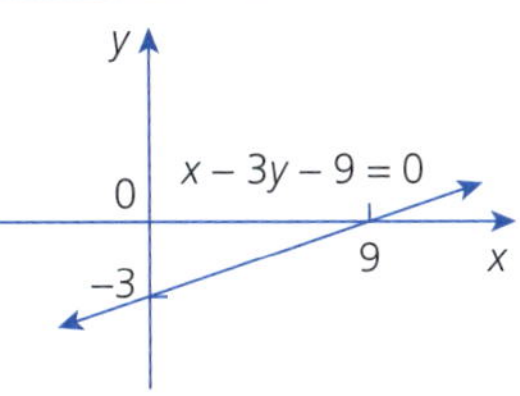

15 a $2x - 3y = 7$

$\therefore \quad 3y = 2x - 7$

$y = \frac{2}{3}x - \frac{7}{3}$

Gradient $= \frac{2}{3}$

$6x - 12y - 5 = 0$

$12y = 6x - 5$

$y = \frac{6}{12}x - \frac{5}{12}$

Gradient $= \frac{1}{2}$

Lines are neither parallel nor perpendicular.

b $x + 2y - 3 = 0$

$\therefore \quad 2y = -x + 3$

$y = -\frac{1}{2}x + \frac{3}{2}$

Gradient $= -\frac{1}{2}$

$y = 2x + 1$

Gradient = 2

Now $2 \times -\frac{1}{2} = -1$

$\therefore$ Lines are perpendicular.

c $2x + 3y = 6$

$3y = -2x + 6$

$y = -\frac{2}{3}x + 2$

Gradient $= -\frac{2}{3}$

$4y = 6x - 5$

$y = \frac{3}{2}x - \frac{5}{4}$

Gradient $= \frac{3}{2}$

Now $-\frac{2}{3} \times \frac{3}{2} = -1$

$\therefore$ Lines are perpendicular.

16 a $y = 4x - 7$, $m = 4$

Line parallel has gradient = 4.

Required line has equation:

$y = mx + b$

$\therefore \quad y = 4x - 4$

$\therefore$ $4x - y - 4 = 0$ is the equation.

b $2x - 3y = 5$

$$3y = 2x - 5$$
$$y = \frac{2}{3}x - \frac{5}{3}$$

Gradient $= \frac{2}{3}$

Line parallel will have gradient $= \frac{2}{3}$.

The required equation will be:

$$y - y_1 = m(x - x_1)$$
$$y + 1 = \frac{2}{3}(x + 1)$$

$\therefore\ 3(y + 1) = 2(x + 1)$

$\therefore\ 3y + 3 = 2x + 2$

$2x - 3y - 1 = 0$ is the equation.

c $x + 3y - 6 = 0$

$\therefore\ 3y = -x + 6$

$$y = -\frac{1}{3}x + 2$$

Gradient $= -\frac{1}{3}$

The perpendicular line will have gradient 3.

Equation is of the form:

$$y - y_1 = m(x - x_1)$$
$$y - 1 = 3(x - 4)$$

$\therefore\ y - 1 = 3x - 12$

$\therefore\ 3x - y - 11 = 0$ is the required equation.

d $2y = 5 - x$

$$y = -\frac{1}{2}x + \frac{5}{2}$$

Gradient $= -\frac{1}{2}$

The line perpendicular will have gradient 2.

The equation will have the form:

$$y - y_1 = m(x - x_1)$$

$\therefore\ y - 3 = 2(x + 2)$

$\therefore\ y - 3 = 2x + 4$

$2x - y + 7 = 0$ is the required equation.

17 Consider $2x - ky = 5$ (1)

$\therefore\ ky = 2x - 5$

$$y = \frac{2}{k}x - \frac{5}{k}$$

The gradient of line (1) $= \frac{2}{k}$.

Consider $3x + 4y - 7 = 0$ (2)

$\therefore\ 4y = -3x + 7$

$$y = -\frac{3}{4}x + \frac{7}{4}$$

The gradient of line (2) $= -\frac{3}{4}$.

As the lines are parallel, their gradients are equal.

$\therefore\ \frac{2}{k} = -\frac{3}{4}$

$-3k = 8$

$$k = -\frac{8}{3}$$
$$= -2\frac{2}{3}$$

18 Call the midpoint of AB, M.

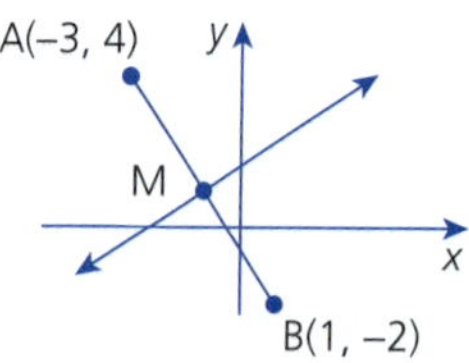

Then the coordinates of M are:

$$\frac{x_1 + x_2}{2}, \frac{y_1 + y_2}{2} = \frac{-3 + 1}{2}, \frac{4 + 2}{2}$$
$$= (-1, 1)$$

The gradient of AB $= \frac{y_2 - y_1}{x_2 - x_1}$

$$= \frac{4 + 2}{-3 - 1}$$
$$= \frac{6}{-4}$$
$$= -\frac{3}{2}$$

The gradient of the perpendicular $= \frac{2}{3}$.

$-\frac{3}{2} \times \frac{2}{3} = -1$

Now the equation of the line ⊥ AB through M is of the form:

$$y - y_1 = m(x - x_1)$$
$$y - 1 = \frac{2}{3}(x + 1)$$

$3(y - 1) = 2(x + 1)$

$\therefore\ 3y - 3 = 2x + 2$

$2x - 3y + 5 = 0$ is the equation of the perpendicular bisector.

19

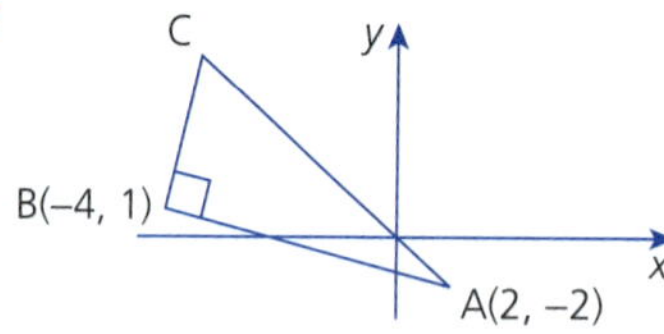

a The equation of AB is of the form:

$$\frac{y - y_1}{x - x_1} = \frac{y_2 - y_1}{x_2 - x_1}$$

$\therefore\ \frac{y - 1}{x + 4} = \frac{-2 - 1}{2 + 4} = \frac{-3}{6}$

$\therefore\ \frac{y - 1}{x + 4} = \frac{-1}{2}$

$\therefore\ 2y - 2 = -x - 4$

$x + 2y + 2 = 0$ is the equation of AB.

b BC ⊥ AB, hence the gradient of BC will be the negative reciprocal of the gradient of AB.

$\therefore\ m_{AB} = -\frac{1}{2}$ [from part **a**],

$\therefore\ m_{BC} = 2$

c The equation of BC has the form

$$y - y_1 = m(x - x_1),$$

That is, $y - 1 = 2(x + 4)$

$$y - 1 = 2x + 8$$

$\therefore\ 2x - y + 9 = 0$ is the equation of BC.

d Distance BC $= \sqrt{(-2 + 4)^2 + (5 - 1)^2}$

$$= \sqrt{4 + 16}$$
$$= \sqrt{20}$$
$$= 2\sqrt{5}$$

e Distance AB $= \sqrt{(2 + 4)^2 + (-2 - 1)^2}$

$$= \sqrt{36 + 9}$$
$$= \sqrt{45}$$
$$= 3\sqrt{5}$$

Area of $\triangle ABC = \frac{1}{2}bh$

$$= \frac{1}{2} \times AB \times BC$$
$$= \frac{1}{2} \times 3\sqrt{5} \times 2\sqrt{5}$$
$$= 15 \text{ units}^2$$

$\sqrt{5} \times \sqrt{5} = 5$

f Midpoint of AB $= \frac{x_1 + x_2}{2}, \frac{y_1 + y_2}{2}$

$$= \frac{-4 + 2}{2}, \frac{1 - 2}{2}$$
$$= -1, -\frac{1}{2}$$

Using $-1, -\frac{1}{2}$ and $(-2, 5)$:

$$\frac{y - y_1}{x - x_1} = \frac{y_2 - y_1}{x_2 - x_1}$$

$\therefore\ \frac{y - 5}{x + 2} = \frac{-\frac{1}{2} - 5}{-1 + 2} = \frac{-\frac{11}{2}}{1}$

$\therefore\ \frac{y - 5}{x + 2} = -\frac{11}{2}$

$\therefore\ 2y - 10 = -11x - 22$

$11x + 2y + 12 = 0$ is the equation of the median.

20

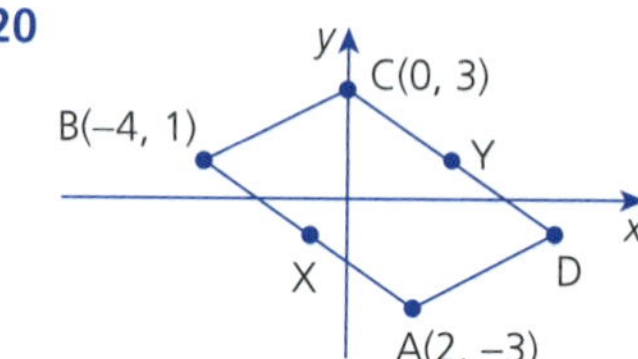

a $m_{AB} = \frac{1 + 3}{-4 - 2}$

$$= \frac{4}{-6}$$
$$= -\frac{2}{3}$$

$\frac{y_2 - y_1}{x_2 - x_1} = m$

The equation of AB is of the form:

$$y - y_1 = m(x - x_1)$$
$$y + 3 = -\frac{2}{3}(x - 2)$$

$3y + 9 = -2x + 4$

$\therefore\ 2x + 3y + 5 = 0$ is the equation of AB.

b CD ∥ AB, gradients are equal.

$m_{CD} = -\frac{2}{3}$ also. Using C(0, 3):

$$y - 3 = -\frac{2}{3}(x - 0)$$

$\therefore\ 3y - 9 = -2x$

$2x + 3y - 9 = 0$ is the equation of CD.

c D will be in the same position relative to A as point C is relative to B, that is, right across 4, up 2.

$D = (2 + 4, -3 + 2)$
$= (6, -1)$

d Midpoint of AB $= \left(\frac{-4+2}{2}, \frac{1-3}{2}\right)$
$= (-1, -1)$

Call it X.

The midpoint of CD $= \left(\frac{0+6}{2}, \frac{3-1}{2}\right)$
$= (3, 1)$

Call it Y.

Using points X and Y, the gradient of XY $= \frac{-1-1}{-1-3} = \frac{-2}{-4} = \frac{1}{2}$

The gradient of BC $= \frac{3-1}{0+4} = \frac{2}{4} = \frac{1}{2}$.

As the gradients are equal, XY ∥ BC.

21 $2x - y + 3 = 0$

$\therefore \quad y = 2x + 3$

$\therefore \quad m = 2$

$y - y_1 = m(x - x_1)$

$y - 1 = 2(x - 2)$

$y - 1 = 2x - 4$

$\therefore 2x - y - 3 = 0$

22 $4x + y + 5 = 0$

$\therefore \quad y = -4x - 5$

∴ Gradient is -4

∴ Gradient of perpendicular is $\frac{1}{4}$

$y - y_1 = m(x - x_1)$

$y - 3 = \frac{1}{4}(x + 1)$

$4y - 12 = x + 1$

$\therefore x - 4y + 13 = 0$

23 Line PQ will have an equation of the form:

$\frac{y+3}{x-7} = \frac{1+3}{-5-7} = \frac{4}{-12}$

$\therefore \quad \frac{y+3}{x-7} = -\frac{1}{3}$

$\therefore \quad 3y + 9 = -x + 7$

$x + 3y + 2 = 0$ is the equation of PQ.

For X to lie on PQ, the coordinates of X must satisfy the equation.

$(-2, 0)$ and $x + 3y + 2 = 0$:

LHS $= -2 + 3(0) + 2 = 0$
$=$ RHS

∴ $(-2, 0)$ lies on line PQ.

24

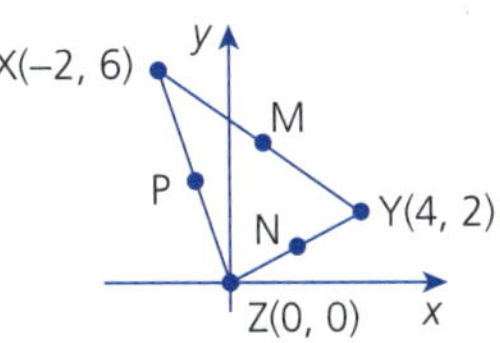

a Using the midpoint formula: $\left(\frac{x_1 + x_2}{2}, \frac{y_1 + y_2}{2}\right)$

The midpoint of XY is:

$M = \left(\frac{-2+4}{2}, \frac{6+2}{2}\right)$
$= (1, 4)$

The midpoint of YZ is:

$N = \left(\frac{4+0}{2}, \frac{2+0}{2}\right)$
$= (2, 1)$

The midpoint of ZX is:

$P = \left(\frac{-2+0}{2}, \frac{6+0}{2}\right)$
$= (-1, 3)$

b For the median MZ, use (1, 4) and (0, 0). The equation is:

$\frac{y-0}{x-0} = \frac{4-0}{1-0}$

$\frac{y}{x} = \frac{4}{1}$

$\therefore \quad y = 4x$ or $4x - y = 0$.

For the median NX, use (2, 1) and (−2, 6). The equation is:

$\frac{y-1}{x-2} = \frac{6-1}{-2-2} = \frac{5}{-4}$

$\therefore \quad \frac{y-1}{x-2} = -\frac{5}{4}$

$4y - 4 = -5x + 10$

$\therefore 5x + 4y - 14 = 0$ is the equation of NX.

For the median PY, use (−1, 3) and (4, 2). The equation is:

$\frac{y-3}{x+1} = \frac{2-3}{4+1} = \frac{-1}{5}$

$\therefore \quad 5y - 15 = -x - 1$

$\therefore x + 5y - 14 = 0$

is the equation of PY.

c For medians to be concurrent they must all pass through the same point. (Call it K.)

Solve MZ and PY to find K.

MZ: $y = 4x$ (1)

PY: $x + 5y = 14$ (2)

Substitute (1) into (2):

$x + 5(4x) = 14$

$21x = 14$

$x = \frac{14}{21}$

$= \frac{2}{3}$

Substitute in (1):

$y = 4\left(\frac{2}{3}\right)$

$= \frac{8}{3}$

K is $\left(\frac{2}{3}, \frac{8}{3}\right)$.

If NX passes through K, all lines pass through K. Test K with NX, $5x + 4y - 14 = 0$.

$\left(\frac{2}{3}, \frac{8}{3}\right)$. LHS $= 5x + 4y - 14$

$= \frac{10}{3} + \frac{32}{3} - 14$

$= \frac{42}{3} - 14$

$= 14 - 14$

$= 0 =$ RHS

That is, NX passes through K, the point of intersection of MZ and PY.

∴ The medians are concurrent.

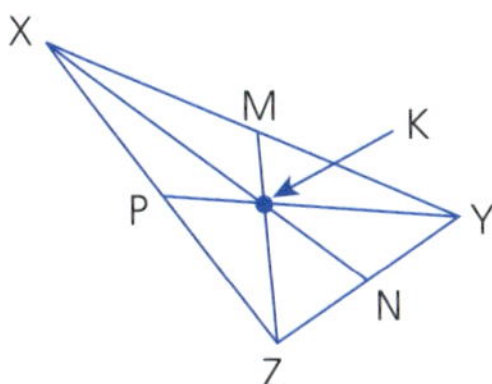

d Using $m = \frac{y_2 - y_1}{x_2 - x_1}$

$m_{XY} = \frac{6-2}{-2-4} = \frac{4}{-6} = -\frac{2}{3}$

$m_{YZ} = \frac{2-0}{4-0} = \frac{1}{2}$

$m_{ZX} = \frac{6-0}{-2-0} = -\frac{3}{1}$

$\therefore m_{XY} \times m_{YZ} \times m_{ZX}$

$= \frac{-2}{3} \times \frac{1}{2} \times \frac{-3}{1} = 1$

e The altitude through Z will be ⊥ to XY and pass through Z.

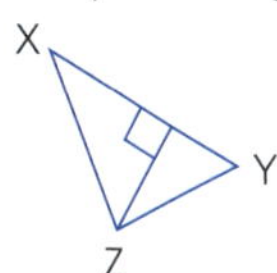

$m_{XY} = -\frac{2}{3}$, $m_{altitude} = \frac{3}{2}$

Using the point–gradient formula with Z(0, 0):

$y - 0 = \frac{3}{2}(x - 0)$ $\quad -\frac{2}{3} \times \frac{3}{2} = -1$

$\therefore 2y = 3x$

$3x - 2y = 0$ is the equation of the altitude through Z.

25 The point $(2d, 5)$ lies on $2x - y - 8 = 0$, so the coordinates must satisfy the equation.

$(2d, 5) \quad 2x - y - 8 = 0$

$2(2d) - 5 - 8 = 0$

$4d = 13$

$d = \frac{13}{4} = 3\frac{1}{4}$

26

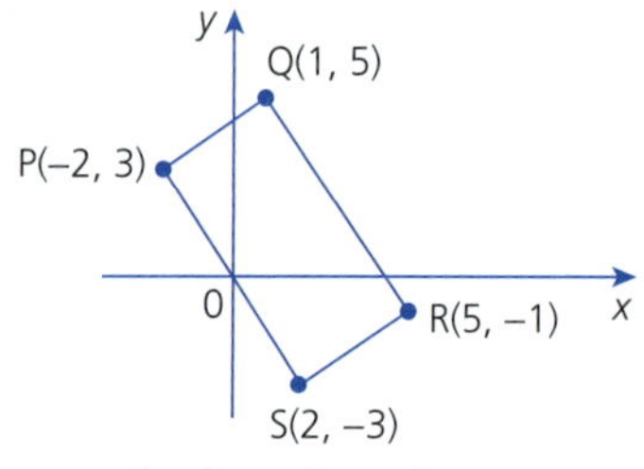

$m_{PS} = \frac{3+3}{-2-2} = \frac{6}{-4} = -\frac{3}{2}$

$m_{QR} = \frac{5+1}{1-5} = \frac{6}{-4} = -\frac{3}{2}$

∴ PS ∥ QR [equal gradients]

$m_{PQ} = \frac{5-3}{1+5} = \frac{2}{3}$

$m_{RS} = \frac{-1+3}{5-2} = \frac{2}{3}$

∴ PQ ∥ RS [equal gradients]

∴ PQRS is a parallelogram.

But PQ ⊥ PS, as:

$m_{PQ} \times m_{PS} = \frac{2}{3} \times -\frac{3}{2} = -1$

∴ PQRS is a rectangle (a parallelogram with one angle a right-angle).

27 $m_{AB} = \frac{5-4}{-3-2} = \frac{1}{-5} = -\frac{1}{5}$

A line ⊥ AB will have a gradient of 5.

$-\frac{1}{5} \times 5 = -1$

The line passes through (0, 0) with a gradient of 5.

The equation is $y - 0 = 5(x - 0)$

∴ $y = 5x$ is the required equation.

28

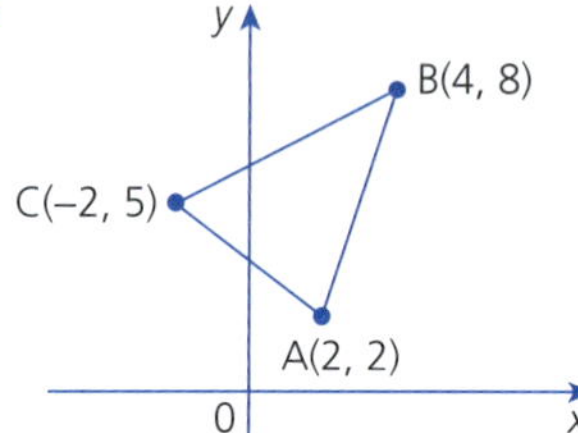

a $BC = \sqrt{(4+2)^2 + (8-5)^2}$
$= \sqrt{36+9}$
$= \sqrt{45}$
$= 3\sqrt{5}$

b The equation is of the form:

$\frac{y - y_1}{x - x_1} = \frac{y_2 - y_1}{x_2 - x_1}$

∴ $\frac{y-8}{x-4} = \frac{5-8}{-2-4} = \frac{-3}{-6}$

∴ $\frac{y-8}{x-4} = \frac{1}{2}$

∴ $2y - 16 = x - 4$

$x - 2y + 12 = 0$ is the required equation.

c A line ∥ BC will have the same gradient as BC, that is, $m = \frac{1}{2}$.

Then the equation will be:

$y - 2 = \frac{1}{2}(x - 2)$ using A

∴ $2y - 4 = x - 2$

$x - 2y + 2 = 0$ is the equation.

d A line ⊥ BC will have a gradient of –2. [Negative reciprocal]

The equation will be:

$y - 2 = -2(x - 2)$

$y - 2 = -2x + 4$

∴ $2x + y - 6 = 0$ is the equation.

e The midpoint of AB $= \frac{4+2}{2}, \frac{8+2}{2}$
$= (3, 5)$

f L(3, 5), C(–2, 5).

The equation is of the form:

$y = b$ as CL is parallel to the x-axis.

(Look at the coordinates.)

∴ $y = 5$ is the equation of the median.

29 The equation joining (–1, 3) and (1, 1) is of the form:

$\frac{y - y_1}{x - x_1} = \frac{y_2 - y_1}{x_2 - x_1}$

i.e. $\frac{y-3}{x+1} = \frac{1-3}{1+1} = \frac{-2}{2} = -1$

∴ $y - 3 = -1(x + 1)$

$y - 3 = -x - 1$ ☒

∴ $x + y - 2 = 0$ is the equation.

Now, for points to be collinear, (4, –2) must lie on this line.

Check (4, –2). $x + y - 2 = 0$

LHS = 4 – 2 – 2 = 0 = RHS

∴ (4, –2) lies on $x + y - 2 = 0$,

i.e. all points are collinear.

Alternative method using gradients

The gradient of the line joining (–1, 3) and (1, 1) is:

$m = \frac{3-1}{-1-1} = \frac{2}{-2} = -1$

The gradient of the line joining (1, 1) and (4, –2) is:

$m = \frac{1+2}{1-4} = \frac{3}{-3} = -1$

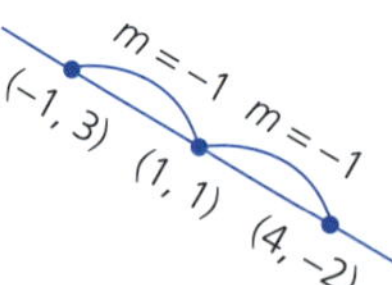

The line must contain all three points.

30

	$2x - 3y = 3$	(1)
	$x + 2y = 5$	(2)
2 × (2)	$2x + 4y = 10$	(3)
(3) – (1)	$7y = 7$	
	$y = 1$	
Subs. in (2)	$x + 2 = 5$	
	$x = 3$	

∴ Point of intersection is (3, 1).

Subs. in $3x - y = 8$

$3(3) - 1 = 8$ (Yes)

∴ All lines pass through (3, 1).

∴ Lines are concurrent.

Chapter 6 . . . . pp. 98–99

Non-linear Relationships

1 a

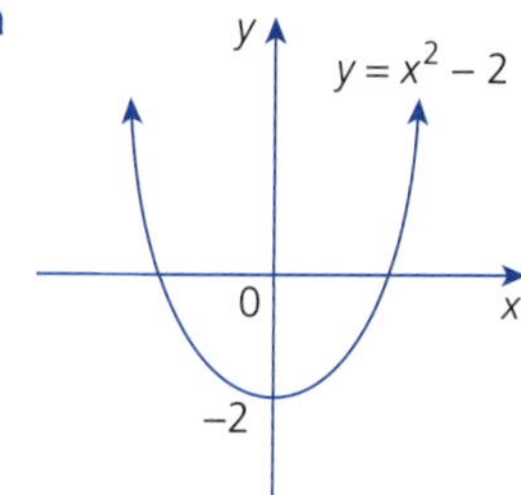

b

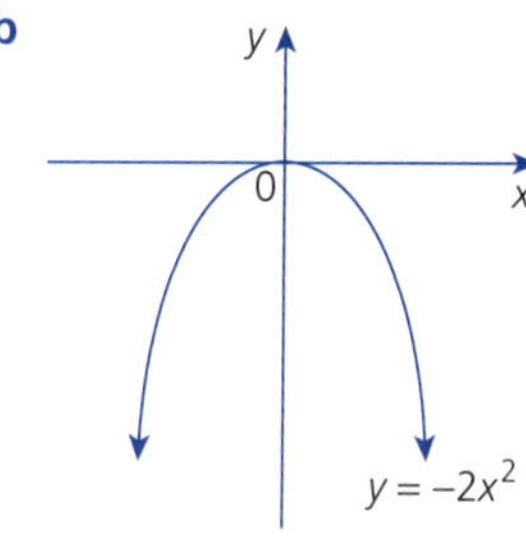

c

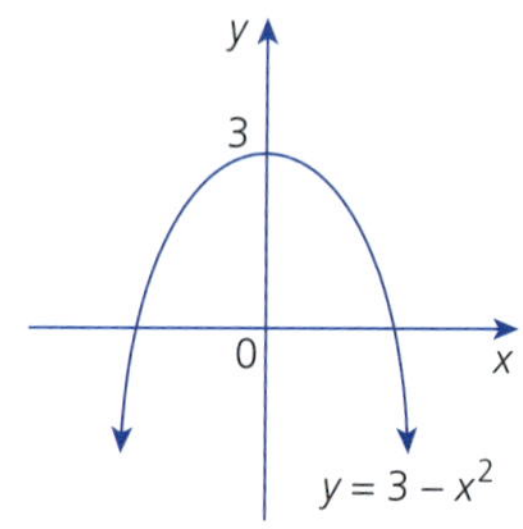

d

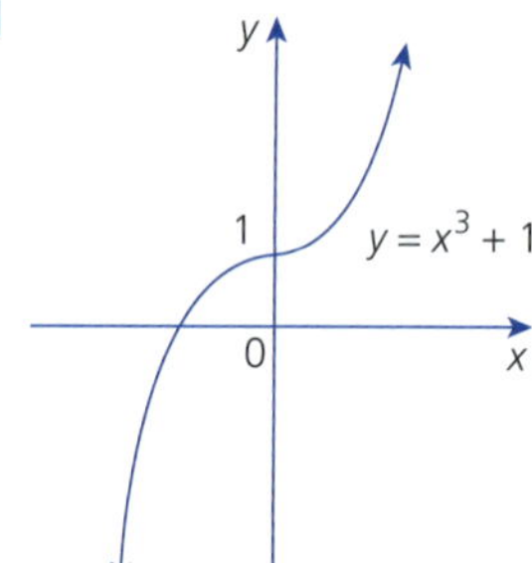

e

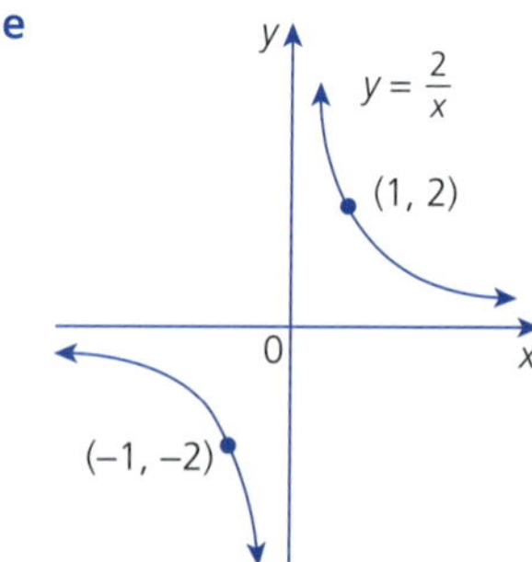

f

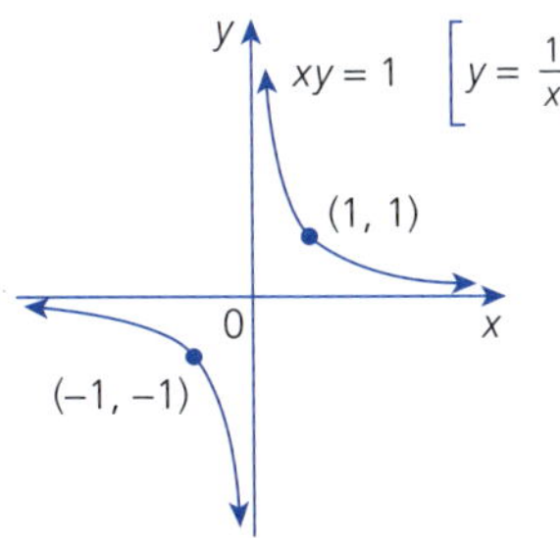

g

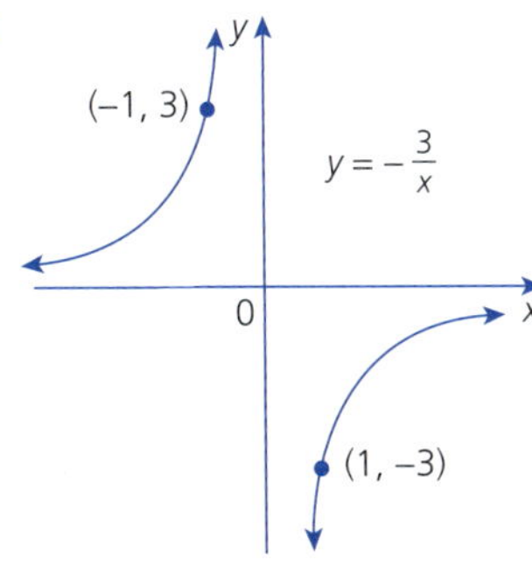

h

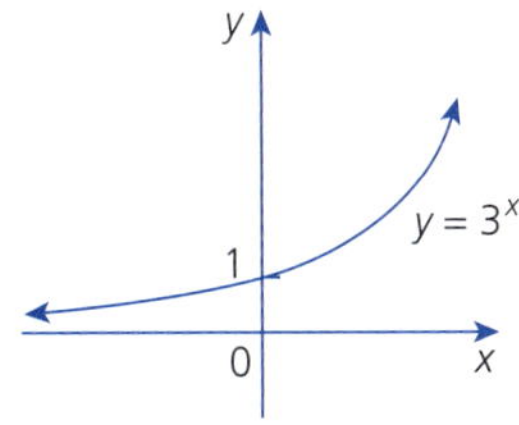

i

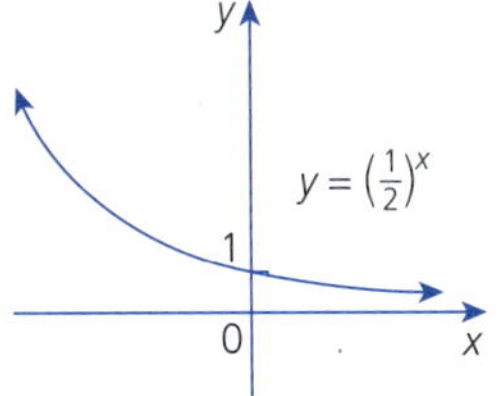

Note: $y = \left(\frac{1}{2}\right)^x$

$\therefore \quad y = (2^{-1})^x$

$= 2^{-x}$

$\therefore \quad y = \left(\frac{1}{2}\right)^x$ is the same as

$y = 2^{-x}$.

j

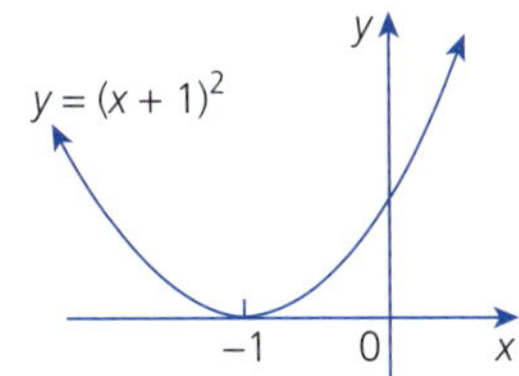

k

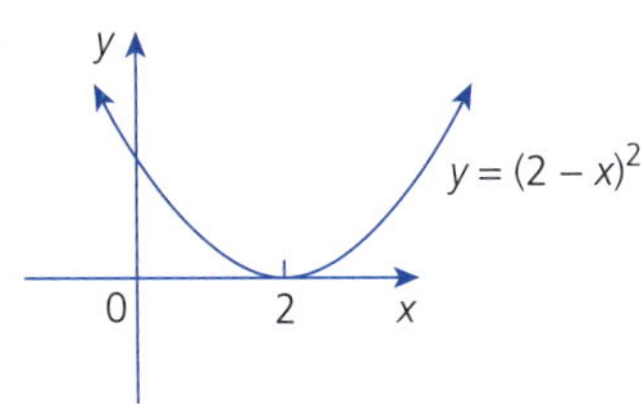

2 a $y = x^2 - 5x + 2$

$y = ax^2 + bx + c$

$\therefore a = 1 > 0$

$\therefore$ Concave up.

b $y = x - x^2$

$y = ax^2 + bx + c$

$\therefore a = -1 < 0$

$\therefore$ Concave down.

c $y = 3 + x^2 - x$

$y = ax^2 + bx + c$

$\therefore a = 1 > 0$

$\therefore$ Concave up.

d $y = 4x^2 - x + 1$

$y = ax^2 + bx + c$

$\therefore a = 4 > 0$

$\therefore$ Concave up.

3 a $y = x^2 + x - 2$

x-intercepts $\qquad \therefore y = 0$

$\therefore 0 = x^2 + x - 2$

i.e. $x^2 + x - 2 = 0$

$(x + 2)(x - 1) = 0$

$x = -2, 1$

$\therefore$ x-intercepts of –2 and 1.

y-intercept $\qquad \therefore x = 0$

$y = 0^2 + 0 - 2$

$y = -2$

$\therefore$ y-intercept of –2.

b $y = x^2 - x$

x-intercepts $\qquad \therefore y = 0$

$\therefore \quad 0 = x^2 - x$

i.e. $x^2 - x = 0$

$x(x - 1) = 0$

$x = 0, 1$

$\therefore$ x-intercepts of 0 and 1.

y-intercepts $\qquad \therefore x = 0$

$\therefore y = 0^2 - 0$

$y = 0$

$\therefore$ y-intercept of 0.

c $y = x^2 - 4$

x-intercepts $\qquad \therefore y = 0$

$\therefore \quad 0 = x^2 - 4$

i.e. $x^2 - 4 = 0$

$x^2 = 4$

$x = \pm 2$

$\therefore$ x-intercepts of 2 and –2.

y-intercepts $\qquad \therefore x = 0$

$\therefore y = 0^2 - 4$

$y = 0 - 4$

$= -4$

$\therefore$ y-intercept of –4.

d $y = 9 - x^2$

x-intercepts $\qquad \therefore y = 0$

$\therefore \quad 0 = 9 - x^2$

i.e. $9 - x^2 = 0$

$x^2 = 9$

$x \pm 3$

$\therefore$ x-intercepts of 3 and –3.

y-intercepts $\qquad \therefore x = 0$

$\therefore y = 9 - 0^2$

$y = 9$

$\therefore$ y-intercept of 9.

4 a $y = 3x^2 - 6x + 9$ $\qquad$ $\boxed{a = -3,\ b = -6,\ c = 9}$

$y = ax^2 + bx + c$

Now, axis of symmetry:

$x = \frac{-b}{2a}$

$= \frac{-(-6)}{2(3)}$

$= \frac{6}{6}$

$= 1$

$\therefore$ Axis of symmetry is $x = 1$.

b $y = 4 - 5x - x^2$ $\qquad$ $\boxed{a = -1,\ b = -5,\ c = 4}$

$y = ax^2 + bx + c$

Now, axis of symmetry:

$x = \frac{-b}{2a}$

$= \frac{-(-5)}{2(-1)}$

$= \frac{5}{-2}$

$= -2\frac{1}{2}$

$\therefore$ Axis of symmetry is $x = -2\frac{1}{2}$.

c $y = 4x^2 - 8x$ $\qquad$ $\boxed{a = 4,\ b = -8,\ c = 0}$

$y = ax^2 + bx + c$

Now, axis of symmetry:

$x = \frac{-b}{2a}$

$= \frac{-(-8)}{2(4)}$

$= \frac{8}{8}$

$= 1$

$\therefore$ Axis of symmetry is $x = 1$.

d $y = 4 - x^2$ $\qquad$ $\boxed{a = -1,\ b = 0,\ c = 4}$

$y = ax^2 + bx + c$

Now, axis of symmetry:

$x = \frac{-b}{2a}$

$= \frac{0}{2(-1)}$

$= 0$

$\therefore$ Axis of symmetry is $x = 0$.

5 a Max. value occurs on axis of symmetry, and $a < 0$.

$y = 4 - x - x^2 \qquad (1)$ $\qquad$ $\boxed{a = -1 < 0,\ b = -1,\ c = 4}$

$y = ax^2 + bx + c$

Axis of symmetry:

$x = \frac{-b}{2a}$

$= \frac{-(-1)}{2(-1)}$

$= \frac{1}{-2}$

$= -\frac{1}{2}$

$\therefore x = -\frac{1}{2}$

Now, substitute $x = -\frac{1}{2}$ in equation (1).

$\therefore y\left(-\frac{1}{2}\right) = 4 - \left(-\frac{1}{2}\right) - \left(-\frac{1}{2}\right)^2$

$= 4 + \frac{1}{2} - \frac{1}{4}$

$= 4\frac{1}{4}$

$\therefore$ Max. value of $4\frac{1}{4}$ (at $x = -\frac{1}{2}$).

b $y = -x^2 + 4x + 2$ (1) $\quad a = -1 < 0,\ b = 4,\ c = 2$

$y = ax^2 + bx + c$

Axis of symmetry:

$x = \frac{-b}{2a}$

$= \frac{-4}{2(-1)}$

$= \frac{-4}{-2}$

$\therefore x = 2$

Now, substitute $x = 2$ in equation (1).

$\therefore y(2) = -2^2 + 4(2) + 2$

$= -4 + 8 + 2$

$= 6$

$\therefore$ Max. value of 6 (at $x = 2$).

6 a Minimum value occurs on axis of symmetry, and $a > 0$.

$y = x^2 + 6x + 12$ (1) $\quad a = 1,\ b = 6,\ c = 12$

$y = ax^2 + bx + c$

Axis of symmetry:

$x = \frac{-b}{2a}$

$= \frac{-6}{2(1)}$

$= \frac{-6}{2}$

$= -3$

Now, substitute $x = -3$ in equation (1).

$\therefore y(-3) = (-3)^2 + 6(-3) + 12$

$= 9 - 18 + 12$

$= 3$

$\therefore$ Minimum value of 3 (at $x = -3$).

b $y = x^2 - 2x + 3$ (1) $\quad a = 1,\ b = -2,\ c = 3$

$y = ax^2 + bx + c$

Axis of symmetry:

$x = \frac{-b}{2a}$

$= \frac{-(-2)}{2(1)}$

$= \frac{2}{2}$

$= 1$

Now, substitute $x = 1$ in equation (1).

$\therefore y(1) = 1^2 - 2(1) + 3$

$= 1 - 2 + 3$

$= 2$

$\therefore$ Min. value of 2 (at $x = 1$).

7 a Passes through (2, 3), and y-intercept of -1.

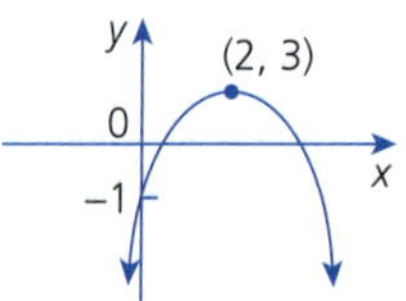

b Passes through (3, −4), and y-intercept of -2.

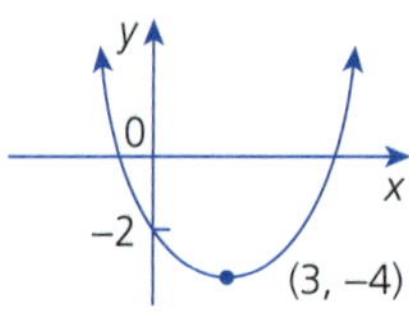

c Passes through (−2, 0), and y-intercept of 3.

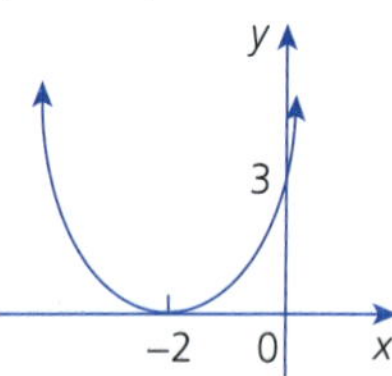

8 a $y = x^2 - 5x - 6$

$\therefore y = (x - 6)(x + 1)$

As $y = 0$

$\therefore (x - 6)(x + 1) = 0$

$x = 6, -1$

Middle of 6 and −1:

$= \frac{6 + -1}{2}$

$= \frac{5}{2} = 2\frac{1}{2}$

Substitute x in $y = (x - 6)(x + 1)$:

$= \left(2\frac{1}{2} - 6\right)\left(2\frac{1}{2} + 1\right)$

$= \left(-3\frac{1}{2}\right)\left(3\frac{1}{2}\right)$

$= -12\frac{1}{4}$

$\therefore$ Vertex $\left(2\frac{1}{2}, -12\frac{1}{4}\right)$.

b $y = x^2 + 2x - 8$

$\therefore \quad y = x^2 + 2x - 8$

this is, $y = x^2 + 2x + 1 - 8 - 1$

$y = (x + 1)^2 - 9$

$\therefore$ Vertex at (−1, −9).

c $y = 2x^2 - 3x + 1$

Find equation of axis of symmetry:

$x = \frac{-b}{2a}$

$= \frac{3}{4}$

That is, $x = \frac{3}{4}$ gives axis of symmetry.

Substitute in y:

$\therefore y = 2\left(\frac{3}{4}\right)^2 - 3\left(\frac{3}{4}\right) + 1$

$= \frac{9}{8} - \frac{9}{4} + 1$

$= -\frac{1}{8}$

$\therefore$ Vertex $= \left(\frac{3}{4}, -\frac{1}{8}\right)$.

d $y = 5 - 2x - x^2$

$\therefore a = -1, b = -2$

$\therefore x = -\frac{b}{2a}$

$= \frac{2}{-2}$

$= -1$

$\therefore x = -1$ is axis of symmetry.

Substitute in y:

$\therefore y = 5 - 2(-1) - (-1)^2$

$= 5 + 2 - 1$

$= 6$

$\therefore$ vertex is (−1, 6).

9 a Let $y = a(x + 1)(x - 3)$

Subs. in (1, −8)

$\therefore -8 = a(1 + 1)(1 - 3)$

$-8 = a(2)(-2)$

$-4a = -8$

$a = 2$

$\therefore y = 2(x + 1)(x - 3)$

b Let $y = ax(x - 4)$.

Subs. in (2, −4)

$\therefore -4 = 2a(2 - 4)$

$-4 = 2a(-2)$

$-4a = -4$

$a = 1$

$\therefore y = x(x - 4)$

c Let $y = a(x + 2)(x - 2)$.

Subs. in (0, 4)

$\therefore 4 = a(0 + 2)(0 - 2)$

$4 = a(2)(-2)$

$-4a = 4$

$a = -1$

$\therefore y = -(x + 2)(x - 2)$

d Let $y = a(x + 2)(x - 4)$.

Subs. in (5, 21)

$\therefore 21 = a(5 + 2)(5 - 4)$

$21 = a(7)(1)$

$7a = 21$

$a = 3$

$\therefore y = 3(x + 2)(x - 4)$

10 a

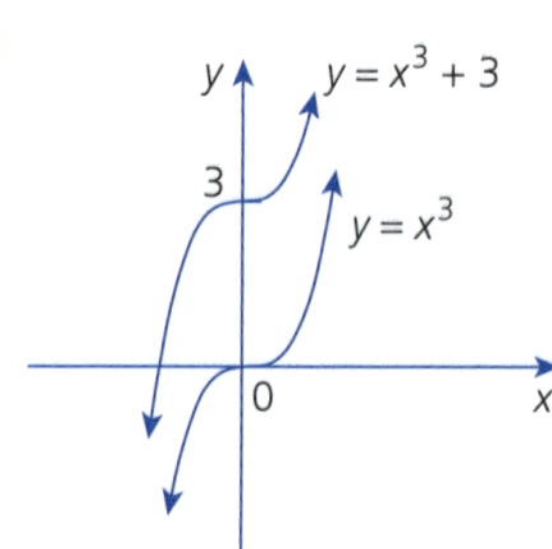

b

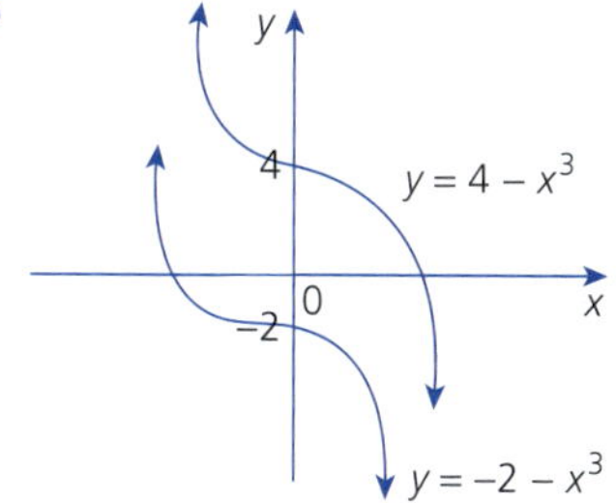

c

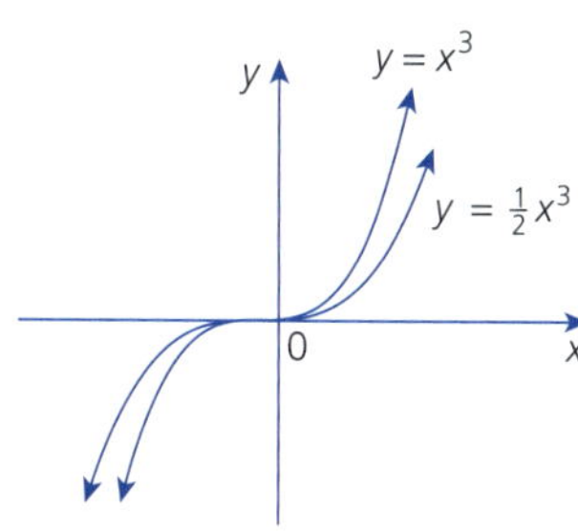

d

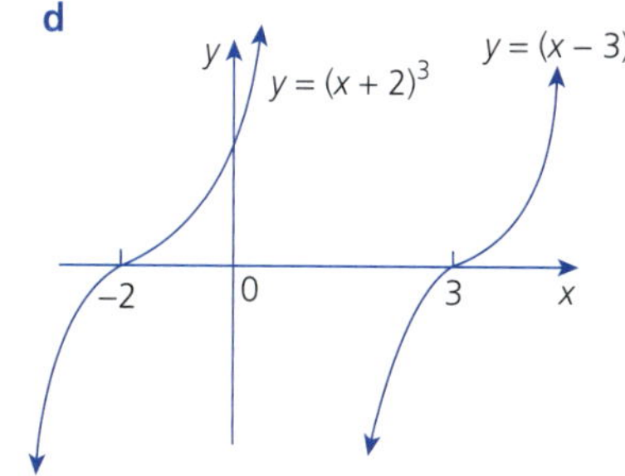

11 a $y = (x - 2)^3$

b $y = 2 - x^3$

12 a

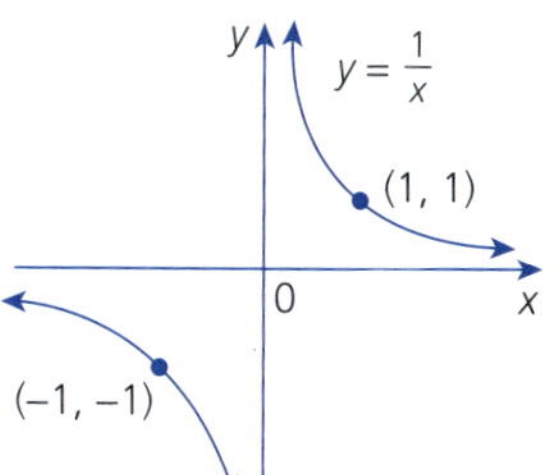

b

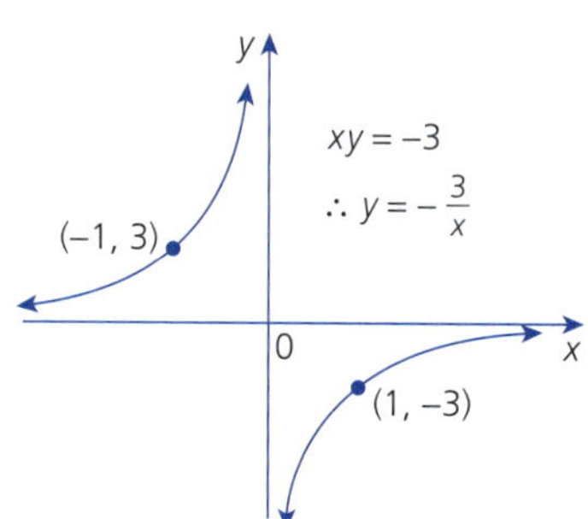

13 a	G	**b**	F
c	J	**d**	B
e	E	**f**	C
g	I	**h**	D
i	A	**j**	H

14 a

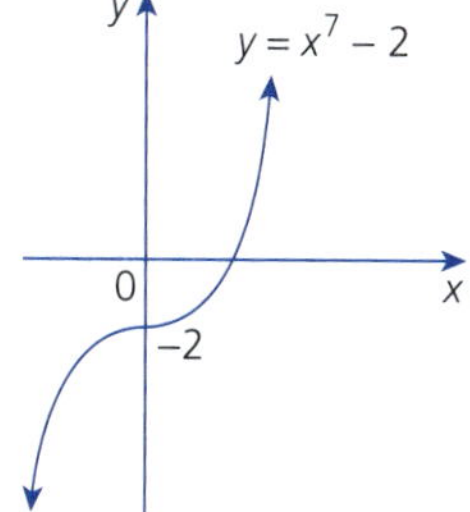

b

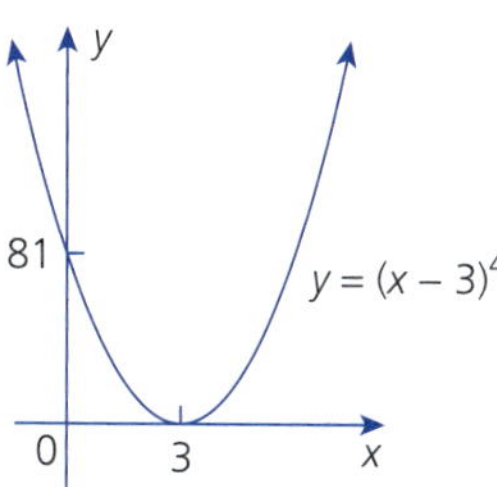

c

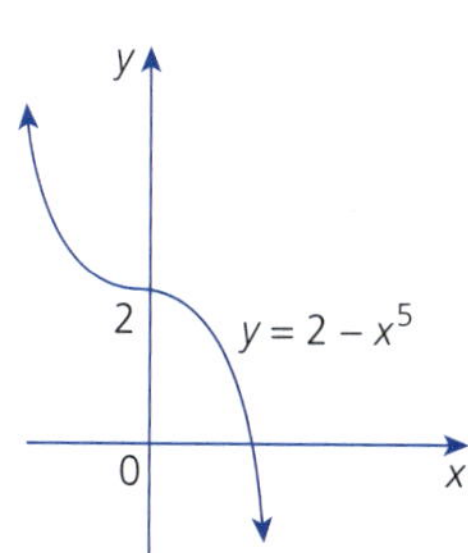

d

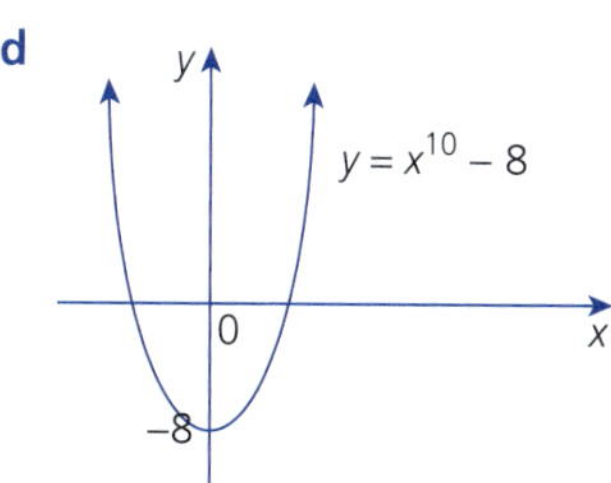

15 a $y = (x - 1)(x - 3)(x + 4)$

x-int: $y = 0 \quad \therefore x = 1, 3, -4$

y-int: $x = 0 \quad \therefore y = (-1)(-3)(4)$

$= 12$

Let $x = 5$.

$\therefore y = (5 - 1)(5 - 3)(5 + 4) > 0$

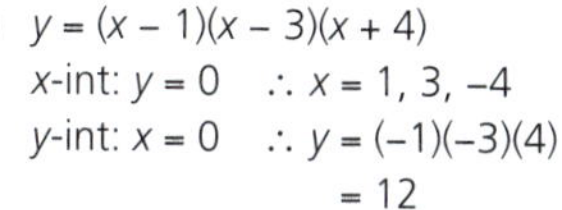

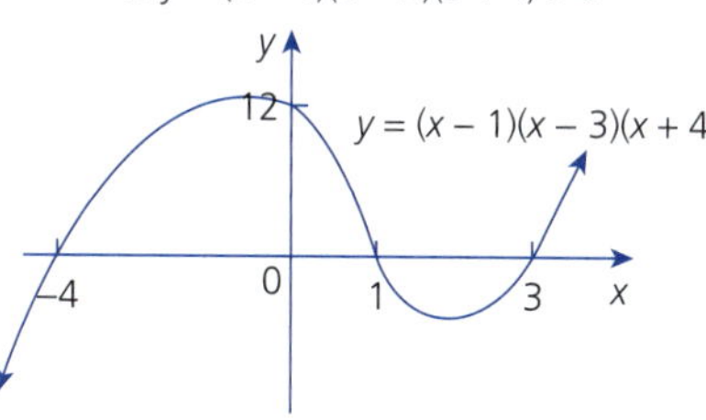

b $y = x(4 - x)(x + 1)$

x-int: $y = 0 \quad \therefore x = 0, 4, -1$

y-int: $x = 0 \quad \therefore y = 0(4 - 0)(0 + 1)$

$= 0$

Let $x = 10$.

$\therefore y = 10(4 - 10)(10 + 1) < 0$

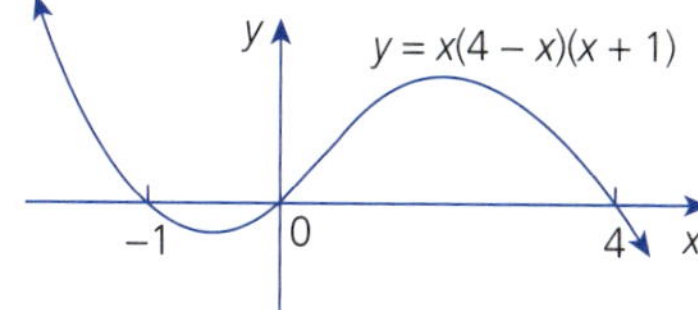

c $y = 2x(x + 3)(x - 1)$

x-int: $y = 0 \quad \therefore x = 0, -3, 1$

y-int: $x = 0 \quad \therefore y = 0(0 + 3)(0 - 1)$

$= 0$

Let $x = 2$.

$\therefore y = 2(2)(2 + 3)(2 - 1) > 0$

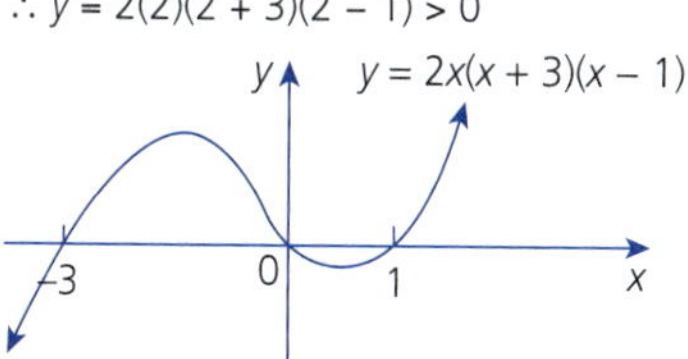

d $y = (4 - x)^2(x + 3)$

x-int: $y = 0 \quad \therefore x = 4, -3$

y-int: $x = 0 \quad \therefore y = (4 - 0)^2(0 + 3)$

$= 48$

Let $x = 5 \therefore y = (4 - 5)^2(5 + 3) > 0$

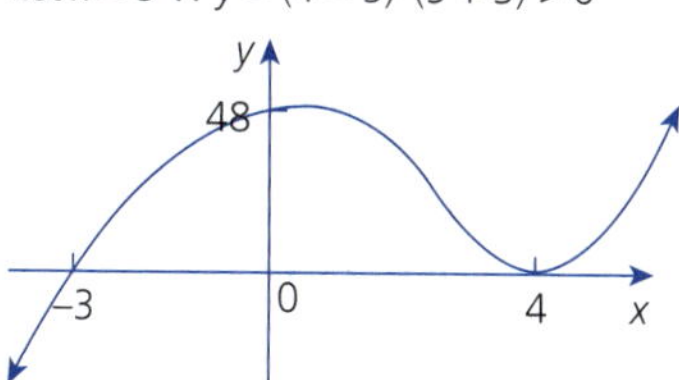

16 a $C(0, 0), r = 7$

$\therefore$ 7 units

b $x^2 + y^2 - 12 = 0$

$x^2 + y^2 = 12$

$\therefore C(0, 0), r = \sqrt{12}$

$= 2\sqrt{3}$

$\therefore 2\sqrt{3}$ units

c $C(0, 0), r = \sqrt{5}$

$\therefore \sqrt{5}$ units

17 a $(x + 3)^2 + (y - 1)^2 = 16$

$C(-3, 1), r = 4$

$\therefore$ Centre $(-3, 1)$ radius 4 units.

b $(x - 1)^2 + (y + 4)^2 = 48$

$C(1, -4), r = \sqrt{48}$

$= 4\sqrt{3}$

$\therefore$ Centre $(1, -4)$, radius $4\sqrt{3}$ units.

c $x^2 + (y + 1)^2 = 24$

$C(0, -1), r = \sqrt{24}$

$= 2\sqrt{6}$

$\therefore$ Centre $(0, -1)$, radius $2\sqrt{6}$ units.

d $x^2 + 8x + y^2 - 10y = 20$

$x^2 + 8x + 16 + y^2 - 10y + 25$

$= 20 + 16 + 25$

$(x + 4)^2 + (y - 5)^2 = 61$

$C(-4, 5), r = \sqrt{61}$

$\therefore$ Centre $(-4, 5)$, radius $\sqrt{61}$ units.

e $x^2 - 6x + y^2 + 16y = 16$

$x^2 - 6x + 9 + y^2 + 16y + 64$

$= 16 + 9 + 64$

$(x - 3)^2 + (y + 8)^2 = 89$

$C(3, -8), r = \sqrt{89}$

$\therefore$ Centre $(3, -8)$, radius $\sqrt{89}$ units.

f $x^2 - 3x + y^2 + 10y = 8$

$x^2 - 3x + \frac{9}{4} + y^2 + 10y + 25$

$= 8 + \frac{9}{4} + 25$

$\left(x - \frac{3}{2}\right)^2 + (y + 5)^2 = 35\frac{1}{4}$

$C\left(1\frac{1}{2}, -5\right), r = \sqrt{\frac{141}{4}}$

$= \frac{\sqrt{141}}{2}$

$\therefore$ Centre $\left(1\frac{1}{2}, -5\right)$, radius $\frac{\sqrt{141}}{2}$ units.

g $2x^2 + 2y^2 + 4y = 36$

$\therefore\ x^2 + y^2 + 2y = 18$

$x^2 + y^2 + 2y + 1 = 18 + 1$

$x^2 + (y + 1)^2 = 19$

$C(0, -1), r = \sqrt{19}$

$\therefore$ Centre $(0, -1)$, radius $\sqrt{19}$ units.

18 a

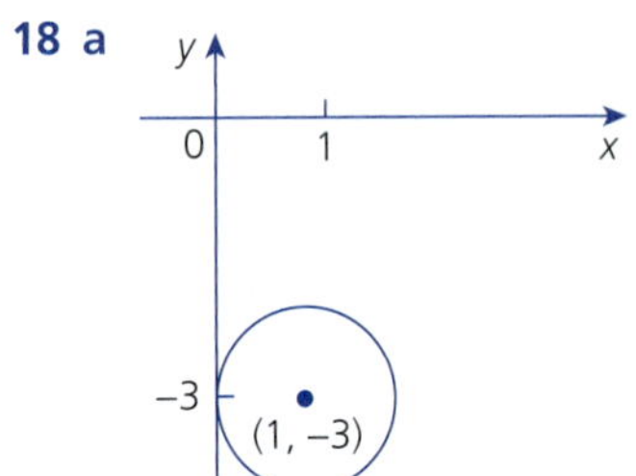

b $x^2 + 4x + y^2 - 6y = 3$

$x^2 + 4x + 4 + y^2 - 6y + 9$

$= 3 + 4 + 9$

$(x + 2)^2 + (y - 3)^2 = 16$

$C(-2, 3), r = 4$

$C(-2, 3)$, radius 4 units.

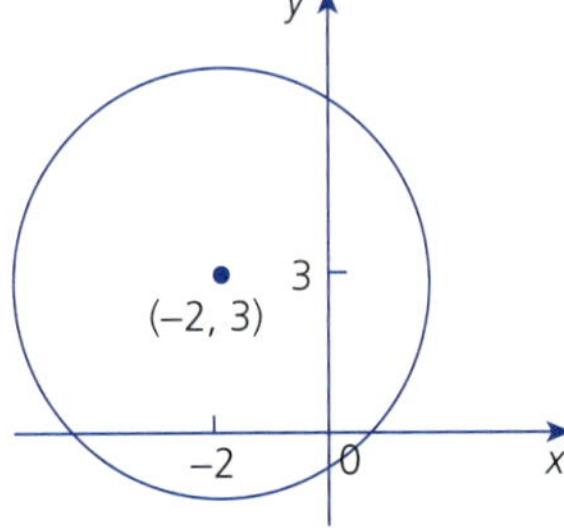

19 a $y = x^2 \quad (1)$

$y = 5x + 24 \quad (2)$

$\therefore\ x^2 = 5x + 24$

$x^2 - 5x - 24 = 0$

$(x - 8)(x + 3) = 0$

$x = 8, -3$

Subs. in (1) $\therefore x = 8, y = 64$

Also, $x = -3, y = 9$

$\therefore (8, 64)$ and $(-3, 9)$

b $y = x^2 + 2 \quad (1)$

$y = 3x \quad (2)$

$\therefore\ x^2 + 2 = 3x$

$x^2 - 3x + 2 = 0$

$(x - 2)(x - 1) = 0$

$x = 2, 1$

Subs. in (2) $\therefore x = 2, y = 6$

Also, $x = 1, y = 3$

$\therefore (2, 6)$ and $(1, 3)$

c $y = x^2 + 2x + 10 \quad (1)$

$y = 2x^2 + 3x + 4 \quad (2)$

$\therefore 2x^2 + 3x + 4 = x^2 + 2x + 10$

$x^2 + x - 6 = 0$

$(x + 3)(x - 2) = 0$

$x = -3, 2$

Subs. in (1):

When $x = -3$,

$y = (-3)^2 + 2(-3) + 10 = 13$

When $x = 2$,

$y = 2^2 + 2(2) + 10 = 18$

$\therefore (-3, 13), (2, 18)$

d $y = \frac{4}{x} \quad (1)$

$y = x + 3 \quad (2)$

$\therefore\ \frac{4}{x} = x + 3$

$\therefore\ 4 = x^2 + 3x$

$x^2 + 3x - 4 = 0$

$(x + 4)(x - 1) = 0$

$x = -4, 1$

Subs. in (2):

When $x = -4, y = -4 + 3$

$= -1$

When $x = 1, y = 1 + 3$

$= 4$

$\therefore (-4, -1)$ and $(1, 4)$

e $x^2 + y^2 = 8 \quad (1)$

$y = 2 \quad (2)$

Subs. (2) in (1):

$\therefore\ x^2 + 4 = 8$

$x^2 = 4$

$x = \pm 2$

$\therefore (2, 2), (-2, 2)$

f $(x + 3)^2 + (y - 1)^2 = 8 \quad (1)$

$y = x \quad (2)$

Subs. (2) into (1):

$(x + 3)^2 + (x - 1)^2 = 0$

$x^2 + 6x + 9 + x^2 - 2x + 1 = 8$

$2x^2 + 4x + 2 = 0$

$x^2 + 2x + 1 = 0$

$(x + 1)^2 = 0$

$\therefore\ x = -1$

Subs. in (2):

$\therefore y = -1$

$\therefore$ Point of intersection $(-1, -1)$.

Chapter 7 . . . pp. 110–112
Rates and Proportion

1 a 100 km = 100 000 m

$100\,000 \div 60 \div 60 = 27\frac{7}{9}$

$\therefore 27\frac{7}{9}$ m/s

b $40 \times 60 \times 60 = 144\,000$

144 000 m = 144 km

$\therefore$ 144 km/h

c 20 km/L = 1 L/ 20 km

Multiply by 5: 5 L/ 100 km

d 10 L/100 km = 100 km/10 L

Divide by 10: 10 km/L

e $28 \times 60 \times 60 = 100\,800$

100 800 cm = 1008 m = 1.008 km

$\therefore$ 1.008 km/h

f 80c/100 g = 100 g/80c

Divide by 80, then multiply by 100:

$100 \div 80 \times 100 = 125$

$\therefore$ 125 g/\$ = 0.125 kg/\$

g 160 c/L = 1 L/160 c

Divide by 160 and multiply by 100:

$1 \div 160 \times 100 = 0.625$

$\therefore$ 0.625 L/\$ = 62.5 L/\$100

h 5 mL/min = 1 min/5 mL

Multiply by 200:

200 min/1000 mL = 200 min/L

$\therefore 3\frac{1}{3}$ h/L

i 2 L/h = 1 h/2 L

As $60 \times 60 = 3600$, then

3600 s/2000 mL

$\therefore$ 1.8 s/mL

2 a $C = 12d$

b increases

c Subs. $d = 35$ in $C = 12d$:

$C = 12 \times 35$

$= 420 \quad \therefore$ \$420

d Subs $C = 504$ in $C = 12d$:

$504 = 12d$

$d = \frac{504}{12}$

$= 42 \quad \therefore$ 42 metres

e

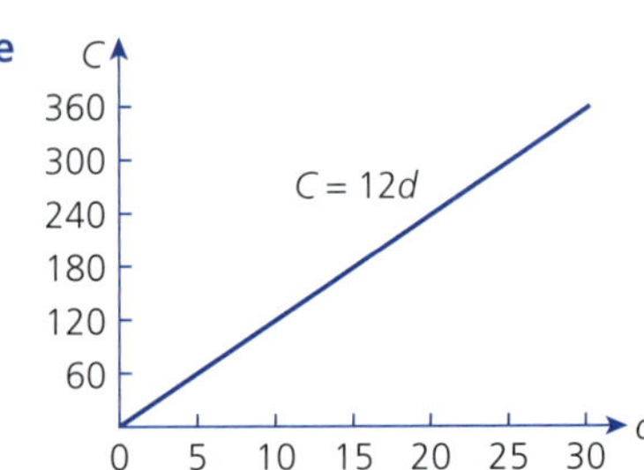

f Gradient $= \frac{\text{Rise}}{\text{Run}}$

$= \frac{360}{30}$

$= 12$

OR: From $C = 12d$, gradient = 12

3 a $W = 16.5n$

b Increases

c Subs. $n = 30$ in $W = 16.5n$:

$W = 16.5 \times 30$

$= 495 \quad \therefore$ \$495

d Subs. $W = 594$ in $W = 16.5n$:
$594 = 16.5n$
$n = \frac{594}{16.5}$
$= 36$ $\therefore$ 36 hours

e

W
700
600
500
400
300
200
100
0 5 10 15 20 25 30 35 40 n

f From $W = 16.5n$, gradient $= 16.5$
OR: Gradient $= \frac{\text{Rise}}{\text{Run}}$
$= \frac{660}{40}$
$= 16.5$

4 a $D = 9n$

b Subs. $n = 35$ in $D = 9n$:
$D = 9 \times 35$
$= 315$ $\therefore$ 315 km

c Subs. $D = 405$ in $D = 9n$:
$405 = 9n$
$n = \frac{405}{9}$
$= 45$ $\therefore$ Used 45 L.
Fraction used $= \frac{45}{60}$
$= \frac{3}{4}$
$\therefore \frac{1}{4}$ of the petrol remains.

5 a

Temperature Conversions

°F
110
100
90
80
70
60
50
40
30
0 10 20 30 40 °C

b i 100 °F

ii 15 °C

6 a i

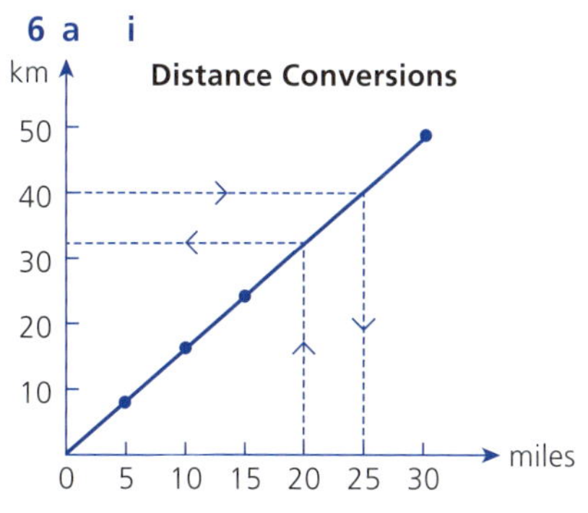

b i 32 km

ii 25 miles

7 a

Kilograms	0	5	10	15	20
Pounds	0	11	22	33	44

b

Mass Conversions

Pounds
50
40
30
20
10
0 5 10 15 20 kg

c i 17.6 pounds

ii As 15 kg = 33 pounds,
then 150 kg = 330 pounds.

8 a i \$A100 = €60

ii €100 = £80

b i \$A60 = €36

ii €42 = \$70

iii €60 = £48

iv £20 = €25

c i Use \$A100 = €60
Multiply by 50:
\$A5000 = €3000

ii Use €60 = \$A100
Multiply by 40:
€2400 = \$A4000

iii Use €100 = £80
Multiply by 7.5:
€750 = £600

iv Use £80 = €100
Multiply by 1.25:
£100 = €125

d i \$A100 = €60 = £48

ii £60 = €75 = \$A125

9 a $y = kx$

b Subs. $y = 12$ and $x = 4$ in $y = kx$:
$12 = 4k$
$k = 3$ $\therefore y = 3x$

c Subs. $x = 20$ in $y = 3x$:
$y = 3 \times 20$
$= 60$

d Subs. $y = 18$ in $y = 3x$:
$18 = 3x$
$x = \frac{18}{3}$
$= 6$

10 a $P = kt$

b Subs. $P = 10$ and $t = 2$ in $P = kt$:
$10 = 2k$
$k = 5$ $\therefore P = 5t$

c Subs. $t = 16$ in $P = 5t$:
$P = 5 \times 16$
$= 80$

d Subs. $P = 30$ in $P = 5t$:
$30 = 5t$
$t = \frac{30}{5}$
$= 6$

11 a $W = kn$
Subs. $W = 468$ and $n = 26$ in $W = kn$:
$468 = 26k$
$k = \frac{468}{26}$
$= 18$ $\therefore W = 18n$

b Subs. $n = 27$ in $W = 18n$:
$W = 18 \times 27$
$= 486$ $\therefore$ \$486

c Subs. $W = 648$ in $W = 18n$:
$648 = 18n$
$n = \frac{648}{18}$
$= 36$ $\therefore$ 36 hours

12 a $P = km$
Subs. $P = 7.8$ and $m = 1.5$ in $P = km$:
$7.8 = 1.5k$
$k = \frac{7.8}{1.5}$
$= 5.2$ $\therefore P = 5.2m$

b Subs. $m = 3.6$ in $P = 5.2m$:
$P = 5.2 \times 3.6$
$= 18.72$ $\therefore$ \$18.72

c Subs. $P = 14.56$ in $P = 5.2m$:
$14.56 = 5.2m$
$m = \frac{14.56}{5.2}$
$= 2.8$ $\therefore$ 2.8 kg

13 a $y = kx$
Subs. $y = 16.2$ and $x = 4.5$ in $y = kx$:
$16.2 = 4.5k$
$k = \frac{16.2}{4.5}$
$= 3.6$ $\therefore y = 3.6x$

b Subs. $x = 1.3$ in $y = 3.6x$:
$y = 3.6 \times 1.3$
$= 4.68$ $\therefore y = 4.68$
Subs. $y = 10.44$ in $y = 3.6x$:
$10.44 = 3.6x$
$x = \frac{10.44}{3.6}$
$= 2.9$ $\therefore x = 2.9$

x	1.3	2.9	4.5
y	4.68	10.44	16.2

14 a $F = km$

Subs. $F = 36$ and $m = 3$ in $F = km$:

$36 = 3k$

$k = 12$ $\quad\therefore F = 12m$

b Subs. $m = 6.2$ in $F = 12m$:

$F = 12 \times 6.2$

$= 74.4$ $\quad\therefore$ 74.4 Newtons

c Subs. $F = 43.2$ in $F = 12m$:

$43.2 = 12m$

$m = \dfrac{43.2}{12}$

$= 3.6$ $\quad\therefore$ 3.6 kg

15 a $y = k\sqrt{x}$

b Subs. $y = 24$ and $x = 9$ in $y = k\sqrt{x}$:

$24 = k\sqrt{9}$

$3k = 24$

$k = 8$ $\quad\therefore y = 8\sqrt{x}$

c Subs. $x = 16$ in $y = 8\sqrt{x}$:

$y = 8 \times \sqrt{16}$

$= 32$

d Subs. $y = 240$ in $y = 8\sqrt{x}$:

$240 = 8\sqrt{x}$

$\sqrt{x} = \dfrac{240}{8}$

$= 30$

$x = 30^2$

$= 900$

16 a $y = kx^2$

b Subs. $y = 48$ and $x = 4$ in $y = kx^2$:

$48 = k \times 4^2$

$16k = 48$

$k = 3$ $\quad\therefore y = 3x^2$

c Subs. $x = 6$ in $y = 3x^2$:

$y = 3 \times 6^2$

$= 108$

d Subs. $y = 588$ in $y = 3x^2$:

$588 = 3x^2$

$x^2 = \dfrac{588}{3}$

$= 196$

$x = 14$ $\quad$ (as $x > 0$)

17

T
20 18 16 14 12 10 8 6 4 2
0 30 60 90 120 150 180 210 240 270 300 s

T is inversely proportional to s because the gradient of the curve is negative and, as s increases, T decreases.

18 a $y = \dfrac{k}{x}$

b Subs. $y = 3$ and $x = 8$ in $y = \dfrac{k}{x}$:

$3 = \dfrac{k}{8}$

$k = 24$

c $\therefore y = \dfrac{24}{x}$

Sub. $x = 16$ in $y = \dfrac{24}{x}$:

$y = \dfrac{24}{16}$

$= 1.5$

d Subs. $y = 80$ in $y = \dfrac{24}{x}$:

$80 = \dfrac{24}{x}$

$80x = 24$

$x = \dfrac{24}{80}$

$= 0.3$

19 a $M = \dfrac{k}{a}$

b Subs. $M = 16$ and $a = 2$ in $M = \dfrac{k}{a}$:

$16 = \dfrac{k}{2}$

$k = 32$

c $\therefore M = \dfrac{32}{a}$

Subs. $a = 8$ in $M = \dfrac{32}{a}$:

$M = \dfrac{32}{8}$

$= 4$

d Subs. $M = 64$ in $M = \dfrac{32}{a}$:

$64 = \dfrac{32}{a}$

$64a = 32$

$a = \dfrac{32}{64}$

$= 0.5$

20 a $T = \dfrac{k}{n}$

Subs. $T = 20$ and $n = 5$ in $T = \dfrac{k}{n}$:

$20 = \dfrac{k}{5}$

$k = 100$ $\quad\therefore T = \dfrac{100}{n}$

b Subs. $n = 2$ in $T = \dfrac{100}{n}$:

$T = \dfrac{100}{2}$

$= 50$ $\quad\therefore$ Takes 50 minutes.

c Subs. $T = 25$ in $T = \dfrac{100}{n}$:

$25 = \dfrac{100}{n}$

$25n = 100$

$n = \dfrac{100}{25}$

$= 4$ $\quad\therefore$ Four tankers used.

21 a $b = \dfrac{k}{a}$

Subs. $b = 0.5$ and $a = 8$ in $b = \dfrac{k}{a}$:

$0.5 = \dfrac{k}{8}$

$k = 4$ $\quad\therefore b = \dfrac{4}{a}$

b Subs. $a = 2$ in $b = \dfrac{4}{a}$:

$b = \dfrac{4}{2}$

$= 2$

Subs. $b = 1$ in $b = \dfrac{4}{a}$:

$1 = \dfrac{4}{a}$

$a = 4$

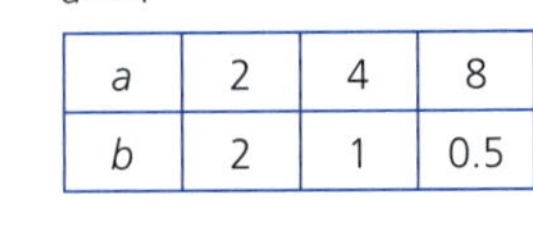

a	2	4	8
b	2	1	0.5

22 a

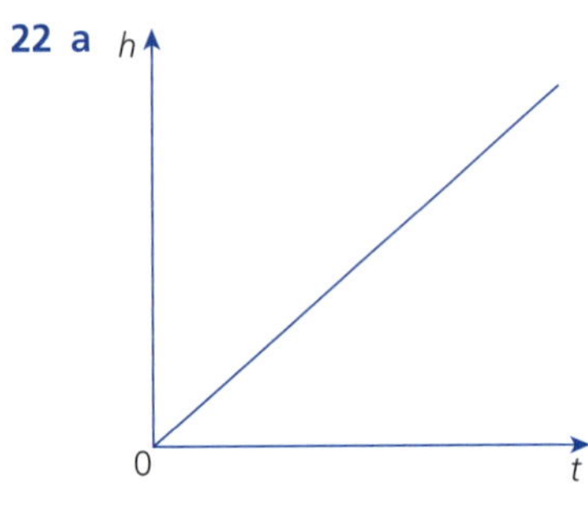

b

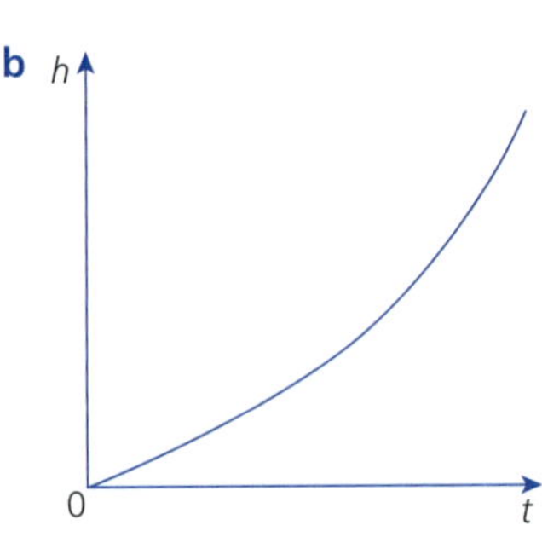

c

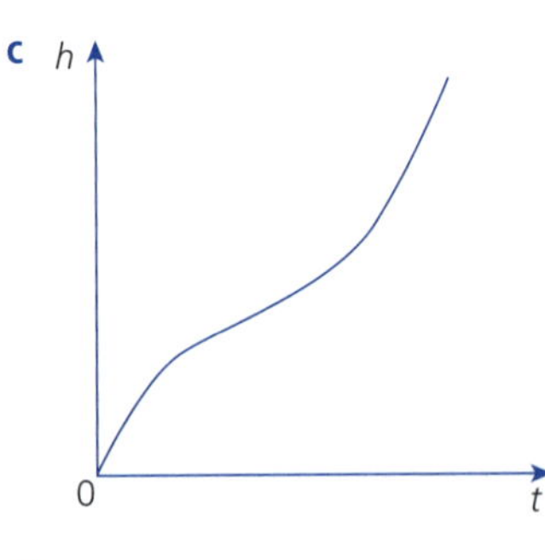

d

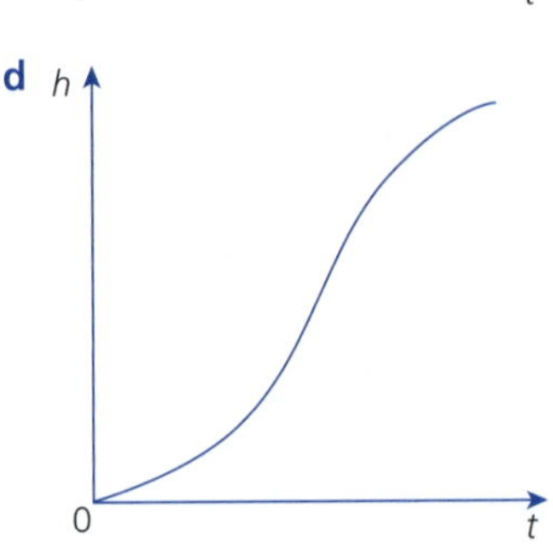

e

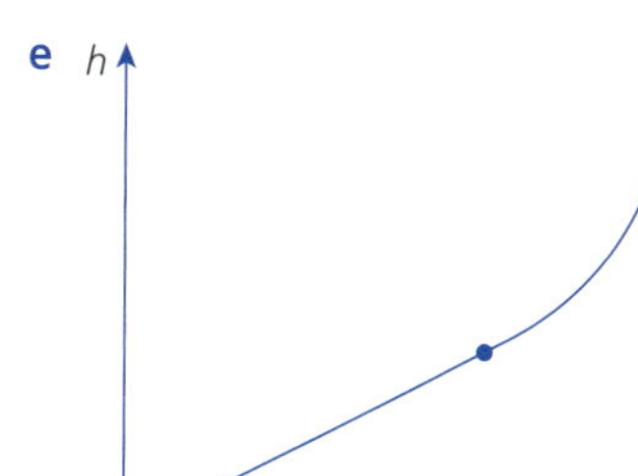

f

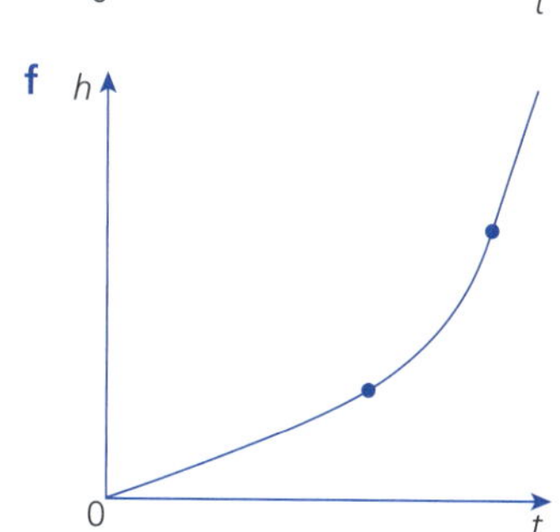

23 a

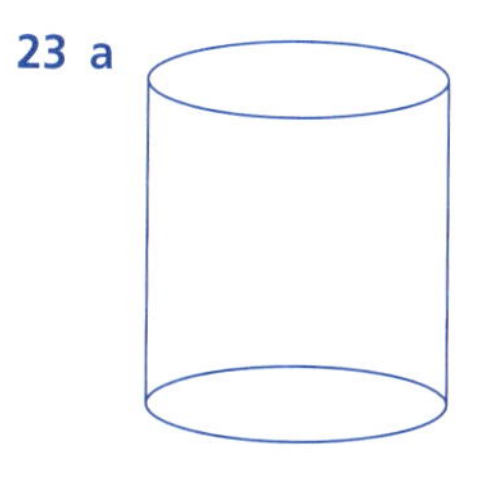

b

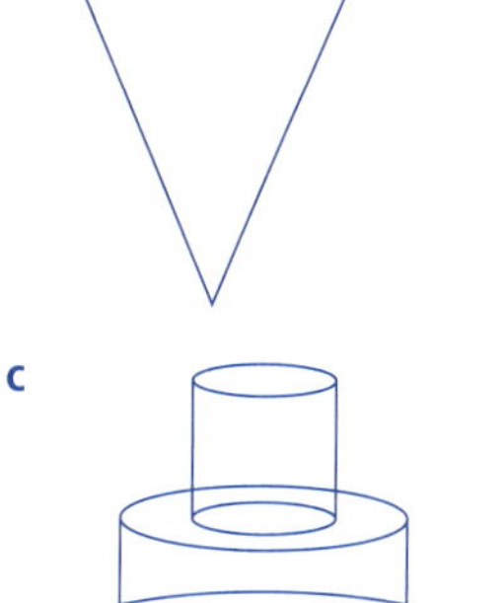

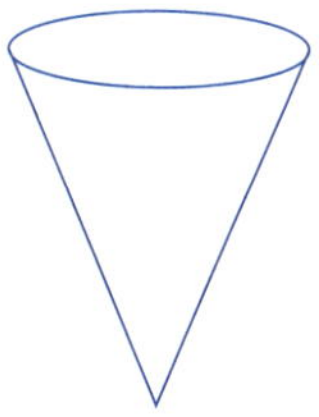

c

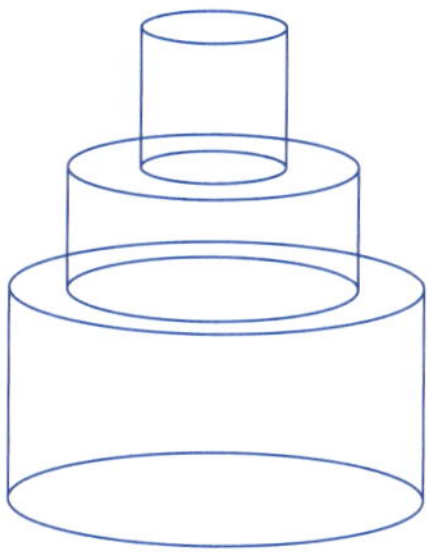

Chapter 8 . . . pp. 134–138

Pythagoras and Trigonometry

1 a $n^2 = 3^2 + 5^2$
$= 9 + 25$
$= 34$
$n = \sqrt{34}$

b $m^2 = 25^2 - 18^2$
$= 625 - 324$
$= 301$
$m = \sqrt{301}$

c $r^2 = 14^2 - 8^2$
$= 196 - 64$
$= 132$
$r = \sqrt{132}$
$= 2\sqrt{33}$

d $s^2 = 19^2 - 8^2$
$= 361 - 64$
$= 297$
$s = \sqrt{297}$
$= 3\sqrt{33}$

2 Let x = length of diagonal.
$x^2 = 16^2 + 20^2$
$= 256 + 400$
$= 656$
$x = \sqrt{656}$
$= 4\sqrt{41}$
∴ The diagonal is $4\sqrt{41}$ cm long.

3 Let x = length of diagonal.
$x^2 = 3^2 + 3^2$
$= 9 + 9$
$= 18$
$x = \sqrt{18}$
$= 4.2$ (1 dec. pl.)
∴ The diagonal is 4.2 cm long.

4 $h^2 = 35^2 - 21^2$
$= 1225 - 441$
$= 784$
$h = \sqrt{784}$
$= 28$
∴ The brace reaches 28 cm.

5 $y^2 = 8^2 - 7.5^2$
$= 64 - 56.25$
$= 7.75$
$y = \sqrt{7.75}$
$= 2.8$ (1 dec. pl.)
∴ Ladder is 2.8 m from wall.

6 Let x = height the wires are attached.
$x^2 = 40^2 - 21$
$= 1600 - 441$
$= 1159$
$x = \sqrt{1159}$
$= 34.044\,089\,06\ldots$
Distance from top $= 45 - 34.044\,089\,06\ldots$
$= 10.96$ (4 sig. fig.)
∴ Wires are attached 10.96 m from the top.

7 Let x = length of side.
$x^2 + x^2 = 18^2$
$2x^2 = 324$
$x^2 = 162$
$x = \sqrt{162}$
$= 12.7$ (1 dec. pl.)
∴ The side length is 12.7 cm.

8 a Check whether $17^2 = 8^2 + 15^2$
LHS $= 17^2$
$= 289$
RHS $= 8^2 + 15^2$
$= 64 + 225$
$= 289$
∴ Triangle is right-angled.

b Check whether $20^2 = 12^2 + 15^2$
LHS $= 20^2$
$= 400$
RHS $= 12^2 + 15^2$
$= 144 + 225$
$= 369$
∴ Triangle is not right-angled.

c Check whether $401^2 = 40^2 + 399^2$
LHS $= 401^2$
$= 160\,801$
RHS $= 40^2 + 399^2$
$= 1600 + 159\,201$
$= 160\,801$
∴ Triangle is right-angled.

9 a $a^2 = 7^2 + 24^2$
$= 49 + 576$
$= 625$
$a = \sqrt{625}$
$= 25$

b $y^2 = 9^2 + 12^2$
$= 81 + 144$
$= 225$
$y = \sqrt{225}$
$= 15$

c $n^2 = 30^2 - 18^2$
$= 900 - 324$
$= 576$
$n = \sqrt{576}$
$= 24$

d $t^2 = 41^2 - 9^2$
$= 1681 - 81$
$= 1600$
$t = \sqrt{1600}$
$= 40$

e $n^2 = 85^2 - 13^2$
$= 7225 - 169$
$= 7056$
$n = \sqrt{7056}$
$= 84$

f $p^2 = 21^2 + 72^2$
$= 441 + 5184$
$= 5625$
$p = \sqrt{5625}$
$= 75$

10 a $\sin 70° = \cos 20°$
$\therefore x = 20$

b $\cos 5° = \sin 85°$
$\therefore y = 85$

11 $\frac{\sin 10Υ}{\cos 80Υ} = \frac{\sin 10Υ}{\sin 10Υ}$
$= 1$

12 a $\frac{y}{48} = \sin 26°$
$y = 48 \times \sin 26°$
$= 21.0$ (1 dec. pl.)

b $\frac{y}{24} = \tan 44°9'$
$y = 24 \times \tan 44°9'$
$= 23.3$ (1 dec. pl.)

c $\frac{y}{4.9} = \cos 62°18'$

$y = 4.9 \times \cos 62°18'$
$= 2.3$ (1 dec. pl.)

d $\frac{18}{y} = \sin 22°$

$y = \frac{18}{\sin 22\Upsilon}$
$= 48.1$ (1 dec. pl.)

e $\frac{18}{y} = \tan 22°$

$\tan = \frac{18}{\tan 22\Upsilon}$
$= 44.6$ (1 dec. pl.)

f $\frac{27}{y} = \cos 46°$

$y = \frac{27}{\cos 46\Upsilon}$
$= 38.9$ (1 dec. pl.)

g $\frac{y}{18} = \sin 42.6°$

$y = 18 \times \sin 42.6°$
$= 12.2$ (1 dec. pl.)

h $\frac{17}{y} = \tan 16.8°$

$y = \frac{17}{\tan 16.8\Upsilon}$
$= 56.3$ (1 dec. pl.)

13 a $\sin\theta = \frac{8}{17}$

$\theta = 28°4'$

b $\tan x = \frac{19}{7}$

$x = 69°47'$

c $\cos\phi = \frac{18}{24}$

$\phi = 41°25'$

d $\tan\theta = \frac{8}{15}$

$\theta = 28°4'$

14 a $\tan\theta = \frac{4}{7}$

$\theta = 30°$ (nearest degree)

b $\cos\theta = \frac{4}{7}$

$\theta = 55°$ (nearest degree)

c $\sin f = \frac{4}{7}$

$f = 34.8°$ (1 dec. pl.)

15 $x^2 + 8^2 = 10^2$

$\therefore \; x^2 = 100 - 64$
$= 36$
$x = \sqrt{36} = 6$

$\sin\theta = \frac{6}{10} = \frac{3}{5}$

$\tan\theta = \frac{6}{8} = \frac{3}{4}$

16 $y^2 = 3^2 + 4^2$
$= 9 + 16 = 25$
$y = \sqrt{25} = 5$

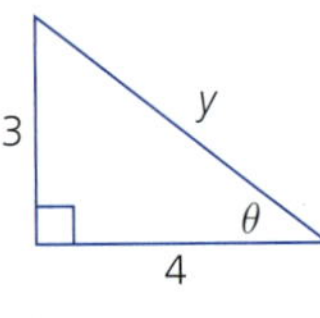

$\sin\theta = \frac{3}{5}$

$\cos\theta = \frac{4}{5}$

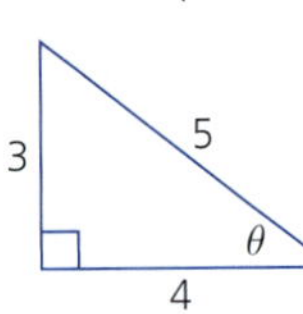

17 Let distance be d metres.
∠ of depression = ∠ of elevation.

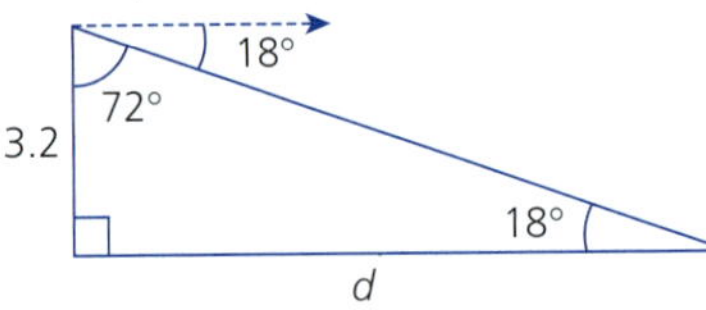

$\frac{6}{3.2} = \tan 72°$

$\therefore \; d = 3.2 \times \tan 72°$
$= 9.848\,5873$
≈ 9.8 (to 1 dec. pl.)

The whale is 9.8 m from the ship.

18 Let the angle be θ.

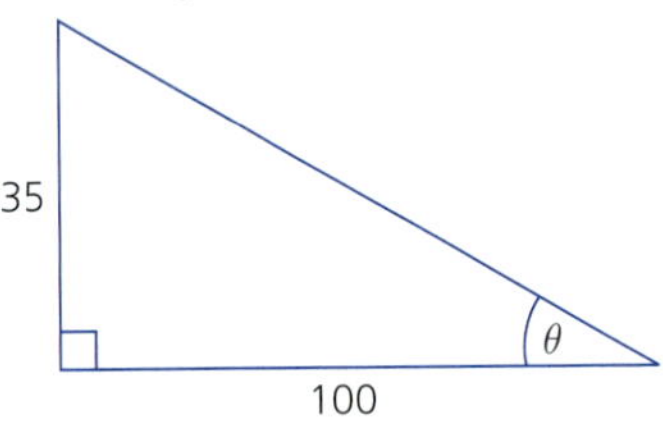

$\tan\theta = \frac{35}{100}$

$\therefore \; \theta = 19°17'$

The angle of elevation is 19°17′.

19 Perpendicular from centre to chord bisects the chord and also the vertical angle. Call the half-chord x metres.

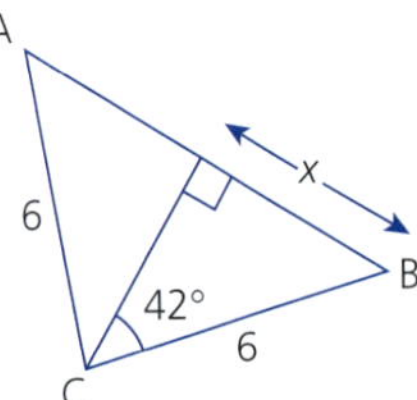

Then $\frac{x}{6} = \sin 42°$

$x = 6 \times \sin 42°$
$= 4.014\,7836,$

$\therefore$ Length of chord $= 2 \times 4.014\,7836$
$= 8.029\,5673$
$= 8.0$ (to 1 dec. pl.)

The chord is 8.0 m long.

20

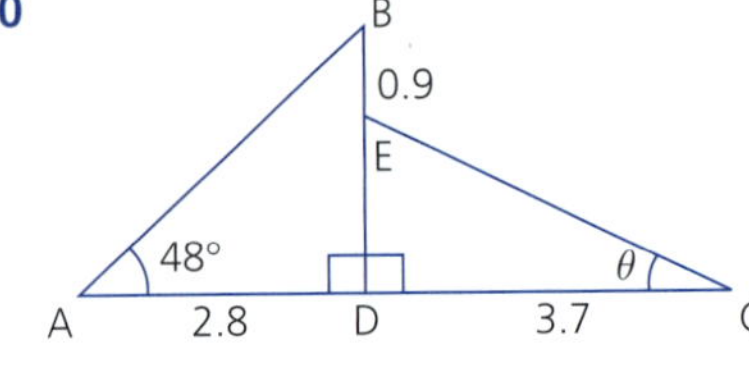

From △ADB, $\frac{BD}{2.8} = \tan 48°$

$\therefore \quad BD = 2.8 \times \tan 48°$
$= 3.109\,715,$

$\therefore \quad DE = 3.109\,715 - 0.9$
$= 2.209\,715$

Let ∠DCE be θ.
From △DCE,

$\tan\theta = \frac{DE}{3.7} = \frac{2.209715}{3.7}$

$\therefore \quad \theta = 30°51'$

DB = 3.1 m (to 1 dec. pl.)
∠DCE = 30°51′

21

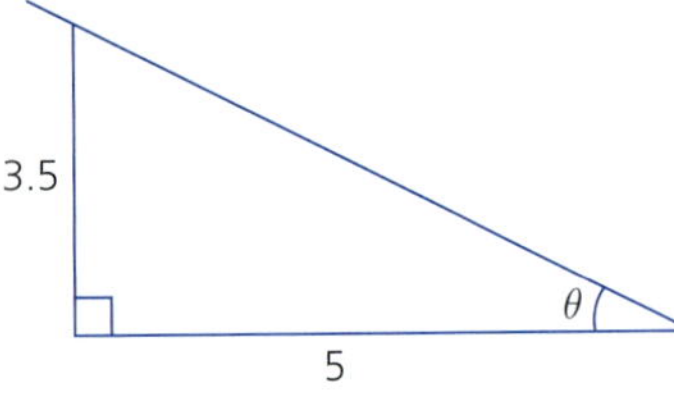

Let the angle of elevation of the sun be θ.

$\tan\theta = \frac{3.5}{5}$

$\therefore \; \theta = 35°$ (to the nearest degree)

The angle of elevation of the sun is 35°.

22 a

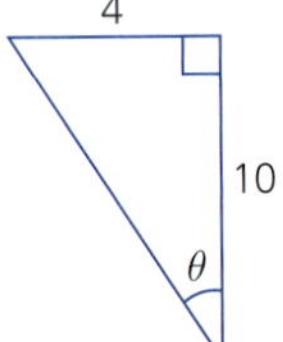

Let θ be the angle that the ball must be kicked to strike the outer can.

$\tan\theta = \frac{4}{10}$

$\theta = 21°48'$.

The ball can be kicked at any angle between 0° and 21°48′ relative to the nearest can.

b As for (**a**), but $\tan\theta = \frac{4}{15}$

$\theta = 14°56'$.

The range of angles in now within 0° and 14°56′.

23

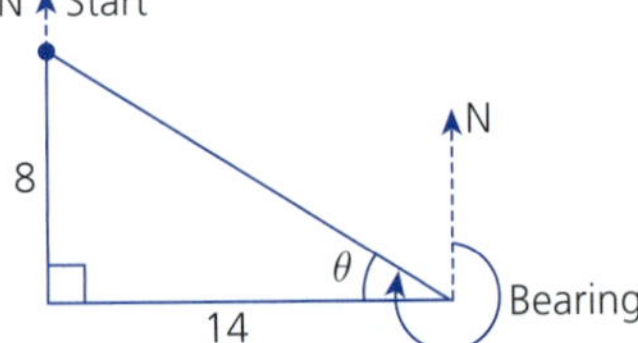

From the △, $\tan\theta = \frac{8}{14}$

$\theta = 29°45'$.

The bearing is measured clockwise from N to the direction of the starting point. This is

$270 + \theta° = 270° + 29°45'$
$= 299°45'$
$= 300°$ (nearest degree)

The bearing of the starting point is 300°.

24 a Using Pythagoras' Theorem, let the length of the support be l metres.

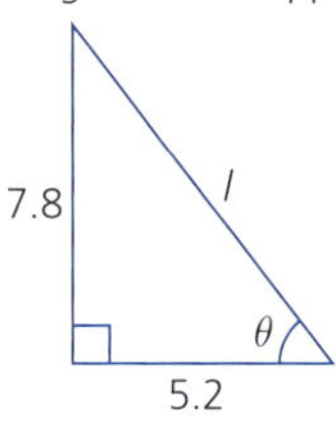

$l^2 = 7.8^2 + 5.2^2$
$= 87.88$
$\therefore\ l = \sqrt{87.88}$
$= 9.374\,4333$
≈ 9.4 [1 dec. pl.].

The length of the support is 9.4 m.

b Let the angle be θ.

$\tan\theta = \dfrac{7.8}{5.2}$

$\therefore\ \theta = 56°19'$.

The support makes an angle of 56°19′ with the ground.

25 $1^2 + x^2 = 2^2$
$x^2 = 4 - 1$
$= 3$
$\therefore\ x = \sqrt{3}$

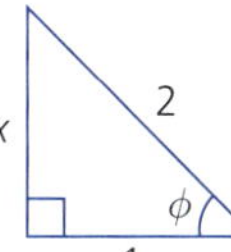

From the diagram:

$\sin\phi = \dfrac{x}{2} = \dfrac{\sqrt{3}}{2}$

$\tan\phi = \dfrac{x}{1} = \sqrt{3}$

$\cos\phi = \dfrac{1}{2}$

$\therefore\ \phi = 60°$

26 $x^2 = 1^2 + 1^2$
$= 2$
$\therefore\ x = \sqrt{2}$

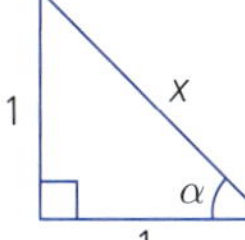

$\sin\alpha = \dfrac{1}{x} = \dfrac{1}{\sqrt{2}}$

$\cos\alpha = \dfrac{1}{x} = \dfrac{1}{\sqrt{2}}$

Now $\tan\alpha = 1$,

$\therefore\ \alpha = 45°$

27 a

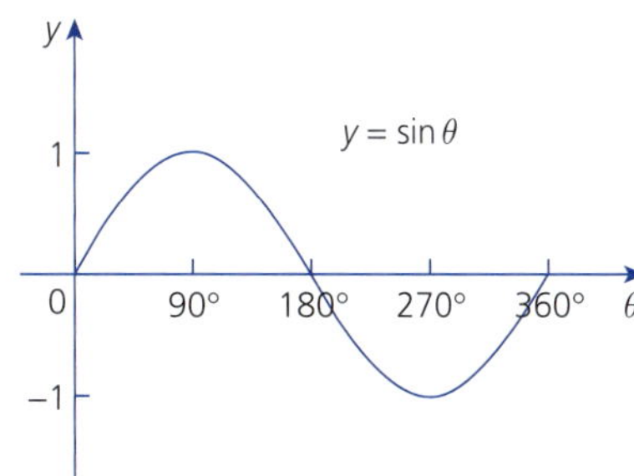

b

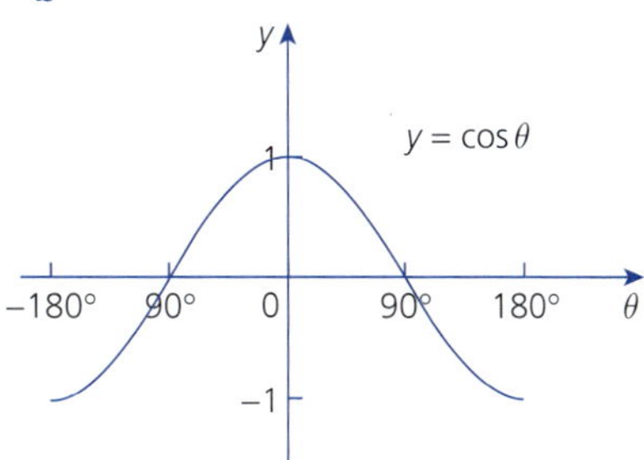

c

y = tan θ

1, −1, 0, 45°, 90°, 135°, 180°, θ

28 a

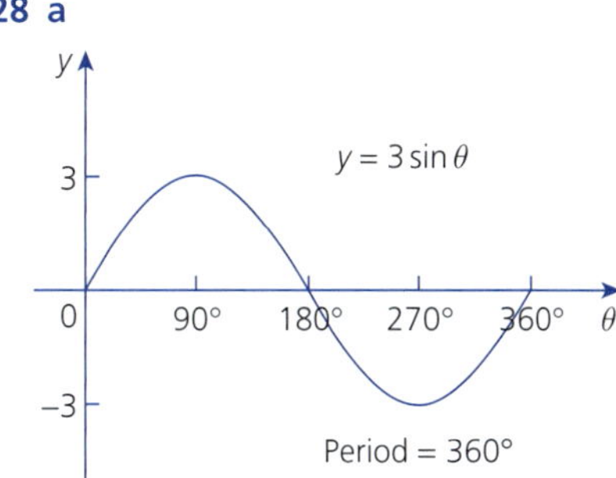

Period = 360°

b

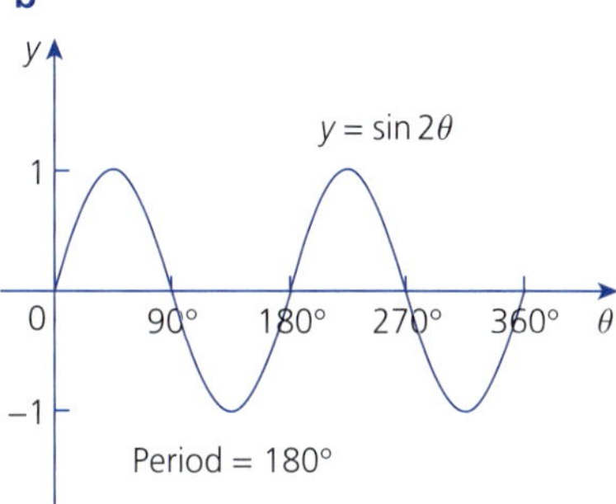

Period = 180°

c

$y = \sin\dfrac{\theta}{2}$

1, 0, 90°, 180°, 270°, 360°, θ

Period = 720°

29

G, 6, D, 8, C

$DG^2 = 8^2 + 6^2$
$= 100$
$DG = 10$

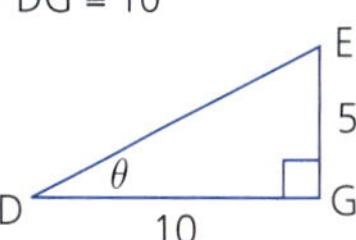

$\tan\theta = \dfrac{5}{10}$

$\theta = 26.565\,05$
$= 26°34'$ (nearest minute)

$\therefore$ Angle of elevation is 26°34′.

30 a $\cos\theta = -0.3165$
$\therefore\ \theta = 108°27'$

b $\tan\theta = -1.2851$
$\therefore\ \theta = 180° - 52°7'$
$= 127°53'$

c $\sin\theta = 0.3896$
$\therefore\ \theta = 180° - 22°56'$
$= 157°4'$

31 a $\dfrac{f}{\sin 18\Upsilon} = \dfrac{64}{\sin 87\Upsilon}$

$\therefore\ f = \dfrac{64\sin 18\Upsilon}{\sin 87\Upsilon}$
$= 19.804\,229$

$\therefore\ f = 19.8$ (to 1 dec. pl.)

b The third angle
$= 180° - (17° + 15°)$
$= 148°$

$\dfrac{f}{\sin 148\Upsilon} = \dfrac{12.1}{\sin 15\Upsilon}$

$\therefore\ f = \dfrac{12.1\sin 148\Upsilon}{\sin 15\Upsilon}$
$= 24.774\,155$
$= 24.8$ (to 1 dec. pl.)

c $\dfrac{f}{\sin 32\Upsilon 16} = \dfrac{40.5}{\sin 86\Upsilon 27}$

$\therefore\ f = \dfrac{40.5\sin 32\Upsilon 16}{\sin 86\Upsilon 27}$
$= 21.662\,919$
$= 21.7$ (to 1 dec. pl.)

32

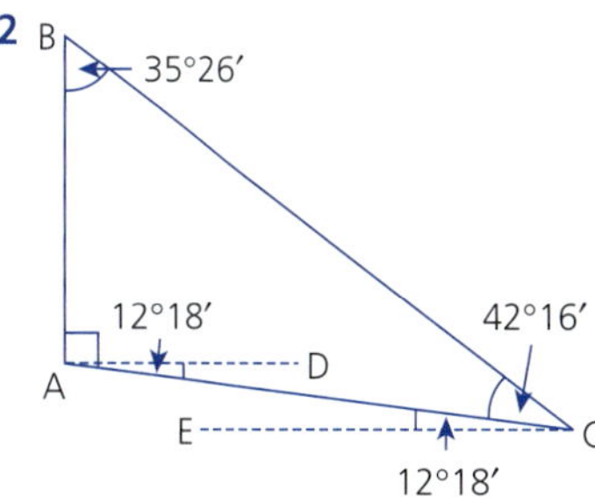

$\angle BCA = 54°34' - 12°18' = 42°16'$
$\angle DAC = 12°18'$
As $\angle DAC = \angle ACE$ [alt. angles AD || EC]
$\therefore\ \angle BAC = 90° + 12°18'$
$= 102°18'$,
$\therefore\ \angle ABC = 180° - (102°18' + 42°16')$
$= 180° - 144°34'$
$= 35°26'$.
Using the sine rule:

$$\frac{AC}{\sin 35Y26} = \frac{AB}{\sin 42Y16}$$

$$\therefore \quad AC = \frac{12\ \sin 35Y26}{\sin 42Y16}$$
$= 10.343813$
$= 10.3$ (to 1 dec. pl.).
The length of the shadow is 10.3 m.

33

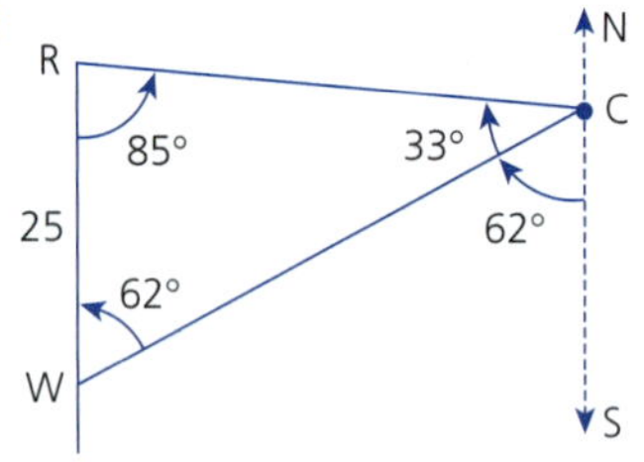

The bearing of W = 242°.
$\therefore\quad \angle SCW = 62°\ (242° - 180°)$
$\angle WCR = 275° - 242° = 33°$,
and $\angle RWC = \angle WCS$
[alt. angles, RW || NS]
$\therefore\quad \angle RWC = 62°$
Then $\angle WRC = 180° - (33° + 62°)$
$= 180° - 95°$
$= 85°$
The shortest distance to a port will be RC, as the shortest side of a triangle is opposite the smallest angle ($\angle RWC < \angle WRC$).
Then, using the sine rule:

$$\frac{RC}{\sin 62Y} = \frac{25}{\sin 33Y}$$

$$\therefore \quad RC = \frac{25 \sin 62Y}{\sin 33Y}$$
$= 40.529026$
$= 40.5$ (to 1 dec. pl.)
The distance to the nearest port (Redhead) is 40.5 km.

34 a i $t^2 = 8^2 + 11^2 - 2(8)(11)\cos 68°$
$= 119.06924$
$\therefore\ t = \sqrt{119.06924}$
$= 10.911885$
$= 10.9$ (to 1 dec. pl.)

ii $t^2 = 6^2 + 13^2 - 2(6)(13)\cos 114°12'$
$= 268.94799$
$\therefore\ t = \sqrt{268.94799}$
$= 16.399634$
$= 16.4$ (to 1 dec. pl.)

iii $t^2 = 4.6^2 + 7.8^2 - 2(4.6)(7.8)\cos 24°8'$
$= 16.512078$
$\therefore\ t = \sqrt{16.512078}$
$= 4.0635056$
$= 4.1$ (to 1 dec. pl.)

b i Area $= \frac{1}{2}(8)(11)\sin 68°$
$= 40.79609$
$= 40.8$ (to 1 dec. pl.)
The area is 40.8 units^2.

ii Area $= \frac{1}{2}(6)(13)\sin 114°12'$
$= 35.572685$
$= 35.6$ (to 1 dec. pl.)
The area is 35.6 units^2.

iii Area $= \frac{1}{2}(4.6)(7.8)\sin 24°08'$
$= 7.3349745$
$= 7.3$ (to 1 dec. pl.)
The area is 7.3 units^2.

35 a $\cos\theta = \dfrac{17^2 + 15^2 - 19^2}{2(17)(15)}$
$= \dfrac{153}{510}$
$= 0.3$
$\theta = 72°33'$

b $\cos\alpha = \dfrac{8^2 + 21^2 - 16^2}{2(8)(21)}$
$= \dfrac{249}{336}$
$= 0.7410714$,
$\therefore\ \alpha = 42°11'$
$\cos\phi = \dfrac{16^2 + 21^2 - 8^2}{2(16)(21)}$
$= \dfrac{633}{672}$
$= 0.9419642$
$\therefore\quad \phi = 19°37'$

36

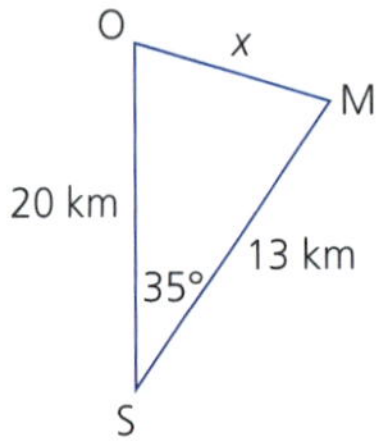

$x^2 = 20^2 + 13^2 - 2(20)(13)\cos 35°$
$= 143.04094$
$\therefore\ x = \sqrt{143.04094}$
$= 11.959972$
$= 12.0$ (to 1 dec. pl.)
Maria is 12.0 km from the start.

37

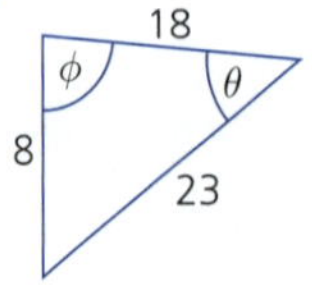

The smallest $\angle$ is opposite the shortest side. The largest $\angle$ is opposite the longest side. Let the smallest angle be θ.

$\cos\theta = \dfrac{18^2 + 23^2 - 8^2}{2(18)(23)}$
$= \dfrac{789}{828}$
$= 0.952899$
$\therefore\ \theta = 17°39'$
Let the largest angle be ϕ.
$\cos\phi = \dfrac{8^2 + 18^2 - 23^2}{2(8)(18)}$
$= \dfrac{-141}{288}$
$= -0.4895833$
$\phi = 119°19'$.
The smallest angle is 17°39′ and the largest angle is 119°19′.
Area $= \frac{1}{2} \times 8 \times 18 \times \sin\phi$
$= 4 \times 18 \times \sin 119°19'$
$= 62.780869$
≈ 62.8 (to 1 dec. pl.)
Area is 62.8 cm^2.

38

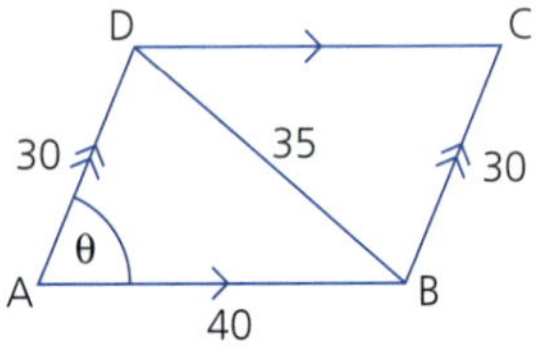

Let one angle of the parallelogram be θ. [As in the diagram.]
$\cos\theta = \dfrac{30^2 + 40^2 - 35^2}{2(30)(40)}$
$= \dfrac{1275}{2400}$
$= 0.53125$
$\therefore\ \theta = 57°55'$
Opposite $\angle$s are equal,
$\therefore\ C = 57°55'$.
Also $\angle ABC = \angle ADC$
$= 180° - 57°55'$
$= 122°05'$
Therefore the angles of the parallelogram are 57°55′ and 122°05′.

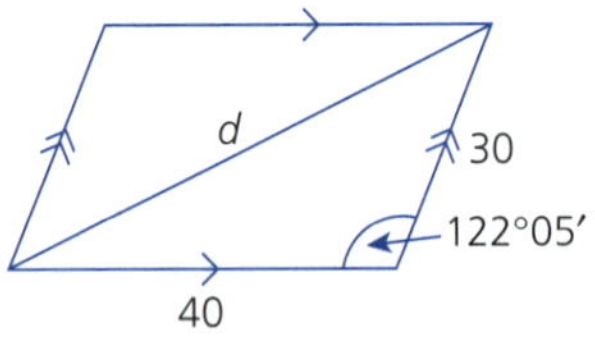

Let the diagonal be d cm.
$d^2 = 40^2 + 30^2 - 2(40)(30)\cos 122°05'$
$= 3775$
$\therefore\ d = \sqrt{3775}$
$= 61.439\ldots$
$= 61.4$ (to 1 dec. pl.)
The other diagonal is 61.4 cm.

39

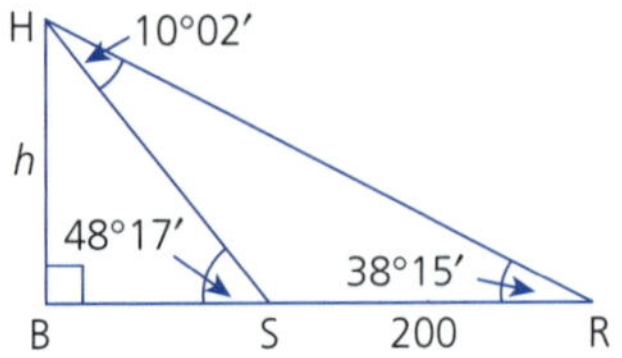

Link method question.
Find the length of HS by using the sine rule in $\triangle$HSR. We need $\angle$SHR for the sine rule.
In $\triangle$HSR,
$\angle SHR = 48°17' - 38°15'$
[$\angle$HSB is the exterior $\angle$ of $\triangle$HSR]
$= 10°02'$.

Then $\dfrac{HS}{\sin 38°15} = \dfrac{200}{\sin 10°02}$

$\therefore \quad HS = \dfrac{200 \sin 38°15}{\sin 10°02}$
$= 710.699\,22$

In $\triangle$HBS,

$\dfrac{h}{HS} = \sin 48°17'$

$\therefore h = HS \times \sin 48°17'$
$= 710.699\,22 \times \sin 48°17'$
$= 530.497\,62$
$= 530$ (to the nearest metre)
The hill is 530 m high.

40

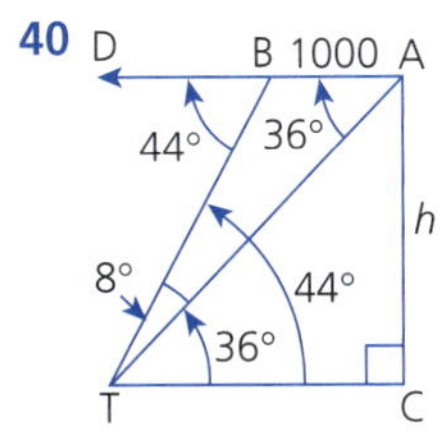

$\angle ATC = 36°$
$\angle BTC = 44°$
[Alt. $\angle$s, AB $\parallel$ TC]
$\therefore \angle BTA = 8°$
Also $\angle TBA = 136°$
[$\angle$ sum of $\triangle$]
Let the altitude AC be h metres.
[Another link method]
Find AT from $\triangle$BAT and use it in $\triangle$TCA.

$\dfrac{AT}{\sin 136°} = \dfrac{1000}{\sin 8°}$

$\therefore \quad AT = \dfrac{1000 \sin 136°}{\sin 8°}$
$= 4991.3264$

In $\triangle$TCA,

$\dfrac{h}{AT} = \sin 36°$

$\therefore h = AT \times \sin 36°$
$= 4991.3264 \times \sin 36°$
$= 2933.828$ metres
$= 2.933\,828$ km
$= 2.9$ km (to 2 sig. fig.)
The altitude of the plane is 2.9 km.

41

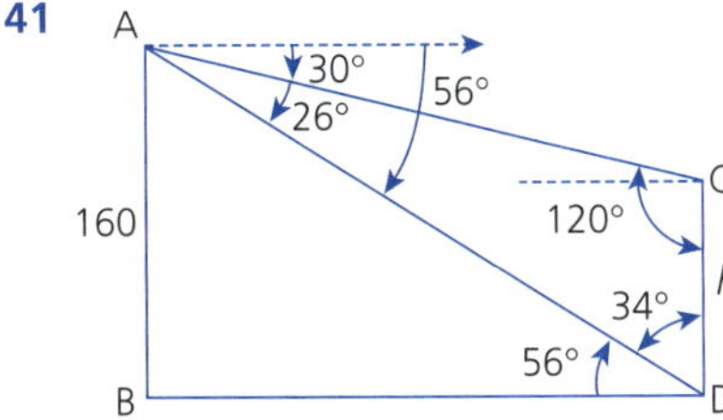

[Another link method]
$\angle ADB = 56°$ [equal to $\angle$ of depression]
$\angle CAD = 26°$
Also $\angle CDA = 90° - 56° = 34°$
$\therefore \angle ACD = 180° - (26° + 34°)$
$= 120°$
Use $\triangle$ABD to find AD and then use AD in $\triangle$ADC.
In $\triangle$ABD,

$\dfrac{160}{AD} = \sin 56°$

$\dfrac{AD}{160} = \dfrac{1}{\sin 56°}$

$\therefore AD = \dfrac{160}{\sin 56°}$
$= 192.994\,87$

In $\triangle$ADC, using the sine rule:

$\dfrac{h}{\sin 26°} = \dfrac{AD}{\sin 120°}$

$\therefore \quad h = \dfrac{192.994\,87 \times \sin 26°}{\sin 120°}$
$= 97.691\,571$
$= 98$ (to the nearest metre)
The height of the building is 98 m.

42

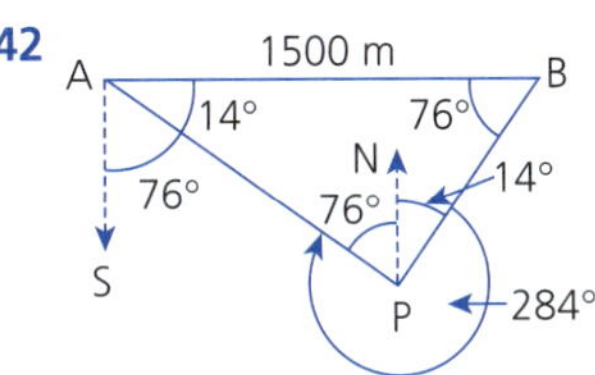

$\angle NPA = 360° - 284°$
$= 76°$
$\therefore \angle SAP = 76°$
[Alt. $\angle$s, AS $\parallel$ NP]
Then $\angle PAB = 90° - 76°$
$= 14°$
[angle between south and east is 90°]
Also $\angle APB = 76° + 14° = 90°$
and $\angle ABP = 76°$

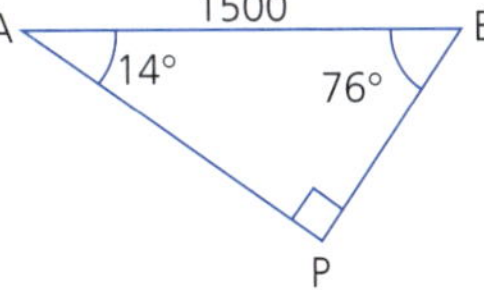

We have a right-angled $\triangle$.

$\dfrac{AP}{1500} = \sin 76°$

$AP = 1500 \times \sin 76°$
$= 1455.4436$
$= 1455$ (to the nearest metre)

$\dfrac{BP}{1500} = \sin 14°$

$BP = 1500 \times \sin 14°$
$= 362.882\,84$
$= 363$ (to the nearest metre)
The distances of the trawler from the ends of the breakwater are 1455 m and 363 m.

43

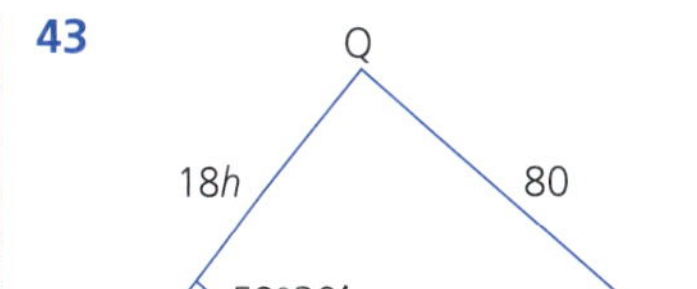

Let the time to be 80 km apart be h hours.
Then PQ = $18h$ $\quad D = S \times T$
PR = $24h$
Now using the cosine rule:
$80^2 = (18h)^2 + (24h)^2$
$- 2(18h)(24h)\cos 58°30'$
$= h^2[18^2 + 24^2 - 2(18)(24)\cos 58°30']$

$\therefore h^2 = \dfrac{80^2}{18^2 + 24^2 - 2(18)(24)\cos 58°30}$
$= 14.267\,84$

$\therefore \quad h = \sqrt{14.267\,84}$
$= 3.777$
That is, the time is 3.777 hours
= 3 hours 47 minutes.

44

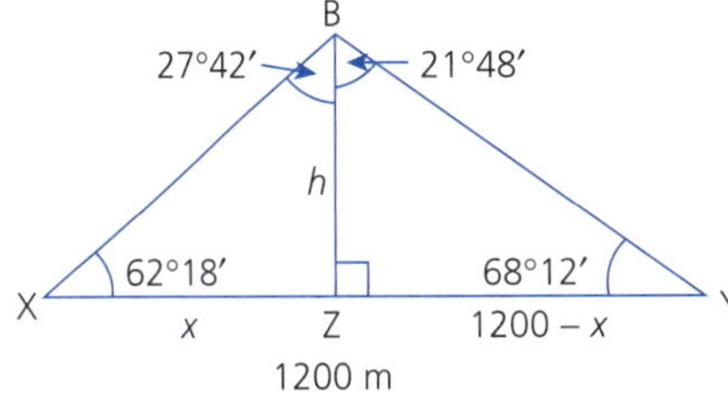

Let XZ be x metres, then
ZY = $(1200 - x)$ metres.
In $\triangle$BXZ,

$\dfrac{x}{h} = \tan 27°42'$.

$\therefore x = h \tan 27°42'$. $\quad$ (1)
In $\triangle$BZY,

$\dfrac{1200 - x}{h} = \tan 27°48'$,

$\therefore 1200 - x = h \tan 21°48'$. $\quad$ (2)
Substitute (1) into (2):
$1200 - h\tan 27°42' = h \tan 21°48'$
$\therefore 1200 = h \tan 27°42' + h \tan 21°48'$
That is,
$h(\tan 27°42' + \tan 21°48') = 1200$,

$\therefore h = \dfrac{1200}{\tan 27°42 + \tan 21°48}$
$= 1297.3209$
$= 1297$ (to the nearest metre)
$= 1300$ (to the nearest 100 m)
The height of the balloon is approximately 1300 m.

45

42° T 54°
120
S
48°
36°
x
y
A
d
B

Monika is at A, Mabel is at B.
$\angle ASB = 90°$
[angle between south and east]
Let the distance required, AB, be d metres.
Let AS = x m, SB = y m.
Then from $\triangle AST$,

$\therefore \frac{x}{120} = \tan 42°$

$x = 120 \tan 42°$.

From $\triangle TSB$,

$\therefore \frac{x}{120} = \tan 54°$

$y = 120 \tan 54°$.

Now in $\triangle ABS$, $d^2 = x^2 + y^2$
[Pythagoras' Theorem]

$\therefore d^2 = (120 \tan 42°)^2 + (120 \tan 54°)^2$
$= 11\,674.475 + 27\,279.752$
$= 38\,954.227$

$\therefore d = \sqrt{38954.227}$
$= 197.368\,25$
$= 197$ (to the nearest metre)

Therefore the distance between Monika and Mabel is 197 metres.

46 a $y = -\frac{1}{2}x + 1$

$m = -\frac{1}{2}$

$\tan \alpha° = -\frac{1}{2}$

$\therefore \alpha° = 153°26'$

b $3y = x + 4$

$y = \frac{1}{3}x + \frac{4}{3}$

$m = \frac{1}{3}$

$\tan \alpha° = \frac{1}{3}$

$\alpha° = 18°26'$

c $3x - 2y = 5$

$3x - 5 = 2y$

$\frac{3}{2}x - \frac{5}{2} = y$

$\therefore m = \frac{3}{2}$

$\therefore \tan \alpha° = \frac{3}{2}$

$\alpha° = 56°19'$

Chapter 9 . . . pp. 151–153
Measurement

1 a $A = bh$
$= 12 \times 8$
$= 96$
$\therefore$ The area is 96 mm^2.

b $A = \frac{1}{2}h(a + b)$

$= \frac{1}{2} \times 3x \times (2x + 6x)$

$= \frac{1}{2} \times 3x \times 8x$

$= 12x^2$

$\therefore$ The area is $12x^2$ units^2.

c $A = \frac{1}{2}xy$

$= \frac{1}{2} \times 12 \times 17$

$= 102$

$\therefore$ The area is 102 cm^2.

d $A = \frac{1}{2}xy$

$= \frac{1}{2} \times 3a \times 2a$

$= 3a^2$

$\therefore$ The area is $3a^2$ units^2.

2 a

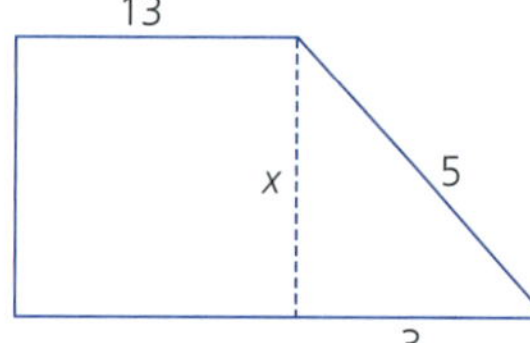

$x^2 = 5^2 - 3^2$
$= 25 - 9$
$= 16$
$x = \sqrt{16}$
$= 4$

$A = \frac{1}{2}h(a + b)$

$= \frac{1}{2} \times 4(13 + 16)$

$= 58$

$\therefore$ The area is 58 cm^2.

Perimeter = 13 + 5 + 16 + 4
= 38

$\therefore$ The perimeter of 38 cm.

b $A = \frac{1}{2}xy$

$= \frac{1}{2} \times 6 \times 8$

$= 24$

$\therefore$ The area is 24 cm^2.

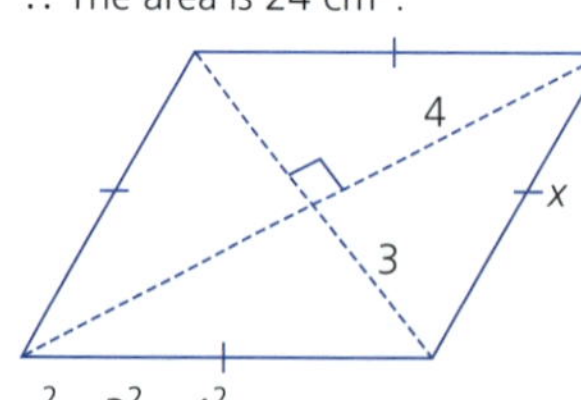

$x^2 = 3^2 + 4^2$
$= 9 + 16$
$= 25$
$x = \sqrt{25}$
$= 5$

Perimeter = 4×5
= 20

$\therefore$ The perimeter is 20 cm.

c

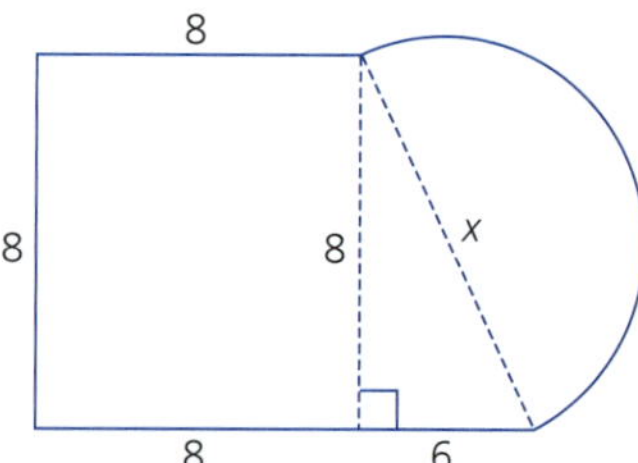

$x^2 = 8^2 + 6^2$
$= 64 + 36$
$= 100$
$x = \sqrt{100}$
$= 10$

$\therefore$ The diameter is 10 cm, the radius is 5 cm.

Area $= \frac{1}{2} \times 8\,(8 + 14) + \frac{1}{2} \times \pi \times 5^2$
$= 127.269\,9082$
$= 127.27$ (to 2 dec. pl.)

$\therefore$ Area is 127.27 cm^2.

Perimeter $= 8 + 8 + 14 + \frac{1}{2} \times 2 \times \pi \times 5$
$= 45.707\,963\,27$
$= 45.71$ (to 2 dec. pl.)

$\therefore$ The perimeter is 45.71 cm.

3 a $V = \frac{1}{3}Ah$

$= \frac{1}{3} \times 12 \times 10 \times 8$

$= 320$

$\therefore$ The volume is 320 cm^3.

b $V = \frac{1}{3}\pi r^2 h$

$= \frac{1}{3} \times \pi \times 10^2 \times 18$

$= 1884.955\,592$
$= 1884.96$ (to 2 dec. pl.)

$\therefore$ The volume is 1884.96 cm^3.

c Diameter = 30 cm
Radius = 15 cm

$\therefore V = \frac{4}{3}\pi r^3$

$= \frac{4}{3} \times \pi \times 15^3$

$= 14\,137.166\,94$
$= 14\,137.17$ (to 2 dec. pl.)

$\therefore$ The volume is 14 137.17 cm^3.

d Volume = Volume of cone + Volume of cylinder

$= \frac{1}{3} \times \pi \times 8^2 \times 10 + \pi \times 8^2 \times 12$

$= 3082.949\,591$
$= 3082.95$ (to 2 dec. pl.)

$\therefore$ The volume is 3082.95 cm^3.

4 Diameter = 25 cm
Radius = 12.5 cm

$V = \frac{4}{3}\pi r^3$

$= \frac{4}{3} \times \pi \times 12.5^3$

$= 8181.230869$

$= 8181.23$ (to 2 dec. pl.)

$\therefore$ The volume is 8181.23 cm^3.

$\therefore$ The capacity is 8181.23 mL.

5 a Diameter PQ = 12 cm
Radius = 6 cm

$\therefore V = \frac{1}{3} \times \pi \times 6^2 \times 10 + \frac{1}{3} \times \pi \times 6^2 \times 8$

$= 678.5840132$

$= 679$ (to 3 sig. fig.)

$\therefore$ The volume is 679 cm^3.

b Diameters 10 cm, 30 cm

$\therefore$ Radii 5 cm, 15 cm.

$V = \frac{1}{3} \times \pi \times 5^2 \times 5 + \frac{1}{3} \times \pi \times 15^2 \times 15$

$= 3665.191429$

$= 3670$ (to 3 sig. fig.)

$\therefore$ The volume is 3670 cm^3.

c Volume = Volume large pyramid − Volume small pyramid

$\therefore V = \frac{1}{3} \times 12 \times 12 \times 8 - \frac{1}{3} \times 3 \times 3 \times 2$

$= 378$

$\therefore$ The volume is 378 cm^3.

d Volume = Volume large cone − Volume small cone

$\therefore V = \frac{1}{3} \times \pi \times 6^2 \times 12 - \frac{1}{3} \times \pi \times 3^2 \times 6$

$= 395.8406743$

$= 396$ (to 3 sig. fig.)

$\therefore$ The volume is 396 cm^3.

6 $V = \frac{1}{2} \times \frac{4}{3}\pi r^3$

$= \frac{1}{2} \times \frac{4}{3} \times \pi \times 12^3$

$= 1152\pi$

$\therefore$ The volume is 1152π cm^3.

7 Volume = Volume of cylinder + 2 × Volume 1 of cones

$= \pi \times 5^2 \times 12 + 2 \times \frac{1}{3} \times \pi \times 5^2 \times 8$

$= 1361.356817$

$= 1361.36$ (to 2 dec. pl.)

$\therefore$ The volume is 1361.36 cm^3.

8 a Diameter = 12 cm
Radius = 6 cm

$V = \frac{4}{3}\pi r^3$

$= \frac{4}{3} \times \pi \times 6^3$

$= 904.7786842$

$= 904.779$ (to 3 dec. pl.)

$\therefore$ The volume is 904.779 cm^3.

b Height of cylinder = 36 cm

$\therefore$ Volume of space

= Volume of cylinder − 3 × Volume of sphere

$= \pi \times 6^2 \times 36 - 3 \times \frac{4}{3} \times \pi \times 6^3$

$= 1357.168026$

$= 1357.168$ (to 3 dec. pl.)

$\therefore$ The volume is 1357.168 cm^3.

9 $V = \frac{4}{3}\pi r^3$

$8 = \frac{4}{3}\pi r^3$

$24 = 4\pi r^3$

$r^3 = \frac{24}{4\neq}$

$r^3 = \frac{6}{\neq}$

$r = \sqrt[3]{\frac{6}{\neq}}$

The radius is $\sqrt[3]{\frac{6}{\neq}}$ cm.

10 $V = \frac{1}{3}\pi r^3 h$

$120 = \frac{1}{3}\pi \times 8^2 \times h$

$360 = 64\pi h$

$\therefore h = \frac{360}{64\pi}$

$= 1.79049311$

$= 1.8$ (1 dec. pl.)

$\therefore$ The height is 1.8 cm.

11 $V = \frac{1}{3}Ah$

$270 = \frac{1}{3} \times 18^2 \times h$

$270 = 108h$

$\therefore h = \frac{270}{108}$

$= 2.5$

$\therefore$ The height is 2.5 cm.

12 a $V = s^3$

$= 18^3$

$= 5832$

$\therefore$ The volume is 5832 cm^3.

b $V = \frac{1}{3}Ah$

$= \frac{1}{3} \times 18^2 \times 18$

$= 1944$

$\therefore$ The volume is 1944 cm^3.

c Diameter = 18 cm
Radius = 9 cm

$V = \frac{1}{3}\pi r^2 h$

$= \frac{1}{3} \times \pi \times 9^2 \times 18$

$= 1526.81403$

$= 1526.81$ (2 dec. pl.)

$= 1526.81$ cm^3

d First cube filled:

$\frac{1944}{5832} \times 100 = 33\frac{1}{3}$

$\therefore 33\frac{1}{3}\%$ filled

Second cube filled:

$\frac{1526.81}{5832} \times 100$

$= 26.17993878$

$= 26.18$ (to 2 dec. pl.)

$\therefore$ 26.18% filled.

13 $SA = 2\pi r^2 + 2\pi rh$

$= 2 \times \pi \times 4^2 + 2 \times \pi \times 4 \times 2$

$= 32\pi + 16\pi$

$= 48\pi$

$\therefore$ The surface area is 48π cm^2.

14 a $SA = 2\pi r^2 + 2\pi rh$

$= 2 \times \pi \times 8^2 + 2 \times \pi \times 8 \times 14$

$= 1105.840614$

$= 1105.84$ (to 2 dec. pl.)

$\therefore$ The surface area is 1105.84 cm^2.

b $SA = 10 \times 10 + 4 \times \frac{1}{2} \times 10 \times 12$

$= 340$

$\therefore$ The surface area is 340 cm^2.

c $SA = \pi r^2 + \pi rl$

$= \pi \times 16^2 + \pi \times 16 \times 20$

$= 1809.557368$

$= 1809.56$ (to 2 dec. pl.)

$\therefore$ The surface area is 1809.56 cm^2.

d $SA = 4\pi r^2$

$= 4 \times \pi \times 10^2$

$= 1256.637061$

$= 1256.64$ (to 2 dec. pl.)

$\therefore$ The surface area is 1256.64 cm^2.

15 a

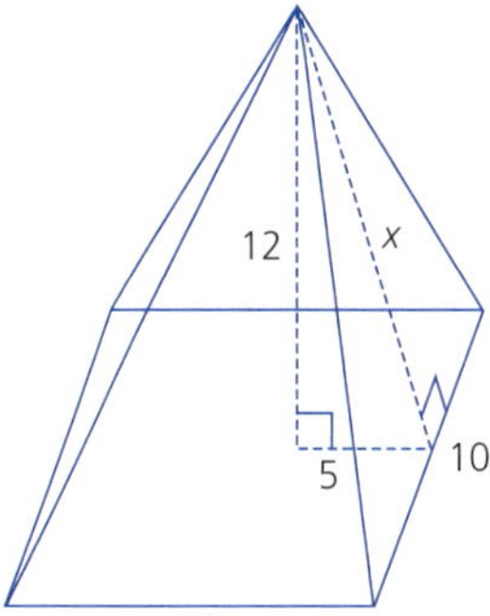

10

$x^2 = 5^2 + 12^2$

$= 25 + 144$

$= 169$

$x = \sqrt{169}$

$= 13$

$SA = 10 \times 10 + 4 \times \frac{1}{2} \times 10 \times 13$

$= 360$

$\therefore$ The surface area is 360 cm^2.

b

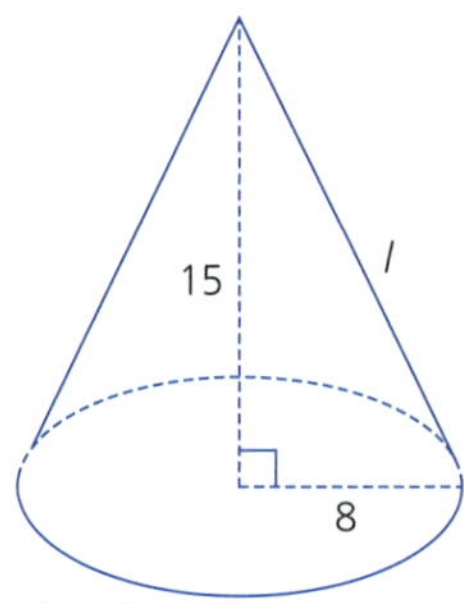

$l^2 = 8^2 + 15^2$
$= 64 + 225$
$= 289$
$l = \sqrt{289}$
$= 17$
$SA = \pi r^2 + \pi rl$
$= \pi \times 8^2 + \pi \times 8 \times 17$
$= 628.3185307$
$= 628.32$ (to 2 dec. pl.)
$\therefore$ The surface area is 628.32 cm^2.

16 a Diameter = 25 cm
Radius = $\frac{25}{2}$ cm
$\therefore SA = 2\pi rh$
$= 2 \times \pi \times \frac{25}{2} \times 24$
$= 600\pi$
$\therefore$ The surface area is 600π cm^2.

b Diameter = 16 cm
Radius = 8 cm
$SA = \pi rl$
$= \pi \times 8 \times 20$
$= 160\pi$
$\therefore$ The surface area is 160π cm^2.

17 $SA = \frac{1}{2} \times 4\pi r^2 + \pi r^2$
$= 3 \times \pi \times 8^2$
$= 192\pi$
$\therefore$ The surface area is 192π cm^2.

18 a $V = \pi r^2 h$
$= \pi \times 4^2 \times 10$
$= 502.6548246$
$= 502.655$ (to 3 dec. pl.)
$\therefore$ The volume is 502.655 cm^3.

b Capacity is 502.654 mL
$\therefore$ 503 mL (to nearest mL)

c $SA = 2\pi r^2 + 2\pi rh$
$= 2 \times \pi \times 4^2 + 2 \times \pi \times 4 \times 10$
$= 351.8583772$
$= 351.86$ (to 2 dec. pl.)
$\therefore$ The surface area is 351.86 cm^2.

d $SA = 2\pi rh$
$= 2 \times \pi \times 4 \times 10$
$= 251.3274123$
$= 251.33$ (to 2 dec. pl.)
$\therefore$ The surface area is 251.33 cm^2.

19 Diameter = 24 cm
Radius = 12 cm
$SA = \pi r^2 + 2\pi rh + \pi rl$
$= \pi \times 12^2 + 2 \times \pi \times 12 \times 30 + \pi \times 12 \times 20$
$= 3468.31829$
$= 3468.32$ (to 2 dec. pl.)
$\therefore$ The label's area is 3468.32 cm^2.

Chapter 10 . . pp. 177–182

Geometry

1 $x = 180 - (90 + 40)$ [$\angle$ sum of straight line]
$= 180 - 130$
$= 50$

2 $a = 180 - (27 + 72)$
$= 180 - 99$
$= 81$ [$\angle$ sum of a straight line]

3 $b + 18 = 90$ [complementary angles]
$\therefore$ $b = 90 - 18$
$\therefore$ $b = 72$

4 $c = 128$ [vertically opposite angles]
also $d = 180 - 128$
[angle sum of a straight line]
$\therefore$ $d = 52$

5 $e + 109 + 112 + 92 = 360$
[angles at a point]
$\therefore$ $e + 313 = 360$
$\therefore$ $e = 360 - 313$
$= 47$

6 $g + 43 + 42 = 180$
[CD is a straight line]
$\therefore$ $g = 180 - 85$
$= 95$
$f + 20 + 95 + 43 = 180$
[AB is a straight line]
That is $f + 158 = 180$
$\therefore$ $f = 180 - 158$
$= 22$

7 $3x + 28 = x + 40$
[vertical opposite angles]
$\therefore$ $2x = 12$
$x = 6$
Then $3x + 28 = 3(6) + 28$
$= 46$
Check $x + 40 = 6 + 40$
$= 46$
The equal angles are 46°.

8 $h = 124$ [alternate angles, parallel lines]

9 $k = 126$ [corresponding angles with parallel lines, then $k°$ and 126° are vertically opposite]

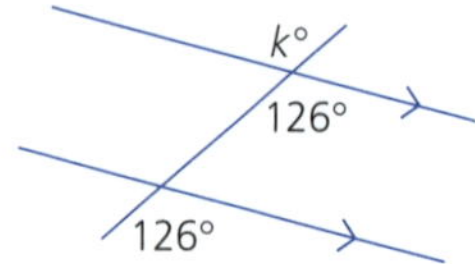

10

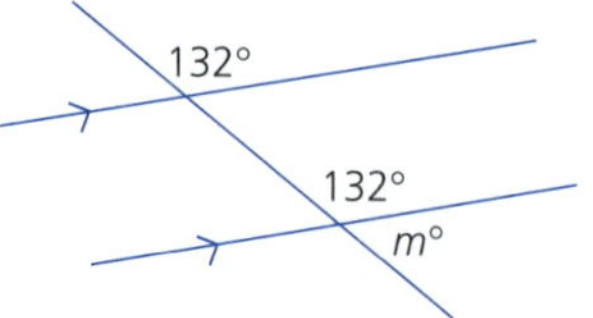

$m + 132 = 180$
$m = 180 - 132$
$= 48$
[corresponding $\angle$s in parallel lines, then $m°$ and 132° are supplementary]

11

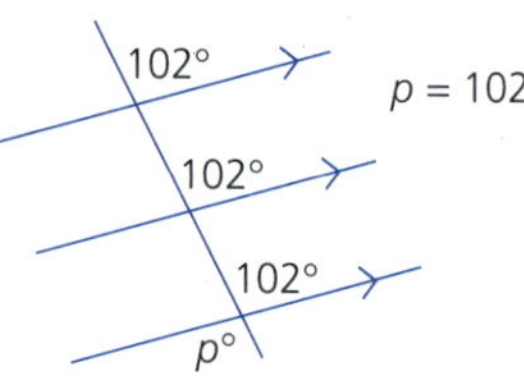

$p = 102$

[corresponding $\angle$s in parallel lines, then $p°$ and 102° are vertically opposite]

12

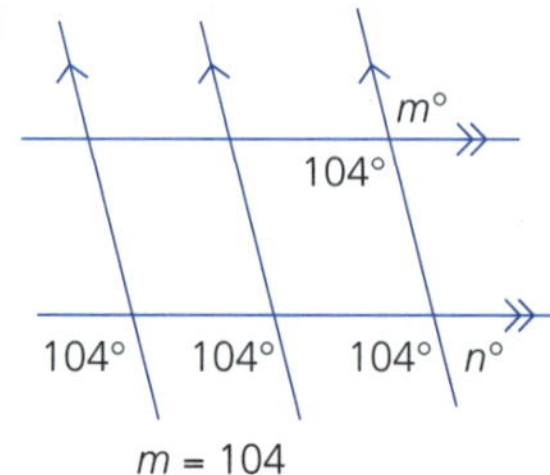

$m = 104$
$n + 104 = 180$
$n = 180 - 104$
$= 76$
For m, corresponding $\angle$s with parallel lines, and $m°$ and 104° are vertically opposite. For n, as for m, but $n + 104 = 180$ [straight line].

13 $x + 78 = 180$
$x = 180 - 78$
$= 102$

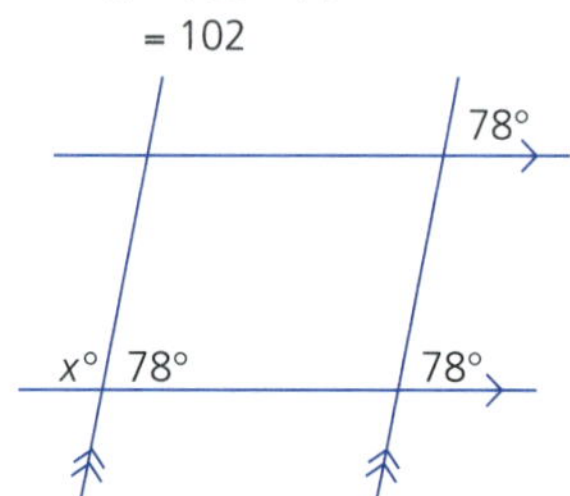

[corresponding $\angle$s with parallel lines, $x°$ and 78° are supplementary]

14 $4y + 120 = 180$
[co-interior angles, parallel lines]
$\therefore$ $4y = 180 - 120$
$4y = 60$
$\therefore$ $y = 15$

15
A
B 110° C
D
a° 110° E
110°
F G
70°

Draw a third line parallel to other two at D [as in the diagram].

$\angle BDE = 110°$ [corr. angles, BC ∥ DE]

$\angle FDE = 110°$ [co-int. angles, DE ∥ FG]

$\therefore \angle ADF = 360° - (110° + 110°)$

$= 360° - 220°$ [angles at a point]

$= 140°$

$\therefore \quad a = 140$

16 $b + 58 + 67 = 180$ [∠ sum of △]

$\therefore \quad b = 180 - 125$

$= 55$

17 $c + c + 56 = 180$ [base angles of isosceles triangle, angle sum of △]

$2c = 180 - 56$

$\therefore \quad 2c = 124$

$\therefore \quad c = 62$

18 $a + a + 2a = 180$ [∠ sum of a △]

$4a = 180$

$a = 45$

19 $d + 59 + 59 = 180$ [base angles of an isosceles △, ∠ sum of a △]

$\therefore \quad d = 180 - 118$

$= 62$

20 $(3x - 9) + (x + 64) + (2x + 35) = 180$

[∠ sum of a △]

$\therefore \quad 6x + 90 = 180$

$\therefore \quad 6x = 90$

$\therefore \quad x = \frac{90}{6}$

That is, $x = 15$

21 $y + 88 + 87 + 57 = 360$

[∠ sum of a quadrilateral]

$\therefore \quad y + 232 = 360$

$\therefore \quad y = 360 - 232$

That is, $y = 128$

22 $(3b + 48) + 2b + b + 2b = 360$

[∠ sum of a quadrilateral]

$\therefore \quad 8b + 48 = 360$

$\therefore \quad 8b = 360 - 48$

$= 312$

$\therefore \quad b = \frac{312}{8}$

That is, $b = 39$

23

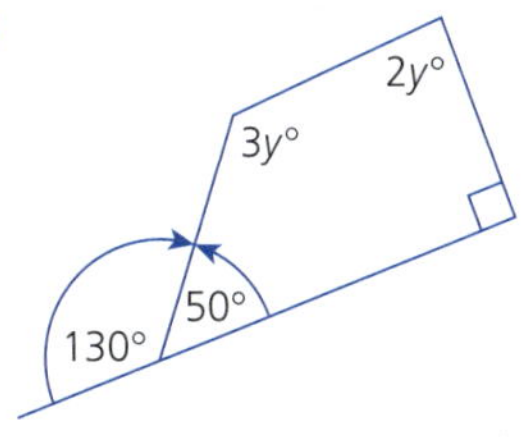

The other angle in the quadrilateral is 50° [supplementary ∠s].

Then $3y + 2y + 90 + 50 = 360$

[∠ sum of a quadrilateral]

$\therefore \quad 5y + 140 = 360$

$\therefore \quad 5y = 360 - 140$

$= 220$

$\therefore y = \frac{220}{5}$

That is, $y = 44$

24 $d + 87 = 132$ [exterior ∠ of a △]

$\therefore \quad d = 132 - 87$

$= 45$

25 $e = 74 + 47$ [exterior ∠ of a △]

$= 121$

26

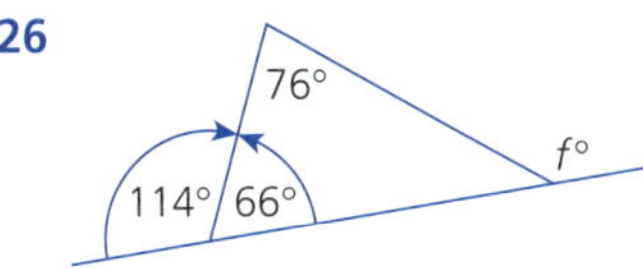

114° and 66° are supplementary.

$f = 76 + 66$ [exterior ∠ of a △]

$= 142$

27 $x = 39$

[alternate angles, parallel lines]

$y + 39 + 54 = 180$ [∠ sum of a △]

$\therefore \quad y + 93 = 180$

$\therefore \quad y = 180 - 93$

$= 87$

28 $5z + 10 = 3z + 50$ [exterior ∠ of a △]

$\therefore \quad 2z = 40$

$z = 20$

29

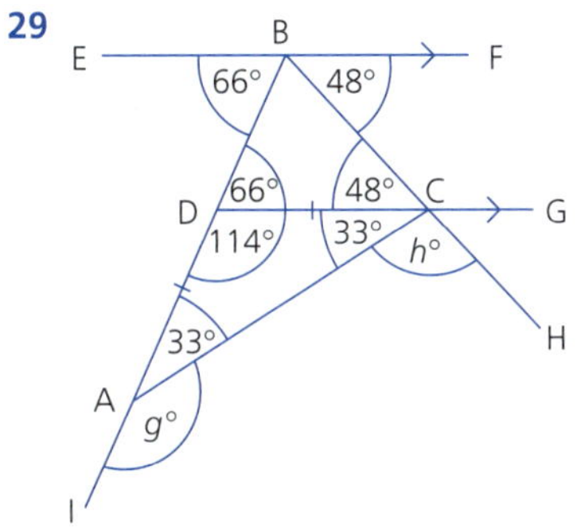

$\angle BDC = 66°$ [alternate angles, EF ∥ DG]

$\therefore \angle ADC = 114°$ [AB is a straight line]

As AD = DC,

$\angle DAC = \angle DCA = 33°$

[base angles of isosceles △, ∠ sum of a △]

Also, $\angle DCB = 48°$

[alternate △, EF ∥ DG]

Then, $g + 33 = 180$ [IB is a straight line]

$\therefore \quad g = 180 - 33$

$= 147$,

and $h + 33 + 48 = 180$

[BH is a straight line]

$\therefore \quad h = 180 - 81$

$= 99$

30

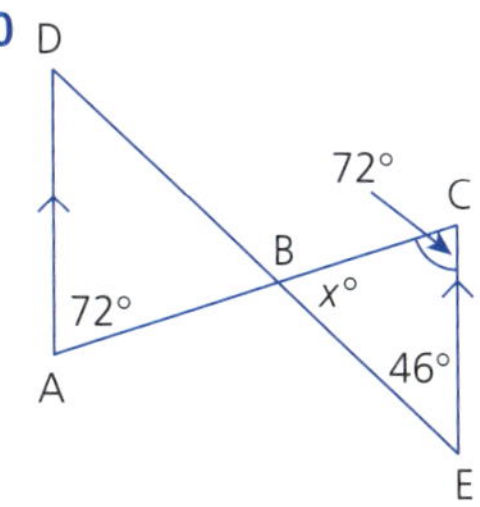

$\angle BCE = 72°$ [alternate ∠s, AD ∥ CE]

$\therefore x + 72 + 46 = 180$ [∠ sum of △]

$\therefore \quad x + 118 = 180$

$\therefore \quad x = 180 - 118$

$= 62$

31

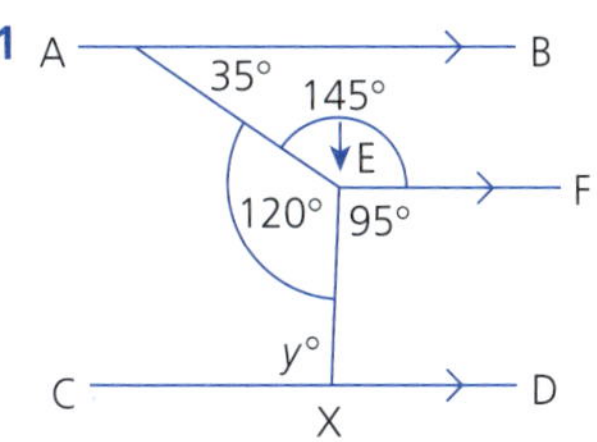

Construct EF ∥ AB and CD.

$\angle AEF = 180° - 35° = 145°$

[co-interior angles, AB ∥ EF]

$\therefore \angle FEX = 360° - (145° + 120°)$

$= 360° - 265°$

$= 95°$,

$\therefore \angle CXE = 95°$ [alternate ∠s, EF ∥ CD]

$\therefore y = 95$

32 **a** Sum $= 180(n - 2)$

$= 180(5 - 2)$

$= 540$

∴ Angle sum is 540°.

b Sum $= 180(12 - 2)$

$= 1800$

∴ Angle sum is 1800°.

c Sum $= 180(20 - 2)$

$= 3240$

∴ Angle sum is 3240°.

33 **a** Sum $= 180(8 - 2)$

$= 1080$

∴ Angle sum is 1080°.

Each angle $= 1080 \div 8$

$= 135$

∴ Each angle is 135°.

b Sum $= 180(12 - 2)$

$= 1800$

∴ Angle sum is 1800°.

Each angle $= 1800 \div 12$

$= 150$

∴ Each angle is 150°.

c Sum $= 180(32 - 2)$

$= 5400$

∴ Angle sum is 5400°.

Each angle $= 5400 \div 32$

$= 168.75$

∴ Each angle is 168.75°.

34 **a** Each angle $= 360 \div 6$

$= 60$

∴ Each exterior angle is 60°.

b Each angle $= 360 \div 10$

$= 36$

∴ Each exterior angle is 36°.

c Each angle $= 360 \div 20$

$= 18$

∴ Each exterior angle is 18°.

35 The polygon has $\frac{360}{n}$ sides.

36

Note: Always mark data on the diagram and then solve the question on the diagram before attempting to write a solution.

In the △AOB and △COD
AB = CD [data]
AO = CO [data]
∠AOB = ∠COD = 90° [AC ⊥ DB]
∴ △AOB ≡ △COD [RHS]

37

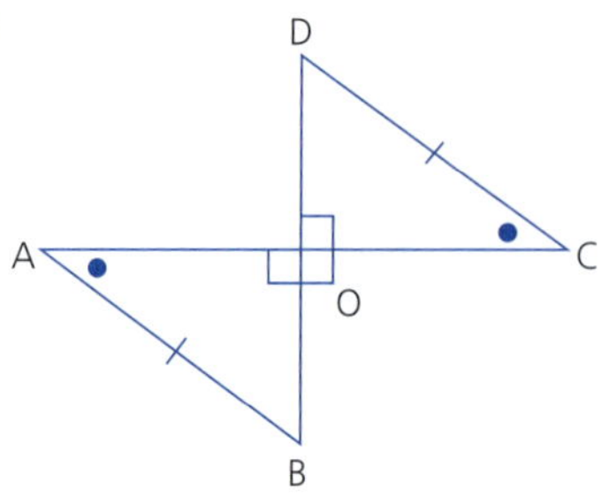

In △s ABO and CDO
AB = CD [data]
∠OAB = ∠OCD [data]
∠AOB = ∠COD [both 90°, AC ⊥ DB]
∴ △AOB ≡ △COD [AAS]

38

In △s ABO and CDO
AO = CO [data]
BO = DO [data]
∠AOB = ∠COD [both 90°, AC ⊥ DB]
∴ △AOB ≡ △COD [SAS]

39

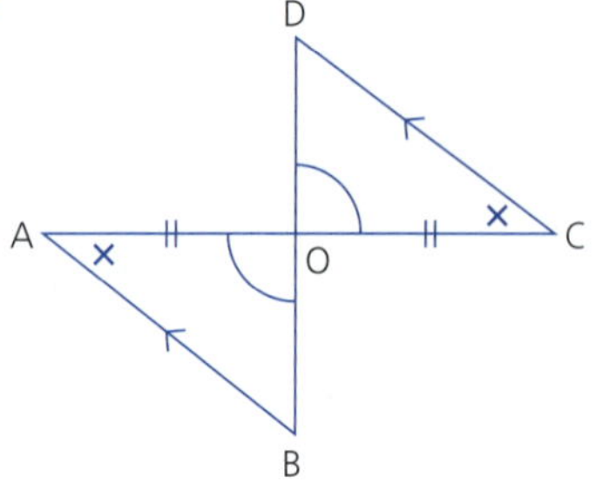

In △s ABO and CDO
AO = CO [data]
∠BAO = ∠DCO [alt. ∠s, AB ∥ CD]
∠AOB = ∠COD [vert opp. angles]
∴ △AOB ≡ △COD [AAS]

40

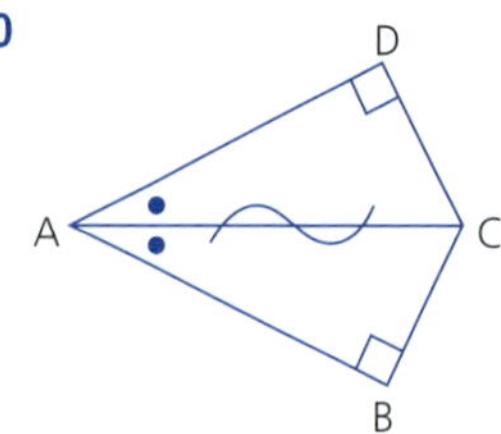

In △s ABC and ADC
AC = AC [common side]
∠BAC = ∠ADC [CA bisects ∠DAB]
∠ABC = ∠ADC [both 90°]
∴ △ABC ≡ △ADC [AAS]

41

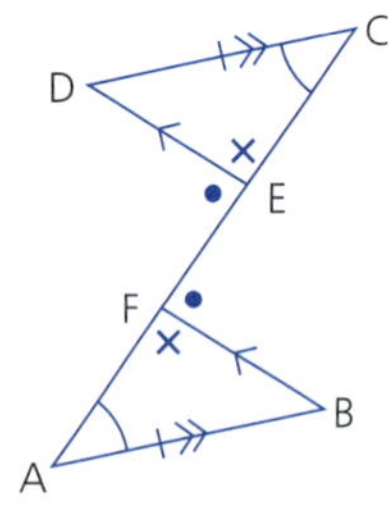

∠BFE = ∠DEF [alternate ∠s, FB ∥ DE]
∴ ∠AFB = ∠CED [supplements of equal angles]
In △s AFB and CED
AB = CD [data]
∠AFB = ∠CED [proved above]
∠FAB = ∠ECD [alternate ∠s, AB ∥ DC]
∴ △ABF ≡ △CDE [AAS]

a Then AF = CE
[corresponding sides in congruent △s]

b Also AF + EF = CE + EF
[adding EF to both]
∴ AE = CF

42

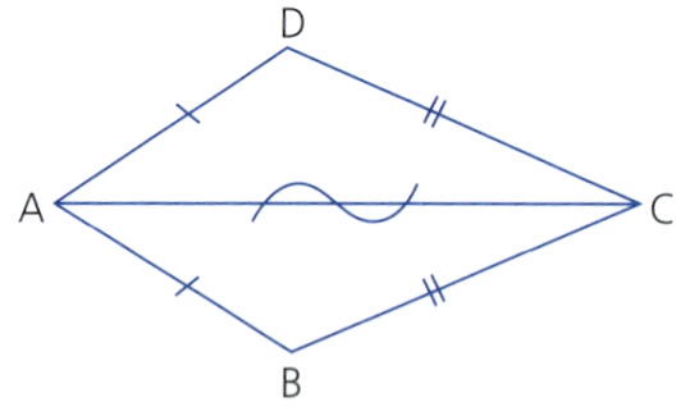

In △s ABC and ADC
AB = AD [data]
BC = DC [data]
AC = AC [common side]
∴ △ABC ≡ △ADC [SSS],
∴ ∠BAC = ∠DAC [corresponding angles in congruent △s]

43

AD + DB = AE + EC [addition of equal sides
∴ AB = AC.
In △s ABE and ADC
AB = AC [proved above]
AE = AD [data]
∠BAE = ∠CAD [common ∠]
∴ △ABE ≡ △ACD [SAS],
∴ ∠ABE = ∠ACD [corresponding angles in congruent △s]

44

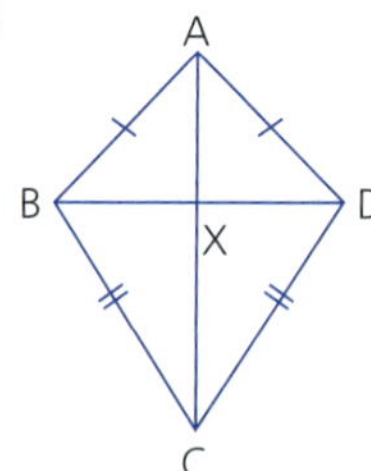

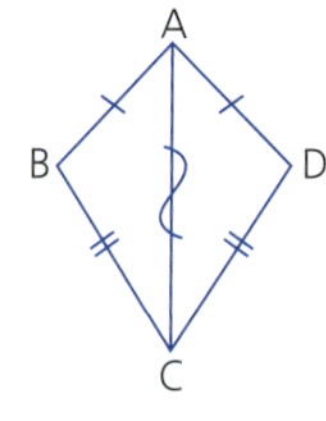

a In △s ABC and ADC
AB = AD [data]
BC = DC [data]
AC = AC [common side]
∴ △ABC ≡ △ADC [SSS]

b Now ∠BAX = ∠DAX [corresponding angles in congruent △s]

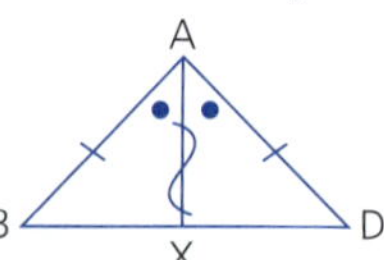

In △s ABX and ADX
AB = AD [data]
AX = AX [common side]
∠BAX = ∠DAX [proved above]
∴ △ABX ≡ △ADX [SAS]

c BX = DX [corresponding sides in congruent △s]
∴ AC bisects BD.
Also ∠BXA = ∠DXA [corresponding angles in congruent △s]
But BD is a straight line
∴ ∠BXA = ∠DXA = 90°
That is, AC ⊥ BD
AC bisects BD at right-angles.

45

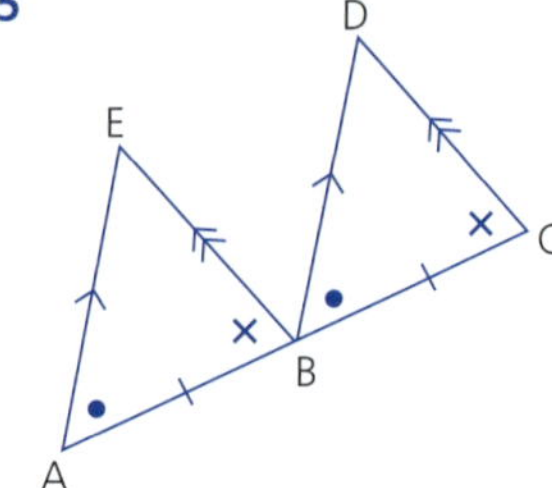

In △s ABE and BCD

AB = BC [B is midpoint of AC]

∠BAE = ∠CBD [corresponding ∠s, AE ∥ BD]

∠ABE = ∠BCD [corresponding ∠s, BE ∥ CD]

∴ △ABE ≡ △BCD [AAS]

46

In △s ABD and CDB

AD = BC [data]

BD = BD [common side]

∠ADB = ∠CBD [alternate ∠s, AD ∥ BC]

∴ △ABD ≡ △CDB [SAS]

a AB = DC [corresponding sides in congruent ∠s]

b ∠ABD = ∠CDB [∠s in corresponding positions in congruent △s]

c As ∠ABD = ∠CDB, AB ∥ DC [a pair of alternate ∠s equal]

47

∠CGF = 116° [corresponding ∠s, AB ∥ CD]

∴ ∠CGH = 180° − 116° [∠ sum of a straight line]

= 64°

48

∠CQE = 78°

∠GSQ = 78° [corr. ∠s, EF ∥ GH]

and ∠RSD = 78° [vert. opposite ∠s]

∴ ∠BRS = 180° − 78° [co-interior ∠s, AB ∥ CD]

= 102°

This is one of a number of equally correct solutions for this diagram.

49

∠BCE + 47° = 141° [exterior ∠ of a △]

∴ ∠BCE = 141° − 47°

= 94°

Then ∠ECD = 180° − 94°

= 86° [AD is a straight line]

50

∠QPR = 47° [alternate ∠s, ST ∥ PQ]

∴ ∠PRQ = 180° − (47° + 48°) [∠ sum of a △]

= 180° − 95°

= 85°

51

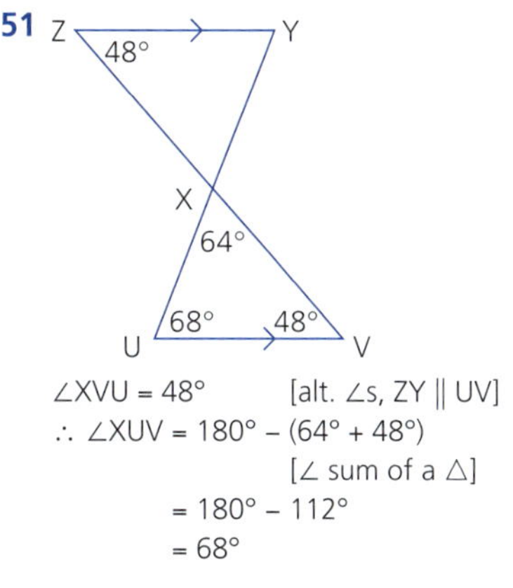

∠XVU = 48° [alt. ∠s, ZY ∥ UV]

∴ ∠XUV = 180° − (64° + 48°) [∠ sum of a △]

= 180° − 112°

= 68°

52

∠CAB = ∠CBA [base ∠s of an isosceles △]

∠CAB + ∠CBA = 180° − 142° [∠ sum of a △]

= 38°

∴ ∠CAB = 19°

Then ∠DAC = 19° [AC bisects ∠DAB]

Then ∠DAB = 38° [∠CAB +∠DAC]

∴ ∠ABD = 38° [base ∠s of an isosceles △]

Then ∠ADB = 180° − (38° + 38°)

= 104° [∠ sum of a △]

53

∠SRP = 70° [base ∠s of an isosceles △]

∴ ∠RPQ = 70° [alt. ∠s, SR ∥ PQ]

∴ ∠PQR = 70° [base ∠s of an isosceles △PQR]

54

∠PQT = 32° [co-int. ∠s, PU ∥ QT]

also ∠RQT = 42° [co-int. ∠s, QT ∥ RS]

∴ ∠PQR = 74° [32° + 42°]

Hence reflex ∠PQR = 360° − 74°

= 286°

55 Construction: Draw CF ∥ AB.

Hence CF ∥ DE [both parallel to AB]

Then ∠BCF = 125° [alt. ∠s, AB ∥ CF]

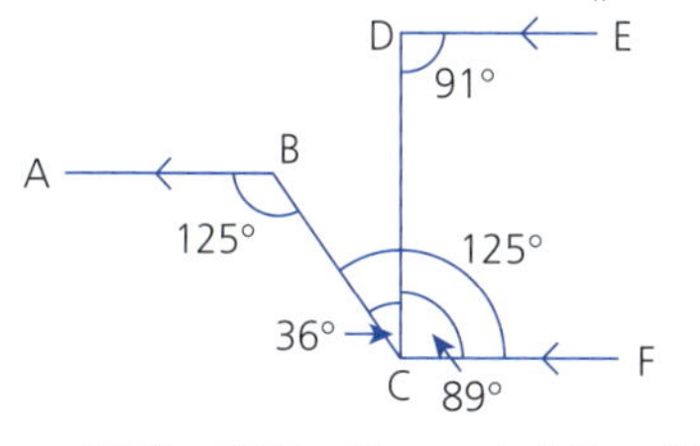

∴ ∠DCF = 125° − 36° [∠BCD = 36°]

= 89°

Then ∠CDE = 180° − 89° [co-int. ∠s, CF ∥ DE]

= 91°

56

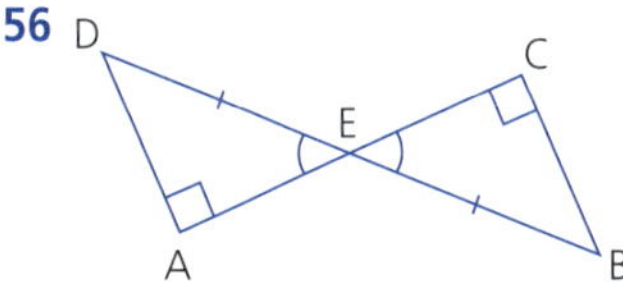

The aim is to prove that AE = CE. Congruent triangles appears to be the method.

In △s ADE and CBE

DE = EB [E is the midpoint of DB]

∠DEA = ∠BEC [vert. opp. ∠s]

∠DAE = ∠BCE [both 90°]

∴ △ADE ≡ △CBE [AAS]

Then AE = CE [corresponding sides in congruent △s]

That is, E is the midpoint of AC, or DB bisects AC.

57

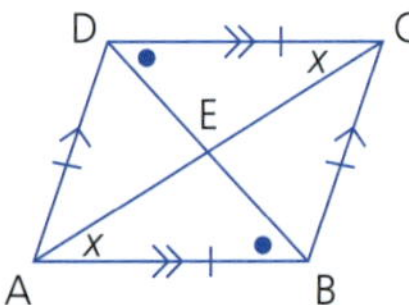

In the $\triangle$s DEC and BEA

DC = AB	[ABCD is rhombus]
$\angle CDE = \angle ABE$	[alternate $\angle$s, DC $\parallel$ AB]
$\angle DCE = \angle BAE$	[alternate $\angle$s, DC $\parallel$ AB]
$\therefore \triangle DEC \equiv \triangle BEA$	[AAS]

Then DE = EB
CE = EA

$\therefore$ The diagonals bisect each other.

Also in $\triangle$s DEC and BEC

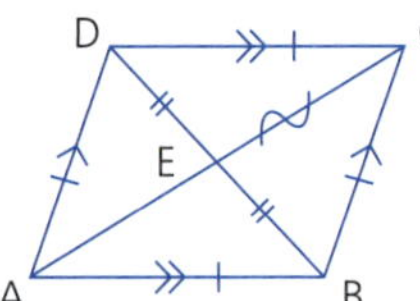

DC = BC	[ABCD is rhombus]
EC = EC	[common side]
DE = EB	[proved above]
$\therefore \triangle DEC \equiv \triangle BEC$	[SSS]

Then $\angle DEC = \angle BEC$
[corresponding angles in congruent triangles]

But $\angle DEC + \angle BEC = 180°$
[DB is a straight line]

$\therefore \angle DEC = \angle BEC = 90°$
[Similarly $\angle DEB = \angle BEA = 90°$]

$\therefore$ The diagonals bisect each other at right-angles.

58

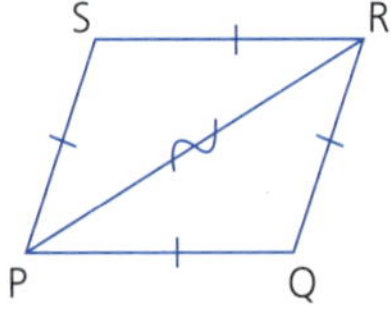

In $\triangle$s PRS and PQR

PS = QR	[given]
SR = PQ	[given]
PR = PR	[common side]
$\therefore \triangle PSR \equiv \triangle RQP$	[SSS]
$\angle SRP = \angle QPR$	[corresponding angles in congruent $\triangle$s]
$\therefore$ SR $\parallel$ PQ	[a pair of alternate angles equal]
Also $\angle SPR = \angle QRP$	[corresponding angles in congruent $\triangle$s]
$\therefore$ SP $\parallel$ RQ	[a pair of alternate angles equal]

$\therefore$ PQRS is a rhombus (parallelogram with all sides equal).

59

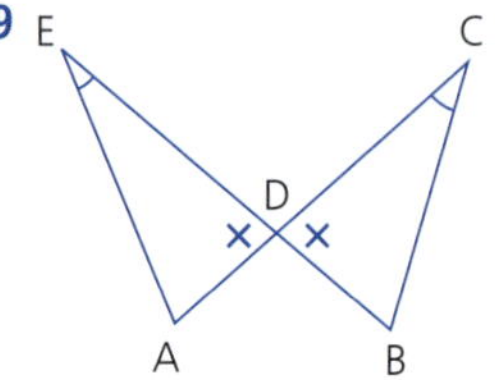

In $\triangle$s ADE and BDC

$\angle AED = \angle BCD$	[data]
$\angle ADE = \angle BDC$	[vertically opposite angles]
$\therefore \triangle ADE \mathbin{\|\|\|} \triangle BDC$	[equiangular]

60

a In $\triangle$s PTQ and RTS

$\angle QPT = \angle SRT$	[alternate angles, SR $\parallel$ PQ]
$\angle PTQ = \angle RTS$	[vertically opposite angles]
$\therefore \triangle PTQ \mathbin{\|\|\|} \triangle RTS$	[equiangular]

b As $\triangle$s are similar, ratios of corresponding sides are equal.

$$\frac{PQ}{\boxed{SR}} = \frac{QT}{\boxed{ST}} = \frac{TP}{\boxed{TR}}$$

Corresponding sides—the sides in each triangle opposite the same equal angle.

61

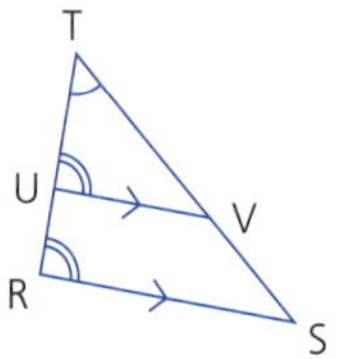

a In $\triangle$s TUV and TRS

$\angle UTV = \angle RTS$	[common angles]
$\angle TUV = \angle TRS$	[corresponding angles, UV $\parallel$ RS],
$\therefore \triangle UVT \mathbin{\|\|\|} \triangle RST$	[equiangular]

b As $\triangle$s are similar, ratios of corresponding sides are equal

$$\frac{TU}{TR} = \frac{UV}{RS} = \frac{TV}{TS}.$$

62

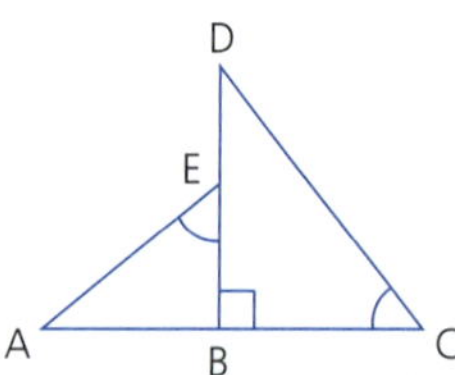

$\angle DBC = 90°$	
$\therefore \angle ABE = 90°$	[angle sum of a straight line]

In $\triangle$s ABE and DBC

$\angle AEB = \angle BCD$	[data]
$\angle ABE = \angle DBC$	[both 90°]
$\therefore \triangle ABE \mathbin{\|\|\|} \triangle DBC$	[equiangular]

63

$$\frac{VW}{WT} = \frac{15}{5} = \frac{3}{1}$$

$$\frac{SW}{WU} = \frac{18}{6} = \frac{3}{1}$$

$$\therefore \frac{VW}{WT} = \frac{SW}{WU}$$

Also $\angle SWV = \angle UWT$ [vertically opposite angles]

$\therefore \triangle SVW \mathbin{\|\|\|} \triangle UTW$ [ratios of two sides in proportion; included angles equal]

64

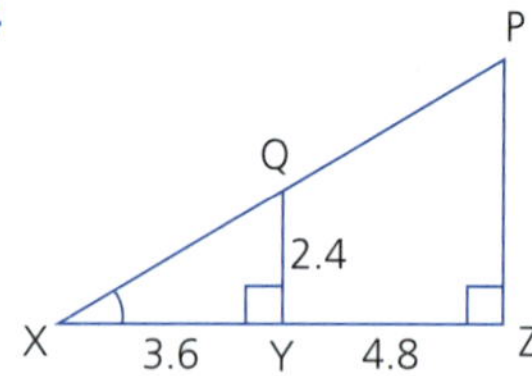

a In $\triangle$s XYQ and XZP

$\angle QYX = \angle PZX$	[both 90°]
$\angle QXY = \angle PXZ$	[common angle]
$\therefore \triangle XYQ \mathbin{\|\|\|} \triangle XZP$	[equiangular]

b Ratios of corresponding sides are equal,

$$\therefore \frac{XY}{XZ} = \frac{XQ}{XP} = \frac{QY}{PZ}$$

Now XY = 3.6, XZ = 3.6 + 4.8, QY = 2.4.

Now using $\frac{XY}{XZ} = \frac{QY}{PZ}$

$$\frac{3.6}{8.4} = \frac{2.4}{PZ}$$

$$\therefore \quad 3.6\,PZ = 2.4 \times 8.4$$

$$\therefore \quad PZ = \frac{2.4 \times 8.4}{3.6} = 5.6$$

PZ is 5.6 m.

65 **a** $P_1 : P_2 = 3 : 6 = 1 : 2$ $\qquad \frac{P_1}{P_2} = \frac{1}{2}$

b $A_1 : A_2 = r_1^2 : r_2^2 = 3^2 : 6^2 = 9 : 36 = 1 : 4$

66 $\frac{A_1}{A_2} = \frac{3^2}{5^2} = \frac{9}{25}$

Now $A_2 = 45$,

$$\therefore \frac{A_1}{45} = \frac{9}{25}$$

$$\therefore A_1 = \frac{9 \times 45}{25} = 16.2.$$

The smaller area is 16.2 cm^2.

67 **a** Ratio $= \frac{32}{40} = \frac{4}{5}$.

Reduction ratio is 4 : 5.

b Call the new circumference C_2.
Then:

$$\frac{C_2}{32} = \frac{4}{5},$$

$$\therefore\ 5C_2 = 4 \times 32$$

$$C_2 = \frac{4 \times 32}{5}$$

$$= 25.6.$$

The circumference 20 m from the base is 25.6 m.

c Let the area at 10 m level be A_2.
Then:

$$\frac{A_2}{750} = \frac{4^2}{5^2} = \frac{16}{25},$$

$$\therefore\ 25\,A_2 = 16 \times 750$$

$$A_2 = \frac{16 \times 750}{25}$$

$$= 480.$$

Let the area at 20 m level be A_3.
Then:

$$\frac{A_3}{480} = \frac{16}{25}$$

$$\therefore\ 25\,A_3 = 16 \times 480$$

$$A_3 = \frac{16 \times 480}{25}$$

$$= 307.2$$

Therefore the cross-sectional areas are 480 m^2 and 307.2 m^2.

68 a **i** Ratio of surface areas = 4 : 25.

ii Ratio volumes = $2^3 : 5^2$
= 8 : 125

b Let the smaller volume be V.
Then

$$\frac{V}{60} = \frac{8}{125},$$

$$\therefore\ 125V = 8 \times 60$$

$$V = \frac{8 \times 60}{125}$$

$$= 3.84$$

Therefore the smaller volume is 3.84 cm^3.

c Let the larger surface area be A.
Then

$$\frac{14}{A} = \frac{4}{25}$$

$$\therefore\ 4A = 14 \times 25$$

$$A = \frac{14 \times 25}{4}$$

$$= 87.5$$

Therefore the larger area is 87.5 cm^2.

69 The ratio of volumes = 32 : 108
= 8 : 27
= $2^3 : 3^3$.

a Diameters of glass balls correspond to widths of cubes,
∴ Ratio of diameters = 2 : 3.

b The ratio of surface areas = $2^2 : 3^2$
= 4 : 9.

70 a Scale = 0.9 : 28.8
= 1 : 32

b Ratio of surface areas = $1^2 : 32^2$
= 1 : 1024.

c Let the quantity of paint required be x litres. Then

$$\frac{x}{48} = \frac{1}{1024}$$

$$\therefore\ x = \frac{48}{1024}\ \text{L}.$$

The quantity of paint

$$= \frac{48}{1024} \times 1000$$

= 46.875
∴ 46.875 mL

d Let the quantity of air be V kL.

$$\frac{V}{8192} = \frac{1}{32^3},$$

$$\therefore\ V = \frac{8192}{32\,768}\ \text{kL}$$

The quantity of air in model

$$= \frac{8192}{32\,768} \times 1000$$

∴ 250 L

Chapter 11 . . pp. 203–206

Statistics

1 a **i** Mean = $\dfrac{8+7+4+8+6+3}{6} = 6$

ii Range = 8 − 3 = 5
Mode = 8

iii 3, 4, (6, 7), 8, 8

Median = $\dfrac{6+7}{2} = 6.5$

b Scores are now 3, 4, 6, 7, 8, 8, 13

Mean = $\dfrac{49}{7} = 7$

Mean has increased by one.

3, 4, 6, (7), 8, 8, 13

Median = 7
Median has increased by 0.5.

c Total no. of goals in eight seasons
= 6.5 × 8 = 52.
Gai scored (52 − 49) goals in the 8th season, that is, three goals.

2 a Mean = $\dfrac{48}{9}$ = 5.3 (one dec. pl.)

3, 4, 4, 4, (5), 6, 7, 7, 8

Median = 5

b Mean = $\dfrac{58}{10} = 5.8$

2, 3, 4, 4, (5, 6), 7, 7, 9, 11
Median = 5.5

3 Total over four exams = 4 × 70 = 280.
Needs to average 75 over five exams, that is, total must be (5 × 75) = 375 after five exams.
Mark needed = 375 − 280 = 95.
Zoltan must score 95 in the 5th exam.

4

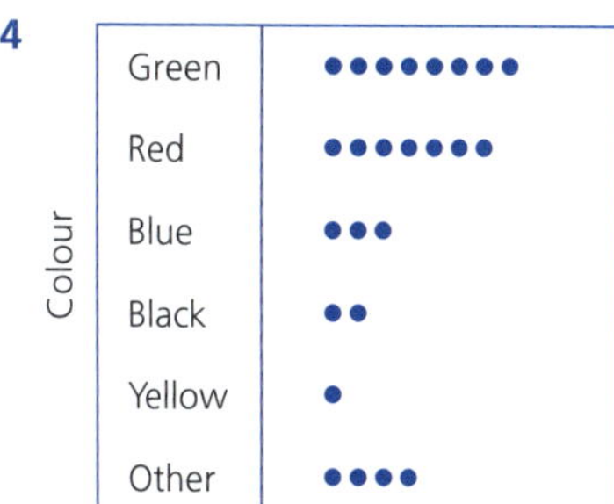

5

Number of students

Shoe size	
4	• •
$4\frac{1}{2}$	•
5	• • • • •
$5\frac{1}{2}$	• •
6	• • •
$6\frac{1}{2}$	• •
7	• • • • •
$7\frac{1}{2}$	•
8	•

→

Shoe size	f
4	2
$4\frac{1}{2}$	1
5	5
$5\frac{1}{2}$	2
6	3
$6\frac{1}{2}$	2
7	5
$7\frac{1}{2}$	1
8	1
Total	22

6 Initial plot:

Stem	Leaf
14	8 9 7 9
15	0 1 5 4 9 2 8 4 0 2
16	3 2 5 7 8 0

Final plot:

Stem	Leaf
14	7 8 9 9
15	0 0 1 2 2 4 4 5 8 9
16	0 2 3 5 7 8

7 Initial plot:

Stem	Leaf
2	6 8
3	9 3 9 9 3
4	7 9 7 6 4 8 3 0 7
5	3 7 2 0 2 4 5 0
6	2 4 0 4 1 2

a Final plot:

Stem	Leaf
2	6 8
3	3 3 9 9 9
4	0 3 4 6 7 7 7 (8 9)
5	0 0 2 2 3 4 5 7
6	0 1 2 2 4 4

b Range = 64 − 26 = 38
Mode = 39, 47
Median between 15th and 16th scores
Median = $\frac{48 + 49}{2}$ = 48.5

8 a Rearranged numbers
4, 5, 6, 6, 7, 8, 8, 9
↑
Median = $\frac{6 + 7}{2}$ = 6.5
(Even number of scores)

b 5, 6, 6, 6, 7, 7, 8, 8, 9, 9, 9
↑
Median = 7
(Odd number of scores)

c 12, 14, 14, 15, 16, 16, 17, 18, 18, 19,
↑
Median = 16

d 7, 7, 7, 8, 8, 9, 9, 10, 11, 11, 12, 12, 12, 13, 14
↑
Median = 10

9 a 23 scores—middle is 12th score.

Stem	Leaf
1	0 7 8
2	1 2 4 6 9
3	7 7 7 (8) 8 9
4	0 0 6 7 8 8 9
5	1 4

◯ Median = 38

b 16 scores—middle is between 8th and 9th scores

Stem	Leaf
10	5 8
11	3 3 6 6 9
12	(4 4) 7
13	5 5 6 8 8
17	7

◯ Median = 124

10 a Rearranged numbers
4, 5, 6, 6, 7, 8, 8, 9
↑
Median = $\frac{6 + 7}{2}$ = 6.5
(Even number of scores)
Consider 4, 5, 6, 6
↑
Lower quartile = $\frac{5 + 6}{2}$ = 5.5
Consider 7, 8, 8, 9
↑
Upper quartile = 8
Interquartile range = 8 − 5.5
= 2.5
Lowest = 4
Highest = 9
Lower quartile = 5.5
Median = 6.5
Upper quartile = 8

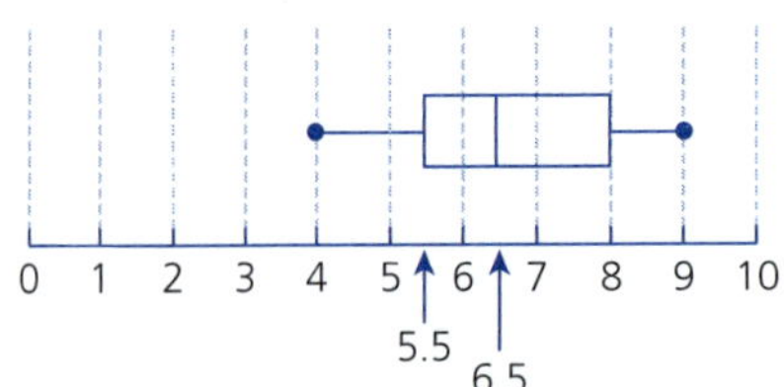

b 5, 6, 6, 6, 7, 7, 8, 8, 9, 9, 9
↑
Median = 7
(Odd number of scores)
Consider 5, 6, 6, 6, 7
↑
Lower quartile = 6
Consider 8, 8, 9, 9, 9
↑
Upper quartile = 9
Interquartile range = 9 − 6 = 3
Highest = 9
Lowest = 5
Lower quartile = 6
Median = 7
Upper quartile = 9

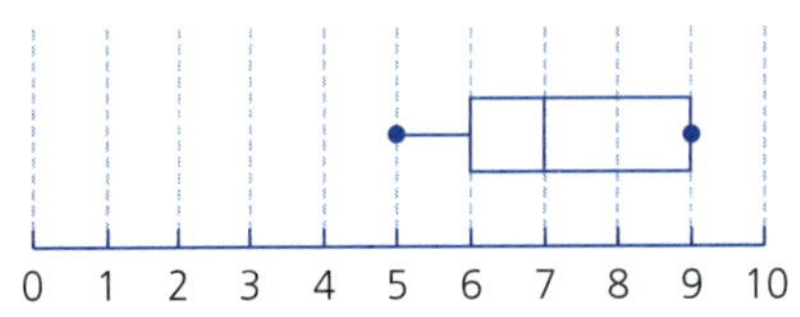

c 12, 14, 14, 15, 16, 16, 17, 18, 18, 19,
↑
Median = 16
Consider 12, 14, 14, 15, 16
↑
Lower quartile = 14
Consider 16, 17, 18, 18, 19
↑
Upper quartile = 18
Interquartile range = 18 − 14
= 4
Lowest score = 12
Highest score = 19
Median = 16
Lower quartile = 14
Upper quartile = 18

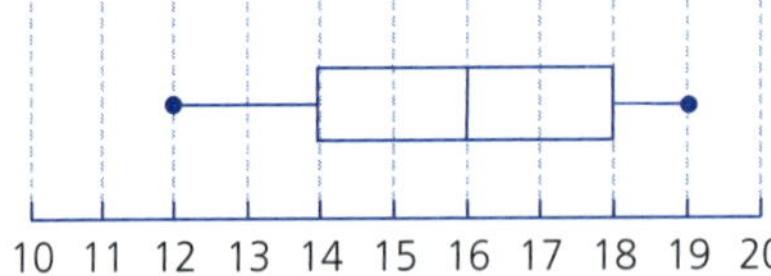

d 7, 7, 7, 8, 8, 9, 9, 10, 11, 11, 12, 12, 12, 13, 14
↑
Median = 10
Consider 7, 7, 7, 8, 8, 9, 9
↑
Lower quartile = 8
Consider 11, 11, 12, 12, 12, 13, 14
↑
Upper quartile = 12
Interquartile range = 12 − 8
= 4
Lowest score = 7
Highest score = 14
Median = 10
Lower quartile = 8
Upper quartile = 12

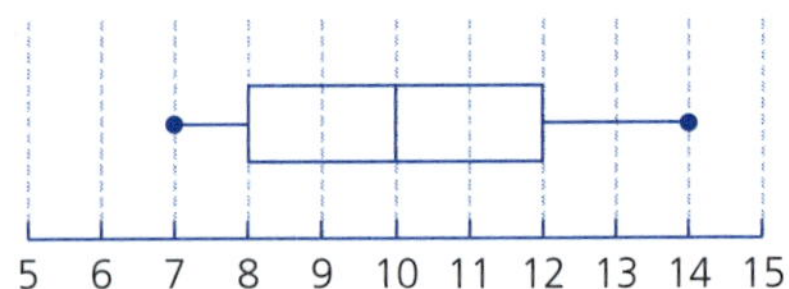

11 a i Median = 16
ii Median = 63

b i Interquartile range = 18 − 15
= 3
ii Interquartile range = 73 − 57
= 16

12 a 23 scores—middle is 12th score.

Stem	Leaf
1	0 7 8
2	1 2 [4] 6 9
3	7 7 7 (8) 8 9
4	0 0 6 [7] 8 8 9
5	1 4

◯ Median = 38
☐ Lower quartile = 24
☐ Upper quartile = 47
Lowest score = 10
Highest score = 54

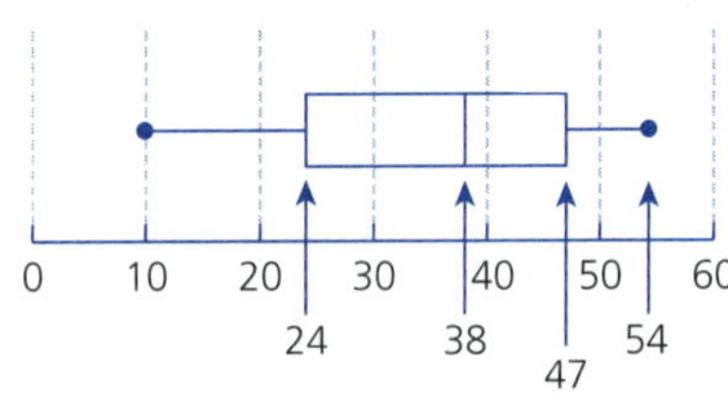

b 16 scores—middle is between 8th and 9th scores

Stem	Leaf
10	5 8
11	3 [3 6] 6 9
12	(4 4) 7
13	5 [5 6] 8 8
14	7

◯ Median = 124
☐ Lower quartile = $\frac{113 + 116}{2}$
= 114.5
☐ Upper quartile = $\frac{135 + 136}{2}$
= 135.5
Lowest score = 105
Highest score = 147

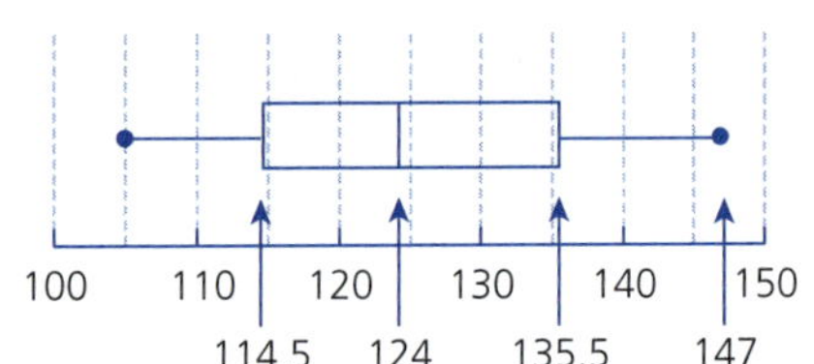

13 a Initial plot:

Stem	Leaf
14	9
15	0 0 4 9 8
16	1 1 7 9 7 8 8 6 2 0 9 6 4
17	2 4 0 5 4 2 0
18	0 2 2 1

Final plot:

Stem	Leaf
14	9
15	0 0 4 8 9
16	0 1 1 2 4 6 6 7 7 8 8 9 9
17	0 0 2 2 4 4 5
18	0 1 2 2

b 30 scores—middle is between 15th and 16th scores

Median = $\frac{167 + 168}{2}$ = 167.5

Lower quartile = 161

Upper quartile = 172

Interquartile range = 172 – 161
= 11

c Lowest score = 149

Highest score = 182

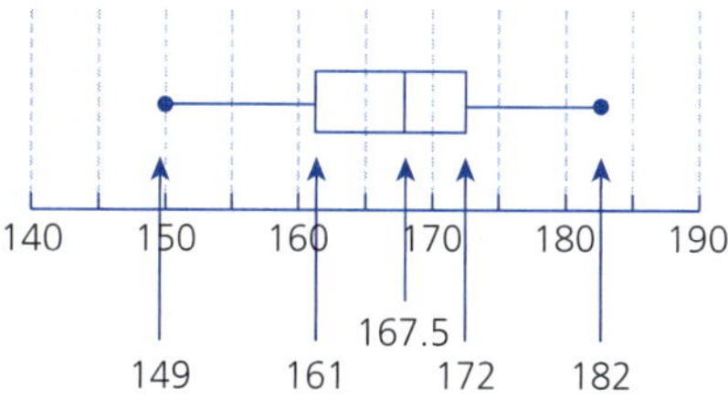

14 a Mean = 12

x	$x - \bar{x}$	$(x - \bar{x})^2$
14	2	4
10	–2	4
18	6	36
15	3	9
9	–3	9
12	0	0
6	–6	36
	Total:	98

b $\sigma = \sqrt{\frac{\Sigma(x - \bar{x})^2}{n}}$

$= \sqrt{\frac{98}{7}}$

= 3.74 (2 dec. pl.) $\therefore \sigma = 3.74$

15 a $\bar{x}$ = 19.5 σ = 3.82

b $\bar{x}$ = 104 σ = 4.46

c $\bar{x}$ = 1.64 σ = 0.0459

16 a Both sets have the same mean of 17. Set A has a lower standard deviation (3.16) than Set B (4.32)

b Both sets have the same standard deviation of 2.83. Set A has a mean (6) lower than Set B (21)

17 a The mean will increase by 2 and the standard deviation will remain the same.

b The mean will double and the standard deviation will double.

c The mean and will remain the same and the standard deviation will decrease.

18 Range = 28 – 19
= 9

19, 19, 23, 23, 25, 25, 28

Interquartile range = 25 – 19
= 6

Standard deviation = 3.04 (2 dec. pl.)

19 a Positively skewed

b Normally distributed (symmetrical)

c Positively skewed

d Negatively skewed

e Negatively skewed

f Symmetrical—without knowing mean and mode we cannot say that it is normally distributed

20 a

Score	Frequency
3	6
4	4
5	3
6	0
7	1
8	1

b Mean = $4\frac{4}{15}$

Mode = 3

Median = 4 (8th score)

c Positively skewed

21 a Negatively skewed

b Positively skewed

22 a Normally distributed

b Median = 8

Mode = 8

Mean = 8 (as the data is normally distributed, it will be the same as the median and mode)

23 The trigonometry test results are positively skewed and the algebra test results are negatively skewed.

24 a 9, 13, 15, 16, 16, 17, 17, 17, 18

Lowest mark: 9

Lower quartile: 14

Median: 16

Upper quartile: 17

Highest mark: 18

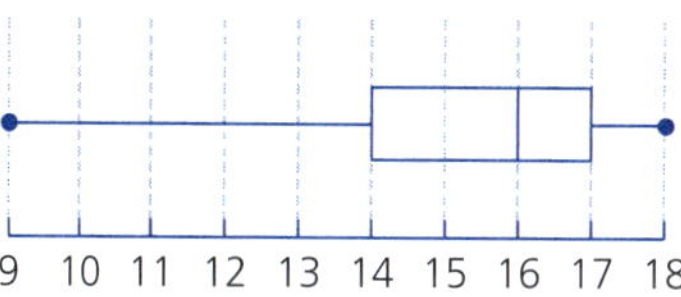

Negatively skewed

b 8, 9, 10, 10, 10, 11, 18

Lowest mark: 8

Lower quartile: 9

Median: 10

Upper quartile: 11

Highest mark: 18

Positively skewed

c 10, 12, 12, 14, 14, 14, 16, 16, 18

Lowest mark: 10

Lower quartile: 12

Median: 14

Upper quartile: 16

Highest mark: 18

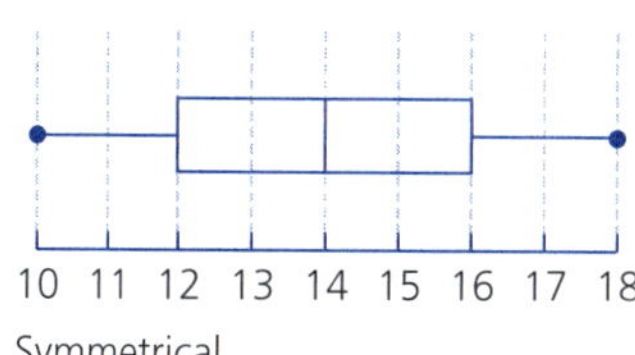

Symmetrical

25 a *r*

b *n*

c *s*

26 a Weak positive

b Strong negative

27 a

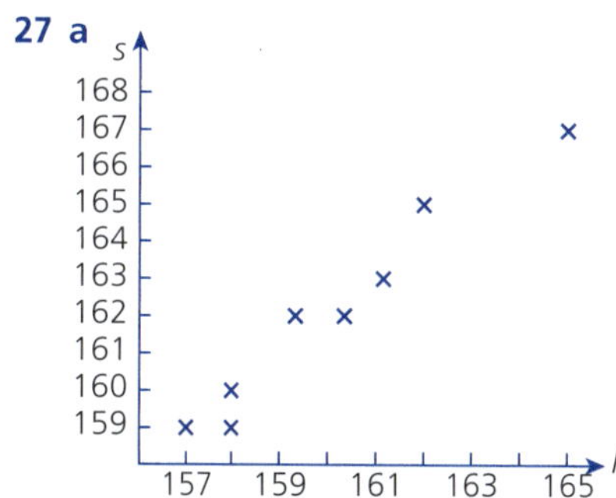

b Strong positive relationship

c About 166–167 cm

d The length of the arm span increases as the height increases.

28 a

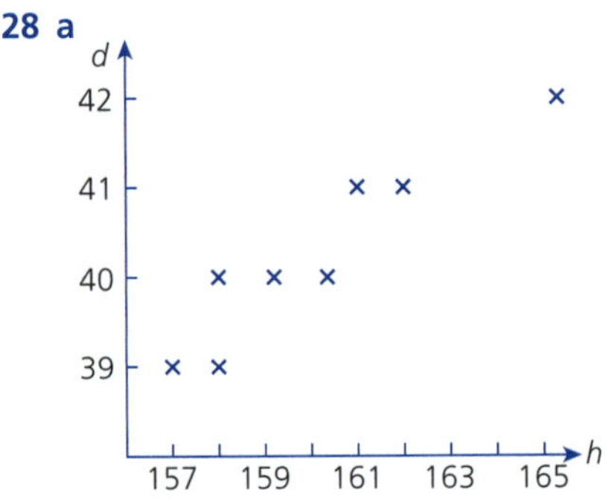

b About 41 cm

29 a

b Negative

c Strong

30 a The line is of the form $y = mx + b$ and passes through (12, 60) and (20, 100)

Gradient $m = \dfrac{100 - 60}{20 - 12}$

$= \dfrac{40}{8}$

$= 5$

$\therefore$ Equation of the form $y = 5x + b$

Now, subs (12, 60) in $y = 5x + b$

$60 = 5(12) + b$

$b = 60 - 60$

$= 0$

$\therefore$ The line is $y = 5x$.

b Subs. $y = 65$ in $y = 5x$

$65 = 5x$

$x = 13$

$\therefore$ The student scored 13 in the topic test.

31 a The line is of the form $y = mx + b$ and passes through (145, 45) and (170, 75).

Gradient $m = \dfrac{75 - 45}{170 - 145}$

$= \dfrac{30}{25}$

$= \dfrac{6}{5}$

$\therefore$ Equation of the form $y = \dfrac{6}{5}x + b$.

Now, subs. (145, 45) in $y = \dfrac{6}{5}x + b$

$45 = \dfrac{6}{5}(145) + b$

$b = 45 - 174$

$= -129$

$\therefore$ The line is $y = \dfrac{6}{5}x - 129$.

b **i** Subs. $x = 160$ in $y = \dfrac{6}{5}x - 129$

$= \dfrac{6}{5}(160) - 129$

$= 63$

$\therefore$ The mass is 63 kg.

ii Subs. $y = 81$ in $y = \dfrac{6}{5}x - 129$

$81 = \dfrac{6}{5}x - 129$

$\dfrac{6}{5}x = 210$

$x = 210 \div \dfrac{6}{5}$

$= 175$

$\therefore$ The student is 175 cm high.

Chapter 12 . . pp. 220–223

Probability

1 a $\frac{4}{52} = \frac{1}{13}$ **b** $\frac{4}{52} = \frac{1}{13}$

c $\frac{13}{52} = \frac{1}{4}$ **d** $\frac{13}{52} = \frac{1}{4}$

e $\frac{26}{52} = \frac{1}{2}$ **f** $\frac{12}{52} = \frac{3}{13}$

g $1 - \frac{3}{13} = \frac{10}{13}$

h Includes 2, 3, 4, 5, 6
[ace may be considered as 1]
$P = \frac{5}{13}$ (No Ace)
or $P = \frac{6}{13}$ (Ace is 1)

i $\frac{1}{52}$

j 26 red cards + 2 black 6s,
$P = \frac{28}{52} = \frac{7}{13}$

k $\frac{2}{52} = \frac{1}{26}$

2 $P(\text{no rain}) = 1 - \frac{7}{10} = \frac{3}{10}$

3 7R, 5B, 3W. Total no. = 15

a $\frac{7}{15}$ **b** $1 - \frac{7}{15} = \frac{8}{15}$

c $\frac{5}{15} = \frac{1}{3}$ **d** $\frac{10}{15} = \frac{2}{3}$

e 0 [There are no greens.]

4 a $\frac{1}{10}$ **b** $\frac{5}{10} = \frac{1}{2}$

c $\frac{6}{10} = \frac{3}{5}$ **d** $\frac{3}{10}$

5 Robyn needs a 50c coin.
$P(50c) = \frac{4}{10} = \frac{2}{5}$

6 Total = 200

a $\frac{36}{200} = \frac{9}{50}$

b As 20 + 14 = 34
then $\frac{34}{200} = \frac{17}{100}$

c As 200 − 112 = 88
then $\frac{88}{200} = \frac{11}{25}$

7 a Number = 1 + 3 + 5 + 8 + 10 + 4 + 2
= 33

b **i** $\frac{8}{33}$

ii As 1 + = 4
then $\frac{4}{33}$

iii As 10 + 4 + 2 = 16
then $\frac{16}{33}$

8 a $\frac{120}{360} = \frac{1}{3}$

b Missing ∠ is 30°
then $\frac{30}{360} = \frac{1}{12}$

c $\frac{130}{360} = \frac{13}{36}$

9

Colour	Frequency	Relative frequency
Red	48	0.48
Green	19	0.19
Blue	33	0.33
Total	100	1

10 a 10 + 6 + 4 = 20
$\therefore$ 20 students

b **i** $\frac{16}{20} = \frac{4}{5}$

ii $\frac{4}{20} = \frac{1}{5}$

iii $\frac{6}{20} = \frac{3}{10}$

11 a Sum = 30

b **i** As 4 + 5 + 2 + 7 = 18
then $\frac{18}{30} = \frac{3}{5}$

ii As 4 + 5 = 9
then $\frac{9}{30} = \frac{3}{10}$

iii $\frac{5}{30} = \frac{1}{6}$

iv As 5 + 3 + 4 = 12
then $\frac{12}{30} = \frac{2}{5}$

12 a

		Die					
		1	2	3	4	5	6
Coin	Head	H1	H2	H3	H4	H5	H6
	Tail	T1	T2	T3	T4	T5	T6

b **i** $P(\text{H4}) = \frac{1}{12}$

ii $P(\text{T1, T3, T5}) = \frac{3}{12} = \frac{1}{4}$

13 See table on page 214.

a $P(9) = \frac{4}{36} = \frac{1}{9}$

b $P(5) = \frac{4}{36} = \frac{1}{9}$

c $P(\text{odd}) = \frac{18}{36} = \frac{1}{2}$

d $P(> 9) = \frac{6}{36} = \frac{1}{6}$

e $P(\leq 9) = 1 - \frac{1}{6} = \frac{5}{6}$

f $P(\neq 9) = 1 - \frac{1}{9} = \frac{8}{9}$

g 6 less than 5 + 16 odd,
$P = \frac{22}{36} = \frac{11}{18}$

h $P = \frac{2}{36} = \frac{1}{18}$ [only the threes]

14 See table on page 215.

a $\frac{2}{36} = \frac{1}{18}$

b $\frac{9}{36} = \frac{1}{4}$

c $\frac{19}{36}$

d $\frac{23}{36}$

e $\frac{20}{36} = \frac{5}{9}$

f $\frac{5}{36}$

g $\frac{2}{3}$

15

	1	2	3	4	5	6
1	1	2	3	4	5	6
2	2	2	3	4	5	6
3	3	3	3	4	5	6
4	4	4	4	4	5	6
5	5	5	5	5	5	6
6	6	6	6	6	6	6

a $\frac{11}{36}$

b $\frac{1}{36}$

c $1 - \frac{11}{36} = \frac{25}{36}$

d $\frac{24}{36} = \frac{2}{3}$

e $\frac{21}{36} = \frac{7}{12}$

f $\frac{17}{36}$

g $\frac{3}{36} = \frac{1}{12}$

h $\frac{35}{36}$

i $\frac{16}{36} = \frac{4}{9}$

j 0

16 a

	Born in Australia	Born overseas	Total
Female	74	46	120
Male	56	24	80
Totals	130	70	200

b **i** $\frac{56}{200} = \frac{7}{25}$

ii $\frac{46}{200} = \frac{23}{100}$

c **i** $\frac{24}{80} = \frac{3}{10}$

ii $\frac{56}{80} = \frac{7}{10}$

17

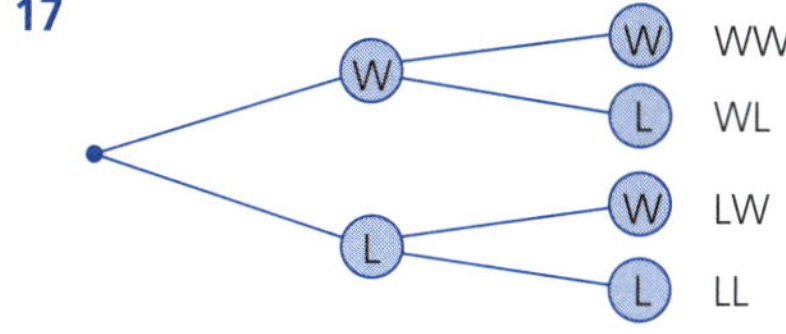

18

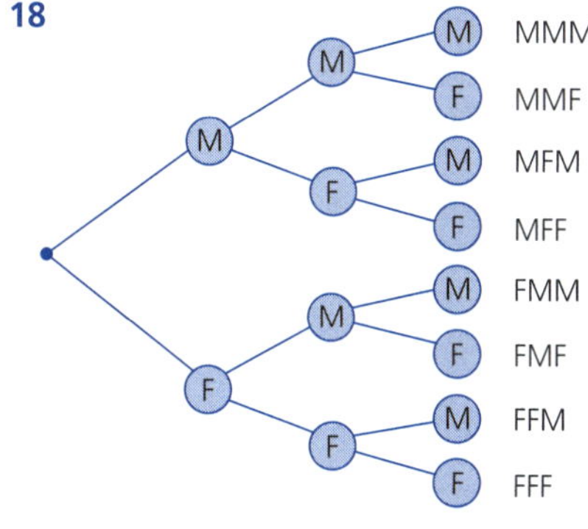

19

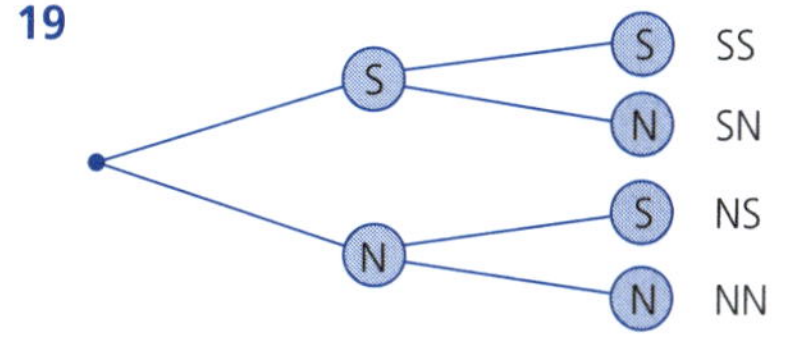

20

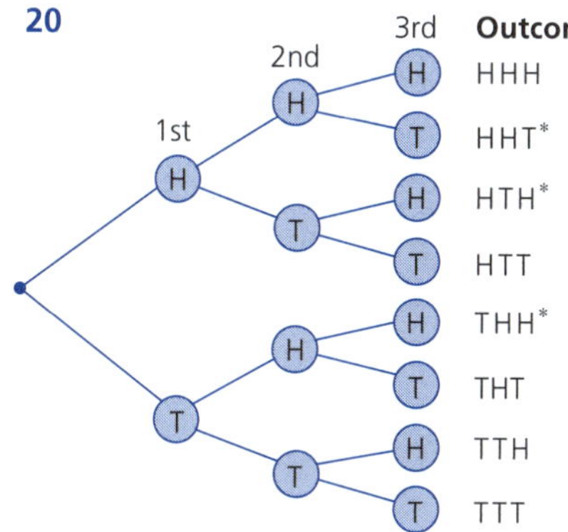

No. of outcomes = 8

a $P(\text{HHH}) = \frac{1}{8}$

b $P(2\text{H}) = \frac{3}{8}$ (marked*)

c $P(1\text{H}) = \frac{3}{8}$

d $P(\text{TTT}) = \frac{1}{8}$

e $P(\text{at least 1H}) = 1 - P(\text{TTT}) = \frac{7}{8}$

21

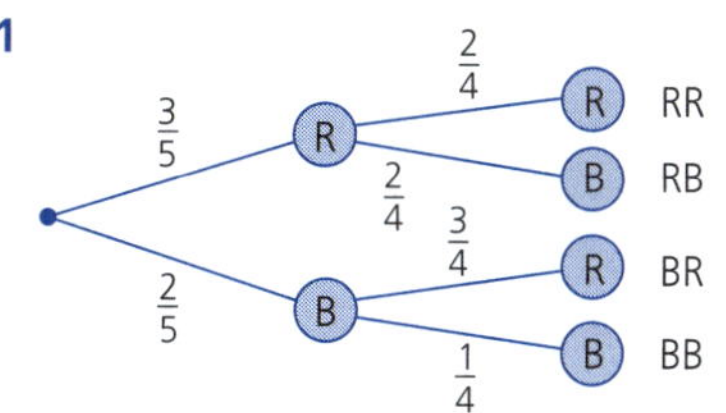

a $P(\text{RR}) = \frac{3}{5} \times \frac{2}{4} = \frac{3}{10}$

b $P(\text{BB}) = \frac{2}{5} \times \frac{1}{4} = \frac{1}{10}$

c $P(\text{RB or BR}) = \frac{3}{5} \times \frac{2}{4} + \frac{2}{5} \times \frac{3}{4} = \frac{3}{5}$

d $P(\text{RR or BB}) = \frac{3}{10} + \frac{1}{10} = \frac{2}{5}$

22

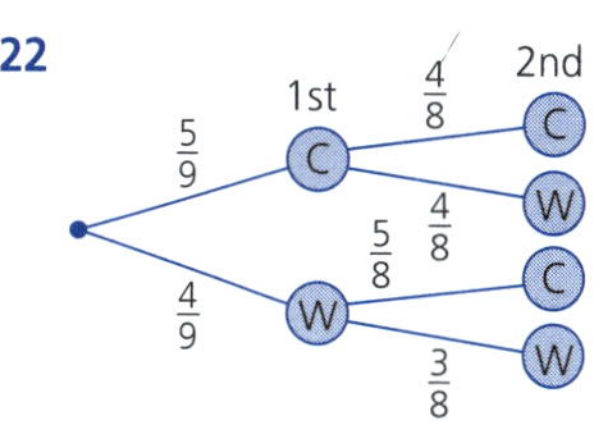

Outcome	Probability
CC	$P(\text{CC}) = \frac{5}{9} \times \frac{4}{8} = \frac{5}{18}$
CW	$P(\text{CW}) = \frac{5}{9} \times \frac{4}{8} = \frac{5}{18}$
WC	$P(\text{WC}) = \frac{4}{9} \times \frac{5}{8} = \frac{5}{18}$
WW	$P(\text{WW}) = \frac{4}{9} \times \frac{3}{18} = \frac{1}{6}$

From the table:

a $P(\text{CC}) = \frac{5}{18}$

b $P(\text{one C}) = P(\text{WC}) + P(\text{CW})$
$= \frac{5}{18} + \frac{5}{18}$
$= \frac{5}{9}$

c $P(\text{CW}) = \frac{5}{18}$

d $P(\text{WW}) = \frac{1}{6}$

e $P(\text{at least one C}) = 1 - P(\text{WW})$
$= 1 - \frac{1}{6} = \frac{5}{6}$

23

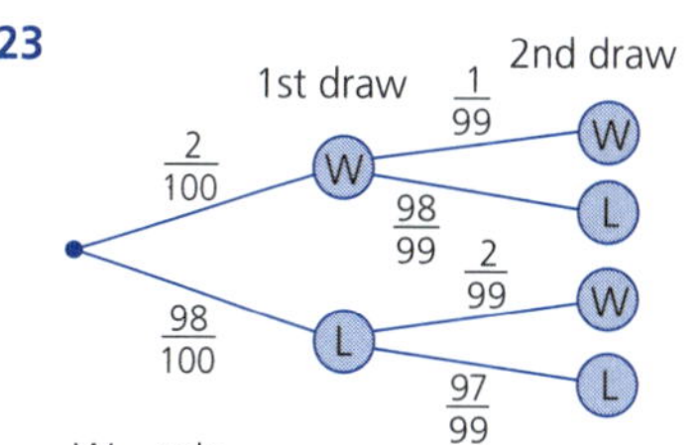

W = win
L = not win

Outcome	Probability
WW	$P(\text{WW}) = \frac{2}{100} \times \frac{1}{99} = \frac{1}{4950}$
WL	$P(\text{WL}) = \frac{2}{100} \times \frac{98}{99} = \frac{98}{4950}$
LW	$P(\text{LW}) = \frac{98}{100} \times \frac{2}{99} = \frac{98}{4950}$
LL	$P(\text{LL}) = \frac{98}{100} \times \frac{97}{99} = \frac{4753}{4950}$

From the diagram:

a $P(\text{first prize}) = \frac{2}{100} = \frac{1}{50}$

b $P(\text{WW}) = \frac{1}{4950}$

c $P(\text{LW}) = \frac{98}{4950} = \frac{49}{2475}$

d $P(\text{a prize}) = P(\text{WW}) + P(\text{WL}) + P(\text{LW})$
or
$1 - P(\text{LL})$
$= 1 - \frac{4753}{4950} = \frac{197}{4950}$

e $P(\text{LL}) = \frac{4753}{4950}$

f $\frac{197}{4950}$ [same as **d**]

24 a $P(4, \text{given even}) = \frac{1}{3}$

b $P(\text{less than } 4, \text{given even}) = \frac{2}{3}$

25 a $P(\text{vowel}) = \frac{5}{26}$

b $P(\text{E, given vowel}) = \frac{1}{5}$

c $P(\text{P, given consonant}) = \frac{1}{21}$

26 a $P(\text{13- or 17-year-old female}) = \frac{12}{50} = \frac{6}{25}$

b $P(\text{18 and over, given female}) = \frac{10}{27}$

c $P(\text{male, given 12 and under}) = \frac{8}{13}$

27 a

	Low	Normal	High	Totals
Men	16	28	26	70
Women	24	22	14	60
Totals	40	50	40	130

b **i** $\frac{28}{130} = \frac{14}{65}$

ii $\frac{24}{130} = \frac{12}{65}$

c **i** $\frac{22}{60} = \frac{11}{30}$

ii As $22 + 14 = 36$,
then $\frac{36}{60} = \frac{3}{5}$

d $\frac{14}{40} = \frac{7}{20}$

28 a $P(\text{diamond, given a red}) = \frac{1}{2}$

b $P(\text{black, given a 7}) = \frac{1}{2}$

c $P(\text{queen, of hearts, given a red queen}) = \frac{1}{2}$

29 a $P(\text{coffee}) = \frac{11}{23}$

b $P(\text{pastry, given a tea drinker}) = \frac{4}{7}$

c $P(\text{coffee drinker, given a pastry eater}) = \frac{6}{12} = \frac{1}{2}$

30 a

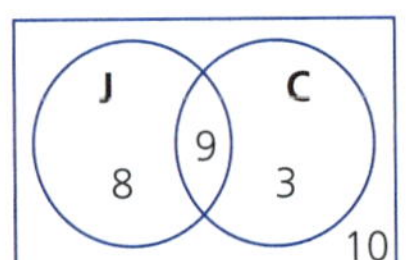

J: Jersey
C: Cap

b $\frac{9}{17}$

c $\frac{8}{18} = \frac{4}{9}$

Chapter 13 . . pp. 237–238

Polynomials

1 a $P(x) = 3x^4 - 7x + 2$
deg: 4, lead t: $3x^4$, lead $c = 3$
const t: 2

b $P(x) = 4 - 2x^5$
deg: 5, lead t: $-2x^5$, lead $c = -2$
const t: 4

c $P(x) = 3x^5 - 2x^3 + 4x$
deg: 5, lead t: $3x^5$, lead $c = 3$
const t: 0

d $P(x) = 7$
deg: 0, lead t: 7, lead $c = 7$
const t: 7

2 a $\frac{4}{x} - 3x + 2 = 4x^{-1} - 3x + 2$
$\therefore$ Not polynomial [power of -1]

b $\frac{5x^4 - x}{x} = 5x^3 - 1$
$\therefore$ Polynomial

c $\frac{3x}{4} + 2 = \frac{3}{4}x + 2$
$\therefore$ Polynomial

d $4x^4 + \sqrt{x} - 2 = 4x^4 + x^{\frac{1}{2}} - 2$
$\therefore$ Not polynomial

e $(x + 4)^2 - 2x + 3$
$= x^2 + 8x + 16 - 2x + 3$
$= x^2 + 6x + 19$
$\therefore$ Polynomial

3 a
$$\begin{array}{r} 4x^4 - 2x^2 + 3 \quad + \\ 3x^4 - 5x^2 - 7 \\ \hline 7x^4 - 7x^2 - 4 \end{array}$$

b
$$\begin{array}{r} 3x^2 - 4x + 1 \quad + \\ 5x^2 + 2x - 5 \\ \hline 8x^2 - 2x - 4 \end{array}$$

c
$$\begin{array}{r} 4x^3 + 12x^2 + 3x + 7 \quad + \\ 5x^3 + 0x^2 + x + 3 \\ \hline 9x^3 + 12x^2 + 4x + 10 \end{array}$$

d
$$\begin{array}{r} 4x^5 - 21x + 2 \quad + \\ 3x^5 + 2x - 1 \\ \hline 7x^5 - 19x + 1 \end{array}$$

e
$$\begin{array}{r} 5x^3 + 2x^2 - x - 1 \quad - \\ 3x^3 - x^2 - x - 5 \\ \hline 2x^3 + 3x^2 \quad + 4 \end{array}$$

f
$$\begin{array}{r} 4x^4 - 2x^2 + 1 \quad - \\ x^4 - 5x^2 - 7 \\ \hline 3x^4 + 3x^2 + 8 \end{array}$$

g
$$\begin{array}{r} 4x^3 + 0x^2 - x + 2 \quad - \\ 3x^3 + 2x^2 + 0x - 1 \\ \hline x^3 - 2x^2 - x + 3 \end{array}$$

h
$$\begin{array}{r} 5x^4 + 0x^3 - 2x^2 + 0x + 1 \quad - \\ 3x^3 + x^2 - 5x + 2 \\ \hline 5x^4 - 3x^3 - 3x^2 + 5x - 1 \end{array}$$

4 a $P(x) \cdot Q(x)$
$= (3x^2 - 4x + 1)(4 - x + x^2)$
$= \ldots + 3x^4 + \ldots$
$\therefore$ Degree is 4.
[only need leading term]

b $P(x) \cdot Q(x)$
$= (5x^4 - 2x - 1)(3x - 5x^2 + x^3)$
$= \ldots + 5x^7 + \ldots$
$\therefore$ Degree is 7.

5 a $(2x^3 - 4)(5x^2 - x + 2)$
$= 10x^5 - 2x^4 + 4x^3 - 20x^2 + 4x - 8$

b $(6x^2 - 4x)(x^3 + x - 1)$
$= 6x^5 + 6x^3 - 6x^2 - 4x^4 - 4x^2 + 4x$
$= 6x^5 - 4x^4 + 6x^3 - 10x^2 + 4x$

6 a
$$\begin{array}{r} x^2 \quad + 7 \\ x - 3 \overline{) x^3 - 3x^2 + 7x - 5} \\ x^3 - 3x^2 \qquad\qquad \\ \hline 0 \quad + 7x - 5 \\ 7x - 21 \\ \hline 16 \end{array}$$
$\therefore x^3 - 3x^2 + 7x - 5$
$= (x - 3)(x^2 + 7) + 16$

b
$$\begin{array}{r} -x^2 + 4x - 13 \\ x + 2 \overline{) -x^3 + 2x^2 - 5x - 1} \\ -x^3 - 2x^2 \qquad\qquad \\ \hline 4x^2 - 5x \qquad \\ 4x^2 + 8x - 26 \\ \hline -13x - 1 \\ -13x - 26 \\ \hline 25 \end{array}$$
$\therefore -x^3 + 2x^2 - 5x - 1$
$= (x + 2)(-x^2 + 4x - 13) + 25$

c
$$\begin{array}{r} x^3 - x^2 + 2x - 2 \\ x^2 + x \overline{) x^5 + 0x^4 + x^3 + 0x^2 - x} \\ x^5 + x^4 \qquad\qquad\qquad \\ \hline -x^4 + x^3 \qquad\qquad \\ -x^4 - x^3 \qquad\qquad \\ \hline 2x^3 + 0x^2 \qquad \\ 2x^3 + 2x^2 \qquad \\ \hline -2x^2 - x \\ -2x^2 - 2x \\ \hline x \end{array}$$
$\therefore x^5 + x^3 - x$
$= (x^2 + x)(x^3 - x^2 + 2x - 2) + x$

d $(x^4 - 1) \div (x^2 - 3)$
$$\begin{array}{r} x^2 \qquad + 3 \\ x^2 - 3 \overline{) x^4 + 0x^3 + 0x^2 + 0x - 1} \\ x^4 \qquad - 3x^2 \qquad\qquad \\ \hline 3x^2 \qquad - 1 \\ 3x^2 \qquad - 9 \\ \hline 8 \end{array}$$
$\therefore x^4 - 1 = (x^2 - 3)(x^2 + 3) + 8$

7 $P(x) = x^3 + 2x^2 - x + 1$
$Q(x) = x - 1$

a $P(x) + Q(x)$
$$\begin{array}{r} \therefore x^3 + 2x^2 - x + 1 \quad + \\ x - 1 \\ \hline x^3 + 2x^2 \qquad\qquad \end{array}$$
$\therefore P(x) + Q(x) = x^3 + 2x^2$

b $P(x) - Q(x)$
$$\begin{array}{r} \therefore x^3 + 2x^2 - x + 1 \quad - \\ x - 1 \\ \hline x^3 + 2x^2 - 2x + 2 \end{array}$$
$\therefore P(x) - Q(x) = x^3 + 2x^2 - 2x + 2$

c $P(x) \cdot Q(x)$
$= (x^3 + 2x^2 - x + 1)(x - 1)$
$= x^4 - x^3 + 2x^3 - 2x^2 - x^2 + x + x - 1$
$= x^4 + x^3 - 3x^2 + 2x - 1$

d $\dfrac{P(x)}{Q(x)}$

$$\begin{array}{r} x^2 + 3x + 2 \\ \therefore\ x - 1 \overline{\big)\, x^3 + 2x^2 - x + 1} \\ \underline{x^3 - x^2} \\ 3x^2 - x \\ \underline{3x^2 - 3x} \\ 2x + 1 \\ \underline{2x - 2} \\ 3 \end{array}$$

$\therefore\ P(x) \div Q(x) = (x^2 + 3x + 2) + \dfrac{3}{x - 1}$

8 a Let $P(x) = x^3 + 4x^2 + x - 1$
Dividing by $(x + 2)$, $\therefore$ find $P(-2)$
$\therefore\ P(-2) = (-2)^3 + 4(-2)^2 + (-2) - 1$
$= -8 + 16 - 2 - 1$
$= 5$
$\therefore$ Remainder is 5.

b Let $P(x) = x^3 - 4x + 2$
Dividing by $(x - 1)$, $\therefore$ find $P(1)$
$\therefore\ P(1) = 1^3 - 4(1) + 2$
$= 1 - 4 + 2$
$= -1$
$\therefore$ Remainder is -1.

9 Let $P(x) = x^3 + 3x^2 - 7x + k$
Dividing by $(x + 2)$, $\therefore\ P(-2) = 5$
$\therefore\ P(-2) = (-2)^3 + 3(-2)^2 - 7(-2) + k = 5$
$-8 + 12 + 14 + k = 5$
$18 + k = 5$
$k = 5 - 18$
$k = -13$
$\therefore$ Value of k is -13.

10 Let $P(x) = 3x^4 - 2x^3 + 2k$
Dividing by $(x + 1)$, $\therefore\ P(-1) = 3$
$\therefore\ P(-1) = 3(-1)^4 - 2(-1)^3 + 2k = 3$
$3 + 2 + 2k = 3$
$5 + 2k = 3$
$2k = 3 - 5$
$2k = -2$
$k = -1$
$\therefore$ Value of k is -1.

11 Let $P(x) = x^2 - 7x + 9$
Dividing by $(x - k)$, $\therefore\ P(k) = 3$
$\therefore\ P(k) = k^2 - 7k + 9 = 3$,
That is $k^2 - 7k + 6 = 0$
$(k - 6)(k - 1) = 0$
$k = 6$ or 1
$\therefore$ k is 6 or 1.

12 Let $P(x) = x^3 + kx - 10$
Dividing by $(x - 5)$, $\therefore\ P(5) = 0$
$\therefore\ P(5) = 5^3 + 5k - 10 = 0$
$125 + 5k - 10 = 0$
$5k + 115 = 0$
$\dfrac{5k}{5} = -\dfrac{115}{5}$
$k = -23$
$\therefore$ Value of k is -23.

13 Let $P(x) = x^3 - 2x^2 - 5x + 6$
If $(x - 1)$ is a factor, then $P(1) = 0$
$\therefore\ P(1) = 1^3 - 2(1)^2 - 5(1) + 6$
$= 1 - 2 - 5 + 6$
$= 0$
$\therefore$ As $P(1) = 0$, then $(x - 1)$ is a factor of $P(x)$.

$$\begin{array}{r} x^2 - x - 6 \\ x - 1 \overline{\big)\, x^3 - 2x^2 - 5x + 6} \\ \underline{x^3 - x^2} \\ -\ x^2 - 5x \\ \underline{x^2 + x} \\ -\ 6x + 6 \\ \underline{-\ 6x + 6} \\ 0 \end{array}$$

$\therefore$ As $x^3 - 2x^2 - 5x + 6 = 0$
$\therefore\quad (x - 1)(x^2 - x - 6) = 0$
$(x - 1)(x - 3)(x + 2) = 0$
$\therefore\quad x = 1, 3, -2$
$\therefore$ The roots of the equation are $x = 1$, 3 and -2.

14 Let $P(x) = 2x^3 - 5x^2 - 4x + 3$
Now find a, where $P(a) = 0$, that is, try
$P(1) = 2(1)^3 - 5(1)^2 - 4(1) + 3$
$= 2 - 5 - 4 + 3$
$\neq 0$
Try
$P(-1) = 2(-1)^3 - 5(-1)^2 - 4(-1) + 3$
$= -2 - 5 + 4 + 3$
$= 0$
$\therefore\ P(-1) = 0$, $\therefore\ (x + 1)$ is a factor.

$$\begin{array}{r} 2x^2 - 7x + 3 \\ x + 1 \overline{\big)\, 2x^3 - 5x^2 - 4x + 3} \\ \underline{2x^3 + 2x^2} \\ -7x^2 - 4x \\ \underline{-7x^2 - 7x} \\ 3x + 3 \\ \underline{3x + 3} \\ 0 \end{array}$$

$\therefore\ 2x^3 - 5x^2 - 4x + 3$
$= (x + 1)(2x^2 - 7x + 3)$
$= (x + 1)(2x - 1)(x - 3)$

15 Let $P(x) = x^3 + 2x^2 - 9x - 18$
Try $P(1) = 1^3 + 2(1)^2 - 9(1) - 18$
$\neq 0$
$P(-1) = (-1)^3 + 2(-1)^2 - 9(-1) - 18$
$\neq 0$
$P(2) = (2)^3 + 2(2)^2 - 9(2) - 18$
$= 8 + 8 - 18 - 18$
$\neq 0$
$P(-2) = (-2)^3 + 2(-2)^2 - 9(-2) - 18$
$= -8 + 8 + 18 - 18$
$= 0$
$\therefore$ As $P(-2) = 0$, $(x + 2)$ is a factor:

$$\begin{array}{r} x^2 - 9 \\ \therefore\ x + 2 \overline{\big)\, x^3 + 2x^2 - 9x - 18} \\ \underline{x^3 + 2x^2} \\ -\ 9x - 18 \\ \underline{-\ 9x - 18} \\ 0 \end{array}$$

As $x^3 + 2x^2 - 9x - 18 = 0$
$(x + 2)(x^2 - 9) = 0$
$(x + 2)(x - 3)(x + 3) = 0$
$\therefore\ x = -2, 3, -3$
Note: $P(x)$ could have been factorised using grouping: $x^2(x + 2) - 9(x + 2)$ $= (x^2 - 9)(x + 2)$ and so on.

16 Let $P(x) = x^3 - 2x^2 - ax + b$
As $x = -3$ is a root, $P(-3) = 0$
$\therefore\ P(-3) = (-3)^3 - 2(-3)^2 - a(-3) + b$
$= 0$
$-27 - 18 + 3a + b = 0$
$-45 + 3a + b = 0$
$\therefore\quad 3a + b = 45.$
Also, as $x = 1$ is a root, $P(1) = 0$.
$\therefore\ P(1) = (1)^3 - 2(1)^2 - a(1) + b$
$= 0$
$1 - 2 - a + b = 0$
$-1 - a + b = 0$
$\therefore\quad a - b = -1$
$\therefore\quad 3a + b = 45 \qquad (1)$
$a - b = -1 \qquad (2)$
(1) + (2)
$4a = 44$
$a = 11$
Substitute $a = 11$ in (2)
$11 - b = -1$
$-b = -1 - 11$
$-b = -12$
$\therefore\quad b = 12,$
$\therefore\ a = 11, b = 12,$
$\therefore\ P(x) = x^3 - 2x^2 - 11x + 12.$
Now as $x = -3$, $x = 1$ are roots,
$\therefore\ (x + 3), (x - 1)$ are factors,
$\therefore\ x^2 - 2x^2 - 11x + 12$
$= (x + 3)(x - 1)(x - a)$
$\therefore\ a = 4$ (by $-3 \times -1 \times 4 = 12$)
$\therefore\ (x + 3)(x - 1)(x - 4)$
$\therefore$ as $P(x) = 0$
$(x + 3)(x - 1)(x - 4) = 0,$
$\therefore\ x = -3, 1, 4$
$\therefore$ The other root is $x = 4$.

17 a $y = (x - 4)(x - 1)(x + 2)$

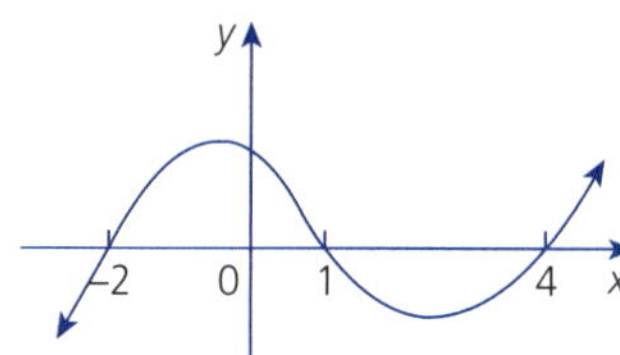

b $y = x(x - 2)^2(4 - x)$

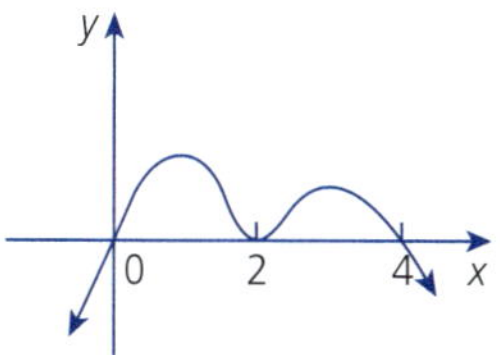

c $y = (3 - x)(x + 1)^3(4 - x)$

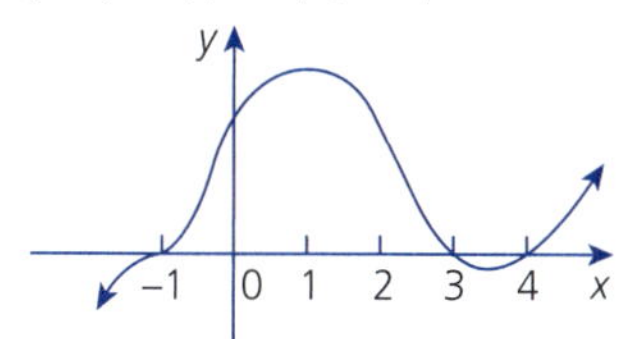

d $y = x^2(1 - x)^2$

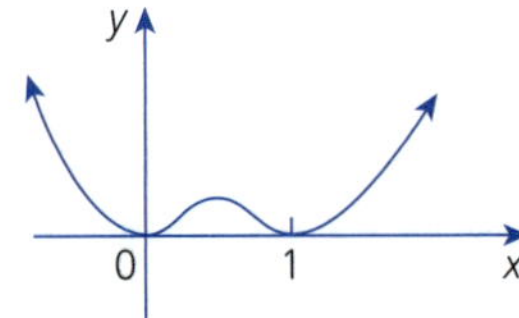

e $y = (1 + x)^3$

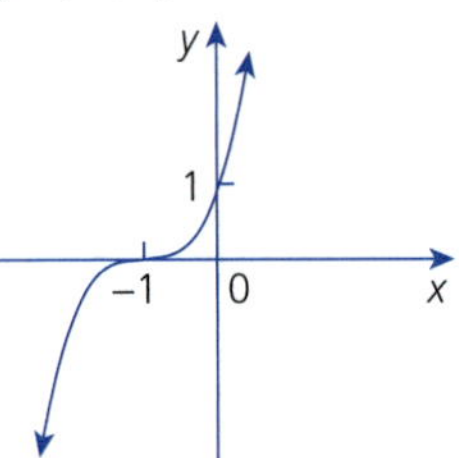

18 a $y = x^3 - 4x^2 + x + 6$

Let $y = P(x)$.

$\therefore P(x) = x^3 - 4x^2 + x + 6.$

Try $P(1) = 1^3 - 4(1)^2 + 1 + 6$
$= 1 - 4 + 1 + 6$
$\neq 0$

$P(-1) = (-1)^3 - 4(-1)^2 - 1 + 6$
$= -1 - 4 - 1 + 6$
$= 0$

$\therefore$ As $P(-1) = 0$, $x + 1$ is a factor,

$$\begin{array}{r} x^2 - 5x + 6 \\ \therefore x + 1\overline{)x^3 - 4x^2 + x + 6} \\ \underline{x^3 + x^2} \\ -5x^2 + x \\ \underline{-5x^2 - 5x} \\ 6x + 6 \\ \underline{6x + 6} \\ 0 \end{array}$$

$\therefore P(x) = x^3 - 4x^2 + x + 6$
$= (x + 1)(x^2 - 5x + 6)$
$= (x + 1)(x - 3)(x - 2)$

$\therefore y = (x + 1)(x - 3)(x - 2)$

Substitute $x = 4$ in y

That is, $y(4) = (4 + 1)(4 - 3)(4 - 1)$
$= (5)(1)(3) > 0$

$\therefore$ Above the x-axis at $x = 4$.

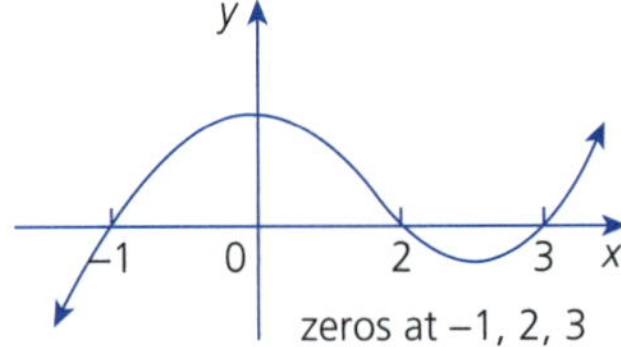

b $y = x^3 - x^2 - x + 1$
$y = x(x^2 - 1) - 1(x - 1)$
$= (x - 1)(x^2 - 1)$
$= (x - 1)(x - 1)(x + 1)$

$\therefore y = (x - 1)^2(x + 1).$

Substitute $x = 2$ in y

That is, $y(2) = (2 - 1)^2(2 + 1)$
$= (1)^2(3) > 0$

$\therefore$ Above x-axis at $x = 4$

That is, $y > 0$ when $x = 4$

(That is, $y > 0$) for $x > 1$.

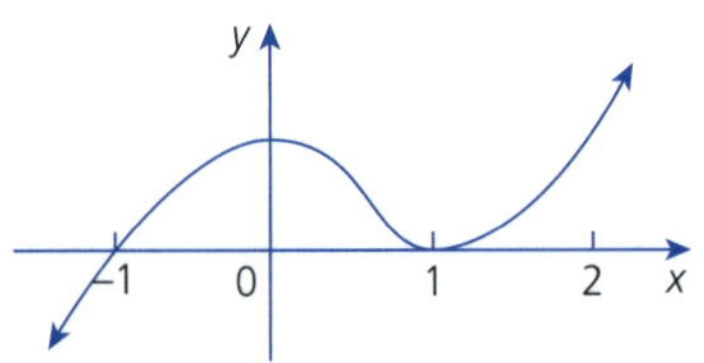

zeros of −1, 1, 2

c $y = x^3 - 3x^2 + 3x - 1$

Let $y = P(x)$.

$\therefore P(x) = x^3 - 3x^2 + 3x - 1$

Try $P(1) = 1^3 - 3(1)^2 + 3(1) - 1$
$= 1 - 3 + 3 - 1$
$= 0$

$\therefore$ As $P(1) = 0$, $(x - 1)$ is a factor,

$$\begin{array}{r} x^2 - 2x + 1 \\ \therefore x - 1\overline{)x^3 - 3x^2 + 3x - 1} \\ \underline{x^3 - x^2} \\ -2x^2 + 3x \\ \underline{-2x^2 + 2x} \\ x - 1 \\ \underline{x - 1} \\ 0 \end{array}$$

$\therefore$ As $P(x) = x^3 - 3x^2 + 3x - 1$
$= (x - 1)(x^2 - 2x + 1)$
$= (x - 1)(x - 1)(x - 1)$
$= (x - 1)^3$

$\therefore y = (x - 1)^3$

Substitute $x = 2$ in y, that is,

$y(2) = (2 - 1)^3 = 1^3 > 0,$

$\therefore$ Above x-axis at $x = 2$.

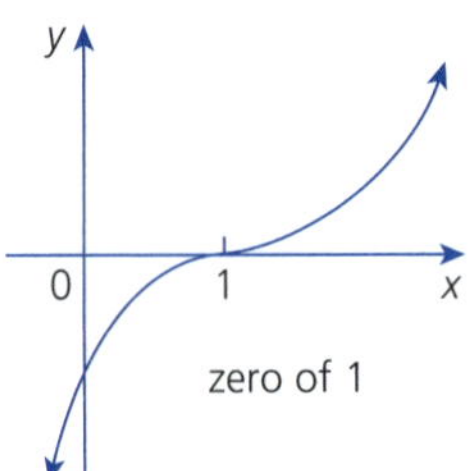

d $y = -x^3 + 12x + 16$

Let $y = P(x)$.

$\therefore P(x) = -x^3 + 12x + 16$

Try $P(1) = -1^3 + 12(1) + 16$
$\neq 0$

$P(-1) = -(-1)^3 + 12(-1) + 16$
$\neq 0$

$P(2) = -2^3 + 12(2) + 16$
$= -8 + 24 + 16$
$\neq 0$

$P(-2) = (-2)^3 + 12(-2) + 16$
$= 8 + 24 + 16$
$= 0$

$\therefore$ As $P(-2) = 0$, $(x + 2)$ is a factor,

$$\begin{array}{r} -x^2 + 2x + 8 \\ \therefore x + 2\overline{)-x^3 + 0x^2 + 12x + 16} \\ \underline{-x^3 + 2x^2} \\ 2x^2 + 12x \\ \underline{2x^2 + 4x} \\ 8x + 16 \\ \underline{8x + 16} \\ 0 \end{array}$$

$\therefore$ As $P(x) = -x^3 + 12x + 16$
$= (x + 2)(-x^2 + 2x + 8)$
$= -(x + 2)(x^2 - 2x - 8)$

$\therefore y = -(x + 2)(x + 2)(x - 4)$
$= -(x + 2)^2(x - 4)$

$\therefore$ Zeros at −2, 4.

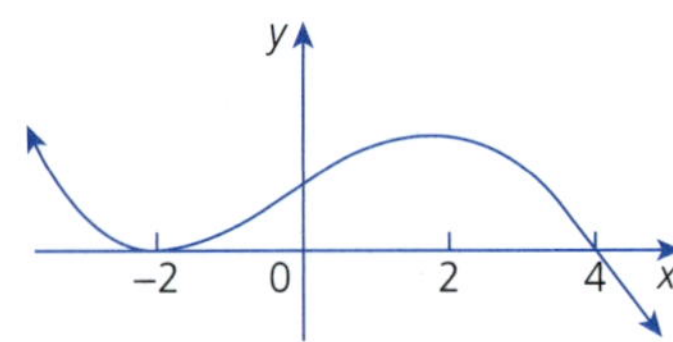

19 a iii

b ii

c iv

d i

Chapter 14 . . pp. 257–259

Functions and Logarithms

1 a Function

b Not function

c Not function

d Function

e Function

f Not function

2 $f(x) = 6x - 11$

a $f(0) = 6(0) - 11 = -11$

b $f(2) = 6(2) - 11 = 1$

c $f(-2) = 6(-2) - 11 = -23$

d $f(5) = 6(5) - 11 = 19$

3 $f(x) = x^2 - 4x$

a $f(0) = 0^2 - 4(0)$
$= 0$

b $f(1) = 1^2 - 4(1)$
$= 1 - 4$
$= -3$

c $f(-1) = (-1)^2 - 4(-1)$
$= 1 + 4$
$= 5$

d $f(3) = 3^2 - 4(3)$
$= 9 - 12$
$= -3$

4 a $f(t) = 1 - t^2$
$f(0) = 1 - 0 = 1$
$f(2) = 1 - (2)^2 = -3$
$f(-2) = 1 - (-2)^2 = -3$

b $f(t) = \dfrac{1+t}{1-t}$

$f(0) = 1$

$f(2) = \dfrac{1+2}{1-2} = -3$

$f(-2) = \dfrac{1+(-2)}{1-(-2)} = \dfrac{-1}{3}$

c $f(t) = t^2 - t^3$

$f(0) = 0$

$f(2) = 2^2 - 2^3 = -4$

$f(-2) = (-2)^2 - (-2)^3$

$= 4 - (-8) = 12$

d $f(t) = (t-3)^2$

$f(0) = (-3)^2 = 9$

$f(2) = (2-3)^2$

$= (-1)^2 = 1$

$f(-2) = (-2-3)^2$

$= (-5)^2 = 25$

e $f(t) = (3-t)^2$

$f(0) = (3)^2 = 9$

$f(2) = (3-2)^2 = 1$

$f(-2) = (3 - -2)^2$

$= 5^2 = 25$

f $f(t) = 3^t$

$f(0) = 3^0 = 1$

$f(2) = 3^2 = 9$

$f(-2) = 3^{-2} = \dfrac{1}{3^2} = \dfrac{1}{9}$

5 $\phi(x) = x^2 + x$

a $\phi(h) = h^2 + h$

b $\phi(x+2) = (x+2)^2 + (x+2)$

$= x^2 + 4x + 4 + x + 2$

$= x^2 + 5x + 6$

c $\phi(x+2) - \phi(2)$

$= (x+2)^2 - (2^2 + 2)$

$= x^2 + 4x + 4 - 6$

$= x^2 + 4x - 2$

d $\phi(-x) = (-x)^2 + (-x)$

$= x^2 - x$

6 a $g(x) = x^4$

$g(-x) = (-x)^4$

$\therefore g(x) = g(-x)$

$(-x) \times (-x) \times (-x) \times (-x) = x^4$

[A function is called an *even* function when $f(x) = f(-x)$.]

b $H(x) = x^3$

$(-x) \times (-x) \times (-x) = x^3$

$H(-x) = (-x)^3$

$= -x^3$

$\therefore H(-x) = -H(x)$

$H(x) = x^3$

[A function is called an *odd* function when $f(-x) = -f(x)$.]

7 a Domain: $x \geq 0$

Range: $y \geq 0$

b Domain: all values of x can be inputs

Range: $y \geq 1$

c Domain: all values of x can be inputs

Range: $y \leq 4$

d Domain: all values of x from -2 to 2 are inputs, i.e. $-2 \leq x \leq 2$.

Range: all values from 0 to 2 may be outputs, that is, $0 \leq y \leq 2$.

8 a

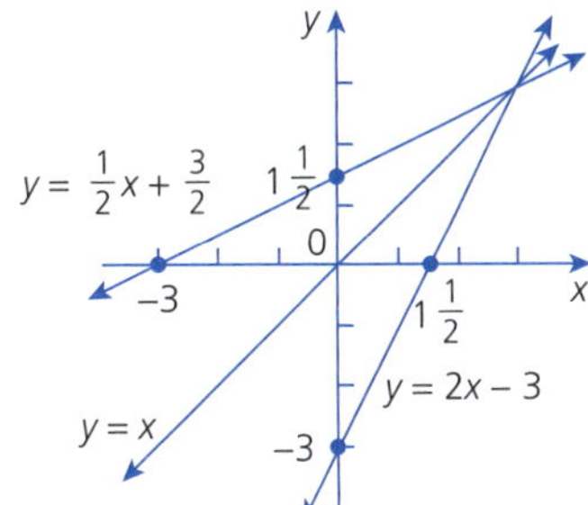

b $y = 2x - 3$

$x = 2y - 3$ ← Interchange x and y

$\therefore 2y = x + 3$

$y = \dfrac{1}{2}x + \dfrac{3}{2}$ ← Make y the subject

Inverse function is $y = \dfrac{1}{2}x + \dfrac{3}{2}$.

c From the diagram it can be seen that the two lines are mirror images through $y = x$.

9 a

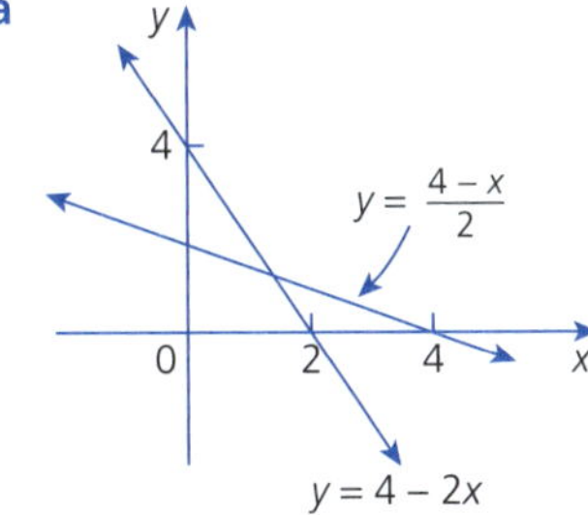

b Swap x and y

$\therefore x = 4 - 2y$

$2y = 4 - x$

$y = \dfrac{4-x}{2}$

10 $y = \sqrt{x}$

Swapping x and y

$\therefore \quad x = \sqrt{y}$

$\therefore \sqrt{y} = x$

$y = x^2$ where $x \geq 0$

11 a $x - 2y = 4$

$\therefore \quad 2y = x - 4$

$y = \dfrac{1}{2}x - 2$ (original function)

Now, $x = \dfrac{1}{2}y - 2$ (interchange x and y)

$\therefore \quad 2x = y - 4$

$y = 2x + 4$ (inverse function)

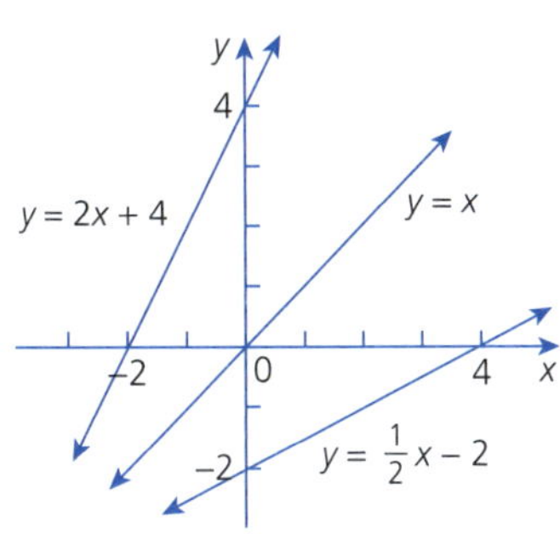

b $y = -x + 2$ (original)

Swap x and y

$x = -y + 2$

$\therefore y = -x + 2$ (inverse)

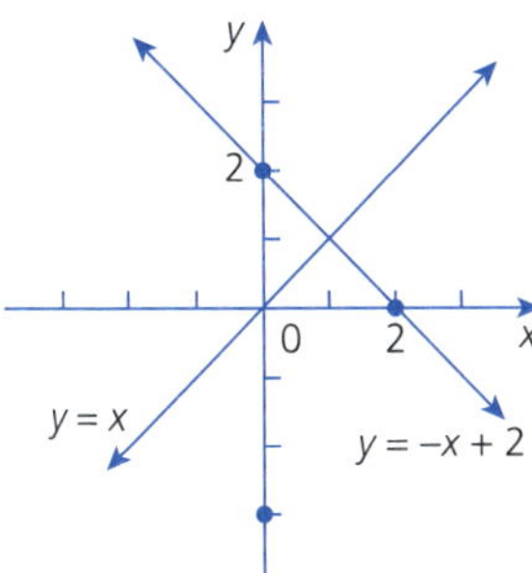

Both original function and inverse function are the same functions, that is, $y = -x + 2$.

$y = -x + 2 \perp y = x$

and hence is its own reflection.

c $y = x + 2$ (original)

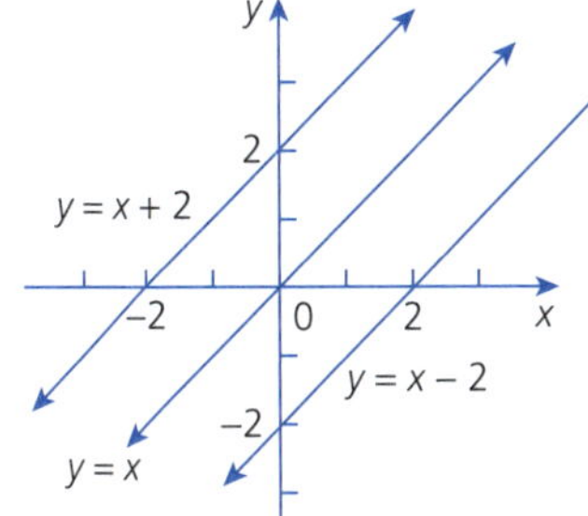

Swap x and y

$x = y + 2$

$\therefore y = x - 2$ (inverse)

Function and inverse function are parallel to and equidistant from $y = x$.

12

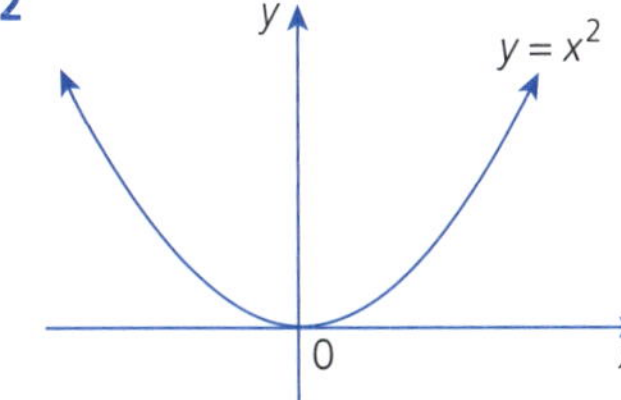

a

y

$y = x^2 - 4$

0

x

-4

b

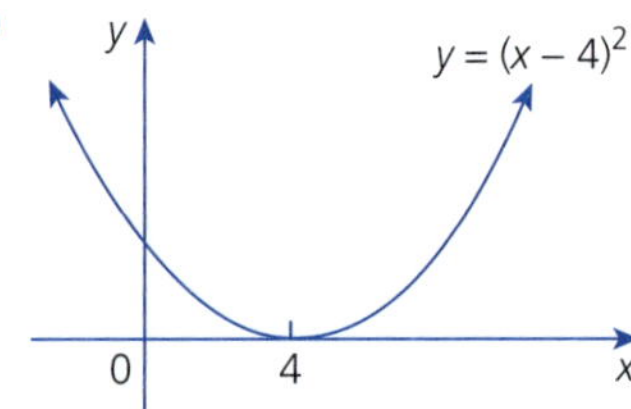

13 a

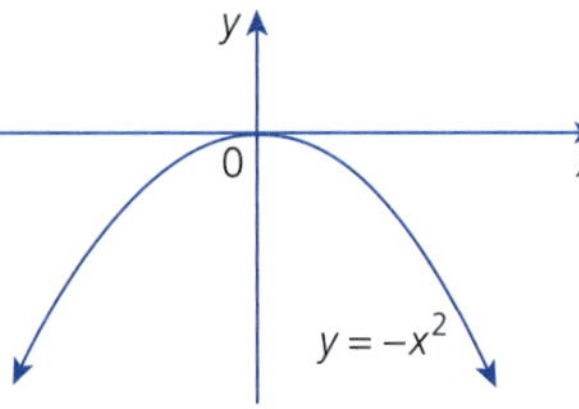

b

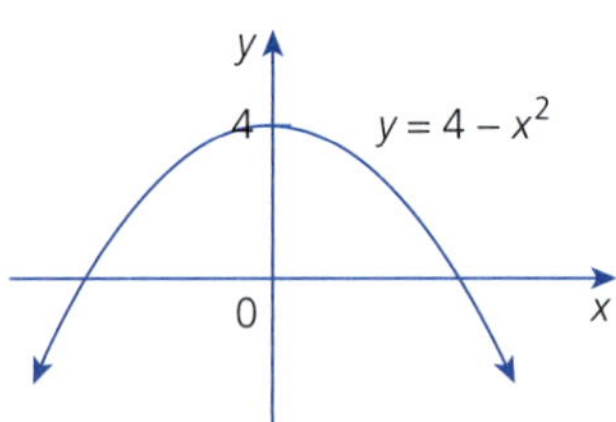

c

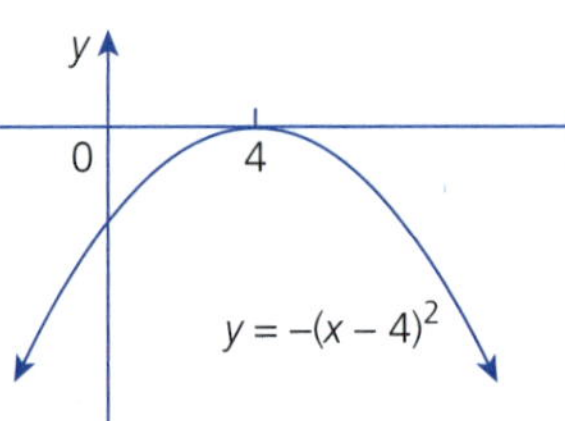

14

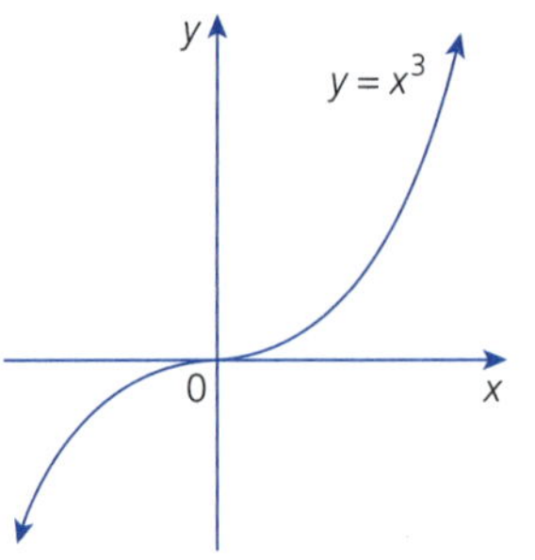

a

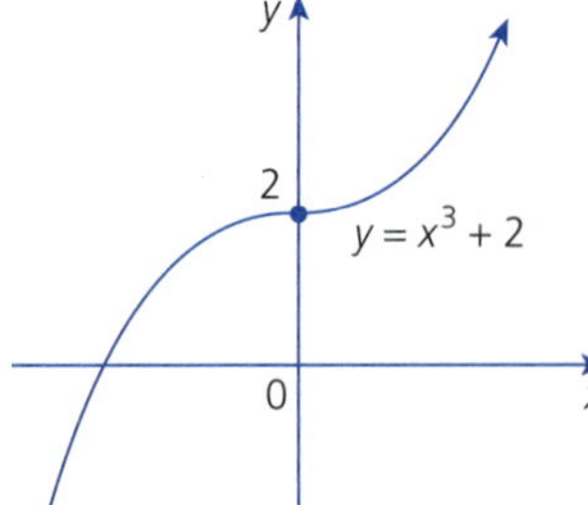

b

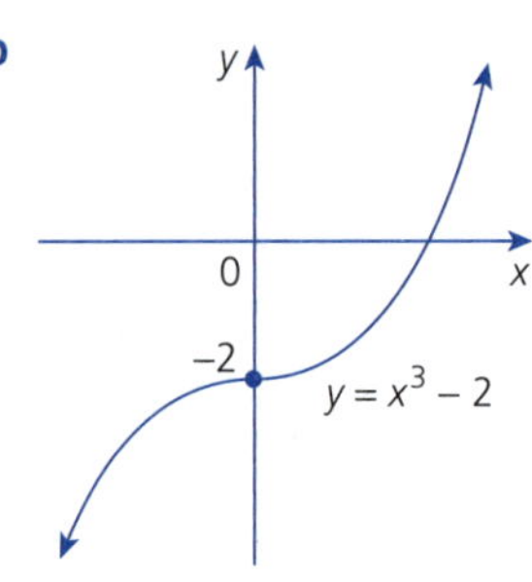

c

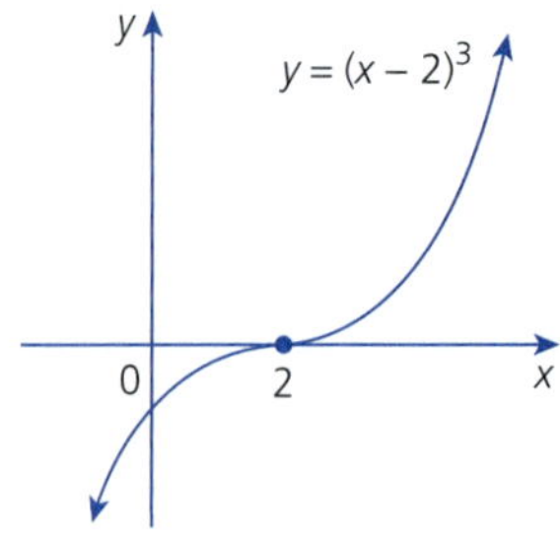

d

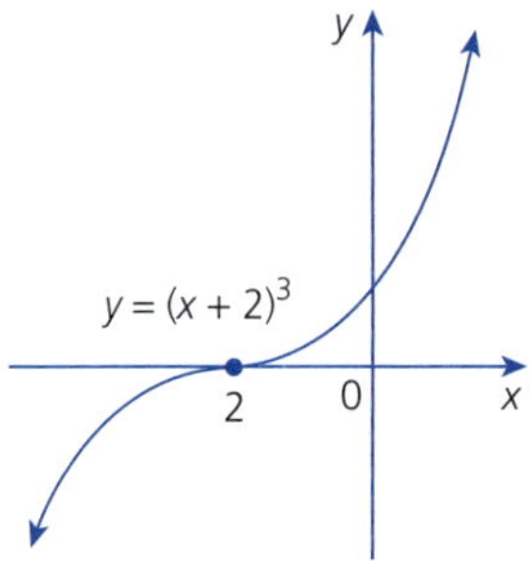

15 a $\dfrac{a^{11}}{a^9} = a^2$

b $\dfrac{15a^{15}}{15a^5} = a^{10}$

c $(2^4)a^{5\times4} = 16a^{20}$

d $\dfrac{(3^4)a^{2\times4}}{(2^2)a^{3\times2}} = \dfrac{81a^8}{4a^6} = \dfrac{81}{4}a^2$

e $\dfrac{a^{\frac{3}{4}-\frac{1}{2}}}{a^{\frac{1}{4}}} = \dfrac{a^{\frac{1}{4}}}{a^{\frac{1}{4}}} = 1$

f $\dfrac{a^{\frac{2}{3}\times4} \times a^{\frac{3}{4}}}{a^{\frac{3}{2}\times3} \times a^{\frac{1}{4}}} = \dfrac{a^{\frac{8}{3}} \times a^{\frac{3}{4}}}{a^{\frac{9}{2}} \times a^{\frac{1}{4}}}$

$= \dfrac{a^{\frac{8}{3}+\frac{3}{4}}}{a^{\frac{9}{2}+\frac{1}{4}}}$

$= \dfrac{a^{\frac{41}{12}}}{a^{\frac{19}{4}}}$

$= a^{-\frac{4}{3}}$

16 a $\sqrt{81} = 9$

b $\sqrt[4]{81} = 3$

c $8 \times \sqrt[6]{64} = 8 \times 2$

$= 16$

d $\sqrt{256} \times \dfrac{1}{\sqrt{16}} = 16 \times \dfrac{1}{4}$

$= 4$

e $(2187^{\frac{1}{7}})^4 = (\sqrt[7]{2187})^4$

$= 3^4$

$= 81$

f $125^{-\frac{2}{3}} = \dfrac{1}{125^{\frac{2}{3}}}$

$= \dfrac{1}{(125^{\frac{1}{3}})^2}$

$= \dfrac{1}{(\sqrt[3]{125})^2}$

$= \dfrac{1}{5^2}$

$= \dfrac{1}{25}$

g $\sqrt[3]{1000} \times (16^{\frac{1}{4}})^{-3} = 10 \times \dfrac{1}{(\sqrt[4]{16})^3}$

$= 10 \times \dfrac{1}{2^3}$

$= 10 \times \dfrac{1}{8}$

$= \dfrac{5}{4}$ or $1\dfrac{1}{4}$

17 a $3^a = 3^4$

$\therefore\ a = 4$

b $a^2 = 1024$

$a = \sqrt{1024}$

$= 32$

c $5^2a = 5^4$

$\therefore\ 2^a = 4$

$a = 2$

d $10^{3x-2} = 10^4$

$\therefore\ 3x - 2 = 4$

$\therefore\ 3x = 6$

$x = 2$

e $2^x = 2^{-2}$

$\therefore\ x = -2$

$\dfrac{1}{4} = \dfrac{1}{2^2} = 2^{-2}$

f $5^{x+1} = 5^{-3}$

$\therefore\ x + 1 = -3$

$\therefore\ x = -4$

$\dfrac{1}{125} = \dfrac{1}{5^3} = 5^{-3}$

g $(3^{\frac{1}{2}})^x = 3^3$

$\therefore\ 3^{\frac{1}{2}x} = 3^3$

$\therefore\ \dfrac{1}{2}x = 3$

That is, $x = 6$.

h $(5^{-\frac{1}{2}})^{4x} = (5^4)^{2x-15}$

$\therefore\ 5^{-2x} = 5^{4(2x-15)}$

$\therefore\ -2x = 8x - 60$

That is, $10x = 60$

$\therefore\ x = 6$

18 a $3^2 \times 3^3 \times 3^3 = 3^8$

$= 6561$

b $\dfrac{2 \times 2^4 \times 2^6 \times 2^7}{2^8} = \dfrac{2^{18}}{2^8}$

$= 2^{10}$

$= 1024$

c $\dfrac{\sqrt{3^6}}{3^2} = \dfrac{3^3}{3^2} = 3$

d $\sqrt{\frac{3^6}{3^2}} = \sqrt{3^4} = 3^2 = 9$

e $\frac{\sqrt{5^5 \times 5^3}}{5^4} = \frac{\sqrt{5^8}}{5^4}$

$= \frac{5^4}{5^4}$

$= 1$

f $\frac{(3^4)^2 \times (3^2)^3}{(3^3)^4} = \frac{3^8 \times 3^6}{3^{12}}$

$= \frac{3^{14}}{3^{12}}$

$= 3^2$

$= 9$

19 a $\log_5 125 = 3$

b $\log_3 \frac{1}{27} = -3$

c $\log_2 \sqrt{8} = \frac{3}{2}$

20 a $10^3 = 1000$

b $2^4 = 16$

c $2^{2.5} = 4\sqrt{2}$

d $5^{-2} = \frac{1}{25}$

21 Change to index form and solve the index equations as in Question **17**.

a $2^x = 64$ (Using table of powers.)

$\therefore 2^x = 2^6$

$\therefore x = 6$

b $5^x = 125$

$\therefore 5^x = 5^3$

$\therefore x = 3$

c $3^x = \sqrt{27}$

$\therefore 3^x = (3^3)^{\frac{1}{2}}$

$= 3^{\frac{3}{2}}$

$\therefore x = \frac{3}{2}$

d $10^x = 1\,000\,000$

This is, $10^x = 10^6$

$\therefore x = 6$

e $2^x = \frac{1}{4}$

$\therefore 2^x = 2^{-2}$

$\therefore x = -2$

f $2^3 = x$

That is, $x = 8$.

g $5^2 = x$

$\therefore x = 25$

h $3^{\frac{1}{2}} = x$

$\therefore x = \sqrt{3}$

i $10^{-1} = x$

$\therefore x = \frac{1}{10}$

j $2^{-3} = x$

$\therefore x = \frac{1}{2^3}$

$= \frac{1}{8}$

k $8^{-\frac{1}{3}} = x$

$\therefore x = \frac{1}{8^{\frac{1}{3}}}$

$= \frac{1}{\sqrt[3]{8}}$

$= \frac{1}{2}$

l $x^8 = 256$

$\therefore x^8 = 2^8$

That is, $x = 2$.

(or $x = (256)^{\frac{1}{8}}$
$= 2$)

m $x^{-2} = \frac{1}{9}$

That is, $x^{-2} = 3^{-2}$

$\therefore x = 3$

(or $x^{-2} = \frac{1}{9}$
$\therefore x^2 = 9$
$\therefore x = \sqrt{9}$
$= 3$)

n $x^{\frac{2}{3}} = 4$

$\therefore x^{\frac{2}{3}} = (8)^{\frac{2}{3}}$

$\therefore x = 8$

(or $x^{\frac{2}{3}} = 4$
$\therefore x = (4)^{\frac{3}{2}}$
$= (4^{\frac{1}{2}})^3$
$= 2^3$
$= 8$)

22 $y = 3^x$

x	-2	-1	0	0.5	1	1.5	2
y	$\frac{1}{9}$	$\frac{1}{3}$	1	≈1.7	3	≈5.2	9

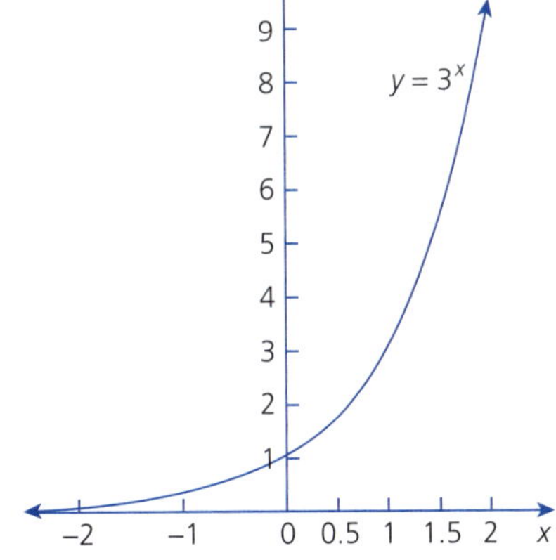

23 $y = 10^x$

x	-1	-0.5	0	0.2	0.5	0.8	1
y	0.1	≈0.3	1	≈1.6	≈3.2	≈6.3	10

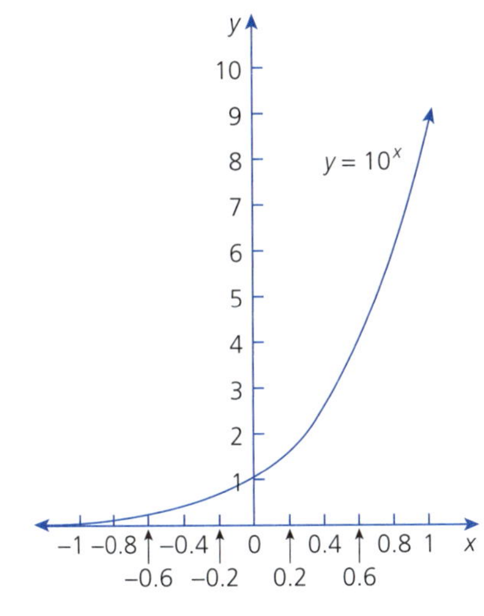

24 $y = \log_3 x$ [Rewrite $x = 3^y$]

x	$\frac{1}{9}$	$\frac{1}{3}$	1	≈1.7	3	≈5.2	9
y	-2	-1	0	0.5	1	1.5	2

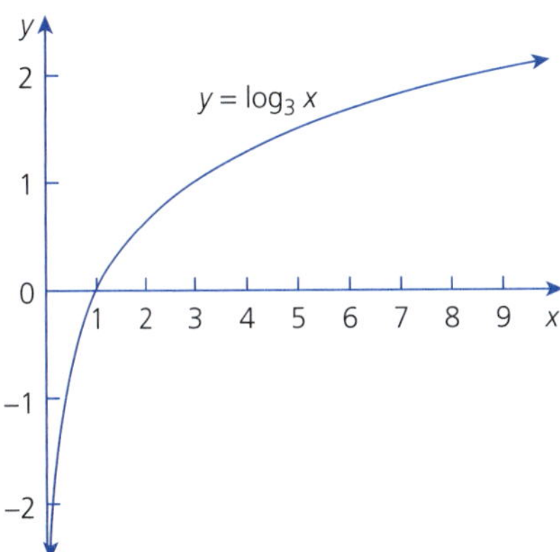

25 $y = \log_{10} x$

x	0.1	≈0.3	1	≈1.6	≈3.2	≈6.3	10
y	-1	-0.5	0	0.2	0.5	0.8	1

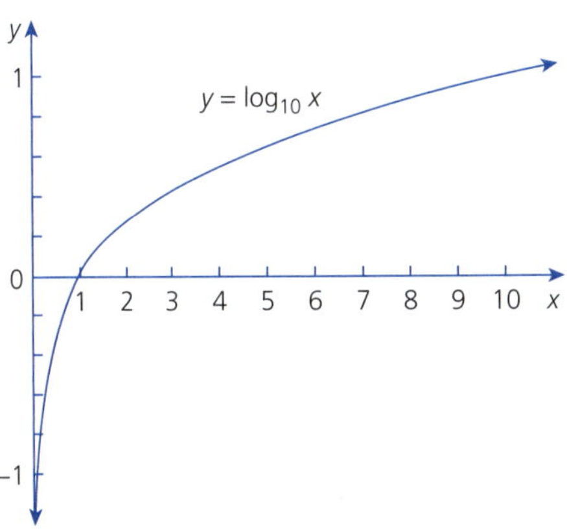

26 a $\log_a NM$

b $\log_a \frac{N}{M}$

c $\log_a N^i$

d $\log_a (N^iM^j)$

e $\log_a \sqrt{N}$

f $\log_a \dfrac{\sqrt{N}}{M^2}$

27 a $\log_a x + \log_a y$

b $\log_a x - \log_a y$

c $\log_a y^{-1} = -\log_a y$

d $\log_a x^{\frac{1}{2}} = \frac{1}{2}\log_a x$

e $\log_a x^2 - \log_a y = 2\log_a x - \log_a y$

f $\log_a x^2 + \log_a \sqrt{y} - \log_a z^4$
$= 2\log_a x + \log_a y^{\frac{1}{2}} - 4\log_a z$
$= 2\log_a x + \frac{1}{2}\log_a y - 4\log_a z$

28 a $\log_2 2^6 = 6\log_2 2$
$= 6$

Log$_a$ $a = 1$

b $\log_4 64 = \log_4 4^3$
$= 3\log_4 4$
$= 3$

c $\log_3 \sqrt{3} = \log_3 3^{\frac{1}{2}}$
$= \frac{1}{2}\log_3 3$
$= \frac{1}{2}$

d $\log_5 5\sqrt{5} = \log_5 5^{1.5}$
$= 1.5\log_5 5$
$= 1.5$

e $\log_5 \dfrac{1}{5^{\frac{1}{2}}} = \log_5 5^{-\frac{1}{2}}$
$= -\frac{1}{2}\log_5 5$
$= -\frac{1}{2}$

f $\log_a a^3 = 3\log_a a$
$= 3$

g $\log_a a^{\frac{1}{2}} = \frac{1}{2}\log_a a$
$= \frac{1}{2}$

h $\log_a a\sqrt{a} = \log_a a^{1.5}$
$= 1.5\log_a a$
$= 1.5$

i $\log_a 1 = 0$ [$\log_2 2 = 1$]

j $\log_x \dfrac{x^2 - 2x}{x - 2} = \log_x \dfrac{x\cancel{(x-2)}}{\cancel{(x-2)}}$
$= \log_x x$
$= 1$

k $\log_5 \dfrac{40}{8} = \log_5 5$
$= 1$

l $\log_3 \dfrac{54}{2} = \log_3 27$
$= \log_3 3^3$
$= 3\log_3 3$
$= 3$

29 a $\log_{10} 3 + \log_{10} 5 = 0.477 + 0.699$
$= 1.176$

b $\log_{10} 5 - \log_{10} 3 = 0.699 - 0.477$
$= 0.222$

c $\log_{10} 3^{\frac{1}{2}} = \frac{1}{2}\log_{10} 3$
$= \frac{1}{2}(0.477)$
$= 0.239$

d $\log_{10} 3\sqrt{3} = \log_{10} 3^{1.5}$
$= 1.5\log_{10} 3$
$= 1.5(0.477)$
$= 0.716$

e $\log_{10} 5 + \log_{10} \sqrt{3}$
$= 0.699 + \frac{1}{2}(0.477)$
$= 0.938$

f $\log_{10} 3^{-1} = -\log_{10} 3$
$= -0.477$

g $\log_{10} \frac{1}{5} = \log_{10} 5^{-1}$
$= -\log_{10} 5$
$= -0.699$

h $\log_{10} 9 + \log_{10} 5$ $9 \times 5 = 45$
$= \log_{10} 3^2 + \log_{10} 5$
$= 2\log_{10} 3 + \log_{10} 5$
$= 2(0.477) + 0.699$
$= 1.653$

i $\log_{10} 25 + \log_{10} 3$
$= \log_{10} 5^2 + \log_{10} 3$
$= 2\log_{10} 5 + \log_{10} 3$
$= 2(0.699) + 0.477$
$= 1.875$

j $\log_{10} 5.4 = \log_{10} \dfrac{27}{5}$
$= \log_{10} 27 - \log_{10} 5$
$= \log_{10} 3^3 - \log_{10} 5$
$= 3\log_{10} 3 - \log_{10} 5$
$= 3(0.477) - 0.699$
$= 0.732$

30 a $\log_a 7 + \log_a 5 = 2.8 + 2.3$
$= 5.1$

b $\log_a 7 - \log_a 5 = 2.8 - 2.3$
$= 0.5$

c $\log_a 1.4 = \log_a \dfrac{7}{5}$
$= 0.5$ [from **b**]

d $\log_a \sqrt{7} = \log_a 7^{\frac{1}{2}}$
$= \frac{1}{2}\log_a 7$
$= \frac{1}{2}(2.8)$
$= 1.4$

e $\log_a \dfrac{1}{7} = \log_a 7^{-1}$
$= -\log_a 7$
$= -2.8$

f $\log_a \dfrac{1}{5^2} = \log_a 5^{-2}$
$= -2\log_a 5$
$= -2(2.3)$
$= -4.6$

g $\log_a \dfrac{1}{7^{\frac{1}{2}}} = \log_a 7^{-\frac{1}{2}}$
$= -\frac{1}{2}\log_a 7$
$= -\frac{1}{2}(2.8)$
$= -1.4$

h $\log_a 175 = \log_a 25 + \log_a 7$
$= \log_a 5^2 + \log_a 7$

$175 = 25 \times 7$

$= 2\log_a 5 + \log_a 7$
$= 2(2.3) + 2.8$
$= 7.4$

i $\log_a 35^{\frac{1}{2}} = \frac{1}{2}\log_a 35$
$= \frac{1}{2}\log_a(5 \times 7)$
$= \frac{1}{2}[\log_a 5 + \log_a 7]$
$= \frac{1}{2}[2.8 + 2.3]$
$= 2.6$ (to 1 dec. pl.)

j $\log_a 9.8 = \log_a 9\frac{4}{5}$
$= \log_a \dfrac{49}{5}$
$= \log_a 49 - \log_a 5$
$= \log_a 7^2 - \log_a 5$
$= 2\log_a 7 - \log_a 5$
$= 2(2.8) - 2.3$
$= 3.3$

31 a $\log_n(3 \times 2) = \log_n 3 + \log_n 2$
$= y + x$ (or $x + y$)

b $\log_n \dfrac{2}{3} = \log_n 3 - \log_n 2$
$= x - y$

c $\log_n \dfrac{3}{2} = \log_n 3 - \log_n 2$
$= y - x$

d $\log_n 2^{\frac{1}{2}} = \frac{1}{2}\log_n 2$
$= \frac{1}{2}x$

e $\log_n 6^{\frac{1}{2}} = \frac{1}{2}\log_n 6$
$= \frac{1}{2}(x + y)$ [from **a**]

f $\log_n(8 \times 3) = \log_n 8 + \log_n 3$
$= \log_n 2^3 + \log_n 3$
$= 3\log_n 2 + 2\log_n 3$
$= 3x + y$

g $\log_n(6 \times \sqrt{2}) = \log_n 6 + \log_n 2^{\frac{1}{2}}$
$= \log_n 6 + \frac{1}{2}\log_n 2$
$= x + y + \frac{1}{2}x$
$= \frac{3}{2}x + y$

h $\log_n 4\frac{1}{2} = \log_n \frac{9}{2}$
$= \log_n 9 - \log_n 2$
$= \log_n 3^2 - \log_n 2$
$= 2\log_n 3 - \log_n 2$
$= 2y - x$

i $\log_n \frac{16}{3} = \log_n 16 - \log_n 3$
$= \log_n 2^4 - \log_n 3$
$= 4\log_n 2 - \log_n 3$
$= 4x - y$

j $\log_n \frac{1}{2} = \log_n 2^{-1}$
$= -\log_n 2$
$= -x$

32 a $N + M = \log_a 10 + \log_a 3$
$= \log_a(10 \times 3)$
$= \log_a 30$

b $N - M = \log_a 10 - \log_a 3$
$= \log_a \frac{10}{3}$

c $\frac{1}{2}N = \frac{1}{2}\log_a 10$
$= \log_a 10^{\frac{1}{2}}$
$= \log_a \sqrt{10}$

d $\frac{1}{2}M - \frac{1}{2}N = \frac{1}{2}[\log_a 3 - \log_a 10]$
$= \frac{1}{2}\log_a \frac{3}{10}$
$= \log_a \sqrt{0.3}$

e $2N + 3M = 2\log_a 10 + 3\log_a 3$
$= \log_a 10^2 + \log_a 3^3$
$= \log_a 100 + \log_a 27$
$= \log_a(100 \times 27)$
$= \log_a 2700$

f $4M - 2N = 4\log_a 3 - 2\log_a 10$
$= \log_a 3^4 - \log_a 10^2$
$= \log_a 81 - \log_a 100$
$= \log_a \frac{81}{100}$
$= \log_a 0.81$

33 a $\log_2 y = \log_2 ab$
$\therefore \quad y = ab$

b $\log_2 y = \log_2 a^{\frac{1}{3}}$
$= \log_2 \sqrt[3]{a}$
$\therefore \quad y = \sqrt[3]{a}$

c $\log_2 y^3 = \log_2 x^{-1}$
$\therefore \quad y^3 = x^{-1}$
$= \frac{1}{x}$
$\therefore \quad y = \sqrt[3]{\frac{1}{x}}$ or $\frac{1}{\sqrt[3]{x}}$

d $\log_2 y = \log_2 2 - \frac{1}{2}\log_2 x$
$= \log_2 2 - \log_2 x^{\frac{1}{2}}$
$= \log_2 2 - \log_2 \sqrt{x}$
$= \log_2 \frac{2}{\sqrt{x}}$
$\therefore \quad y = \frac{2}{\sqrt{x}}$

34 a $\log_{10} a = \log_{10}(3 \times 8)$
$\therefore \quad a = 3 \times 8$
$= 24$

b $\log_{10} a = \log_{10} 81^{\frac{1}{2}}$
$= \log_{10} \sqrt{81}$
$\therefore \quad a = \sqrt{81}$
$= 9$

c $\log_{10} a = \log_{10} \frac{72}{9}$
$= \log_{10} 8$
$\therefore \quad a = 8$

d $\log_{10} a = \log_{10} 2^3 - \log_{10} 5^2$
$= \log_{10} 8 - \log_{10} 25$
$= \log_{10} \frac{8}{25}$
$\therefore \quad a = \frac{8}{25}$

e $\log_{10} a = \log_{10} 10 - \log_{10} 5^2$
$= \log_{10} 10 - \log_{10} 25$
$= \log_{10} \frac{10}{25}$
$\therefore \quad a = \frac{10}{25}$
$= \frac{2}{5}$

f $\log_{10} \frac{16a}{a+6} = \log_{10} 10$
$\therefore \quad \frac{16a}{a+6} = 10$
$\therefore \quad 16a = 10(a + 6)$
$= 10a + 60$
$\therefore \quad 6a = 60$
$a = 10$

g $\log_2 \frac{a}{a-2} = 3 \times 1$
$= 3\log_2 2$
$= \log_2 2^3$
$= \log_2 8$
$\therefore \quad \frac{a}{a-2} = 8$
$\therefore \quad a = 8(a - 2)$
$= 8a - 16$
$\therefore \quad 7a = 16$
$a = 2\frac{2}{7}$

Note the use of $1 = \log_{10} 10$ and $1 = \log_2 2$.

35 a $3^x = 5$ — Taking $\log_{10}$ on each side.
$\log_{10} 3^x = \log_{10} 5$
$\therefore x\log_{10} 3 = \log_{10} 5$
$\therefore \quad x = \frac{\log_{10} 5}{\log_{10} 3}$
$= 1.4649735,$
$= 1.46$ (to 2 dec. pl.)

b $3^x = 10$ — Take logs of both sides.
$\log_{10} 3^x = \log_{10} 10$
$\therefore x\log_{10} 3 = 1$
$x = \frac{1}{\log_{10} 3}$
$= 2.0959033$
$= 2.10$ [to 2 dec. pl.]

c $3^{2x} = 5$
$\log_{10} 3^{2x} = \log_{10} 5$
$\therefore 2x\log_{10} 3 = \log_{10} 5$
$2x = \frac{\log_{10} 5}{\log_{10} 3}$
$x = \frac{\log_{10} 5}{2\log_{10} 3}$
$= 0.7324867$
$= 0.73$ (to 2 dec. pl.)

d $3^x = \sqrt{5}$
$\log_{10} 3^x = \log_{10} \sqrt{5}$
$= \log_{10} 5^{\frac{1}{2}}$
$\therefore x\log_{10} 3 = \frac{1}{2}\log_{10} 5$
$\therefore \quad x = \frac{\frac{1}{2}\log_{10} 5}{\log_{10} 3}$
$= 0.73$ (to 2 dec. pl.)

e $3^x = 5^{-1}$
$\log_{10} 3^x = \log_{10} 5^{-1}$
$\therefore x\log_{10} 3 = -\log_{10} 5$
$\therefore \quad x = \frac{-\log_{10} 5}{\log_{10} 3}$
$= -1.46$ [from **a**]

f $\log_{10} 3^x = \log_{10} 2^{3x-5}$

$\therefore x\log_{10} 3 = (3x - 5)\log_{10} 2$

$= 3x\log_{10} 2 - 5\log_{10} 2$

$\therefore 3x\log_{10} 2 - x\log_{10} 3 = 5\log_{10} 2$

$\therefore x(3\log_{10} 2 - \log_{10} 3) = 5\log_{10} 2$

$\therefore x = \dfrac{5\log_{10} 2}{3\log_{10} 2 - \log_{10} 3}$

$= 3.533\,4753$

$= 3.53$ (to 2 dec. pl.)

g $2^{3-x} = 5^{2x+1}$

$\therefore \log_{10} 2^{3-x} = \log_{10} 5^{2x+1}$

$\therefore (3 - x)\log_{10} 2 = (2x + 1)\log_{10} 5$

$\therefore 3\log_{10} 2 - x\log_{10} 2$

$= 2x\log_{10} 5 + \log_{10} 5$

$\therefore 3\log_{10} 2 - \log_{10} 5$

$= 2x\log_{10} 5 + x\log_{10} 2$

$= x(2\log_{10} 5 + \log_{10} 2)$

$\therefore x = \dfrac{3\log_{10} 2 - \log_{10} 5}{2\log_{10} 5 + \log_{10} 2}$

$= 0.120\,1433$

$= 0.12$ (to 2 dec. pl.)

Chapter 15 . . pp. 277–281

Circle Geometry

1 Data: AB = CD, OE ⊥ AB, OF ⊥ CD, O is the centre of the circle.

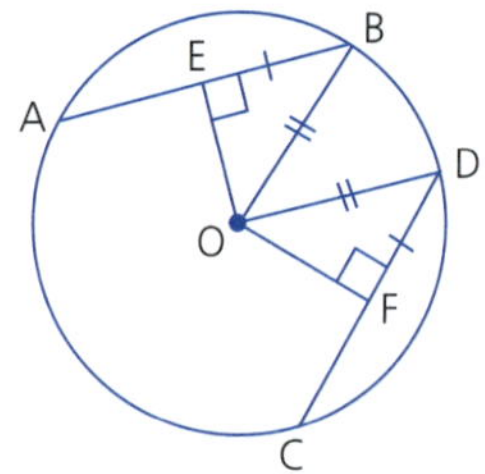

Aim: To prove that OE = OF.

Construction: Join OB, OD.

Proof: AE = EB [perpendicular from centre to chord]

CF = FD [perpendicular from centre to chord]

∴ EB = FD [halves of equal chords AB and CD]

In △s EBO and FDO

EB = FD [proved above]

OB = OD [equal radii]

∠OEB = ∠OFD

= 90° [OE ⊥ AB, OF ⊥ CD]

∴ △EBO ≡ △FDO [RHS]

Then EO = FO [corresponding sides in congruent △s]

Hence, equal chords are equidistant from the centre.

2 a $x = 7$ [perpendicular from centre to chord]

b $x = 90$ [line from centre to midpoint of chord]

c $y = 8.4$ [equal chords subtend equal ∠s at the centre]

d $\theta = 110$ [equal chords subtend equal ∠s at centre]

3 a

CB = 8 as the perpendicular from the centre bisects the chord.

Let OC = x cm

$10^2 = x^2 + 8^2$

$\therefore x^2 = 100 - 64$

$= 36$

$x = \sqrt{36}$

$= 6$

OC = 6 cm

b

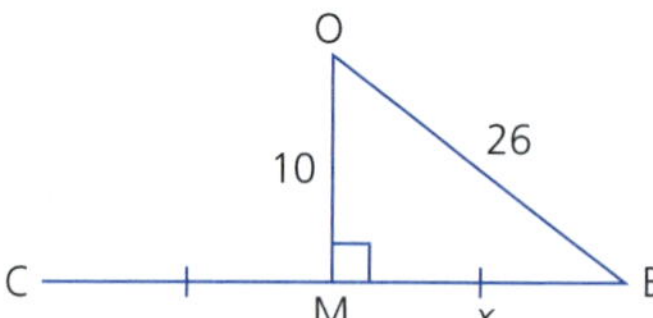

Line from centre to midpoint of chord is perpendicular to chord,

∴ ∠OMB = 90°.

Join the radius OB.

Let MB = x cm.

$26^2 = 10^2 + x^2$

$\therefore x^2 = 26^2 - 10^2$

$= 576$

$x = \sqrt{576}$

$= 24$

MB = 24 cm

∴ AB = 2 × 24 cm

= 48 cm

c

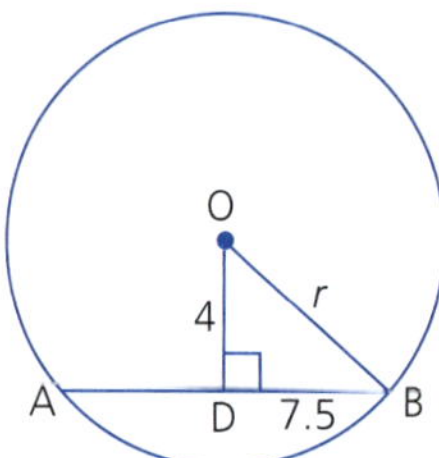

DB = 7.5 cm

Perpendicular from the centre bisects the chord.

AB = 15 cm, OD = 4 cm

Join the radius OB. Let OB = r cm.

$r^2 = 4^2 + 7.5^2$

$= 72.25$

$\therefore r = \sqrt{72.25}$

$= 8.5$

The radius is 8.5 cm.

d

OB = OD = 20 cm

CD = 24 cm

∴ ND = 12 cm

AB = 32 cm

∴ MB = 16 cm

Perpendicular from the centre bisects the chord.

In △OMB, let OM = x cm.

Then $20^2 = x^2 + 16^2$

$\therefore x^2 = 20^2 - 16^2$

$= 144$

$\therefore x = \sqrt{144} = 12$

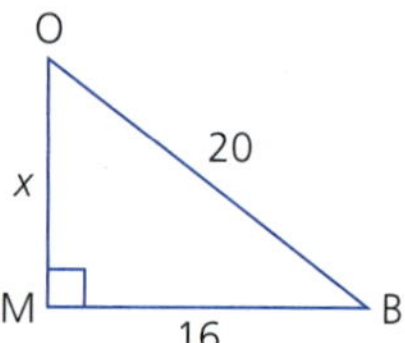

OM = 12 cm

In △OND, let ON = y cm.

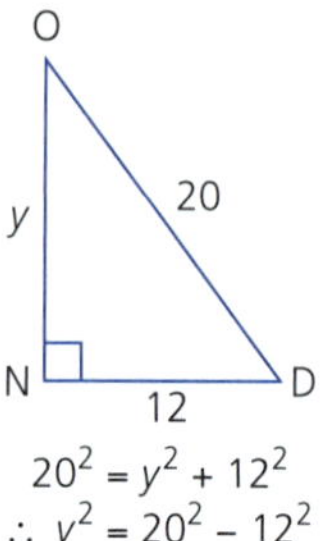

$20^2 = y^2 + 12^2$

$\therefore y^2 = 20^2 - 12^2$

$= 256$

$\therefore y = \sqrt{256} = 16$

ON = 16 cm

Now MN = ON − OM

= 16 − 12 cm

= 4 cm

The distance between the chords is 4 cm.

4

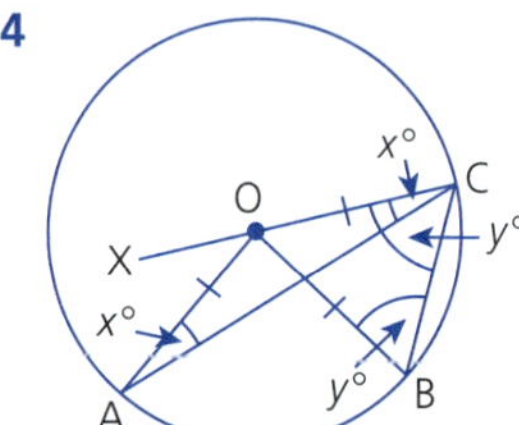

Data: O is centre.

Aim: To prove that ∠AOB = 2 × ∠ACB.

Construction: Join CO and produce to X.

Proof: Let ∠OAC = x°, ∠OBC = y°.

∠OCA = x° [OA = OC, base ∠s of an isosceles △]

∴ ∠XOA = 2x° [exterior ∠ of △OCA]

Also ∠OCB = y° [OB = OC, base ∠s of an isosceles △]

Then ∠XOB = 2y° [exterior ∠ of △BOC]

Then ∠AOB = ∠XOB − ∠XOA

= 2y° − 2x°

= 2(y° − x°)

and ∠ACB = ∠OCB − ∠OCA

= y° − x°.

Then ∠AOB = 2 × ∠ACB

5 Data: PQRS is a cyclic quadrilateral.
Aim: To prove that $\angle SPQ = \angle QRT$.

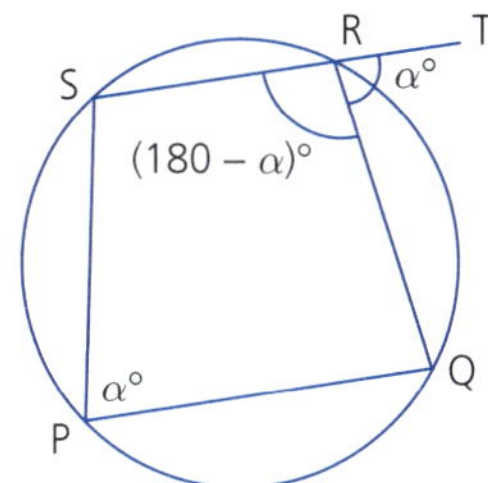

Proof: Let $\angle SPQ = \alpha°$.
Then $\angle SRQ = (180 - \alpha)°$
[opposite $\angle$s of a cyclic quadrilateral]
Then $\angle SRQ + \angle TRQ = 180°$
[ST is a straight line]
$\therefore \angle TRQ = \alpha°$
[$\angle TRQ + 180° - \alpha° = 180°$]
That is, $\angle SPQ = \angle TRQ$ [both $\alpha°$].

6 a $a = \frac{1}{2} \times 86 = 43$
[$\angle$ at the circumference is half $\angle$ at the centre on the same arc]

b $\angle PRQ = 90°$ [$\angle$ in semicircle]
$\therefore b + 52 = 90$ [$\angle$ sum of a $\triangle$]
$b = 38$

c $y = 38$ [$\angle$s in the same segment]

d $x = 37$ [$\angle$s in the same segment]
$\therefore y = 180 - (42 + 37)$ [$\angle$ sum of a $\triangle$]
$\therefore y = 101$

e $g = 180 - 84$
[opposite $\angle$s of a cyclic quadrilateral]
$= 96$

f $h = 114$
[exterior $\angle$ of a cyclic quadrilateral]

g $a = 114$
[corresponding $\angle$s, AD $\parallel$ BC]
$b = 114$
[exterior $\angle$ of a cyclic quadrilateral]
$c + 114 = 180$
[opposite $\angle$s of a cyclic quadrilateral]
$\therefore c = 180 - 114$
$= 66$

h OB = OC [equal radii]
$\therefore y = 48$
[base $\angle$s of an isosceles $\triangle$]
But $\angle ABC = 90°$
[angle in a semicircle],
$\therefore x = 42$ [90 − 48]
Also OB = OA [equal radii]
$\therefore z = 42$
[base $\angle$s of an isosceles $\triangle$]

i Reflex $\angle AOC = 360° - 138°$
$= 222°$ [$\angle$s at a point]
$\therefore \theta = \frac{1}{2}(222)$
[$\angle$ at circumference is half $\angle$ at the centre on the same arc]
That is $\theta = 111$.

j Obtuse $\angle POR = 2 \times 88°$
$= 176°$
[$\angle$ at the centre is twice $\angle$ at the circumference on the same arc]
$\therefore \phi = 360 - 176$
$= 184$ [$\angle$s at a point].

k $y = 124$
[alternate angles, AB $\parallel$ OD]
Reflex $\angle AOC = 248°$
[$\angle$ at the centre is twice $\angle$ at the circumference on the same arc]
$\therefore x = 360 - 248$
$= 112$ [$\angle$s at a point]
Then $z = 180 - 112$
[co-interior angles, AB $\parallel$ OD]
$= 68$
Note: ABCO is not a cyclic quadrilateral.

l $a = 2 \times 48$
[$\angle$ at the centre is twice $\angle$ at the circumference on the same arc]
$= 96$
Then $b = 180 - 48$
[opposite $\angle$s of a cyclic quadrilateral PQRS]
$\therefore b = 132$

m $x = 42$
[$\angle$s in the same segment]
$\therefore y = 42$
[alternate $\angle$s, AD $\parallel$ BC]
Then
$\angle ATD = 180° - (42° + 42°)$
$= 96°$ [$\angle$ sum of a $\triangle$]
$\therefore z = 96$ [vertically opposite $\angle$s]

n $y = 44$ [$\angle$ in alternate segment]

o $x = 52$ [$\angle$ in alternate segment]
$\therefore y = 2 \times 52$
[$\angle$ at the centre is twice $\angle$ at the circumference on the same arc]
$= 104$

p AT = AS
[tangents from external point]
Then $\angle TSA = a°$
[base $\angle$s of an isosceles $\triangle$]
$\therefore a + a + 82 = 180$ [$\angle$ sum of a $\triangle$]
$2a = 98$
$a = 49$

q $x + 46 = 90$
[radius drawn to the point of contact of a tangent]
$\therefore x = 90 - 46$
$= 44$
Also OT = OA [equal radii]
$\therefore \angle OAT = 44°$
[base $\angle$s of an isosceles $\triangle$]
$\therefore y + 44 + 44 = 180$
[$\angle$ sum of a $\triangle$]
That is $y = 180 - 88$
$= 92$

r $x = 38$
[$\angle$ in the alternate segment]
$\angle TOA = 2 \times 38°$
[$\angle$ at the centre is twice $\angle$ at the circumference on the same arc]
$= 76°$
OA = OT [equal radii],
$\therefore \angle OTA = y°$
[base $\angle$s of an isosceles $\triangle$]
Then $y + y + 76 = 180$
[$\angle$ sum of a $\triangle$]
$2y = 180 - 76$
$= 104$
$\therefore y = 52$

s $\angle OTP = 90°$
[radius drawn to the point of contact]
$\therefore \phi + 26 = 90$ [$\angle$ sum of a $\triangle$]
$\phi = 90 - 26$
$= 64$

t $\angle OTP = 90°$
[radius drawn to the point of contact]
$\therefore a + 68 = 90$ [$\angle$ sum of a $\triangle$]
$a = 90 - 68$
$= 22$
OA = OT [equal radii]
Then $\angle ATO = b°$
[base $\angle$s of an isosceles $\triangle$]
$\therefore b + b = 68$ [external $\angle$ of $\triangle AOT$]
$\therefore 2b = 68$
$\therefore b = 34$

7 a

$\angle ACB = 90°$ [$\angle$ in a semicircle]
AB = 20 [radius = 10 cm]
$\therefore h^2 + 16^2 = 20^2$
[Pythagoras' Theorem]
$\therefore h^2 = 20^2 - 16^2$
$= 144$
$h = \sqrt{144}$
$= 12$
That is, h is 12.

b

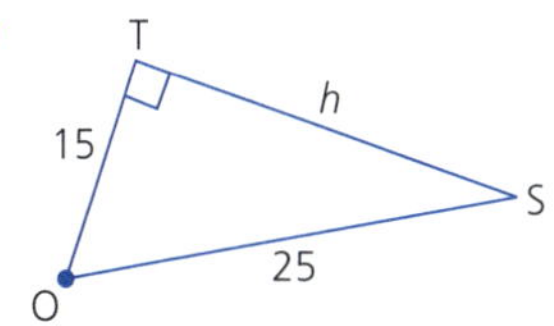

$\angle OTS = 90°$
[radius drawn to the point of contact]
$\therefore h^2 + 15^2 = 25^2$
[Pythagoras' Theorem]
$\therefore h^2 = 25^2 - 15^2$
$= 400$
$h = \sqrt{400}$
$= 20$

c Let OS = x cm
$\angle OTS = 90°$
[radius drawn to the point of contact]

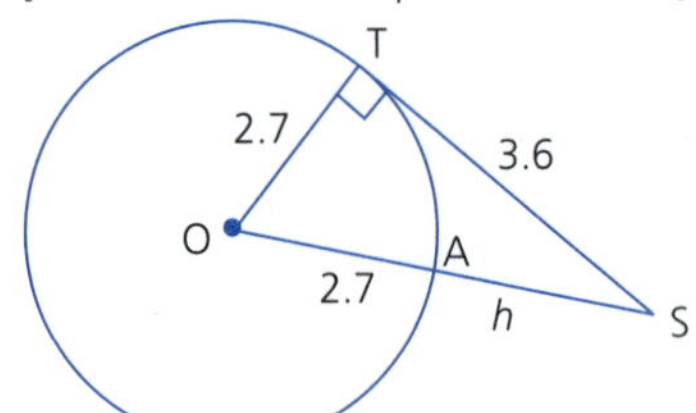

Then $x^2 = 2.7^2 + 3.6^2$
$= 20.25$
$\therefore \quad x = \sqrt{20.25}$
$= 4.5.$
Now $\quad h = OS - OA$
$= 4.5 - 2.7$
$= 1.8$

d

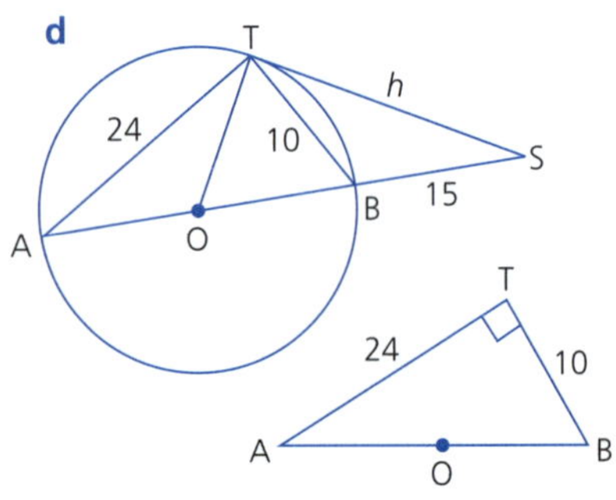

In $\triangle ABT$
$\angle ATB = 90°$ [angle in a semicircle]
Let AB = x cm.
$x^2 = 24^2 + 10^2$
$= 676$
$\therefore \quad x = \sqrt{676} = 26$
Then OB $= \frac{1}{2} \times 26 = 13$
[OB is a radius]

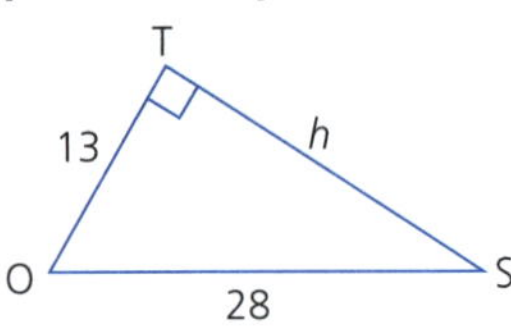

In $\triangle OTS$
OT = 13 [radius]
OS = 13 + 15 = 28.
Then $h^2 + 13^2 = 28^2$
$h^2 = 28^2 - 13^2$
$= 615$
$\therefore \quad h = \sqrt{615}$
$= 24.799\,194$
$= 24.8$ [1 dec. pl.]

8 a $4x = 5 \times 8$ [DE × EC = AE × EB]
$= 40$
$\therefore x = 10$

b $t^2 = 15 \times 8$ [$ST^2 = AS \times SB$]
$\therefore t^2 = 120$
$t = \sqrt{120}$
$= 10.954\,451$
$= 11.0$ (to 1 dec. pl.)

c $6(y + 6) = 18 \times 8$ [AC × BC = EC × CD]
$y + 6 = 3 \times 8$
$= 24$
$\therefore \quad y = 18$

d $8(a + 8) = 12^2$ [$AS \times SB = ST^2$]
$8(a + 8) = 144$
$\therefore \quad a + 8 = 18$
$\therefore \quad a = 10$
Then $y(7 + y) = 12^2$ [$CS \times SD = TS^2$]
$7y + y^2 = 144$
That is,
$y^2 + 7y - 144 = 0$
$(y + 16)(y - 9) = 0$
$\therefore \quad y = 9$ or -16
Ignore the negative value as we are finding a length, that is, $y = 9$.

e $3x = 5 \times 9$ [AE × EB = DE × EC]
$= 45$
$x = 15$
Then AB = 18.
Also $t^2 = 36 \times 18$ [$TS^2 = AS \times SB$]
$\therefore \quad t = \sqrt{648}$ [$\sqrt{18^2 \times 2}$]
$= 18\sqrt{2}$

9 a

$\angle QRP = 37°$
[∠ at the circumference is half ∠ at the centre on the same arc]
Also, reflex $\angle QOP = 286$ [∠ at a point]
$\angle RPQ = 360 - (37 + 286 + 28)$
$= 9$ [angle sum of quadrilateral]

b

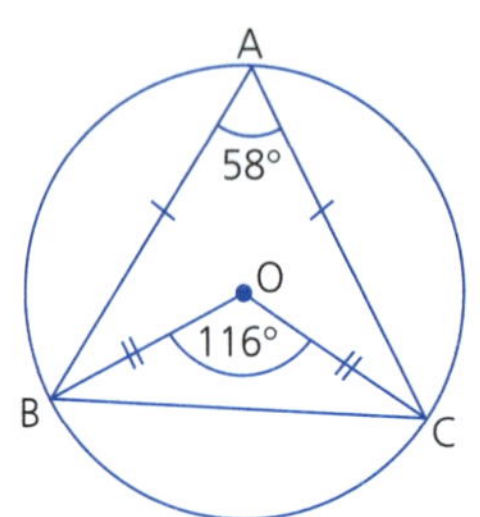

$\angle BAC = 58°$
[∠ at the circumference is half ∠ at the centre on the same arc]
Now $\angle ABC = \angle ACB$
[base ∠s of an isosceles △].
But $\angle ABC + \angle ACB + 58° = 180°$
$\therefore \quad 2\angle ABC = 180° - 58°$
$= 122°$
$\therefore \quad \angle ABC = 61°$
Also BO = OC [equal radii]
$\therefore \angle OBC = \angle OCB$
[base ∠s of an isosceles △]
But $\angle OBC + \angle OCB + 116° = 180°$
[∠ sum of a △]
$\therefore \quad 2\angle OBC = 180° - 116°$
$= 64°$
$\therefore \quad \angle OBC = 32°$
But $\angle ABO = \angle ABC - \angle OBC$
$= 61° - 32°$
$= 29°$

c

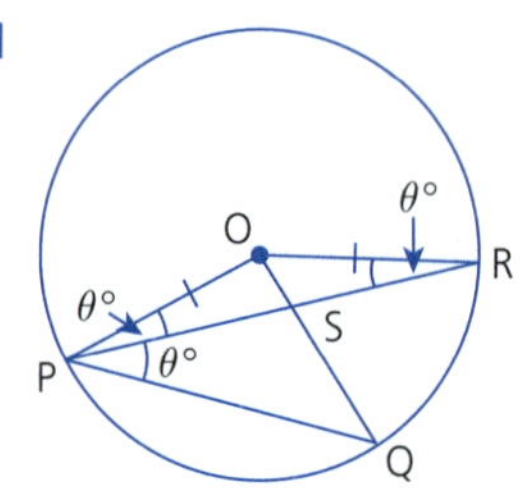

Let $\angle ORQ = \alpha°$.
But OR = OQ [equal radii]
$\therefore \angle OQR = \alpha°$
$\angle ROQ + 142° + 94° = 360°$
[angles at a point]
$\therefore \angle ROQ = 360° - 236°$
$= 124°$
In $\triangle ROQ$, $\alpha + \alpha + 124 = 180$
[∠ sum of a △]
$\therefore \quad 2\alpha = 180 - 124$
$= 56$
$\therefore \quad \alpha = 28$
That is, $\angle ORQ = 28°$.

d

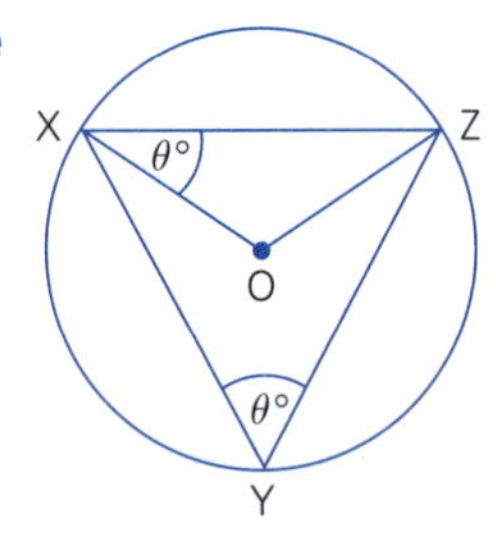

$\angle ORS = \theta°$
OP = OR [equal radii]
$\therefore \angle OPR = \theta°$
[base ∠s of an isosceles △]
$\therefore \angle RPQ = \theta°$
[given $\angle OPR = \angle QPR$]
Then $\angle ROQ$ is twice $\angle RPQ$
[∠ at the centre is twice ∠ at the circumference on the same arc]
That is, $\angle ROS = 2\theta°$
[$\angle ROS$ is the same as $\angle ROQ$]

e

OX = OZ [equal radii]
$\therefore \angle OZX = \theta°$
[base ∠s of an isosceles △]
Also, $\angle XOZ = 2\theta°$
[∠ at the centre is twice ∠ at the circumference on the same arc]
$\therefore \theta + \theta + 2\theta = 180$ [∠ sum of △XOZ]
$4\theta = 180$
$\therefore \quad \theta = 45$
$\angle XOZ = 2\theta° = 90°$

f

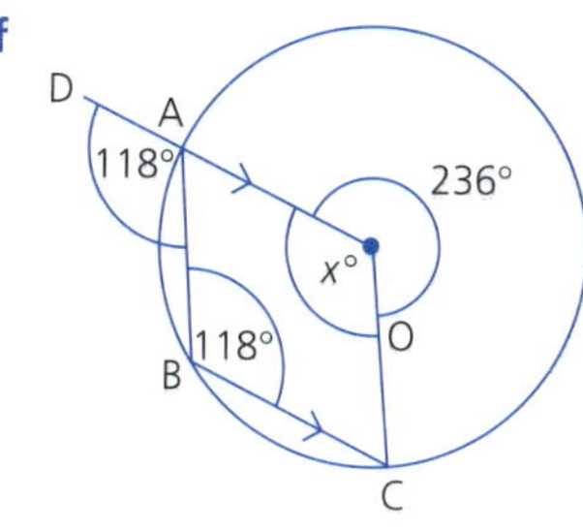

∠ABC = 118°

[alternate ∠s, DO ∥ BC]

Then reflex ∠AOC is twice 118°

[∠ at the centre is twice ∠ at the circumference on the same arc]

That is, reflex ∠AOC = 236°

Then $x + 236 = 360$

[angles at a point]

∴ $x = 360 - 236$

$= 124$

g

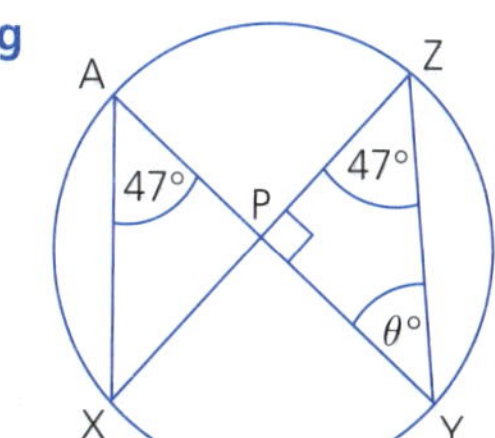

∠XZY = 47°

[∠s in the same segment]

∴ $\theta + 47 + 90 = 180$ [∠ sum of a △]

∴ $\theta = 180 - 137$

$= 43$

h

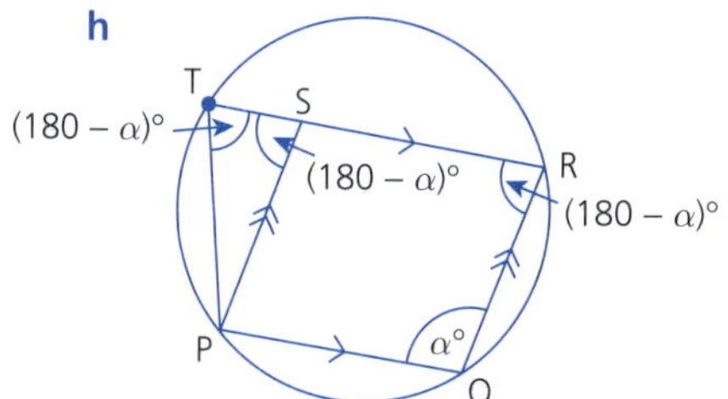

Let ∠PQR = α°.

∠PTR = $(180 - \alpha)$°

[opposite angles of a cyclic quadrilateral]

∠SRQ = $(180 - \alpha)$°

[co-interior angles, PQ ∥ SR]

∴ ∠TSP = $(180 - \alpha)$°

[corresponding angle, SP ∥ RQ]

In △PST

∠PTS = ∠PST [both $(180 - \alpha)$°]

Then △PST is isosceles

[base angles equal]

∴ PT = PS

i

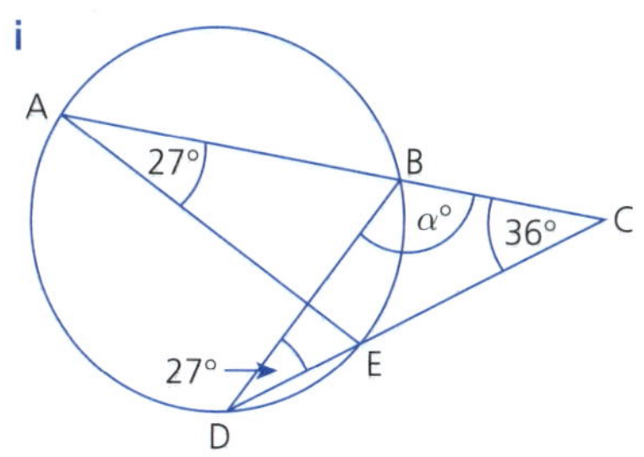

Let ∠DBC = α°.

Now ∠BDE = 27°

[angles in the same segment]

In △BDC

$\alpha + 27 + 36 = 180$

[angle sum of a △]

∴ $\alpha + 63 = 180$

$\alpha = 180 - 63$

$= 117$

∠DBC = 117°

j

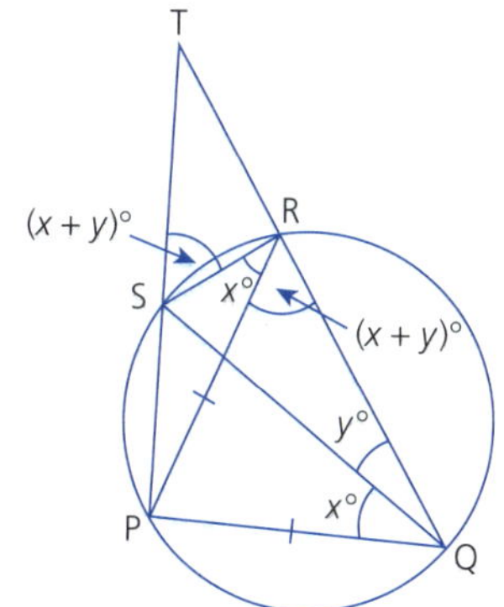

∠PRQ = $(x° + y°)$

[base angles of isosceles △RPQ]

Also ∠SRP = x°

[angles in the same segment]

and ∠TSR = $(x° + y°)$

[exterior angles of a cyclic quadrilateral PQRS].

Now ∠SRQ = $x° + (x + y)°$

$= (2x + y)°$

From △SRT

∠STR + ∠TSR = ∠SRQ

[exterior ∠ = sum of interior opposite angles]

∴ ∠STR + $(x + y)° = (2x + y)°$

That is, ∠STR = $(2x + y)° - (x + y)°$

$= x°$

k

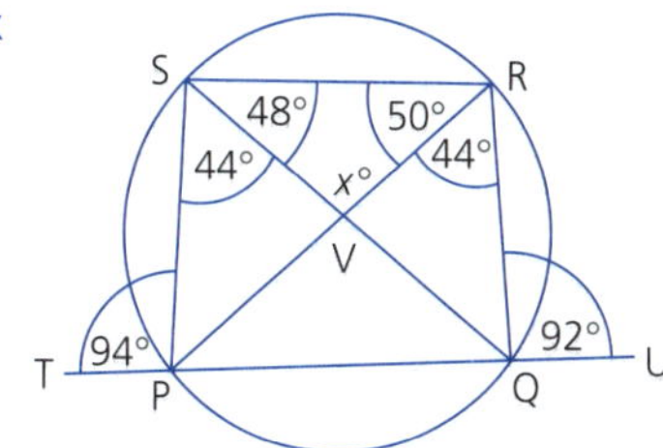

Let ∠SVR = x°.

Now ∠PSR = 92°

[exterior ∠ of a cyclic quadrilateral]

But ∠PSV = ∠PSR – ∠QSR

= 92° – 48°

= 44°

∴ ∠PRQ = 44°

[angles in the same segment]

But ∠SRQ = 94°

[exterior angle of a cyclic quadrilateral]

Now ∠SRV = ∠SRQ – ∠PRQ

= 94° – 44°

= 50°

In △SVR

$x + 50 + 48 = 180$

[angle sum of a triangle]

∴ $x = 180 - 98$

$= 82$

∴ ∠SVR = 82°

l

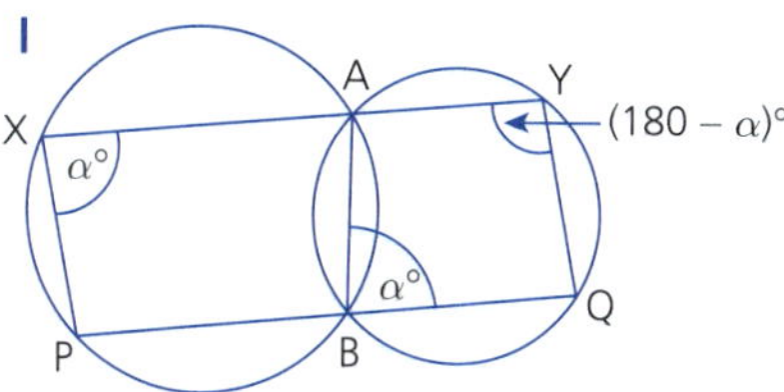

Join PX, BA and QY.

Call ∠PXA = α°

∴ ∠ABQ = α°

[exterior angle of a cyclic quadrilateral PBAX]

Also, ∠AYQ = 180° – α°

[opposite angles of a cyclic quadrilateral BQYA]

Then ∠PXA + ∠AYQ = α° + 180° – α°

= 180°

∴ PX ∥ YQ

[a pair of co-interior angles are supplementary]

m

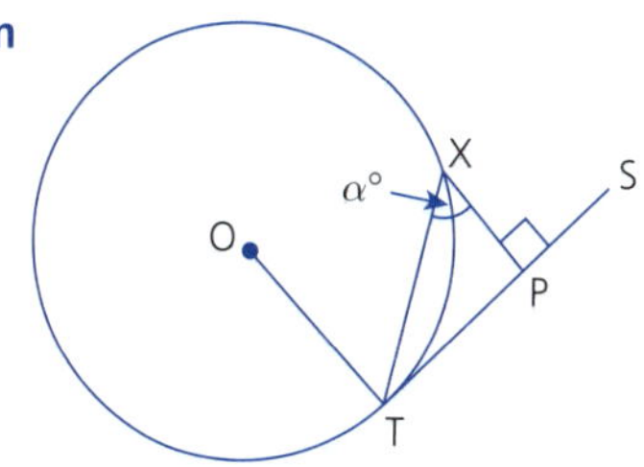

Let ∠TXP = α°.

Now ∠OTP = 90°

[angle from the centre to the point of contact]

Then ∠OTP = ∠XPS [both 90°]

∴ OT ∥ XP

[pair of corresponding ∠s equal]

∴ ∠OTX = α°

[alternate angles, OT ∥ XP]

That is, ∠OTX = ∠TXP.

n

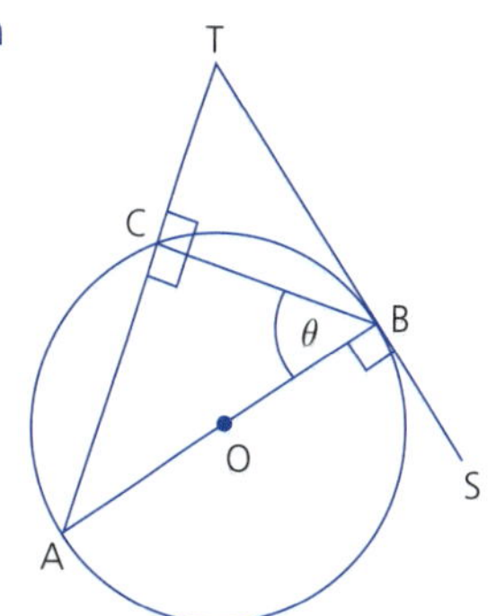

∠ACB = 90°

[angle in a semicircle]

∴ ∠TCB = 90° [AT is a straight line]

∠ABS = 90°

[radius to a point of contact]

∴ ∠CBS = ∠CBA + ∠ABS

= 90° + θ°

Then in $\triangle$TCB
$\angle$ATB + $\angle$TCB = $\angle$CBS
[exterior angle of a triangle]
That is, $\angle$ATB + 90° = 90° + θ°
$\therefore$ $\angle$ATB = θ°

o

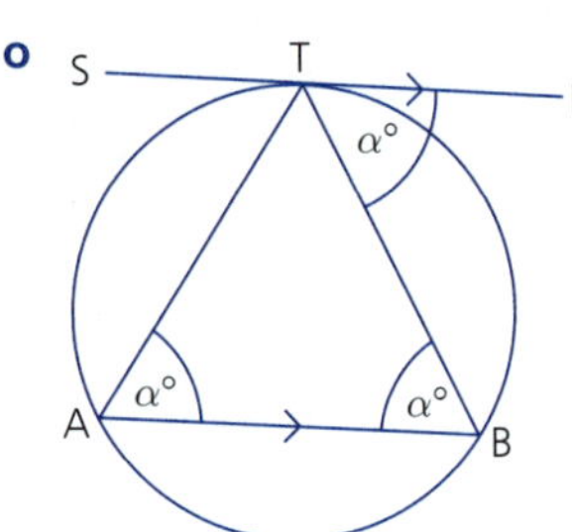

$\angle$ABT = α°
[alternate angles, SP || AB]
Also $\angle$TAB = α°
[angle in the alternate segment]
$\therefore$ $\angle$TAB = $\angle$TBA [both α°]
That is, $\triangle$ABT is isosceles
[base angles equal]

p

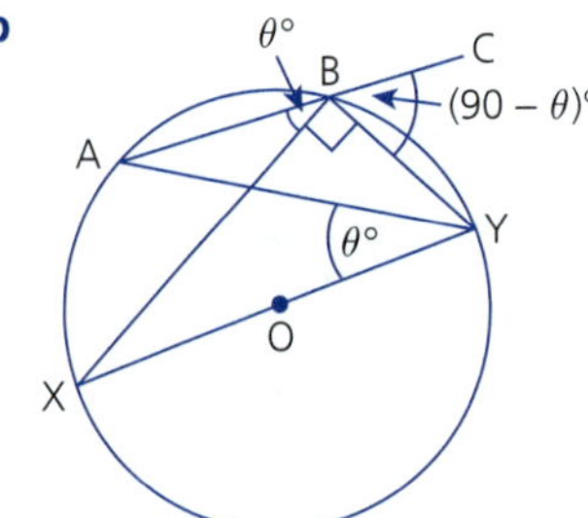

$\angle$XYA = θ°
$\angle$ABX = θ°
[angles in the same segment]
Also $\angle$XBY = 90°
[angle in a semicircle]
Then $\angle$YBC + 90° + θ° = 180°
[AC is a straight line]
$\therefore$ $\angle$YBC = 180° – 90° – θ°
= 90° – θ°
Then
$\angle$XYA + $\angle$YBC = θ° + 90° – θ°
= 90°

q

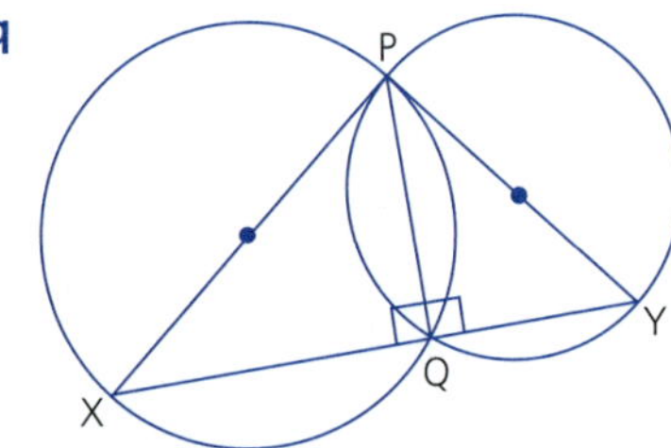

Join XQ, QY and PQ.
Now $\angle$PQY = 90°
[angle in a semicircle]
Also $\angle$XQP = 90°
[angle in a semicircle]
Then $\angle$PQY + $\angle$XQP = 180°
$\therefore$ XY is a straight line
That is, X, Q, Y lie on a straight line.
X, Q, Y are collinear.

Worked Solutions to Tests

Chapter 1
Financial Mathematics

Intermediate Test

p. 12

1 a Tax = 3572 + 0.325 × (38400 − 37000)
= 4027
∴ Tax of \$4027. ✓

b i Taxable income = 100000 − 10000
= 90000
∴ Taxable income of \$90000. ✓

ii Tax = 17547 + 0.37 × (90000 − 80000)
= 21247
∴ Tax of \$21247. ✓

2 $I = Prn$
= 700 × 0.09 × 3
= 189
∴ Interest of \$189. ✓

3 $I = Prn$
= 6400 × 0.075 × 5
= 2400
∴ Interest of \$2400. ✓

4 $I = Prn$
= 4200 × 0.05 × 3
= 2400
∴ Interest of \$630.
4200 + 630 = 4830
∴ Amount repaid is \$4830. ✓

5 $I = Prn$
= 650 × 0.0075 × 12
= 58.5
∴ Interest of \$58.50. ✓

6 a $I = Prn$
= 12000 × 0.08 × 4
= 3840
∴ Interest of \$3840. ✓

b Total repayments = 12000 + 3840
= 15840
Monthly repayments = 15840 ÷ 48
= 330
∴ Repayments of \$330 each month. ✓

7 New price = 249 × 0.85 × 0.85 ✓
= 179.9025
= 179.90 (2 dec. pl.)
∴ New price is \$179.90. ✓

8 a Deposit = 0.1 × 46000
= 4600
∴ Deposit is \$4600. ✓

b Balance owing = \$41400
Interest = 41400 × 0.06 × 4
= 9936
Total repayments = 41400 + 9936
= 51336 ✓
Monthly repayment = 51336 ÷ 48
= 1069.5
∴ Instalment of \$1069.50. ✓

9 $A = P(1 + r)^n$
$= 6000(1.05)^3$
= 6945.75 ✓
Interest = 6945.75 − 6000
= 945.75
∴ Compound interest of \$945.75. ✓

10 $A = P(1 + r)^n$
$= 4500(1.07)^{10}$
= 8852.181108… ✓
= 8852.18 (2 dec. pl.)
∴ Compound interest of \$8852.18. ✓

11 $A = P(1 - r)^n$
$= 20000(0.84)^6$
= 7025.960632… ✓
= 7025.96 (2 dec. pl.)
∴ Value of car is \$7025.96. ✓

12 $I = Prn$
= 2000 × 0.05 × 15
= 1500 ✓
$A = P(1 + r)^n$
$= 2000(1.04)^{15}$
= 3601.887011…
= 3601.89
Comp. int. = 3601.89 − 2000
= 1601.89 ✓
Difference = 1601.89 − 1500
= 101.89
∴ Compound interest account by \$101.89. ✓

13 $A = P(1 - r)^n$
$8352.1 = P(0.85)^4$ ✓
$P = \dfrac{8352.1}{0.85^4}$
= 16000
∴ The jet ski cost \$16000. ✓

Advanced Test

p. 13

1 a \$4742 tax means the taxable income is between \$37000 and \$80000.
Amount over 3572 = 4742 − 3572
= 1170
As 1170 ÷ 0.325 = 3600, the income is \$3600 more than \$37000.
∴ Taxable income is \$40600. ✓

b \$35011 tax means the taxable income is between \$80000 and \$180000.
Amount over 17547 = 35011 − 17547
= 17464
As 17464 ÷ 0.37 = 47200, the income is \$47200 more than \$80000.
∴ Taxable income is \$127200. ✓

2 $I = Prn$
1536 = 6400 × r × 4 ✓
= 25600r
$r = \dfrac{1536}{25600}$
= 0.06
∴ Interest rate of 6% p.a. ✓

3 $I = Prn$
1071 = 3600 × 0.0425 × n ✓
= 153n
$n = \dfrac{1071}{153}$
= 7
∴ Olivia invested for 7 years. ✓

4 New price = 200 × 0.75 × 0.8
= 120 ✓
Total discount = 200 − 120
= 80
∴ Total discount of \$80. ✓

5 As 0.7 × 0.8 = 0.56, and 1 − 0.46 = 0.44, the discounts is 44%, ∴ P is 44
OR: Let original price be \$100
New price = 100 × 0.7 × 0.8
= 56 ✓
Total discount = 100 − 56
= 44
Total % discount = 44%
∴ P is 44. ✓

6 Deposit = 0.25 × 32000
= \$8000 ✓
Balance owing = \$24000
Interest = 24000 × 0.05 × 3
= 3600 ✓
Total repayments = 24000 + 3600
= 27600
Monthly repayment = 27600 ÷ 36
= 766.666666…
= 766.67 (2 dec. pl.)
∴ Instalment of \$766.67. ✓

7 4% p.a. is 1% per quarter, for 20 quarters
$A = P(1 + r)^n$
$= 8000(1.01)^{20}$
= 9761.52032…
= 9761.52 (2 dec. pl.) ✓
Interest = 9761.52 − 8000
= 1761.52
∴ Interest of \$1761.52. ✓

8 $12\,000 + 6416 = 18\,416$

$18\,416 = 12\,000(1 + r)^8$

$(1 + r)^8 = \frac{18\,416}{12\,000}$ ✓

$1 + r = \frac{18\,416}{12\,000}^{\frac{1}{8}}$

$= 1.054\,998\,294\ldots$

$r = 0.054\,998\,294\ldots$

$= 0.055$ (3 dec. pl.)

∴ Interest rate of 5.5% p.a. ✓

9 Kaylee:

$A = P(1 + r)^n$

$= 20\,000(1.06)^{10}$ ✓

$= 35\,816.953\,93\ldots$

$= 35\,816.96$ (2 dec. pl.)

Interest $= 35\,816.96 - 20\,000$

$= 15\,816.96$

∴ Interest of \$15 816.96. ✓

Joseph:

$I = Prn$

$15\,816.96 = 20\,000 \times 0.06 \times n$ ✓

$= 1200n$

$n = \frac{15\,816.96}{12\,000}$

$= 13.1808$

$13.1808 - 10 = 3.1808$ ✓

∴ It would take 3.1808 years longer (or 4 years as interest would be added at the end of the year).

10 $360\,000 - 117\,936 = 242\,064$

$A = P(1 - r)^n$

$242\,064 = 360\,000(0.82)^n$ ✓

$0.82^n = \frac{242\,064}{360\,000}$

$= 0.6724$

Use 'guess and check' strategy:

$n = 2$

∴ It will take 2 years. ✓

11 Pop. in 2005: $A = 500\,000(1.02)^{10}$

$= 609\,497.21$ ✓

Pop. in 2015: $A = 609\,497.21(0.97)^{10}$

$= 449\,457.9479\ldots$

$= 449\,458$ (nearest whole)

∴ The population was 449 458. ✓

Chapter 2

Algebraic Techniques

Intermediate Test

p. 32

1 a $3x + 5x = 8x$ ✓

b $6x \times y = 6xy$ ✓

c $\$4x = 400x$ cents ✓

2 a $2 \times -6 \times 4 = -48$ ✓

b $5 + (-6) + 4 = 3$ ✓

c $(-6)^2 - 4^2 = 36 - 16 = 20$ ✓

d $\frac{-6+4}{-6-4} = \frac{-2}{-10} = \frac{1}{5}$ ✓

3 a $10m$ ✓

b $36y^2$ ✓

c $5ab - 12ab = -7ab$ ✓

d $16y^2$ ✓

e $\frac{{}^{3}\cancel{21x}y}{{}_{1}\cancel{7x}} = 3y$ ✓

f $\frac{{}^{1}\cancel{9}\,{}^{1}\cancel{a}b}{{}^{2}\cancel{18}\cancel{a^2}^{1}} = \frac{b}{2a}$ ✓

4 a $\frac{{}^{6}\cancel{42}\cancel{x}\cancel{y}}{{}_{1}\cancel{7}\cancel{x}\cancel{y}y} = \frac{6x}{y}$ ✓

b $\frac{2(2m) - 3(3m)}{6} = \frac{4m - 9m}{6}$

$= \frac{-5m}{6}$ ✓

c $\frac{3(11) - 2(7)}{6x} = \frac{33 - 14}{6x}$

$= \frac{19}{6x}$ ✓

d $\frac{2x}{5} \times \frac{10x}{4} = \frac{20x^2}{20} = x^2$ ✓

e $\frac{2x}{5} \times \frac{4}{10x} = \frac{8x}{50x} = \frac{4}{25}$ ✓

5 a $15x - 10y$ ✓

b $5 - 6x + 10 = 15 - 6x$ ✓✓

c $12x - 9 - 12x + 16 = 7$ ✓✓

6 a $\frac{4x + 3(2x - 1)}{12} = \frac{4x + 6x - 3}{12}$ ✓

$= \frac{10x - 3}{12}$ ✓

b $\frac{3(4x - 3) + 2(2x - 3)}{6}$

$= \frac{12x - 9 + 4x - 6}{6}$ ✓

$= \frac{16x - 15}{6}$ ✓

7 a $x^2 - 3x + 5x - 15$

$= x^2 + 2x - 15$ ✓

b $4x^2 - 25$ ✓

c $4x^2 + 2(2x)(5) + 25$

$= 4x^2 + 20x + 25$ ✓

8 a $7(y - 3)$ ✓

b $5m(m + 3n)$ ✓

c $(m + 3)(m - 3)$ ✓

d $(a + 8)(a - 5)$ ✓

e $(y - 9)(y - 2)$ ✓

f $y(y + 2x) + 9(y + 2x)$ ✓

$= (y + 2x)(y + 9)$ ✓

g $\frac{{}^{x+5}\cancel{(2x+10)}(2x-1)}{\cancel{2}_1} - 10$ ✓

$= (x + 5)(2x - 1)$ ✓

h $7(x^2 - 4)$ ✓

$= 7(x - 2)(x + 2)$ ✓

9 a $\frac{{}^{3}\cancel{27}}{{}_{1}\cancel{9}(x + y)} = \frac{3}{x + y}$ ✓

b $\frac{{}^{1}\cancel{(m-2n)}(m + 2n)}{4\cancel{(m-2n)}_1} = \frac{m + 2n}{4}$ ✓✓

c $\frac{\cancel{5}x}{{}_{1}\cancel{(x-5)}(x + 5)} \times \frac{{}^{1}\cancel{5}x\cancel{(x-5)}}{\cancel{25}_{\cancel{5}_1}}$ ✓✓

$= \frac{x^2}{x + 5}$ ✓

d $\frac{5}{x + 1} + \frac{3x + 1}{x(x + 1)}$ ✓

$= \frac{5x + 3x + 1}{x(x + 1)}$ ✓

$= \frac{8x + 1}{x(x + 1)}$ ✓

Advanced Test

p. 33

1 a $8y - 5y = 3y$ ✓

b $8a \times 3a = 24a^2$ ✓

c $\$\frac{3x}{100}$ ✓

2 a $-2t$ ✓

b $-42mn$ ✓

c $18m^2 - 35m^2 = -17m^2$ ✓

d $\frac{{}^{3}\cancel{42a}b}{{}_{1}\cancel{14a}} = 3b$ ✓

e $27m^2n$ ✓

f $10a - 10c$ ✓

3 a $\frac{{}^{5}\cancel{35}\,\cancel{ab}bb}{\cancel{7}\,\cancel{ab}} = 5b^2$ ✓

b $\frac{25a - 6a}{10} = \frac{19a}{10}$ ✓

c $\frac{21 - 5}{3m} = \frac{16}{3m}$ ✓

d $\frac{{}^{3}\cancel{15}}{{}_{1}\cancel{x}y} \times \frac{3\cancel{x}}{\cancel{5}_1} = \frac{9}{y}$ ✓

e $\frac{{}^{5}\cancel{25}\cancel{y}y}{{}_{2}\cancel{14}\cancel{x}} \times \frac{{}^{1}\cancel{7}\cancel{x}\cancel{x}}{{}_{1}\cancel{5}\cancel{x}\cancel{y}} = \frac{5y}{2}$ ✓

4 a $35m - 21m^2$ ✓

b $15t - 16t + 40$ ✓

$= -t + 40$ ✓

c $12a + 30 - 15a + 10$ ✓

$= -3a + 40$ ✓

5 a $\frac{39x - 4(4x - 1)}{12}$ ✓

$= \frac{39x - 16x + 4}{12}$

$= \frac{23x + 4}{12}$ ✓

b $\frac{5(5m - 2) - 2(2m - 5)}{10}$

$= \frac{25m - 10 - 4m + 10}{10}$ ✓

$= \frac{21m}{10}$ ✓

6 a $3x^2 + 15x - 2x - 10$

$= 3x^2 + 13x - 10$ ✓

b $25x^2 - 9$ ✓

c $25x^2 - 2(5x)(3) + 9$

$= 25x^2 - 30x + 9$ ✓

d $2x^2 - x - 6 - 2x^2 + 6$ ✓

$= -x$ ✓

e $4x^2 - 4x + 1 - (4x^2 + 4x + 1)$ ✓

$= 4x^2 - 4x + 1 - 4x^2 - 4x - 1$

$= -8x$ ✓

7 a $9(6 - a)$ ✓

b $(m - 8)(m + 8)$ ✓

c $9y(3y - 1)$ ✓

d $(p + 7)(p - 6)$ ✓

e $(p - 9)(p + 8)$ ✓

f $(3x - 2)(2x + 1)$ ✓

g $m(m - 8) - 7t(m - 8)$ ✓

$= (m - 8)(m - 7t)$ ✓

h $(a^2 + b^2)(a^2 - b^2)$ ✓

$= (a^2 + b^2)(a + b)(a - b)$ ✓

8 a $\frac{42x - 35}{6x^2 - 5x} = \frac{7(6x - 5)}{x(6x - 5)}$ ✓

$= \frac{7}{x}$ ✓

b $\frac{5}{2x + 1} - \frac{5}{2x - 1}$

$= \frac{5(2x - 1) - 5(2x + 1)}{(2x + 1)(2x - 1)}$ ✓

$= \frac{10x - 5 - 10x - 5}{(2x + 1)(2x - 1)}$

$= \frac{-10}{(2x + 1)(2x - 1)}$ ✓

c $\frac{15ab}{9a^2 - 1} \times \frac{9a + 3}{5b^2}$

$= \frac{{}^{3}\cancel{15}a\cancel{b}^{1}}{(3a - 1)\cancel{(3a + 1)}_{1}} \times \frac{3\cancel{(3a + 1)}^{1}}{{}^{1}\cancel{5}b^{\cancel{2}1}}$ ✓

$= \frac{9a}{b(3a - 1)}$ ✓

d $\frac{3y - 2}{8y} \div \frac{9y^2 - 4}{6y + 4}$

$= \frac{{}^{1}\cancel{3y - 2}}{{}_{4}\cancel{8}y} \times \frac{{}^{1}\cancel{2}\cancel{(3y + 2)}^{1}}{{}_{1}\cancel{(3y + 2)}\cancel{(3y - 2)}_{1}}$ ✓

$= \frac{1}{4y}$ ✓

e $\frac{2}{x^2 - 16} - \frac{1}{x^2 - 4x}$

$= \frac{2}{(x + 4)(x - 4)} - \frac{1}{x(x - 4)}$

$= \frac{2x - (x + 4)}{x(x + 4)(x - 4)}$ ✓

$= \frac{2x - x - 4}{x(x + 4)(x - 4)}$

$= \frac{x - 4}{x(x + 4)(x - 4)}$

$= \frac{1}{x(x + 4)}$ ✓

Chapter 3
Surds and Indices

Intermediate Test

p. 47

1 a $\sqrt{\frac{9}{25}} = \frac{\sqrt{9}}{\sqrt{25}}$

$= \frac{3}{5}$ ✓

b $\sqrt{\frac{\cancel{12}^{\sqrt{4}}}{\cancel{3}_{1}}} = 2$ ✓

2 a $3\sqrt{2}(\sqrt{2} - 3)$

$= 6 - 9\sqrt{2}$ ✓

b $(\sqrt{2} - 1)^2$

$= 2 - 2\sqrt{2} + 1$ ✓

$= 3 - 2\sqrt{2}$ ✓

3 $\sqrt{18} + \sqrt{32} = 3\sqrt{2} + 4\sqrt{2}$

$= 7\sqrt{2}$ ✓

$\therefore \quad a = 7$ ✓

4 $\frac{1}{2 + \sqrt{3}} \times \frac{2 - \sqrt{3}}{2 - \sqrt{3}}$ ✓

$= \frac{2 - \sqrt{3}}{4 - 3}$

$= 2 - \sqrt{3}$ ✓

5 $\frac{4\sqrt{3} - \sqrt{2}}{3\sqrt{3}} \times \frac{\sqrt{3}}{\sqrt{3}}$ ✓

$= \frac{12 - \sqrt{6}}{9}$ ✓

6 $\frac{4x^4 \times 6x^2}{12x^9} = \frac{24x^6}{12x^9}$

$= \frac{2}{x^3}$ ✓✓

7 $27^{-\frac{2}{3}} = 3^{-2}$

$= \frac{1}{9}$ ✓

8 $3^{x - 1} = 27$

$3^{x - 1} = 3^3$ ✓

$\therefore x - 1 = 3$

$x = 4$ ✓

9 a $\sqrt{x^4y^6} = x^2y^3$ ✓

b $(8x)^{\frac{1}{3}} = 2x^{\frac{1}{3}}$

$= 2\sqrt[3]{x}$ ✓

10 Let $x = 0.47474747\ldots$

$100x = 47.474747474\ldots$

$\therefore 99x = 47$

$x = \frac{47}{99}$ $\qquad \therefore 0.\dot{4}\dot{7} = \frac{47}{99}$ ✓

11 $\sqrt{\frac{3.9812 - 0.0012}{0.0063^2}} = 316.6656721\ldots$

$= 3.17 \times 10^2$ ✓

Advanced Test

p. 48

1 $\sqrt{a^3} + a\sqrt{a} = \sqrt{a^2 \times a} + a\sqrt{a}$

$= a\sqrt{a} + a\sqrt{a}$ ✓

$= 2a\sqrt{a}$ ✓

2 $y = 2\sqrt{3}$

$y^2 = 12$

$y^4 = 144$ ✓

$\therefore y^4 - 2y^2 + 1 = 144 - 24 + 1$

$= 121$ ✓

3 $\frac{3 - 2\sqrt{2}}{3 + 2\sqrt{2}} \times \frac{3 - 2\sqrt{2}}{3 - 2\sqrt{2}}$

$= \frac{9 - 12\sqrt{2} + 8}{9 - 8}$ ✓

$= 17 - 12\sqrt{2}$

$\therefore a = 17, b = -12$ ✓

4 $\frac{6}{\sqrt{5} + \sqrt{2}} \times \frac{\sqrt{5} - \sqrt{2}}{\sqrt{5} - \sqrt{2}}$

$= \frac{6\sqrt{5} - 6\sqrt{2}}{5 - 2}$ ✓

$= \frac{3(2\sqrt{5} - 2\sqrt{2})}{3}$

$= \sqrt{20} - \sqrt{8}$

$\therefore x = 20, y = 8$ ✓

5 As $(2\sqrt{3} - \sqrt{11})(2\sqrt{3} + \sqrt{11})$
$= 12 - 11$
$= 1$
And, $(2\sqrt{2} - 3)(2\sqrt{2} + 3)$
$= 8 - 9$
$= -1$
$\therefore 1 \times -1 = -1$ ✓✓

6 $8^{\frac{2}{3}} \times 9^{-\frac{3}{2}} = \left(8^{\frac{1}{3}}\right)^2 \times \left(9^{\frac{1}{2}}\right)^{-3}$
$= 2^2 \times 3^{-3}$ ✓
$= 4 \times \frac{1}{27}$
$= \frac{4}{27}$ ✓

7 $\frac{{}^{1}\not{4}a^{-2} \times {}^{1}\not{3}a^{-4}}{\not{12}_{1}a} = \frac{a^{-6}}{a}$
$= \frac{1}{a^7}$ ✓

8 a $4^{x+2} = 32$
$\therefore (2^2)^{x+2} = 2^5$ ✓
$2^{2x+4} = 2^5$
$\therefore 2x + 4 = 5$
$2x = 1$
$x = \frac{1}{2}$ ✓

b $3^{x-1} = 0$
$3^x = 1$
$3^x = 3^0$
$x = 0$ ✓

9 $(2x^m)^3 \times (2x)^{m-2} = nx^2$
$2^3x^{3m} \times 2^{m-2}x^{m-2} = nx^2$ ✓
Now $x^{3m} \times x^{m-2} = x^2$
$\therefore x^{4m-2} = x^2$
$\therefore 4m - 2 = 2$
$4m = 4$
$m = 1$ ✓
Also $2^3 \times 2^{m-2} = n$
$\therefore 2^3 \times 2^{-1} = n$
$2^2 = n$
$\therefore n = 4$
$\therefore m = 1, n = 4$ ✓

10 $\frac{5}{\sqrt[3]{1 - 3x + x^2}} = 5(1 - 3x + x^2)^{-\frac{1}{3}}$ ✓

Chapter 4
Equations

Intermediate Test

p. 67

1 a $4x - 11 = x + 17$
$4x - x = 17 + 11$ ✓
$3x = 28$
$x = \frac{28}{3} = 9\frac{1}{3}$ ✓

b $x - 2 = 5 \times 8$ ✓
$= 40$
$x = 40 + 2$
$= 42$ ✓

2 $-1 < x \le 2$ ✓

3 $1 - 4x > 9$
$-4x > 8$ ✓
$x < -2$ ✓

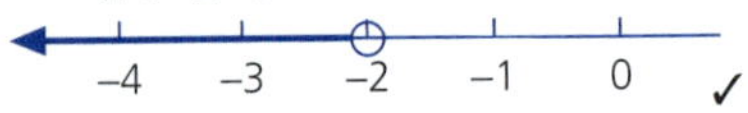

✓

4 $V = 7(4) + \frac{1}{2}(5)(4^2)$ ✓
$= 68$ ✓

5 $7x - 2y = 9$
$7x = 9 + 2y$
$7x - 9 = 2y$
$y = \frac{7x - 9}{2}$ ✓✓

6 $4x^3 = 32$
$x^3 = 8$
$x = 2$ ✓

7 $2x - y = 11$ (1)
$3x + y = 4$ (2)
(1) + (2) $5x = 15$ ✓
$x = 3$
Subst. in (2) $9 + y = 4$ ✓
$y = -5$
Solution is $x = 3, y = -5$. ✓

8 a $(x - 8)(x + 1) = 0$ ✓
$\therefore x = 8$ or -1 ✓

b $5x^2 - 4x = 0$
$x(5x - 4) = 0$ ✓
$\therefore x = 0$ or $x = \frac{4}{5}$ ✓

9 $x = \frac{-b \pm \sqrt{b^2 - 4ac}}{2a}$
$= \frac{5 \pm \sqrt{(-5)^2 - 4(1)(-2)}}{2}$ ✓
$= \frac{5 \pm \sqrt{25 + 8}}{2}$
$= \frac{5 \pm \sqrt{33}}{2}$ ✓

10 $x^2 - 8x = -12$
$\therefore x^2 - 8x + 16 = -12 + 16$ ✓
$(x - 4)^2 = 4$ ✓
$\therefore x - 4 = \pm 2$
$\therefore x = 4 + 2$ or $4 - 2$
$= 6$ or 2 ✓

11 $2x + y = 14$ (1)
$y = x^2 + 3x$ (2)
Substitute (2) into (1)
$2x + (x^2 + 3x) = 14$ ✓
$\therefore x^2 + 5x - 14 = 0$
$(x + 7)(x - 2) = 0$ ✓
$\therefore x = -7$ or 2 ✓
Substitute into (2)
$x = -7, y = (-7)^2 + 3(-7) = 28$
$x = 2, y = 2^2 + 3(2) = 10$
$\therefore$ Points of intersection $(-7, 28)$, $(2, 10)$. ✓

12 Width $= x$ cm
Length $= 2x + 3$ cm ✓
$A = x(2x + 3) = 2$
$\therefore 2x^2 + 3x - 2 = 0$
$(2x - 1)(x + 2) = 0$ ✓
$\therefore x = \frac{1}{2}$ or -2 ✓
As we cannot have side of negative length then:
$x = \frac{1}{2}$ ✓
Width $= \frac{1}{2}$ cm
Length $= 2x + 3$
$= 2\,\frac{1}{2} + 3$
$= 4$ cm
Dimensions are $\frac{1}{2}$ cm and 4 cm. ✓✓

Advanced Test

p. 68

1 a $7 - 4x = 13 - 7x$
$7x - 4x = 13 - 7$ ✓
$3x = 6$
$x = 2$ ✓

b $7x(3) - 4(3x - 5) = 2 \times 12$ ✓
$21x - 12x + 20 = 24$
$9x + 20 = 24$ ✓
$9x = 4$
$x = \frac{4}{9}$ ✓

2

✓

3 $6x - 15 - 8x + 32 < 13$ ✓
$-2x + 17 < 13$
$-2x < -4$
$x > 2$ ✓

4 $b^2 - 4ac = (2ax + b)^2$
$5^2 - 4(2)(2) = (2 \times 2 \times x + 5)^2$ ✓
$25 - 16 = (4x + 5)^2$
$9 = (4x + 5)^2$
$\therefore 4x + 5 = \pm 3$ ✓
$4x = 3 - 5$ or $-3 - 5$
$4x = -2$ or -8
$x = -\frac{1}{2}$ or -2 ✓

5 a From the quadratic formula, consider $b^2 - 4ac$
As $(-5)^2 - 4(1)(2) \ge 0$
$\therefore$ The roots are real. ✓

b Again, from the quadratic formula, consider $b^2 - 4ac$
As $(1)^2 - 4(1)(3) < 0$
$\therefore$ The roots are not real. ✓

6 a $M = x(y - b)$

$M = xy - bx$ ✓

$xy = M + bx$

$y = \frac{M + bx}{x}$ ✓

[Or $\frac{m}{x} = y - b$

$\therefore y = \frac{m}{x} + b$]

b $P = \frac{x + y}{m} + \frac{x - y}{n}$

$Pmn = n(x + y) + m(x - y)$ ✓

$Pmn = nx + ny + mx - my$

$my - ny = nx + mx - Pmn$ ✓

$y(M - n) = nx + mx - Pmn$

$y = \frac{nx + mx - Pmn}{m - n}$ ✓

7 $3x - 2y = 1$ (1)

$x + 3y = 15$ (2)

(2) × 3 $3x + 9y = 45$ (3) ✓

(3) − (1) $11y = 44$

$y = 4$ ✓

Subst. in (2) $x + 12 = 15$

$x = 3$

Solution is $x = 3, y = 4$. ✓

8 a $6x^2 - 11x - 10 = 0$

$(2x - 5)(3x + 2) = 0$ ✓

$\therefore x = \frac{5}{2}$ or $-\frac{2}{3}$ ✓

b $x - 2 = \frac{35}{x}$

$\therefore x^2 - 2x = 35$

$\therefore x^2 - 2x - 35 = 0$ ✓

$\therefore (x - 7)(x + 5) = 0$

$\therefore x = 7$ or -5 ✓

9 $y = \frac{5 \pm \sqrt{5^2 - 4(1)(1)}}{2}$

$= \frac{5 \pm \sqrt{21}}{2}$ ✓

$= \frac{5 \pm \sqrt{21}}{2}$ or $\frac{5 - \sqrt{21}}{2}$

$= 4.79$ or 0.21 ✓

10 $y = 2x - 1$ (1)

$y = \frac{3}{x}$ (2)

Substitute (2) into (1)

$2x - 1 = \frac{3}{x}$

$\therefore 2x^2 - x = 3$ ✓

$2x^2 - x - 3 = 0$

$\therefore (2x - 3)(x + 1) = 0$ ✓

$\therefore x = \frac{3}{2}$ or -1 ✓

Substitute into (1)

$x = \frac{3}{2}, y = 2\ \frac{3}{2}\ - 1$

$= 2$

$x = -1, y = 2(-1) - 1$

$= -3$

Points of intersection are

$\frac{3}{2}, 2$ and $(-1, -3)$. ✓

11 Let one integer be x then next integer is $(x + 1)$.

Then $x^2 + (x + 1)^2 = 221$ ✓

$\therefore x^2 + x^2 + 2x + 1 = 221$

$2x^2 + 2x + 1 = 221$

$2x^2 + 2x - 220 = 0$

$\therefore x^2 + x - 110 = 0$ ✓

$(x + 11)(x - 10) = 0$

$\therefore x = -11$ or 10 ✓

But $x > 0$

$\therefore x = 10$

Then $(x + 1) = 11$

The integers are 10 and 11. ✓

Chapter 5

Linear Relationships

Intermediate Test

p. 85

1 a Distance $= \sqrt{(5 + 3)^2 + (-4 - 1)^2}$ ✓

$= \sqrt{8^2 + (-5)^2}$

$= \sqrt{89}$ ✓

b Midpoint $= \frac{-3 + 5}{2}, \frac{1 - 4}{2}$ ✓

$= 1, \frac{-3}{2}$ ✓

c Gradient $= \frac{-4 - 1}{5 + 3}$ ✓

$= -\frac{5}{8}$ ✓

2 Gradient $(m) = \frac{\text{Rise}}{\text{Run}}$

$= -\frac{4}{3}$ (↘) ✓

Also y-intercept $= 4$ ✓

Using $y = mx + b$

$y = \frac{-4}{3}x + 4$

Or $4x + 3y - 12 = 0$ ✓

3 $y = \frac{2}{4}x - \frac{5}{4}$

$= \frac{1}{2}x - \frac{5}{4}$ ✓

$m = \frac{1}{2}, b = -\frac{5}{4}$ ✓

4 y-intercept $= -4$ ✓

Gradient $= 2$

Using $y = mx + b$

$y = 2x - 4$ ✓

5 Using two-point formula:

$\frac{y - 5}{x + 2} = \frac{-2 - 5}{5 + 2} = \frac{-7}{7} = \frac{-1}{1}$ ✓

$\therefore y - 5 = -(x + 2)$

$y - 5 = -x - 2$

$\therefore x + y - 3 = 0$ ✓

6 Take (3, −5), (9, 7) $m_1 = \frac{-5 - 7}{3 - 9} = 2$ ✓

Take (9, 7), (−2, − 15) $m_2 = \frac{7 + 15}{9 + 2} = 2$

As gradients are equal and (9, 7) is on both lines, the 3 points must be collinear. ✓

7 $m = \frac{1}{2}$, through (0, 3) ✓

Use point–gradient formula:

$y - 3 = \frac{1}{2}(x - 0)$

$\therefore 2y - 6 = x$

i.e. $x - 2y + 6 = 0$ ✓

8 a $AB = \sqrt{(3 + 3)^2 + (1 + 2)^2}$

$= \sqrt{6^2 + 3^2}$

$= \sqrt{45}$

$= 3\sqrt{5}$ ✓

b $m_{AC} = \frac{-2 - 4}{-3 + 1} = \frac{-6}{-2} = 3$ ✓

c $MP_{AC} = \frac{-3 - 1}{2}, \frac{-2 + 4}{2} = (-2, 1)$ ✓

d $y - 4 = 3(x + 1)$ ✓

$y - 4 = 3x + 3$

$\therefore 3x - y + 7 = 0$ ✓

e Perpendicular bisector passes through (−2, 1).

Gradient $= -\frac{1}{3}$ (negative reciprocal) ✓

Equation is $y - 1 = -\frac{1}{3}(x + 2)$

$3y - 3 = -x - 2$

$\therefore x + 3y - 1 = 0$ ✓

f Gradient OC $= -4$ ✓

Midpoint of AB: $0, -\frac{1}{2}$

$y + \frac{1}{2} = -4(x - 0)$

$2y + 1 = -8x$

$8x + 2y + 1 = 0$ ✓

9 (2, k) $x + 3y = 8$

$2 + 3k = 8$ ✓

$\therefore 3k = 6$

$k = 2$ ✓

Advanced Test

p. 86

1 a Distance $= \sqrt{(4+3)^2 + (-3-1)^2}$
$= \sqrt{7^2 + (-4)^2}$
$= \sqrt{65}$ ✓✓

b $m_{AC} = \dfrac{1+5}{-3+2} = \dfrac{6}{-1} = -6$ ✓✓

c MP $= \dfrac{4-2}{2}, \dfrac{-3-5}{2}$
$= (1, -4)$ ✓✓

d $\frac{1}{4}$ way will be midpoint of C and midpoint (1, −4)
$\dfrac{1}{4}$ pt $= \dfrac{-2+1}{2}, \dfrac{-5+4}{2}$
$= -\dfrac{1}{2}, -4\dfrac{1}{2}$ ✓✓

2 Gradient $= \dfrac{\text{Rise}}{\text{Run}}$
$= -\dfrac{4}{2}$
$= -2$ ✓
y-intercept $= -4$ ✓
Using $y = mx + b$
$y = -2x - 4$ ✓

3 $2y = 6x + 7$
$y = 3x + \dfrac{7}{2}$
Gradient $= 3$ ✓
y-intercept $= 3\dfrac{1}{2}$ ✓

4 Using point–gradient form:
$y + 3 = \dfrac{3}{4}(x - 2)$ ✓
$\therefore \quad 4y + 12 = 3x - 6$
or $3x - 4y - 18 = 0$ ✓

5 Using two-point formula:
$\dfrac{y-1}{x+3} = \dfrac{-5-1}{3+3} = \dfrac{-6}{6} = \dfrac{-1}{1}$ ✓
$\therefore \quad y - 1 = -(x + 3)$
$y - 1 = -x - 3$
$\therefore x + y + 2 = 0$ ✓

6 $x + 2y - 5 = 0$
$2y = -x + 5$
$y = \dfrac{-1}{2}x + \dfrac{5}{2}$
$m = \dfrac{-1}{2}$ ✓
$\therefore$ Gradient perpendicular $= 2$ ✓
Equation of perpendicular is
$y - 4 = 2(x + 1)$
$y - 4 = 2x + 2$
$\therefore 2x - y + 6 = 0$ ✓

7 $2(3m) - 3(-5) - 3 = 0$ ✓
$6m + 15 - 3 = 0$
$6m = -12$
$m = -2$ ✓

8 a X(−4, 3)
centre
(−1, −1)
Y(p, q)

Midpoint XY = (−1, −1)
(−4, 3) (p, q)
MP $= \dfrac{-4+p}{2}, \dfrac{3+q}{2}$
$\therefore \dfrac{-4+p}{2} = -1$ and $\dfrac{3+q}{2} = -1$
$-4 + p = -2 \qquad 3 + q = -2$
$p = 2 \qquad q = -5$ ✓✓

b Distance X → centre
$= \sqrt{(-1+4)^2 + (-1-3)^2}$
$= \sqrt{3^2 + (-4)^2}$
$= 5$ ✓

9 Solve $x + 2y - 6 = 0$ (1)
with $y = 3x - 4$ (2)
Substitute (2) into (1)
$x + 2(3x - 4) = 6$
$7x - 8 = 6$
$7x = 14$
$x = 2$ ✓
In (2) $y = 3(2) - 4 = 2$
Lines intersect at (2, 2) ✓
Substitute (2, 2) into $x = 2y - 2$
$2 = 4 - 2$
(2, 2) lies on $x = 2y - 2$
Then all lines pass through (2, 2)
Lines are concurrent. ✓

10 a AB $= \sqrt{(3+4)^2 + (-1+2)^2}$
$= \sqrt{7^2 + 1^2} = \sqrt{50} = 5\sqrt{2}$ ✓

b $m_{AB} = \dfrac{-2+1}{-4-3} = \dfrac{-1}{-7} = \dfrac{1}{7}$ ✓

c $y + 2 = \dfrac{1}{7}(x + 4)$ ✓
$7y + 14 = x + 4$
$\therefore x - 7y - 10 = 0$ ✓

d DC ∥ AB $\therefore m_{DC} = \dfrac{1}{7}$
Using D, $y - 2 = \dfrac{1}{7}(x + 2)$ ✓
$7y - 14 = x + 2$
$\therefore x - 7y + 16 = 0$ ✓

e $MP_{AD} = \dfrac{-4-2}{2}, \dfrac{-2+2}{2}$
$= (-3, 0)$ ✓

f A → B
(−4, −2) → (3, −1) $(x + 7, y + 1)$
(−4 → 3, −2 → −1)
Then D → C
(−2 + 7, 2 + 1)
C = (5, 3) ✓

g CD $= \sqrt{50}$
$= 5\sqrt{2}$
Opposite sides of parallelograms are equal. ✓

11 Midpoint AC = (2, 0) = midpoint BD ✓
Gradient AC $= \dfrac{-1-1}{3-1}$
$= \dfrac{-2}{2}$
$= -1$ ✓
Gradient BD $= \dfrac{-1-1}{1-3}$
$= \dfrac{-2}{-2}$
$= 1$
As $-1 \times 1 = 1$, then perpendicular.
$\therefore$ Diagonals bisect at right-angles. ✓

Chapter 6

Non-linear Relationships

Intermediate Test

p. 101

1 Parabola concave down, with x-intercepts of ±3
$\therefore y = 9 - x^2$
$\therefore$ D ✓

2 x-intercepts of −2 and 3
$\therefore y = (x + 2)(x - 3)$ could be equation
$\therefore$ B ✓

3

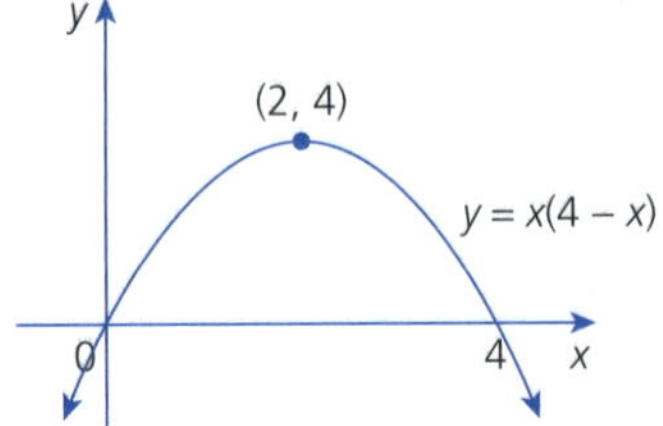

✓

4 a $y = (x + 3)(x - 1)$
As x-intercepts are −3 and 1
$\therefore x = -1$ is eqn of axis of symmetry ✓

b Substitute $x = -1$ in y
$\therefore y(-1) = (-1 + 3)(-1 - 1)$
$= 2 \times (-2)$
$= -4$
$\therefore$ Minimum value of −4. ✓

c

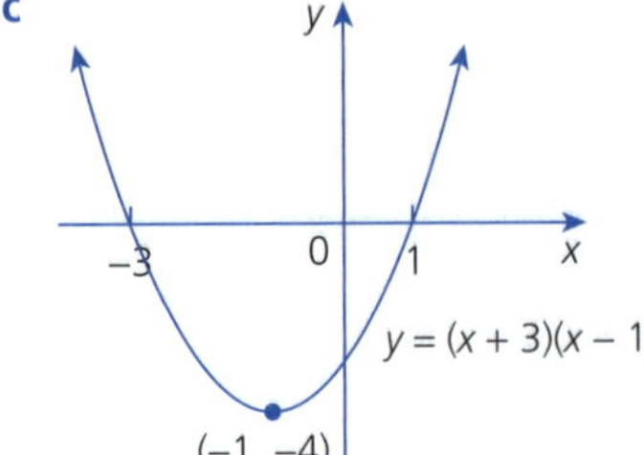

✓

5 $y = 4 - 2x - x^2$

Now, $x = \frac{-b}{2a}$

$= \frac{-(-2)}{2(-1)}$

$= \frac{2}{-2}$

$= -1$

$\therefore$ Axis of symmetry is $x = -1$. ✓

Also substitute $x = -1$ in y

$\therefore y = 4 - 2(-1) - (-1)^2$

$= 4 + 2 - 1$

$= 5$

$\therefore$ Maximum value of 5. ✓

6 x-intercept of -2, 4

$\therefore y = a(x + 2)(x - 4)$ ✓

Substitute in $(1, -9)$

$\therefore -9 = a(1 + 2)(1 - 4)$

$-9 = a(3)(-3)$

$-9a = -9$

$a = 1$

$\therefore y = (x + 2)(x - 4)$ ✓

7 a

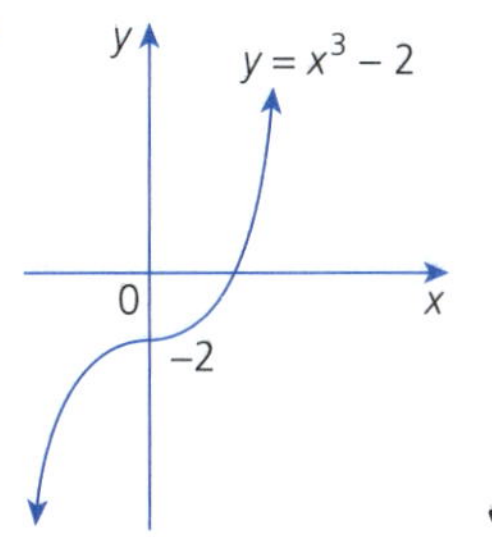

✓

b

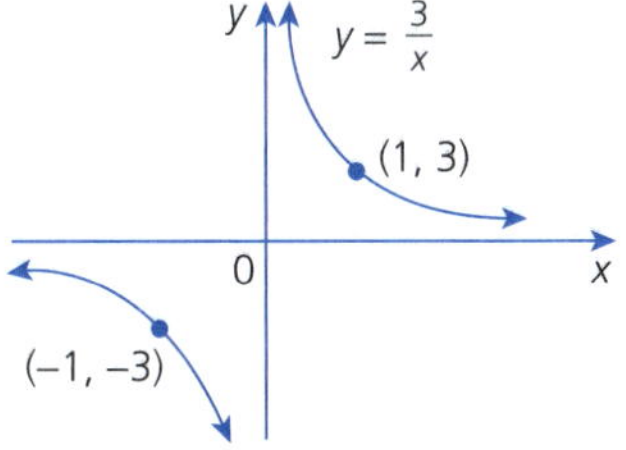

✓

c

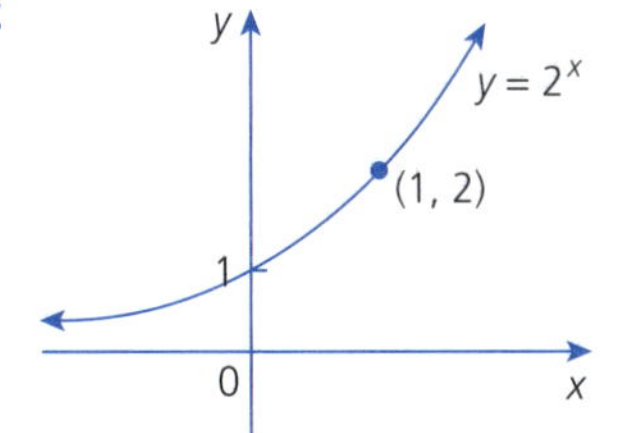

✓

8 $x^2 + y^2 = r^2$

$\therefore x^2 + y^2 = 16$ ✓

9 $y = x^2 - x - 20$

x-intercepts $\therefore y = 0$

$\therefore x^2 - x - 20 = 0$

$(x - 5)(x + 4) = 0$

$x = 5, -4$ ✓

y-intercepts $\therefore x = 0$

$\therefore y = 0 - 0 - 20$

$y = -20$ ✓

10 $y = 25 - x^2$

As $x = \frac{-b}{2a}$

$= 0$

$\therefore x = 0$ is equation of axis of symmetry. ✓

11 a

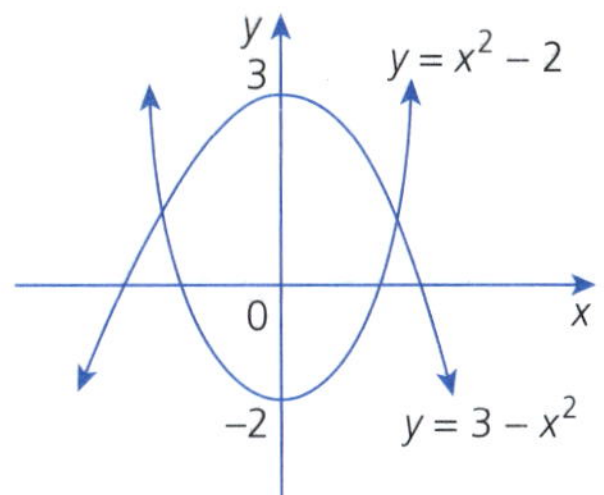

✓

b

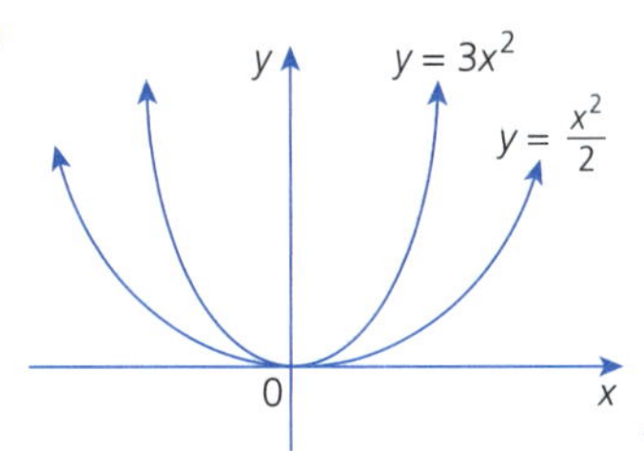

✓

12 a As $(3 - x)(x + 2) = 0$

$x = 3, -2$

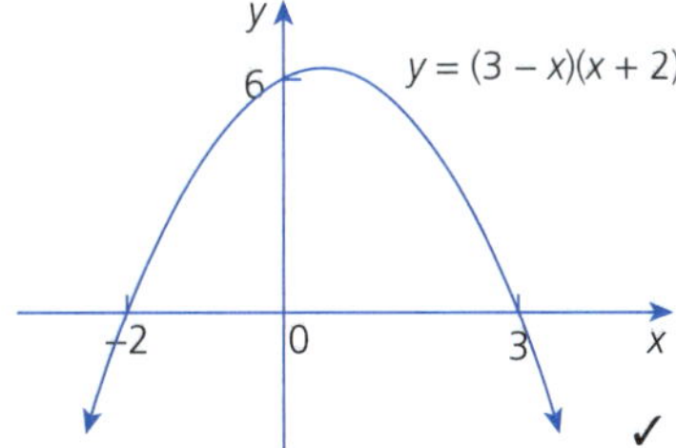

✓

b As $x(x + 2) = 0$

$x = 0, -2$

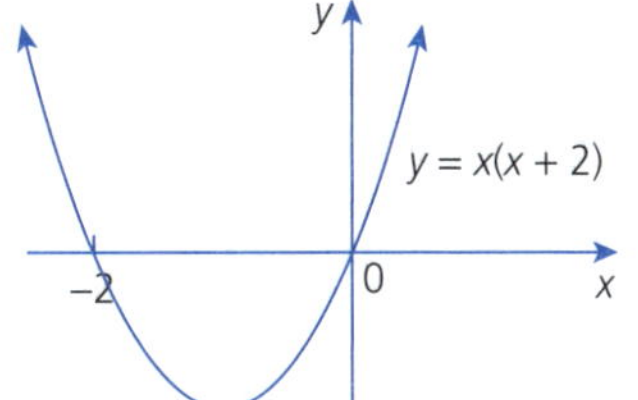

✓

c As $x^2 - 5x - 14 = 0$

$(x - 7)(x + 2) = 0$

$x = 7, -2$ ✓

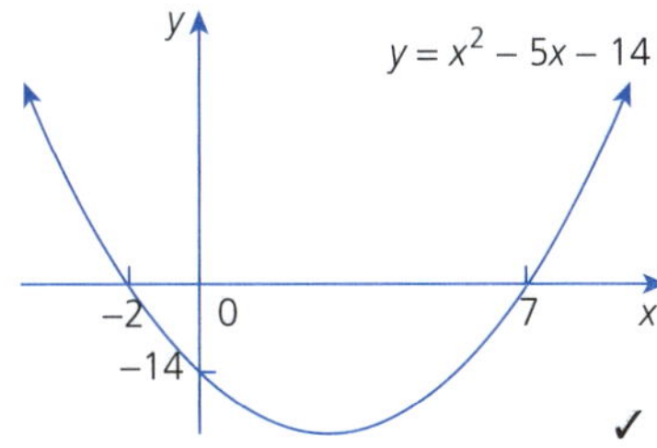

✓

13 a $y = 3 - x^2$ ✓

b $y = (x + 2)(x - 4)$ ✓

c $y = x(x - 3)$ ✓

14 a

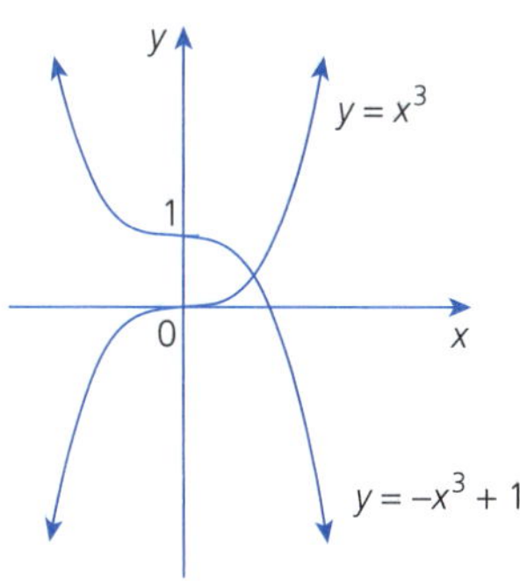

✓✓

b

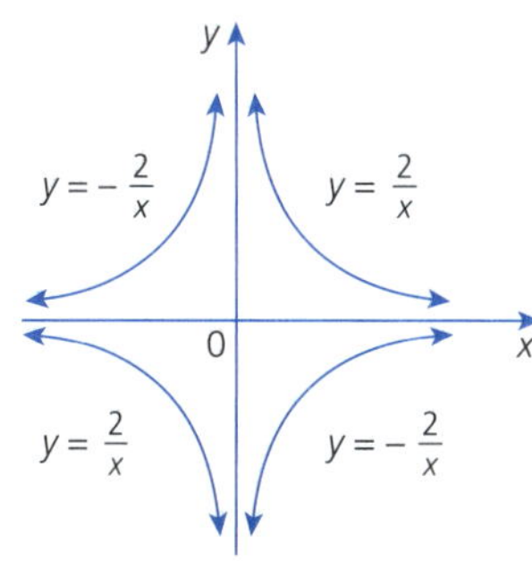

✓

15 a $(x - h)^2 + (y - k)^2 = r^2$

$(x - 5)^2 + (y + 2)^2 = 16$ ✓

b $(x - h)^2 + (y - k)^2 = r^2$

$x^2 + y^2 = 17$ ✓

c C(2, 2), radius: 2 units ✓

$(x - h)^2 + (y - k)^2 = r^2$

$(x - 2)^2 + (y - 2)^2 = 4$ ✓

16 a Let $y = 0$

$\therefore (x + 2)(x - 1)(x + 3) = 0$

$x = -2, 1, -3$

x-intercepts of -2, 1, -3 ✓

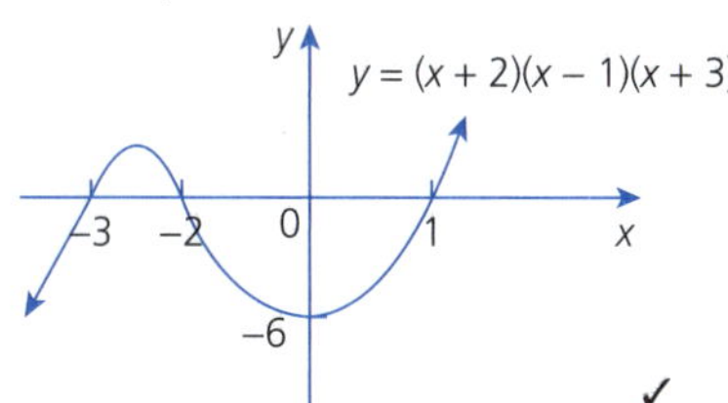

✓

b Let $y = 0$

$\therefore 2x(x + 1)(x - 2) = 0$

$x = 0, -1, 2$

x-intercepts of 0, -1, 2 ✓

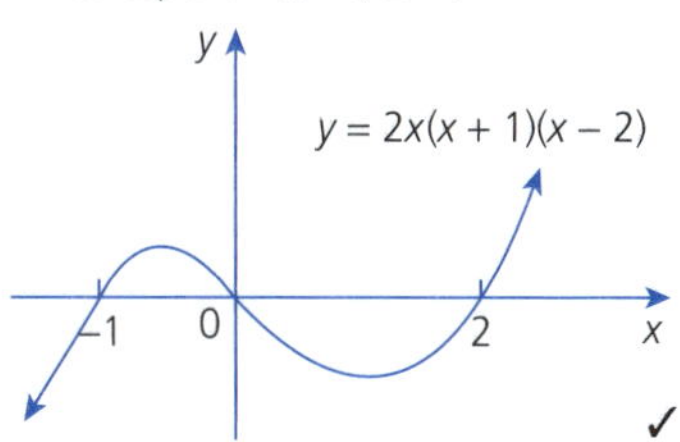

✓

17 $y = x^2$, $y = 2x + 3$
$\therefore \quad x^2 = 2x + 3$ ✓
$x^2 - 2x - 3 = 0$
$(x - 3)(x + 1) = 0$
$x = 3, -1$ ✓
Substitute into $y = x^2$
$\therefore x = 3, y = 9$
$x = -1, y = 1$
$\therefore$ The points of intersection are (3, 9), (−1, 1). ✓

Advanced Test

p. 102

1 Only one x-intercept of $x = 3$
$\therefore y = (x - 3)^2$
$\therefore$ C ✓

2

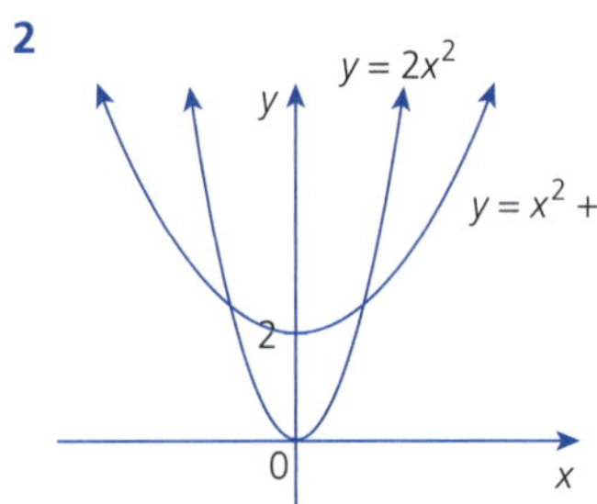

Both parabolas are concave up
$\therefore$ B ✓

3 $y = x^2 - 4x - 5$
$\therefore y = (x - 5)(x + 1)$
Letting $y = 0$
$\therefore \quad (x - 5)(x + 1) = 0$
$x = 5, -1$ ✓
Also, letting $x = 0 \quad \therefore y = -5$

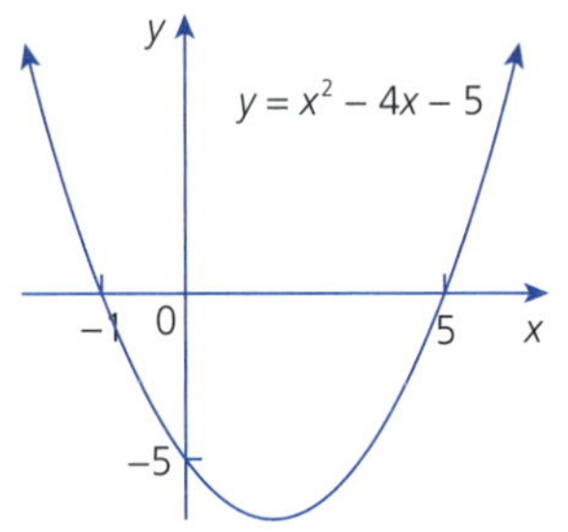

✓

4 a As x-intercepts of 5, 2 and middle of 5 and 2 is $3\frac{1}{2}$
$\therefore$ Axis of symmetry is $x = 3\frac{1}{2}$. ✓

b Substitute $x = 3\frac{1}{2}$ in y
$\therefore y \; 3\frac{1}{2} = 3\frac{1}{2} - 5 \;\; 2 - 3\frac{1}{2}$
$= -1\frac{1}{2} \times -1\frac{1}{2}$
$= 2\frac{1}{4}$
$\therefore$ Maximum value of $2\frac{1}{4}$. ✓

5 $y = 2x^2 - 4x + 5$
Now, as $x = \dfrac{-b}{2a}$
$= \dfrac{4}{2(2)}$
$= 1$
$\therefore$ The equation of axis of symmetry is $x = 1$. ✓
Substitute x = 1 in y
$\therefore y(1) = 2(1)^2 - 4(1) + 5$
$= 2 - 4 + 5$
$= 3$
$\therefore$ Minimum value of 3. ✓

6 a Let $y = ax(x - 4)$
Substitute in (2, 12)
$\therefore \; 12 = 2a(2 - 4)$ ✓
$12 = 2a(-2)$
$-4a = 12$
$a = -3$
$\therefore y = -3x(x - 4)$ ✓

b x-intercepts of −4 and 6
$\therefore$ Middle of −4 and 6 is 1
$\therefore$ (1, −50) lies on parabola
Now, let $y = a(x + 4)(x - 6)$ ✓
$-50 = a(1 + 4)(1 - 6)$
$-50 = a(5)(-5)$
$-25a = -50$
$a = 2$
$\therefore y = 2(x + 4)(x - 6)$ ✓

c y-intercept of −64
$\therefore$ (0, −64) lies on parabola
x-intercept of −2 and 8:
$\therefore$ Let $y = a(x + 2)(x - 8)$ ✓
$-64 = a(0 + 2)(0 - 8)$
$-64 = a(2)(-8)$
$-16a = -64$
$a = 4$
$\therefore y = 4(x + 2)(x - 8)$ ✓

7 Checking alternatives:
In A, for $y = x^2 - 6x + 7$ axis of sym.: $x = 3$
And substitute (1, 2) into equation:
$2 = 1^2 - 6(1) + 7$
$2 = 1 - 6 + 7$
$2 = 2$
$\therefore$ Passes through (1, 2)
$\therefore$ The equation is $y = x^2 - 6x + 7$.
$\therefore$ A ✓

8

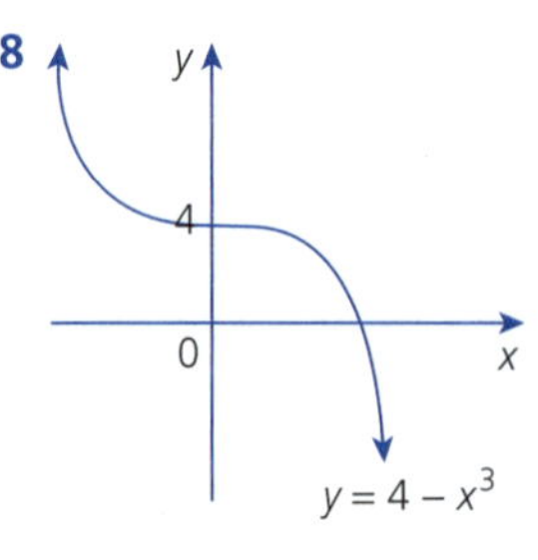

✓

9 $2x^2 + 2y^2 = 96$
$\therefore x^2 + y^2 = 48$
$\therefore \quad r^2 = 48$
$r = \sqrt{48}$
$r = 4\sqrt{3}$
$\therefore$ The radius is $4\sqrt{3}$ units. ✓

10 a If $y = 0$, then $x(x + 4) = 0$
$x = 0, -4$
$\therefore$ x-intercepts of 0, −4
$\therefore x = -2$ is axis of sym. ✓
Now, substitute $x = -2$ into y:
$y(-2) = -2(-2 + 4) = -4$
$\therefore$ Vertex (−2, −4). ✓

b If $y = 0$ then $(x + 3)(x - 5) = 0$
$x = -3, 5$
$\therefore$ x-intercepts of −3, 5
$\therefore x = 1$ is axis of sym. ✓
Now, substitute $x = 1$ into y:
$y(1) = (1 + 3)(1 - 5)$
$= -16$
$\therefore$ Vertex (1, −16). ✓

c $y = 2x^2 - 4x + 1$
Axis of sym: $x = \dfrac{-b}{2a}$
$= \dfrac{4}{4}$
$= 1$
$\therefore x = 1$ is axis of sym. ✓
Now, substitute $x = 1$ into y:
$y(1) = 2(1)^2 - 4(1) + 1 = -1$
$\therefore$ Vertex (1, −1). ✓

11 a As x-intercepts of −1, 3
$\therefore y = a(x + 1)(x - 3)$ ✓
Substitute (1, −12):
$\therefore \quad -12 = a(1 + 1)(1 - 3)$
$-12 = -4a$
$\therefore \quad a = 3$
$\therefore y = 3(x + 1)(x - 3)$ ✓

b As vertex is (3, 18) and an x-intercept of $x = 0$, then other x intercept is $x = 6$
$\therefore y = ax(x - 6)$ ✓
Substitute (3, 18):
$\therefore \quad 18 = a \times 3(3 - 6)$
$18 = -9a$
$a = -2$
$\therefore y = -2x(x - 6)$ ✓

c As x-intercepts of ±3,
$y = a(x - 3)(x + 3)$ ✓
Substitute (0, −36):
$-36 = a(0 - 3)(0 + 3)$
$-9a = -36$
$a = 4$
$\therefore y = 4(x - 3)(x + 3)$ ✓

12 a

y
$y = (x - 2)^3$
0
2
x
−8

✓

b

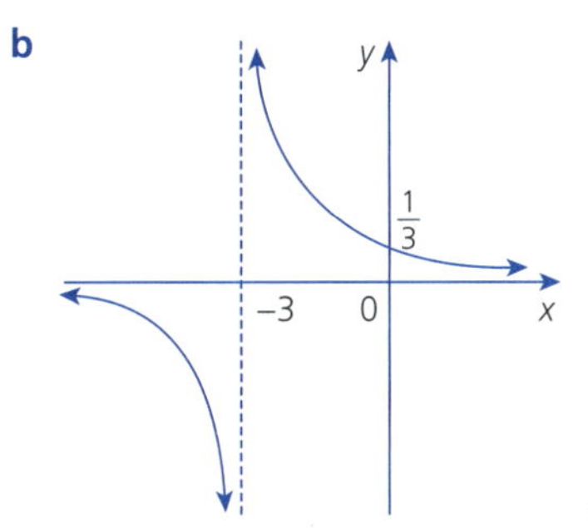

✓

13
$$x^2 - 4x + y^2 - 12y = 9$$
$$x^2 - 4x + 4 + y^2 - 12y + 36 = 9 + 4 + 36$$
$$(x - 2)^2 + (y - 6)^2 = 49 \quad ✓$$
$\therefore C(2, 6)$, radius 7 units. ✓

14
$$x^2 + y^2 + 6y = 0$$
$$x^2 + y^2 + 6y + 9 = 9$$
$$x^2 + (y + 3)^2 = 9 \quad ✓$$
$\therefore C(0, -3)$, radius 3 units.

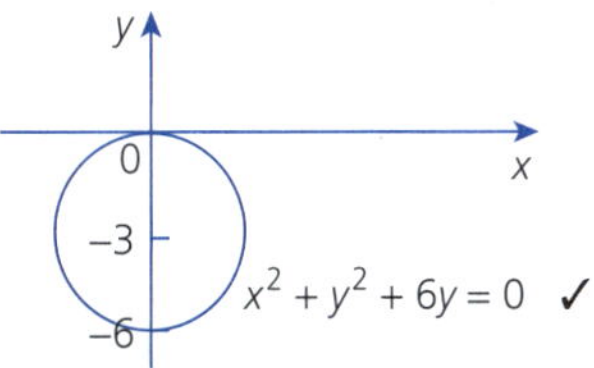

15 If $y = 0$, then
$$-(x + 2)^2(x - 1) = 0$$
$$x = -2, 1$$
$\therefore$ x-intercepts of −2 and 1. ✓

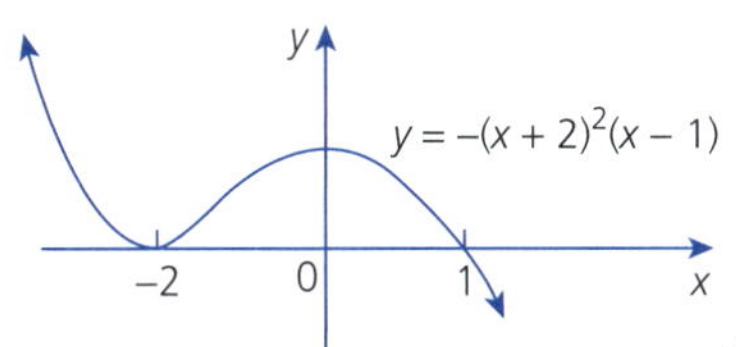

✓

16 As $y = \frac{12}{x}$ and $y = 3x$

$\therefore \quad \frac{12}{x} = 3x$

$\therefore \quad 12 = 3x^2$

$\therefore \quad x^2 = 4$

$x = \pm 2$ ✓

Now, subs. in $y = 3x$

$\therefore x = 2, y = 6$

$x = -2, y = -6$

$\therefore$ The points of intersection (2, 6), (−2, −6). ✓

17 As $y = 4x - 1$ and $y = x^2 + 3$

$\therefore \quad x^2 + 3 = 4x - 1$

$\therefore \; x^2 - 4x + 4 = 0$

$(x - 2)^2 = 0$

$x = 2$

Now, as only one point of intersection $\therefore$ they touch at one point. ✓

Chapter 7

Rates and Proportion

Intermediate Test

p. 114

1 $0.4 \times 365 = 146$

$\therefore$ 146 mm/year = 14.6 cm/year ✓

2 a $10\,000 \text{ m}^2 = 1$ ha

$20 \times 10\,000 = 200\,000$

$\therefore$ 200 000 g/ha = 200 kg/ha ✓

b $100 \times 100 \times 100 = 1\,000\,000$

$\therefore 1\,000\,000 \text{ cm}^3 = 1 \text{ m}^3$

$0.08 \times 1\,000\,000 = 80\,000$

$\therefore 80\,000 \text{ c/m}^3 = \$800 \text{ /m}^3$ ✓

3 Bolt: 100 m/9.58 s

$100 \div 9.58 \times 60 \times 60$

= 37 578 (nearest whole)

$\therefore$ 37 578 m/h = 37.578 km/h ✓

Cheetah: 100 m/5.95 s

$100 \div 5.95 \times 60 \times 60$

= 60 504 (nearest whole)

$\therefore$ 60 504 m/h = 60.504 km/h ✓

Difference = 22.926

$\therefore$ Cheetah is faster by 23 km/h ✓

4 a

h	0	1	2	3	4
D	0	90	180	270	360

✓

b $D = 90h$ ✓

c 90 km/h ✓

d 75 minutes = $1\frac{1}{4}$ hours = 1.25 hours

Subs. $h = 1.25$ in $D = 90h$:

$D = 90 \times 1.25$

$= 112.5$

$\therefore$ 112.5 km ✓

e Second car: $D = 60h$

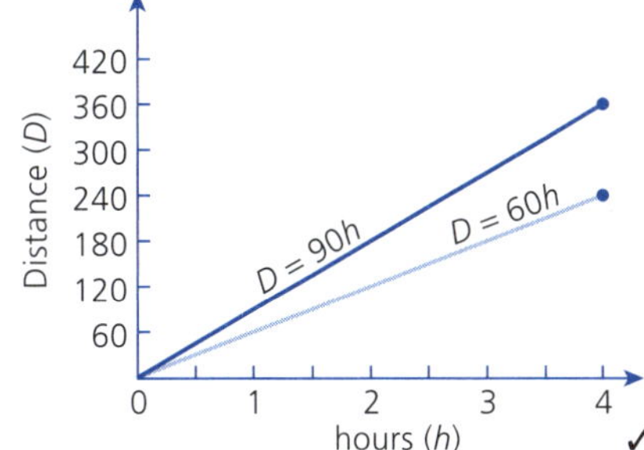

✓

f Subs. $h = 3.5$ in $D = 90h$:

$D = 90 \times 3.5$

$= 315$

$\therefore$ First car travelled 315 km.

Subs. $h = 3.5$ in$D = 60h$:

$D = 60 \times 3.5$

$= 210$

$\therefore$ Second car travelled 210 km.

Difference = 105

$\therefore$ First car travelled 105 km more. ✓

g Subs. $D = 180$ in $D = 90h$:

$180 = 90h$

$h = 2$

$\therefore$ First car takes 2 hours. ✓

Subs. $D = 180$ in $D = 60h$:

$180 = 60h$

$h = 3$

$\therefore$ Second car takes 3 hours.

Difference = 1

$\therefore$ Second car takes 1 hour longer. ✓

5 a $C = kA$ ✓

Subs. $C = 235\,000$ and $A = 200$ in $C = kA$:

$235\,000 = 200k$

$k = 1175$

$\therefore C = 1175A$ ✓

b Subs. $A = 310$ in $C = 1175A$:

$C = 1175 \times 310$

$= 364\,250$

$\therefore$ \$364 250 ✓

6 a $d = k\sqrt{h}$ ✓

Subs. $d = 4.16$ and $h = 1.69$ in $d = k\sqrt{h}$:

$4.16 = k \times \sqrt{1.69}$

$4.16 = 1.3k$

$k = \frac{4.16}{1.3}$

$= 3.2$

$\therefore d = 3.2\sqrt{h}$ ✓

b Subs. $d = 8$ in $d = 3.2\sqrt{h}$:

$8 = 3.2\sqrt{h}$

$\sqrt{h} = \frac{8}{3.2}$

$= 2.5$

$h = 6.25$

$\therefore$ 6.25 metres ✓

c Subs. $h = 18$ in $d = 3.2\sqrt{h}$:

$d = 3.2\sqrt{18}$

$= 13.58$ (2 dec. pl.)

As 13.58 > 12, then the person can see the island. ✓

7 a

h

0

t ✓

b 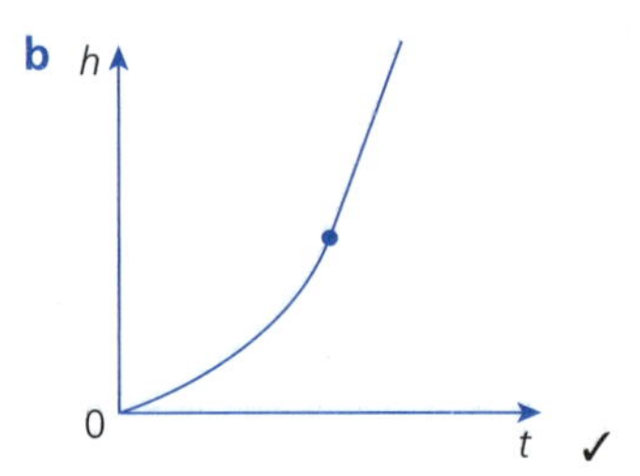

✓

8 a 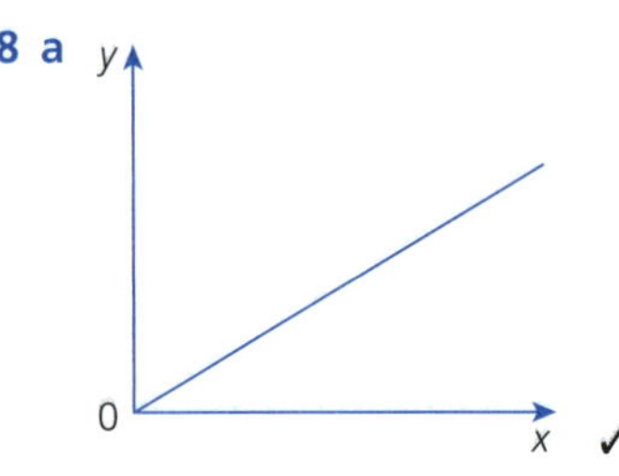

✓

b 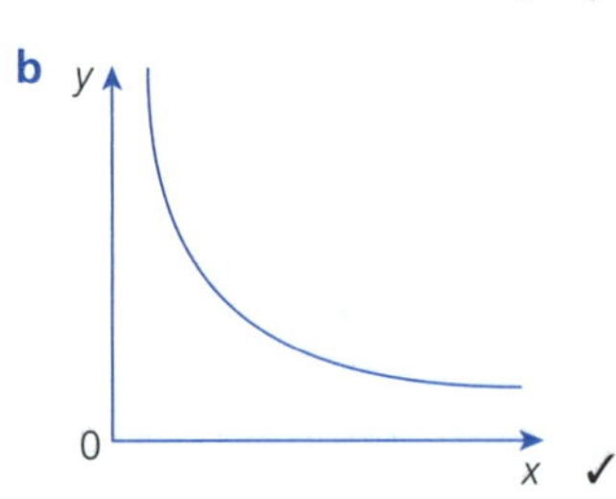

✓

Advanced Test

1 20g/125 mL
$20 \div 125 \times 1000 = 160$
$\therefore$ 160 grams/1000 mL = 160 g/L ✓

2 $y = kx$
If $x = 1$, then $y = k$
If $x = 2$, then $y = 2k$
$\therefore$ If x is doubled, then y is doubled. ✓

3 $y = kx^2$
If $x = 1$, then $y = k$
If $x = 2$, then $y = 4k$
$\therefore$ If x is doubled, then y is multiplied by 4. ✓

4 a $y = k\sqrt{x}$ ✓
Subs. $x = 16$ and $y = 8$ in $y = k\sqrt{x}$:
$8 = k \times \sqrt{16}$
$8 = 4k$
$k = 2$ ✓
$\therefore y = 2\sqrt{x}$

b Subs. $x = 4$ in $y = 2\sqrt{x}$:
$y = 2\sqrt{4}$
$= 4$
Subs. $x = 49$ in $y = 2\sqrt{x}$:
$y = 2\sqrt{49}$
$= 14$
Subs. $y = 24$ in $y = 2\sqrt{x}$
$24 = 2\sqrt{x}$
$\sqrt{x} = 12$
$x = 144$

x	4	16	49	144
y	4	8	14	24

✓✓✓

5 a Truck A:
$\text{Gradient} = \frac{\text{Rise}}{\text{Run}}$
$= \frac{480}{6}$
$= 80$
$\therefore D = 80t$ ✓
Truck B:
$\text{Gradient} = \frac{600}{6}$
$= 100$
$\therefore D = 100t$ ✓

b Consider the difference in the two equations:
As $100t - 80t = 20t$, then to find differences between the 2 trucks use $D = 20t$.
Subs. $t = 4$ in $D = 20t$:
$D = 20 \times 4$
$= 80$
$\therefore$ 80 km more ✓✓

c Subs. $D = 60$ in $D = 20t$:
$60 = 20t$
$t = 3$
$\therefore$ After 3 hours. ✓✓

6 a $C = kn^2$ ✓
Subs. $n = 3$ and $C = 64.8$ in $C = kn^2$:
$64.8 = k \times 3^2$
$64.8 = 9k$
$k = 7.2$
$\therefore C = 7.2n^2$ ✓

b Subs. $C = 45$ in $C = 7.2n^2$:
$45 = 7.2n^2$
$n^2 = \frac{45}{7.2}$
$= 6.25$
$n = 2.5$
$\therefore$ 2.5 cm ✓

c Subs. $n = 4.3$ in $C = 7.2n^2$:
$C = 7.2 \times (4.3)^2$
$= 133.13$ (2 dec. pl.)
$\therefore$ Costs \$133.13. ✓

7 Let time = T and p = power
$\therefore T = \frac{k}{p}$
Subs. $T = 6$ and $p = 360$ in $T = \frac{k}{p}$:
$6 = \frac{k}{360}$
$k = 2160$
$\therefore T = \frac{2160}{p}$ ✓

Subs. $p = 500$ in $T = \frac{2160}{p}$:
$T = \frac{2160}{500}$
$= 4.32$
$\therefore$ Takes 4.32 minutes. ✓

8 a 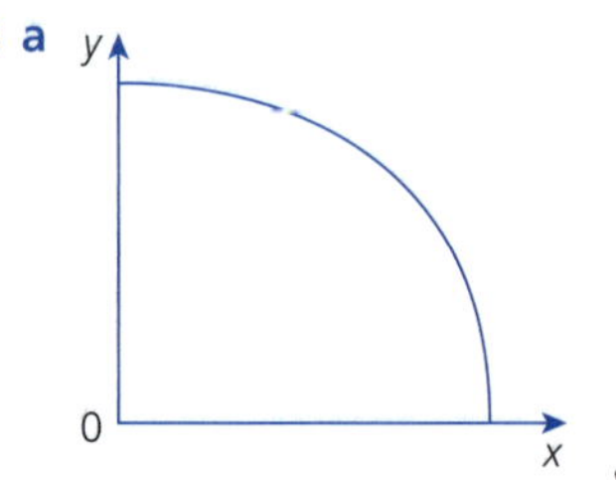

✓

b 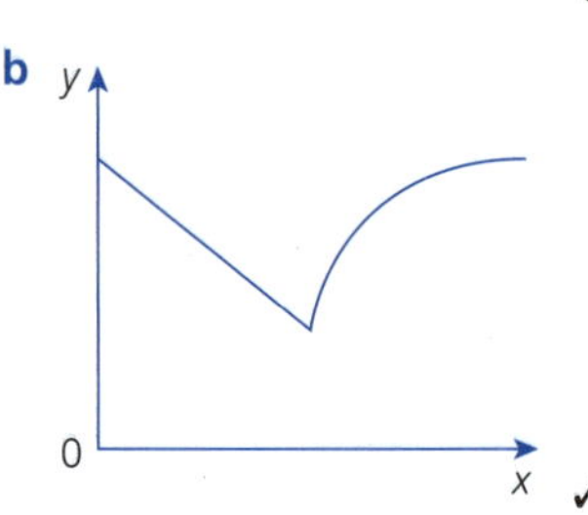

✓

9 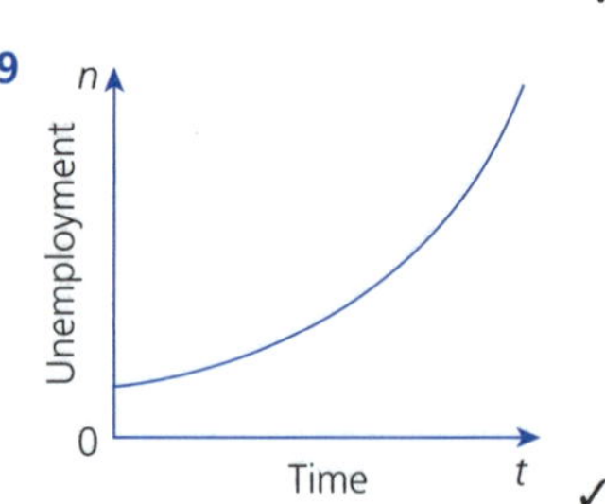

✓

10 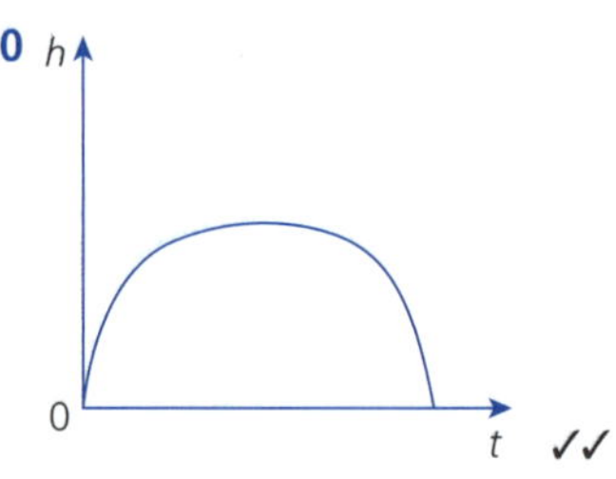

✓✓

Chapter 8

Pythagoras and Trigonometry

Intermediate Test

1 a $x^2 = 7^2 + 9^2$ ✓
$= 49 + 81$
$= 130$
$x = \sqrt{130}$
$= 11.401\,754\,25\ldots$
$= 11.40$ (2 dec. pl.) ✓

b $x^2 = 8.6^2 - 6.3^2$ ✓
$= 73.96 - 39.69$
$= 34.27$
$x = \sqrt{34.27}$
$= 5.854\,058\,421\ldots$
$= 5.85$ (2 dec. pl.) ✓

2 Determine whether $91^2 = 35^2 + 84^2$ ✓
LHS $= 91^2$
$= 8281$
RHS $= 35^2 + 84^2$
$= 1225 + 7056$
$= 8281$
∴ The triangle is right-angled. ✓

3

5 x θ 12

$x^2 = 5^2 + 12^2$
$= 169$
$x = \sqrt{169}$
$= 13$ ✓

$\tan\theta = \frac{5}{12}$ ✓

$\sin\theta = \frac{5}{13}$ ✓

4 $\sin 70° = \cos 20°$
$x = 20$ ✓

5 $\sin 65° = \cos 25°$ ✓
∴ $2y + 1 = 25$
$2y = 24$
$y = 12$ ✓

6 a $\frac{m}{18} = \tan 42°14'$ ✓
$m = 18 \times \tan 42°14'$
$= 16.3$ (to 1 dec. pl.) ✓

b $\frac{m}{27} = \sin 39°15'$ ✓
$m = 27 \times \sin 39°15'$
$= 17.1$ (1 dec. pl.) ✓

c $\frac{16.7}{m} = \cos 56°$ ✓
$m = \frac{16.7}{\cos 56\Upsilon}$
$= 29.9$ (1 dec. pl.) ✓

7 a $\tan x° = \frac{17}{5}$ ✓
$x° = 73°37'$ ✓

b $\cos y° = \frac{19}{24}$ ✓
$y° = 37°40'$ ✓

8

18°15′ h 5.2

$\frac{h}{5.2} = \cos 18°15'$ ✓
$h = 5.2 \times \cos 18°15'$
$= 4.9$ (to 1 dec. pl.)
The ladder reaches 4.9 m up the wall. ✓

9 a $\frac{x}{\sin 42\Upsilon 18} = \frac{16.4}{\sin 54\Upsilon}$ ✓
$\therefore x = \frac{16.4 \sin 42\Upsilon 18}{\sin 54\Upsilon}$
$= 13.6$ (to 1 dec. pl.) ✓

b $y^2 = 8.7^2 + 9.5^2 - 2(8.7)(9.5)\cos 87°18'$ ✓
$= 158.1533\ldots$
$y = \sqrt{158.153\ldots}$
$= 12.6$ (to 1 dec. pl.) ✓

10 Area $= \frac{1}{2}ab\sin C$
$= \frac{1}{2} \times 17 \times 15 \times \sin 16°40'$ ✓
$= 36.567\ldots$
≈ 36.6 cm^2 (to 1 dec. pl.) ✓

11 $m = \frac{3}{4}$

$\tan\alpha = \frac{3}{4}$ ✓
$\therefore \alpha = 36°52'$ ✓
The line makes an angle of 36°52′.

12

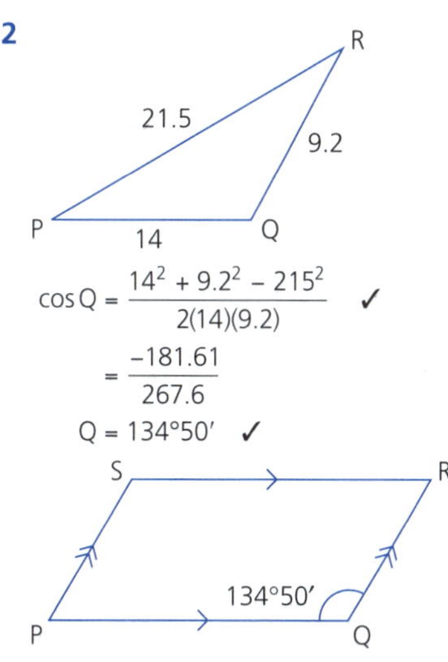

$\cos Q = \frac{14^2 + 9.2^2 - 215^2}{2(14)(9.2)}$ ✓
$= \frac{-181.61}{267.6}$
$Q = 134°50'$ ✓

$Q + R = 180°$
(co-interior angles, PQ ∥ SR)
$\therefore \hat{R} = 180° - 134°50'$
$= 45°10'$ ✓
Angles of the parallelogram are 134°50′ and 45°10′.

Advanced Test

pp. 142–143

1 Let x = missing side.
$x^2 = 25^2 - 7^2$ ✓
$= 625 - 49$
$= 576$
$x = \sqrt{576}$
$= 24$
∴ Length is 24 m. ✓

2 Let $AC = x$
$x^2 = 12^2 + 16^2$
$= 144 + 256$
$= 400$
$x = 20$ ✓
∴ Radius is 10 cm.

Area $= \frac{1}{2}\pi r^2$
$= \frac{1}{2} \times \pi \times 10^2$
$= 50\pi$
∴ Area is 50π cm^2. ✓

3

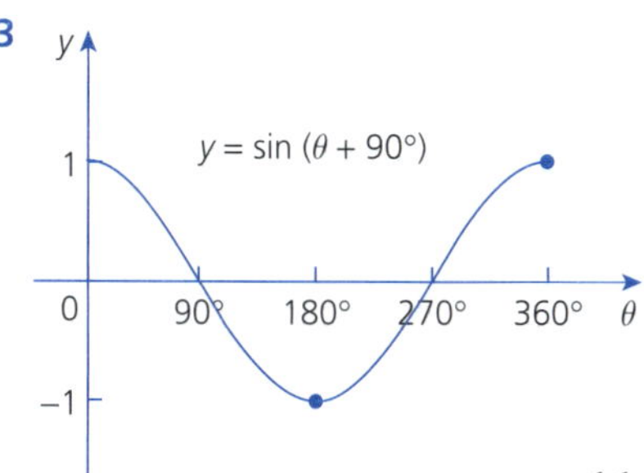

✓✓

4

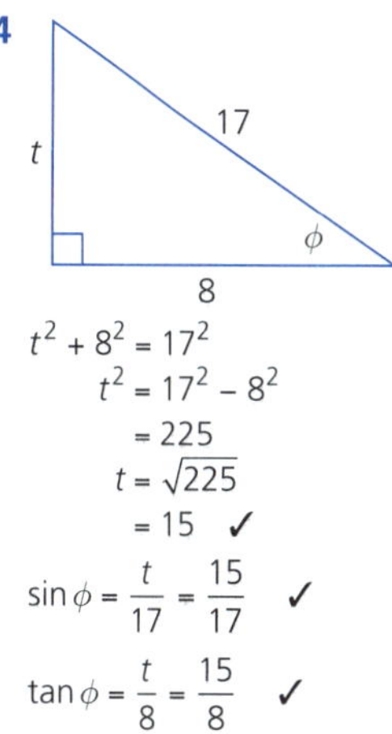

$t^2 + 8^2 = 17^2$
$t^2 = 17^2 - 8^2$
$= 225$
$t = \sqrt{225}$
$= 15$ ✓

$\sin\phi = \frac{t}{17} = \frac{15}{17}$ ✓

$\tan\phi = \frac{t}{8} = \frac{15}{8}$ ✓

5 a $\cos 45° = \frac{1}{\sqrt{2}}$ ✓

b $\tan 150° = -\tan 30° = \frac{-1}{\sqrt{3}}$ ✓

c $\sin 60° = \sin 120° = \frac{\sqrt{3}}{2}$

$\cos 120° = -\cos 60° = \frac{-1}{2}$

$(\sin 120°)^2 + (\cos 120°)^2$
$= \left(\frac{\sqrt{3}}{2}\right)^2 + \left(\frac{-1}{2}\right)^2$ ✓
$= \frac{3}{4} + \frac{1}{4}$
$= 1$ ✓

6 a $\frac{BD}{4.9} = \sin 49°17'$ ✓
$BD = 4.9 \times \sin 49°17'$
$= 3.71393$
$BD = 3.7$ ∴ 3.7 cm ✓

b $\dfrac{BF}{3.8} = \sin 52°$ ✓

$BF = 3.8 \times \sin 52°$

$= 2.99444$

$BF = 3.0$ (to 1 dec. pl.) ∴ 3.0 cm ✓

c $FD = BD - BF$

$= 0.7194$

$= 0.7$ (to 1 dec. pl.) ∴ 0.7 cm ✓

7 a $\theta = 65°$ ✓

b $m° = 51°20'$ ✓

8 $\dfrac{21}{t} = \cos 37°15'$ ✓

$\therefore t = \dfrac{21}{\cos 37°15'}$

$= 26.4$ (to 1 dec. pl.) ✓

9 

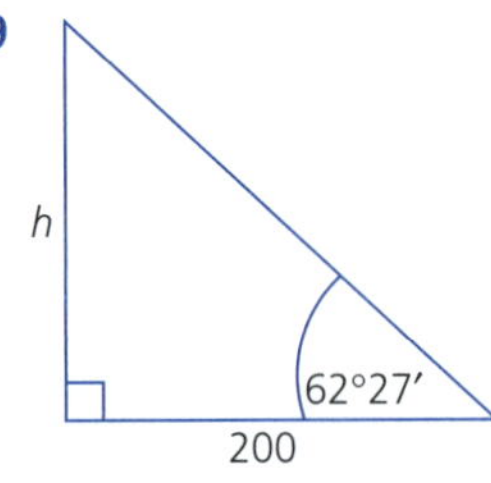

$\dfrac{h}{200} = \tan 62°27'$ ✓

$h = 200 \times \tan 62°27'$

$= 383.4$ (to 1 dec. pl.)

The height of aerial is 383.4 metres. ✓

10 a Third angle $= 180° - 112° - 27°15'$

$= 40°45'$

$\therefore \dfrac{y}{\sin 40°45'} = \dfrac{16.9}{\sin 112°}$ ✓

$y = \dfrac{16.9 \sin 40°45'}{\sin 112°}$

$= 11.9$ (to 1 dec. pl.)

Length is 11.9 cm. ✓

b $x^2 = 5.9^2 + 7.8^2 - 2(5.9)(7.8)\cos 110°27'$ ✓

$\therefore x = \sqrt{127.8078...}$

$= 127.8078...$

$= 11.3$ (to 1 dec. pl.)

Length is 11.3 cm. ✓

11 Largest ∠ is opposite longest side:

$\cos\theta = \dfrac{8^2 + 12^2 - 17^2}{2 \times 8 \times 12}$ ✓

$= \dfrac{-81}{192}$

$\therefore \theta = 114°57'$

The largest angle is 114°57′. ✓

12

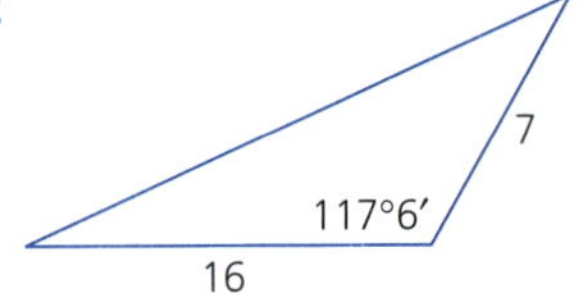

$A = \dfrac{1}{2}(16)(7)\sin 117°6'$ ✓

$= 49.8519$

Area of parallelogram $= 2 \times 49.8519$

$= 99.703...$

$= 99.7\text{ cm}^2$ ✓

13 At A, angle between line and → x-axis is given by

$\tan\alpha = -2$

$\alpha = 116°34'$

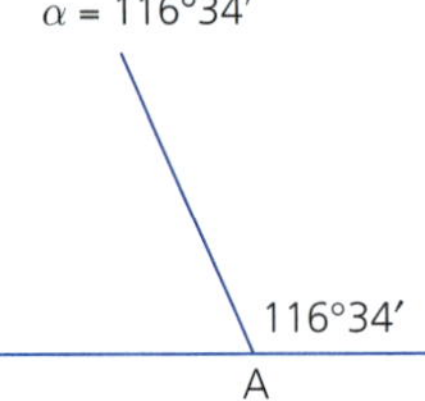

∴ Angle in △

$\angle BAC = 180° - 116°34'$

$= 63°26'$ ✓

At B, $4x - 3y + 12 = 0$

$4x + 12 = 3y$

$\dfrac{4}{3}x + 4 = y$

$m = \dfrac{4}{3}$

$\therefore \tan\theta = \dfrac{4}{3}$

$\theta = 53°8'$

$\therefore \angle CBA = 53°8'$ ✓

Then at C

$\angle BCA = 180° - 53°8' - 63°26'$

$= 63°26'$

∴ The angles of the triangle are 63°26′, 53°8′ and 63°26′. ✓

14 a

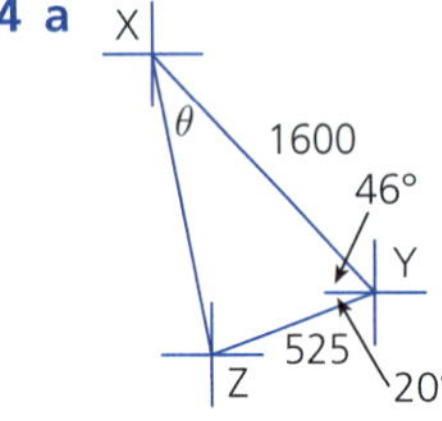

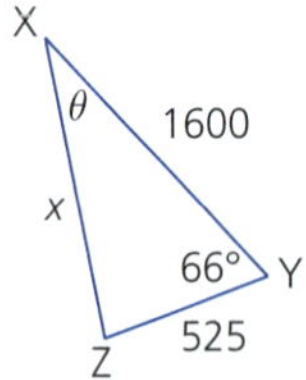

✓

b $x^2 = 1600^2 + 525^2 - 2(1600)(525)\cos 66°$ ✓

$= 215\,307$

$x = 1467$ (nearest whole)

∴ 1467 km ✓

c $\dfrac{\sin\theta}{525} = \dfrac{\sin 66°}{1467}$

$\sin\theta = \dfrac{525 \sin 66°}{1467}$ ✓

$= 0.3269$

$\theta = 19°$ (nearest whole)

$136 + 19 = 155$

∴ Bearing of Z from X is 155°. ✓

Chapter 9

Measurement

Intermediate Test

pp. 155–156

1 a $V = \dfrac{1}{3}Ah$

$= \dfrac{1}{3} \times 12 \times 10 \times 8$

$= 320$

∴ The volume is 320 cm³. ✓

b $V = \dfrac{1}{3}\pi r^2 h$

$= \dfrac{1}{3} \times \pi \times 9^2 \times 12$ ✓

$= 1017.876\,02$

$= 1017.88$ (to 2 dec. pl.)

∴ The volume is 1017.88 cm³. ✓

c $V = \dfrac{4}{3}\pi r^3$

$= \dfrac{4}{3} \times \pi \times 6^3$ ✓

$= 288\pi$

∴ The volume is 288π cm³. ✓

d Volume = Volume of hemisphere + Volume of cone

$V = \dfrac{1}{2} \times \dfrac{4}{3} \times \pi \times 10^3 + \dfrac{1}{3} \times \pi \times 10^2 \times 26$ ✓✓

$= 4817.108\,736$

$= 4817.11$ (to 2 dec. pl.)

∴ The volume is 4817.11 cm³. ✓

2 a $SA = 2\pi r^2 + 2\pi rh$

$= 2 \times \pi \times 6^2 + 2 \times \pi \times 6 \times 24$ ✓

$= 1130.973\,355$

$= 1130.97$ (to 2 dec. pl.)

∴ The surface area is 1130.97 cm². ✓

b SA = Area of square + 4 × Area of triangle

$= 10 \times 10 + 4 \times \dfrac{1}{2} \times 10 \times 8$ ✓

$= 260$

∴ The surface area is 260 cm². ✓

c $SA = \pi r^2 + \pi rl$

$= \pi \times 12^2 + \pi \times 12 \times 18$ ✓

$= 1130.973\,355$

$= 1130.97$ (to 2 dec. pl.)

∴ The surface area is 1130.97 cm². ✓

d $SA = 4\pi r^2$

$= 4 \times \pi \times 8^2$ ✓

$= 256\pi$

∴ The surface area is 256π cm². ✓

3 a Volume
= Volume of cylinder +
2 × Volume of cone

$= \pi \times 4^2 \times 10 + 2 \times \frac{1}{3} \times \pi \times 4^2 \times 10$ ✓
$= 837.758041$
$= 838$ (to 3 sig. fig.)
∴ The volume is 838 cm^3. ✓

b Volume of top cone $= \frac{1}{3}\pi r^2 h$

Volume of solid $= \pi r^2 h + 2 \times \frac{1}{3}\pi r^2 h$
$= \frac{5}{3}\pi r^2 h$

∴ The cone is $\frac{1}{5}$ of solid. ✓

4 $l^2 = 3^2 + 12^2$
$= 9 + 144$
$= 153$
$l = \sqrt{153}$ ✓

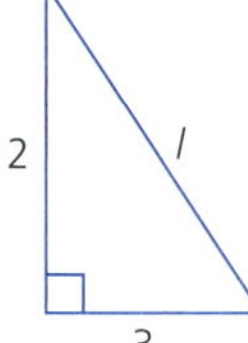

Curved SA $= \pi r l$
$= \pi \times 3 \times \sqrt{153}$
$= 116.5780651$
$= 116.58$ (to 2 dec. pl.)
∴ Surface area is 116.58 cm^2. ✓

5 $\pi r^2 h = 80\pi$
$16h = 80$
$h = 5$ ✓
$2\pi rh = 2 \times \pi \times 4 \times 5$
$= 40\pi$
∴ Curved surface area of 40π cm^2. ✓

6 a $\frac{1}{3}\pi r^2 h = 16\pi$

$\frac{1}{3}\pi r^2 \times 3 = 16\pi$
$r^2 = 16$ ✓
$r = 4$
∴ Radius of 4 cm. ✓

b Let slant height $= l$
$l^2 = 3^2 + 4^2$
$= 25$
$l = 5$ ✓
SA $= \pi r^2 + \pi r l$
$= \pi \times 4^2 + \pi \times 4 \times 5$ ✓
$= 36\pi$
∴ Surface area is 36π cm^2. ✓

7 Vol. of cone $= \frac{1}{3}\pi r^2 h$,

Vol. of sphere $= \frac{4}{3}\pi r^3$

$\frac{1}{3}\pi \times 15^2 \times h = \frac{4}{3}\pi \times 15^3$ ✓
$75h = 4500$
$h = 60$
∴ Height of 60 cm. ✓

Advanced Test

pp. 157–158

1 a In △ABC
$h^2 = 13^2 - 5^2$
$= 169 - 25$
$= 144$
$h = \sqrt{144}$
$= 12$ ✓

Now, $V = \frac{1}{3}Ah$
$= \frac{1}{3} \times 10 \times 10 \times 12$
$= 400$
∴ The volume is 400 cm^3. ✓

b $h^2 = 10^2 - 6^2$
$= 100 - 36$
$= 64$
$h = \sqrt{64}$
$= 8$ ✓
∴ $V = \frac{1}{3}\pi r^2 h$
$= \frac{1}{3} \times \pi \times 6^2 \times 8$
$= 96\pi$
∴ The volume is 96π cm^3. ✓

c $V = \frac{1}{2} \times \frac{4}{3}\pi r^3$
$= \frac{1}{2} \times \frac{4}{3} \times \pi \times 16^3$ ✓
$= 8578.642339$
$= 8578.64$ (to 2 dec. pl.)
∴ The volume is 8578.64 cm^3. ✓

2 a $V = \frac{4}{3}\pi r^3$
$= \frac{4}{3} \times \pi \times 8^3$
$= 2144.660585$
$= 2144.66$ (to 2 dec. pl.)
∴ The volume is 2144.66 cm^3. ✓

b Volume = Volume of cylinder
− 2 × Volume of sphere
$= \pi r^2 h - 2 \times \frac{4}{3}\pi r^3$
$= \pi \times 8^2 \times 32 - 2 \times \frac{4}{3} \times \pi \times 8^3$ ✓
$= 2144.660585$
$= 2144.66$ (to 2 dec. pl.)
∴ The volume is 2144.66 cm^3. ✓

3 SA $= 2\pi rh$
$= 2 \times \pi \times 8 \times 10$ ✓
$= 160\pi$
∴ The surface area is 160π cm^2. ✓

4 a Firstly,
$x^2 = 6^2 + 8^2$
$= 36 + 64$
$= 100$
$x = \sqrt{100}$
$= 10$
∴ AM is 10 cm. ✓

Also,
$y^2 = 5^2 + 6^2$
$= 25 + 36$
$= 61$
$y = \sqrt{61}$ ∴ AN is $\sqrt{61}$ cm ✓
Now,
SA $= 2 \times \frac{1}{2} \times 10 \times 10 + 2 \times \frac{1}{2} \times 16 \times \sqrt{61} + 16 \times 10$
$= 384.9639948$
$= 384.96$ (to 2 dec. pl.)
∴ The surface area of 384.96 cm^2. ✓

b $l^2 = 12^2 + 10^2$
$= 144 + 100$
$= 244$
$l = \sqrt{244}$ ✓

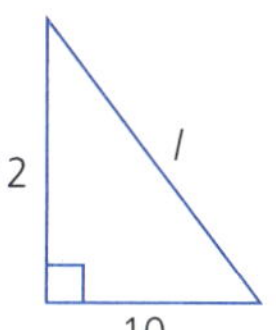

SA $= \pi r^2 + \pi r l$
$= \pi \times 10^2 + \pi \times 10 \times \sqrt{244}$
$= 804.8917254$
$= 804.89$ (to 2 dec. pl.)
∴ The surface area is 804.89 cm^2. ✓

c SA $= \frac{1}{2} \times 4\pi r^2 + \pi r^2$
$= \frac{1}{2} \times 4 \times \pi \times 13^2 + \pi \times 13^2$ ✓
$= 1592.787475$
$= 1592.79$ (to 2 dec. pl.)
∴ The surface area is 1592.79 cm^2. ✓

5 Diameter = 2 cm
Radius = 1 cm
Height = 2 cm

a Volume of cylinder $= \pi \times 1^2 \times 2$
$= 2\pi$
Total volume
$= \pi \times 1^2 \times 2 + \frac{1}{2} \times \frac{4}{3} \times \pi \times 1^3$
$= 2\pi + \frac{2\pi}{3}$
$= \frac{8\pi}{3}$ ✓
∴ Fraction $= 2\pi \div \frac{8\pi}{3}$
$= \frac{3}{4}$ ✓

b SA hemisphere $= \frac{1}{2} \times 4 \times \pi \times 1^2 = 2\pi$

Total SA $= \frac{1}{2} \times 4 \times \pi \times 1^2 + 2 \times \pi \times 1 \times 2 + \pi \times 1^2$
$= 7\pi$ ✓
∴ Fraction $= \frac{2\pi}{7\pi}$
$= \frac{2}{7}$ ✓

6 Let radius = 1 cm
$\therefore$ SA – $4\pi r^2$
$= 4\pi$ ✓
Now, let radius = 2 cm
$\therefore$ SA = $4\pi r^2$
$= 4\pi \times 2^2$
$= 16\pi$
$\therefore$ The surface area is four times the original. ✓

7 Let length of each side of square = 2 cm.
Volume of cube = 2^3
$= 8$
$\therefore$ Volume is 8 cm^3. ✓
As diameter of ball is 2 cm, then radius = 1 cm

Volume of ball = $\frac{4}{3} \times \pi \times 1^3$

$= \frac{4\neq}{3}$

$\therefore$ Volume is $\frac{4\neq}{3}$ cm^3. ✓

Percentage used by ball

$= \frac{4\neq}{3} \div 8 \times 100\%$ ✓

= 52.359 877 56…
= 52 (nearest whole)
$\therefore$ 52% of cube taken by basketball. ✓

8 Volume of shaded cone:

$V = \frac{1}{3}\pi r^2 h$

$240\pi = \frac{1}{3} \times \pi \times r^2 \times 20$

$720\pi = \pi r^2 \times 20$

$\therefore 20r^2 = 720$

$r^2 = 36$

$r = 6$ ✓

Now, using similar triangles:

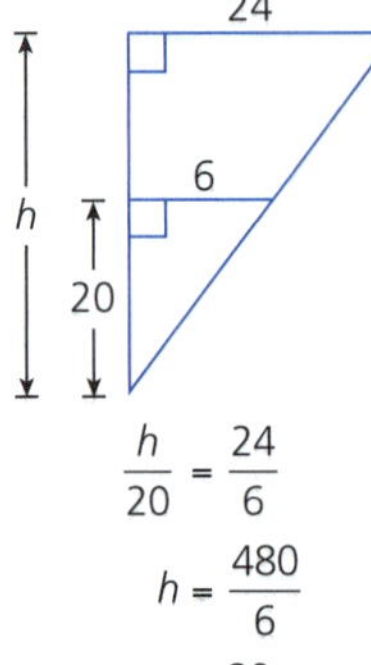

$\frac{h}{20} = \frac{24}{6}$

$h = \frac{480}{6}$

$= 80$

$\therefore$ The height is 80 cm. ✓

Chapter 10
Geometry

Intermediate Test

p. 184

1 a $a° = 180° - 88° - 18°$ ✓
$a = 74$ ($\angle$ sum of $\triangle$)

b $b = 69 + 57$ (exterior $\angle$ of $\triangle$)
$= 126$ ✓

c $c = 180 - 2 \times 47$
(equal $\angle$s of isosceles $\triangle$)
$= 86$ ✓

d $d + 79 + 58 + 117 = 360$
($\angle$ sum of quadrilateral)
$d = 360 - 254$
$= 106$ ✓

e $e = 78$
(opposite $\angle$s of parallelogram) ✓

f $f = 39$
(corresponding $\angle$s parallel lines) ✓

2

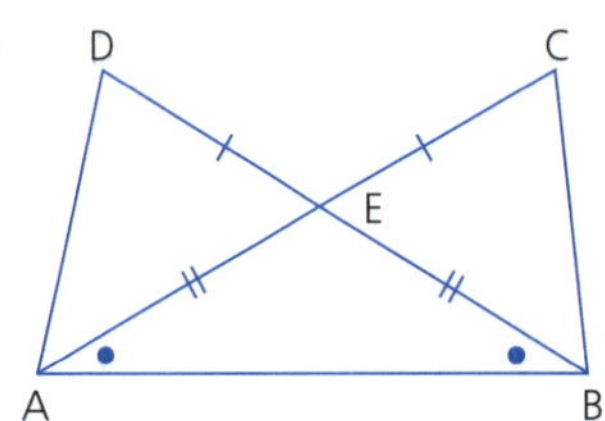

In $\triangle$s ABD and BAC
BD = AC (sum of equal sides)
AB = AB (common)
$\angle$ABD = $\angle$BAC (base $\angle$s isosceles $\triangle$)
$\therefore$ $\triangle$ABD $\equiv$ $\triangle$BAC (SAS) ✓✓✓
$\therefore$ AD = BC
(corresponding sides in congruent $\triangle$s) ✓

3

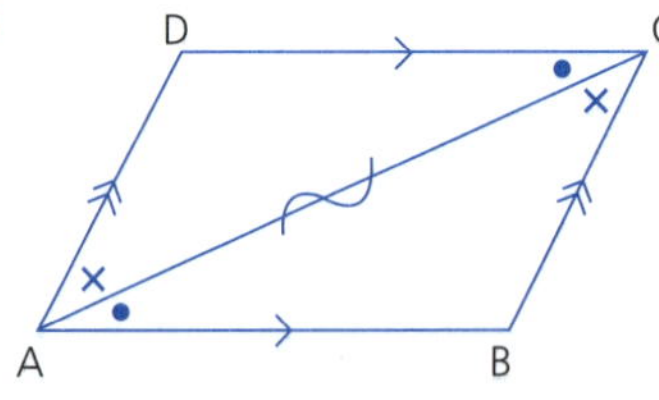

Consider $\triangle$s ABC and CDA
In $\triangle$s ABC and CDA
$\angle$CAB = $\angle$ACD (alternate $\angle$s, AB || DC)
$\angle$BCA = $\angle$DAC (alternate $\angle$s, AD || BC)
AC = AC (common)
$\therefore$ $\triangle$ABC = $\triangle$CDA (AAS) ✓✓
Then $\hat{B} = \hat{D}$ ($\angle$s corresponding angles in congruent $\triangle$s)
Similarly using diagonal BD it can be shown that $\hat{A} = \hat{C}$
$\therefore$ Opposite angles of the parallelogram are equal. ✓✓

4

In $\triangle$s QPR and SPT
$\angle$RQP = $\angle$TSP
(corresponding $\angle$s, QR || ST)
$\angle$QRP = $\angle$STP
(corresponding $\angle$s, QR || ST)
$\therefore$ $\triangle$QPR ||| $\triangle$SPT (equiangular) ✓✓
Then ratios of corresponding sides are equal.

$\frac{PQ}{PS} = \frac{PR}{PT} = \frac{QR}{ST}$

We don't need $\frac{PR}{PT}$

$\frac{PQ}{PS} = \frac{QR}{ST}$

$\therefore \frac{PQ \times ST}{PS} = QR$ ✓✓

5

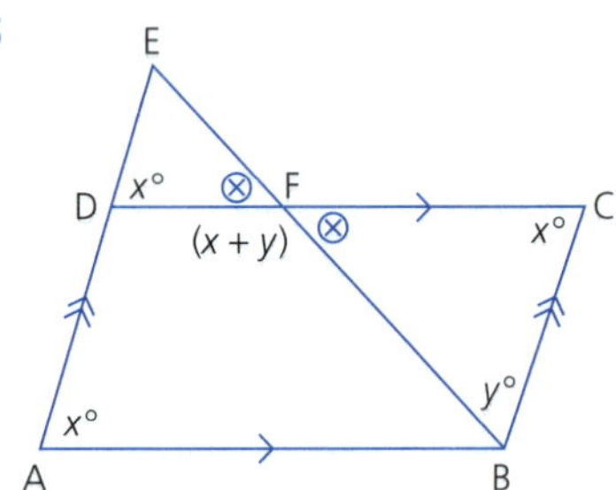

a Let $\angle$DAB = $x°$, $\angle$FBC = $y°$
Then $\angle$BCF = $x°$
(opposite $\angle$s of parallelogram)
Then $\angle$DFB = $x° + y°$
(exterior $\angle$ of $\triangle$FBC)
$\therefore$ $\angle$DFB = $\angle$DAB + $\angle$FBC ✓✓✓
also $\angle$EDF = $x°$
(corresponding $\angle$s, DC || AB)

b In $\triangle$s DFE and CFB
$\angle$EDF = $\angle$BCF (Both $x°$)
$\angle$DFE = $\angle$CFB (vertically opp. $\angle$s)
$\therefore$ $\triangle$DFE ||| $\triangle$CFB (equiangular)
✓✓

c Ratios of corresponding are equal

$\therefore \frac{DE}{BC} = \frac{EF}{BF} = \frac{DF}{FC}$ ✓

We don't need $\frac{EF}{BF}$

$\frac{4}{BC} = \frac{6}{8}$

$\therefore 6 \times BC = 32$

$BC = \frac{32}{6}$

$= 5\frac{1}{3}$ cm ✓

6 Sum = 180(10 − 2)
= 1440
∴ Angle sum is 1440°.
Each angle = 1440 ÷ 10
= 144
∴ Each angle is 144°. ✓

7 a i Ratio = 3 : 4 ✓

ii Ratio of areas = $3^2 : 4^2$
= 9 : 16 ✓

iii Ratio of volumes = $3^3 : 4^3$
= 27 : 64 ✓

b Volume of large cone = 640
∴ Vol. of small cone = $\frac{640}{64} \times 27$
= 270
∴ The volume is 270 cm^3. ✓

Advanced Test

pp. 185–186

1 a $2x + 3x + x = 180$ (∠ sum of △)
$6x = 180$
$x = 30$ ✓

b $y = 56 + (180 - 114)$
$= 56 + 66$ (ext. ∠ of △)
$= 122$ ✓

c $2c + 48 = 180$ (equal ∠s in isosc. △)
∴ $2c = 132$
$c = 66$ ✓

d $m + 112 = 180$
(opp. ∠s of parallelogram)
$m = 68$ ✓

e

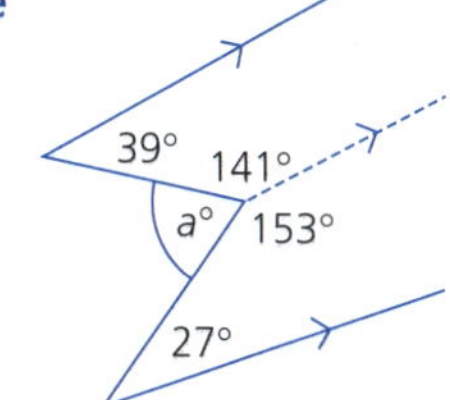

$a + 141 + 153 = 360$
$a = 66$ ✓

2 a

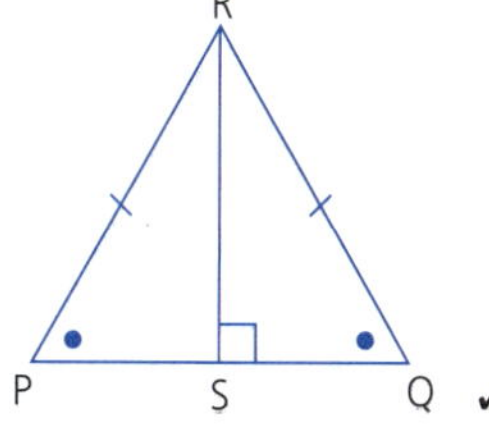

✓

b In △s PSR and QSR
∠RPS = ∠RQS (base ∠s isosceles △)
∠PSR = ∠QSR (RS ⊥ PQ)
PR = QR (given)
∴ △PSR ≡ △QSR (AAS) ✓✓

c PS = QS
(corresponding sides in congruent △s)
∠PRS = ∠QRS (corresponding angles in congruent △s) ✓✓

3 a

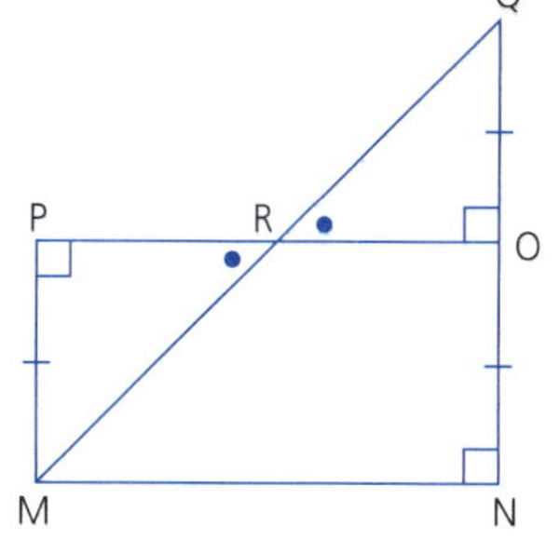

✓

b All ∠s of rectangle are 90°
∴ $R\hat{O}Q$ = 90° (NQ is straight line)
Opposite sides of rectangle are equal
∴ PM = ON
In △s PMR and OQR
PM = OQ (both equal NO)
∠MPR = ∠ROQ (both 90°)
∠MRP = ∠QRO (vertically opposite)
∴ △PMR ≡ △OQR (AAS) ✓✓

c PR = RO (corresponding sides in congruent △s)
∴ MQ bisects PO ✓

4

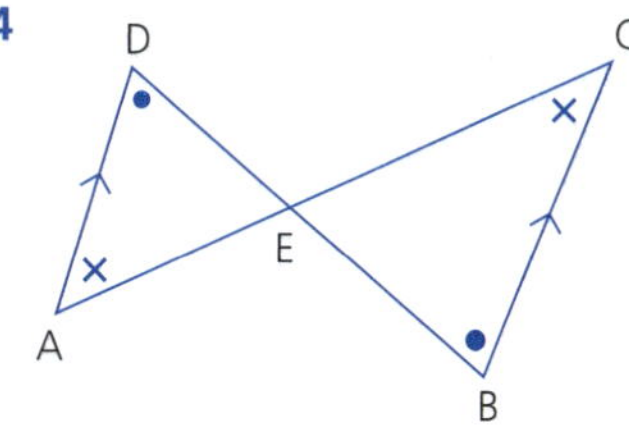

a In △s AED and CEB
∠DAE = ∠BCE
(Alternate ∠s, AD ∥ BC)
∠ADE = ∠EBC
(Alternate ∠s AD ∥ BC)
∴ △AED ||| △CEB (equiangular) ✓✓

b Ratios of corresponding sides are equal
∴ $\frac{DE}{EB} = \frac{AE}{EC} = \frac{AD}{BC}$
i.e. $\frac{DE}{EB} = \frac{AE}{EC}$ ✓

5

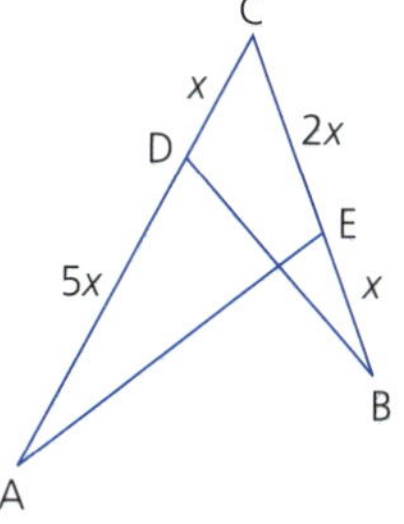

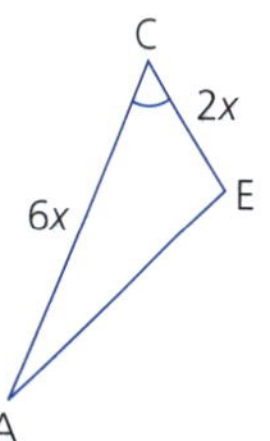

a In △s AEC and BDC
$\frac{AC}{BC} = \frac{6x}{3x} = \frac{2}{1}$ ✓
$\frac{EC}{DC} = \frac{2x}{1} = \frac{2}{1}$ ✓
∴ $\frac{AC}{BC} = \frac{EC}{DC}$
Also ∠ACE = ∠BCD
(common angle at C)
∴ △AEC ||| △BDC ✓
(ratios of corresponding sides equal, included angles equal)

b $\frac{AE}{BD} = \frac{2}{1}$
If BD = y
Then AE = $2y$ ✓

6

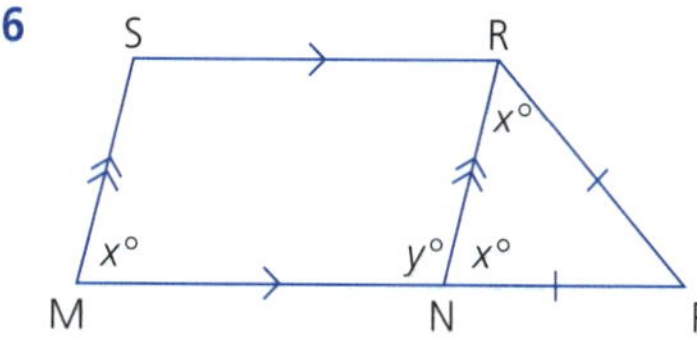

a Let ∠SMN = x°
∴ ∠RNP = x°
(corresponding ∠s, SM ∥ RN)
∴ ∠NRP = x°
(∠s opposite equal sides)
∴ ∠SMN = ∠NRP (Both x°) ✓✓

b Let ∠MNR = y°
Then ∠NPR + ∠NRP = ∠MNR
∠NPR = y° − ∠NRP
= y° − x°
= ∠MNR − ∠SMN ✓✓

7 Exterior angle = 18
Number of sides = 360 ÷ 18
= 20
Angle sum = $180(n - 2)$
= 180 × 18
= 3240
∴ Angle sum is 3240°. ✓

8 a Ratio of volumes = 1 : 8
∴ Ratio of lengths = 1 : 2 ✓
As diameter small cylinder = 3 cm
∴ Radius small cylinder = 1.5 cm
∴ Radius of large cylinder = 3 cm ✓

b Ratio of areas = 1 : 4 ✓
∴ Surface area = 40 × 4
= 160
∴ The surface area of large cylinder is 160 cm^2. ✓

Chapter 11
Statistics

Intermediate Test

p. 208

1 C as it has two modes: 3 and 4 ✓

2 a Mean $= \frac{26 + 32 + 28 + 39 + 30}{5}$
$= 31$
∴ Mean number of points is 31. ✓

b No change means the team needs to score the mean
∴ Team needs to score 31 points. ✓

c Mean after 6 games $= \frac{31 \times 5 + 37}{6}$
$= 32$
∴ Mean after 6 games is 32. ✓

d Mean after 7 games = 30
Points scored in 7th game
$= 7 \times 30 - 6 \times 32$
$= 18$ ✓
Score was 28 − 18
∴ Team lost by 10 points. ✓

3 a Mean = 4 ✓
Mode = 5 ✓
Range = 6 − 1
= 5 ✓

b 1, 2, 3, 3, 3, 4, 4, 4, 5, 5, 5, 5, 6, 6
Median = 4 ✓
Lower quartile = 3
Upper quartile = 5
∴ Interquartile range = 2 ✓

c

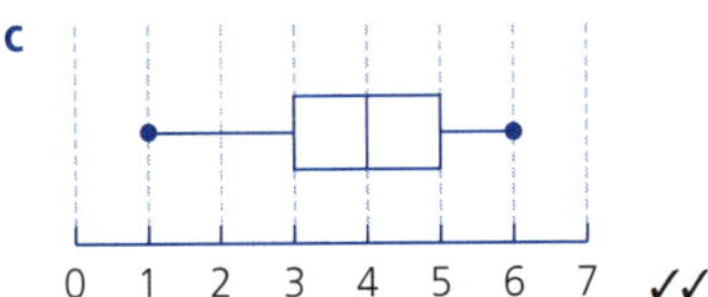

✓✓

4 a Mean = 31.5 ✓
Stand dev. = 4.61 (2 dec. pl.) ✓

b Mean = 2.8 ✓
Stand dev. = 2.48 (2 dec. pl.) ✓

5 a Highest was 24 out of 25; ✓
lowest was 3 out of 25. ✓

b i 50% ✓

ii 75% ✓

iii 88% is 22 out of 25
∴ 25% ✓

c 25% of 28 students = 7
∴ 7 students ✓

d 72% is 18 out of 25
∴ bottom half ✓

e Negatively skewed ✓

6 a i Wildfires: 51 − 8 = 43
Diamonds: 57 − 19 = 38
∴ Wildfires has bigger range. ✓

ii Wildfires: 19; Diamonds: 36
∴ Diamonds has bigger mode. ✓

iii Wildfires: 20.5; Diamonds: 41
∴ Diamonds has bigger median. ✓

b Wildfires is positively skewed;
Diamonds is negatively skewed. ✓

7 a

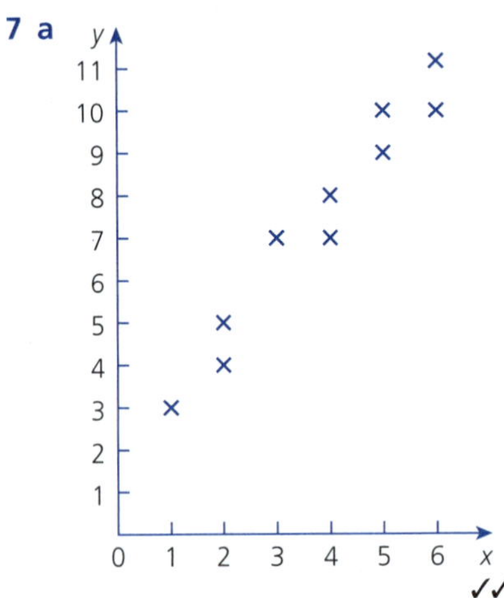

✓✓

b y ✓

c Positive ✓

d Strong ✓

8 Subs. $x = 50$ in $y = 2x + 20$
$= 2(50) + 20$
$= 120$ ✓

Advanced Test

pp. 209–210

1 a Normally distributed ✓

b Bi-modal ✓

c Negatively skewed ✓

d Positively skewed ✓

2 a Mode = 10; range = 35 ✓✓

b 18 ✓

c Lower quartile = 10;
Upper quartile = 31 ✓
∴ Interquartile range = 21 ✓

d

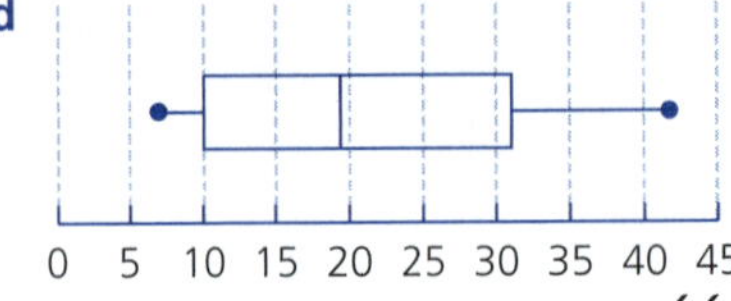

✓✓

3 a Negatively skewed ✓

b Standard deviation = 1.04 (2 dec. pl.) ✓

c The class mean is 8.7. This means that this class' mean is higher than the year mean. ✓✓

4 a $10 ✓

b 75% ✓

c 100 books ✓

d Shop A had IQR of 25, which is higher than Shop B's IQR of 15. ✓✓

5 Mean is 30 and standard deviation is unchanged. ✓✓

6 a

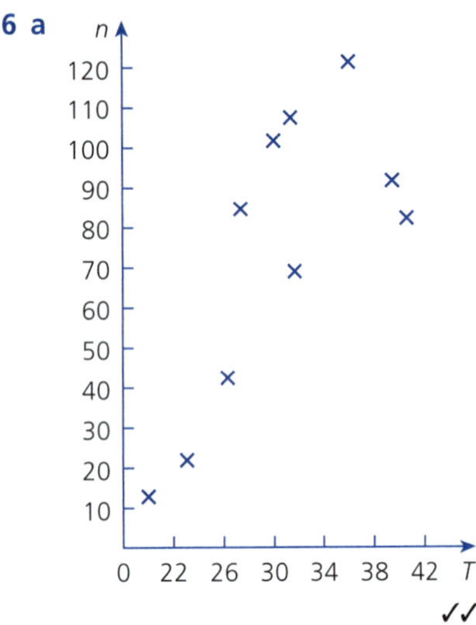

✓✓

b Positive ✓

c Strong relationship ✓
The hotter the temperature, the more people want to go to the beach. ✓

7 Lower quartile = 7.4;
Upper quartile = 10.1
∴ Interquartile range = 10.1 − 7.4
= 2.7 ✓

8 a Weak ✓

b 50 ✓

c $a = 45$; $b = 40$ ✓✓

9 a The line is of the form $y = mx + b$ and passes through (20, 50) and (30, 400)
Gradient $m = \frac{400 - 50}{30 - 20}$
$= \frac{350}{10}$
$= 35$ ✓
∴ Equation of the form $y = 35x + b$
Now, subs. (20, 50) in
$y = 35x + b$
$50 = 35(20) + b$
$50 = 700 + b$
$b = -650$
∴ The line is $y = 35x - 650$. ✓

b Subs. $x = 37$ in $y = 35(37) - 650$
$= 645$
∴ There would be about 645 at the pool. ✓

Chapter 12
Probability

Intermediate Test

pp. 225–226

1 $P(\text{head}) = \frac{1}{2}$

a $P(\text{HH}) = \frac{1}{2} \times \frac{1}{2}$
$= \frac{1}{4}$ ✓

b $P(\text{at least are head}) = 1 + P(\text{TT})$
$= 1 - \frac{1}{4}$
$= \frac{3}{4}$ ✓

2 a

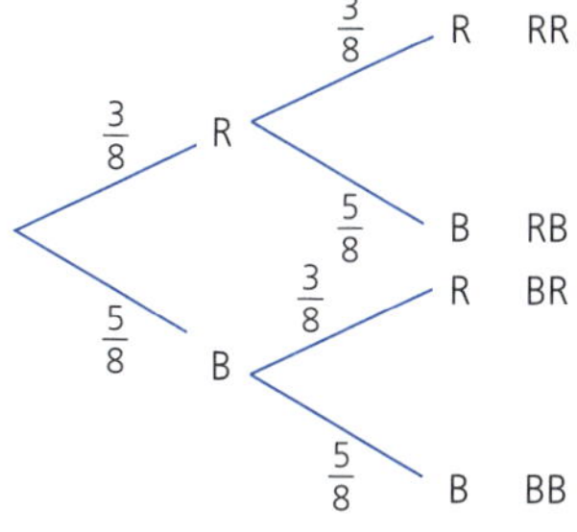

✓

b **i** $P(\text{BB}) = \frac{5}{8} \times \frac{5}{8}$
$= \frac{25}{64}$ ✓

ii $P(\text{RB}) + P(\text{BR}) = \frac{3}{8} \times \frac{5}{8} + \frac{5}{8} \times \frac{3}{8}$
$= \frac{15}{64} + \frac{15}{64}$
$= \frac{15}{32}$ ✓

3 a As $P(3) = \frac{1}{4}$, $P(\text{green}) = \frac{1}{3}$
$\therefore P(3 \text{ and green}) = \frac{1}{4} \times \frac{1}{3}$
$= \frac{1}{12}$ ✓

b $P(\text{even}) = \frac{1}{2}$, $P(\text{not blue}) = \frac{2}{3}$
$\therefore P(\text{even and not blue}) = \frac{1}{2} \times \frac{2}{3}$
$= \frac{1}{3}$ ✓

4 a As 100 males,
$P(\text{male smoker}) = \frac{27}{100}$
$= 0.27$ ✓

b As 100 females,
$P(\text{female non-smoker}) = \frac{64}{100}$
$= 0.64$ ✓

c As 200 people,
$P(\text{non-smoker}) = \frac{137}{200}$ ✓

5 80 commuters involved.

a $P(\text{bus}) = \frac{37}{80}$ ✓

b $P(\text{train only}) = \frac{43}{80}$ ✓

6 a $P(\text{head and 5}) = \frac{1}{12}$ ✓

b $P(\text{tail and odd}) = \frac{3}{12} = \frac{1}{4}$ ✓

c $P(\text{head and less than 3}) = \frac{2}{12} = \frac{1}{6}$ ✓

7 a

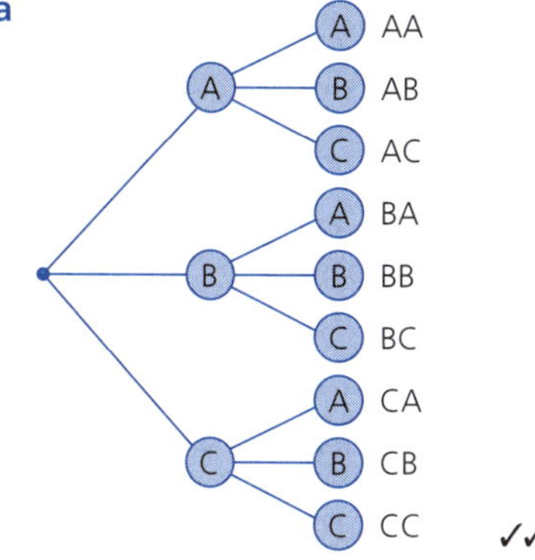

✓✓

b **i** $P(\text{BB}) = \frac{1}{9}$ ✓

ii $P(\text{BA}) = \frac{1}{9}$ ✓

iii $P(\text{at least 1 B}) = \frac{5}{9}$ ✓

iv $P(\text{AA, BB or CC}) = \frac{3}{9} = \frac{1}{3}$ ✓

v $P(\text{different letters}) = 1 - P(\text{same})$
$= 1 - \frac{1}{3}$
$= \frac{2}{3}$ ✓

8 a

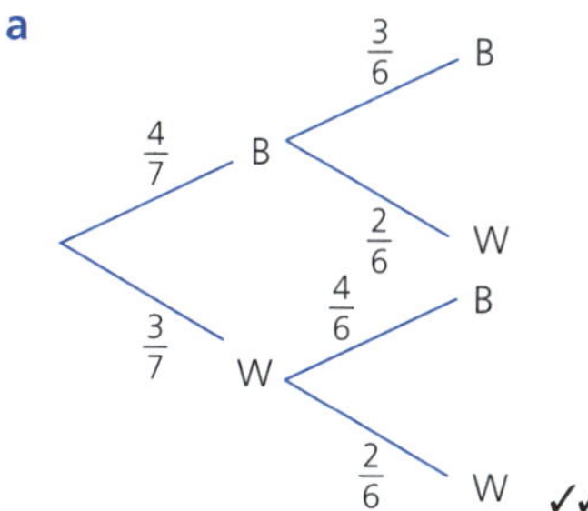

✓✓

b **i** $P(\text{BW}) = \frac{4}{7} \times \frac{3}{6}$
$= \frac{2}{7}$ ✓

ii $P(\text{WB}) = \frac{3}{7} \times \frac{4}{6}$
$= \frac{2}{7}$ ✓

iii $P(\text{WW}) = \frac{3}{7} \times \frac{2}{6}$
$= \frac{1}{7}$ ✓

9 a $P(\text{less than 6}) = \frac{5}{20} = \frac{1}{4}$ ✓

b $P(\text{even, given that more than 10})$
$= \frac{5}{10} = \frac{1}{2}$ ✓

c $P(\text{factor of 24, given that multiple of 4}) = \frac{3}{5}$ ✓

Advanced Test

pp. 227–228

1 a

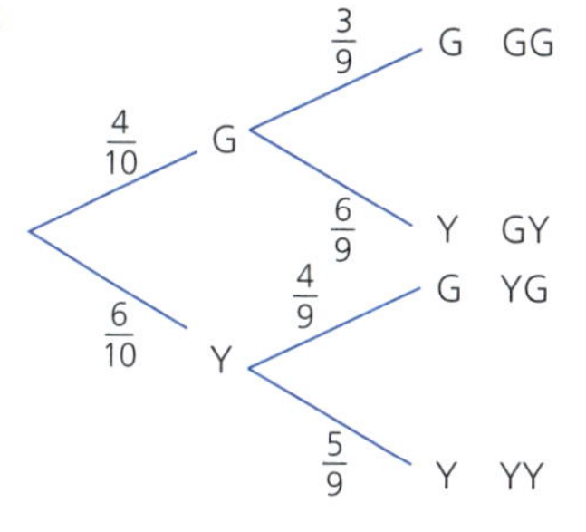

✓

b **i** $P(\text{GY}) + P(\text{YG})$
$= \frac{4}{10} \times \frac{6}{9} + \frac{6}{10} \times \frac{4}{9}$
$= \frac{24}{90} + \frac{24}{90}$
$= \frac{48}{90}$
$= \frac{8}{15}$ ✓

ii $P(\text{at least 1 yellow})$
$= 1 - P(\text{no yellow})$
$= 1 - P(\text{GG})$
$= 1 - \frac{4}{10} \times \frac{3}{9}$
$= 1 - \frac{12}{90}$
$= \frac{78}{90}$
$= \frac{13}{15}$ ✓

2 a 100 women surveyed.
$P(\text{women voted Lib–NP}) = \frac{45}{100}$
$= 0.45$ ✓

b 200 people surveyed.
85 people voted ALP.
$\therefore P(\text{voted ALP}) = \frac{85}{200}$
$= 0.425$ ✓

3 a **i** As $18 + 16 + 12 + 15 = 61$
$\therefore P(\text{like PDHPE}) = \frac{61}{100}$
$= 0.61$ ✓

ii As $13 + 18 = 31$
$\therefore P(\text{like English, not Science})$
$= \frac{31}{100}$
$= 0.31$ ✓

b As $14 + 16 + 12 + 12 = 54$
there are 54 students who like Science.

i As $14 + 16 = 30$
$\therefore P(\text{like English}) = \frac{30}{54}$
$= \frac{5}{9}$ ✓

ii $P(\text{like all 3}) = \frac{16}{54}$
$= \frac{8}{27}$ ✓

4 a

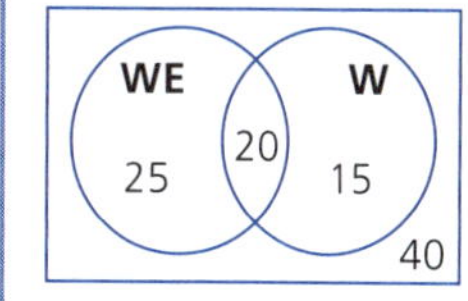

WE: Weekend Sport
W: Week Sport

✓

b $P(\text{both}) = \frac{20}{100} = \frac{1}{5}$ ✓

c $P(\text{week, not weekend}) = \frac{15}{55} = \frac{3}{11}$ ✓

5 a $P(\text{bat and bowler left hand})$
$= \frac{6}{30} = \frac{1}{5}$ ✓

b $P(\text{right-hand bowler, given left-hand batsman}) = \frac{4}{10} = \frac{2}{5}$ ✓

6 a

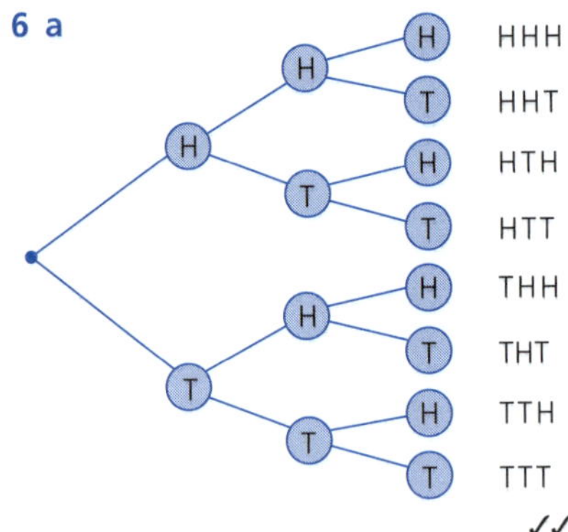

✓✓

b i $P(\text{TTT}) = \frac{1}{8}$ ✓

ii $P(\text{HTT, THT, TTH}) = \frac{3}{8}$ ✓

iii $P(\text{at least one H})$
$= 1 - P(\text{no heads})$
$= 1 - P(\text{TTT})$
$= 1 - \frac{1}{8}$
$= \frac{7}{8}$ ✓

7 a

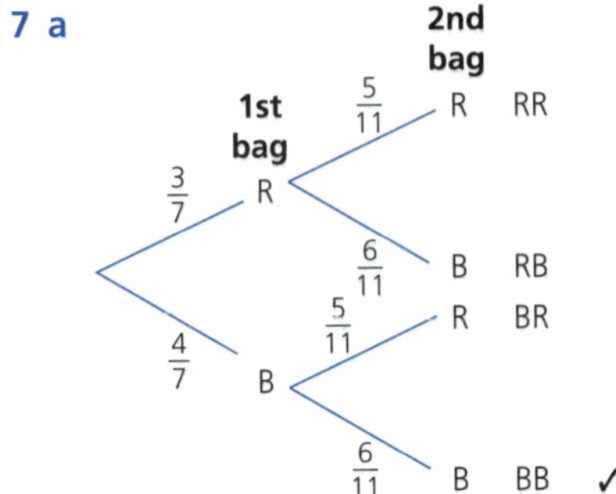

✓

b i $P(\text{RR}) = \frac{3}{7} \times \frac{5}{11}$
$= \frac{15}{77}$ ✓

ii $P(\text{RR or BB}) = \frac{3}{7} \times \frac{5}{11} + \frac{4}{7} \times \frac{6}{11}$
$= \frac{39}{77}$ ✓

iii $P(\text{RB or BR}) = 1 - P(\text{same})$
$= 1 - \frac{39}{77}$
$= \frac{38}{77}$ ✓

iv $P(\text{at least a blue}) = 1 - P(\text{no blue})$
$= 1 - P(\text{RR})$
$= 1 - \frac{15}{77}$
$= \frac{62}{77}$ ✓

8 a

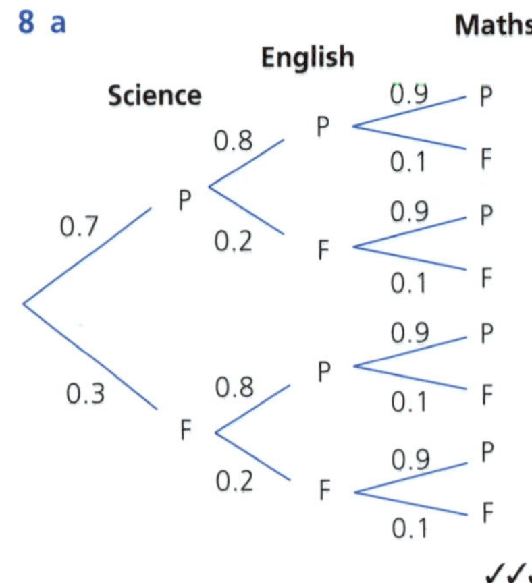

✓✓✓

b i $P(\text{PPP}) = 0.7 \times 0.8 \times 0.9$
$= 0.504$ ✓

ii $P(\text{FFP}) = 0.3 \times 0.2 \times 0.9$
$= 0.054$ ✓

iii $P(\text{FPP, PFP or PPF})$
$= 0.3 \times 0.8 \times 0.9 + 0.7 \times 0.2 \times 0.9 + 0.7 \times 0.8 \times 0.1$
$= 0.398$ ✓

Chapter 13
Polynomials

Intermediate Test

p. 240

1 a 4 ✓

b 2 ✓

2 a $(3x - 1)(2x^2 + x - 1)$
$\therefore$ degree = 3, constant term = 1 ✓✓

b $(4x^3 - 1)(2 - 7x^2)$
$\therefore$ degree = 5, constant term = −2 ✓✓

3
$$\begin{array}{r} x^2 - x + 5 \\ x - 1 \overline{) x^3 - 2x^2 + 6x - 1} \\ \underline{x^3 - x^2 \quad\quad\quad} \\ -x^2 + 6x \quad \\ \underline{-x^2 + x \quad} \\ 5x - 1 \\ \underline{5x - 5} \\ 4 \end{array}$$

$\therefore x^3 - 2x^2 + 6x - 1$
$= (x - 1)(x^2 - x + 5) + 4$ ✓✓

4 $P(x) = 2x^3 - 4x^2 + 5x - 1$

a $P(0) = 2(0)^3 - 4(0)^2 + 5(0) - 1$
$= -1$ ✓

b $P(3) = 2(3)^3 - 4(3)^2 + 5(3) - 1$
$= 54 - 36 + 15 - 1$
$= 32$ ✓

c $P(-2) = 2(-2)^3 - 4(-2)^2 + 5(-2) - 1$
$= -16 - 16 - 10 - 1$
$= -43$ ✓

5 Let $P(x) = x^3 - 4x^2 + x + 6$.
$\therefore P(-1) = (-1)^3 - 4(-1)^2 + (-1) + 6$ ✓
$= -1 - 4 - 1 + 6$
$= 0$
$\therefore (x + 1)$ is factor ✓

6 Let $P(x) = x^3 + 3x^2 - 6x - 8$.
Consider $P(1) = 1^3 + 3(1)^2 - 6(1) - 8$
$= 1 + 3 - 6 - 8$
$\neq 0$
Consider $P(-1) = (-1)^3 + 3(-1)^2 - 6(-1) - 8$
$= -1 + 3 + 6 - 8$
$= 0$
$\therefore (x + 1)$ is factor ✓

$$\begin{array}{r} x^2 + 2x - 8 \\ x + 1 \overline{) x^3 + 3x^2 - 6x - 8} \\ \underline{x^3 + x^2 \quad\quad\quad\quad} \\ 2x^2 - 6x \quad \\ \underline{2x^2 + 2x \quad} \\ -8x - 8 \\ \underline{-8x - 8} \\ 0 \end{array}$$

$\therefore (x + 1)(x^2 + 2x - 8)$ ✓
$\therefore (x + 1)(x + 4)(x - 2)$ ✓

7 a

$y = (x - 2)(x + 1)(x - 3)$

y; 6; −1; 0; 1; 2; 3; x ✓

b

$y = x^2(2 - x)(x + 1)^3$

y; −3; −2; −1; 0; 1; 2; 3; x ✓

8 $y = a(x + 1)(x - 2)(x - 5)$ ✓
Subs. (0, 20)
$\therefore 20 = a(1)(-2)(-5)$
$\therefore 10a = 20$
$a = 2$
$\therefore y = 2(x + 1)(x - 2)(x - 5)$ ✓

Advanced Test

p. 241

1

$$
\begin{array}{r}
x^2 - \ x + 2 \\
x - 3\overline{)x^3 - 4x^2 + 5x + 2} \\
\underline{x^3 - 3x^2} \\
-x^2 + 5x \\
\underline{-x^2 + 3x} \\
2x + 2 \\
\underline{2x - 6} \\
8
\end{array}
$$

$\therefore x^3 - 4x^2 + 5x + 2$
$= (x - 3)(x^2 - x + 2) + 8$ ✓✓

2

$$
\begin{array}{r}
2x^2 + 2x + 1 \\
x - 1\overline{)2x^3 + 0x^2 - \ x - 1} \\
\underline{2x^3 - 2x^2} \\
2x^2 - \ x \\
\underline{2x^2 - 2x} \\
x - 1 \\
\underline{x - 1} \\
0
\end{array}
$$

$\therefore 2x^3 - x - 1$
$= (x - 1)(2x^2 + 2x + 1)$ ✓✓

3 $P(x) = x^3 - 6x^2 - x + 30$
$P(5) = 5^3 - 6(5)^2 - 5 + 30$ ✓
$= 125 - 150 - 5 + 30$
$= 0$
$\therefore (x - 5)$ is factor ✓

4 Let $P(x) = x^3 - 3x^2 + kx - 2$
$\therefore P(-1) = (-1)^3 - 3(-1)^2 + k(-1) - 2$
$= -1 - 3 - k - 2$
$= -6 - k$ ✓
As remainder is 5, then $P(-1) = 5$:
$-6 - k = 5$
$k = -6 - 5$
$= -11$ ✓

5 $P(x) = x^3 + ax^2 + bx - 3$
$\therefore P(1) = 1^3 + a(1)^2 + b(1) - 3$
$= 1 + a + b - 3 = -1$
$a + b - 2 = -1$
$a + b = 1$ ① ✓

Also, $P(-1) = (-1)^3 + a(-1)^2 + b(-1) - 3$
$= -1 + a - b - 3 = -9$
$a - b - 4 = -9$
$a - b = -5$ ②
① + ②: $2a = -4$
$a = -2$
Subs. in ①: $-2 + b = 1$
$b = 3$
$\therefore a = -2, b = 3$ ✓

6 Let $P(x) = x^3 - x^2 - 14x + 24$.
Consider $P(1) = 1^3 - 1^2 - 14(1) + 24$
$= -14 + 24$
$\neq 0$
Consider $P(2) = 2^3 - 2^2 - 14(2) + 24$
$= 8 - 4 - 28 + 24$
$= 0$
$\therefore (x - 2)$ is factor ✓

$$
\begin{array}{r}
x^2 + x - 12 \\
x - 2\overline{)x^3 - x^2 - 14x + 24} \\
\underline{x^3 - 2x^2} \\
x^2 - 14x \\
\underline{x^2 - 2x} \\
-12x + 24 \\
\underline{-12x + 24} \\
0
\end{array}
$$
✓

$\therefore (x - 2)(x^2 + x - 12)$
$\therefore (x - 2)(x + 4)(x - 3)$
$\therefore 2, -4, 3$ ✓

7 Let $P(x) = x^3 - 7x + 6$.
Consider $P(1) = (1)^3 - 7(1) + 6$
$= 0$
$\therefore (x - 1)$ is factor ✓

$$
\begin{array}{r}
x^2 + \ x - 6 \\
x - 1\overline{)x^3 + 0x^2 - 7x + 6} \\
\underline{x^3 - \ x^2} \\
x^2 - 7x \\
\underline{x^2 - \ x} \\
-6x + 6 \\
\underline{-6x + 6} \\
0
\end{array}
$$

$\therefore (x - 1)(x^2 + x - 6)$ ✓
$\therefore (x - 1)(x + 3)(x - 2) = 0$
$\therefore x = 1, -3, 2$ ✓

8 a $y = x^3 + x^2 - 12x$
$= x(x^2 + x - 12)$
$= x(x + 4)(x - 3)$ ✓

y
$y = x^3 + x^2 - 12x$
-4 0 3 x
✓

b $y = x^3 + x^2 - 16x + 20$
Consider $y(2) = 2^3 + 2^2 - 16(2) + 20$
$= 8 + 4 - 32 + 20$
$= 0$
$\therefore (x - 2)$ is factor ✓

$$
\begin{array}{r}
x^2 + \ 3x - 10 \\
x - 2\overline{)x^3 + \ x^2 + 16x + 20} \\
\underline{x^3 - 2x^2} \\
3x^2 - 16x \\
\underline{3x^2 - \ 6x} \\
-10x + 20 \\
\underline{-10x + 20} \\
0
\end{array}
$$

$\therefore y = (x - 2)(x^2 + 3x - 10)$
$\therefore y = (x - 2)(x - 2)(x + 5)$
$\therefore x = (x - 2)^2(x + 5)$

y
$y = x^3 + x^2 - 16x + 20$
20
-5 0 2 x
✓

Chapter 14

Functions and Logarithms

Intermediate Test

p. 261

1

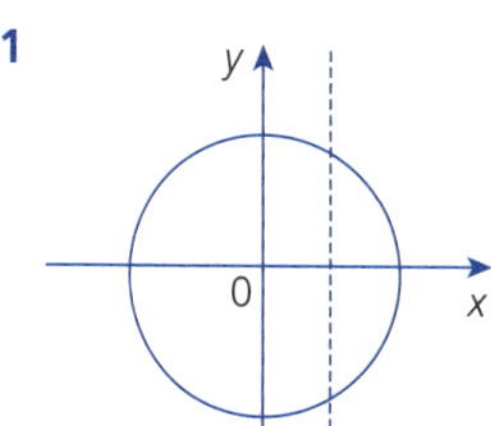

D, as vertical line cuts more than once,
$\therefore$ not function. ✓

2 $f(x) = 2x^2 - 4x$
$f(-3) = 2(-3)^2 - 4(-3)$ ✓
$= 18 + 12$
$= 30$ ✓

3 $g(x) = 4 - x$
$g(x + 2) = 4 - (x + 2)$ ✓
$= 4 - x - 2$
$= 2 - x$ ✓

4 a $y = \frac{1}{x}$
$x \neq 0$ ✓

b $y = \sqrt{x}$
$x \geq 0$ ✓

5 a $y = 2x + 1$

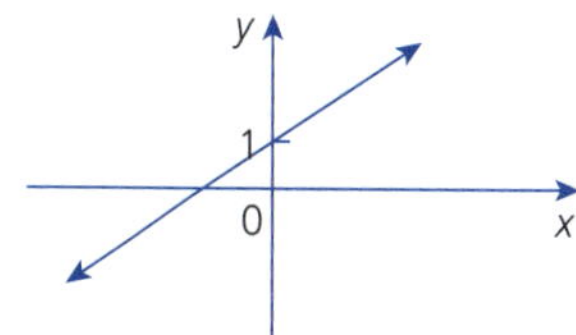

Domain: all real x ✓
Range: all real y ✓

b $y = x^2 + 3$

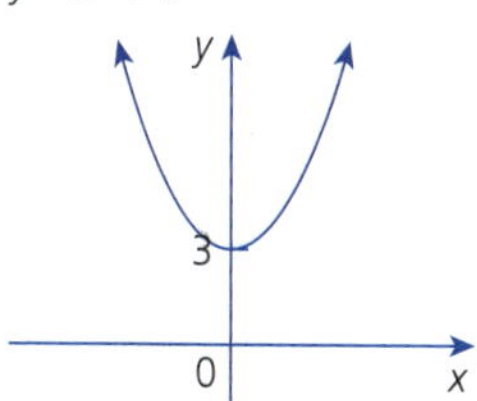

Domain: all real x ✓
Range: $y \geq 3$ ✓

6 a $y = 3x - 2$

swap x with y:

$\therefore x = 3y - 2$

$3y = x + 2$

$y = \dfrac{x+2}{3}$ ✓

b

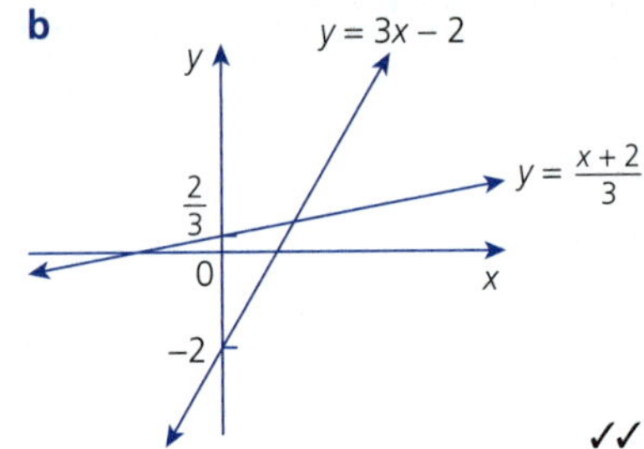

✓✓

7

✓✓

8 $\log_2 x = 3$

$\therefore \quad x = 2^3$

$= 8$ ✓

9

✓✓

10 a $\log_a 12 = \log_a (4 \times 3)$

$= \log_a 4 + \log_a 3$

$= \log_a 2^2 + \log_a 3$

$= 2\log_a 2 + \log_a 3$ ✓

$= 2(0.4176) + 0.6413$

$= 1.4765$ ✓

b $\log_a 1.5 = \log_a \dfrac{3}{2}$

$= \log_a 3 - \log_a 2$ ✓

$= 0.6413 - 0.4176$

$= 0.2237$ ✓

c $\log_a \sqrt{6} = \log_a 6^{\frac{1}{2}}$

$= \dfrac{1}{2}\log_a 6$

$= \dfrac{1}{2}\log_a (3 \times 2)$

$= \dfrac{1}{2}[\log_a 3 + \log_a 2]$ ✓

$= \dfrac{1}{2}[0.6413 + 0.4176]$

$= 0.52945$ ✓

11 a $\log_5 15 - \log_5 3 = \log_5 \dfrac{15}{3}$

$= \log_5 5$

$= 1$ ✓

b $2\log_2 8 + \log_2 4$

$= 2\log_2 2^3 + \log_2 2^2$

$= 6\log_2 2 + 2\log_2 2$

$= 8\log_2 2$

$= 8 \times 1$

$= 8$ ✓

c $\dfrac{\log_3 27}{\log_3 3} = \dfrac{\log_3 3^3}{\log_3 3}$

$= \dfrac{3\log_3 3}{\log_3 3}$

$= 3$ ✓

12 a $\log_{10} 10\,000 = \log_{10} 10^4$

$= 4\log_{10} 10$

$= 4 \times 1$

$= 4$ ✓

b $3^{x+2} = 81$

$3^{x+2} = 3^4$

$\therefore x + 2 = 4$

$x = 4 - 2$

$x = 2$ ✓

Advanced Test

p. 262

1 a $f(x) = x^2 - 4x - 6$

$f(-2) = (-2)^2 - 4(-2) - 6$

$= 4 + 8 - 6$

$= 6$ ✓

b $f(x + 2) = (x + 2)^2 - 4(x + 2) - 6$

$= x^2 + 4x + 4 - 4x - 8 - 6$

$= x^2 - 10$ ✓

c $f(x) = 6$

$\therefore x^2 - 4x - 6 = 6$

$x^2 - 4x - 12 = 0$ ✓

$(x - 6)(x + 2) = 0$

$x = 6, -2$ ✓

2 a $y = \dfrac{1}{x - 2}$

Domain is $x \neq 2$. ✓

b $y = \dfrac{1}{2x^2 + 5x - 12}$

Check $2x^2 + 5x - 12 = 0$

$(2x - 3)(x + 4) = 0$

$x = \frac{3}{2}, -4$

Domain is $x \neq \frac{3}{2}$ or -4. ✓

c $y = \sqrt{3 - x}$

Check $3 - x \geq 0$

$x \leq 3$

Domain is $x \leq 3$. ✓

d $y = \sqrt{9 - x^2}$

Check $9 - x^2 \geq 0$

Now, use $y = 9 - x^2$ to find $y \geq 0$

This means $-3 \leq x \leq 3$

Domain is $-3 \leq x \leq 3$. ✓✓

3 a $y = 4 - x^2$

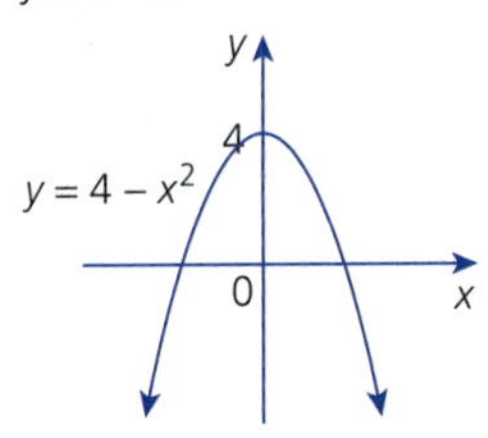

Domain: all real x ✓

Range: $y \leq 4$ ✓

b $y = \dfrac{1}{2x + 1}$

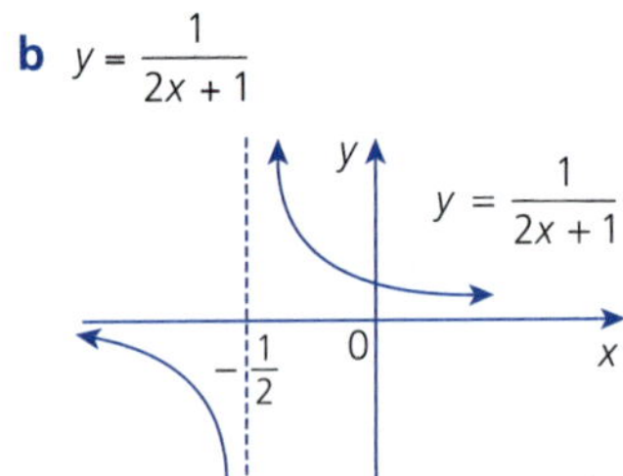

Check $2x + 1 = 0$

$2x = -1$

$x = -\dfrac{1}{2}$

$\therefore$ Domain: all real x, except

$x = -\dfrac{1}{2}$ ✓

Range all real y, except $y = 0$. ✓

c $y = 3^x + 1$

Domain: all real x ✓

Range: $y > 1$ ✓

4

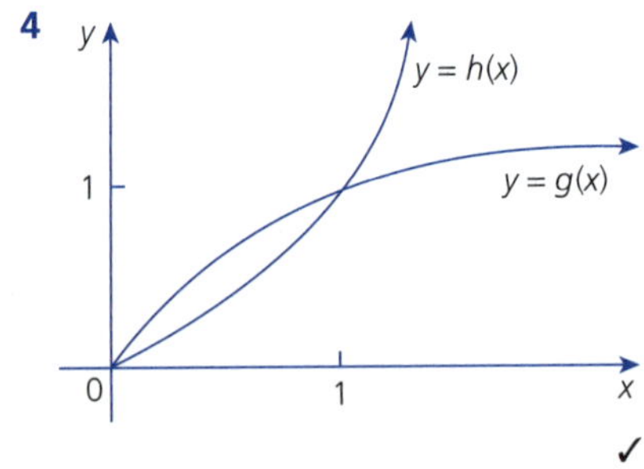

✓

5 a $\log_2 4\sqrt{2} = \log_2(2^2 \times 2^{\frac{1}{2}})$

$= \log_2 2^{2\frac{1}{2}}$

$= 2\frac{1}{2} \times \log_2 2$

$= 2\frac{1}{2} \times 1$

$= 2\frac{1}{2}$ ✓

b $\log_5 b + \log_5 \frac{1}{b} = \log_5 \ b \times \frac{1}{b}$

$= \log_5 1$

$= 0$ ✓

c $\log_2 36 - \log_2 9 = \log_2 \frac{36}{9}$

$= \log_2 4$

$= \log_2 2^2$

$= 2\log_2 2$

$= 2 \times 1$

$= 2$ ✓

6 a $x + y = \log_b 3 + \log_b 4$

$= \log_b (3 \times 4)$

$= \log_b 12$ ✓

b $2x - y = 2\log_b 3 - \log_b 4$

$= \log_b 3^2 - \log_b 4$ ✓

$= \log_b 9 - \log_b 4$

$= \log_b \frac{9}{4}$ ✓

7 a $\log_2 3x - \log_2 4 = \log_2 6$

$\log_2 \frac{3x}{4} = \log_2 6$ ✓

$\therefore \quad \frac{3x}{4} = 6$

$3x = 24$

$x = 8$ ✓

b $\log_2(x + 2) + \log_2(x - 5) = 3$

$\log_2(x + 2)(x - 5) = 3\log_2 2$

$\log_2(x + 2)(x - 5) = \log_2 2^3$ ✓

$\therefore \quad (x + 2)(x - 5) = 8$

$x^2 - 3x - 10 = 8$

$x^2 - 3x - 18 = 0$

$(x - 6)(x + 3) = 0$

$x = 6, -3$ ✓

But $x \neq -3$ as $\log_2(-3 + 2)$ is impossible, $\therefore x = 6.$ ✓

8 $\log_3(\log_4 64) = \log_3(\log_4 4^3)$

$= \log_3[3\log_4 4]$

$= \log_3(3 \times 1)$

$= \log_3 3$

$= 1$ ✓

9 $4^{x+2} = \frac{1}{2\sqrt{2}}$

$4^{x+2} = \frac{1}{\sqrt{8}}$

$4^{x+2} = 8^{-\frac{1}{2}}$

$(2^2)^{x+2} = (2^3)^{-\frac{1}{2}}$

$2^{2x+4} = 2^{-\frac{3}{2}}$

$\therefore \ 2x + 4 = -\frac{3}{2}$ ✓

$4x + 8 = -3$

$4x = -11$

$x = -2\frac{3}{4}$ ✓

Chapter 15

Circle Geometry

Intermediate Test

pp. 283–284

1 a Passes through the centre of the circle ✓

b Equal angles at the centre ✓

c Are supplementary ✓

d A right-angle ✓

e Is a right-angle ✓

2 a Radius ✓

b Chord ✓

c Minor segment ✓

3 a $x = 4$ (perpendicular from centre bisects chord) ✓✓

b $y = 44$ (∠ at circumference is half ∠ at centre on same arc) ✓✓

c $a = 12$ (tangents from external point) ✓✓

d $b + 23 = 90$

$b = 67$ (angle in semicircle) ✓✓

e $c = \frac{1}{2}(360 - 116)$

$= 122$ ✓

(reflex angle at centre is twice ∠ at circumference) ✓

f $d + 116 = 180$

$d = 64$ ✓

(opposite angles of cyclic quadrilateral) ✓

4

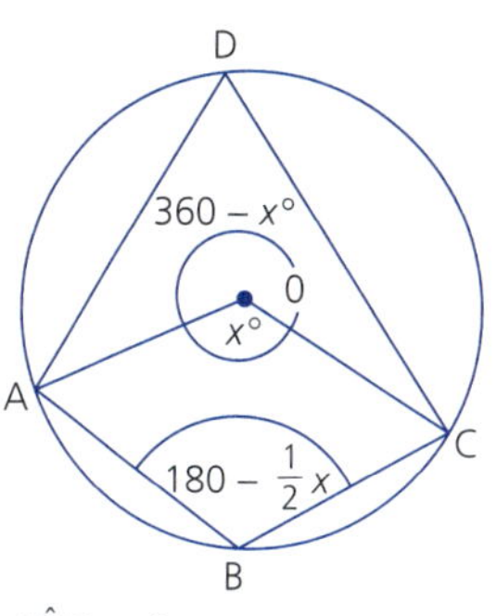

$A\hat{O}C = x°$

$\therefore$ Reflex $= A\hat{O}C = 360° - x°$

(angle sums of revolution)

Then $\angle ABC = \frac{1}{2}(360° - x°)$

(∠ at circumference is half ∠ at centre on same arc)

$\therefore \angle ABC = 180° - \frac{1}{2}x$ϒ ✓✓✓

5 $8(y + 8) = 12^2$ (product of intercepts = square of tangent)

$8y + 64 = 144$

$8y = 80$

$y = 10$ ✓✓

6

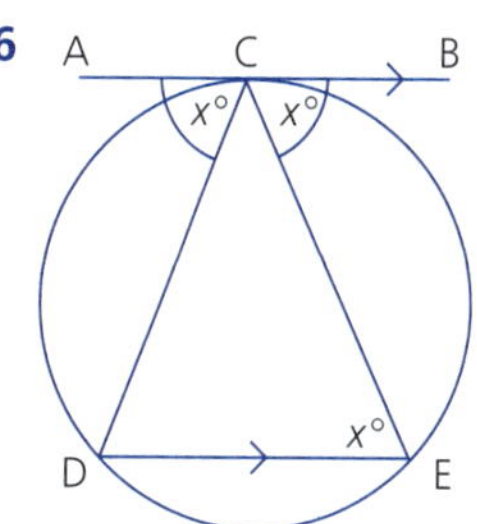

a Let $\angle ACD = x°$

Then $\angle DEC = x°$

(angle in alternate segment)

Also $\angle BCE = x°$

(alternate angles, DE ∥ AB) ✓✓✓

$\therefore \angle ACD = \angle BCE$ (both $x°$)

b $\angle CDE = x°$

(alternate to ∠ACD, AB ∥ DE)

Then $\angle CDE = \angle CED$ (both $x°$)

$\therefore$ △DEC is isosceles

(base ∠s equal) ✓✓

7

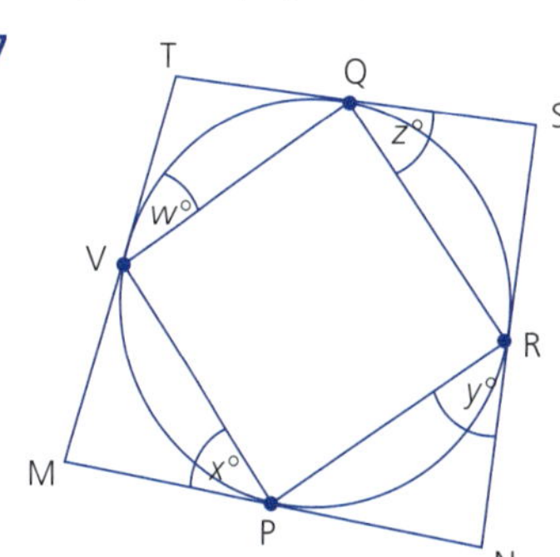

MP = MV	(tangents from external point)
∴ ∠PVM = $x°$	(∠s opposite equal sides in △PVM)
PN = NR	(tangents from external point)
∴ ∠RPN = $y°$	(∠s opposite equal sides in △PNR)

Similarly $\angle QRS = z°$ (in $\triangle QRS$)
$\angle TQV = w°$ (in $\triangle TVQ$)
Then $\angle VMP = 180° - 2x°$
($\angle$ sum of $\triangle PVM$)
Similarly $\angle PNR = 180° - 2y°$
($\angle$ sum of $\triangle PNR$)
$\angle QSR = 180° - 2z°$ ($\triangle QRS$)
$\angle QTV = 180° - 2w°$ ($\triangle TVQ$)
Then $180° - 2x° + 180° - 2y°$
$+ 180° + 2z° + 180° - 2w° = 360$
(angle sum of quadrilateral MNST)
$\therefore\ 720° - 2(x + y + z + w)° = 360°$
$2(x + y + z + w) = 360$
$\therefore\quad x + y + z + w = 180$
✓✓✓✓✓

Advanced Test

pp. 285–286

1 a … the product of the intercepts of any secants from the point. ✓

b … and the point of contact are collinear. ✓

c … equal to the angle in the alternate segment. ✓

d … bisects their common chord at right-angles. ✓

e … is twice the angle at the circumference standing on the same arc. ✓

2 a Arc ✓

b Sector ✓

c Tangent ✓

3 a $a = 46$ (equal chords subtend equal $\angle$s at centre) ✓✓

b $b = 37$ ($\angle$s on same arc at circumference) ✓✓

c $c + 54 = 90$ ($\angle$ in a semicircle)
$c = 36$ ✓✓

d $d = 102$ (exterior $\angle$ of cyclic quadrilateral equals interior opposite $\angle$) ✓✓

e $e = 68$ (angle in alternate segment) ✓✓

f $f + 32 = 90$ ($\angle$ between tangent and radius is right angle)
$f = 58$ ✓✓

4 $\angle NPM = 90°$ (angle in a semicircle)
$\therefore\ \angle PMO = 90° - 43°$ ($\angle$ sum of $\triangle MNP$)
$= 47°$
Now OM = OP (radii)
Then $\angle MPO = \angle PMO = 47°$
($\angle$s opposite equal sides)
$\therefore\ \angle MPO = 47°$
Then $\angle MOP = 180° - 47° - 47°$
($\angle$ sum of $\triangle MOP$)
$\therefore\quad \angle MOP = 86°$ ✓✓✓✓

5 $x(7 + x) = 12^2$
(square of tangent = product of intercepts cut from secant)
$7x + x^2 = 144$
$\therefore\ x^2 + 7x - 144 = 0$
$(x + 16)(x - 9) = 0$
$\therefore\quad x = 9$ or -16
i.e. $\quad x = 9$
(cannot have negative length) ✓✓✓

6

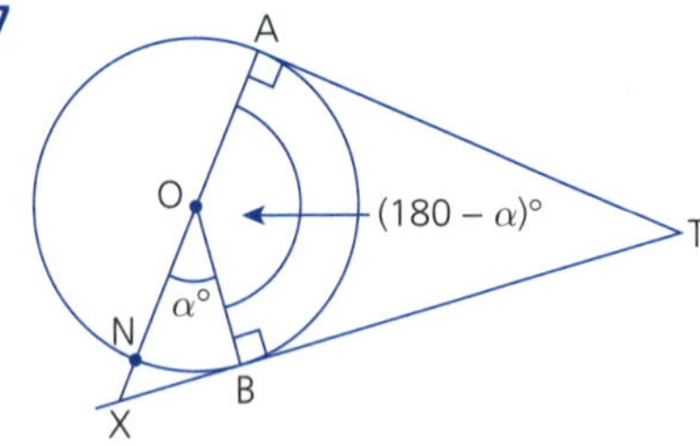

MN is common tangent to each circle at T.
Join OT and QT.
Then $M\hat{T}Q = 90°$
($\angle$ between radius and tangent = 90°)
Also $O\hat{T}M = 90°$ (same reason)
$\therefore\ O\hat{T}Q = 180°$ ($\angle OTM + \angle MTQ$)
$\therefore$ OTQ is straight line
i.e. O, T, Q are collinear ✓✓✓

7

Let $\angle BOX = \alpha°$
$\therefore\ \angle BOA = 180° - \alpha°$
[AX is a straight line]
Also $\angle OBT = 90°$
[radius to a point of contact]
and $\angle OAT = 90°$
[radius to a point of contact]
From quadrilateral OBTA
$180° - \alpha° + 90° + 90° + \angle ATB = 360°$
[angle sum of a quadrilateral]
$\therefore\ \angle ATB = 360° - 360° + \alpha°$
$= \alpha°$
$\therefore\ \angle BOX = \angle ATB$ [both $\alpha°$] ✓✓✓

$\angle OAB = \frac{1}{2}\alpha°$
[$\angle$ at the circumference = $\frac{1}{2}\angle$ at the centre on the same arc NB] ✓
Then $\angle ATB = 2 \times \angle OAB$
[$\alpha° = 2 \times \frac{1}{2}\alpha°$] ✓

Worked Solutions to Sample Examinations

Examination Paper 1

pp. 287–292

1 a Tax = 3572 + 0.325 × (72 340 − 37 000)
= 15 057.50
∴ \$15 057.50 ✓

b $I = Prn$
$1134 = 4200 \times 0.045 \times n$ ✓
$189n = 1134$
$r = \frac{1134}{189}$
$= 6$
∴ Invested for 6 years. ✓

c Price = 24 000 × 0.85 × 0.95
= 19 380
∴ Car cost \$19 380. ✓

d $A = P(1 + r)^n$
$A = 6500(1.0375)^4$
= 7531.227 698…
= 7531.23 (2 dec. pl.) ✓
∴ Interest = 7531.23 − 6500
= 1031.23
∴ Interest of \$1031.23. ✓

e $A = P(1 - r)^n$
$A = 52\,000 \times 0.88^5$ ✓
= 27 442.059 67…
= 27 442.06 (2 dec. pl.)
∴ Value of caravan is \$27 442.06. ✓

2 a $(x + 5)^2 - (x + 5)(x - 5)$
$= x^2 + 10x + 25 - (x^2 - 25)$ ✓
$= 10x + 50$ ✓

b i $(x - 3)(x + 2)$ ✓

ii $(2x - 3)(x + 1)$ ✓

c i $\frac{2x - 2}{x^2 - 1} = \frac{2(x - 1)}{(x - 1)(x + 1)}$ ✓
$= \frac{2}{x + 1}$ ✓

ii $\frac{2a - 5}{3} - \frac{4 - a}{2}$
$= \frac{2(2a - 5) - 3(4 - a)}{6}$ ✓
$= \frac{4a - 10 - 12 + 3a}{6}$
$= \frac{7a - 22}{6}$ ✓

3 a $\sqrt{12} + \sqrt{27} = 2\sqrt{3} + 3\sqrt{3}$
$= 5\sqrt{3}$ ✓

b $\frac{2 - \sqrt{3}}{2 + \sqrt{3}} \times \frac{2 - \sqrt{3}}{2 - \sqrt{3}} = \frac{4 - 4\sqrt{3} + 3}{4 - 3}$
$= 7 - 4\sqrt{3}$ ✓
∴ $a = 7$, $b = -4$ ✓

c $(\sqrt{2} - 1)^2 - (\sqrt{2} + 1)^2$
$= (2 - 2\sqrt{2} + 1) - (2 + 2\sqrt{2} + 1)$ ✓
$= 3 - 2\sqrt{2} - 3 - 2\sqrt{2}$
$= -4\sqrt{2}$ ✓

d $8^{-\frac{2}{3}} \times \left(\frac{1}{2}\right)^{-2} = \frac{1}{4} \times 4$
$= 1$ ✓

e $4^{x - 2} = 8^x$
$2^{2(x - 2)} = 2^{3x}$ ✓
$2^{2x - 4} = 2^{3x}$
$2x - 4 = 3x$
$3x - 2x = -4$
$x = -4$ ✓

4 a $3x - 4y = 5$
$3x - 5 = 4y$
$\frac{4y}{4} = \frac{3x - 5}{4}$
$y = \frac{3x - 5}{4}$ ✓

b $x^2 - 6x - 16 = 0$
$(x - 8)(x + 2) = 0$
$x = 8, -2$ ✓

c $3x^2 - 5x - 1 = 0$
$\therefore x = \frac{-b \pm \sqrt{b^2 - 4ac}}{2a}$
$= \frac{5 \pm \sqrt{25 - 4(3)(-1)}}{2(3)}$ ✓
$= \frac{5 \pm \sqrt{25 + 12}}{6}$
$= \frac{5 \pm \sqrt{37}}{6}$ ✓

d If $2x - y = 5$ ∴ $y = 2x - 5$ (1)
$y = x^2 - 4x$ (2)
Let (1) = (2)
∴ $2x - 5 = x^2 - 4x$ ✓
∴ $x^2 - 6x + 5 = 0$
$(x - 5)(x - 1) = 0$
$x = 5, 1$ ✓
Substitute in (1)
if $x = 5$, $y = 5$
if $x = 1$, $y = -3$
∴ (5, 5), (1, −3) ✓

e $x^4 - 2x^2 + 1 = 0$
let $m = x^2$
∴ $m^2 - 2m + 1 = 0$
$(m - 1)^2 = 0$
$m = 1$
∴ $x^2 = 1$
i.e. $x = \pm 1$ ✓

5 a i $AB = \sqrt{(-1 - 3)^2 + (2 + 4)^2}$
$= \sqrt{16 + 36}$
$= \sqrt{52}$
$= 2\sqrt{13}$
∴ $2\sqrt{13}$ units ✓

ii $MP = \left(\frac{-1 + 3}{2}, \frac{2 - 4}{2}\right)$
$= (1, -1)$ ✓

iii $m = \frac{-4 - 2}{3 + 1}$
$= -\frac{6}{4}$
$= -\frac{3}{2}$ ✓

b i $3x - 2y = 4$
$3x - 4 = 2y$
$\frac{2y}{2} = \frac{3x - 4}{2}$
$y = \frac{3x - 4}{2}$
$\therefore m = \frac{3}{2}$ ✓

ii $b = -2$ ✓

c As $3x - 4y = 2$
∴ $\frac{4y}{4} = \frac{3x - 2}{4}$
∴ Gradient $= \frac{3}{4}$
∴ $m = -\frac{4}{3}$ (negative reciprocal) ✓
∴ $y - 4 = -\frac{4}{3}(x + 2)$
$3y - 12 = -4x - 8$
$4x + 3y - 4 = 0$ ✓

d Substitute $(m + 2, 4)$ into equation:
∴ $2(m + 2) - 3(4) = 4$
$2m + 4 - 12 = 4$
$2m = 12$
$m = 6$ ✓

6 a i as $(x - 4)(2 - x) = 0$
$x = 4, 2$
∴ Axis of sym.: $x = 3$ ✓

ii Substitute $x = 3$ into y
∴ $y(3) = (3 - 4)(2 - 3)$
$= -1 \times -1$
$= 1$
∴ Maximum value of 1. ✓

iii

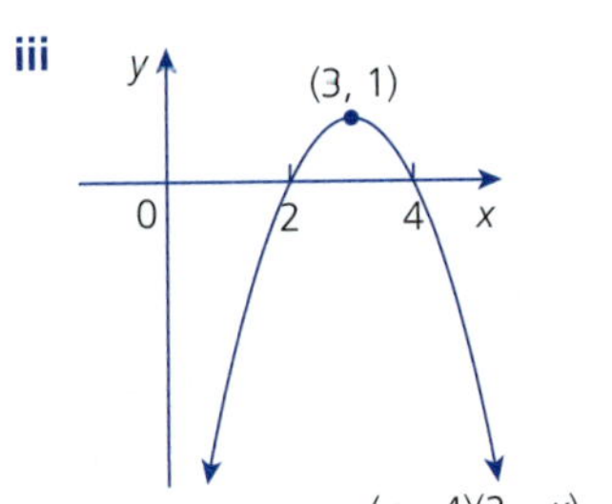

✓

b Let $y = a(x + 2)(x - 4)$
Substitute in (0, 24)
$\therefore 24 = a(0 + 2)(0 - 4)$ ✓
$24 = a(2)(-4)$
$-8a = 24$
$a = -3$
$\therefore y = -3(x + 2)(x - 4)$ ✓

c $x^2 + 4x + y^2 - 6y = 3$
$x^2 + 4x + 4 + y^2 - 6y + 9 = 3 + 4 + 9$ ✓
$(x + 2)^2 + (y - 3)^2 = 16$
C(–2, 3) and radius is 4 units. ✓

d

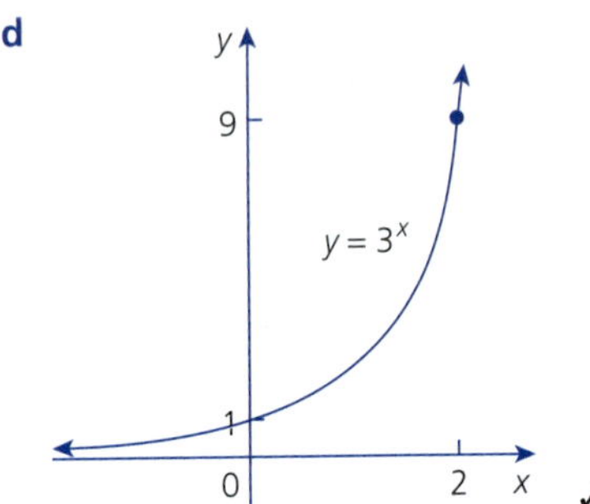

✓

7 a 1000 g = 1 kg and 10 000 m^2 = 1 ha.
As 30 × 10 000 = 300 000,
300 000 grams/ha = 300 kg/ha ✓

b **i** $y = kx$ ✓

ii Sub. $y = 18$ and $x = 6$ in $y = kx$:
$18 = 6k$
$k = \frac{18}{6}$
$= 3$ $\therefore y = 3x$ ✓

iii Sub. $x = 20$ in $y = 3x$
$= 3 \times 20$
$= 60$ ✓

iv Sub. $y = 96$ in $y = 3x$
$96 = 3x$
$x = \frac{96}{3}$
$= 32$ ✓

c $W = kn$ ✓
Sub. $W = 528$ and $n = 24$ in $W = kn$:
$528 = 24k$
$k = \frac{528}{24}$
$= 22$ $\therefore W = 22n$ ✓

d

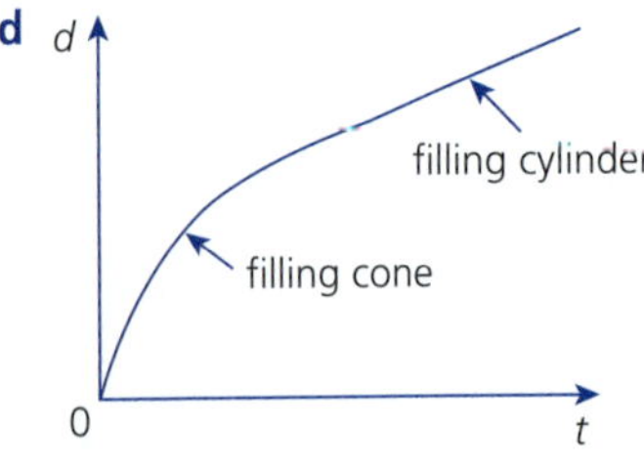

∴ **B** ✓

8 a

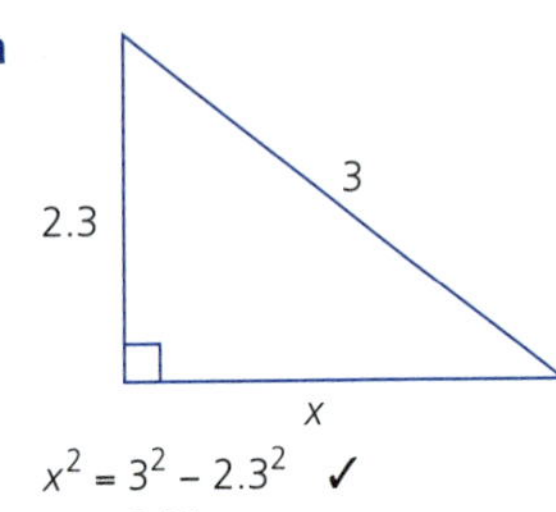

$x^2 = 3^2 - 2.3^2$ ✓
$= 3.71$
$x = 1.926\,136\,028\ldots$
$= 1.93$ (3 sig. fig.) ✓
∴ 1.93 m

b $\tan 150° = -\tan 30°$
$= -\frac{1}{\sqrt{3}}$ ✓

c

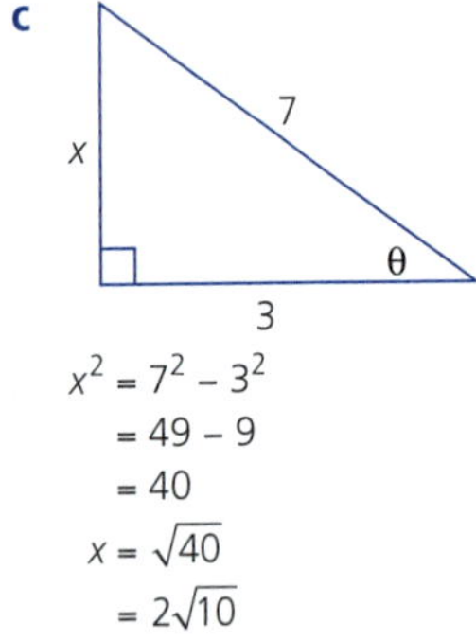

$x^2 = 7^2 - 3^2$
$= 49 - 9$
$= 40$
$x = \sqrt{40}$
$= 2\sqrt{10}$

$\therefore \tan\theta = \frac{2\sqrt{10}}{3}$ ✓

d $\frac{\frac{1}{2}}{x} = \cos 30°$

$\therefore x = \frac{1}{2} \div \cos 30°$ ✓
$= \frac{1}{2} \div \frac{\sqrt{3}}{2}$
$= \frac{1}{2} \times \frac{2}{\sqrt{3}}$
$= \frac{1}{\sqrt{3}}$ ✓

e

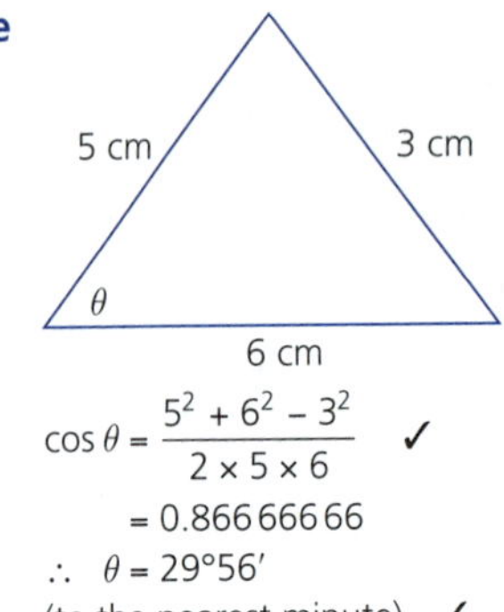

$\cos\theta = \frac{5^2 + 6^2 - 3^2}{2 \times 5 \times 6}$ ✓
$= 0.866\,666\,66$
$\therefore \theta = 29°56'$
(to the nearest minute) ✓

9 a

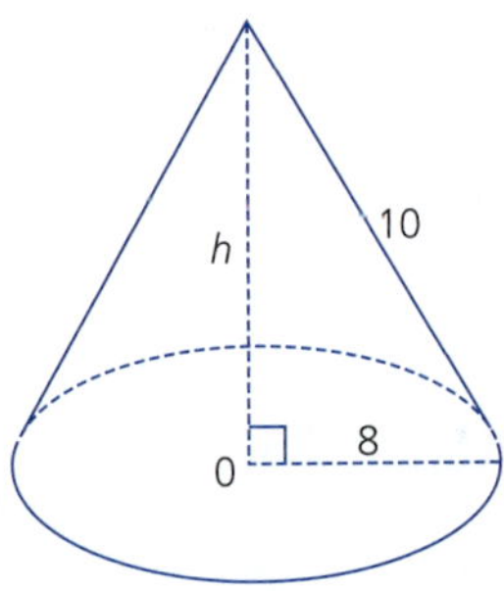

$h^2 = 10^2 - 8^2$
$= 100 - 64$
$= 36$
$h = 6$ ✓
$\therefore V = \frac{1}{3} \times \pi \times 8^2 \times 6$
$= 128\pi$
∴ The volume is 128π cm^3. ✓

b $V = \frac{1}{2} \times \frac{4}{3} \times \pi \times 12^3$ ✓
$= 3619.114\,737$
$= 3619.11$ (to 2 dec. pl.)
∴ The volume is 3619.11 cm^3. ✓

$SA = \frac{1}{2} \times 4 \times \pi \times 12^2 + \pi \times 12^2$ ✓
$= 1357.168\,026$
$= 1357.17$ (to 2 dec. pl.)
∴ The surface area is 1357.17 cm^2. ✓

c

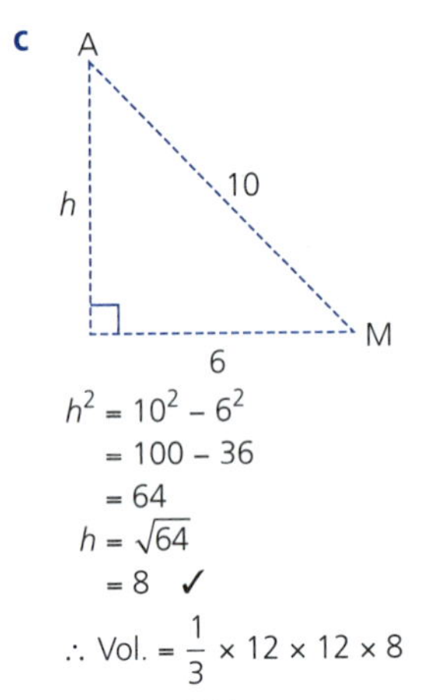

$h^2 = 10^2 - 6^2$
$= 100 - 36$
$= 64$
$h = \sqrt{64}$
$= 8$ ✓
$\therefore \text{Vol.} = \frac{1}{3} \times 12 \times 12 \times 8$
$= 384$
∴ The volume is 384 cm^3. ✓

10 a **i**

A

C M B

CM = BM (given)
AC = AB (given)
AM is common
∴ △ACM ≡ △ABM
(SSS test) ✓✓

ii ∴ ∠CAM = ∠BAM
(corresponding angles of congruent △s equal) ✓

b i

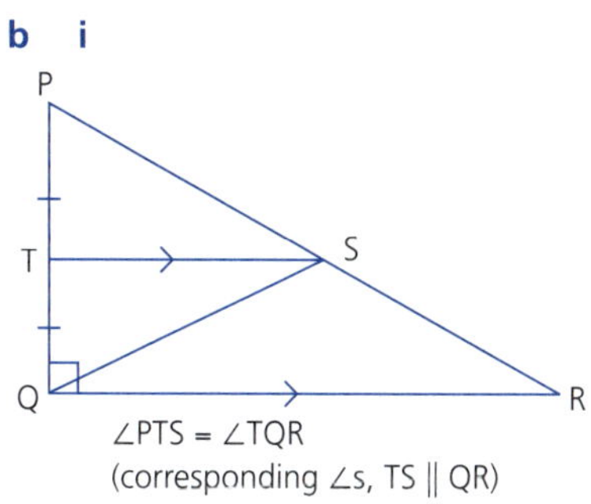

$\angle PTS = \angle TQR$
(corresponding $\angle$s, TS || QR)
$\therefore \angle PTS = 90°$
$\therefore \angle QTS = 90°$
$\therefore \angle PTS = \angle QTS$
PT = QT (given)
TS is common
$\therefore \triangle PTS \equiv \triangle QTS$ (SAS) ✓✓

ii Let $\angle RQS = x°$
$\therefore$ $\angle QST = x°$ (alternate $\angle$s, TS || QR)
And $\angle PST = x°$ (corresponding $\angle$s of congruent $\triangle$s equal)
$\therefore$ $\angle SRQ = x°$ (corresponding $\angle$s, TS|| QR)
$\therefore$ $\angle RQS = \angle SRQ$
$\therefore \triangle SQR$ is isos. $\triangle$ (base $\angle$s equal)
$\therefore$ QS = SR (sides of isosceles $\triangle$) ✓✓

c Ratio of areas = 25.2 : 56.7
= 4 : 9
$\therefore$ Ratio of sides = 2 : 3
$\therefore$ Ratio of diameters is 2 : 3. ✓

11 a i 25% ✓

ii 50% is 10 out of 20
$\therefore$ 25% of students = 0.25 × 24
= 6
$\therefore$ Six students. ✓

iii Negatively skewed ✓

b i 3.13 (2 dec. pl.) ✓

ii 1.03 (2 dec. pl.) ✓

c i The line is of the form $y = mx + b$ and passes through (25, 40) and (50, 65).

Gradient $m = \dfrac{65 - 40}{50 - 25}$
$= \dfrac{25}{25}$
$= 1$ ✓

$\therefore$ Equation of the form $y = x + b$
Now, sub. (25, 40) in $y = x + b$

$40 = 25 + b$
$b = 40 - 25$
$= 15$
$\therefore$ The line is $y = x + 15$. ✓

ii Sub. $x = 52$ in $y = x + 15$
$y = 52 + 15$
$= 67$
$\therefore$ Estimated history mark of 67. ✓

12 a i

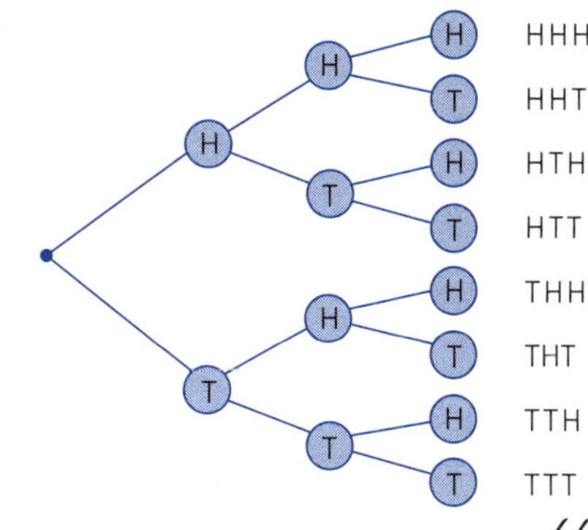

✓✓

ii 1 $P(\text{HHH}) = \dfrac{1}{8}$ ✓

2 $P(\text{head and 2 tails}) = \dfrac{3}{8}$ ✓

3 $P(\text{at least one tail})$
$= 1 - P(\text{no tails})$
$= 1 - \dfrac{1}{8}$
$= \dfrac{7}{8}$ ✓

b i $P(\text{less than 10}) = \dfrac{9}{15}$
$= \dfrac{3}{5}$ ✓

ii $P(\text{even, given more than 5})$
$= P(6, 8, 10, 12, 14)$
$= \dfrac{5}{10}$
$= \dfrac{1}{2}$ ✓

iii $P(\text{prime, given factor of 15})$
$= P(3, 5)$
$= \dfrac{2}{4} = \dfrac{1}{2}$ ✓

13 a

$$\begin{array}{r} 2x^2 - 3x \phantom{{}+3x-2} \\ x - 1\overline{)2x^3 - 5x^2 + 3x - 2} \\ \underline{2x^3 - 2x^2 \phantom{{}+3x-2}} \\ -3x^2 + 3x \phantom{{}-2} \\ \underline{-3x^2 + 3x \phantom{{}-2}} \\ -2 \end{array}$$ ✓

$(2x^3 - 5x^2 + 3x - 2)$
$= (x - 1)(2x^2 - 3x) - 2$ ✓

b $P(x) = x^3 - 2x^2 - 5x + 6$
$P(3) = (3)^3 - 2(3)^2 - 5(3) + 6$ ✓
$= 27 - 18 - 15 + 6$
$= 0$
$\therefore (x - 3)$ is factor ✓

c Let $P(x) = x^4 + kx^2 - 2x + 1$
$P(-2) = (-2)^4 + k(-2)^2 - 2(-2) + 1$ ✓
$= 16 + 4k + 4 + 1 = 9$
$4k + 21 = 9$
$4k = 9 - 21$
$4k = -12$
$k = -3$ ✓

d $y = x^3 - 6x^2 + 9x$
$y = x(x^2 - 6x + 9)$
$y = x(x - 3)^2$ ✓

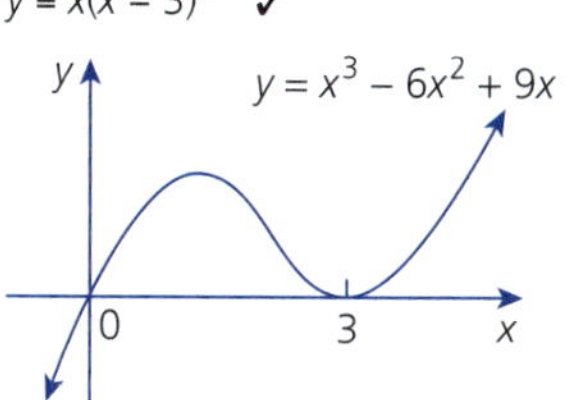

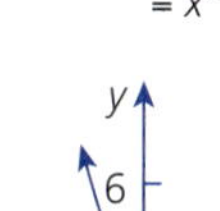

✓

14 a $f(x) = x^2 - 4x + 5$
$f(x - 2) = (x - 2)^2 - 4(x - 2) + 5$ ✓
$= x^2 - 4x + 4 - 4x + 8 + 5$
$= x^2 - 8x + 17$ ✓

b

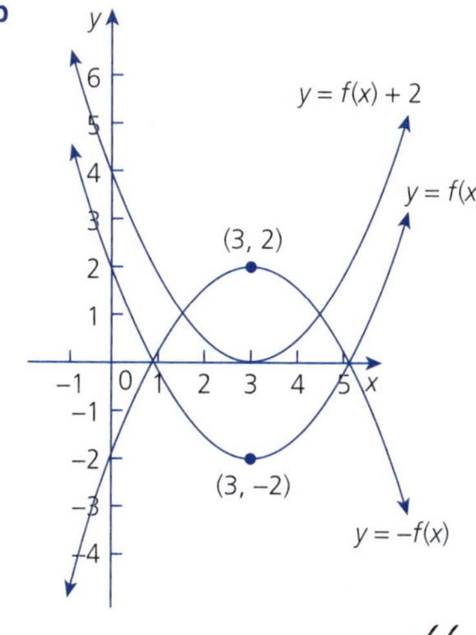

✓✓

c $\log_5 x = 3$ $\quad \therefore 5^3 = x$
$x = 125$ ✓

d $\log_a 18 = \log_a(3^2 \times 2)$
$= \log_a 3^2 + \log_a 2$ ✓
$= 2\log_a 3 + \log_a 2$
$= 2(0.3416) + 0.1476$
$= 0.8308$ ✓

e $2^x = 11$
$\log_{10} 2^x = \log_{10} 11$
$x\log_{10} 2 = \log_{10} 11$
$x = \dfrac{\log_{10} 11}{\log_{10} 2}$
$= 3.459431619\ldots$
$= 3.46$ (2 dec. pl.) ✓

15 a

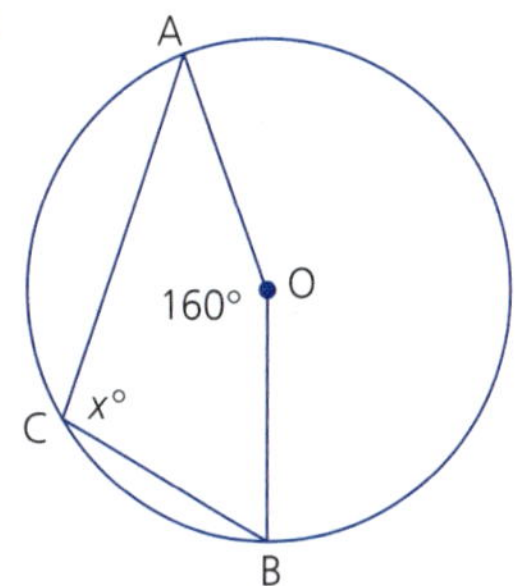

Reflex $\angle AOB = 200°$
(angles at a point) ✓
$x = 100$
($\angle$ at circumference is half $\angle$ at centre on same arc) ✓

b

Let $\angle PCA = x$

$\therefore \angle TPA = x$ ($\angle$s in alt. segment) ✓

$\therefore \angle PBC = x$ (alt. $\angle$s, TP $\|$ BC) ✓

$\therefore \angle PCA = \angle PBC$

c Let $BC = x$

$x(x + 6) = 5^2$ ✓

$x^2 + 6x - 25 = 0$

(square of the tangent to a circle from an external point equals the product of the intercepts of any secants from the point)

$$x = \frac{-6 \pm \sqrt{6^2 - 4(1)(-25)}}{2(1)}$$

$$= \frac{-6 \pm \sqrt{136}}{2}$$

$= 2.83$ (2 dec. pl.) ($x > 0$) ✓

$\therefore$ BC is 2.83 cm.

d

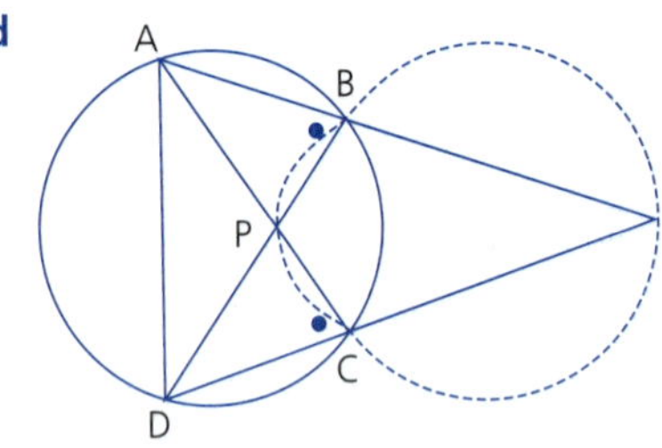

$\angle ABD = \angle ACD = x°$ (angles in same segment)

$\angle QBP = \angle QCP = (180 - x)°$ (supplementary angles)

But $\angle QBP + \angle QCP = 180°$ (opp. angles of cyclic quad. PBQC supplementary) ✓

$\therefore 180 - x + 180 - x = 180$

$2x = 180$

$x = 90$

$\therefore \angle ABD = 90°$ ✓

Examination Paper 2

Part A . . . . pp. 293–296

1 $(x - 2)(3x + 1) = 0$

$x = 2, -\frac{1}{3}$

$\therefore$ B ✓

2 $3x + y = 2$

$y = 2 - 3x$

$\therefore m = -3$

$\therefore$ C ✓

3 $\dfrac{1}{\sqrt{11} - 3} \times \dfrac{\sqrt{11} + 3}{\sqrt{11} + 3} = \dfrac{\sqrt{11} + 3}{11 - 9}$

$= \dfrac{\sqrt{11} + 3}{2}$

$\therefore$ A ✓

4 10 cm/20 s

$\therefore$ 30 cm/1 min

$\therefore$ 1800 cm/1 h

$\therefore$ 18 m/h

$\therefore$ 0.018 km/h

$\therefore$ A ✓

5 $P(\text{same colour}) = P(BB) + P(RR)$

$= \frac{2}{4} \times \frac{1}{3} + \frac{2}{4} \times \frac{1}{3}$

$= \frac{1}{6} + \frac{1}{6}$

$= \frac{1}{3}$

$\therefore$ B ✓

6 $4 - 2(5 - 3x) = 4 - 10 + 6x$

$= -6 + 6x$

$\therefore$ D ✓

7 $y = \dfrac{4x + 3}{2x}$

$\therefore \quad 2xy = 4x + 3$

$4x - 2xy = -3$

$x(4 - 2y) = -3$

$x = \dfrac{-3}{4 - 2y}$

$\therefore \quad x = \dfrac{3}{2y - 4}$

$\therefore$ C ✓

8 $y = ax(x - 1)(x - 3)$

Substitute in (4, 12)

$\therefore 12 = 4a(4 - 1)(4 - 3)$

$12 = 12a$

$a = 1$

$\therefore y = x(x - 1)(x - 3)$

$\therefore$ A ✓

9 $\dfrac{AB}{BC} = \sin 53°$

$= 0.7986$ (to 4 dec. pl.)

$\therefore$ A ✓

10 $a^{-\frac{2}{3}} = \dfrac{1}{a^{\frac{2}{3}}}$

$= \dfrac{1}{\sqrt[3]{a^2}}$

$\therefore$ D ✓

11 $2x^2 - 3x - 4 = 0$

$$x = \frac{-(-3) \pm \sqrt{(-3)^2 - 4(2)(-4)}}{2(2)}$$

$$= \frac{3 \pm \sqrt{41}}{4}$$

$\therefore$ B ✓

12 $\triangle ABC \,|||\, \triangle DEC$ (2 $\angle$s equal)

Also AC = 5 cm (by Pythagoras' Theorem)

$\therefore \dfrac{CD}{5} = \dfrac{8}{4}$

$CD = \dfrac{40}{4}$

$= 10$

$\therefore$ CD = 10 cm

$\therefore$ A ✓

13

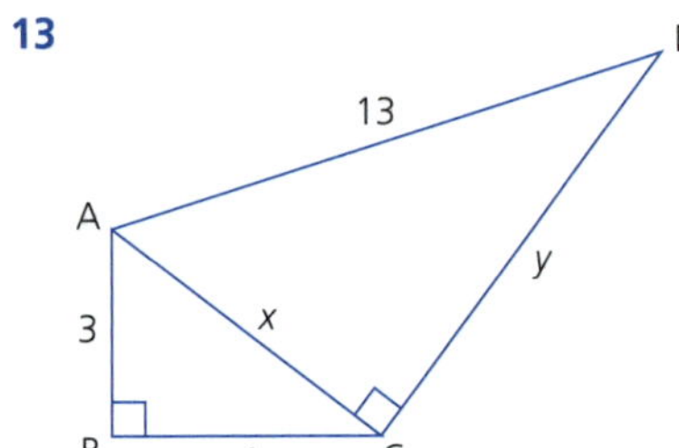

$x^2 = 3^2 + 4^2$

$= 9 + 16$

$= 25$

$x = 5$

$y^2 = 13^2 - 5^2$

$= 169 - 25$

$= 144$

$y = 12$

$\therefore$ D ✓

14 Let $P(x) = x^3 - 2x^2 - x + 2$

$P(2) = 2^3 - 2(2)^2 - 2 + 2$

$= 0$

$\therefore$ A ✓

15

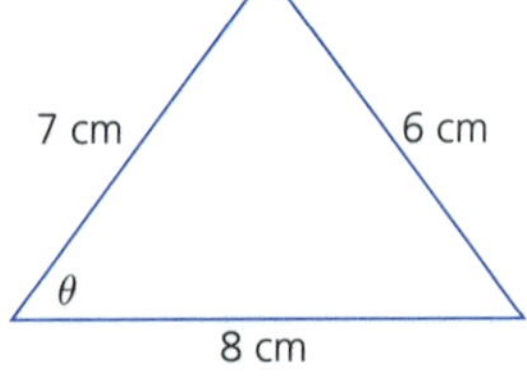

$\cos\theta = \dfrac{7^2 + 8^2 - 6^2}{2 \times 7 \times 8}$

$= 0.6875$

$\theta = 47°$ (to the nearest degree)

$\therefore$ B ✓

16 $\dfrac{3x}{5} - \dfrac{x}{2} = \dfrac{6x - 5x}{10} = \dfrac{x}{10}$

$\therefore$ A ✓

17 $f(x) = 3x^2 - x$

$f(-2) = 3(-2)^2 - (-2)$

$= 12 + 2$

$= 14$

$\therefore$ C ✓

18 Area $= \frac{1}{2} ab\sin C$

$= \frac{1}{2} \times 4.2 \times 4.2 \times \sin 60°$

$= 7.64$ (to 2 dec. pl.)

$\therefore$ Area is 7.64 cm^2

$\therefore$ B ✓

19 $4 - 2x \geq 6$

$-2x \geq 2$

$x \leq -1$

−3 −2 −1 0 1

$\therefore$ B ✓

20

$3x - 2y = 0 \quad (1)$

$x + y = 5 \quad (2)$

$\therefore \quad 2x + 2y = 10 \quad (3)$

$(1) + (3) \quad 5x = 10$

$x = 2$

$\therefore$ D ✓

21

35° 55° 70° 55°

$\therefore$ B ✓

22 $x^2 - 4x + 4 + y^2 + 6y + 9 = 2 + 4 + 9$

$(x - 2)^2 + (y + 3)^2 = 15$

$\therefore$ C(2, −3)

$\therefore$ A ✓

23 $\frac{\sin\theta}{r} = \frac{\sin 34\Upsilon}{q}$

$\sin\theta = \frac{r\sin 34\Upsilon}{q}$

$\therefore$ C ✓

24 $\sqrt{8} + \sqrt{18} = 2\sqrt{2} + 3\sqrt{2}$

$= 5\sqrt{2}$

$= \sqrt{25} \times \sqrt{2}$

$= \sqrt{50}$

$\therefore k = 50$

$\therefore$ D ✓

25 $\frac{a^3b^2}{(a^2b)^2} = \frac{a^3b^2}{a^4b^2}$

$= \frac{1}{a}$

$\therefore$ C ✓

26 Method A: P = $2n$

Method B: P = $n + 1$

$\therefore$ Difference: $2n - n - 1 = n - 1$

$\therefore$ A ✓

27 Axis of symmetry: $x = -2$

y-intercept: 2

Checking alternatives:

$\therefore y = x^2 + 4x + 2$

$\therefore$ D ✓

28 Ratio of volumes = 64 : 287.496

$\therefore$ Ratio of sides = 4 : 6.6

= 2 : 3.3

Ratio of areas = 4 : 10.89

$\therefore$ Surface area = $\frac{16}{4} \times 10.89$

= 43.56

$\therefore$ Surface area is 43.56 cm^2.

$\therefore$ B ✓

29 $V = \frac{1}{2} \times \frac{4}{3}\pi r^3$

$= \frac{2}{3} \times \pi \times 3^3$

$= 18\pi$

$\therefore$ The volume is 18π units^3.

$\therefore$ B ✓

30 $p = 16, q = 4$

$\therefore$ C ✓

Part B . . . . pp. 296–298

31 a $d = \sqrt{(x_2 - x_1)^2 + (y_2 - y_1)^2}$

$= \sqrt{(-3 - 2)^2 + (4 + 1)^2}$

$= \sqrt{25 + 25}$

$= \sqrt{50}$

$= 5\sqrt{2}$

$\therefore 5\sqrt{2}$ units ✓

b $4x^3 = 32$

$x^3 = 8$

$x = 2$ ✓

c $^{2\times}3x = \frac{^{2\times}x}{2} + 5^{\times 2}$

$6x = x + 10$

$6x - x = 10$

$5x = 10$

$x = 2$ ✓

d $\frac{x^2 - 5x + 6}{x^2 - 9} = \frac{(x - 3)(x - 2)}{(x - 3)(x + 3)}$

$= \frac{x - 2}{x + 3}$ ✓

e $P(x) \cdot Q(x)$

$= (3x^2 - 4x^3 + 7)(2x^4 - 7x^2 + 1)$

$= \ldots - 8x^7 + \ldots$

$\therefore$ Degree of 7. ✓

f $v = u + at$

$60 = 24 + 6t$

$6t + 24 = 60$

$6t = 60 - 24$

$6t = 36$

$t = 6$ ✓

g SA = $2 \times x \times 2 + 2 \times x \times 3 + 2 \times 2 \times 3$

$= 4x + 6x + 12$

$= 10x + 12$

$\therefore 10x + 12 = 62$

$10x = 62 - 12$

$10x = 50$

$x = 5$ ✓

h 80% of 60 = 0.8×60

= 48

From box plot 50% scored 48 or less.

$\therefore$ 12 students. ✓

i Gradient of $y = x$ is 1

$\therefore$ gradient of perpendicular is −1

Now, l has gradient of −1 and passes through (3, 0)

$\therefore \quad y - 0 = -1(x - 3)$

$y = -x + 3$

$\therefore x + y - 3 = 0$ ✓

j $2x^2 = 4x$

$2x^2 - 4x = 0$

$2x(x - 2) = 0$

$x = 0$ or $x = 2$ ✓

k $6x^2 - 5x - 4$

$= (3x - 4)(2x + 1)$ ✓

l ∠OCD = 32° (alt. ∠s, DC ∥ AB)

∠COD = 64° (∠ at centre is twice ∠ at circum. on same arc)

∠CDO = 84° (∠ sum of △)

$\therefore$ ∠ BDA = 84° (vertically opp. ∠s) ✓

m $M = kn$

$128 = 8n$

$n = \frac{128}{8}$

$= 16$

$\therefore M = 16n$ ✓

n $\log_a 12 = \log_a(3 \times 4)$

$= \log_a 3 + \log_a 4$

$= \log_a 3 + \log_a 2^2$

$= \log_a 3 + 2\log_a 2$

$= 1.4 + 2(0.9)$

$= 1.4 + 1.8$

$= 3.2$ ✓

o Numbers could be 6, 0. ✓

32 a

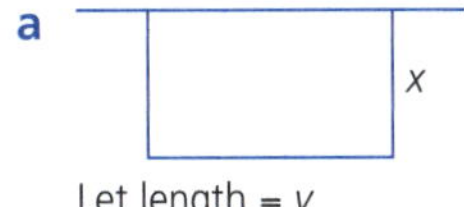

Let length = y

$\therefore y + 2x = 80$

$y = 80 - 2x$

$\therefore$ The length is $(80 - 2x)$ m. ✓

b $80 - 2x = 50$

$\therefore \quad 2x = 30$

$x = 15$

$\therefore$ Width is 15 m. ✓

Area = 50×15

= 750

$\therefore$ The area is 750 m^2. ✓

33 a The line is of the form $y = mx + b$ and passes through (20, 10) and (40, 90).

Gradient $m = \frac{90 - 10}{40 - 20}$

$= \frac{80}{20}$

$= 4$ ✓

$\therefore$ Equation of the form $y = 4x + b$.

Now, sub. (20, 10) in $y = 4x + b$

$10 = 4(20) + b$

$b = 10 - 80$

$= -70$

$\therefore$ The line is $y = 4x - 70$. ✓

b Sub. $x = 38$ in $y = 4x - 70$:
$y = 4(38) - 70$
$= 82$
∴ The shop might sell 82 icecreams. ✓

34 a $y = x^2 - 3x$
$y = x(x - 3) = 0$
$x = 0, 3$
Using $x = 3$, then A(3, 0). ✓

b $y = x^2 - 3x$ (1)
$y = 2x$ (2)
(1) = (2) ∴ $x^2 - 3x = 2x$
$x^2 - 5x = 0$
$x(x - 5) = 0$
$x = 5$ ✓
Substitute in (2) $y = 2 \times 5$
$= 10$
∴ B(5, 10) ✓

35

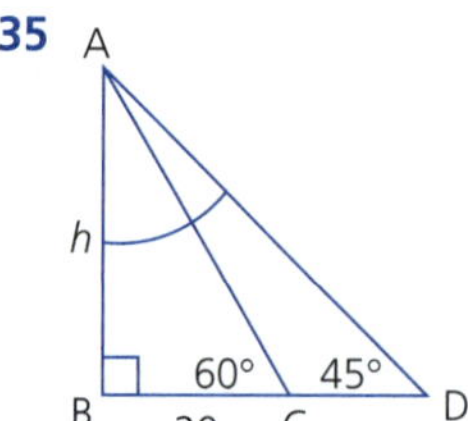

a $\frac{h}{20} = \tan 60$
∴ $h = 20 \tan 60$
$= 34.64$
∴ Height of 34.64 m. ✓

b As △ABD is isosceles ✓
∴ BD = 34.64
∴ CD = 34.64 – 20
= 14.64
That is, CD is 14.64 m. ✓

36 a Let $h = 0$
∴ $32 + 4x - x^2 = 0$
That is, $x^2 - 4x - 32 = 0$
$(x - 8)(x + 4) = 0$
∴ $x = 8, -4$
∴ B is 8 metres from the cliff. ✓

b Using axis of symmetry formula:
$x = \frac{-b}{2a}$
$= \frac{-4}{2(-1)}$
$= 2$ ✓
Substitute $x = 2$ in h
∴ $h = 32 + 4(2) - 2^2$
$= 36$
∴ The maximum height is 36 m. ✓

37

Let ∠CAB = x°
∴ ∠ACB = x°
(base ∠s of isosceles △ equal)
∴ ∠CBD = $2x$°
(ext. ∠ of △ equals sum of 2 int. ∠s)
But ∠CBE = ∠EBD
(BE bisects ∠CBD)
∴ ∠CBE = ∠EBD = x°
∴ ∠CAB = ∠EBD
∴ AC ∥ BE
(corresponding angles equal) ✓✓✓

38 ∠DAB = 40° ✓
(angle in alternate segment)
∠DAC = 15° (angle in same segment) ✓
∴ ∠CAB = 40° – 15°
= 25° ✓

39 a

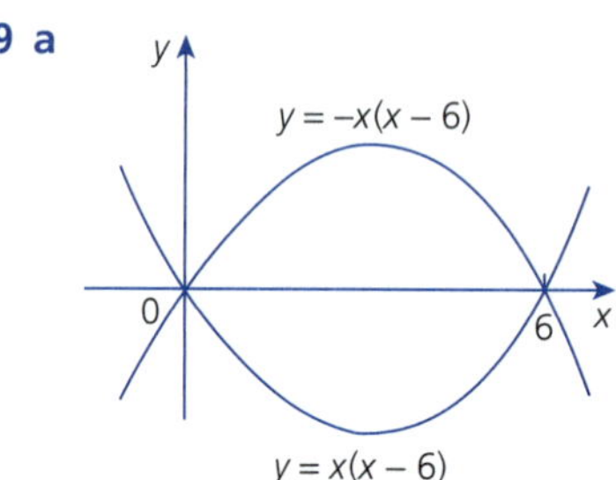

Any parabola with equation of form $y = ax(x - 6)$, for example, $y = x(x - 6)$, $y = -x(x - 6)$. ✓

b Minimum value occurs at $x = 3$
∴ (3, –18) lies on parabola.
∴ Substitute in $y = ax(x - 6)$
∴ $-18 = 3a(3 - 6)$ ✓
∴ $-18 = -9a$
$a = 2$
∴ $y = 2x(x - 6)$ ✓

40 a

Shape A

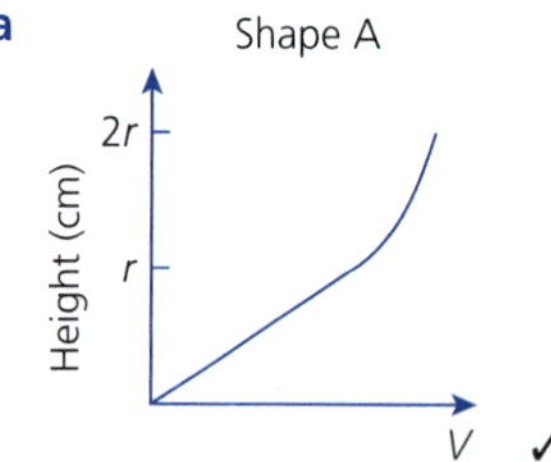

✓

Shape B

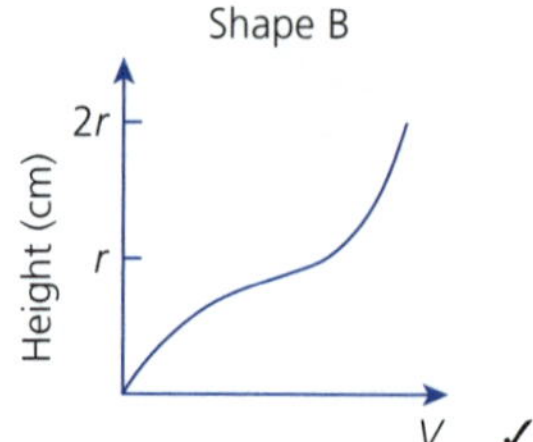

✓

b Shape A:
Volume of cylinder + Cone
(radius = r, height = r)
$= \pi \times r^2 \times r + \frac{1}{3} \times \pi \times r^2 \times r$
$= \pi \times r^3 + \frac{1}{3}\pi r^3$
$= \frac{4}{3} \neq r^3$

Shape B:
Volume of sphere (radius = r)
$= \frac{4}{3} \times \pi \times r^3$
$= \frac{4}{3} \neq r^3$

∴ As volumes are the same, their capacities are the same. ✓

41 a Substitute $x = 3$ in
$x^2 + y^2 = 25$
∴ $9 + y^2 = 25$
∴ $y^2 = 16$
$y = \pm 4$
∴ Obviously $y = 4$
∴ $a = 4$ ∴ (3, 4) ✓

b Substitute (3, 4) in
$3x + 4y + k = 0$
That is, $3(3) + 4(4) + k = 0$
$9 + 16 + k = 0$
$k = -25$ ✓
∴ $3x + 4y - 25 = 0$ is the equation of the line.

c x: substitute $y = 0$
∴ $3x - 25 = 0$
$3x = 25$
$x = 8\frac{1}{3}$
y: substitute $x = 0$
∴ $4y - 25 = 0$
$4y = 25$
$y = 6\frac{1}{4}$
∴ $= \frac{1}{2} \times 8\frac{1}{3} \times 6\frac{1}{4}$
$= 26\frac{1}{24}$
∴ The area is $26\frac{1}{24}$ units^2. ✓

Index

Index

Index

Index

NOTES

NOTES